D1243015

Organometallics in Synthesis

Organometallics in Synthesis

A Manual

Edited by

M. Schlosser
Université de Lausanne, Lausanne, Switzerland

JOHN WILEY & SONS
Chichester · New York · Brisbane · Toronto · Singapore

Other Wiley Editorial Offices

John Wiley & Sons, Inc., 605 Third Avenue,
New York, NY 10158-0012, USA

Jacaranda Wiley Ltd, G.P.O. Box 859, Brisbane,
Queensland 4001, Australia

John Wiley & Sons (Canada) Ltd, 22 Worcester Road,
Rexdale, Ontario M9W 1L1, Canada

John Wiley & Sons (SEA) Pte Ltd, 37 Jalan Pemimpin #05-04,
Block B, Union Industrial Building, Singapore 2057

Library of Congress Cataloging-in-Publication Data

Organometallics in synthesis : a manual / edited by M. Schlosser.
p. cm.
Includes bibliographical references and index.
ISBN 0 471 93637 5
1. Organic compounds—Synthesis. 2. Organometallic compounds.
I. Schlosser, M. (Manfred)
QD262.0745 1994
547'.05—dc20 93-2193
CIP

British Library Cataloguing in Publication Data
A catalogue record for this book is available from the British Library

ISBN 0 471 93637 5

Typeset in Times 11/13pt by Techset Composition Ltd, Salisbury, Wiltshire, UK.
Printed and bound in Great Britain by Bookcraft (Bath) Ltd

Contents

List of Contributors

L. S. HEGEDUS — *Department of Chemistry, Colorado State University, Fort Collins, CO 80523, USA*

B. H. LIPSHUTZ — *Department of Chemistry, University of California, Santa Barbara, CA 93106, USA*

H. NOZAKI — *Department of Applied Chemistry, Okayama University of Science, 1–1 Ridai-cho, Okayama 700, Japan*

M. T. REETZ — *Max-Plank-Institut für Kohlenforschung, Kaiser-Wilhelm-Platz 1, D-45470 Mülheim Ruhr, Germany*

P. RITTMEYER — *Chemetall GmbH, Reuterweg 14, Postfach 10 15 01, D-60271 Frankfurt am Main 1, Germany*

M. SCHLOSSER — *Institut de Chimie Organique, Université de Lausanne, Rue de la Barre 2, 1005 Lausanne, Switzerland*

K. SMITH — *Department of Chemistry, University College, Singleton Park, Swansea SA2 8PP, UK*

F. TOTTER — *Chemetall GmbH, Reuterweg 14, Postfach 10 15 01, D-60271 Frankfurt am Main 1, Germany*

H. YAMAMOTO — *Department of Applied Chemistry, Faculty of Engineering, Nagoya University, Furocho, Chikusa, Nagoya 464-01, Japan*

Preface

The origin of this Manual, was a series of three post-graduate workshops which I had the privilege to host in the late eighties. These five-day seminars provided ample opportunity for the participants to talk to the lecturers, solve exercises and, in 1987 at least, to practise their new skills in the laboratory. One common feature became apparent in all such interactions between tutors and novices: the psychological and practical barrier for newcomers to enter the field of organometallic chemistry is still very high

Therefore, this Manual is meant primarily for those researchers who have not yet had a chance to familiarize themselves with the basic concepts and techniques of organometallic reactions. It is a nuts-and-bolts text: full of useful hints, rules of thumb and, last but not least, carefully selected working procedures. Thus, it summarizes what organometallic reagents can do for modern organic synthesis and at the same time explains how they are conveniently employed.

However, as Ludwig Boltzmann recognized, 'nothing is more practical than a good theory'. The message applied to our case is that sound mechanistic insight is required if we wish not only to document the course of known reactions but also predict the outcome of future experiments. Therefore, despite its down-to-earth approach, this Manual puts great emphasis on the presentation of first principles that can provide a rational basis for understanding organometallic reactivity, a fascinating and at the same time still widely mysterious subject.

In order not to discourage the reader by too voluminous a book, we had to restrict the coverage to a selection of the most popular metals and methods. Hence, many important reagents and topics had to be neglected for the moment. They may, however, be included in a future, more comprehensive edition.

It is now my pleasure to express by profound gratitude to my co-authors who, notwithstanding their numerous other commitments, have agreed to share their expertise with the reader and thus to contribute to the propagation of organometallics in synthesis. I wish also to acknowledge the help of my coworkers who have checked many working procedures. Finally, I am indebted to my wife Elsbeth, who has made the Chem-Art china ink drawings (contributions Nos 1, 2, 3 and 8) with artistic skill and aesthetic taste.

Lausanne, Spring 1994 — Manfred Schlosser

1

Organoalkali Reagents

MANFRED SCHLOSSER
Université de Lausanne, Lausanne, Switzerland

Organometallics in Synthesis—A Manual. Edited by M. Schlosser

1.1 THE ORGANOMETALLIC APPROACH: A LOOK BACK

The cradle of polar organometallic chemistry stood some 150 years ago in Northern Hesse [1]. There, in 1831, Wöhler became a teacher at the newly founded Higher Industrial School (Höhere Gewerbeschule) in Cassel. Previously, he had held an appointment at a similar institution in Berlin, where he had succeeded in the preparation of the 'natural product' urea from the undisputably inorganic precursors ammonia and silver or lead cyanate [2, 3]. Thus, he had defeated the dogma of the indispensable *vis vitalis* and had fired the starting shot for the start of the race towards organic synthesis. An outburst of cholera prompted him to leave the Prussian capital and to return to the proximity of his native town. Among his numerous noteworthy accomplishments in Cassel was the preparation of diethyltellurium [4] as the first Main Group organometallic compound. This discovery, however, remained without aftermath.

In 1836, Wöhler occupied the vacant chair of chemistry at the University of Göttingen while the young Privatdozent Robert Wilhelm Bunsen became his successor in Cassel. Scientifically ambitious, Bunsen decided to tackle one of the greatest experimental challenges of his time. He was determined to solve the mysteries of the dreadful Cadet's liquor, an atrociously smelling, poisonous distillate that forms when arsenic and potassium acetate are exposed to red heat. At the risk of his health, he managed to isolate the main component and to identify its elementary composition as $C_4H_{12}As_2O$ [5]. While he initially called the new substance alcarsine oxyde, he soon adopted the more revealing name of cacodyl oxyde (*κακόζ* = stinking, *όδηζ* = odor).

Nowadays, we can unambiguously assign the structure of bis(dimethylarsanyl) oxide to this compound. However, in the middle of last century, the notions of chemical bonding, valency and connectivity had not yet seen the light of day, although a forerunner concept had just emerged. The great Swedish chemist Jons Jacob Berzelius, the dominant authority of the time, conceived organic molecules to consist of two electrostatically matched subunits, for example an alcohol of an alkyl and a hydroxy moiety, exactly as mineral salts are made up from cations and anions. If the organic world really were nothing but a reproduction of the inorganic, should one not be able to identify highly reactive molecular fragments in just the same way as metallic sodium and elementary chlorine had been obtained from sodium chloride? Indeed, there was a general consensus that the isolation of organic radicals in the free state would constitute compelling evidence for the correctness of Berzelius's 'electrostatic theory,' which postulated a 'dualistic' constitution of *all* matter. Unfortunately, all experiments undertaken in this respect so far had failed.

Bunsen was the man to create a sensation. He treated cacodyl oxide with concentrated hydrochloric acid to obtain cacodyl chloride (dimethylarsanyl chloride). This substance was painstakingly purified and filled into glass ampoules which contained activated zinc sheets and which had beforehand been purged with carbon dioxide before they were flame sealed and heated on a water-bath. After 3 h, the vessel was opened, the zinc chloride formed extracted and the remaining oil collected. When exposed to air, the compound spontaneously ignited. It was, as we know, tetramethyldiarsane. Bunsen, however, believed he had isolated the free cacodyl radical, and many of his contemporaries shared his conviction.

$$(H_3C)_2As{-}O{-}As(CH_3)_2 \xrightarrow{HCl} (H_3C)_2As{-}Cl \xrightarrow{Zn} \left((H_3C)_2As\cdot\right) \longrightarrow (H_3C)_2As{-}As(CH_3)_2$$

Otherwise feared as a merciless critic, Berzelius wholeheartedly praised Bunsen. The latter moved as an associate (1839) and later full professor (1841) to Marburg before, after a short interlude in Breslau, he climbed in 1852 to the then most prestigious position in Heidelberg. His rising renown attracted scholars from all over the world. One of them, Edward Frankland, hoped to obtain the free ethyl radical by applying his mentor's method, i.e. reduction with zinc to ethyl iodide. He prepared the first organozinc compounds instead [6]. A few years later, John A. Wanklyn, another English research fellow in Bunsen's laboratory, treated dialkylzincs with sodium and produced the first, though ill-defined, organosodium species [7].

$$H_5C_2\text{-}I \xrightarrow{Na} H_5C_2\text{-}Na$$

$$H_5C_2\text{-}I \xrightarrow{Zn} H_5C_2\text{-}ZnI$$

$$(H_5C_2)_2Zn \xrightarrow{Na} H_5C_2\text{-}Na$$

The work of Wanklyn was followed up by Paul Schorigin (from 1906) and Wilhelm Schlenk (from 1922). Frankland inspired Sergei Nikolaijevitch Reformatzky (from 1887), whose α-bromo ester-derived zinc enolates became the first synthetically useful organometallic reagents. Philippe Barbier (from 1899) and Victor Grignard (from 1900) changed from the organozinc to the organomagnesium series. The latter compounds, being universally accessible and at the same time more reactive, soon became favourite tools for synthesis. It still needed the ingenuity and perseverance of pioneers such as Morris S. Kharasch, Avery A. Morton, Henry Gilman, Georg Wittig and Karl Ziegler, who systematically developed, complemented and refined the methods that became the foundations of the organometallic approach to synthesis. The seeds planted by such outstanding researchers in the 1930s and 1940s began to grow in the years after the Second World War. A few decades later, organic synthesis has undergone a complete metamorphosis. The genealogy of polar organometallic chemistry is outlined in Figure 1.1.

What is so special and is so novel about the organometallic approach to organic synthesis? To gain a perspective, let us first consider the most classical and typical organic reaction, the S_N2 process (route a_2 in the scheme overleaf): a standard nucleophile M—*Nu* (iodide, cyanide, enolate, etc.) encounters in a bimolecular collision a suitable carboelectrophile —C—X from the rear with respect to the nucleofugal leaving group X (halide, sulfonate, etc.) and promotes the replacement of the latter under Walden inversion of configuration. In general, such a type of nucleophilic substitution proceeds with great ease if the carboelectrophilic substrate has a methyl or primary alkyl group as the organic moiety. With a *sec*-alkyl carbon as the reaction centre, generally only a poor yield is obtained. *tert*-Alkyl, 1-alkenyl, 1-alkynyl, aryl, hetaryl and cyclopropyl halides or sulfonates are completely S_N2 inactive. In other words, the concerted bimolecular nucleophilic substitution has a fairly narrow scope of applicability.

Organometallic chemistry now offers a way out of this dilemma. Say we wish to convert an aryl bromide into the corresponding benzoic acid. It would be more than naive to treat it with potassium cyanide in the hope of obtaining in this way the aryl nitrile which subsequently could be hydrolysed. However, if it is allowed to react with magnesium or lithium (butyllithium may also be used advantageously, as we shall see

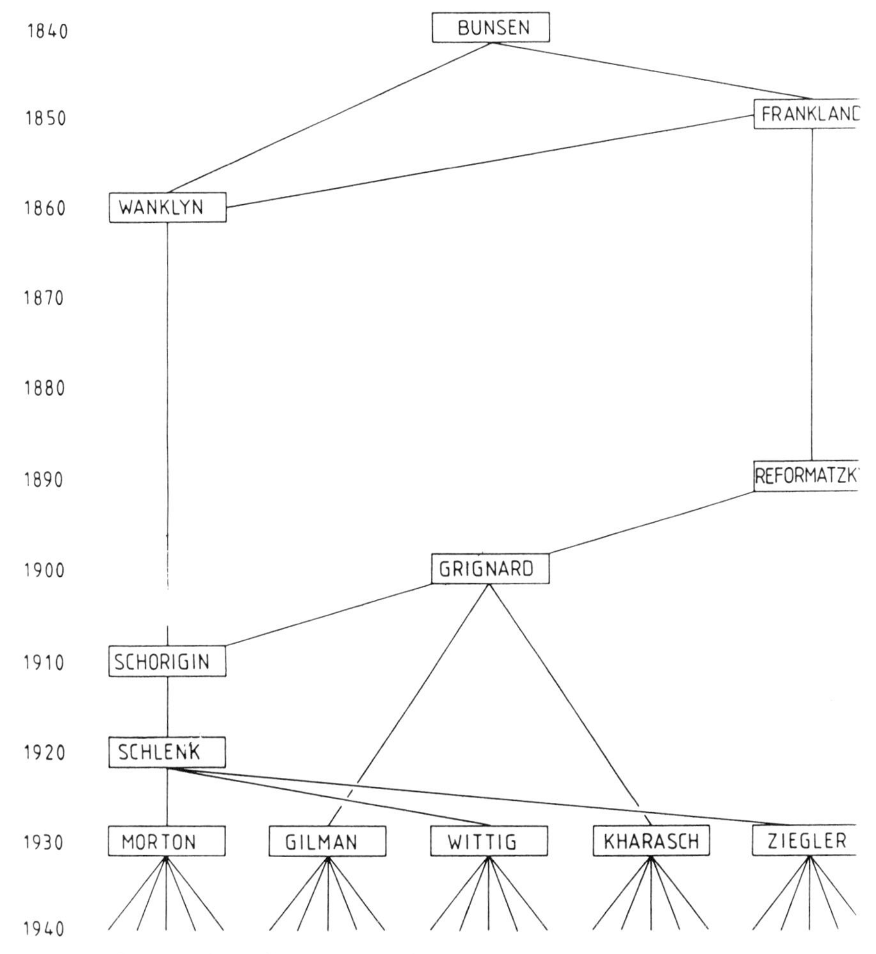

Figure 1.1. Genealogy of polar organometallic chemistry

later), a highly reactive organometallic species is formed (route b), which can be readily condensed with cyanogen bromide or cyanogen to give the expected nitrile (route c_2). Alternatively, the carboxylic acid may be directly obtained by nucleophilic addition of the organometallic intermediate to carbon dioxide.

Switching from a halide to an organometallic compound implies an *Umpolung* [8]. The term was coined by Wittig and later popularized by Seebach. By the reversal of the familiar product polarities, the experimentalist is rewarded with a second option: he may now select from a fresh set of electrophiles *El*—X his preferred building block, the choice again being embarrassingly large. It is like reshuffling a pack of cards and starting the game over again.

The reductive or permutational replacement of halogen by a metal M (route b) can hardly fail and subsequent electrophilic replacement of the metal (route c_2) rarely causes trouble. On the other hand, the selective access to the required starting material (route a_1) may give us a true headache. In general, neither such a halide nor the corresponding alcohol exists as such in nature. Ultimately, most of them have to be prepared from a suitable saturated or unsaturated hydrocarbon. The site-selective introduction of the heteroatom is far from being a trivial task. There is a great temptation to try a short-cut and to convert the hydrocarbon directly into an organometallic intermediate (route c_1).

Obviously, this is easier said than done. Simple hydrocarbons are only minimally acidic and hence extremely reluctant to undergo deprotonation. The second obstacle is that several or even many different hydrogen atoms, not just one, are present in ordinary substrates (omitting the few high-symmetry structures known, such as benzene, neo-

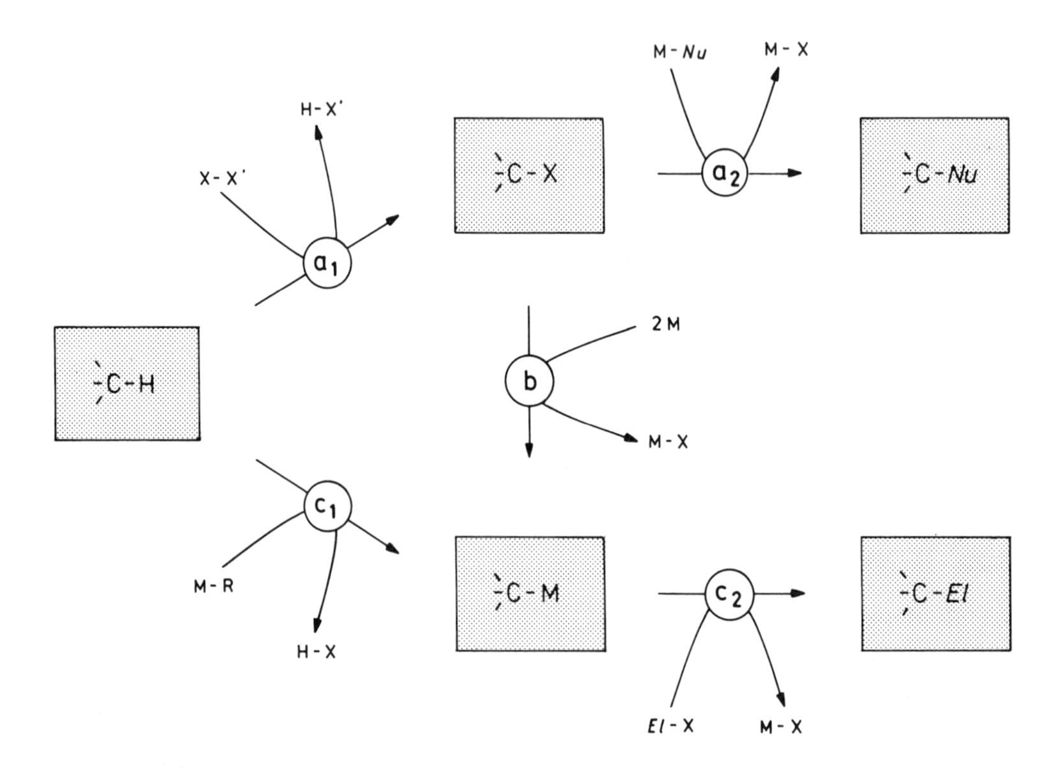

pentane and cyclohexane). We have to learn how to exploit subtle differences in their environment in order to discriminate between them chemically. In concrete terms this means that we have to conceive metallating reagents that unite two seemingly incompatible reactivity profiles: maximum reactivity and maximum selectivity. A substantial part of this contribution will be devoted to this fascinating problem. It will be shown how the two antagonistic features can be reconciled and what impressive practical results can be achieved in this way.

In summary, the present contribution assigns a higher priority to the *generation* of organometallic reagents than to their transformation, their reactivity or other properties. When we consider the various possibilities that exist for the preparation of key organometallic intermediates, emphasis will be put on the most direct method, permutational hydrogen–metal exchange.

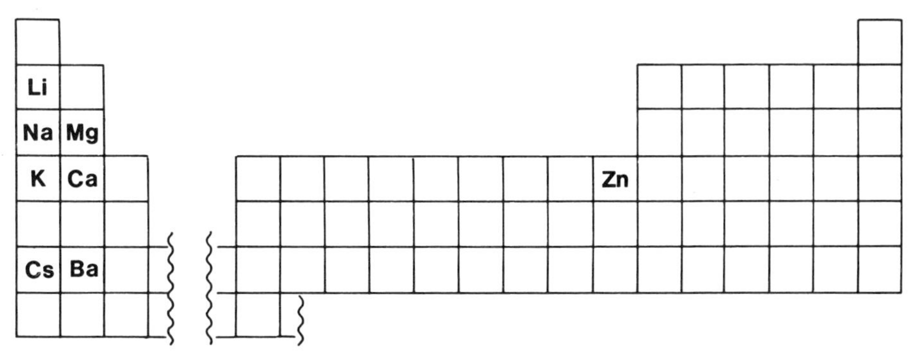

Figure 1.2. A Periodic Table featuring the major elements used as constituents of 'polar' organometallic reagents

The coverage will be restricted to the so-called *polar organometallics* [9]. This class of compounds includes the derivatives of the most common alkali and alkaline earth metals and also organozinc compounds, which deserve to be assimilated with their organomagnesium analogues (Figure 1.2). Although all of these species are characterized by a polar carbon–metal bond, the degree and pattern of polarity vary significantly with the element. While nucleophilicity and basicity are the absolutely dominant features of organic derivatives of potassium, caesium and barium, the reactions of lithium, magnesium and zinc compounds are, in increasing order, triggered by the electrophilicity (Lewis acidity) of the metal. As will be stressed over and over again, the individuality of the metal involved is the most crucial parameter for orienting and 'fine-tuning' organometallic reactions.

This contribution neglects organic derivatives of transition elements, no matter whether produced in a stoichiometric reaction or as a transient species in a catalytic cycle. Of course, one cannot ignore either the interdependence or the complementarity of main group and transition block chemistry in modern organic synthesis. For example, organotitanium and organocopper compounds are most conveniently prepared through organolithium or organomagnesium reagents. Organocopper and organopalladium compounds allow us to escape completely from the S_N2 restrictions even in such cases as when, despite *Umpolung*, no suitable electrophile can be matched with a polar organometallic reagent. The area of transition element chemistry (like that of boron, aluminium and tin or the technical implications of lithium reagents) will be treated in the following chapters by somc of the most competent workers in the field.

1.2 STRUCTURES

Two decades ago, the term 'polar organometallics' [1] was coined as a common designation for the organic derivatives of magnesium, zinc, lithium, sodium and potassium. While their first appearance was inconspicuous, they have in the meantime become key reagents for modern organic synthesis. At first sight they are all endowed with the same high basicity and nucleophilicity. Such behaviour is readily compatible with a bond model in which the metal-bearing carbon carries a negative charge and the inorganic counterpart is a cation. Therefore, such reactive intermediates have frequently been called 'carbanions' [2]. This primitive description was very helpful indeed when, in the years after the Second World War, Wittig and other pioneers began to popularize the rapidly developing branch of organometallic chemistry. Nevertheless, the conceptual reduction of organometallic species to carbanions is an oversimplification which must lead to misjudgments. In fact, no difference in the reactivity pattern of given organometallic reagents can be rationalized unless the metal and its specific interactions with the accompanying carbon backbone, the surrounding solvent and the substrate of the reaction are explicitly taken into account.

In other words, in order to understand *reactivity* we need a detailed knowledge of the *structures* involved. Only if we have a realistic idea about the nature of an organometallic bond can we dare to predict what changes it may undergo under the influence of a suitable reaction partner until a thermodynamically more stable entity will emerge from such a molecular reorganization. For this reason, this section is devoted to the unique architecture of polar organometallic compounds.

Many of us may have heard in high school that metals want to get rid of their valence electrons in order to become cations having the same electron shell as noble gases. Although common, this belief is entirely wrong, of course. We just have to look up the ionization potentials of monoatomic metals in order to convince ourselves of the contrary. Stripping off an electron from even the most 'electropositive' metals requires energies of 124 (lithium), 118 (sodium), 100 (potassium), 98 (rubidium) and 90 (caesium) kcal mol^{-1} (1 kcal = 4.187 kJ) [3]. Why, then, should a metal want to become a cation? Actually it does not. As with most other elements also, it merely seeks to form a chemical *compound*. In the case of metals this is, however, not a trivial matter.

Let us consider a metal atom, say lithium, in the gas phase. When it is allowed to combine with another radical, binding energy is gained. If a lithium chloride molecule is formed, this amounts to 112 kcal mol^{-1}. In order to evaluate its heat of formation from the elements, one now has to deduct half of the dissociation energy of elemental chlorine (58 kcal mol^{-1}) and, in addition, the heat of fusion, vaporization and atomization of bulk lithium (all together some 70 kcal mol^{-1}). In other words, the formation of lithium chloride from the elements is an endothermic process as long as monomeric species are produced in the gas phase. This conclusion holds also for other alkali metal halides. They become thermodynamically strongly favoured only in the solid state. What structural features make the crystal lattice so advantageous?

The answer can be found in any textbook of general chemistry. Owing to the dense packing of the ball-like ions, each of them is surrounded by several, e.g. six, counter ions. In this way, the attraction between oppositely charged particles increases steeply while the repulsion of ions having the same sign remains moderate because of the longer internuclear distances. The electrostatic model can be used to describe smaller molecular packages such as aggregates or clusters. As an extension, solvation may be portrayed as a charge–dipole interaction.

Although nothing is wrong with these ideas, we wish to adopt a different point of view or, perhaps to put it more properly, to use a different vocabulary. We prefer to conceive of the sodium chloride crystal being held together by 'partial bonds' (electron-deficient bonds) rather than 'ionic bonds'. This approach has the merit of universality; it creates a continuum of binding interactions spanning from perfectly covalent to highly polar bonds as the extremes.

The reader may immediately wish to object. How can somebody dare to deny the ionic nature of the sodium chloride crystal? Has it not been proved that within the radii of 0.95 and 1.81 Å around sodium and chloride nuclei precisely 10 and 18 electrons, respectively, can be found and that, most revealingly, the electron density between such spheres drops to zero? [4] Yet, electron population analyses may be fallacious [5]. A lithium atom in the gas phase would have to be visualized as an 'electride' if treated in the same way as solid lithium fluoride: at a 0.65 Å distance from the nucleus the electron density is zero; inside this 'ion radius' confinement we find just two 1s electrons, but outside is the 2s valence electron! [6, 7]

To feel reassured, the reader should not forget how imperfect any description of the physical reality using man-made equations, images and terminology notoriously is; hence the liberty, if not necessity, to choose in a given case the model which appears to be the most appropriate one. For example, the τ ('banana bond') and the σ–π depiction of ethylene are equivalent. Depending on the situation, one or the other representation may be more suitable. In the same way, it is a matter of convenience whether to deal with metal derivatives in terms of ionic and partial bonds. Since all molecular ensembles are tied together by the electrostatic attraction between protons and electrons, ultimately both models must converge and lead to the same conclusions.

With all these precautions in mind, we shall now attempt to rationalize the most characteristic features of metal binding, in particular aggregation and coordination. Our selection of examples is focused entirely on carbon–metal compounds, although metal amides, metal alkoxides or aroxides and metal halides exhibit the same structural peculiarities.

1.2.1 FIRST-ORDER PRINCIPLES OF METAL COORDINATION

Subunits can assemble to afford larger chemical entities when the attractive forces overcompensate the repulsive forces. In order to minimize repulsion, electrons must be able to spread out over a maximum of space. In order to maximize attraction, the electrons have to approach the atomic nuclei as closely as possible. These seemingly contradictory requirements can best be reconciled if the electrons surround the atomic nuclei spherically. Within a diatomic molecule (**1**) such as lithium chloride, electrons will inevitably concentrate between the two poles of attraction, the metal and halogen nuclei, and thus create an energetically unfavourable asymmetric charge distribution. In a crystal lattice (**2**), however, each metal atom has four, six or eight heteroatomic neighbours with which to share the valence electrons. The phenomenon of solvation can be treated in a similar manner. As we have already seen, a naked metal ion (**3**) is extremely energetic if kept in empty space. It can, however, be efficiently stabilized by solvent molecules (*Sv*, **4**) or other ligands which surround it spherically and share their non-bonding electrons with the metal.

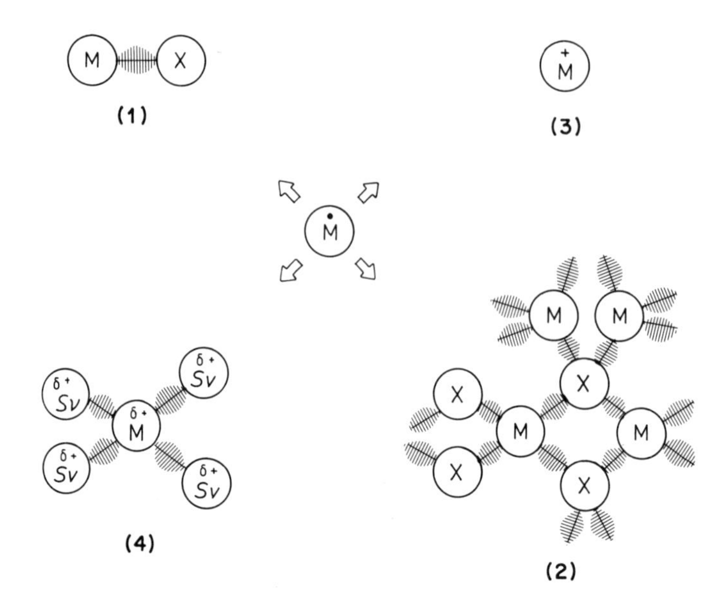

The typical coordination number of lithium [8], magnesium [9] and zinc [10] is four and of larger cations such as sodium [11] or potassium [12] six. Special effects, in particular steric hindrance, may deplete the coordination sphere and reduce it to two ligands (e.g. lithium perchlorate dietherate, **5** [13]), bis[tris(trimethylsilyl)methyl]magnesium [9], while with crown ether- or cryptand-like complexands exceptionally high coordination numbers of six (lithium tetrafluorborate bis[*cis,cis*-1,2;3,4;5,6-triepoxycyclohexane], **6** [14]), magnesium dibromide tetrakis(tetrahydrifuran) dihydrate [15]), seven (sodium perchlorate benzo-15-crown-5 [16]), potassium 2-phenylethenolate 18-crown-6 [17]), eight (diethylmagnesium 18-crown-6 [18], diethylzinc 18-crown-6 [18], sodium and potassium thiocyanate nonactin [19]), potassium ethyl acetoacetate 18-crown-6 [20]) and ten (barium bis[1,1,1,5,5,5-hexafluoro-1,3-pentanedione anion] 18-crown-6, **7** [21]).

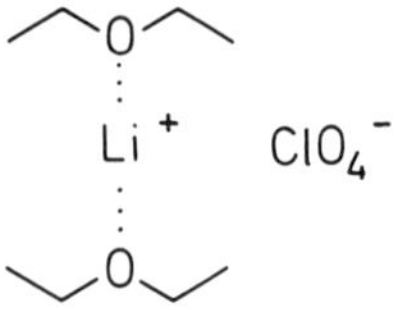

Li^+ ClO_4^-

(5)

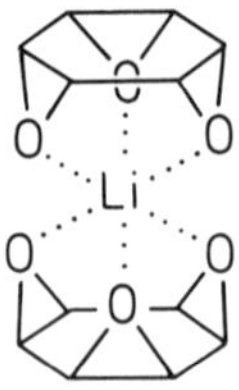

(6)

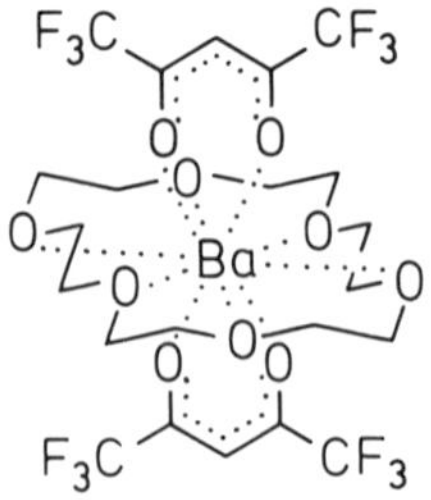

(7)

1.2.2 AGGREGATION IN THE GAS PHASE, IN THE CRYSTALLINE STATE AND IN SOLUTION

The fascinating structures of organometallic compounds become immediately intelligible if we keep in mind that metals in organic as in inorganic derivatives always strive for maximum coordination numbers. Partial or electron-deficient bonds are the instruments with which this goal is materialized. This trick allows organometallic species to cluster into oligomeric or polymeric superstructures. These so-called aggregates exist in the vapour phase, in solution as in the solid state.

1.2.2.1 Gas-Phase Structures

The prototype of all electron-deficient bonds is the double hydrogen bridge that keeps the diborane molecule together. The monomeric borane, being isoelectronic with the methyl cation, has only six electrons in the valence sphere. In order to attain tetracoordination, the metalloid atom abandons one ordinary boron–hydrogen bond in favour of two electron-deficient bonds, which then allow two BH_3 units to be combined.

Although only a total of four electrons is available to construct the four bridging B—H bonds, the resulting dimer (**8**) is thermodynamically very stable, the aggregation enthalpy amounting to 35 kcal mol^{-1} [22]. Boron–hydrogen, boron–boron and, where appropriate, boron–carbon electron-deficient bonds are also at the origin of the stability of polyboranes [23], such as pentaborane (**9**), and carboranes [23], such as the 1,2-carborane (**10**); open and filled circles represent BH and CH groups, respectively.

Unlike dimethylborane (**11**), trimethylborane is no longer capable of dimerization [24]. This is to some extent a consequence of the inferior strength of boron–carbon compared with boron–hydrogen electron-deficient bonds. The increased steric hindrance is another crucial, even determining, factor. Despite their substantially longer

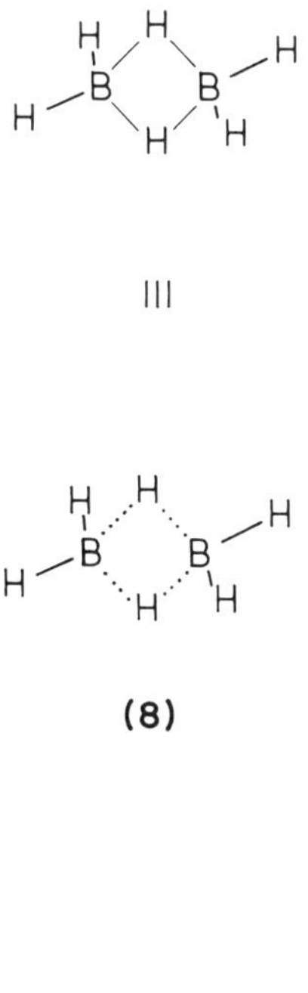

(8) 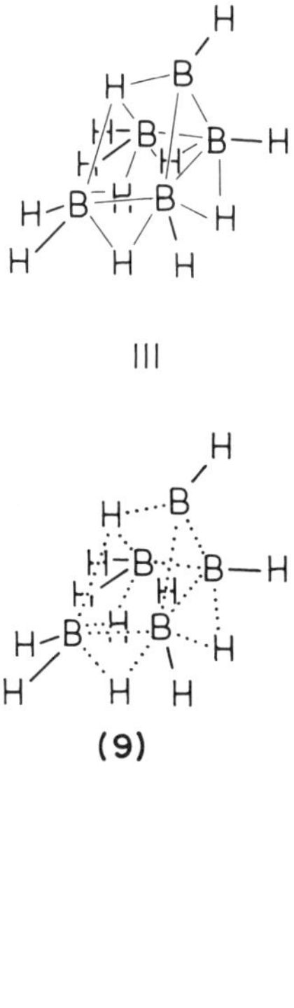 (9) 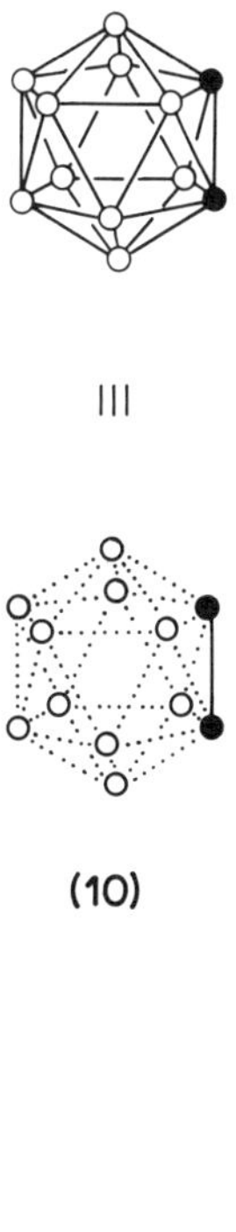 (10)

metal–carbon bonds, triisopropylaluminium and trineopentylaluminium are also monomeric [25]. Unbranched trialkylaluminiums can better accommodate two bridging carbon moieties and therefore dimerize. The binding forces being relatively moderate ($\Delta H^\circ = 20\ \text{kcal mol}^{-1}$) [26], monomers and dimers of trimethylaluminium (**12**) are concomitantly observed in the gas phase [27], whereas benzene solutions contain only the dimer [27]. An X-ray diffraction crystal analysis [28] shows the electron-deficient Al—C bonds to be markedly longer (2.24 Å) than the ordinary ones (2.00 Å). The exocyclic C—Al—C angle is found slightly wider (124°) than in the monomer (120°) whereas the endocyclic angle within the bridge approximates tetrahedral geometry (110°). The Al—C—Al angle is compressed to 70°.

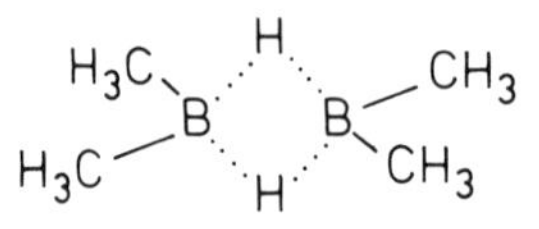

(**11**) 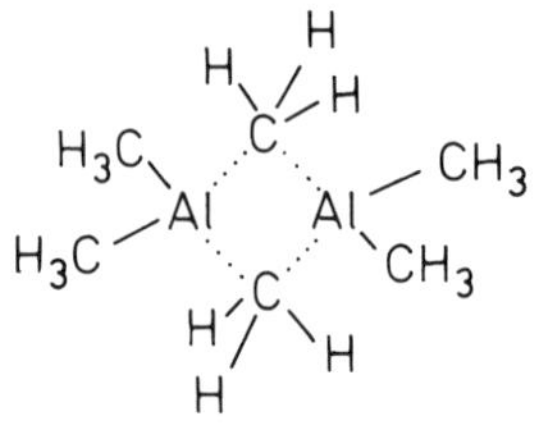 (**12**)

Dimethylberyllium vapour consists of an equilibrium mixture of monomers, dimers (**13**) and trimers (**14**) [29]. On slow condensation colourless needles form. The X-ray analysis of the latter reveals a polymer (**15**) having Be—C—Be—C rhombs alternating in two perpendicular planes. The bond angles around the metal atoms are close to tetrahedral (114°) while the carbon-centred Be—C—Be angles are only 66° wide [30]. Obviously, beryllium has to contribute both of its valencies to the construction of electron-deficient bonds in order to attain tetracoordination (as in **15**). If only one classical metal–carbon bond is sacrificed and converted into two half-bonds, the metal will be placed in a trigonal environment (as in **13**).

For the monovalent methyllithium it should be particularly difficult, if not impossible, to reach tetracoordination. On the other hand, its driving force for oligomer formation

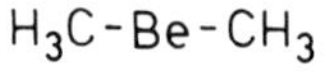

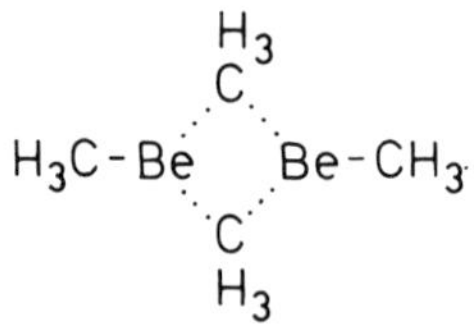

$H_3C\text{-}Be\text{-}CH_3$

(**13**) 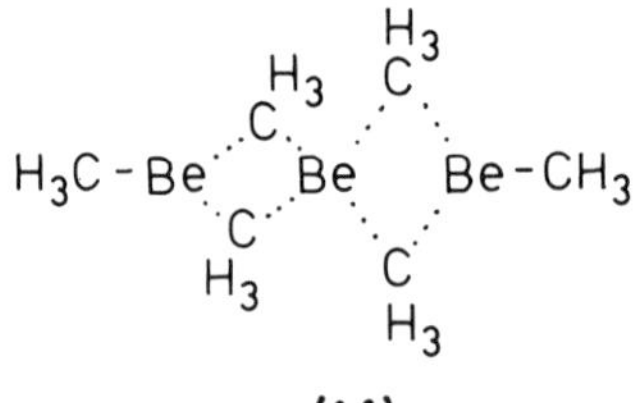 (**14**)

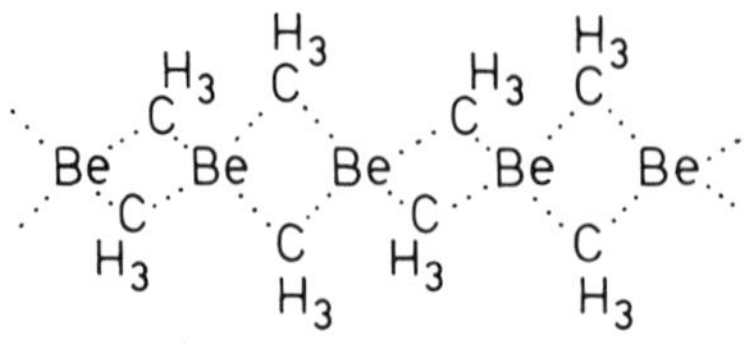

(**15**)

is remarkably high. The lightest alkali metal allows fairly reliable *ab initio* calculations to be performed if large basis sets are used and electron correlation is included. Thus, the aggregation enthalpy was evaluated to approximate 20, 25 and 30 kcal mol^{-1} per $LiCH_3$ unit in the methyllithium dimer, trimer and tetramer, respectively [31]. Actually no monomers, only oligomers, can be observed in the gas phase [32]. The monomer can be generated, of course, but it must be trapped in an argon matrix in order to be preserved [33]. On the other hand, bis(trimethylsilyl)methyllithium was found to be cleanly monomeric in the gas phase, although it is polymeric in the crystalline state [34].

Unfortunately, gas-phase studies of organometallic species are often hampered by practical problems such as insufficient volatility of the sample. Even the most powerful techniques for structure elucidation of vapours, infrared spectrometry and electron diffraction [35], do not always provide unambiguous and conclusive answers. Therefore, the available information is scarce in comparison with crystal and solution data.

1.2.2.2 X-Ray Crystallography

The first-single crystal X-ray analysis of a solvent-free organometallic compound was that of ethyllithium reported in 1963 and refined two decades later [36]. It revealed tetrameric subunits having a core built up from Li_4 tetrahedra with four methyl-bearing methylene moieties each of which is bridging an Li_3 face. Thus, every α-carbon atom is hexacoordinated and every lithium site has three α-carbon nearest neighbours. In addition, however, short Li—H contacts and elongated α-C—H bonds suggest that one of the proximal hydrogen atoms acts as a fourth pseudo-ligand of the metal. Since then, coordination of C—H or B—H bonds to lithium has frequently been observed (e.g., in the polymeric lithium tetramethylborate **16** [37], the monomeric lithium tetrakis(di-*tert*-butylmethyleneimidoaluminate **17** [38] and the dimeric lithium boronate–*N*,*N*,*N'*,*N'*-tetramethylethylenediamine adduct **18** [39]) and the ‘agostic’ C—H—Li interactions have become a popular concept [40].

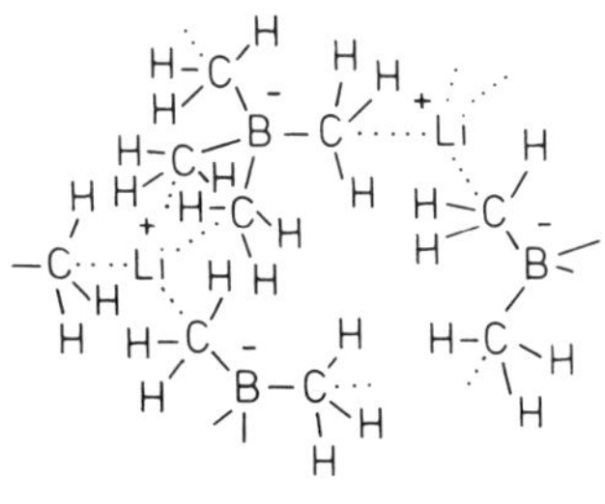

(16) 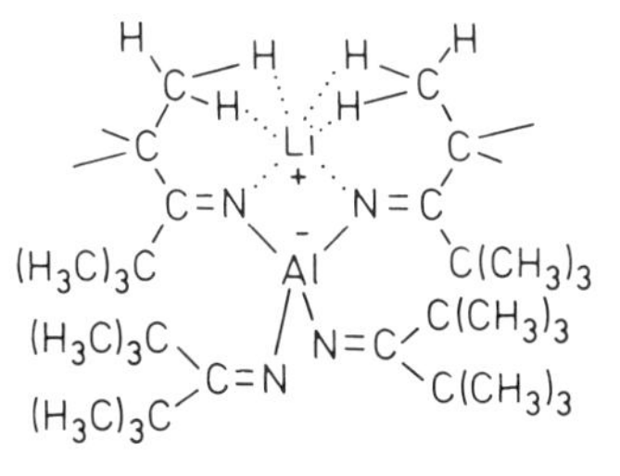 **(17)** 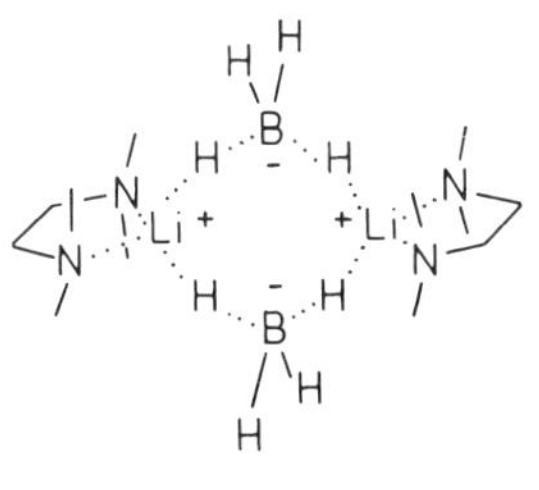 **(18)**

The trimethylsilylmethyllithium (**19**) crystal [41] is composed of hexameric units exhibiting a chair-like arrangement of the metal atoms. The structure may be visualized as a distorted octahedron having two small- and four medium-sized triangular faces which are each bridged by a trimethylsilylmethyl group. Thus, every lithium atom is tricoordinated.

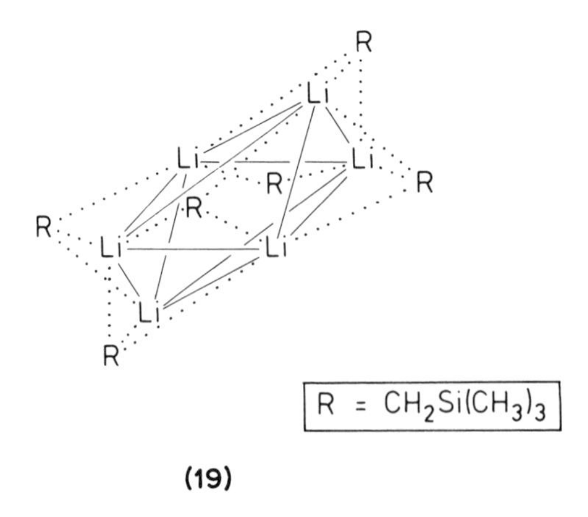

(19)

Methylalkali metal compounds, the smallest organometallics, can only be isolated as powders. Nevertheless, it was possible to collect accurate and meaningful data by submitting the perdeuteriated species to a combination of neutron and synchroton radiation diffraction techniques at 1.5 and 290 K (Figures 1.3–1.5). Methyllithium [42] was found to contain tetrameric units. A carbon and a smaller lithium tetrahedron penetrate each other to form a distorted Li_4C_4 cube with methyl groups adopting a staggered position relative to the adjacent lithium triplet (elementary cell **20**). Methylpotassium [43] shows a monotonously repeating orthorhombic unit cell (**21**) containing

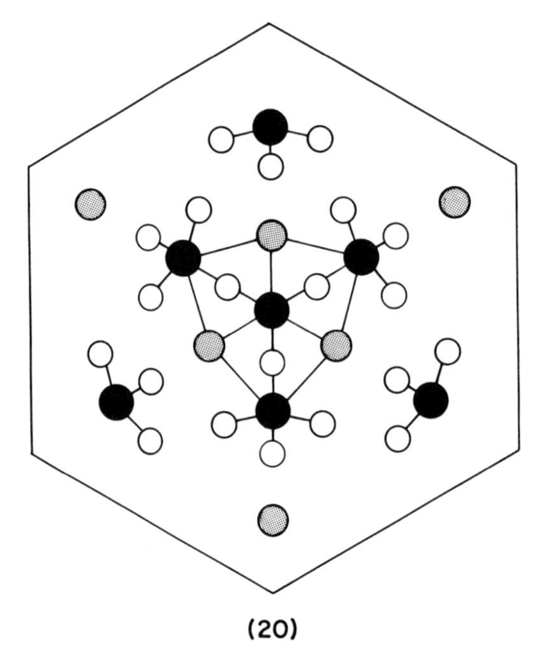

(20)

Figure 1.3. Crystal lattice of methyllithium (**20**)

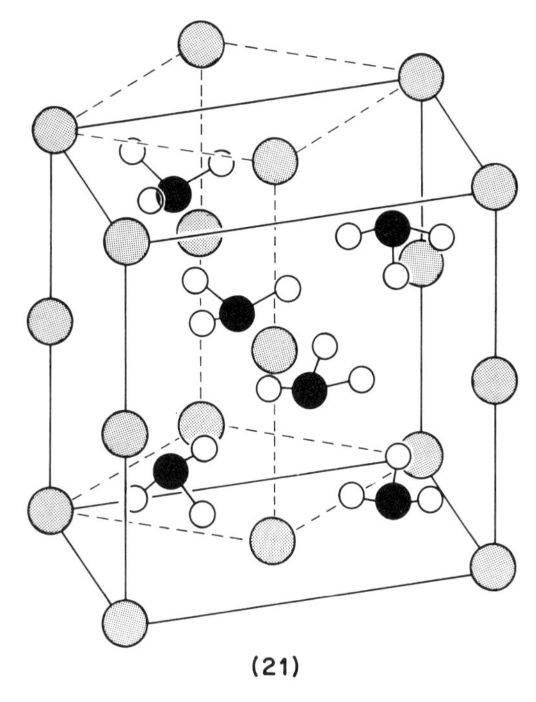

(21)

Figure 1.4. Crystal lattice of methylpotassium (**21**)

four methylpotassium species in a trigonal-prismatic array. The methyl groups have alternating orientations, each one facing with their electron pair three potassium atoms at a close distance (3.0 Å) while turning their hydrogen-covered rear side ($\angle$HCH 105°) to three others at a longer distance (3.3 Å). The structures of methylrubidium [44] and methylcaesium [44] are similar and again resemble an ion lattice of the NiAs type. Methylsodium **22** [45] presents features which are characteristic of the methyllithium and others which are characteristic of the methylpotassium structure. The space group is orthorhombic and contains 16 molecules. Half of them are clustered together to give tetramers which are linked by the remaining 8 molecules arranged in Na zig-zag chains.

Solvent- or ligand-free crystals of organometallic compounds are rare. In general, they incorporate ether or amine molecules as, for example, monomeric phenylmagnesium bromide dietherate (**23**) [46], dimeric ethylmagnesium bromide bis(diisopropyletherate) [47], monomeric bis(pentafluorophenyl)zinc di(tetrahydrofuranate) [48], dimeric mesityllithium di(tetrahydrofuranate) (**24**) [49], dimeric 2,4,6-triisopropylphenyllithium etherate [50], tetrameric 3,3-dimethyl-1-butynyllithium tetrahydrofuranate [51] (**25**) and tetradecameric 3,3-dimethyl-1-butynyllithium [51] (**26**) carrying just four tetrahydrofuran molecules at the cluster periphery or dimeric ethylmagnesium bromide trimethylamine [52] and the phenyllithium–TMEDA [53] and the 1-bicyclobutyllithium–TMEDA [54] 1:1 adducts.

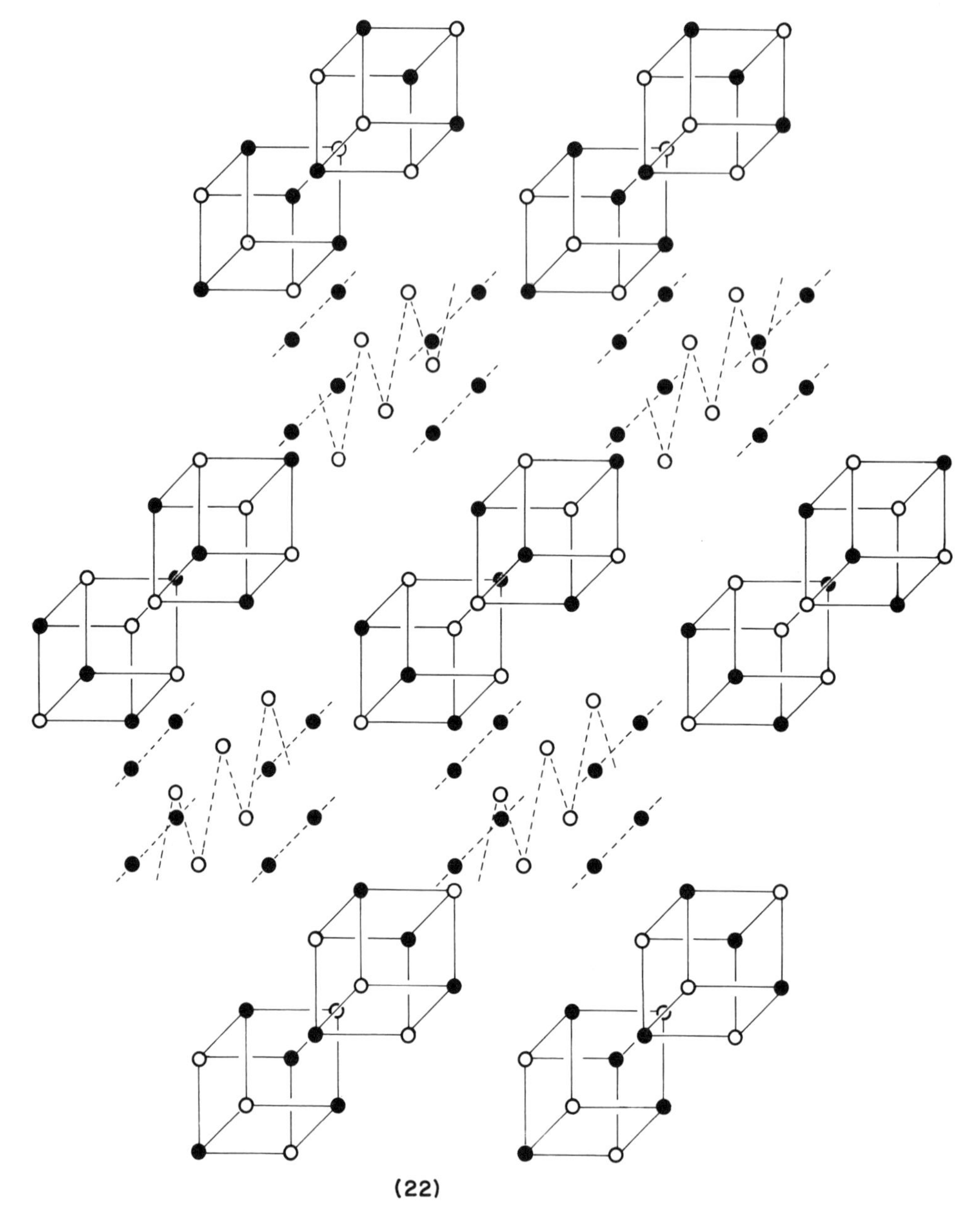

Figure 1.5. Crystal lattice of methylsodium (**22**)

1.2.2.3 Composition in Solution

As Hein and Schramm [55] already recognized in 1930, organolithium compounds tend to form oligomers. In the meantime, numerous systematic investigations have been carried out, the principal techniques of investigation being cryoscopy, ebullioscopy or

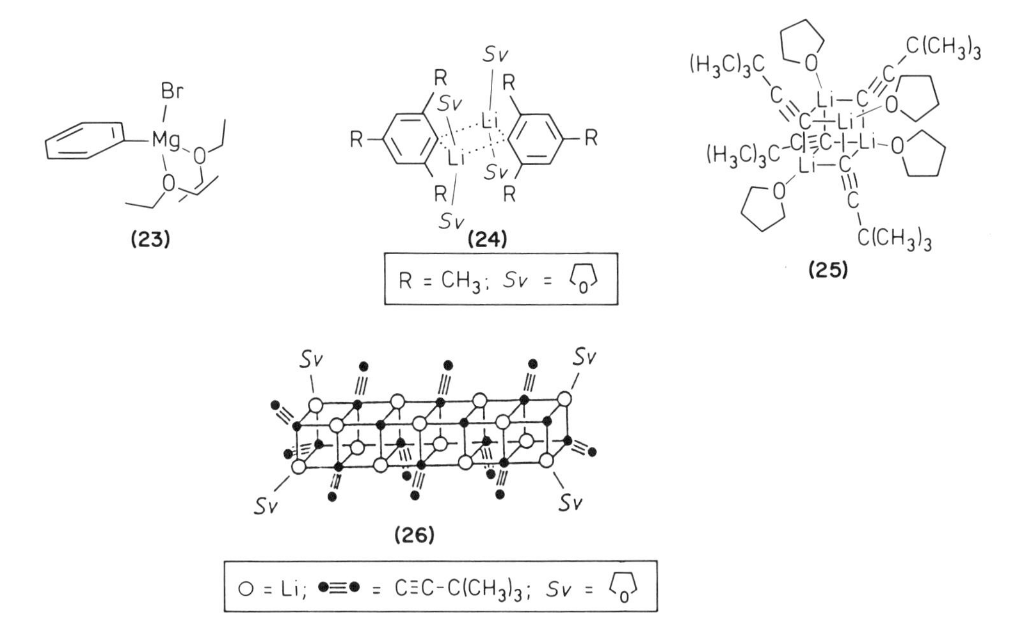

vapour pressure measurements and nuclear magnetic resonance spectrometry. A comparison of existing data reveals a few clear trends:

1. The aggregation number may, although need not always, be higher in paraffinic or aromatic solvents when compared with ethereal solvents. Ethyllithium [56–58] and butyllithium [56–60] are hexameric in hexane, but tetrameric in diethyl ether or tetrahydrofuran; dilute solutions of trimethylsilylmethyllithium in hexane are hexameric but only tetrameric in benzene [58, 61].
2. Steric bulk hinders aggregation. In contrast to primary alkyllithiums, isopropyllithium [58] *sec*-butyllithium [62] and *tert*-butyllithium [58, 63] are not hexameric, but rather tetrameric, in hexane. *tert*-Butyllithium appears to be dimeric in diethyl ether and monomeric in tetrahydrofuran, at least at low temperatures [64].
3. Hybridization towards increasing s-character disfavours aggregation. Whereas, for example, isopropyllithium [58] is always tetrameric in ethers, phenyllithium [65–67] and 3,3-dimethyl-1-butynyllithium [63, 68] exist as equilibrium mixtures in which, depending on temperature and concentrations, dimers or tetramers dominate.
4. Charge delocalization also disfavours aggregation. Polyisoprenyllithium is dimeric in light petroleum and monomeric in diethyl ether or tetrahydrofuran [69]; allyl-type species such as 2-methylallyllithium [70] and 2-methylallylpotassium [71] in tetrahydrofuran are apparently monomeric [72].
5. Low temperatures cause deaggregation. Below −100 °C in tetrahydrofuran, the butyllithium tetramer disintegrates to a large extent to form a dimer whereas the phenyllithium dimer establishes an equilibrium with its monomer [67].

Aggregate statics and dynamics are of prime importance for the reactivity of organometallic reagents. Since, in general, the rate efficient species is the monomer [1] and only occasionally a dimer [73] or higher aggregate fragment, total or partial deaggregation has to take place before a reaction can occur. Organolithium reagents that combine to give particularly tight oligomers are handicapped in this respect. As a monomer, methyllithium should be more basic than phenyllithium by more than 10 pK units (see below). At a 0.01 M concentration in tetrahydrofuran methyllithium is, however, only three times more reactive than phenyllithium and at a 0.5 M concentration it is even of inferior reactivity [65].

Alkyllithium reagents have a high tendency to combine with other organolithium [74], organomagnesium, organozinc [75] and organosodium compounds [76]. The mixed aggregates (heterooligomers) thus obtained are in general thermodynamically more stable than the corresponding 'pure blood' homooligomers. Mixed aggregate formation can promote rapid intracomplex hydrogen–metal exchange processes and thus explain the ease with which certain dimetallation reactions can be brought about [1].

1.3 REACTIVITY AND SELECTIVITY

The way in which a metal atom and a carbon moiety establish a chemical bond between themselves is the contrary of a Romeo and Juliet story! The two elements *can* get together, but actually they do not want to. What is the reason for this lack of affinity? No matter whether we look upon the sodium chloride crystal as a three-dimensional inorganic polymer kept together by electron-deficient bonds or as a cubic lattice of electrostatically attracted cations or anions, we immediately realize how ideally the metal and halide fit together owing to their spherical symmetries and complementary electronegativities. The matching of an alkali or alkaline earth metal with an organic moiety is energetically far less favourable. The two unlike binding partners have to enter into a compromise: the metal has to content itself with an imperfectly designed ligand and low coordination numbers and the carbon counterpart has to tolerate hypervalency and a fractional negative imposed by the high polarity of the organometallic bonds.

Lack of thermodynamic stability means a high reactivity potential. The conversion of an organometallic compound into an essentially covalent hydrocarbon and a salt-like metal derivative will inevitably be accompanied by a substantial gain in free reaction enthalpy and should therefore provide an important driving force for the chemical transformation.

1.3.1 CHEMICAL POTENTIAL OF ORGANOMETALLIC REAGENTS

A crude estimate of reaction enthalpies can be based on a comparison between the bond strengths of starting materials and reaction products. As a compilation of literature data [1, 2] reveals, the homolytic dissociation energies of metal species vary over a wide range of numbers (see Figure 1.6). Metal–metal bonds as present in small clusters [3] or in the bulk state are notoriously weak. At the other end of the scale we find molecular entities formed between metals and electronegative counterparts. Metal halides or metal

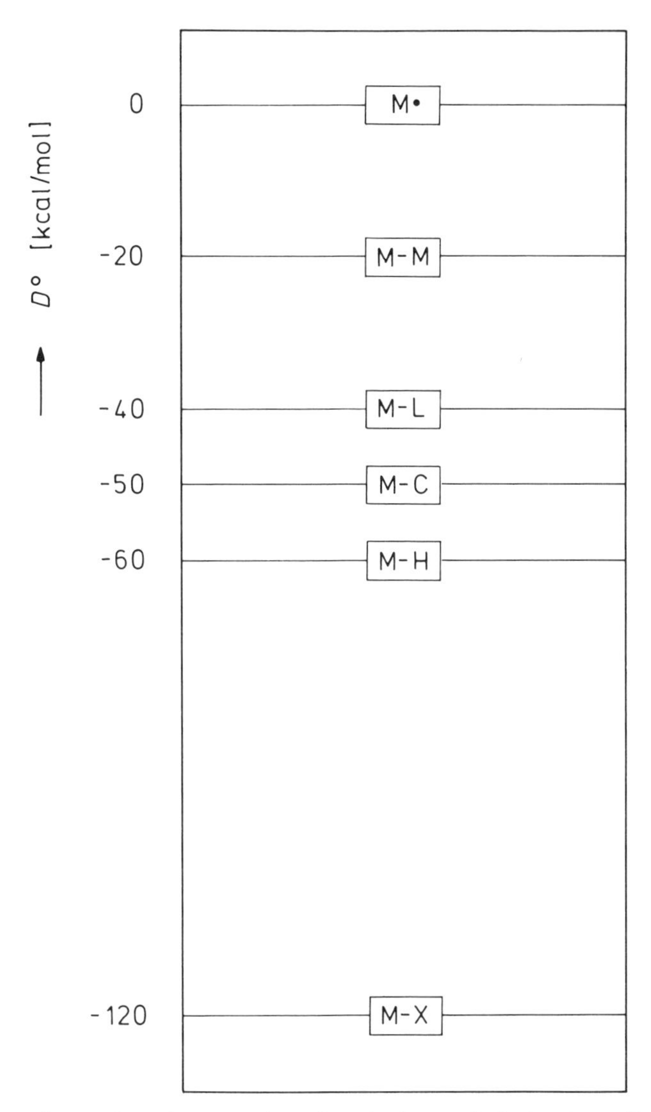

Figure 1.6. Approximate bond strengths of typical metal derivatives (including elemental metal)

alkoxides, for example, are held together by very strong forces. Half-way between these extremes are located metal hydrides, metal-ligand complexes and finally organometallic compounds. The respective bond strengths are moderate and may be compared with that of a carbon–iodine bond.

If we want to select the most suitable reagent for a given transformation, it is not enough to know that organometallics are globally reactive entities. What we need is a kind of a yardstick that will allow us to evaluate the *relative* chemical potential of any individual organometallic reagent. A simple, non-computational method would be most welcome provided that it leads to predictions that are at least semiquantitatively reliable.

Our attempt to solve this problem begins with the wrong assumption that naked carbanions were involved in our organometallic reactions. Then we would expect the reactivity roughly to parallel the proton affinity of those intermediates. Since in reality we deal with organometallic contact species, we have to find out how and to what extent the metal counterpart attenuates the carbanion basicity. This can easily be done if we just consider that the polar organometallic bond to be a hybrid between a covalently bonded and an ionic non-bonded limiting structure. With a highly electropositive metal such as potassium the charge-separated resonance form will become preponderant and the organometallic compound will conserve most of the carbanionic reactivity. At the other extreme, a weakly electropositive metal such as mercury will favour the homeopolar resonance form and make the reagent almost as inert as a pure hydrocarbon. We may even try to assign 'polarity coefficients' to various metals and metalloids: hydrogen (as the standard) 0%, lithium 35%, sodium 50%, potassium 65%, caesium 75%, and so on [4]. Although these numbers are arbitrarily chosen, they can serve to illustrate the principle.

$$\overset{\delta+}{M} \blacktriangleleft \overset{\delta-}{C}{<} \quad \equiv \quad M{-}C{<} \quad \leftrightarrow \quad \overset{+}{M} \quad :\overset{-}{C}{<}$$

The reactivity potential of an organometallic reagent not only depends on the metal but also on the organic moiety. If, on the basis of our polarity coefficients, we postulate methyllithium to have retained 35% of the methyl anion basicity, we do not know yet what chemical potential to attribute to, say, vinyllithium. To answer this question we have to know the reactivity of the vinyl anion and, in a wider sense, of all typical carbanions relative to the methyl anion. A very suitable measure is gas-phase proton affinities, which have recently become available for numerous anions [3]. The deprotonation of ethylene turns out to be less endothermic than that of methane by *ca* 10 kcal mol^{-1}. This difference may shrink to *ca* 3 kcal mol^{-1} if the corresponding lithium compounds are compared.

The chemical potential of typical organometallic compounds can now be represented as a function of both the metal and the organic counterpart in a simple graph (Figure 1.7) in which carbanion basicities are plotted against metal-dependent polarity coefficients. When the metal is replaced with hydrogen, the polarity and at the same time the chemical potential of all species vanish.

A two-parameter set can hardly cope with all the complexity of organometallic reagents and reactions. Hence it is obvious that the fit with experimental results will be approximate at best. But does one not have to challenge the approach as such? Let us just examine the most important of possible objections.

1. How can one hope to correlate 'basicity' via an ill-defined 'chemical potential' with 'reactivity'? First of all, this has already been attempted before. In the Brønsted equation [5], proton transfer equilibrium constants are correlated with the corresponding transfer rates. The problem with organometallic reactions is that not only are protons (or other electrophiles) transferred, but also metal atoms which are

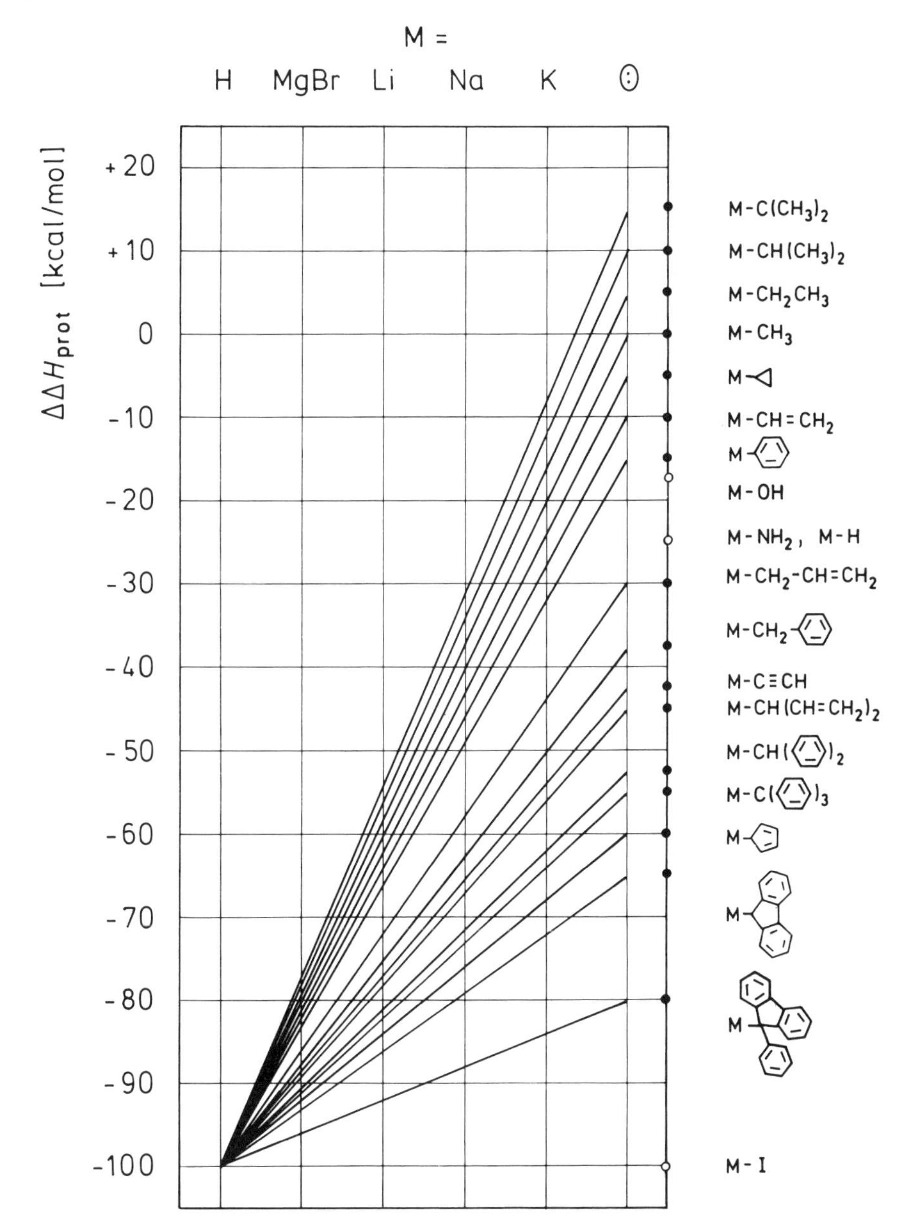

Figure 1.7. Gas-phase basicities of free carbanions (listed on the right) and their attenuation as a function of the metal electropositivity

delivered in the countercurrent sense. This may require extremely complicated mechanisms. Nevertheless, a proportionality between reactivity and basicity appears probable as long as acid–base-type processes or reversible reactions are investigated.

2. Why are only straight lines and no curves drawn in the diagram? Does a given metal really affect the carbanion basicity always in the same way irrespective of the nature

of the organic moiety? Certainly not, especially as organometallic compounds exist in a variety of structural patterns which differ in the aggregation state (monomers, oligomers, polymers) and the coordination type (η^1, η^3, η^5 bonding). As already emphasized, however, we seek simplicity and generality even at the expense of accuracy. As will be demonstrated later (Section 1.3.2), the approach outlined above allows one to rationalize many fundamental findings in a qualitative and even semiquantitative manner.

3. What makes gas-phase data more meaningful than dissociation constants obtained in or extrapolated to aqueous media? In protic media, hydrogen bonding may completely override intrinsic carbanion stabilities. In aqueous solution, for example, the enormous hydration enthalpy [6] of the hydroxy anion makes water (pK_a 15.6) considerably more acidic than, say, toluene ($pK \geqslant 41$). The gas-phase acidities are reversed: the deprotonation of water is more endothermic than that of toluene by *ca* 12 kcal mol^{-1}! Hydroxide or alkoxide anions are strong and amines moderately strong whereas carbanions are relatively weak hydrogen bond acceptors. With the last class of species it makes a substantial difference, however, whether the negative charge is localized at just one carbon atom or spread out over an entire area of unsaturation. Although the conjugatively stabilized [50] anions of the allyl and benzyl type may simultaneously interact with a multitude of protic solvent molecules, the total gain in binding energy is small because of the low local charge density. Geometrically stabilized [50] carbanions of the methyl, cyclopropyl, vinyl or acetylide type can form much stronger hydrogen bonds and hence their basicity is far more sensitive to solvent effects. The pK_a values of acetylene and fluorene differ by only 5 units in aqueous or alcoholic solution whereas the gas-phase acidities are at least 15 units apart ($\Delta\Delta H^{\circ}_{deprot} \geqslant 20$ kcal mol^{-1}). In ethereal or paraffinic solvents, the standard reaction media for strongly basic organometallic intermediates, no hydrogen bonding can occur. Therefore, the relative thermodynamic stabilities of such reactive species should be better reflected by gas-phase rather than aqueous solution data.

1.3.2 METAL EFFECTS ON REACTION ENTHALPIES

Our key postulate is that a metal counter ion inevitably attenuates the differences in free carbanion basicities. The effect should be the more pronounced the less electropositive the metal is. Let us present some experimental evidence for the correctness of this affirmation.

The first example features a set of irreversible reactions. More than a decade ago, carefully measured heats of neutralization were reported for typical organomagnesium [7] and organolithium [8] compounds (see Table 1.1). If these species had not retained any carbanion character, the reaction enthalpies should be virtually the same. On the other hand, if they were completely ionized, the differences in the heats of neutralization would not depend on the metal and would be identical with the corresponding gas-phase data. As the comparison demonstrates (see Table 1.1), the difference in the thermo-

Table 1.1. Relative heats of neutralization (kcal mol^{-1}) as measured for representative organomagnesium [7] and organolithium [8] compounds in comparison with the proton affinities [3] of the corresponding carbanions in the gas phase

R	M = MgBr	M = Li	M = :
$C(CH_3)_3$	+9	+13	+15
$CH(CH_3)_2$	+6	+9	+10
$CH_2C_3H_7$	+3	+7	+5
CH_3	0	0	0
C_6H_5	−3	−3	−15
$CH_2CH{=}CH_2$	−5	−4	−30
$CH_2C_6H_5$	−5	—[a]	−38

[a] Not determined.

dynamic stabilities of a pair of carbanions is two or three times larger than that of their magnesium and lithium derivatives.

A strict proportionality between the three series was, of course, not to be expected. For instance, methyllithium forms a particularly tight tetrameric aggregate (see Section 1.2) and hence is far less basic than it would if it were a monomer as most of the other species are.

The metal effect can also be probed under the conditions of a reversible reaction. The hydrogen–metal exchange between a variety of toluene derivatives and benzyl-type organometallics **1** in tetrahydrofuran solution was followed kinetically until equilibrium was attained by approaching it from both sides. If potassium was involved as the metal, the equilibrium positions were significantly more extreme than in the analogous reactions with organolithium compounds. A slope of $pK^{M=Li}/pK^{M=Li} = 0.7$ was obtained when the metal-specific equilibrium constants were plotted against each other (see Figure 1.8) [9, 60].

The investigation of an intramolecular reaction has provided a most elegant and conclusive illustration of how metals can effect equilibria positions. In general, the so-called cyclopropylcarbinyl–homoallyl rearrangement operates in one direction, producing the ring-opened product exclusively. This holds not only for cyclopropylmethyl-type cations [10] and radicals [11] but also for simple organometallic species such as

(1)

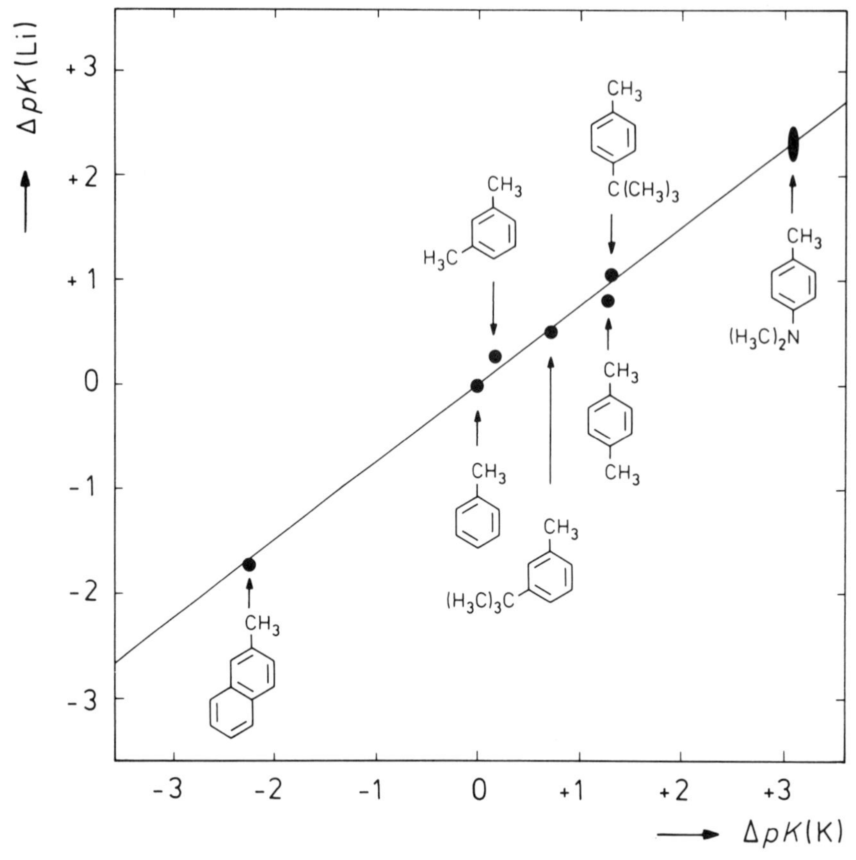

Figure 1.8. Proton and metal transfer between two benzyl-type moieties: metal effect on the equilibrium constants [10, 11]

cyclopropylmethylmagnesium bromide [12] or cyclopropylmethyllithium [13]. Also cyclopropyldiphenylmethylmagnesium bromide [14] (**2**-MgBr), as soon as generated, is instantaneously converted into 4,4-diphenyl-3-butenylmagnesium bromide (**3**-MgBr). In contrast, the reaction takes place in the opposite sense if the corresponding potassium compounds are studied. The reaction is self-monitoring: the colourless 4,4-diphenyl-3-butenylpotassium (**3**-K) disappears rapidly to re-emerge as the bright red cyclopropyldiphenylmethylpotassium (**2**-K) [14].

How can one rationalize this metal-mediated reversal of relative stabilities? Let us

M···C (2) ⇄ (3) C=C(H)–CH2–CH2–M

M = MgBr (→)

M = K (←)

$\delta+$ M, $\delta-$ CH_2

(2) **(3)**

first consider the extreme cases, the hydrocarbons **2** and **3** (M = H) and the carbanions **2** and **3** (M = **:**). The olefinic hydrocarbon should be thermodynamically more stable by about 15 kcal mol^{-1} than its three-membered ring isomer. The main contributions in favour of the ring-opened form are relief of ring strain (cyclopropane strain 27 kcal mol^{-1}; olefinic strain, i.e. the difference in the homolytic dissociation energies between two ethane and one ethylene carbon–carbon bonds, 22 kcal mol^{-1}), relief of steric strain due to angle widening ($\angle$CCC of *ca* 112° and 122° in the cyclic and acyclic derivative, respectively), double bond conjugation and free rotation around one extra carbon–carbon bond after ring opening. At the carbanion level, the same parameters are effective. In addition, however, delocalization of the negative charge into the two phenyl rings comes into play and proves to be a dominant factor. Despite the uncertainty of how to assess the steric and electronic effects of the three-membered ring, resonance can be safely assumed to stabilize the cyclopropyldiphenylmethyl anion (**2**, M = **:**)) relative to the isomeric primary carbanide (**3**, M = **:**) by at least 45 kcal mol^{-1}.

The 'morphology bonus' of *ca* 15 kcal mol^{-1} in favour of the open form is metal invariant and always remains the same. It is largely overcompensated for by the delocalization term not only when the free carbanion is considered but also in the case of the potassium derivative (**2**, M = K). If we suppose, for the sake of argument, that the latter species has still retained two thirds of the full carbanion character, the resonance stabilization will still amount to *ca* 30 kcal mol^{-1}. On the other hand, the magnesium analogue (**2**, M = MgBr) is carbanion-like only to the extent of, say, one fifth and consequently benefits little from charge delocalization, probably by 10 kcal mol^{-1} at best. This is not enough to counterbalance the driving force of 15 kcal mol^{-1} for the ring opening.

The corresponding lithium compounds (**2** and **3**, M = Li) behave as structural chameleons: in diethyl ether they exist in the open form (**6**-Li, colourless) whereas in tetrahydrofuran they prefer the ring-closed shape (**2**-Li, red) [14]. Obviously, new factors now have to be taken into account. The open form should benefit from substantial extra stabilization due to the aggregation phenomenon which is so typical of alkyllithium compounds. Nevertheless, when diethyl ether is replaced with tetrahydrofuran, this superior donor solvent coordinates so strongly to the metal that the oligomers break up. The ionicity of the organometallic bond is increased or loose ion pairs are even formed. In either case, the gain in resonance energy rises considerably.

1.3.3 METAL EFFECTS ON REACTION RATES

An equilibrium constant is the ratio of two reaction rates, one referring to the forward and the other to the reverse process. Since a change in the metal was found to cause a shift in equilibrium positions, it should also affect irreversible reactions by either accelerating or retarding them. This has indeed been observed in numerous substitution, addition, elimination and isomerization reactions.

The metal dependence of the Grovenstein–Zimmerman rearrangement [15] is particularly impressive. In the course of this process, an alkylmetal intermediate is converted into a resonance-stabilized 1,1-diarylalkyl species. Bis(2,2,2-triphenylethyl)mercury is a perfectly stable compound which can serve as the precursor to all analogous metal derivatives **4** when submitted to metal–metal cleavage. Whereas the magnesium compound requires reflux temperatures to undergo slowly 1,2-phenyl migration leading to the 1,1,2-triphenylethylmetal isomer **5**, the rearrangement of the corresponding lithium, sodium, potassium and caesium species occurs at about 0, −25, −50 and −75 °C, respectively [16].

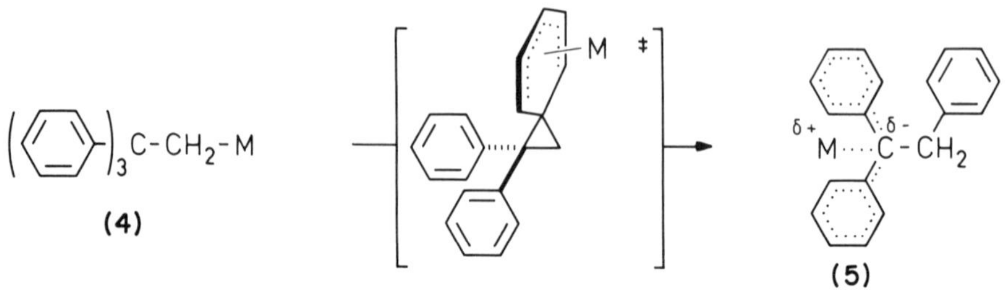

The gain in resonance energy constitutes the driving force for the reaction. It increases with increasing electropositivity of the metal, as we have already seen. For the reaction kinetics, however, the extent of the resonance stabilization at the pentadienylmetal-like 1,2-aryl bridged transition state becomes the crucial issue. Metal η^6-complexation [17] may also contribute.

The transient species can evolve into an interceptable intermediate (**6**) if the charge delocalization in the bridged species is further improved by replacing the migrating phenyl with a *p*-biphenyl moiety [18]. At the same time, any resonance stabilization of the starting material or the rearrangement product must be avoided (hence R = H, CH_3 or CD_3 in **6**).

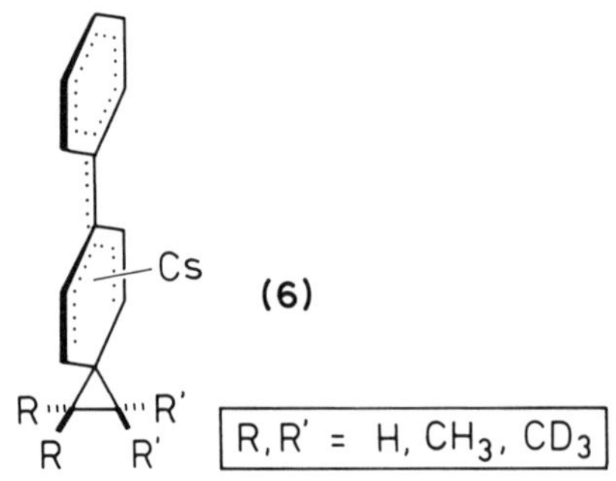

A different isomerization mechanism is observed if one of the three phenyl groups is replaced by a benzyl moiety. This time the benzyl rather than an aryl group migrates and hence again an extensively delocalized 1,1-diarylalkylmetal compound **8** is formed. As crossover experiments with isotope-labelled material have unequivocally demonstrated, the rearrangement is brought about in a two-step elimination–readdition sequence with a benzylmetal species **7** as the crucial intermediate. However, only lithium and sodium derivatives give a clean reaction. Benzyl and phenyl migration compete with each other if potassium or caesium analogues are used [19].

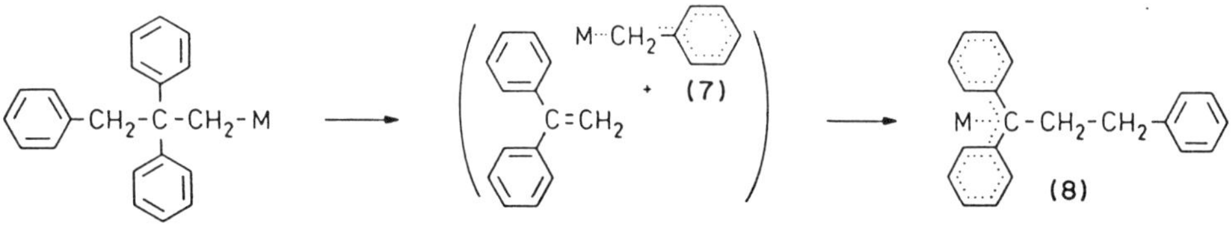

Readdition must not necessarily take place. 3,3-Diphenylpropyllithium (**9**) was found to decompose irreversibly into diphenylmethyllithium and ethylene [20]. The corresponding magnesium derivative is stable under the same conditions.

Astonishingly, however, the opposite behaviour was found with 3-methyl-3-(4-pyridyl)butylmagnesium bromide (**10**) and the corresponding lithium analogue. Whereas the latter proved to be perfectly stable in tetrahydrofuran, the former rapidly decomposes with evolution of ethylene [21].

Triphenylmethylmagnesium bromide readily reacts with tetrahydrofuran to form the adduct bromomagnesium 5,5,5-triphenylpentanolate whereas triphenylmethyllithium, -sodium and -potassium are stable under the same conditions [22]. Evidently, the ring opening of the cyclic ether requires electrophilic participation. The alkaline earth metal is, of course, a stronger Lewis acid than any alkali metal ion.

Inverse metal effects, that is, decreasing reactivity with increasing metal electropositivity, have also been identified in connection with simple elimination reactions. Although cyclohexadienyllithium [23] and -sodium [24] rapidly lose hydride to afford benzene, cyclohexadienylpotassium [25] and alkylated derivatives [26] thereof were found to be perfectly stable in tetrahydrofuran at −60 °C. Evidently it is this time the metal–hydride bond strength rather than the more or less efficient interaction of the metal with the pentadienyl-type organic backbone that dictates the course of the reaction.

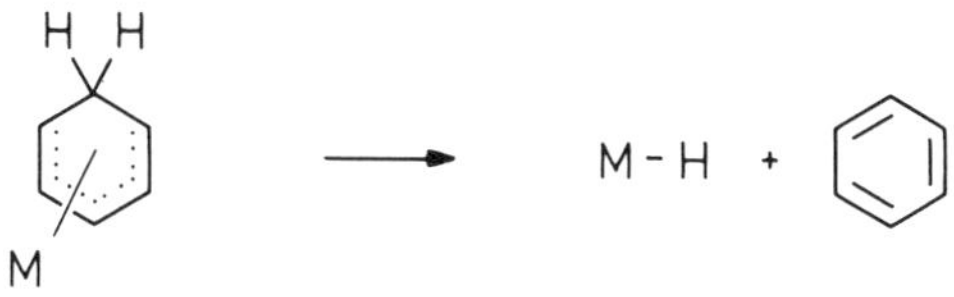

Inverse metal effects are generally observed when the reaction pathway leads through carbocationic intermediates. While being inert towards Grignard reagents, *tert*-butyl chloride rapidly and quantitatively undergoes β-elimination of hydrogen chloride when treated with butyllithium. With organozinc compounds, however, a substitution reaction occurs producing a quaternary carbon derivative with high yield [27]. Organozinc compounds are neither particularly basic nor nucleophilic, but have a pronounced Lewis acid character. Hence one may assume the reaction to be initiated by the formation of a carbenium zincate intermediate (**11**).

$$(H_3C)_3C\text{-}Cl \xrightarrow{LiC_4H_9} \left[H_2C\cdots H\cdots C_4H_9 ;\; H_3C\text{-}C(CH_3)\cdots Cl\cdots Li \right]^{\ddagger} \longrightarrow (H_3C)_2C{=}CH_2$$

$$(H_3C)_3C\text{-}Cl \xrightarrow{BrMgC_4H_9} \text{(no reaction)}$$

$$(H_3C)_3C\text{-}Cl \xrightarrow{ClZnC_4H_9} \left(H_3C\text{-}\overset{+}{C}(CH_3)_2 \quad Cl_2\overset{-}{Zn}C_4H_9 \right) \longrightarrow (H_3C)_3C\text{-}C_4H_9$$

(11)

Organomagnesium compounds can also give rise to the paradox of organometallic reactions operating through carbocationic mechanisms. Whereas orthoformates and other orthoesters are not attacked by organolithium compounds, they condense smoothly with ordinary Grignard reagents to give acetals [28]. The inverse metal effect suggests again an electrophilically assisted ionization generating a *gem*-dialkoxycarbenium ion (**12**) to be the crucial step of the reaction.

$$HC(OCH_3)_3 \xrightarrow{BrMg\text{-}R} \left(H\overset{+}{C}(OCH_3)_2 \quad H_3CO\text{-}\overset{-}{Mg}(R)\text{-}Br \right) \longrightarrow (H_3CO)_2CH\text{-}R$$

(12)

R = alkyl, aryl

The reactivity behaviour of two diastereomeric cyclic orthoformates corroborates this hypothesis. When treated with ethereal methylmagnesium bromide, the *trans*-2-methoxy-*cis*-4,6-dimethyl-1,3-dioxane (*trans*-**13**) rapidly exchanges the methoxy substituent against a methyl group with retention of its configuration [29]. The leaving group in the axial position can fully benefit from the neighbouring group assistance provided by the two ring heteroatoms. The all-*cis* stereoisomer (*cis*-**13**) lacks this possibility since, for steric reasons, the triaxial form is virtually unpopulated in the conformational equilibrium and the relevant orbitals occupy *gauche* positions in the triequatorial form. Hence no reaction can take place.

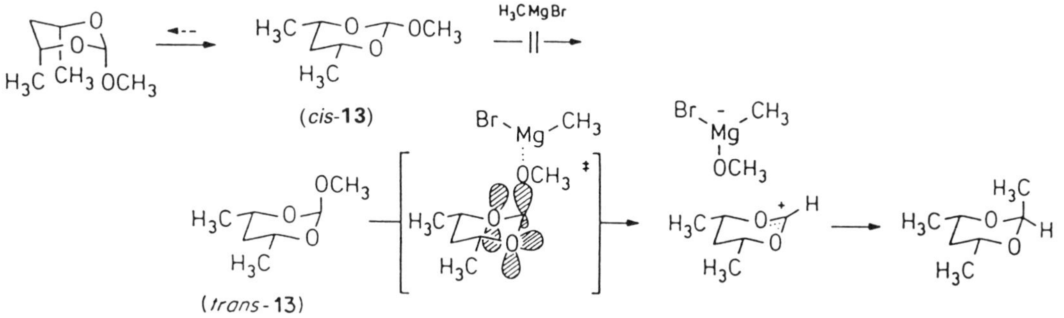

(*cis*-**13**)

(*trans*-**13**)

The smooth replacement of the oxygen functional group in an α-alkoxy- [30] or α-silyloxyamine [31] by the organic moiety of a Grignard reagent can again be rationalized in terms of a cationic mechanism. The postulated carbenium–immonium intermediate **14** enjoys a particularly efficient resonance stabilization.

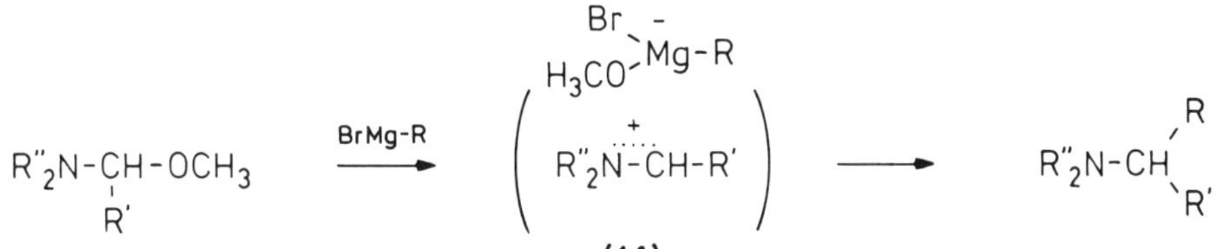

(14)

R'' = alkyl, $(H_3C)_3Si$; R' and R = alkyl, aryl

Finally, even organosodium compounds have conserved sufficient electrophilic power to be capable of generating cationic intermediates if the latter are only stable enough. This is apparently the case when *N*,*N*-dimethyl(2-chloropropyl)amine is treated with sodium diphenylketenimide, prepared by deprotonation of diphenylacetonitrile with phenylsodium. A secondary amine was isolated rather than the expected primary isomer [32], which would have been the result of an S_N2-like direct substitution. Obviously the product forms through an intermediate aziridinium ion **15**, which must have emerged from a cation triggered, β-amino group-assisted departure of a chloride ion.

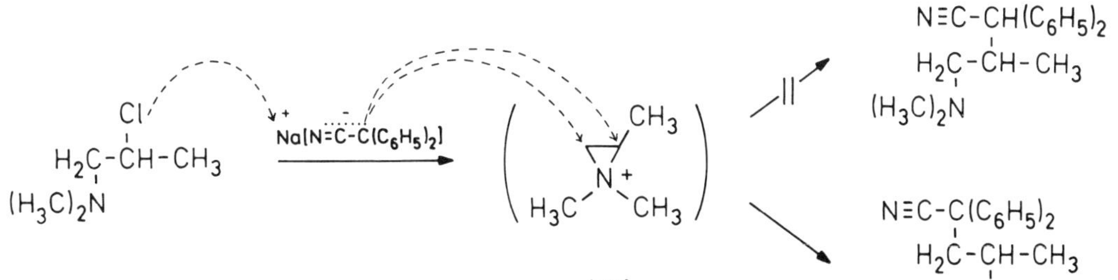

(15)

Radical mediated mechanisms become more frequently operative than cationic ones. They invariably intervene when halides get exposed to 'radical-anions' (metal–arene 1:1 adducts) or bulk metal, evidence existing in the latter case for both freely diffusing and surface attached species [33]. Radicals may also be generated when organometallics are treated with oxygen or other electron acceptors and may give rise to rearrangement cyclizaton or hydrogen transfer [34].

Acetylides [35] are known to have a characteristic reactivity profile. Only moderately basic, they are relatively nucleophilic owing to the high polarity of the carbon–metal bond. Therefore, one would expect them to exhibit the normal metal effect and indeed this is frequently found. For example, phenylethynylmagnesium bromide, -lithium, -sodium and -potassium combine in diethyl ether with acetonitrile or benzonitrile to form the imide-type adduct **16** with increasing ease [36].

$$R{-}C{\equiv}N + M{-}C{\equiv}C{-}C_6H_5 \longrightarrow R{-}C({=}N{-}M){-}C{\equiv}C{-}C_6H_5 \; (16) \longrightarrow R{-}C({=}O){-}C{\equiv}C{-}C_6H_5$$

R = H_3C, H_5C_6; M = BrMg, Li, Na, K

The same rate enhancement as a function of the metal electropositivity is observed when metal 3,3-dimethyl-1-butynides are allowed to react with methyl iodide in diethyl ether. The electrophilic substitution reaction shows normal metal dependence: Li < Na < K < Cs [37]. However, changing the solvent from diethyl ether to tetrahydrofuran produces discontinuity in the otherwise monotonous series. The reactivity order now becomes Li ≫ Na < K < Cs [38]. Why does this breaking away of the smallest alkali metal occur? Presumably, all intermediates, regardless of the nature of the metal, exist as contact species in the less polar ethereal medium and hence have to employ a multi-centre type of mechanism [39] (transition state **17**, X = iodide or acetylide). On the other hand, the higher donor capacity of tetrahydrofuran may offer the opportunity to the lithium derivative to form a (solvent-separated) ion pair, and thus a carbanion mechanism [39] (transition state **18**) may become operative in the latter solvent.

$$\overset{+}{M} \;\; :\overset{-}{C}{\equiv}C{-}C(CH_3)_3 \longrightarrow \left[\overset{+}{M} \;\; \overset{\delta-}{I}\cdots CH_3\cdots C{\equiv}C{-}C(CH_3)_3^{\;\delta-}\right]^{\ddagger} (18) \longrightarrow H_3C{-}C{\equiv}C{-}C(CH_3)_3$$

$$\updownarrow$$

$$M{-}C{\equiv}C{-}C(CH_3)_3 \longrightarrow \left[\begin{array}{c} X\cdots M \\ M \;\; H \;\; C{\equiv}C{-}C(CH_3)_3 \\ I\cdots C{-}H \\ H \end{array}\right]^{\ddagger} (17) \longrightarrow H_3C{-}C{\equiv}C{-}C(CH_3)_3$$

The remarkable solvent and metal dependence of acetylide reactivity can be exploited to achieve optional transformation of ω-hydroxy-1-alkynes at either the acetylenic or the hydroxylic site. The dilithio compounds **19** react with alkyl halides in tetrahydrofuran exclusively at the carbanionic centre. In contrast, the dipotassio analogue **20** undergoes predominantly *O*-alkylation [37].

$$\underset{(19)}{Li{-}C{\equiv}C{-}(CH_2)_n{-}OLi} \xrightarrow{RCH_2Br} RCH_2{-}C{\equiv}C{-}(CH_2)_n{-}OLi$$

$$\underset{(20)}{K{-}C{\equiv}C{-}(CH_2)_n{-}OK} \xrightarrow{RCH_2Br} K{-}C{\equiv}C{-}(CH_2)_n{-}OCH_2R$$

$$n = 4 - 8$$

Control over the typoselective outcome of a reaction can be commonly achieved by the appropriate choice of the organometallic component. The only prerequisite is that the competing sites show a marked difference in their susceptibility to metal effects. The nucleophilic addition to enolizable ketones represents a typical example. Thus, acetophenone is almost quantitatively converted into the corresponding tertiary alcoholate if a Grignard reagent such as phenylmagnesium bromide is employed. On the other hand, more polar reagents such as phenyllithium, phenylsodium and phenylpotassium favour the formation of the enolate **21** at the expense of the adduct **22** [38].

$H_3C{-}C{=}O$ (acetophenone) + M–C_6H_5 → $H_2C{=}C(C_6H_5){-}O\cdots M$ **(21)** + $H_3C{-}C(C_6H_5)_2{-}O{-}M$ **(22)**

M	(21)	(22)
MgBr	0 %	97 %
Li	75 %	14 %
Na	60 %	4 %
K	67 %	0 %

Analogously, complete α-deprotonation is ensured when aliphatic nitriles are treated with organosodium [34] or sometimes even organolithium [40] reagents. The resulting ketene-imides (**23**) readily undergo electrophilic substitution, in particular alkylation, at the electron-rich carbon centre. In contrast, ethereal organomagnesium derivatives cleanly add to the carbon–nitrogen triple bond to afford metal imides (**24**), which, on treatment with water, are rapidly hydrolysed to ketones [41]. Proton abstraction occurs only if the Grignard reagent is employed in a highly polar solvent such as hexamethylphosphoric triamide [42].

$R-CH_2-C{\equiv}N \xrightarrow{M-R'}$ $R-CH(M)-C{\equiv}N$ **(23)** $\xrightarrow{R'CH_2Br}$ $R-CH(CH_2R')-C{\equiv}N$

$R-CH_2-C{\equiv}N \xrightarrow{M-R'}$ $R-CH_2-C(R')=N-M$ **(24)** $\xrightarrow{H_2O}$ $R-CH_2-C(R')=O$

It is a common feature that the electropositivity of the metal affects the basicity more strongly than the nucleophilicity of an electron-rich species. Thus, potassium hydride is capable of converting aliphatic ketones by α-proton abstraction into the corresponding enolates *ca* 1000 times faster than sodium hydride and at least 10^6 times faster than lithium hydride does [43]. All three alkali metal hydrides are poor reducing agents; in other words, all of them lack major nucleophilic activity.

The nature of the metal is particularly crucial for the regioselectivity exhibited by delocalized entities. As a juxtaposition of prenyl type (3-methyl-2-butenyl) organometallics clearly reveals (Table 1.2), the magnesium compounds react almost exclusively at the inner, alkyl-substituted reactive centre, producing vinyl-branched derivatives **25**, whereas the potassium analogues combine with electrophiles preferentially at the unsubstituted terminus of the allyl moiety, thus giving rise to chain-elongated products **26** [44].

Similarly, electrophiles may attack benzyl-type organometallics at the exocyclic α- or at the *ortho*- and occasionally [45] even the *para*-position. The regioselectivities are

Table 1.2. Reaction between prenyl-type organomagnesium and -potassium compounds and a variety of electrophiles: ratio of branched vs chain-elongated products **25** and **26**

	R = H		R = H_3C	
X-*El*	M = MgBr	M = K	M = MgBr	M = K
$ClSi(CH_3)_3$	1:99	1:99	0:100	0:100
$FB(OCH_3)_2$	80:20	5:95	80:20	2:98
ICH_3	—	—	92:8	6:94
IC_3H_7	75:25	15:85	70:30	10:90
$BrCH_2CH{=}CH_2$	—	—	73:27	19:81
$O(CH_2)_2$	98:2	43:57	98:2	20:80
$O{=}CH_2$	99:1	28:72	—	30:70
$O{=}C{=}O$	99:1	10:90	—	4:96
$H_2C{=}CH_2$	—	—	—	100:0

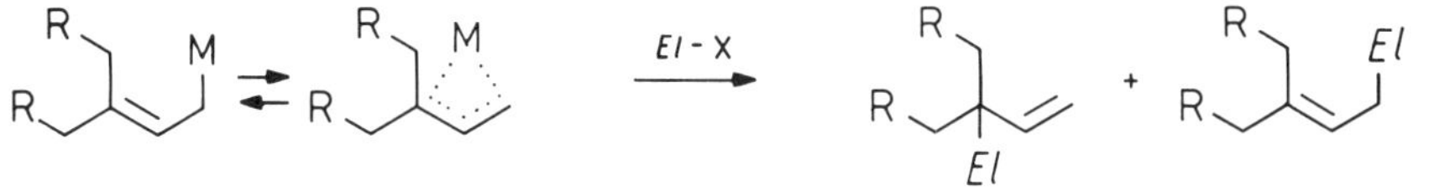

(25) (26)

again high enough to be of practical utility. For instance, 4-*tert*-butylbenzylpotassium reacts with formaldehyde to produce the 2-(4-*tert*-butylphenyl)ethanolate, whereas the corresponding magnesium derivative gives an *ortho*-substituted adduct **27** which spontaneously tautomerizes to the rearomatized 5-*tert*-butyl-2-methylbenzyl alcoholate [46].

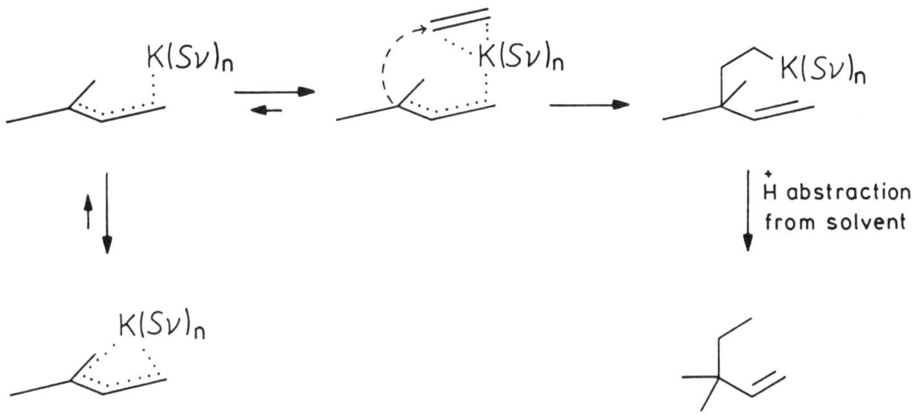

H abstraction from solvent

(27)

How does the metal steer the electrophile to its particular position? At the moment, any attempt to answer this question must be speculative. Nevertheless, a few facets of the problem have been clarified in the past two decades. Organomagnesium compounds are tightly σ-bonded species. Neither in the front nor in the back lobe of the carbon–metal linking orbital is enough charge density available to capture the electrophile directly (transition states **28**). Therefore, the electrophile has to approach the sterically more accessible vinylogous position of the allyl moiety and eventually becomes attached there, a steady flow of electrons to that centre being secured at the expense of the gradually disappearing organometallic bond. The product emerging from such a 'vicarious' S_E' reaction can be unequivocally distinguished from the one resulting from a normal S_E attack by the dislocated double bond, which has undergone an 'allyl shift.' What is unknown yet, however, is whether the electrophile attacks supra- or antarafacially, i.e. on the same or on the opposite side of the metal (transition states **29**^{supra} and **29**^{antara}). For a long time the former hypothesis was favoured since it suggests the operation of an aesthetically appealing multi-centred concerted cyclic mechanism (**29**^{supra}). Such a process should involve retention of the configuration of the electrophile. With oxiranes, however, the addition reaction of crotylmagnesium bromide was demonstrated to occur with inversion of configuration [47]. Hence 'open-circuit' mechanisms (**29**^{antara}) are now generally accepted as plausible alternatives at least.

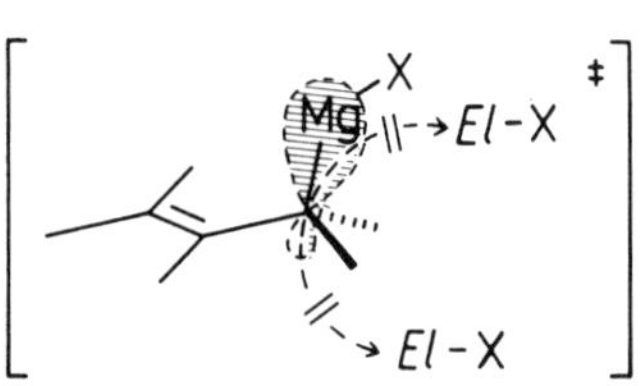

(28) 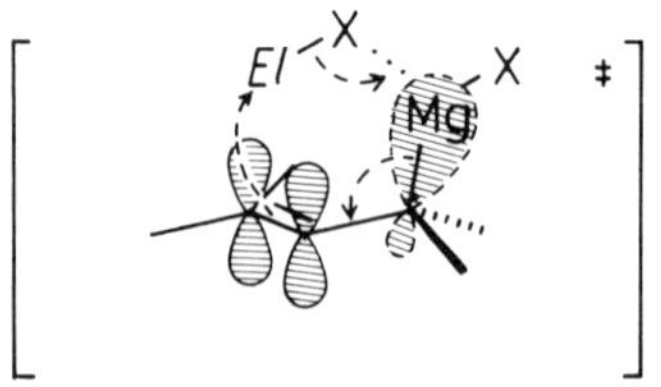 (29^supra) 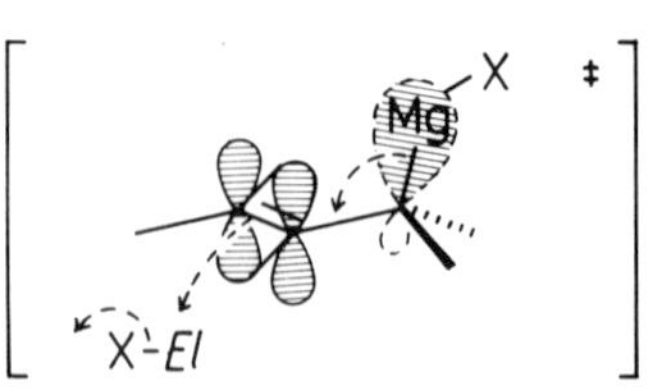 (29^antara)

Alkali metal cations combine with allyl anions [48], and also with many benzyl moieties [49], to form more or less symmetrical π-complexes having a trihapto (η^3) interaction between the metal and the carbon backbone. Consequently, the electrophile encounters at both termini of the allyl entity roughly the same electronic situation. Under such circumstances, the reagent will be directed to the sterically least hindered site. It will presumably enter on the metal-bearing face (transition state **30**) since retention is the ordinary mode of direct electrophilic substitution.

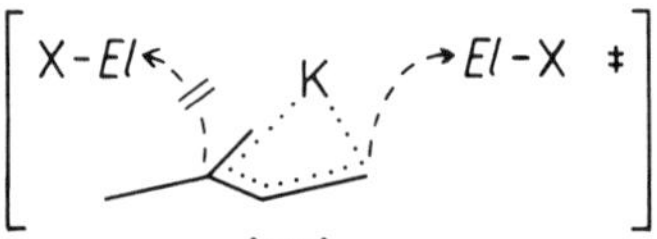

(30)

The electrophile must, however, not necessarily react with the η^3 ground state species but rather may intercept the η^1 metallomer which coexists with the former in finite although minute equilibrium concentrations. This course of events becomes particularly plausible if the coordinative gap that opens at the η^1-bonded metal is immediately occupied by complexation of the electrophile. In a subsequent rapid step, the latter will be delivered by suprafacial intramolecular transfer to the vinylogous alkyl-substituted allyl terminus. Such a mechanism takes place with a high degree of certainty when ethylene [50] acts as the electrophile (intermediate **31**; *Sv* means a solvent molecule) and it may also be operative, partially at least, with aldehydes, oxiranes and oxetanes.

(31)

Enolates [51], phenolates [52], deprotonated Schiff bases [53, 54], *N*-metalloanilines [55], pyrroles [56] and indoles [56] are oxa or aza analogues of allyl- and benzyl-type organometallics. Once again, the proper choice of the metal is of prime importance for achieving regiocontrol over the chemical transformations of such intermediates.

Finally, also the stereoselectivity of organometallic reactions is prone to element effects. Nucleophilic additions to carbonyl compounds are particularly prone to such influences. Thus, the polar allylalkali metal reagents show some preference for approaching 4-*tert*-butylcyclohexanone on the sterically less hindered *cis* face (affording adduct *eq*-**32**), whereas allylzinc bromide becomes mainly attached on the *trans* face (affording adduct *ax*-**32**) [57]. The change in orientation (see Table 1.3) may be a consequence of preceding complexation of the less electropositive metal to the carbonyl oxygen.

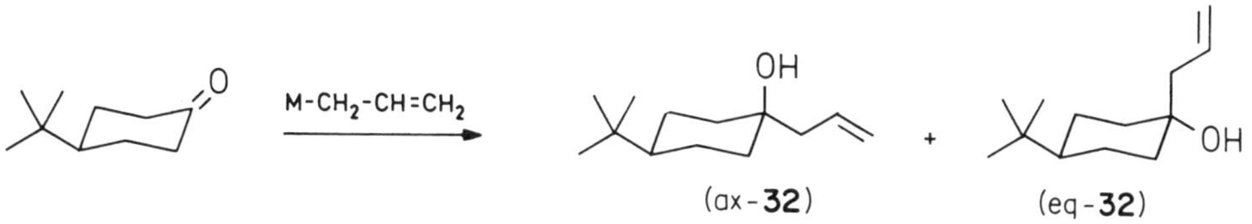

M-CH_2-CH=CH_2

(*ax*-**32**) (*eq*-**32**)

Table 1.3. Addition of allylmetals to 4-*tert*-butylcyclohexanone: face selectivity as a function of the metal

M	axial vs equatorial OH
ZnBr	85:15
MgBr	45:55
Li	35:65
Na	35:65
K	37:63

Frequently, steric accessibility favours one mode of reaction and metal complexation (or chelate formation) over the other [58]. The chemistry of α-hetero-substituted aldehydes offers many striking examples for the struggle between two conflicting parameters. Among them we find the addition of diethylzinc, ethylmagnesium bromide and ethyllithium to 2-(3,4-dihydro-2*H*-pyranyl)carbaldehyde ('acrolein dimer'). The strongly coordinating bivalent metals afford essentially the *threo* adduct (*threo*-**33**) whereas the alkali metal derivative produces large proportions of the *erythro* isomer (*erythro*-**33**), particularly in the presence of the polar cosolvent hexamethylphosphoric triamide (Table 1.4) [59].

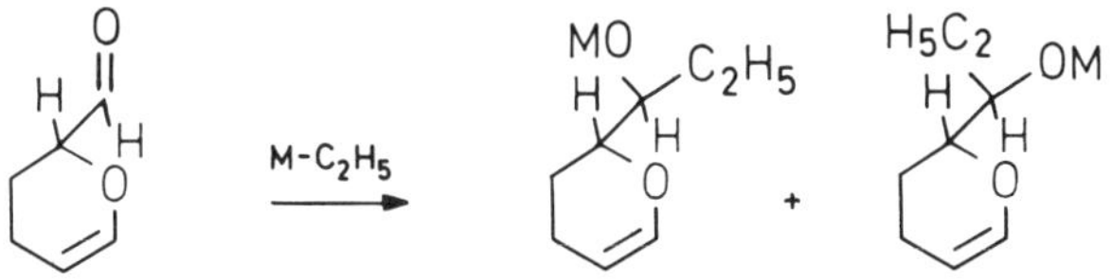

M-C_2H_5

(*erythro*-**33**) (*threo*-**33**)

The conformational preferences of allyl- and pentadienyl-type organometallics belong to the most fascinating stereochemical phenomena [11]. The metal-dependent differences are pronounced enough to have great potential for stereocontrolled synthesis. They begin to level out with heptatrienyl and have practically completely vanished with nonatetraenyl species.

2-Alkenylmetal compounds (**34**) may be generated by metal insertion into the carbon–hetero bond of allyl halides and ethers or, more conveniently, by metallation

Table 1.4. Addition of ethylmetals to acrolein dimer: diastereoselectivity as a function of the metal

M	*erythro*- vs *threo*- **33**
ZnC_2H_5	15:85
MgBr	30:70
Li	72:28
Li + HMPT	88:12

of a suitable alkene using a strong organometallic base. If the configuration of the original double bond is conserved, a *trans*-alkene, or a derivative thereof, will give rise to an *exo*- or (*E*)- and a *cis* precursor will lead to an *endo*- or (*Z*)- organometallic intermediate (**34**). On trapping with an electrophile, *trans*- and *cis*-isomers, respectively, should be produced selectively. Allyl-type magnesium and lithium compounds, however, undergo rapid torsional isomerization (via transition state **35**) at temperatures below −50 °C, typical energies of activation falling in the range 10–15 kcal mol^{-1} [60]. Sodium and notably potassium and caesium analogues are conformationally far more stable [61]. Without catalysis it requires several hours until their *endo–exo* equilibria are established.

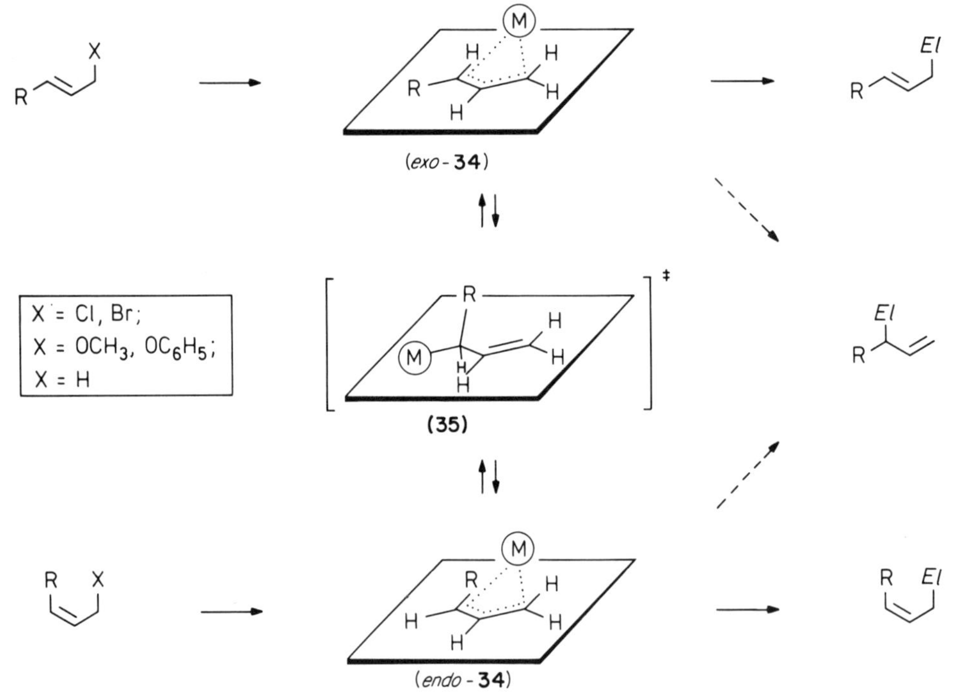

X = Cl, Br;
X = OCH_3, OC_6H_5;
X = H

(*exo*-**34**)

(**35**)

(*endo*-**34**)

Typical *endo–exo* distributions of allylmagnesium or allyllithium derivatives do not deviate very much from a 1:1 ratio. However, remarkably enough, the heavier alkali metals show a strong preference for the sterically more congested *endo* conformation (Table 1.5) [61]. The *endo*/*exo* equilibrium ratios reach their maximum with 2-butenylpotassium (124 : 1) and 2-butenylcaesium (> 500 : 1) and diminish on replacement of the methyl with a primary alkyl group and even more on introduction of a secondary alkyl group. If the substituent is a tertiary alkyl group, the *exo* species becomes the thermodynamically favoured isomer. Nevertheless, as a comparison with the corresponding hydrocarbon 4,4-dimethyl-2-pentene reveals, also this organometallic species must display some electronic effect that offers a substantial bonus (of *ca* 3 kcal mol^{-1}) to the conformer carrying the doubly branched alkyl substituent in the *endo* rather than the *exo* position.

Table 1.5. *endo*/*exo* Ratios of 2-alkenylmetal compounds **35** (M = Li, Na, K, Cs) after torsional equilibration and, for comparison, *cis*/*trans* equilibrium ratios of the corresponding 2-alkenes (M = H)

	M = H	M = Li	M = Na	M = K	M = Cs
R = H_3C	1 : 5	2 : 1	10 : 1	100 : 1	500 : 1
R = $H_3C—CH_2$	1 : 5			15 : 1	
R = $(H_3C)_2CH$	1 : 5			5 : 1	
R = $(H_3C)_3C$	1 : 10000			1 : 10	

If a vinyl group is connected to the allyl moiety, the area of delocalization will be extended provided all five carbon atoms lie in the same plane. The vinyl tail may occupy an *endo* or an *exo* position. In either case, it may be twisted inward or outward. In this way, a set of three privileged, since coplanar, and hence perfectly resonance stabilized species results: a horseshoe-like *U*, a sickle-shaped *S* and a zig-zag lined *W* conformer. Pentadienyllithium (**36**, M = Li) can be readily generated by treating 1,4-pentadiene with *sec*-butyllithium at −75 °C in tetrahydrofuran. It rapidly and completely adopts the sterically and electrostatically optimized *W* form as evidenced by NMR spectroscopy

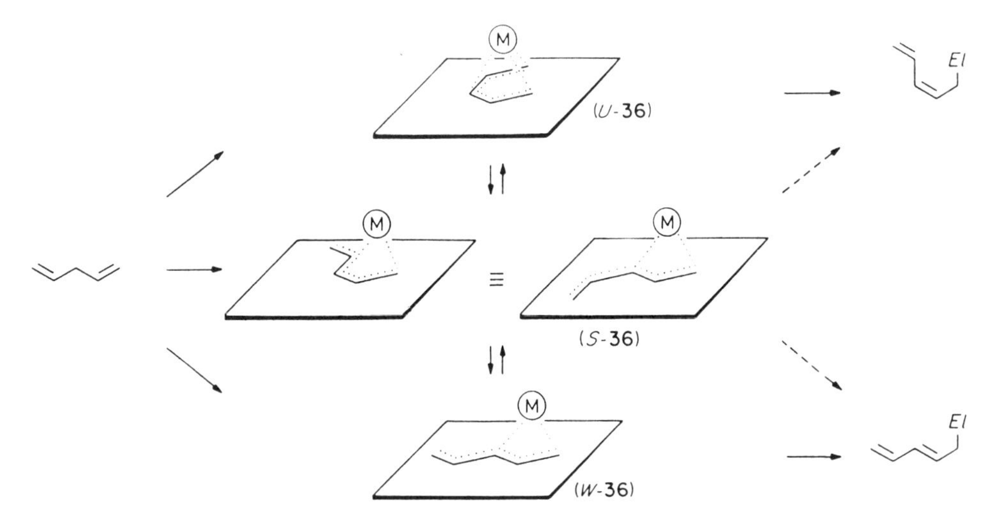

[62] or by chemical transformation to derivatives having the *trans* configuration [63]. Pentadienylpotassium (**36**, M = K) in hexane suspension [63, 65] exhibits the same structural preference. When dissolved in tetrahydrofuran, however, it chooses the *U*-shaped conformation, the driving force being apparently the formation of a pentahapto (η^5)-coordinated open sandwich [63]. Trapping with electrophiles produces *cis* isomers.

The introduction of an alkyl group into a terminal position of the pentadienyl backbone first of all removes the degeneracy of the *S* form. The substituent will be either located at the end of the sickle handle or at the tip of the sickle blade. Moreover, it may occupy the *endo* or the *exo* position. This formally gives rise to a new pattern of eight privileged, maximally delocalized conformers (**37**). In reality, steric crowding prohibits the coplanarity of the *endo-U* and of one of the *endo-S* forms. Three of the remaining six species can be selectively 'cultured' under the conditions of conformational competition. The *endo-W* form becomes preponderant to the extent of >97% if an organolithium reagent is used to deprotonate a *cis*-1,4-alkadiene and the *exo-U* shape results if an organopotassium compound is allowed to react with a *trans*-1,4-diene [66]. When generated from the same *trans*-isomer, organolithium species exist mainly (*ca* 90%), although not exclusively, in the *exo-W* shape.

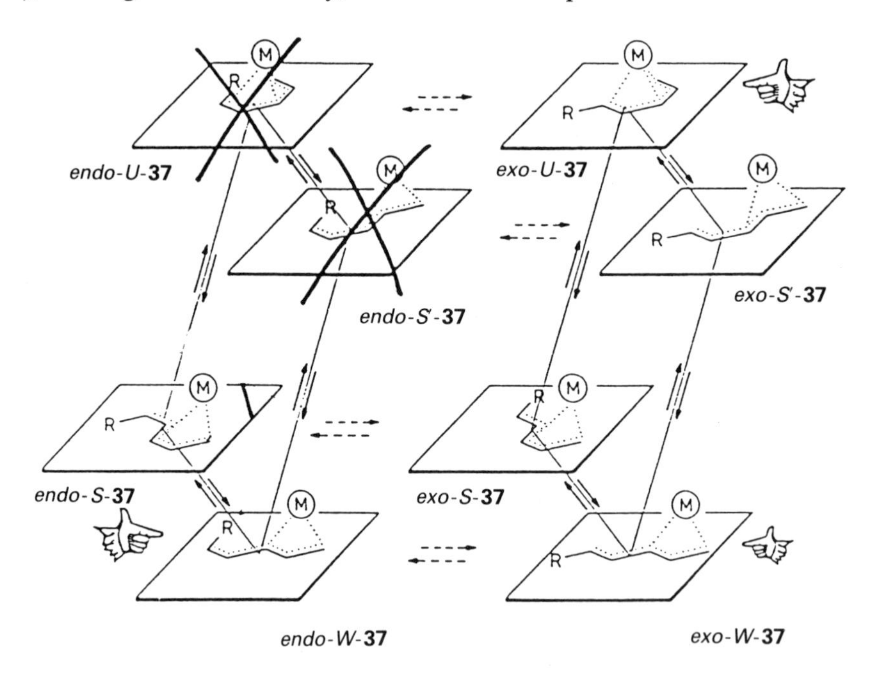

Once again, the analogy between allyl- or pentadienylmetal compounds and their nitrogen and oxygen analogues should be stressed. The conformational dynamics and preferences of deprotonated, α,β-unsaturated Schiff bases have been extensively studied [67]. Enolate reactions take a different stereochemical course depending on whether they carry an alkyl substituent in the *endo* or *exo* position [68].

1.4 PREPARATION OF ORGANOMETALLIC REAGENTS AND INTERMEDIATES

There are two principal options for the introduction of a metal atom into an organic moiety [1]. One of them is to use another organometallic reagent as a carrier for the metal M, the latter being exchanged against a mobile electrofugal group Z which is suitably located in the target molecule and which is transferred countercurrent to the reagent. Such a permutational interconversion is a most versatile, convenient and still powerful means for the generation of 'carbanionic' intermediates.

$$\geq C{-}Z \;+\; M{-}R \;\rightleftharpoons\; \geq C{-}M \;+\; Z{-}R$$

The use of a ready-made organometallic reagent only defers the problem of the *de novo* creation of carbon–metal bonds. The ultimate source of the metal can only be the element itself. Therefore, the most straightforward and frequently also the most rational approach is the direct reductive replacement of the leaving group Z by the insertion of elemental metal.

$$\geq C{-}Z \;+\; 2M\cdot \;\longrightarrow\; \geq C{-}M \;+\; Z{-}M$$

In either category, the favourite leaving groups are hydrogen, halogens (notably bromine) and metalloids (notably mercury and tin). The standard procedures for the preparation of organometallic species are based on the replacement of such elements. They are complemented by methods which have a more restricted area of application and which imply the cleavage of carbon–oxygen, carbon–sulfur or carbon–carbon bonds (Table 1.6).

Table 1.6. Synopsis of general methods for the preparation of organometallic species; the arrows indicate in which chronological order the various methods will be dealt with and their relative importance is represented by one to three plus signs

Reductive replacement	Z	Permutational interconversion
+ + +	X (halogen)	+ + +
+ +	Y (oxygen, sulfur . . .)	(+)
+	H (hydrogen)	+ + +
+	C (carbon)	+
+ +	Q (mercury, tin . . .)	+ +

1.4.1 REDUCTIVE REPLACEMENT (OR DIRECT INSERTION) METHODS

The ease with which the various types of metal insertions occur parallels fairly closely the respective enthalpy gains. As a comparison of the respective bond strengths reveals, the reductive replacement of a halogen by a metal should be a highly exothermic process (see p. 19). The new metal halide or metal alkoxide interaction overcompensates the sacrificed carbon–heteroelement bond and the strength of the carbon–metal linkage exceeds the energy required for the atomization of bulk metal. The driving force and hence the enthalpy gain diminishes when less polar metal species, such as metal hydrides or metal derivatives of organometalloids (e.g., tributylstannyllithium), are formed at the expense of carbon–hydrogen or carbon–metalloid bonds. Finally, the imaginary cleavage of ethane by elementary lithium to afford two equivalents of methyllithium would have to be an endothermic reaction.

1.4.1.1 Reductive Replacement of Halogen by Metal

In a formal sense, the reductive insertion of a metal always follows the same scheme, irrespective of the nature of the elements involved. The practical execution, however, has to adjust for huge differences in absolute and relative conversion rates. Therefore, individually distinct and optimized protocols have to be elaborated.

The preparation of organomagnesium [2] and organozinc [3] compounds is notoriously sluggish. Sometimes an acceleration can be achieved by merely replacing the organic chloride, if this is the precursor, with the corresponding bromide or iodide. More frequently, one attempts to overcome the inertness of the bulk metal by mechanically dividing it to very fine dispersions.

Metal clusters of atomic dimensions can be prepared if metal salts are reduced with elemental potassium [4] or, even more rigorously, if metal vapours are condensed into a solvent matrix at the temperature of liquid nitrogen [5]. Alternatively, rather than to increase the surface of the metal by pulverization, one can activate it by etching with an aggressive substance. In the case of zinc, washing with mineral acid [6] or incrustation of a heavy metal (e.g. producing a zinc–copper couple [7]) are equally recommended. If a reluctant organic halide has to be converted into the organomagnesium species, mechanical *in situ* grinding or ultrasonication [8] will generally provide sufficient activation. Otherwise the ‘entrainment method’ [9] may be used to solve the problem. The trick is continuously to add small quantities of a very reactive halide, say ethyl bromide or 1,2-dibromoethane or iodine, which then drags along the poorly reactive substrate.

For a long time, the ‘pacemaker’ was believed to act merely as a surface sweeper, corroding the metal even when it becomes covered with oxide deposits and thus always supplying new areas of high activity. More recent evidence, however, points at a more specific role [10]. The reductive insertion of magnesium into a carbon–halogen bond inevitably involves the intermediacy of free radicals **1**. They may be immediately generated as such ($R^{\bullet}$) by homolytic scission of a freely diffusing radical anion produced by single electron transfer from the metal to the organic halide [11]. Alternatively, they may, at the moment of their formation, remain sequestered by adsorption on the metal surface [12]. Even if this were the case, they might eventually manage to become detached and to enter into the solution. In order to be processed to the organometallic compound (RM), the radical has to return to the metal surface and to absorb a second electron. While in the ethereal phase, however, the radical may abstract an atomic hydrogen from the solvent and thus be lost for the production of an organometallic reagent. This is especially the fate of secondary or tertiary and allylic or benzylic radicals. Such endangered species, however, can be efficiently trapped by metal transfer from a primary alkylmagnesium halide which has been formed from the entrainment promoter. The concomitantly resulting primary alkyl radical appears to find its way back to the metal surface with particular ease and thus becomes metal loaded again before it can be destroyed by solvent attack.

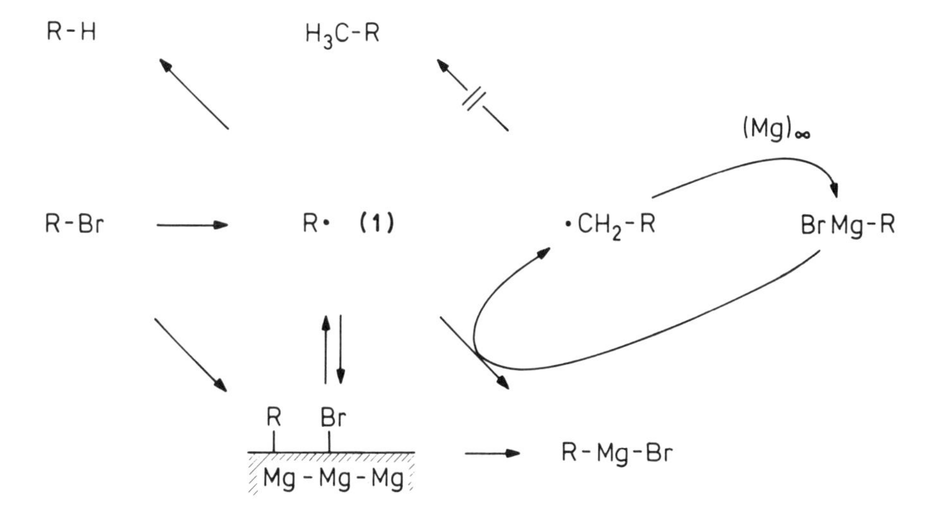

With increasing electropositivity of the metal, the transfer of the second electron and, as a corollary, the interception of the radical become more and more efficient. Now, however, new obstacles arise. If we replace magnesium or zinc with lithium or sodium, not only the reactivity of the metal towards the organic halide is enhanced but also, and even to a much greater extent, the reactivity of the resulting organometallic derivative towards its own precursor. Depending on the organic moiety, the latter reaction mainly proceeds according to S_N2, *E*2 or radical pair mechanisms [13], leading to a mixture of coupling and dismutation products.

R - X

>C=C<

R-H ←--- (R·)

(R· + ·R) ---→ R-H

M-R

R-R

Fortunately, the nature of the halide employed affects much less the rate with which it is reduced by the metal than the rate with which it reacts with the organometallic reagent formed. In 1930, a landmark semikinetic study [14] was undertaken in which butyl chloride was recognized to be almost inert towards butyllithium, irrespective of the solvent. On the other hand, butyl bromide was found to react rapidly in diethyl ether, unless the mixture was cooled below 0 °C, and the iodide did so even in benzene (Table 1.7). The practical advice is to prepare butyllithium from the chloride rather than the bromide and never from the iodide.

Table 1.7. Approximate half-life periods, $\tau_{1/2}$, of butyllithium in the presence of butyl chloride, bromide or iodide, all of them having a 0.5 M initial concentration in benzene or diethyl ether at 25 °C [14]

	$\tau_{1/2}$ (h)	
$Li\text{-}C_4H_9 + H_9C_4\text{-}X$	Benzene	Diethyl ether
X = Cl	>100	40
X = Br	40	0.5
X = I	3	<0.1

Special precautions are required in order to make tertiary alkyllithium compounds. The only convenient starting material for the preparation of *tert*-butyllithium is the chloride, and hexane is the solvent of choice. Moreover, sodium–lithium alloy [15] or a sodium–lithium codispersion [16] have to be used unless standard quality lithium is activated by abrasion [17].

Aryl halides are, of course, much less prone than their aliphatic counterparts to coupling with the corresponding organolithium reagents. Phenyllithium can be readily prepared from bromobenzene in ethereal medium [18]. Chlorobenzene reacts with lithium in diethyl ether relatively slowly, but rapidly in tetrahydrofuran [19].

The Krotos reverse behaviour (the children swallow their procreator) becomes more acute than ever when the reductive replacement of exceptionally 'mobile' halogen atoms is attempted. The conversion of chloromethyl methyl ether into methoxymethyllithium [20] is typical. The use of dimethoxymethane (formaldehyde dimethyl acetal) as an excellent donor solvent and of finely dispersed metal together with careful temperature control allows the problems to be overcome. In other cases, for example when allylic chlorides are involved as the precursors, an *in situ* interception of the organometallic intermediate may give satisfactory results [21]. More often, however, the use of 'dissolved metals' is the best measure to be taken in order to ensure a sufficiently rapid electron transfer. All alkali metals dissolve when stirred with ethereal solutions of electron acceptor arenes with which they form deeply green or blue so-called 'radical anions' [22]. These are 1:1 adducts such as naphthalene–lithium (lithium dihydronaphthylide) or biphenyl–lithium (lithium dihydrobiphenylylide). Naphthalene–lithium has been successfully applied to convert a host of alkyl chlorides carrying hydroxy, amide, acetal or dithioacetal functional groups into the corresponding organolithium species [23]. The 1 : 1 adduct between lithium and 4,4-di-*tert*-butylbiphenyl (Freeman's reagent [24]) offers the advantage of improved chemical stability of the hydrocarbon. It was found to be very convenient for the generation of 7-norbornadienyllithium (**2**) [25] from its labile chloro precursor.

(2 equiv.) **(2)**

The preparation of organo*sodium* reagents (**3**) from the corresponding halides is particularly troublesome. Only chlorides were found to sustain the attack of such organometallic species and even then only for a short period of time. Therefore, the reaction should be completed as quickly as possible. Once again, the particle size of the metal turns out to be the crucial factor. With the less critical chlorobenzene or *p*-chlorotoluene it suffices to employ sodium 'sand' [26, 27]. However, alkylsodium compounds such as propyl- [29], butyl- [29], pentyl- [30, 31], octyl- [32] and decyl- [33] sodium can be obtained with fair to good yields only if a very detailed protocol is scrupulously followed. An ultrafine dispersion ($\leqslant 25$ µm average particle diameter) of the metal must be prepared; high-speed stirring (5000–20 000 rpm) and vortical fluid motion in special glass equipment ('greased flasks') have to be used to ensure efficient mixing and the reaction temperature must be kept at -10 °C.

$$\text{R-Cl} \xrightarrow{\text{2 Na}} \text{R-Na} \quad \textbf{(3)}$$

Despite all such efforts, alkyl chlorides always afford the corresponding *potassium* compounds in poor yields (e.g. pentylpotassium [31] 35%; dodecylpotassium [33] 10%). In contrast, the preparation of aryl- and 1-alkenyl-type potassium derivatives gives acceptable results (e.g., phenylpotassium 55% [34]; vinylpotassium 73% [35]; 1-dodecenylpotassium 47% [36]).

1.4.1.2 Reductive Cleavage of Ethers and Thioethers

Metals undergo the insertion into carbon–oxygen or carbon–sulfur bonds much less readily than into carbon–chlorine, carbon–bromine or carbon–iodine bonds [37]. Obviously, chalcogens do not pick up electrons as efficiently as halogens do when exposed to a metal. Thus, saturated ethers are perfectly stable against reductive cleavage under ordinary conditions. Nevertheless, at reflux temperatures 1,4-dioxane is slowly decomposed by potassium–sodium alloy [38]. Even more astonishingly, magnesium can intrude into the carbon–oxygen bond of tetrahydrofuran and produce a six-membered magnesia cycle (**4**). The process, however, requires catalysis by titanium tetrachloride or cobalt dichloride unless oligoatomic metal ('Rieke magnesium') is employed [39]. In a similar way, ring-opened products (e.g. **5**), are obtained when tetrahydrofuran, 2,2-dimethyltetrahydrofuran or 2-vinyltetrahydrofuran are simultaneously treated with the 4,4′-di-*tert*-butylbiphenyl–lithium radical-ion (Li-DBBP) and boron trifluoride [40].

Mg; O–Mg **(4)**; Li · DBBP + BF_3; O Li; ⊕ Li F_3B **(5)**

Small rings can gain additional driving force by the relief of angular strain. Therefore, oxiranes and oxetanes in tetrahydrofuran do not require electrophilic assistance to undergo reductive ring opening. However, while exclusively primary alkyllithium derivatives (e.g. **6a** and **7a**) are formed under such conditions, isomers with secondary or tertiary organometallic bonds (e.g. **6b** and **7b**) result when the reaction is carried out in the presence of Lewis acid catalysts such as boron trifluoride or triethylaluminium [41].

Li · DBPP + LiBr, -75°C; Li O Li **(6a)**; -75°C, BF_3 + Li · DBPP; O $\bar{B}F_3$ Li Li^+ **(6b)**

Li · DBPP, 0°C; Li O Li **(7a)**; O $\bar{A}l(C_2H_5)_3$ Li Li^+ **(7b)**

The reductive ring scission of α-lactones can be brought about with potassium in tetrahydrofuran in the presence of 1,4,7,10,13,16-hexaoxacyclooctadecane (18-crown-6) at −20 °C. This results in a (2-enolato)oxyalkylpotassium species of type **8** [42].

R–(β-lactone)=O —18-crown-6→ R–CH(K)–O–C(=O)–O–K

(8)

Diaryl ethers and alkyl aryl ethers undergo reductive cleavage with remarkable ease. In the case of an unequally substituted ether, two pairs of possible reaction products have to be envisaged. In general, the less basic organometallic compound is formed preferentially or exclusively. Thus, phenylpotassium and potassium methoxide are obtained when anisole is treated with potassium in refluxing heptane [34]. However, if applied in liquid ammonia, the same metal produces potassium phenolate and methane, the latter obviously via methylpotassium [43].

C_6H_5–OCH_3 —in NH_3→ C_6H_5–O–K + K–CH_3

C_6H_5–OCH_3 —in C_7H_{16}→ C_6H_5–K + K–O–CH_3

How can one rationalize this dichotomy in product composition? It may reflect a dichotomy in mechanism [44]. Apparently, the aromatic moiety of the anisole serves as an electron sink where negative charge can be accumulated. The uptake of the first electron leads to a radical anion (**9**). There is ample evidence that carbon–oxygen bond scission may directly occur at that stage [45]. The dianion **10**, having an antiaromatic decet of delocalized π-electrons, is not likely to be involved. Nevertheless, a second mode of reductive ether cleavage can be conceived of. Under certain conditions, the transfer of the second electron may not merely *follow* the carbon–oxygen bond scission in a posterior step but rather may occur *simultaneously* and actively support the cleavage of the radical anion. Since the latter process would by-pass any transient radical intermediate, the relative thermodynamic stabilities of the possible fragment combinations (phenylpotassium plus potassium methoxide vs methylpotassium plus potassium phenolate and methyl radical plus potassium phenolate vs phenyl radical plus potassium methoxide) could be reversed.

C_6H_5–O–CH_3 ⇌ K^+[(7π)–O–CH_3]$^{\cdot-}$ **(9)** ⇌ K^+ K^+[(8π)–O–CH_3]$^{2-}$ **(10)**

(9) → C_6H_5–O–K + $\cdot CH_3$

(9) ⇸ $C_6H_5\cdot$ + K–O–CH_3

(9) → C_6H_5–K + K–O–CH_3

Allylsodium (**11**, M = Na) [46] and allylpotassium (**11**, M = K) [47] are readily produced by the treatment of allyl phenyl ether with the alkali metal in ethereal or even paraffinic solvents. However, the preparation of allyllithium (**11**, M = Li) [48], crotyllithium [49], prenyllithium (3-methyl-2-butenyllithium) [44] and other homologues by ether cleavage with lithium requires tetrahydrofuran as the solvent. Invariably the corresponding metal phenolate is formed as a by-product. Owing to its relative inertness it does not compromise subsequent reactions.

OC_6H_5 — M (2 equiv.), $-[MOC_6H_5]$ → M (**11**)

Methyl or ethyl benzyl ether, bis(1-phenylethyl) ether and butyl 2-methoxybenzyl ether react smoothly with lithium in tetrahydrofuran to afford benzyllithium [50, 51], 1-phenylethyllithium [50] and 2-methoxybenzyllithium [52], respectively. Remarkably enough, the heavier alkali metals cleave benzylic ethers only sluggishly if at all unless the latter carry alkyl or aryl substituents in the α-position. The generation of the ether-soluble, fairly stable 2-phenyl-2-propylpotassium ('cumylpotassium') [53, 54] by treatment of cumyl methyl ether with potassium–sodium alloy became a milestone in the development of the preparative organometallic chemistry. The disappearance of its deep red colour allows the progress of its transformations to be monitored visually.

OCH_3, $H_3C\text{-}C\text{-}CH_3$ — K (2 equiv.), $-[KOCH_3]$ → K, $H_3C\text{-}C\text{-}CH_3$

The scission of symmetrical ethers is advantageous since it avoids any ambiguity about the cleavage site. In this way, α-methoxydiphenylmethylpotassium [55, 56] and -sodium [56], diphenylmethylpotassium [56], α,α-dibenzylbenzylpotassium [53], 9-fluorenylsodium [57], 9-phenylfluorenylsodium [57] and 9-methoxyfluorenylsodium [58], among others, have been made accessible.

Compared with their oxygen analogues, thioethers have considerably higher electron affinities and weaker carbon–heteroelement bonds. Moreover, thiolates are much less basic than alcoholates or even phenolates. All these factors should favour the reductive cleavage of sulfides. Actually, lithium insertion into carbon–sulfur bonds is such a facile process that even heterocycles such as thiophene [59] and *N*-alkylphenothiazines [60] can be cleaved in this manner.

From a preparative point of view, the reductive cleavage of alkyl phenyl sulfides deserves attention. It can be accomplished with bulk metal or, more conveniently, with radical anions such as di-*tert*-butylbiphenyl–lithium and yields alkylmetal compounds in addition to thiophenolate [61, 62]. The reduction of allyl phenyl sulfides is of particular practical importance [63]. In an elegant study, *exo*-3-methoxyallylpotassium (*exo*-**12**, M = K) was generated by treatment of the thioether with naphthalene–potassium at −120 °C. While the potassium compound underwent only slow equilibration to afford the thermodynamically more stable *endo* isomer (*endo*-**12**, M = Li), stereomutation occurred very rapidly after replacement of the heavier alkali metal with lithium.

H_3CO SC_6H_5 M (*exo*-**12**) H_3CO M (*endo*-**12**)

Mercaptals (dithioacetals) [65], thioorthoformates [66], α-(phenylthio)amines [67] and α-(phenylthio)ethers [68] can again be very efficiently cleaved. The last type of substrate gives access to the very versatile α-alkoxyalkyllithium reagents (e.g. **13**). The favourite reducing agent for such reactions is the 1-dimethylaminonaphthalene–lithium 1:1 adduct ('LIDMAN') [69, 70], despite its instability in tetrahydrofuran above −50 °C. It offers the great advantage of being easily removed from the crude reaction mixture by simple acid extraction.

O O SC_6H_5 O Li **(13)**

Not only aryl derivatives of chalcogens, as described, but also those of pnictogens are prone to reductive cleavage by alkali metals. Shaking a solution of triphenylamine, -phosphine, -arsane or -stiborane in tetrahydrofuran with elemental lithium produces, in addition to phenyllithium, lithium diphenylamide [71, 72], diphenylphosphide [72, 73], diphenylarsanide [72] and diphenylstiboride [72], respectively.

1.4.1.3 Reductive Replacement of Hydrogen by Metal

The conversion of acetylene into sodium acetylide is the oldest [74] and most extensively studied reaction of this type. It is generally carried out in liquid ammonia [75], the evaporation of which leaves behind the solid material [76]. The reducing agent sodium may be replaced with lithium, potassium, rubidium, caesium or an alkaline-earth metal [77]. The transformation of higher 1-alkynes or arylacetylenes into the corresponding lithium or sodium acetylides is equally possible [78]. In all these cases, the replaced hydrogen is in part evolved as the diatomic molecule. The rest is consumed by a reduction of the acetylene to the olefin.

A reductive replacement can also be expediently brought about between cyclopentadiene or methylcyclopentadiene with potassium in refluxing benzene [79] or with sodium in refluxing xylene [81]. The dimeric cycloadduct can be used directly if higher boiling solvents such as decalin [81] or polyethylene glycol ethers [82] are used and reaction temperatures in the range 170–200 °C are applied. Under such conditions, half an equivalent of molecular hydrogen is produced per cyclopentadienylsodium or potassium formed. If, however, the reaction is carried out in liquid ammonia [80, 81], half an equivalent of cyclopentene is obtained instead as the by-product. Obviously, the ammonia intercepts both negatively charged intermediates, the radical anion **14** and the allylmetal species **15** by protonation. The concomitantly resulting sodium amide then reacts immediately with cyclopentadiene under proton abstraction [82]. In paraffinic or ethereal solvents the reaction has to take a different route. Presumably via a diene–metal (1 : 2) adduct (**16**), sodium hydride is eliminated, which is basic enough to abstract a proton from cyclopentadiene. In either reaction step one equivalent of cyclopentadienylsodium is formed.

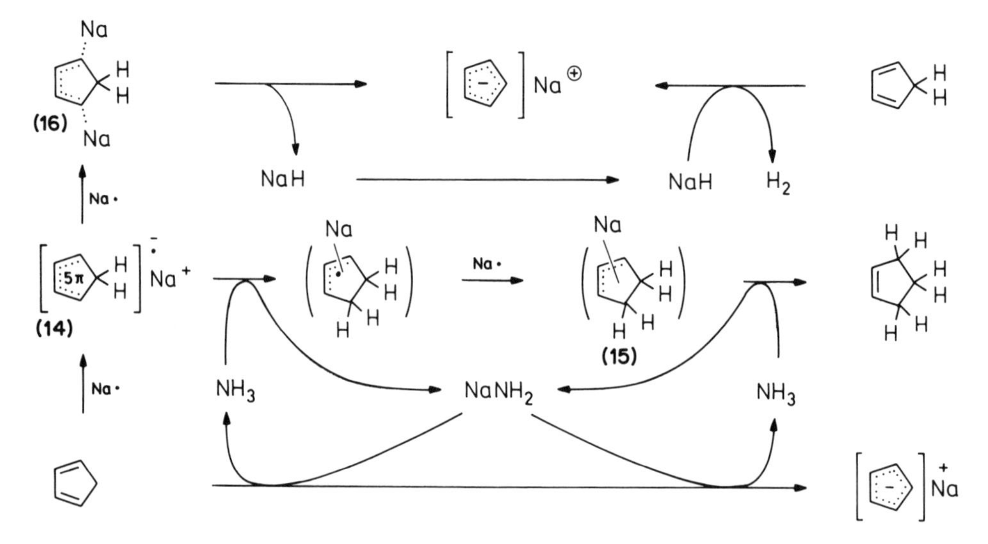

Indenylsodium [82] and -potassium [82] and fluorenyllithium [83], -sodium [84] and -potassium [85] can be prepared under similar conditions. Even in aprotic solvents reduced hydrocarbons (e.g. tetrahydrofluorene) are unavoidable by-products.

Both triphenylmethylsodium and -potassium can be easily generated by reductive hydrogen replacement in liquid ammonia. On evaporation of the solvent, however, the sodium derivative decomposes to triphenylmethane and sodium amide [86]. Only the potassium compound can be isolated intact and crystalline [86]. Alternatively, it may be prepared in ethylene glycol dimethyl ether ('glyme') [87].

All hydrocarbons studied so far belong to the most acidic ones known. Attempts to extend the method to moderately or poorly acidic substrates were only sporadically successful. Toluene [88] and allylbenzene [88] do react with caesium reasonably well. On the other hand, the reaction of toluene with potassium requires hexamethylphosphoric triamide as a cosolvent [89], which unfortunately is rapidly attacked by the benzylpotassium thus formed. Benzylpotassium and naphthylmcthylpotassium and derivatives carrying alkyl substituents in the exocyclic α-position or on the ring can be obtained with moderate to good yields if the alkyl arene and potassium sand are heated in the presence of sodium oxide to temperatures around or above 110 °C [90]. Furan [91], 2-methylfuran [91], a variety of thiophenes [92] and benzothiophene [93] do react with sodium or potassium under reductive replacement, but the process again is not clean and is without practical value.

Under favourable circumstances, even saturated hydrocarbons or simple olefins can interact with elemental metals. Atomic calcium in the vapour phase was found to insert smoothly into the C—H bonds of cyclohexane [94]. Benzene has been claimed to give phenylcaesium when treated with metallic caesium [95]. As a careful reinvestigation has revealed, however, a temperature-dependent mixture of products is obtained containing the black 1 : 1 adduct (radical anion **17**), the yellow dimer (**18**) and the black biphenyl–caesium 1 : 2 adduct (**19**) as the principal components [96].

Cs, THF, −75 °C → **(17)** → (−45 °C) **(18)** → (+35 °C) biphenyl → (Cs) **(19)**

Allylpotassium is formed if, in an autoclave at 80 atm pressure, propene is heated together with a potassium dispersion to 150 °C [97]. Much milder conditions can be applied with 1,4-pentadiene as the substrate. It suffices to treat it in tetrahydrofuran at 0 °C in the presence of triethylamine with elemental potassium to produce pentadienylpotassium (**20**) [98].

(20)

Finally, ethylene can be cleanly converted into vinyllithium if a solution in tetrahydrofuran is vigorously stirred at −25 °C with finely dispersed lithium in the presence of a cocktail of arene-type electron carriers [99]. Apparently, 1,2-dilithioethane (**21**) is involved as a key intermediate in the reaction sequence. At −25 °C, it cannot be observed, however, since it immediately loses lithium hydride. In contrast, the corresponding caesium compound appears to be stable. It forms spontaneously when caesium is in contact with ethylene [100].

$$H_2C{=}CH_2 \rightleftharpoons (LiCH_2{-}\dot{C}H_2) \rightleftharpoons (LiCH_2{-}CH_2Li)\ \mathbf{(21)} \xrightarrow{-[LiH]} H_2C{=}CH{-}Li$$

An extension of the method to terminal alkenes as substrates appears to be possible if 2,5-diphenyl-1,6,6aλ^4-trithiapentalene is employed as a catalyst. The latter compound can bind several lithium atoms [101]. With this electron carrier, 1-octene can be efficiently converted into isomerically almost pure (*E*)-1-octenyllithium (**22**) [102].

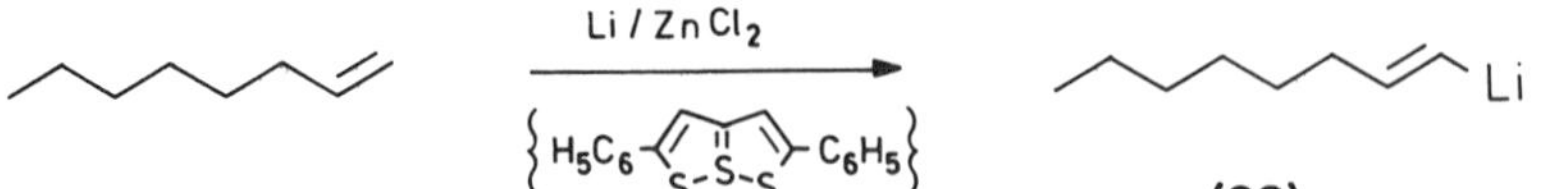

Li / ZnCl_2

{ H_5C_6 ... C_6H_5 }

Li

(22)

1.4.1.4 Reductive Replacement of Carbon by Metal

The reaction of ethylene with lithium affording vinyllithium and lithium hydride not only involves a carbon–hydrogen cleavage but also the scission of the carbon–carbon π-bond. Olefins may be conceived of as two-membered carbocycles. The relief of the corresponding ring strain of *ca* 22 kcal mol^{-1} provides the driving force to promote metal addition to the olefin. The reductive opening of three- and four-membered rings (e.g. dicyclopropylidene [103] and dihydrobenzobutene [104]) can be rationalized in the same way.

Alkynes are far better electron captors than alkenes. The stereoselective reduction of alkynes to *trans*-alkenes in liquid ammonia passes through transient alkyne–metal 1:2 adducts [105]. It is possible to generate such a dimetal species (**23**) under aprotic conditions and thus to make it survive by treating the strained cyclooctyne with lithium. A 2:2 adduct (**24**) is obtained as a by-product [106].

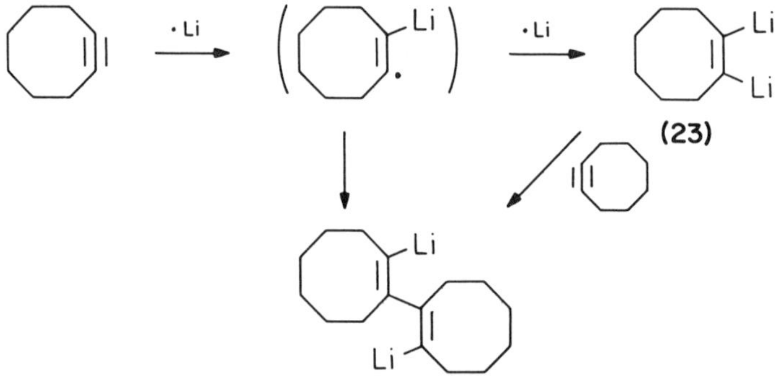

·Li Li ·Li Li Li

(23)

Li

Li

(24)

The reductive cleavage of ethane to give two equivalents of methyllithium or of higher alkanes to alkyllithium fragments has never been observed. Such processes would be endothermic or thermoneutral at best. Moreover, no reasonable mechanism for the required electron transfer can be operative. Phenyl substituents may attenuate or remove both shortcomings. They can serve as electron-acceptor moieties and thus improve the kinetics of the envisaged reaction and they can make it thermodynamically feasible by providing resonance stabilization to the resulting organometallic products.

The oldest example of this type is the preparation of triphenylmethylsodium from the triphenylmethyl dimer, at that time erroneously considered to be hexaphenylethane

[107]. 1,1,2,2-Tetraphenyl-1,2-(phenylethynyl)ethane [108], 9,9′-dixanthyl [109] and 9,9′-di-9-methylxanthyl [109] again react smoothly with sodium under reductive C—C bond scission. Rapid cleavage of 1,1,2,2-tetraphenylethane [109], however, occurs only with potassium whereas 1,2-di(phenoxy)-1,1,2,2-tetraphenylethane [110] may be cleaved with either lithium, sodium or potassium. With the corresponding dimethyl ether as the substrate, the carbon–carbon bond scission (affording **25**) has to compete with a reductive methoxide elimination. The latter process leads, presumably through tetraphenylethylene, to the dilithio compound **26** [111].

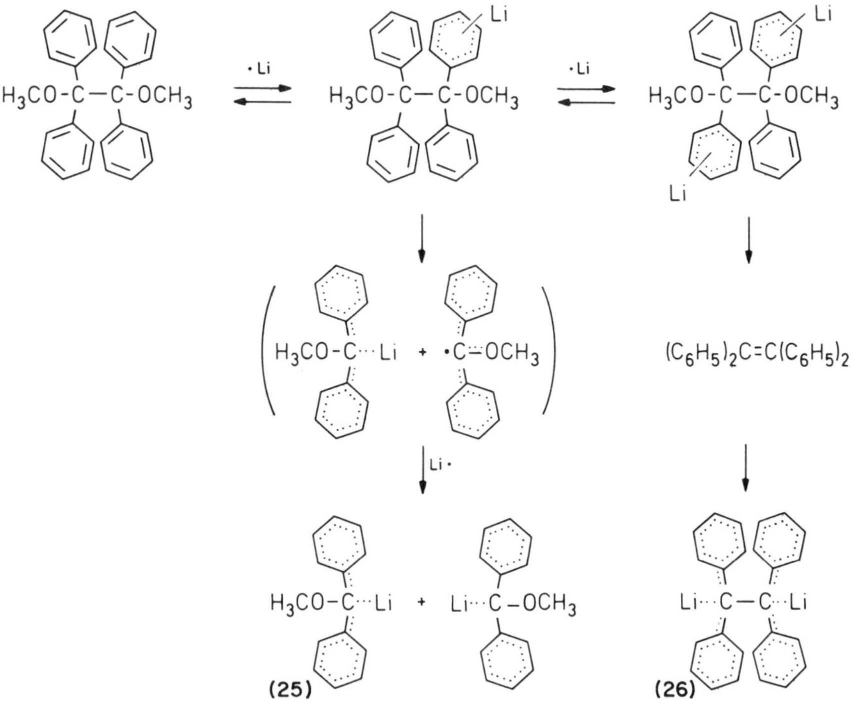

$H_3CO\text{-}C\text{—}C\text{-}OCH_3$ ⇌ (·Li) $H_3CO\text{-}C\text{—}C\text{-}OCH_3$ (Li) ⇌ (·Li) $H_3CO\text{-}C\text{—}C\text{-}OCH_3$ (Li, Li)

$H_3CO\text{-}C\cdots Li$ + $\cdot C\text{=}OCH_3$ → (Li·) $H_3CO\text{-}C\cdots Li$ + $Li\cdots C\text{-}OCH_3$ (25)

$(C_6H_5)_2C{=}C(C_6H_5)_2$ → $Li\cdots C\text{—}C\cdots Li$ (26)

Potassium is again required for the cleavage of 1,2-di-*p*-biphenylylethane [112], 1,1,6,6-tetraphenyl-1,5-hexadiene [112] and 2,2,7,7-tetramethyl-3,5-octadiyne [113], whereas only caesium is capable of promoting the cleavage of 1,2-diphenylethane [114]. Lithium in tetrahydrofuran converts 9,9-diphenylfluorene into a mixture of 9-phenyl-9-fluorenyllithium and phenyllithium [115]. The central bicyclic unit is split when triptycene (**27**) is treated with potassium–sodium alloy [116].

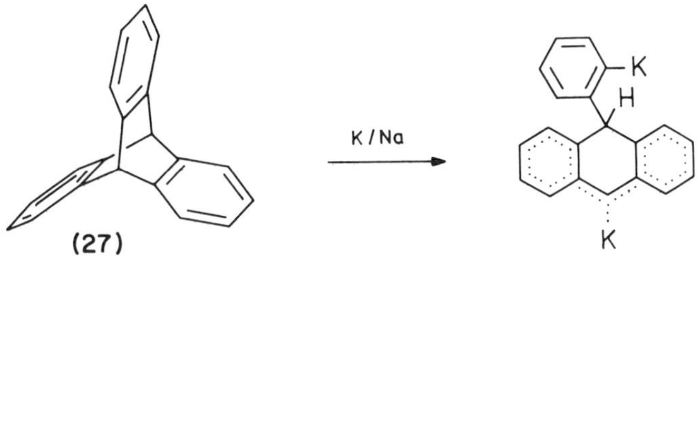

(27) —K/Na→ K, H, K

1.4.1.5 Reductive Replacement of Another Metal or Metalloid by Metal

The controlled preparation of organic alkali metal derivatives was first achieved by the reductive cleavage of dialkylzinc compounds [117]. A short time later the use of diorganomercury compounds [118] was found to be preferable since the reaction is much cleaner and, despite its potential reversibility, it goes to completion if an excess of the alkali metal is employed. In this way numerous organolithium, -sodium, -potassium and -caesium compounds have been conveniently prepared, e.g. methyllithium [119], ethyllithium [119], propyllithium [119], butyllithium [120], 2,2-diphenylpropyllithium [121], benzyllithium [122], cyclopropyllithium [123], vinyllithium [124], phenyllithium [125], *o*-phenylenedilithium [126], 2,2′-biphenylidenedilithium [127], meththylsodium [119, 128], ethylsodium [119, 129–130], propylsodium [119, 131], butylsodium [132], octylsodium [119], benzylsodium [119], phenylsodium [119], methylpotassium [119, 129, 133], ethylpotassium [130], butylpotassium [132], phenyl-potassium [134], trimethylsilylmethylpotassium [135] and trimethylsilylmethylcaesium [136].

As a rule, the less electronegative ('more electropositive') metal drives the more electronegative one out of its bonding interaction with carbon. This makes sense if one conceives of the conversion of the element into an organometallic derivative as a partial oxidation of the metal. If we compare the ionization potentials of potassium (99 kcal mol^{-1}) and lithium (126 kcal mol^{-1}), we realize immediately that the former has a greater reductive potential than the latter. In agreement with this relationship, the lighter alkali metal is replaced by the heavier one if solutions of organolithium reagents are stirred with bulk potassium or, more efficiently, potassium–sodium alloy. This method has been applied to the preparation of triphenylmethylpotassium [137], ethylpotassium [138], butylpotassium [138], *sec*-butylpotassium [138] and *tert*-butylpotassium [138]. The samples thus obtained are, however, impure. In addition to finely dispersed lithium and excess of potassium, they appear to contain small amounts of the organolithium precursor, which coprecipitates with the organopotassium compound to form mixed aggregates.

The mechanism of this reductive metal–metal replacement process is still obscure. Radicals (such as **28**) may be involved since *sec*-butylbenzene was identified as one of

$(H_3C)(H_5C_2)CH\text{-}Li \xrightarrow{\cdot K} \left((H_3C)(H_5C_2)CH\cdot\right)$ **(28)** $\xrightarrow{\cdot K} (H_3C)(H_5C_2)CH\text{-}K \dashrightarrow$ mixed aggregate

(28) + C_6H_6 ⇢ $H_3C\text{-}CH(C_6H_5)\text{-}C_2H_5$

$(H_3C)(H_5C_2)CH\text{-}K$ + C_6H_6 → C_6H_5K

the products when a benzene solution of *sec*-butyllithium was allowed to react with potassium [139].

1.4.2 PERMUTATIONAL INTERCONVERSION (OR INDIRECT EXCHANGE) METHODS

It is often more convenient to perform a reaction of the permutational interconversion type rather than the reductive replacement type. Such methods, of course, rely on the availability of a highly reactive organometallic auxiliary MR′. In most cases, one of the commercial reagents, notably butyllithium, hexyllithium, *sec*-butyllithium, *tert*-butyllithium, methyllithium or phenyllithium, is used for this purpose.

$$\text{R-Z} + \text{M-R}' \rightleftharpoons \text{R-M} + \text{Z-R}'$$

Formally at least, all permutation reactions are reversible. In a first approximation, the position of the equilibrium established in a given case depends on the relative basicities of the two organometallic species, MR and MR′, involved. The reaction tends to produce a less basic at the expense of a more basic reagent. For example, dimethylmercury and ethyllithium are almost completely converted into diethylmercury and methyllithium when they are allowed to react in a paraffinic or aromatic, non-ethereal solvent [140].

When evaluating basicities, one has to keep in mind, of course, what has been explained earlier (see pp. 20–22). It is an oversimplification to correlate the chemical potential of a series of organometallic compounds merely with the relative thermodynamic stabilities of the corresponding carbanions without taking other factors into account. Differences in aggregation energies may attenuate or even reverse the natural basicity order which one would observe with the monomers. An equilibrium may be even more drastically affected if one of its components precipitates.

1.4.2.1 Permutational Metal–Metal Exchange

The interconversion between alkyllithium reagents and organic derivatives of 'soft' and fairly electronegative metals, so-called 'metalloids', proceeds particularly smoothly. When treated with butyllithium, dimethylmercury sets free methyllithium [141]. Bis(1-methylpropyl)mercury and bis(*N*,*N*-diethylcarbomoyl)mercury were found to be suitable precursors for the generation of optically active *sec*-butyllithium [142] and *N*,*N*-diethylcarbonyllithium [143], respectively. A large excess of *tert*-butyllithium converts tetrakis(chloromercurio)methane (**29**) into tetralithiomethane [144], previously

$$CCl_4 \xrightarrow[\text{Li}]{ClB(OCH_3)_2} C[B(OCH_3)_2]_4 \longrightarrow C(HgCl)_4 \longrightarrow CLi_4$$

(29)

identified as one of the products formed upon the reaction between tetrachloromethane and lithium vapor [145].

Tetravinyllead may be used to prepare vinyllithium [146]. (3,3-Dichloroallyl)triphenyllead, when treated with butyllithium, sets free *gem*-dichloroallyllithium (**30**) [147].

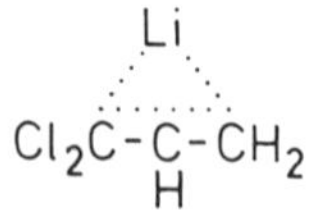

$$\underset{\text{H}}{Cl_2C{-}\overset{\text{Li}}{C}{-}CH_2}$$

(30)

The most versatile components for organometal–organometalloid interconversions are organotin compounds. This is not only because of the ease with which the permutational exchange occurs. A second crucial advantage is that the starting materials are accessible by a variety of different, complementary methods. There is on the one hand the classical route, which implies the reaction between a Grignard reagent and an organic tin chloride. Alternatively, the polar organometallic precursor may be condensed with a (halomethyl)tin derivative. Finally, one can reverse the polarities and couple an organic halide or tosylate with trimethyl- or tributylstannyllithium.

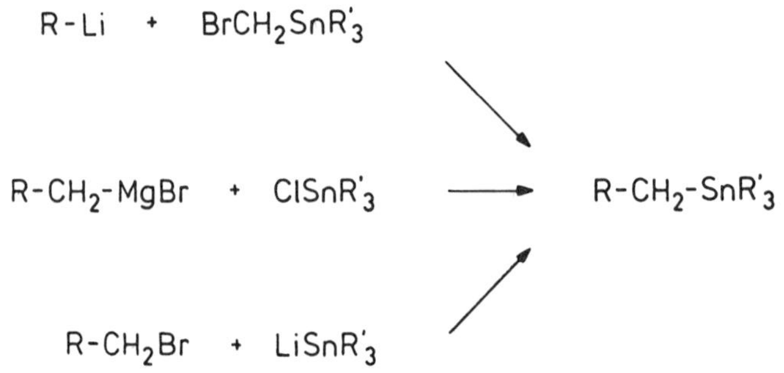

$$R{-}Li + BrCH_2SnR'_3 \longrightarrow R{-}CH_2{-}SnR'_3$$

$$R{-}CH_2{-}MgBr + ClSnR'_3 \longrightarrow R{-}CH_2{-}SnR'_3$$

$$R{-}CH_2Br + LiSnR'_3 \longrightarrow R{-}CH_2{-}SnR'_3$$

Organotin compounds allow the efficient preparation of organolithium reagents which are otherwise troublesome to obtain: cyclopropyllithium [148], vinyllithium [149], (*Z*)- or (*E*)-1-propenyllithium [150], trifluorovinyllithium [151], (*Z*)- or (*E*)-1,2-diphenylvinyllithium [142], (*Z*)- or (*E*)-2-*p*-chloro-1,2-diphenylvinyllithium [142], 1-(2-chloroethyl)vinyllithium [152], allyl-, 2-methylallyl- and crotyllithium [153, 154], benzyllithium [153, 155], 1-phenylethyllithium [154] and *o*-bromobenzyllithium [156]. Allyl- and benzyl-type organopotassium and organocaesium compounds are again readily accessible by the interconversion method [157]. Bis(trialkyltin) precursors such as the olefin **31** [158] and the diene **32** [159] can be submitted to a unilateral [158] or bilateral [157, 159] exchange.

Tin reagents open up a very convenient entry to α-alkoxyalkyllithium species [160] which may subsequently undergo a Wittig rearrangement [161] or can act as synthetic equivalents of α-deprotonated alkanols [162] if they carry an acetal function or a benzyloxy group which may hydrolytically or reductively be removed at a later stage. Among the numerous α-alkoxyorganolithium compounds prepared in this way are

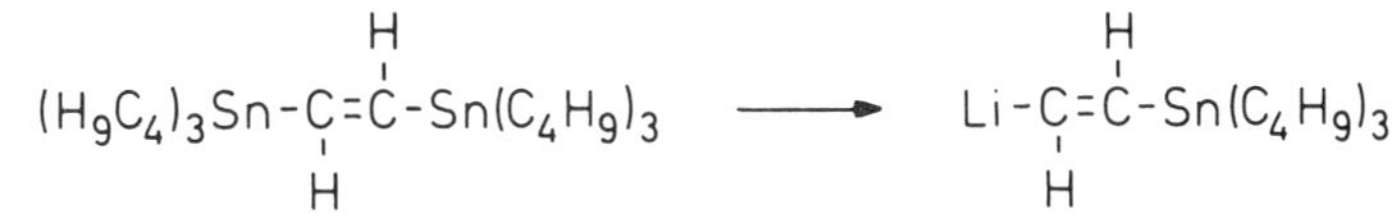

$$(H_9C_4)_3Sn-CH=CH-Sn(C_4H_9)_3 \longrightarrow Li-CH=CH-Sn(C_4H_9)_3$$

(31)

$$(H_3C)_3Sn-C(CH_3)=CH-CH=C(CH_3)-Sn(CH_3)_3 \longrightarrow Li-C(CH_3)=CH-CH=C(CH_3)-Li$$

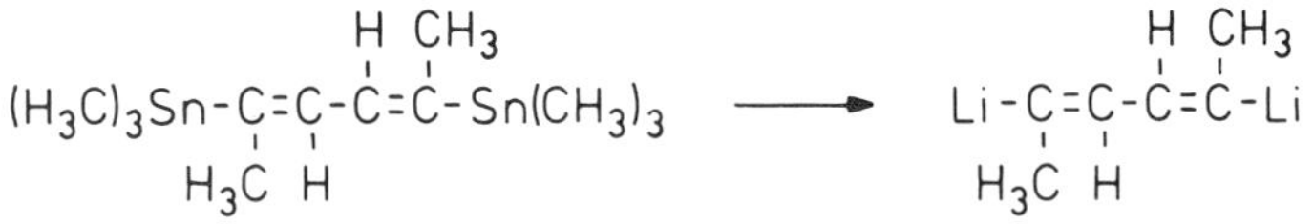

(32)

benzyloxymethyllithium [163], methoxymethoxymethyllithium (**33**) [164], (1-ethoxyethoxy)methyllithium [165], (2-tetrahydropyranyloxy)methyllithium [166], (benzyloxymethoxy)methyllithium (**34**) [167], 1-(methoxymethoxy)cyclohexyllithium [168], 2-tetrafuryllithium [168], 2-tetrahydropyryllithium [168] and lithiated sugar derivatives such as **35** [169].

$Li-CH_2OCH_2OCH_3$ **(33)**

$Li-CH_2OCH_2OCH_2C_6H_5$ **(34)**

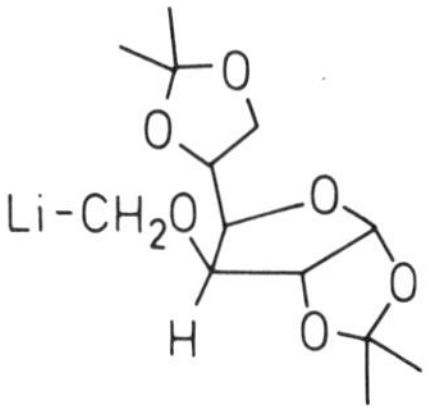

(35)

Permutational interconversions are not restricted to organometallic reagents and organometalloid compounds, but can also be carried out between organometallic reagents and metal (or metalloid) salts [170]. In general, the exchange proceeds in the direction where the more electropositive metal combines with the less basic organic moiety. Thus, treatment with ethereal lithium bromide converts 2-naphthylmethylpotassium [171] and allylpotassium [172] into the lithium compounds. In the same way allylsodium [172] can be prepared if sodium tetraphenylborate in tetrahydrofuran is used. Nevertheless, the exchange can also produce the more basic organometallic species. This is notably the case when organolithium compounds are allowed to interact with soluble sodium [173], potassium [173] or caesium [174] alkoxides. As the equilibrium is driven to the side of the exceptionally strong oxygen–lithium bond, the carbon has to accept the heavier alkali metal as a binding partner.

1.4.2.2 Permutational Halogen–Metal Exchange

The interconversion process between an organic halide and an organometallic reagent belongs to the synthetically most useful reactions [175]. The driving force lies in a partial neutralization: the more strongly basic organometallic species is converted into the less basic component of the equilibrium. For example, when phenyl iodide is added to an ethereal solution of propyllithium, the latter disappears almost completely from the equilibrium, which contains almost 10 000 times more of phenyllithium [176]. On the other hand, vinyllithium and phenyllithium coexist under similar conditions in an approximate ratio of 250:1 [176]. As the phenyl anion has a smaller protonation enthalpy than vinyl anion (see Figure 1.7), the basicity gradient does not explain why vinyllithium should be favoured. Other factors such as strong aggregation and poor solubility have to be invoked.

$$C_6H_5\text{-I} + \text{Li-R} \rightleftharpoons C_6H_5\text{-Li} + \text{I-R}$$

The dynamic nature of the exchange equilibrium can be demonstrated by 'up-hill' (endoenergetic) metal transfers. When phenyllithium is added to an alkyl halide, only infinitely small amounts of the alkyllithium species are formed. However, if an efficient mode of interception exists, this intermediate will be continuously transformed as supplied in the equilibrium until all material is consumed. In this way, *endo*-tricyclo[4.2.1.0^{2,5}]non-7-ene (**36**) [177] was readily obtained. The analogous preparation of 6,7-dihydrodibenzo[*b,f*]oxepin (**37**) [178] energetically runs 'down-hill' all the way, of course, since the treatment of bis(2-bromomethylphenyl) ether with phenyllithium generates a benzylic, hence resonance stabilized organolithium intermediate.

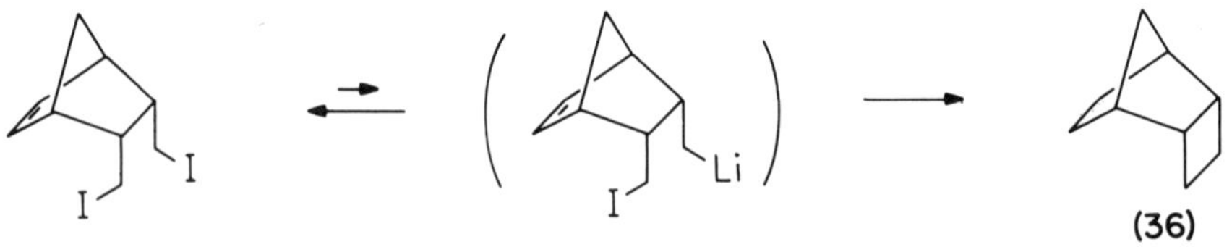

(36)

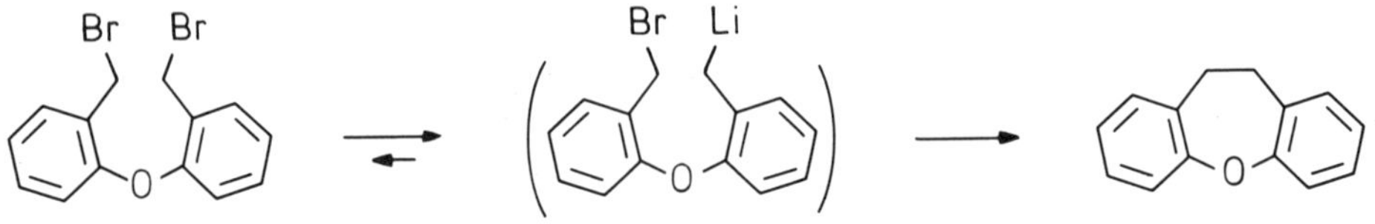

(37)

Being intramolecular, such cyclizations are fast. Nevertheless, they can be suppressed at low temperature. Treatment of 2-(2-bromoethyl)bromobenzene with butyllithium generates 2-(2-bromoethyl)phenyllithium (**38**), which can be trapped if the temperature is maintained at -100 °C [179].

(38)

From a practical point of view, the halogen–metal interconversion mode can almost entirely be narrowed down to bromine–lithium and iodine–lithium exchange. Chlorine–lithium exchange gives satisfactory results only when polyhalogenated substrates are involved. Thus, for example, trichloromethyllithium [180] can be obtained from tetrachloromethane, dichloroallyllithium [181] from 3,3,3-trichloropropene, *endo*- and *exo*-7-(7-chloro)bicyclo[4.1.0]heptyllithium (**39**) [180] from 7,7-dichloronorcarane, 1-chloro-2,2-diphenylvinyllithium [180] from 1,1-dichloro-2,2-diphenylethylene, trifluorovinyllithium (**40**) [182] from trichlorofluoroethylene and 1,1-dichloro-2,2,2-trifluoroethyllithium [183] from 1,1,1-trichloro-2,2,2-trifluoroethane.

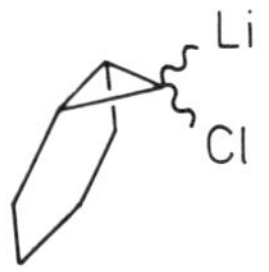

(39) 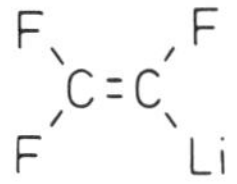 **(40)**

Alkaline earth or heavy alkali metals rarely participate in halide interconversion processes. Methyl iodide [184], 1- or 2-dodecenyl bromide [185], phenyl bromide [186], *p*-tolyl bromide [186] and 1-naphthyl bromide [186] do form the corresponding sodium derivatives, although in moderate yields, when treated with butylsodium or pentylsodium. 3-Thienylpotassium [187], from 3-bromothiophene and cumylpotassium, appears to be the sole potassium species so far prepared by halogen–metal exchange.

The element effects can be readily understood on the basis of the exchange mechanism. The intermediacy of ate complexes first postulated by Wittig and Schöllkopf [188] is nowadays generally accepted. A lithium diphenyliodate (**41**) has been identified as a reaction intermediate in tetrahydrofuran [189]. While the heavy halogen centre can accommodate the negative charge most efficiently, the lithium cation will benefit from a particularly high solvation energy. The crucial role of solvation in ion-pair formation is reflected by the thermal stability of lithium bis(pentafluorophenyl)iodate, which can be isolated in crystalline form. It undergoes vigorous decomposition above −75 °C if the metal is surrounded by tetrahydrofuran molecules (**42**; $Sv = C_4H_8O$; $n = 4$), but the complex is stable up to 25 °C if coordinated with two equivalents of *N,N,N′,N′*-tetramethylethylenediamine [**42**; $Sv = (H_3C)_2NCH_2CH_2N(CH_3)_2$; $n = 2$] [190].

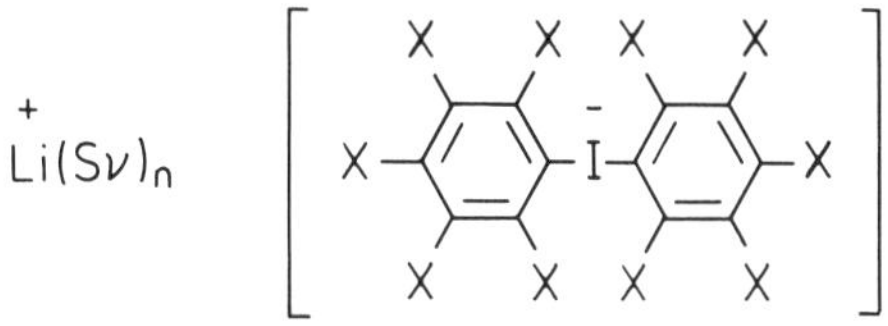

41: X = H
42: X = F

The halogen–metal exchange is the method of choice to introduce specifically a lithium atom into a given, non-activated position. Thus, it allows the selective preparation of *o*-, *m*- and *p*-anisyllithium whereas the hydrogen–metal exchange method, i.e. the direct 'metallation' of anisole (see p. 86), can only give access to the *ortho* isomer (see page 58).

The mild reaction conditions make halogen–metal interconversion processes particularly attractive for the preparation of labile organolithium species. A few α-, β-, γ-, δ- or ε-heterofunctional organometallic compounds, all of them generated by permutational bromine–lithium exchange, will be mentioned as typical examples: heptafluoropropyllithium [191], (*Z*)-1-chloro-2-phenylvinyl chloride (**43**) [192], *endo*- and *exo*-7-fluoro-7-bicyclo[4.1.0]heptyllithium (**44**) [193], 1-bromo-2,2-dimethyl-1-cyclopropyllithium [194] and 1-methoxy-1-cyclopropyllithium [195]; (*Z*)-2-ethoxyvinyllithium [196] (**45**, OR = OC_2H_5), (*Z*)-2-(trimethylsilyloxy)vinyllithium [197] [**45**, OR = $OSi(CH_3)_3$], 1-(ethoxymethyl)vinyllithium [198], 6-ethoxy-1-cyclohexenyllithium [198], 1-(diethoxymethyl)vinyllithium [199], 3,3,6,6-tetramethoxy-1,4-cyclohexadienyllithium (**46**) [200], 1-(lithiooxymethyl)vinyllithium [201], (*Z*)-1-lithiooxymethyl-1-propenyllithium [202], (*Z*)-3-lithiooxy-1-methyl-1-propenyllithium [202] (**47**) and 2-lithiooxy-6-methyl-1-cyclohexenyllithium [203]; 2-(2-methyl-2-dioxolanyl)propyllithium [204] (**48**), 3,3-diethoxy-1-propenyllithium [205], (*E*)-3-methoxymethoxy-1-octenyllithium [206],

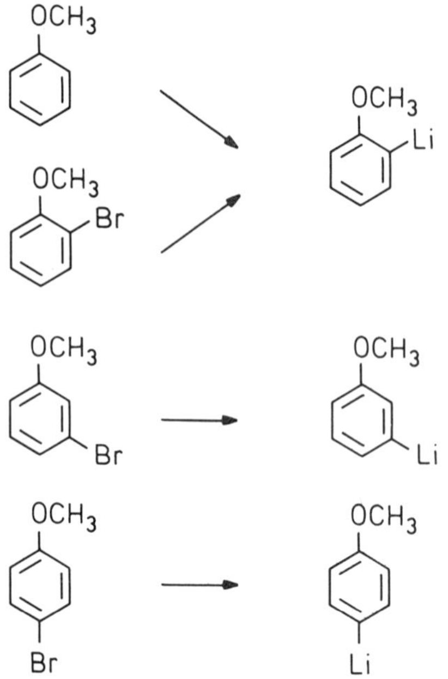

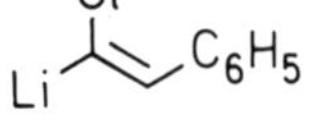

(43) 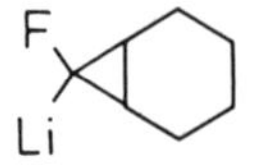 (44) 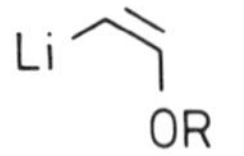 (45) 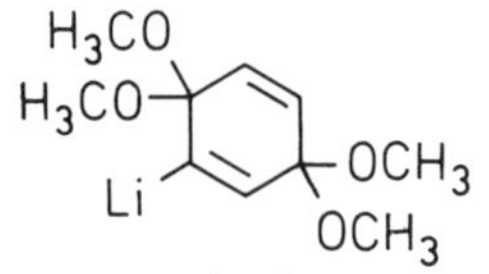 (46) 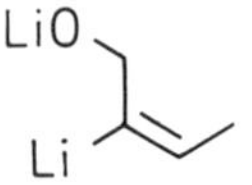 (47)

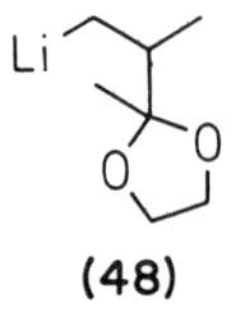

(48) 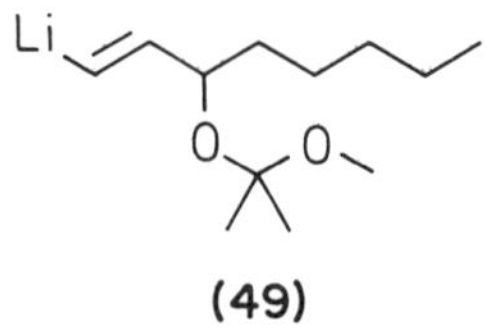 (49) 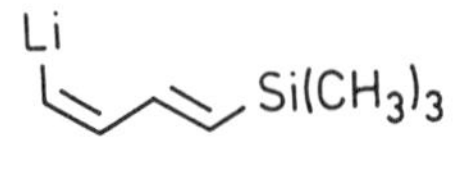 (50)

(*S*)-3-(1-methoxy-1-methylethoxy)-(*E*)-1-octenyllithium [206] (**49**) and 3-trimethylsilyloxy-4,4-dimethyl-1-octenyllithium [207]; 4,4,4-trimethoxybutyllithium [208]; a convincing (1*Z*,3*E*)-4-trimethylsilyloxy-1,3-butadienyllithium [209] (**50**) and (1*Z*,3*E*/1*E*,3*E*)-6,6-diethoxy-1,3-hexadienyllithium [210].

The halogen–metal interconversion also offers a reliable entry to a great variety of metal-bearing heterocyclic compounds. Again, we list just a few: 2-pyridyllithium [211], 3-pyridyllithium [211], 4-pyridyllithium [212], 3,4,6-triphenyl-2-pyridyllithium [213], 5,6,7,8-tetrahydro-2-quinolyllithium [214], 2-quinolyllithium [215] (**51**), 3-quinolyllithium [215], 4-quinolyllithium [215], 1-isoquinolyllithium [215] and 4-isoquinolyllithium (**52**) [215], 3-furyllithium (**53**) [216], 4-methyl-2-thienyllithium [217], 1-triisopropylsilyl-3-pyrryllithium (**54**, R = isopropyl) [218] and -*H*14-chloro-5,7-dilithio-7-(pyrrolo[2,3-*d*]pyrimidyl)lithium (**55**) [219].

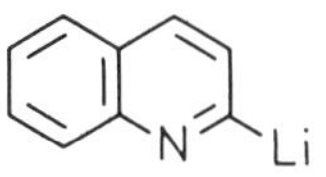

(51) 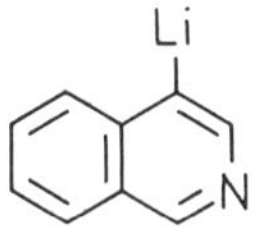 (52) 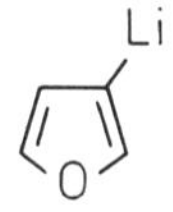 (53)  (54) 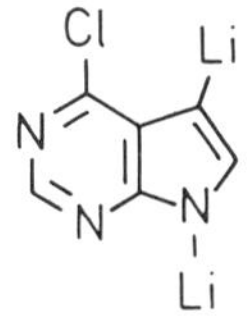 (55)

If several halogen atoms are present in the substrate, iodide is exchanged in preference to bromine and chlorine. If several like halogen atoms are present in a substrate, the one which occupies the most acidic site is first replaced. As a convincing demonstration of this principle, 2,4,5-tribromo-1-imidazoles (**56**; R = CH_3, $CH_2C_6H_5$) was submitted to three consecutive electrophilic substitutions at positions 2, then 4 and finally 5 [220]. Similarly, lithium was consecutively introduced into 2,4-dibromothiazole first at position 2 and next at position 4 [221].

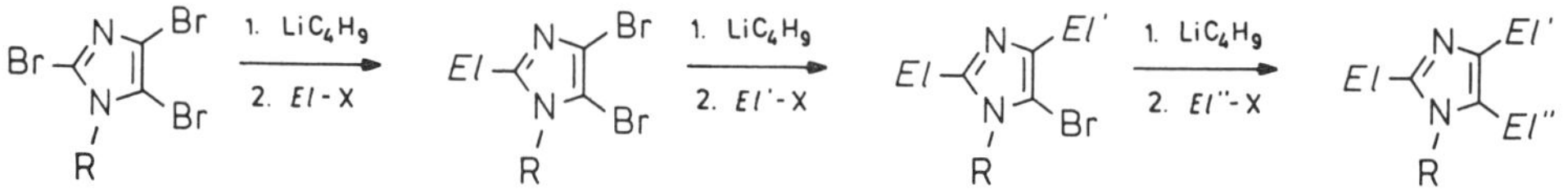

(56)

The organolithium species initially generated may gradually be transformed to a less basic isomer, if such a possibility exists. Thus, 3-bromo-2-chlorothiophene and butyllithium react to afford 2-chloro-3-thienyllithium (**57**), which is slowly converted into 5-chloro-2-thienyllithium (**58**) [222]. Analogously, 1-benzyloxymethyl-4,5-diiodoimidazole gives first 1-benzyloxymethyl-5-iodo-4-imidazolyllithium (**59**), which subsequently 'metamorphoses' to 1-benzyloxymethyl-5-iodo-2-imidazolyllithium (**60**) [223].

Such isomerization processes involve hydrogen–metal interconversions. It suffices to produce incidentally, for example by deprotonation of 3-bromo-2-chlorothiophene with 2-chloro-3-thienyllithium, a small amount of 2-chlorothiophene. This compound now can enter an exothermic acid–base reaction in which 5-chloro-2-thienyllithium is

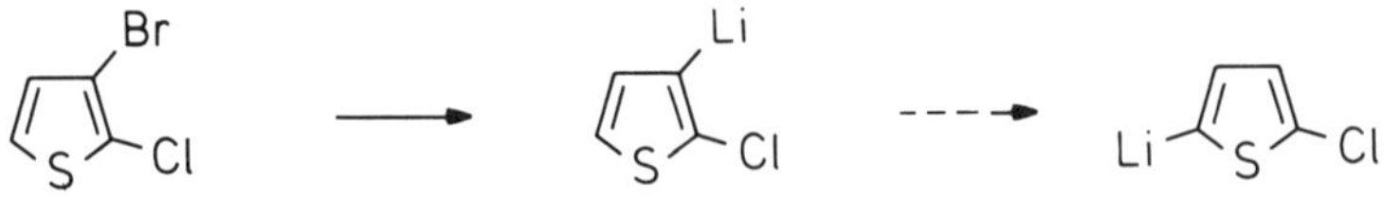

(57) (58)

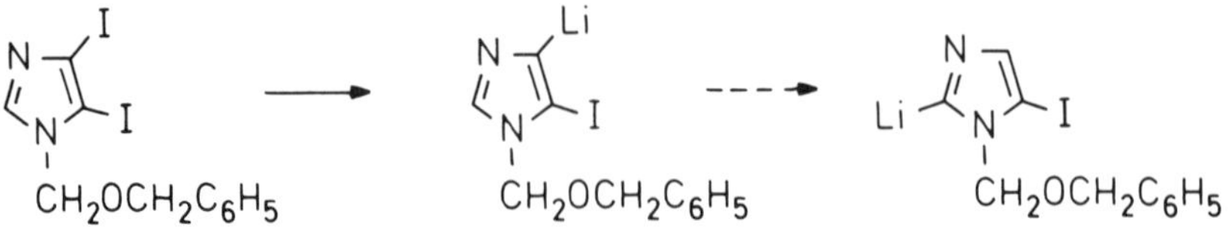

(59) (60)

continuously produced at the expense of the 3-chloro isomer. The 2-chlorothiophene, acting like a turntable, is continuously recovered as consumed.

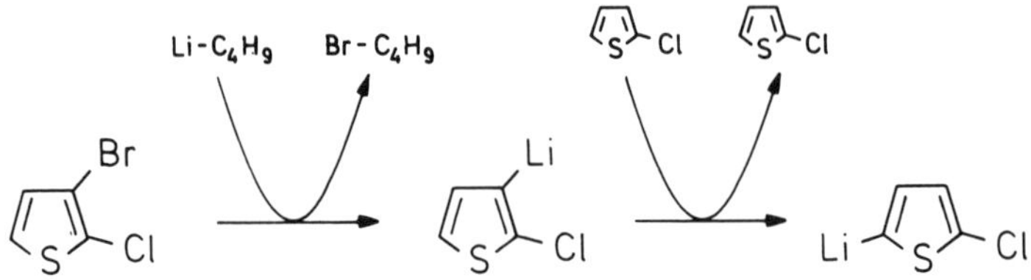

Bromine–lithium and iodine–lithium exchange reactions occur instantaneously, even at −75 °C, if an ethereal solvent is employed [224]. Prolonged exposure times are not only unnecessary but have to be avoided because of threatening side-reactions. The newly generated organolithium species may in particular undergo a condensation reaction with butyl bromide or butyl iodide. There are, however, tricks for removing such inevitable by-products of the halogen–metal exchange. If the interconversion is carried out in a hydrocarbon medium such as benzene or toluene, most aryllithium and 1-alkenyllithium compounds are sufficiently insoluble to permit removal of the butyl halide by washing and filtration [225]. The same goal can be achieved more conveniently if two equivalents of *tert*-butyllithium are used as the exchange reagent [226]. The *tert*-butyl bromide or iodide formed is immediately destroyed by the unconsumed *tert*-butyllithium, which promotes a β-elimination of hydrogen halide thus producing lithium bromide or iodide the presence of which generally does affect the outcome of

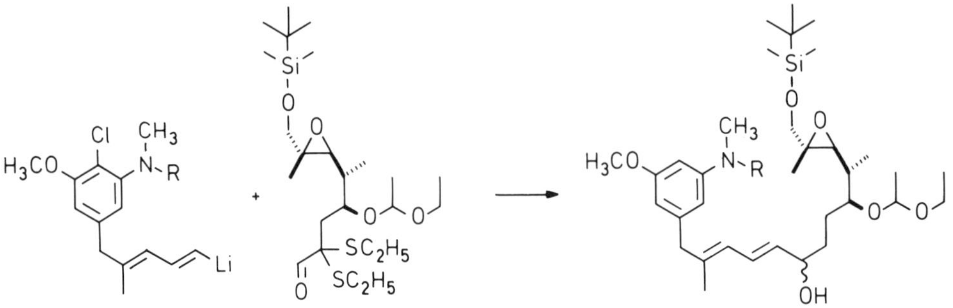

(61)

R = $COOCH_2CH_2Si(CH_3)_3$

the subsequent reaction. In this way, it was possible to generate the organolithium intermediate **61** at $-120\,^{\circ}C$ from a ω-bromodiene precursor and to use it as a key building block in the first enantioselective synthesis of maysine [227].

tert-Butyllithium may also be advantageously employed to shift exchange equilibria towards a dimetallated species. Thus 1,5-diiodonaphthalene was found to give cleanly the monolithio and dilithio derivatives **62** and **63** when treated with two and four equivalents of *tert*-butyllithium, respectively [228].

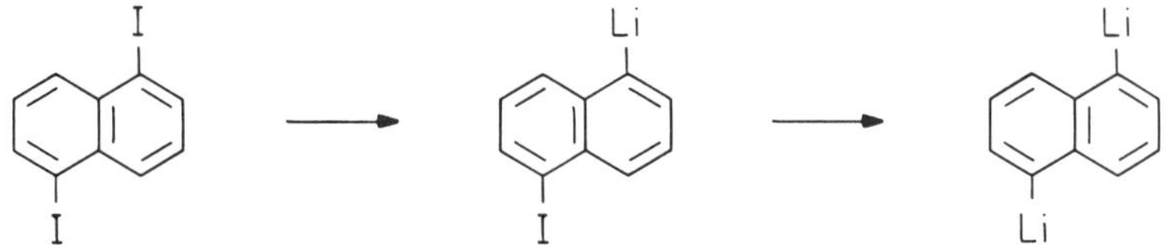

(62) **(63)**

Imperceivable halogen–metal exchange processes may give rise to products which, at first glance, could be mistaken as resulting from an S_N2-forbidden reaction. On closer inspection, of course, one realizes that the amide **64** obtained after neutralization was not formed by direct coupling of (*E*)-1-iodo-1-hexene with *N*,2-dilithio-*N*-benzoylethylamine but rather via the two components generated by halogen–metal interconversion [229].

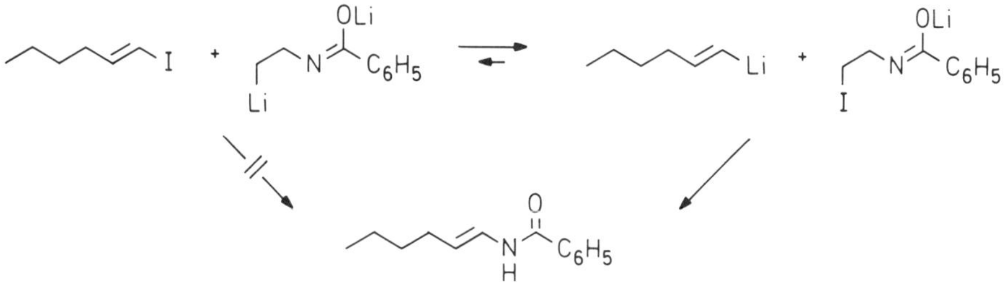

(64)

1.4.2.3 Permutational Chalcogen–Metal Exchange

According to all available evidence, first-row heteroelements do not participate at all in interconversion reactions. In the second row, silanes [230] undergo a rapid and phosphines [231] a slow exchange of substituents when they are treated with alkyllithium reagents and diphenyl sulfide [232] is practically inert.

The aryl–aryl replacement can, however, be accelerated by a factor of *ca* 10^9 if one switches from the inter- to an intramolecular reaction. Deuteriated 2-(2′-lithiophenyl)dibenzothiophene (see formula scheme on p. 62) needs less than 1 min at $-75\,^{\circ}C$ in tetrahydrofuran to achieve complete isotope scrambling between positions 9 and 3′ [233]. The ate complex **65**, which acts as a transient intermediate in the exchange process, is present in undetectable concentrations. However, the corresponding selenium compound (**65**, Se instead of S) becomes the preponderant, spectroscopically observable species if hexamethylphosphoric triamide is added to the solution in tetrahydrofuran in order to improve the solvation of the lithium cation [233].

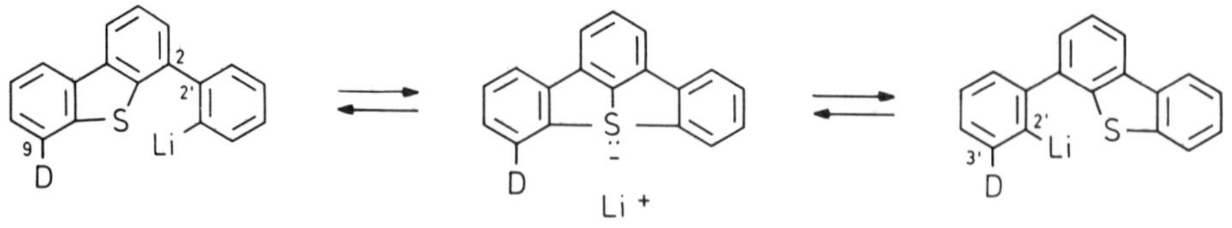

(65)

Selenides [233] and tellurides [234] are 10^4 and 10^{13} times, respectively, more reactive than sulfides towards organolithium reagents. While the synthetic potential of tellurides is just beginning to be explored [235], organoselenium compounds have attracted wide attention as intermediates in organic synthesis [236]. Diseleno acetals, carrying at least one hydrogen atom at the α-carbon position, can be deprotonated with alkali metal dialkylamides whereas, unlike dithio acetals [237], they react with alkyllithium reagents with selenium–lithium exchange [238]. This reaction mode opens up a wide field for structural modifications, as illustrated by the synthesis of the spiro compound **66** [239].

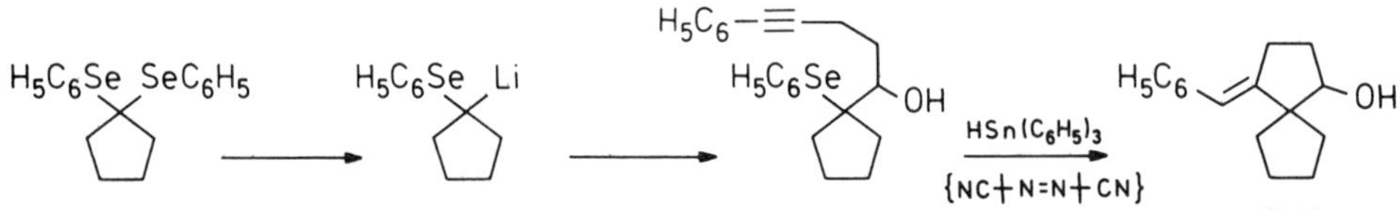

$HSn(C_6H_5)_3$ {NC–N=N–CN}

(66)

1.4.2.4 Permutational Carbon–Metal Exchange

Ordinary carbon–carbon bonds are completely resistant to the attack of organolithium reagents. However, the latter do undergo nucleophilic addition to many olefins. This process can be conceived of as the replacement of one carbon moiety by another, the metal being simultaneously transferred from the entering to the departing fragment.

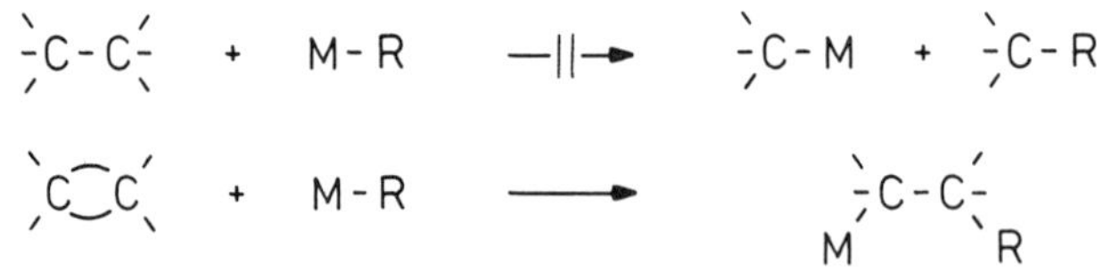

$$\geq C-C \leq \; + \; M-R \; \not\longrightarrow \; \geq C-M \; + \; \geq C-R$$

$$>C{=}C< \; + \; M-R \; \longrightarrow \; M-\overset{|}{\underset{|}{C}}-\overset{|}{\underset{|}{C}}-R$$

Butyllithium combines with ethylene not only at elevated, but also at ordinary temperatures. Since the adducts are as reactive as their precursors, the addition can continue and linear polyethylene of low molecular weight is produced [240]. On the other hand, the reaction stops cleanly after the first addition step if isopropyllithium or *tert*-butyllithium is used, since the resulting primary alkyllithiums are less basic than the secondary or tertiary compounds [241].

$$H_2C{=}CH_2 \; + \; Li{-}C(CH_3)_3 \; \longrightarrow \; Li{-}CH_2{-}CH_2{-}C(CH_3)_3$$

Relief or ring strain facilitates the organometallic addition reaction and also contributes to control its outcome. Norbornene [242] and cyclopropene [243] rapidly react with a variety of organolithium and, in the latter case, even organomagnesium compounds. Finally, neighbouring group assistance may play an important role. Allyl ethers and alcoholates do react under conditions where simple olefins are completely inert. When treated with Grignard reagents, butyl cinnamyl ether undergoes a vinylogous substitution either via the adduct **67** or, more likely, in a concerted fashion [244]. Allylmagnesium bromide combines with homoallyl alcoholates to give adducts of type **68** (R = e.g. C_6H_5) [245] while organolithium reagents add smoothly to allyl alcoholates producing branched derivatives such as **69** (R = alkyl, aryl) [246].

H_9C_4O ... C_6H_5 → (H_9C_4O, C_7H_{15}, C_6H_5, MgBr) **(67)** → $-[H_9C_4O-MgBr]$ → C_7H_{15}, C_6H_5

BrMgO, R, R → O — Mg, R, R **(68)** → $H^{\oplus}$ / H_2O → HO, R, R

LiO → LiO Li, R **(69)** → CO_2 → O, =O, R

Allylic and benzylic sodium and potassium compounds are particularly efficient nucleophiles that readily combine with ethylene although the only remaining driving force is the transformation of a π to a σ bond (in other words, the relief of strain inherent in the double bond which may be visualized as a 'two-membered ring'). Actually, adducts (e.g. **70** [247] and **71** [248]) are far more basic than their organometallic precursors. The newly formed intermediates are actually so reactive that they cannot

K → K **(70)** →

K → K **(71)** →

be trapped as such but rather are destroyed by proton abstraction from the solvent, if ethereal, or from another C,H acid.

The high propensity of delocalized organometallics for nucleophilic addition to carbon–carbon double bonds is at the origin of the so-called anionic polymerization of butadiene, isoprene and styrene [249]. In general, an alkyllithium reagent is used to initiate the reaction. This first adduct is formed relatively slowly but its subsequent reaction is fast. Since each addition step to the diene or styrene restores an allyl- or benzyl-type structure, the polymer chain grows rapidly until all monomer has been consumed. The chain growth can be started again if another monomer is introduced into the batch. In other words, the polymer remains 'living' [25] until the reactive ends are quenched by solvolysis. When butyllithium is allowed to react with 1,1-diphenylethylene rather than styrene, one encounters the reverse situation: any subsequent reaction is too slow to compete with the relatively fast first addition step. In this way, 1,1-diphenylhexyllithium (**72**) is obtained virtually quantitatively.

$$C_6H_5(H)C{=}CH_2 \xrightarrow[\text{slow}]{Li\text{-}C_4H_9} Li\cdots CH(C_6H_5)\text{-}CH_2\text{-}C_4H_9 \xrightarrow[\text{fast}]{C_6H_5\text{-}CH{=}CH_2} Li\cdots CH(C_6H_5)\text{-}CH_2\text{-}CH(C_6H_5)\text{-}CH_2\text{-}C_4H_9$$

$$(C_6H_5)_2C{=}CH_2 \xrightarrow[\text{fast}]{Li\text{-}C_4H_9} Li\cdots C(C_6H_5)_2\text{-}CH_2\text{-}C_4H_9 \;(\mathbf{72}) \;\not\longrightarrow\; Li\cdots C(C_6H_5)_2\text{-}CH_2\text{-}C(C_6H_5)_2\text{-}CH_2\text{-}C_4H_9$$

(72)

Organolithium reagents add with particular ease to fulvenes [251] and at more moderate rates to polycyclic aromatic or quasi-aromatic ring systems such as anthracene [252] or azulene [253]. Phenyllithium undergoes slow addition to the triple bond of diphenylacetylene [254]. All such reactions are of only limited value for the preparation of organometallic reagents. Since they are of principal importance, they nevertheless have deserved appropriate coverage.

1.4.2.5 Permutational Hydrogen–Metal Exchange

The terms 'metallation' and 'transmetallation' are widely used to designate the transfer of metal from an organometallic reagent to a hydrocarbon substrate in exchange for a carbon-bound hydrogen atom. This reaction mode was accidentally discovered by Schorigin [255] when he submitted diethylmercury, dissolved in benzene, to reductive cleavage with sodium metal. Rather than the expected ethylsodium, he obtained phenylsodium as the reaction product, without doubt the result of a subsequent interconversion process.

$$Hg(C_2H_5)_2 \xrightarrow{Na} Na\text{-}C_2H_5 \xrightarrow[-[C_2H_6]]{C_6H_5} Na\text{-}C_6H_5$$

In the following decades, systematic investigations have established the universality of this transmetallation route. In accordance with their formal relationship to acid–base reactions [256], these interconversion processes inevitably generate a less basic organometallic compound at the expense of a more basic compound. For example, fluorene was found to react with ethyllithium to afford ethane and 9-fluorenyllithium [257]. A convenient working procedure for the preparation of benzylsodium [258] uses chlorobenzene, sodium and toluene as the ingredients; obviously the method relies on the *in situ* transformation of phenylsodium by transmetallation with toluene.

Scope and Limitations

Gradually it was also realized that the ease with which a hydrogen–metal exchange reaction occurs depends not only on the acidity gradient established between the two hydrocarbons involved but also on the electropositivity of the metal employed (see Figure 1.9). While alkylpotassium reagents are claimed to be powerful enough to metallate even pentane, in other words to promote thermoneutral reactions [259], organolithium, unlike organosodium, compounds are unable to deprotonate benzene or even toluene if no special activation is provided. Finally, organomagnesium reagents do not participate in transmetallation processes unless a very acidic substrate such as an acetylene or cyclopentadiene is present as the proton source.

Being soluble in ethereal and frequently also in paraffinic solvents, alkyllithium reagents are most convenient to work with. Moreover, many of them are commercially available. Therefore, it is of great practical importance to define the scope and limitations of alkyllithium-promoted metallation reactions. Generally, only substrates

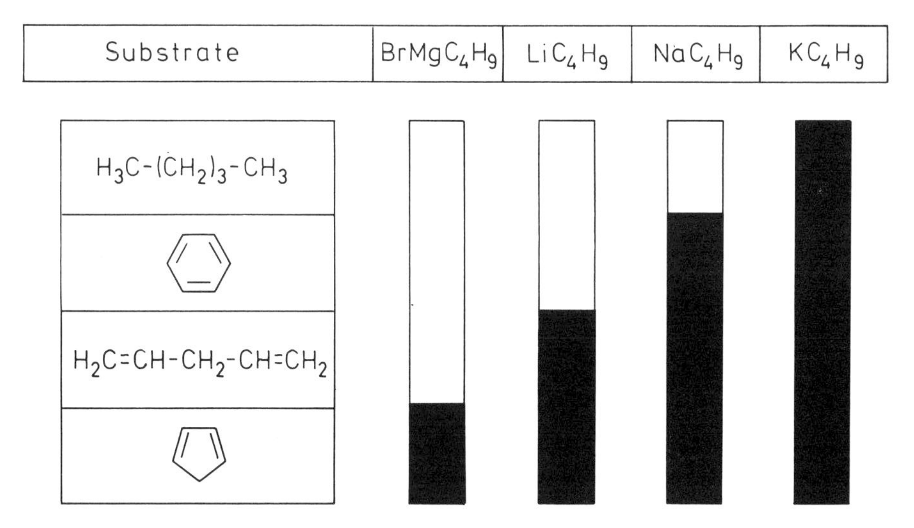

Figure 1.9. Pictorial representation of the interdependence between acidity gradients and metal effects in prototype transmetallation reactions (employing butylmagnesium bromide and butyllithium in an ethereal and butylsodium and butylpotassium in a paraffinic solvent). Open boxes mean no reaction and shaded boxes mean the reaction is taking place

with enhanced C,H acidity are amenable to hydrogen–lithium exchange. One can differentiate between several fundamental carbanion-stabilizing and thus acidity-increasing factors, as follows.

Ring Tension The gas-phase protonation enthalpy of cyclohexanide is at least 10 kcal mol^{-1} higher than that of cyclopropanide (see Figure 1.7). The relationship between geometry ('hybridization') and acidity can be understood on the basis of the Gillespie–Nyholm model (valence shell electron pair repulsion concept, VSEPR) [260, 261]. Cyclopropane reacts only sluggishly with organosodium (see pp. 72–73) and not at all with organolithium compounds. More severely strained three-membered rings such as tricyclo[3.1.1.0^{6,7}]heptane [262] and 1,3,3-trimethylcyclopropene [263] are, however, readily deprotonated by butyllithium, giving rise to intermediates **73** and **74**.

Li Li

(73) **(74)**

Following Pauling's advice that 'bent bonds are better' [264], ethene and ethyne can be conceived as 'cycloethane' and 'bicyclo[0.0.0]ethane' [261]. The enhanced proton mobility of olefinic, allenic and acetylenic C,H bonds can then again be attributed to C—C—C angle compression [261].

Finally, ring strain contributes significantly to the C,H acidity of five-membered carbo- and heterocyclic rings. Pentylsodium metallates spiro[4.4]nona-1,3-diene at position 2 (intermediate **75**) [265] under conditions where an ordinary open-chain and conjugated diene would simply become the prey of polymerization. 1-Trimethylsilylpyrrole undergoes metallation with butyllithium in hexane at the nitrogen adjacent position 2 (see p. 85), but with *tert*-butyllithium at the sterically more accessible position 3 (intermediate **76**) [266].

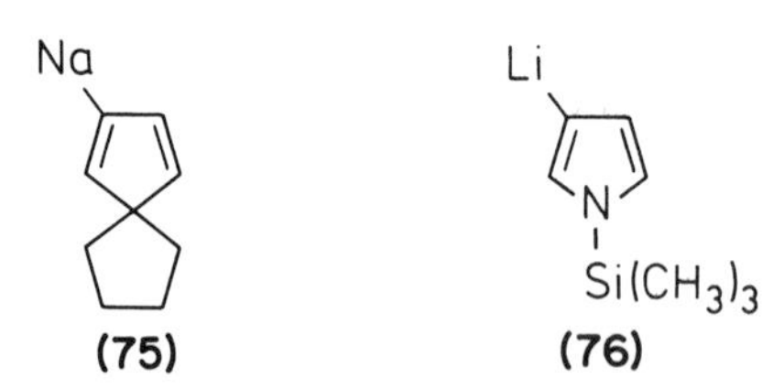

(75) **(76)**

Polarization In the last two cases a second factor has to be taken into consideration. Conjugated or homoconjugated double bonds can interact with each other in a way that helps to attenuate a charge excess by deformation ('polarization') of their electron clouds [267]. In these terms it can be explained why norbornadiene (bicyclo[2.2.1]-hepta-2,5-diene) [268, 269], bicyclo[3.2.0]hepta-2,6-diene [268] and cycloheptatriene

[268], when metallated with pentylsodium to afford the intermediates **77**, **78** and **79**, react one or two orders of magnitude more rapidly than the monounsaturated analogues. Polarization phenomena may also be invoked to explain why butyllithium deprotonates 4*H*-pyran [270] (generating intermediate **80**) more rapidly than 3,4-dihydro-2*H*-pyran [271] and to rationalize certain reactivity patterns of substituted arenes.

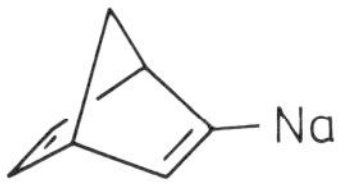

(77) 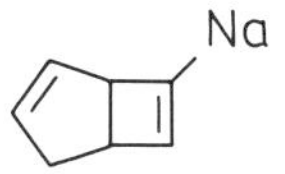 **(78)** 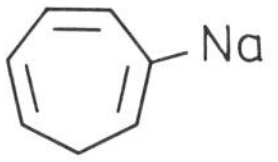 **(79)** 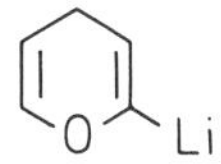 **(80)**

p-Orbital Resonance Effects Charge delocalization offers a more efficient mechanism for the attenuation of electron excess than mere bond polarization. The proton affinities (see Figure 1.7) of the allyl and benzyl anions are roughly 35 and 40 kcal mol^{-1}, respectively, smaller than that of the propyl anion (1-propanide ion). These energy differences reflect the resonance energy which is gained by charge delocalization within a molecular framework having alternating loops (antinodal points) and nodes.

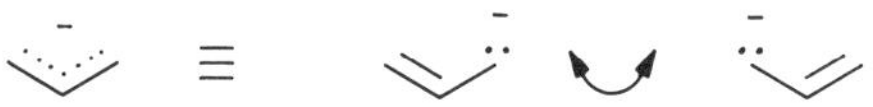

Butyllithium is not reactive enough to metallate simple alkenes or methylarenes unless it is activated by complexation with *N*,*N*,*N′*,*N′*-tetramethylethylenediamine (TMEDA). Under such circumstances allyllithium [272], 2-methylallyllithium [272], 2-butenyllithium [273] and benzyllithium [274] were prepared with acceptable yields. 1,4-Dienes [275], 1,3,6-trienes [276], diphenylmethane [277] and triphenylmethane undergo a rapid hydrogen–lithium interconversion with butyllithium in ethereal solvents. The deprotonation of fluorene [275, 278], indene [279] and cyclopentadiene [280] occurs instantaneously, all of them generating quasi-aromatic organometallic entities.

Resonance effects become even more pronounced if they allow the negative charge to be displaced from a carbon to a hetero atom. Enamide and enolate ions accumulate most of the excess electron density at the nitrogen or, in particular, oxygen atom. This is the reason why aliphatic carbonyl compounds are more acidic than azomethines (Schiff's bases) and the latter for their part more acidic than alkenes.

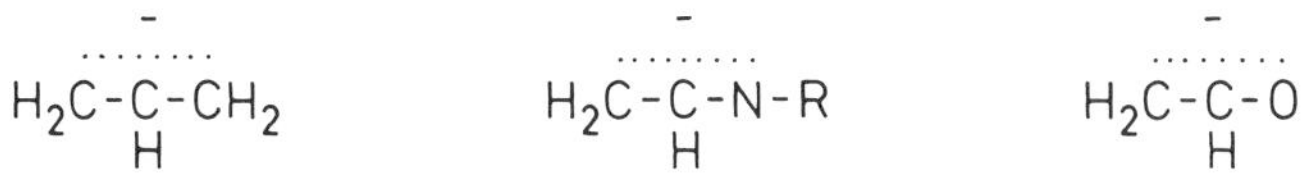

$H_2C{-}C(H){-}CH_2$ $H_2C{-}C(H){-}N{-}R$ $H_2C{-}C(H){-}O$

Inductive Electron Withdrawal The polarization effects discussed under one of the preceding headings could also have been ascribed to inductive electron withdrawal caused by the olefinic carbon centres which are arguably more electronegative than saturated centres. We prefer to speak of inductive effects only in conjunction with heteroelements.

The electrostatic attraction exerted by a heteroatom becomes especially strong if the latter carries a positive charge. Thus, tetramethylammonium bromide is rapidly deprotonated to afford a so-called nitrogen ylid (**81**), more precisely the lithium bromide adduct of trimethylammoniomethanide [281]. The metal-free ylid is unstable; it decomposes instantaneously by methyl migration (Stevens rearrangement) or methylene transfer [281]. *N*-Ylid-related structures are diazomethane (**82**) and lithiated isocyanides (**83**) [282].

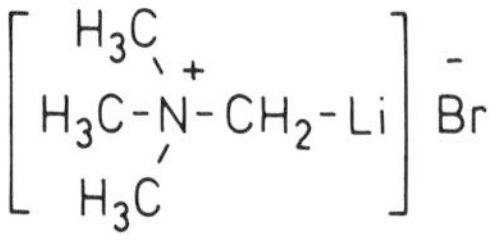

$$\left[(H_3C)_3\overset{+}{N}-CH_2-Li\right]\overset{-}{Br} \qquad :N\equiv\overset{+}{N}-\overset{-}{C}H_2 \qquad :\overset{-}{C}\equiv\overset{+}{N}-CH_2-Li$$

(81) 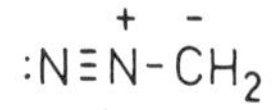 **(82)** 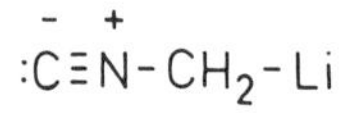 **(83)**

We do not expect trimethylamine and dimethyl ether to be much more acidic than methane in the gas phase ($\Delta pK \leqslant 5$) [283]. In other words, uncharged nitrogen and oxygen substituents should only modestly stabilize carbanions and organometallics derived therefrom. They facilitate, however, very markedly hydrogen–metal exchange processes occurring in their vicinity by providing neighbouring group assistance (see pp. 81–88).

d-Orbital Resonance Effects Like trimethylsulfonium salts, tetramethylphosphonium bromide was found to be many orders of magnitude more acidic than the corresponding ammonium salt [284]. On the basis of electrostatic forces alone, the first-row element should outperform its higher isologue because of the shorter distance between the oppositely charged centres in the zwitterionic ylids originating from deprotonation. The second-row element must hence offer extra stabilization to the carbanion and this was attributed to an expansion of its valence shell beyond the usual electron octet. Hence the true bonding situation in the ylid could be represented by two resonating limiting structures, an 'ylid' and an 'ylene' form (**84a** and **84b**, respectively) [284, 285].

$$(H_3C)_3\overset{+}{P}-\overset{-}{C}H_2 \longleftrightarrow (H_3C)_3P=CH_2$$

(84a) **(84b)**

The capacity of phosphorus to accommodate ten or even twelve valence electrons has been amply demonstrated. The syntheses of pentaphenylphosphorane [286]

(**85**) and bis(*o,o'*-biphenylene)phosphonium tris(*o,o'*-biphenylene)phosphate [287] (**86**) illustrate this in a particularly convincing manner.

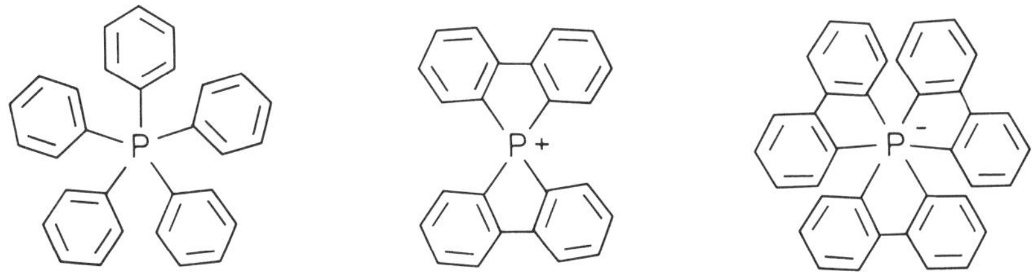

(85) (86)

Although the concept of d-orbital resonance remains controversial [288], there can be no doubt about the existence of a special acidifying effect of second- and higher row elements. The heteroatom must not carry a positive charge to become effective. Methyldiphenylphosphine is readily deprotonated at the α-position by TMEDA activated butyllithium giving (diphenylphosphino)methyllithium (**87**) [289]. The corresponding phosphine oxide reacts, of course, much faster to afford (diphenylphosphinoyl)methyllithium (**88a**) [290], which presumably exists preferentially in a metallomeric ylid form (**88b**) [291].

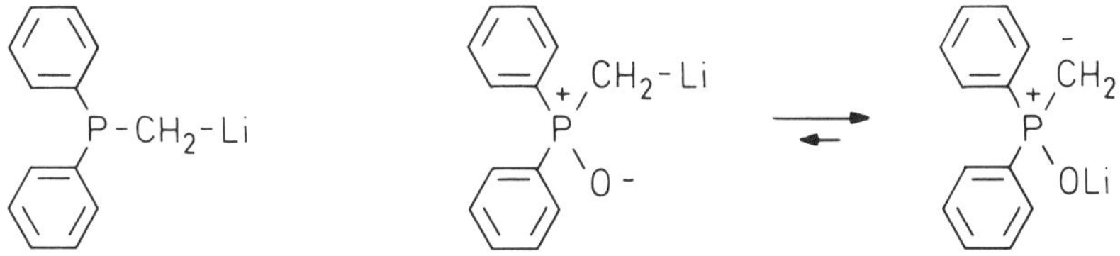

P-CH_2-Li | CH_2-Li, P, O- | CH_2, P, OLi

(87) (88a) (88b)

Analogously, methyl phenyl sulfide (thioanisole) can be converted into phenylthiomethyllithium [292], 2-methyl-1,3-dithiane into 2-methyl-2-(1,3-dithianyl)lithium [293] and trimethyl trithioorthoformate into tri(methylthio)methyllithium [294]. Tetramethylsilane and ethoxytrimethylsilane can be metallated to give trimethylsilyllithium [295] and (ethoxydimethylsilyl)methyllithium [296], respectively.

Unlike the corresponding fluoro analogues, dichloromethyllithium and trichloromethyllithium can be easily generated and can be conserved at low temperatures (e.g. −110 °C) [297]. Tribromomethane and trichloromethane are far more acidic than trifluoromethane, despite the very short carbon–halogen bond in the latter [298]. A rationalization of these differences must again take into account d-orbital resonance effects.

Second-row and heavier elements can exert their special carbanion-stabilizing interaction without being subject to any geometrical restriction. At olefinic centres they display an unimpaired acidifying effect: 1-alkenyl chlorides [299] are much more readily deprotonated than 1-alkenyl fluorides and enesulfides [300] than enethers and thiophene [301] is more reactive than furan (overleaf).

The most striking feature is the formal violation of Bredt's rule by the second-row element sulfur. While the tricyclic enolate **89** is by at least 15 pK_a units more basic than an open-chain analogue [302], the rigid trisulfone anion **91** [303] is only slightly more basic ($\Delta pK_a \approx 3$) and the trisulfide **90** [304] is even less basic than the corresponding acyclic counterpart.

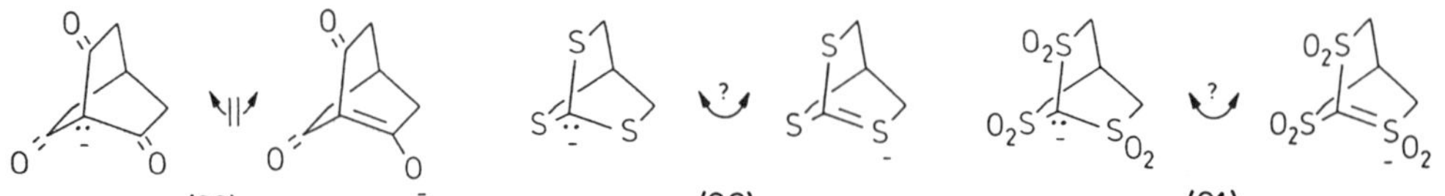

(89) (90) (91)

The d-orbital resonance effect appears to originate from a deformation of the carbanion electron density in the direction toward the heavy heteroatom. In a wider sense, the strong acidifying effect exerted by transition elements on complexed cyclopentadienyl or arene ligands may be considered as a related phenomenon. For example, ferrocene [305] and di(benzene)chromium [306] can both be submitted to efficient monometallation (species **92** or **93**) and dimetallation.

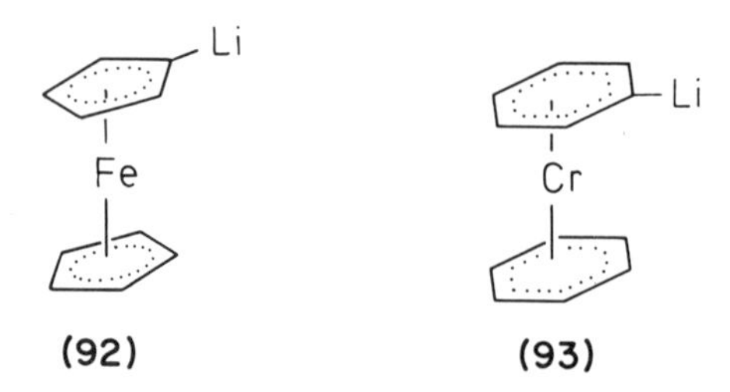

(92) (93)

As we have seen above, a wide variety of substrates undergo smooth hydrogen–lithium exchange with butyllithium alone. In the case of a relatively inert substrate, the reagent may require activation by added TMEDA. What can be done, if under such circumstances the reaction still fails to proceed in the desired way?

Let us try to gain some insight into the action of such a chelating complexand as TMEDA before considering what options for improvements exist. The perfect transition state (**94**) for the proton transfer from one carbon moiety to another would be characterized by a linear C—H—C bridge. This could only be accomplished, however, if the metal were not directly involved in the interconversion process but rather were to be separated from the carbon backbone as a persolvated ion (YR_2 means a solvent molecule) and wait until the new binding site has been uncovered. As we have seen previously (p. 30), such a carbanion mechanism can only be realized if exceptionally stable carbanions are involved. The general exchange mode implies a four-centre-type transition state (**95**) which suffers from compressed bond angles at the crucial carbon centres and an angular C—H—C bridge. Moreover, the reorganizing carbon–metal bonds are weak. In order to compensate for the latter defect, the solvation of the metal has to be improved. This can be achieved by using a chelating complexand such as TMEDA (schematized transition state **96**) or even a macrocyclic polyether.

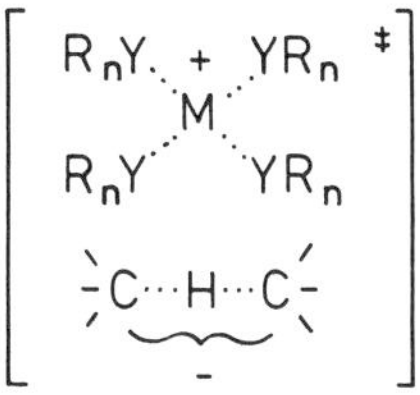

(94) 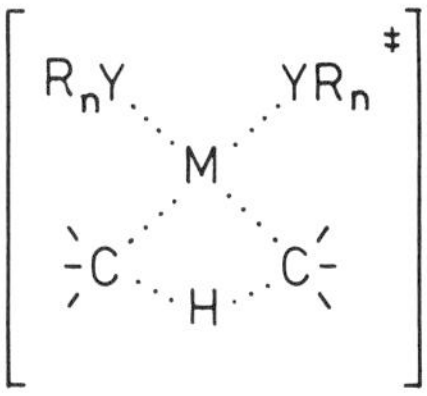 (95) 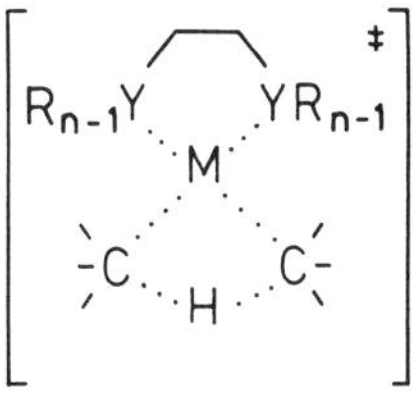 (96)

Actually, TMEDA-type diamines and crown ethers exert a twofold effect. As described above, they polarize, i.e. weaken, carbon–metal bonds and thus facilitate four-centre bonded transition states (as depicted in **96**). Before doing that, they break up the alkyllithium hexamers or tetramers into far more reactive smaller subunits, mainly dimers or monomers. These two effects taken together provide powerful activation to metallating reagents. Nevertheless, the application of such complexands has narrow practical limits. Although the activation may not yet be strong enough to cope with a reluctant substrate, it may already be too much to allow undamaged survival of the fragile complexand. TMEDA undergoes slow metallation at one of its methyl groups [307] whereas crown ethers are fairly rapidly degraded by β-elimination [308]. If we wish to optimize activation, we have to identify a potential ligand of the metal being endowed with a superior donor capacity and, at the same time, being itself completely resistant against attack of an organometallic reagent.

There are two independent solutions to the problem. One can use a relatively weak electron-donor moiety but covalently attach it to the substrate so that its complexing power is potentiated by the entropy-saving proximity effect. This is the approach based on *neighbouring group assistance* to metallation (transition state **97**). Alternatively, one can add potassium *tert*-butoxide or another soluble alcoholate having a heavy alkali metal as a counter ion. In this way, one disposes of an extremely electron-rich complexand (schematized transition state **98**: MYR represents, e.g., potassium *tert*-butoxide) without incurring the risk of losing it by destructive collision with the organometallic reagent. This is the *superbase approach*. Let us first deal with the latter aspect. We can restrict ourselves to a brief outline, since 'LIC–KOR' [309] mixtures of alkyllithium (LIC) and potassium alcoholates (KOR) have gained in the meantime a solid reputation as powerful tools for modern organic synthesis. Moreover, the superbase chemistry has been reviewed recently [310, 311].

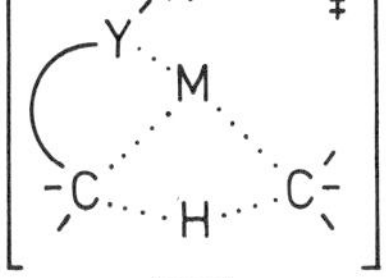

(97) 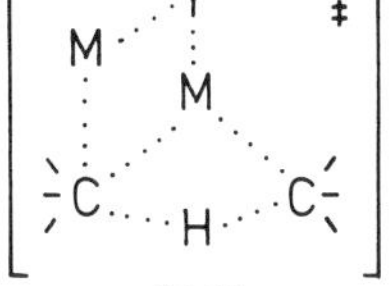 (98)

Superbase Approach

The combination of butyllithium and potassium *tert*-butoxide should be conceived of as a mixed metal reagent. It is not simply butylpotassium, as has sometimes been claimed. The alleged identity of the two reagents can be disproved on the basis of fairly sophisticated arguments [309, 312], but also by a simple empirical comparison of their behaviour. Butylpotassium can be very conveniently generated from dibutylmercury [313]. It reacts with tetrahydrofuran at temperatures below −100 °C [314], whereas the LIC–KOR reagent can be conserved for some time in the same solvent at −60 °C [314].

The increased solvent stability is doubtless one of the key factors that can explain why metallation protocols employing superbases usually outperform similar reactions carried out with neat organosodium or organopotassium reagents. Typical results featuring three cyclopropanes, three olefins without resonance-active allyl positions and three ordinary alkenes are listed in Table 1.8.

Superbase reactions are easy to perform, even on a ton scale. Moreover, the reagents are not only extremely powerful but at the same time show a high degree of typo-, regio- and stereocontrol. Thus they combine brute force with subtlety or, in chemical terms, reactivity with selectivity, two attributes that are difficult to match and rather tend mutually to exclude each other. In order to establish a collection of leading references, we shall now present a concise but systematic overview of the different classes of organometallic intermediates that have been generated by means of superbasic reagents with preparatively useful yields and selectivities.

Cyclopropylmetal Compounds Bicyclo[4.1.0]heptane ('norcarane') [265] and spiro-[2.2]pentane [324] are the least acidic hydrocarbons that have so far been successfully metallated under stoichiometrically controlled conditions. Tricyclo[2.2.1.0^{2,6}]heptane ('nortricyclene') [265, 325] is considerably more reactive owing to increased ring strain.

Functional groups at cyclopropane rings accelerate the interconversion process by neighbouring group assistance and direct the metal to their closest neighbourhood. This principle is illustrated by the metallation of 3-methoxytricyclo[2.2.1.0^{2,6}]heptane [318].

1-Alkenylmetal Compounds Not only ethylene [326] but also bicyclo[2.2.1]heptene [268], bicyclo[3.2.0]heptene [268] and bicyclo[2.2.2]octene [265] can be efficiently metallated. With 3,3-dimethyl-1-butene [265] and camphene [265, 318] as substrates the metal enters exclusively into the sterically unhindered *E*-position (to afford species **99** and **100**).

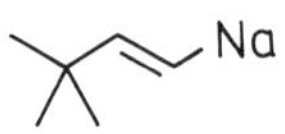

Na

(99)

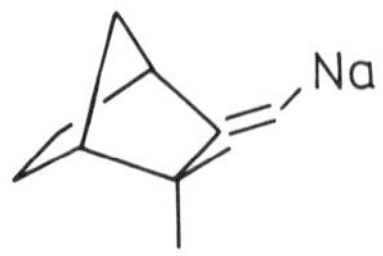

Na

(100)

Table 1.8. Metallation of model hydrocarbons either with standard organosodium and -potassium reagents or with mixed metal superbases (reaction conditions and yields obtained after trapping with the electrophiles being specified in brackets)

Metallation reaction[a]	Previous work	Superbase results
 → M	NaC_5H_{11} HEX, 100 h, 25 °C, 3% [CO_2] [265]	NaC_5H_{11}–$KOC(CH_3)_3$[b] HEX, 50 h 25 °C, 30% [CO_2] [265]
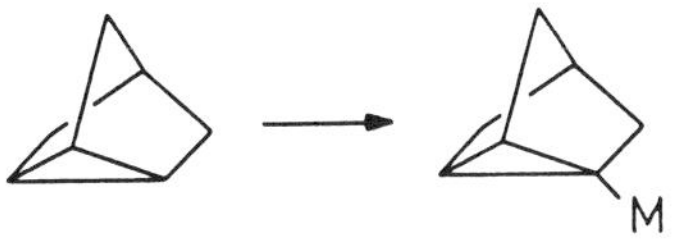 → M	NaC_5H_{11} HEX, 22 days, 25 °C, 5% [CO_2] [315]	NaC_5H_{11}–$KOC(CH_3)_3$[b] PEN, 24 h, 25 °C, 60% [$ClSi(CH_3)_3$] [265]
 → M	NaC_5H_{11} HEX, 7 days, 25 °C, 20%[c] [CO_2] [316]	—[d]
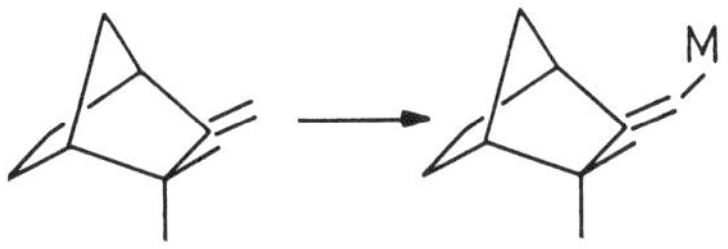 → M	NaC_5H_{11} HEX, 69 days, 25 °C, 18%[e] [CO_2] [317]	NaC_5H_{11}–$KOC(CH_3)_3$[b] PEN, 1 h, 25 °C, 68% [CH_3I] [318]
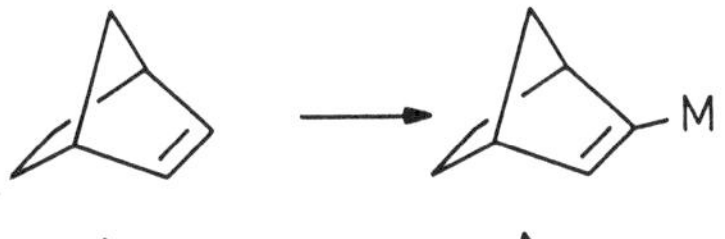 → M	NaC_5H_{11} HEX, 17 days, 25 °C, 60% [CO_2] [315]	LiC_4H_9–$NaOC(CH_3)_3$[b] THF, 15 h, −50 °C, 88% [$ClSi(CH_3)_3$] [268]
 → M	NaC_5H_{11} HEX, 5 h 25 °C, 0%[f] [CO_2] [315]	LiC_4H_9–$NaOC(CH_3)_3$[b] THF, 15 h[g,i], −50 °C, 92% [$ClSi(CH_3)_3$] [268]
 → M	NaC_5H_{11} HEX, 24 h, 25 °C, 0% [CO_2] [319]	LiC_4H_9–$KOC(CH_3)_3$ THF, 1 h, −55 °C, 74% [CO_2] [320, 321]
 → M	NaC_5H_{11} HEX, 30 days, 25 °C, 38%[h] [CO_2] [322]	$Li^sC_4H_9$–$KOC(CH_3)_3$[b] PEN, 24 h, 25 °C, 77% [$O(CH_2)_2$] [318]
 $H_{17}C_8$ → $H_{17}C_8$ M	KC_4H_9[j] HEX, 22 h, 25 °C, 52%[k] [CO_2] [323]	LiC_4H_9–$KOC(CH_3)_3$ THF, 6 h, −50 °C[k] [324] 70% [$FB(OCH_3)_2$–H_2O_2]

[a] Abbreviations: M = metal, HEX = hexane, PEN = pentane, THF = tetrahydrofuran, $Li^sC_4H_9$ = *sec*-butyllithium.
[b] Similar results, although with lower yields, with LiC_4H_9–$KOC(CH_3)_3$.
[c] Cyclopropane used as a cosolvent.
[d] Not attempted.
[e] Impure material.
[f] Only fragmentation products were identified.
[g] Virtually the same yield was obtained after 2 h.
[h] With 'aged' (e.g. 4 weeks old) pentylsodium and sodium isopropylalcoholate about 80% yield.
[i] With pentylsodium (HEX, 18 h, 25 °C) and a fourfold excess of the olefin about 45% yield.
[j] Mixtures of 2-nonyl-3-butenoic acid (main component), (*E*)-2-tridecenoic acid and (*Z*/*E*)-3-tridecenoic acid.

Vinyl ethers [327] and enesulfides [300] can already undergo a hydrogen–metal exchange at the α-position when conventional organolithium reagents are used. 3,4-Dihydro-2*H*-pyran requires at least organosodium reagents [271]. The LIC–KOR mixture offers distinct advantages, in particular when the heterocyclic ring carries labile substituents (e.g. intermediates **101** [328] and **102** [329]).

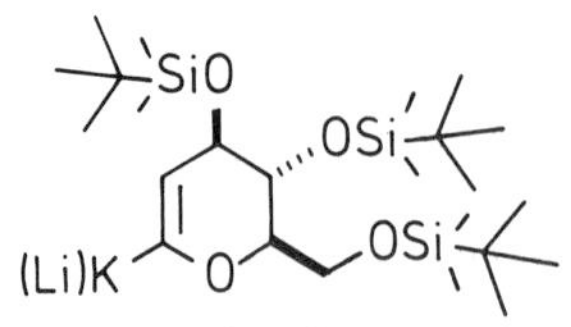

(101) 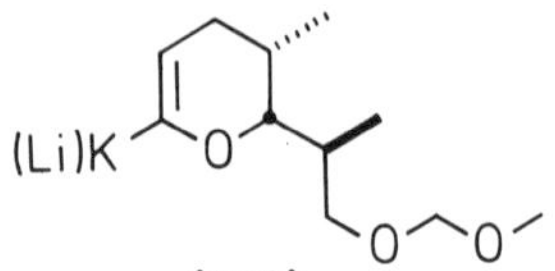 (102)

Arylmetal Compounds The instantaneous and quantitative metallation of excess of benzene by butyllithium and potassium *tert*-butoxide was the first revelation of the enormous reactivity potential of the LIC–KOR superbase [277]. Subsequently, conditions were disclosed that allow stoichiometric amounts of benzene to be submitted to mono- or dimetallation in tetrahydrofuran or hexane medium, respectively [312]. 1-(Methylcyclopropyl)benzene [312] and *tert*-butylbenzene [312, 330] react roughly five times more slowly, producing a *ca* 1:1 mixture of *meta*- and *para*-isomers. 1,3-Di(*tert*-butyl)benzene, however, gives a single intermediate (**103**) [312].

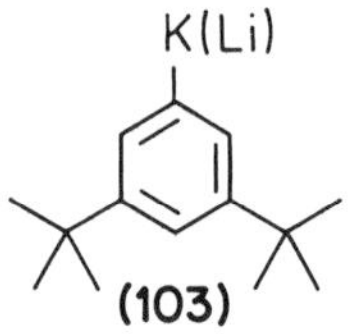

(103)

Alkinyl substituents enhance the reactivity of the arene. Lithium or potassium phenylacetylide undergoes clean hydrogen–metal exchange at the *ortho*-position [331]. On the other hand, (3,3-dimethyl-1-butynyl)benzene and (3-hydroxy-3-methyl-1-butynyl)benzene form mixtures of *ortho*-, *meta*- and *para*-isomers in the ratios of 9:78:13 and 69:25:6, respectively [331].

Heterosubstituted arenes such as fluorobenzene, anisole and aniline derivatives are particularly reactive. The *ortho*-metallated intermediates which can be readily generated from such substrates is discussed in detail later (pp. 85–103).

Benzylmetal Compounds Toluene can be cleanly α-metallated with the LIC–KOR superbase [277] whereas TMEDA-activated [332] butyllithium gives rise to some contamination by ring-metallated products. α-Alkyl substituents diminish the reactivity of benzylic positions. 1-Ethyl-4-methylbenzene is selectively deprotonated at the methyl group and 1-ethyl-4-isobutylbenzene at the uncrowded methylene group (leading to intermediates **104** and **105**) [333].

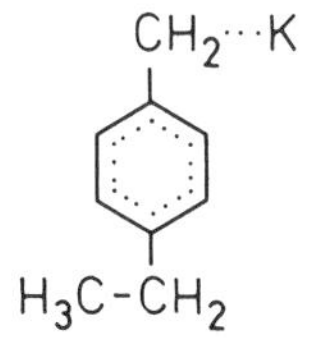

CH2···K / H3C-CH2 **(104)**   H3C-CH···K / (H3C)2CH-CH2 **(105)**

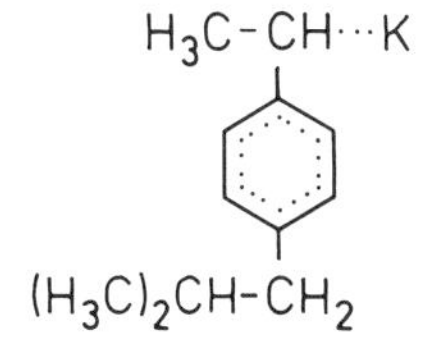

Cyclopropylbenzene [248] and isopropylbenzene (cumene) [318] react only sluggishly. The LIC–KOR superbase metallates the latter hydrocarbon mainly at the *meta*- and *para*-positions regardless of whether a paraffinic or ethereal solvent is employed. Clean α-deprotonation is only accomplished with trimethylsilylmethylpotassium as the metallating reagent [318].

Smooth monometallation [334] but only partial dimetallation [335] occurs with *o*- and *p*-xylene. *m*-Xylene [335], however, can be readily converted into a dimetallated and 1,3,5-trimethylbenzene (mesitylene) [336] into a trimetallated species.

The LIC–KOR base promotes the hydrogen–metal exchange selectively at the benzylic position of *N*,*N*,4-trimethylaniline (50 h at −75 °C in tetrahydrofuran, then addition of dimethyl sulphate; 65% alkylation product) [337] whereas TMEDA-activated butyllithium exclusively attacks the *ortho*-position [338]. Depending on the choice of the reagent and metallation conditions, 2-, 3- and 4-methoxytoluene (cresyl methyl ethers) [339], 2-, 3- and 4-fluorotoluene [340] and 3-(trifluoromethyl)toluene [340] can be selectively deprotonated at the benzylic methyl group or at a position adjacent to the hetero moiety. This ambiguity is, of course, removed with 2,6-dimethylanisole [341] and 1,4-bis(methoxymethoxy)-2,3,5,6-tetramethylbenzene [342] as substrates, since proton abstraction can now occur only at a methyl group (producing intermediates **106** and **107**, the latter being a potential α-tocopherol building block).

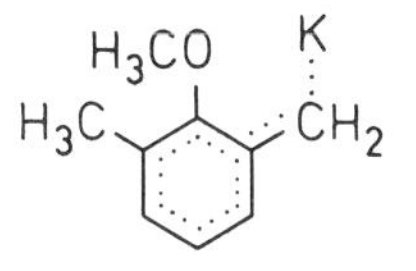

H3CO, K, H3C, CH2 **(106)**   H3C, K, O, O, CH2, H3C, O, O, H3C **(107)**

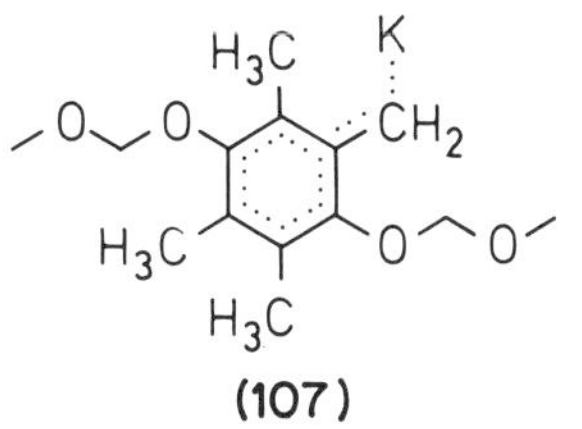

Allylic Organometal Compounds Allyl-type organoalkali species can be prepared with particular ease: allylpotassium [343], allylcaesium [136], 2-methylallylpotassium [171, 344], 2-ethylallylpotassium [345], 2-isopropylallylpotassium [346], 2-*sec*-butylallylpotassium [321], 2-*tert*-butylallylpotassium [346, 347], 2-(1,1-dimethylpropyl)-allylpotassium [344], 2-(1-ethyl-1-methylpropyl)allylpotassium (**108**) [344], (1-cyclopropenyl)methylpotassium (**109**, from methylenecyclopropane) [348], pinenylpotassium(**110**) [350, 351], 2-phenylallylpotassium [347], 2-[2-(1-methylvinyl)]allylpotassium [347], 2-(2-biphenylyl)allylpotassium [347] and bicyclo[3.2.1]octa-3,6-dien-3-ylpotassium (**11**) [349]. As the deprotonation of allylic methylene or methine centres is in general more difficult to accomplish than that of corresponding methyl groups, the metallation of cyclohexene [318, 320] and cycloheptene [320] requires optimized reaction conditions. The same is true for the generation of 2,3-dimethylallylpotassium (**112**) [320] from 2,3-dimethyl-2-butene ('tetramethylethylene'), 3-methyl-2-butenylpotassium ('prenylpotassium) from 3-methyl-1-butene [318] and 3-ethyl-2-pentenylpotassium (**113**) from 3-ethyl-1-pentene [352].

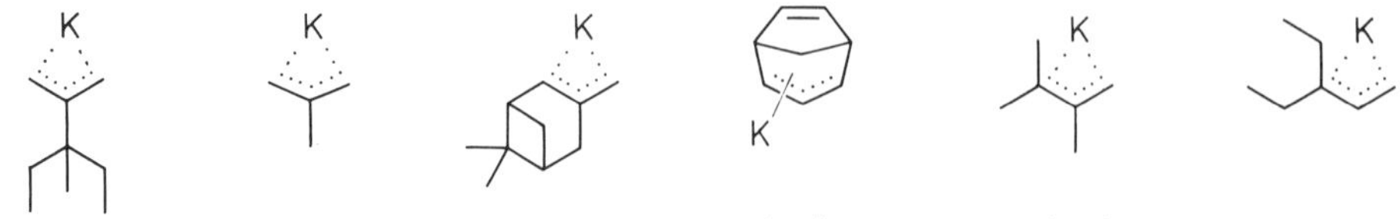

(**108**) (**109**) (**110**) (**111**) (**112**) (**113**)

Allyl-type organopotassium compounds are most attractive species for organic synthesis. They do not only offer the desired typo- and regioselectivity, but provide stereocontrol as an extra. The stereocontrol can be brought about in two different ways. Either the original configuration of the alkene precursor may be delivered via the allylpotassium intermediate unchanged to the final product ('stereoconservative' or 'stereodefensive' approach), or one may allow for torsional equilibration at the stage of the organometallic intermediate which generally will then exhibit a strong preference for one of the possible conformations ('stereoselective approach') [353]. For example, (*Z*)- and (*E*)-2-butenylpotassium (**114**) can be selectively generated by superbase deprotonation of *cis*- and *trans*-2-butene, respectively [354–356]. If, however, the organometallic solutions or suspensions are kept for hours before the electrophile is added, progressively an equilibration takes place until a 124 : 1 *Z*/*E* ratio is attained [354, 355]. It is also possible to catalyse this isomerization process [353].

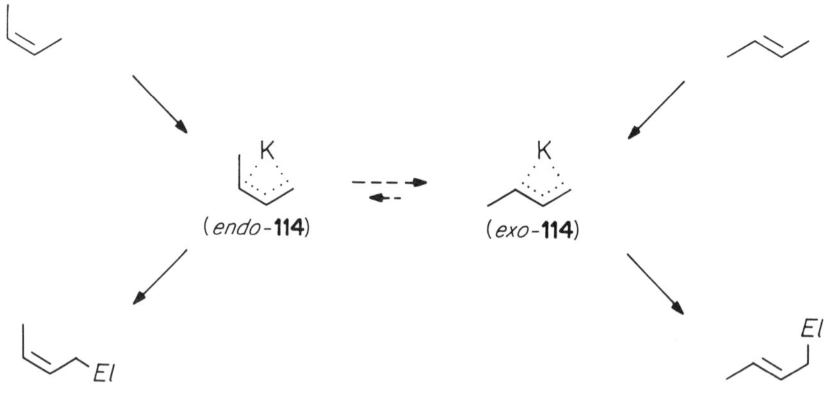

(*endo*-**114**) (*exo*-**114**)

The same principles apply to all other allylpotassium compounds that carry a primary alkyl group at a terminal position. Starting with the 2-alkene of the desired configuration the *endo* (*Z*) or *exo* (*E*) organopotassium species can be selectively generated and subsequently transformed into a stereochemically homogeneous derivative. Alternatively, any mixture of *endo*- and *exo*-allylpotassium species may be prepared, from either the 2-alkenes or the corresponding 1-alkene, and submitted to torsional isomerization. Again the *endo* (*Z*) component turns out to be clearly favoured, although this time only to the extent of approximately 94:6 [353]. Pertinent studies have been performed with (*Z*)- and (*E*)-2-pentenylpotassium [357], (*Z*)- and (*E*)-2-hexenylpotassium [354, 355, 358], (*Z*)- and (*E*)-2-octenylpotassium [359], (*Z*)-2,7-octadienylpotassium [360], (*Z*)-2-dodecenylpotassium [357], (*Z*)-2,13-tetradecadienylpotassium [360] and (*Z*,*Z*)-2,12-heptadecadienylpotassium (**115**) [360].

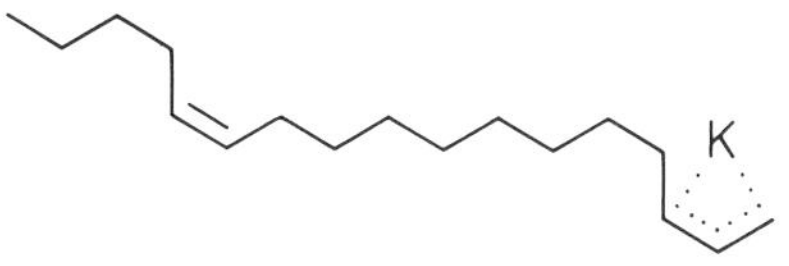

K

(115)

When allylpotassium species **116** carrying a methyl and a primary alkyl substituent (e.g. R = C_2H_5 [321], C_3H_7 [361]), $(H_3C)_2CHCH_2CH_2CH_2$ [361] at the same terminal position were allowed to undergo torsional isomerization, amazingly high stereopreferences were observed again. Apparently for conformational reasons the methyl group proves to be more successful in conquering the *endo* site, thus giving rise to *Z*/*E* equilibrium ratios of typically 9:91 [361].

R K ⇌ R K

(*Z*-**116**) (*E*-**116**)

The *endo* preference reaches maximum values if the allylpotassium compound has alkyl substituents at both positions 1 and 2. Disadvantaged anyway, the *exo* isomer now suffers from the additional handicap of steric repulsion between the two alkyl moieties. This brings the *Z*/*E* ratios in the range 50:1 to 500:1 [362]. Relevant systematic investigations have focused on 1-alkyl-2-methylallyl, 1-alkyl-2-ethylallyl and 1-alkyl-2-isopropylallyl species **117** (R = methyl [350], ethyl [321], butyl [354], 2-(2,3-dimethyltricyclo[2.2.1.0^{2,6}]hept-3-yl)ethyl [320]), **118** (R = methyl [362], ethyl [362], butyl [362]) and **119** (R = methyl [321], ethyl [321]). The 1,2,3-trialkyl-substituted allyl species **120** [363] shows the same *endo*-selective behaviour.

R K R K R K R K

H_5C_6 C_6H_5

(117) 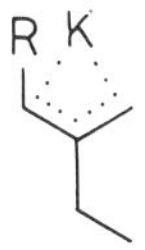 **(118)**  **(119)** 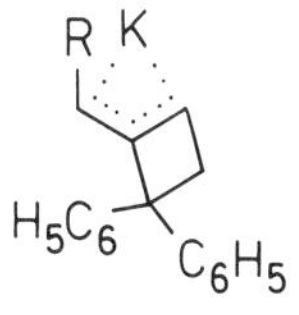 **(120)**

Functionalized allylmetal compounds have also captured much attention. Metallated enamines [364], allylamines [365], enethers [343] and allyl ethers [343, 366–370] have been used as versatile intermediates in organic synthesis. Hydroxy- and carboxy-substituted alkenes have been converted into doubly *O*,*C*-deprotonated species such as

121 [372], **122** [343, 373], **123** [343], **124** [374], **125** [359, 374] and **126** [374] (M = Li, K, etc.).

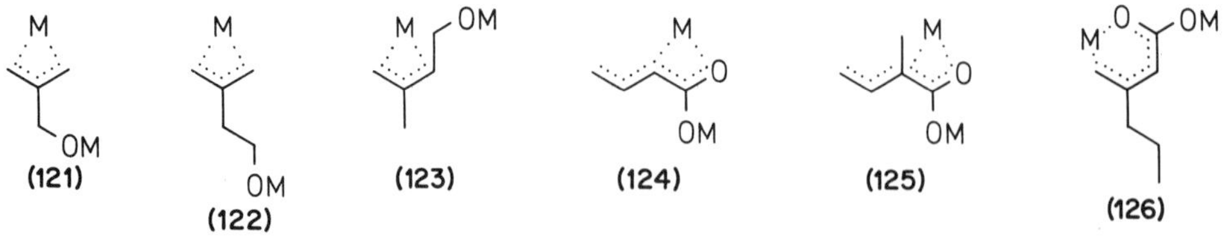

M M M OM M M M O OM
O O
OM OM OM
OM
(121) **(122)** **(123)** **(124)** **(125)** **(126)**

The superbase proved to be the reagent of choice for the metallation of a series of allylic silanes such as allyltrimethylsilane [375], *cis*- and *trans*-2-butenyltrimethylsilane [376], *cis*- and *trans*-2-hexenyltrimethylsilane [376], trimethyl-2-methylallylsilane [377], butoxydimethyl-2-methylallylsilane [377] and (diisopropylamino)dimethyl-2-methylallylsilane [377]. The last three substrates gave organometallic intermediates **127** ($X = CH_3$, OC_4H_9, $N(i\text{-}C_3H_7)_2$; M = Li or K) which covered a wide range of α/γ-selectivities when reacting with alkyl halides [377].

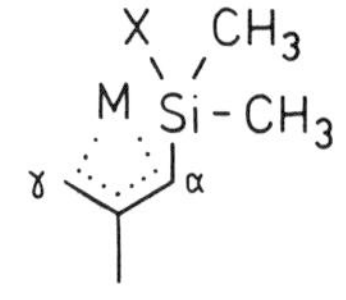

X CH_3
M $Si-CH_3$
γ α

(127)

Dehydroallylmetal Compounds 2-Alkynes and allenes can be conveniently deprotonated with the LIC–KOR reagent producing the dehydroalkenylpotassium intermediates **128** and **129** [375], neither being subject to stereoisomerism. Unless the substituent R is a bulky *tert*-alkyl group, electrophiles bind preferentially, if not exclusively, to the inner end of the delocalized moiety [348].

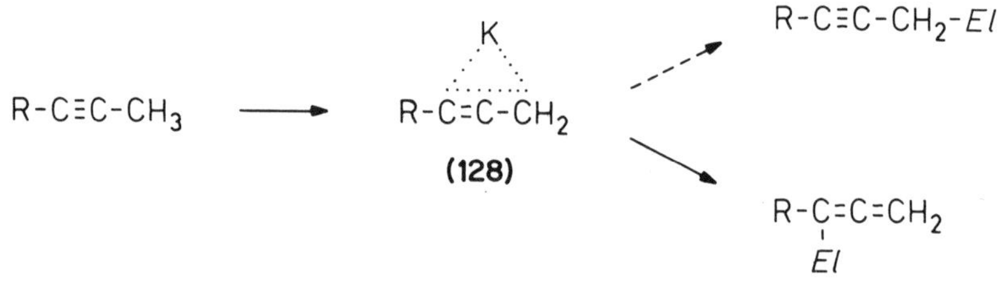

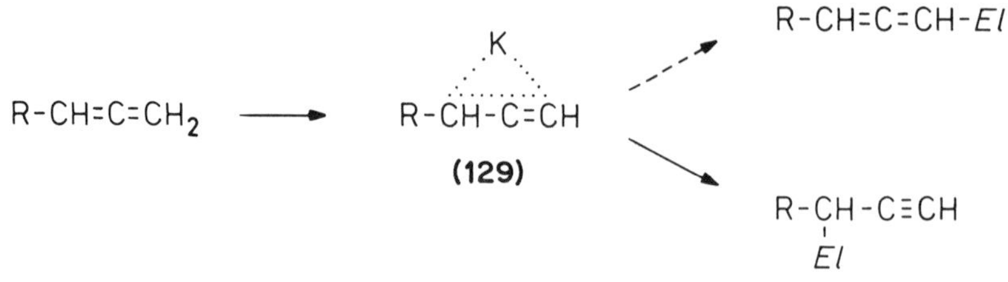

$R-C{\equiv}C-CH_3$ ⟶ K / $R-C{=}C-CH_2$ **(128)** ⇢ $R-C{\equiv}C-CH_2-El$; ⟶ $R-C(El){=}C{=}CH_2$

$R-CH{=}C{=}CH_2$ ⟶ K / $R-CH-C{=}CH$ **(129)** ⇢ $R-CH{=}C{=}CH-El$; ⟶ $R-CH(El)-C{\equiv}CH$

Replacement of the alkyl or aryl group by a heterofunctional substituent affects the outcome of such a metallation–derivatization sequence in several respects. First, an ordinary organolithium reagent will generally suffice to abstract a proton from a heteroallene. Furthermore, it is no longer the free terminal position but rather the position adjacent to the heteroatom which preferentially undergoes deprotonation. Thus, diethyl-amino-1,2-propadiene generates the same organometallic intermediate (**130**) which results from the treatment of 1-diethylamino-1-propyne with the superbase [378].

Finally, the regioselectivity of the electrophilic attack may change. While protonation and alkylation are oriented towards the γ-position of the 2-alkynyl ether-derived species **131** [379, 380], all electrophiles prefer to become linked to the α-position of the allenyl ether-derived species **132** [380, 381].

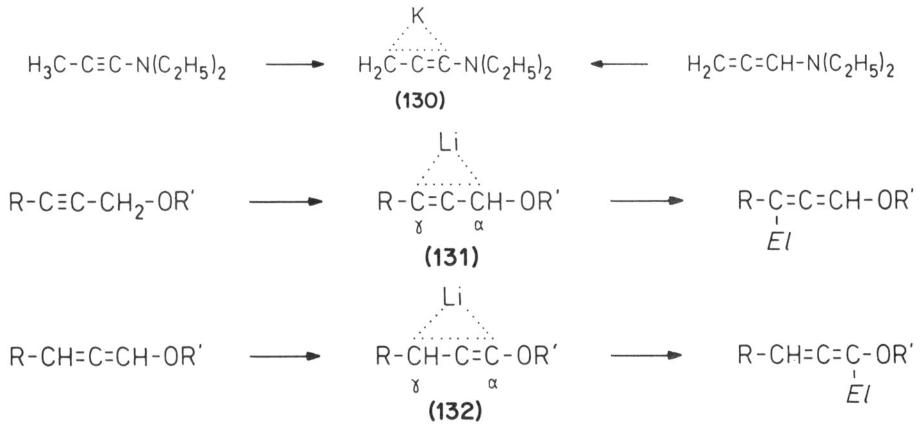

$H_3C\text{-}C{\equiv}C\text{-}N(C_2H_5)_2$ → $H_2C\text{-}C{=}C\text{-}N(C_2H_5)_2$ (K) **(130)** ← $H_2C{=}C{=}CH\text{-}N(C_2H_5)_2$

$R\text{-}C{\equiv}C\text{-}CH_2\text{-}OR'$ → $R\text{-}C{=}C\text{-}CH\text{-}OR'$ (Li; γ, α) **(131)** → $R\text{-}C(El){=}C{=}CH\text{-}OR'$

$R\text{-}CH{=}C{=}CH\text{-}OR'$ → $R\text{-}CH\text{-}C{=}C\text{-}OR'$ (Li; γ, α) **(132)** → $R\text{-}CH{=}C{=}C(El)\text{-}OR'$

Pentadienylmetal Compounds Like allylarenes [277, 382] and alk-1-en-4-ynes [383], 1,4-alkadienes also undergo with butyllithium or, better, *sec*-butyllithium very rapid deprotonation at the position flanked by the two unsaturated moieties. Hydrogen–metal exchange can still be accomplished at an allylic position of a conjugated diene such as 2,4-dimethyl-1,3-pentadiene [384], 1-isopropyl-4-methyl-1,3-cyclohexadiene (α-terpinene) [385] or 7,8-dehydrocholesterol (pro-vitamin D_3, leading to the intermediate **133**) [386].

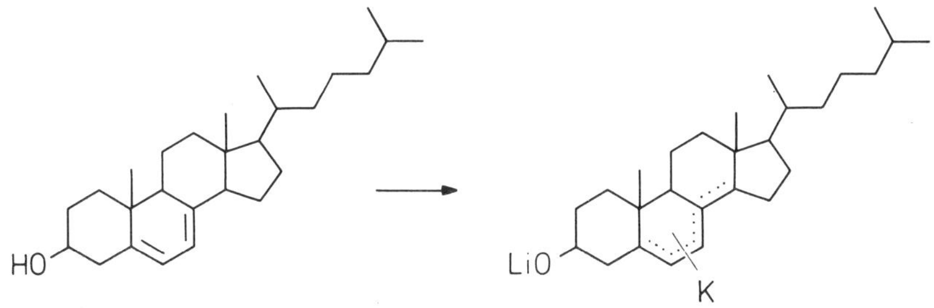

HO → LiO, K

(133)

The fascinating stereochemistry of pentadienyl-type organometallics [352] will be summarized only briefly. A resonance-stabilized pentadienyl moiety can choose from among three coplanar structures: a horseshoe-like *U* shape, a sickle-like *S* shape and a zig-zag band-like *W* shape. Pentadienylpotassium in tetrahydrofuran exists in the *U* shape (*U*-**134**), as evidenced by the formation of (*Z*)-2,4-pentadien-1-ol upon treatment with hydrogen peroxide [384]. In contrast, the corresponding lithium derivative favours the *W* shape (*W*-**134**), and when submitted to the borylation–oxidation sequence, is converted into the (*E*)-2,4-pentadienol [382].

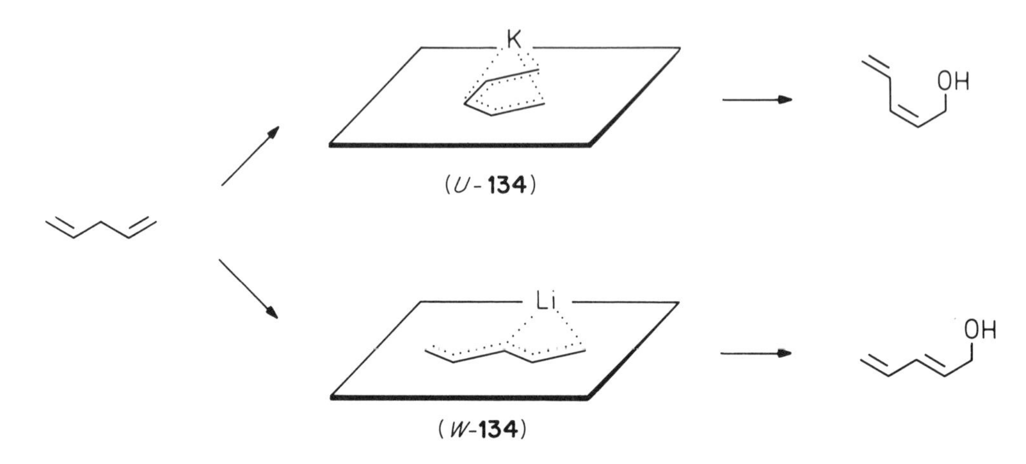

The sensitivity of the parent system **134** to metal and solvent effects can be plausibly explained [384]. Additional, notably steric, factors may, however, become dominant. Therefore, 3-methylpentadienyl and 2,4-dimethylpentadienyl species opt for the *W* and *U* conformations, respectively (*W*-**135** and *U*-**136**), regardless of what metal (M = Li or K) or solvent (hexane, tetrahydrofuran, etc.) is involved [384].

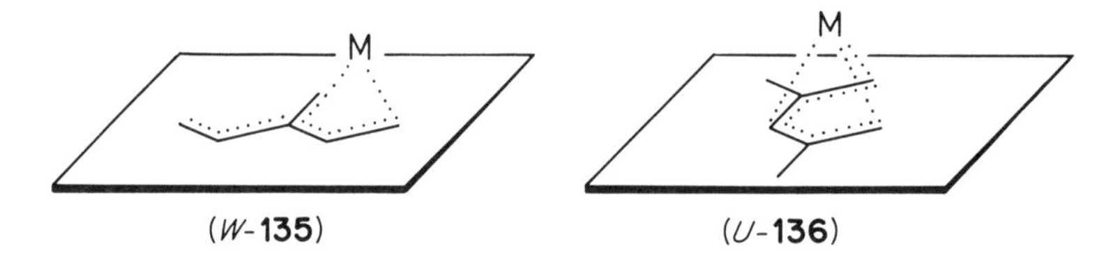

The number of conformers increases to eight when a single substituent is attached to the C_5 backbone, provided that it is not located at the centre (3-position) but in one of the wings. From a practical point of view, unbranched 2,4-alkadienylmetals are most important. Depending on the geometry of the hydrocarbon precursor, the metal, the solvent and the metallation temperature, either of three structures can be generated: the (*Z*)-*W*, the (*E*)-*W* and the (*E*)-*U* shape (*endo*-*W*-**137**, *exo*-*W*-**137** and *exo*-*U*-**137**; R = methyl, pentyl, hexyl, etc.). After borylation and oxidation, three dienols having the (2*E*,4*Z*), the (2*E*,4*E*) and the (2*Z*,4*E*) configurations can be isolated with high isomeric purity [387].

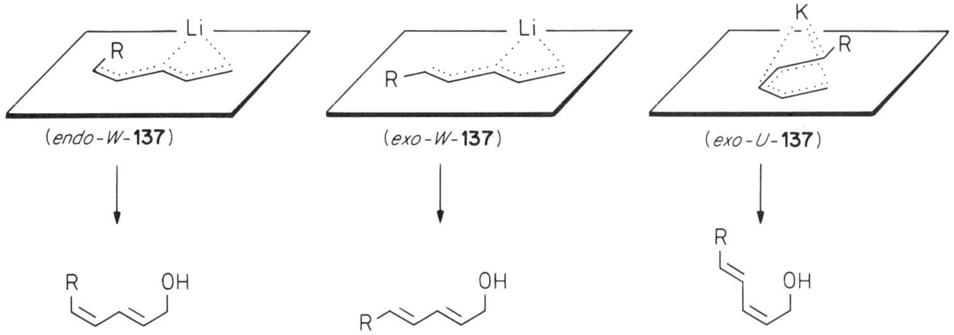

(*endo*-*W*-**137**) (*exo*-*W*-**137**) (*exo*-*U*-**137**)

Heptatrienylmetal Compounds The number of coplanar, and hence optimally resonance-stabilized, structures increases exponentially with increasing areas of delocalization. Thus, even unsubstituted heptatrienylmetal species can exist in ten different, at least formally, coplanar conformations. In reality, however, only two of them are populated in the proportions of 1:4. The major component has a zig-zag band like a 'triple-V' shape (*W*-*W*-**138**), while the serpent-like semicoiled conformation (*U*-*W*-**138**) can only be tentatively assigned to the minor conformer [348].

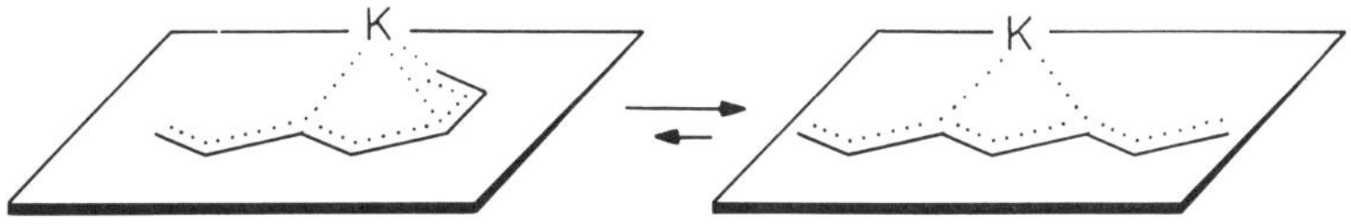

(*U*-*W*-**138**) (*W*-*W*-**138**)

Methyl or primary alkyl substituents at antinodal points favour the *W*-*W* conformation to the extent of exclusivity. Irrespective of the nature of their metal (M = Li, Na or K), 3,7-dimethyl-2,4-6-octatrienyl (**139**) and β-ionylideneethyl (**140**) species occupy only the completely outstretched *W*-*W* shapes and consequently produce all-(*E*) derivatives when trapped with an electrophile [388].

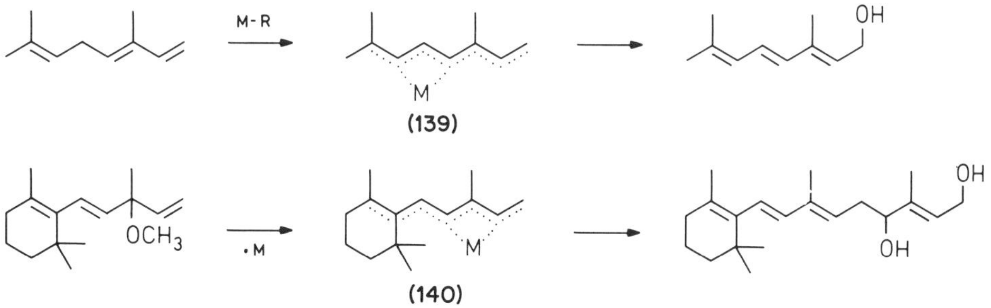

(139)

(140)

Neighbouring Group Assistance

The successful replacement of a hydrogen atom at the *ortho* position of anisole with sodium [389] or lithium [390] was one of the landmark events in the evolution of preparative organometallic chemistry. The consequence of this achievement was obvious: a method began to emerge that allows one to carry out electrophilic substitutions in an *ortho*-selective manner, thus avoiding the regioisomeric mixtures notoriously obtained in Brønsted- or Lewis acid-catalysed aromatic substitution reactions. Since then the *ortho*-selective metallation of heterofunctionalized aromatic substrates has been systematically explored and has opened up new entries to a variety of important natural and synthetic products [391, 392].

Despite the progress already made, '*ortho*'-directed metallations continue to attract much attention. One tends to forget, however, that this subject is only one manifestation of a more general phenomenon. Heteroatoms are capable of accelerating hydrogen–metal interconversion not only at aromatic but also at heterocyclic, olefinic, cyclopropanic and aliphatic sites. Before summarizing the present state of the art in the aromatic field, we shall take a few glimpses at these other areas. When doing so, we shall focus on the first-row elements fluorine, oxygen and nitrogen. Otherwise, d-orbital resonance effects (see pp. 68–70) may turn out to be preponderant when elements of the higher periods such as chlorine, bromine, sulfur, selenium, phosphorus or silicon are involved.

Trifluoromethane can be readily deprotonated by ethereal methyllithium at -75 °C, but the organometallic intermediate **141** is too labile to be trapped with electrophiles [393]. Dimethyl ether [314], *tert*-butyl methyl ether [314, 394] and tetrahydrofuran [314] react rapidly with butylpotassium at -75 or -100 °C to afford methoxymethylpotassium (**142a**), *tert*-butyloxymethylpotassium (**142b**) and 2-tetrahydrofuranylpotassium, respectively. Tetrahydropyran and oxepane react more sluggishly [314]. Organolithium or mixed metal reagents convert simple amines such as trimethylamine (**143**, $R = R' = CH_3$) [395], *N*-methylpiperidine (**143**, $R + R' = (CH_2)_5$) [396], *N*,*N*,*N'*,*N'*-tetramethylethylenediamine (TMEDA) [397], *N*-(*tert*-butyloxycarbonyl) pyrrolidine [398] and *N*-lithiooxycarbonyl-1,2,3,4-tetrahydroquinoline [399] into the corresponding metallomethyl or metallomethylene derivatives.

$Li{-}CF_3$ (**141**)

$K{-}CH_2{-}OR$

142a : $OR = OCH_3$

142b : $OR = OC(CH_3)_3$

$M{-}CH_2{-}NRR'$

143 : R, R' = Alkyl

Allylic or benzylic halides, ethers or amines are doubly activated and hence particularly reactive. Lithium diisopropylamide (LIDA) and lithium 2,2,6,6-tetramethylpiperidide (LITMP) are basic enough to abstract a proton from allyl chloride [400] or benzyl chloride [401]. Alkyl allyl ethers [343, 366–370] and alkyl benzyl ethers [402] react instantaneously with *sec*-butyllithium or even phenyllithium to give organometallic compounds which can be submitted to electrophilic substitution at temperatures below -50 °C or, when warmed up, undergo the Wittig rearrangement. Finally, versatile reactive intermediates can be generated by the metallation of enamines and allylamines [364, 403].

Owing to the ring strain effect, cyclopropane is not only considerably more acidic than cyclohexane but even more than methane [404]. Whereas alkyl-substituted cyclopropanes undergo a hydrogen–metal exchange only with the most powerful sodium and potassium reagents known [265], the oxygen functional derivative **144** can be readily prepared by treatment of the sterically hindered cyclopropanecarboxylate with *sec*-butyllithium in the presence of TMEDA [405].

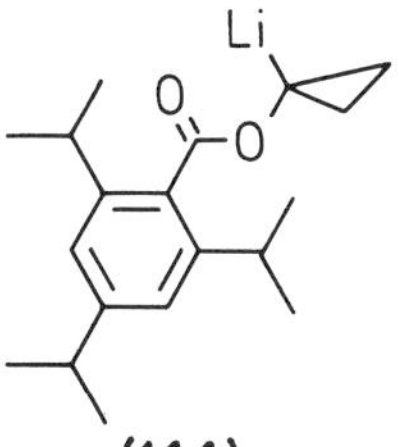

(144)

The incorporation of an oxygen atom into the three-membered ring has a particularly strong acidifying effect. Oxiranes are readily deprotonated by organolithium reagents or metal amides. Under favourable circumstances, the α-metallated species (e.g. **145**) can be intercepted by electrophilic attack [406]. Generally, however, it tautomerizes by metallotropic migration to become a carbene, which evolves into final stable products by a β-hydrogen shift or by insertion into more distant C—H bonds [407]. Alternatively or competitively, proton abstraction from a β-position and simultaneous C^{α}—O bond scission may occur, leading to an allyl alcoholate [407, 408].

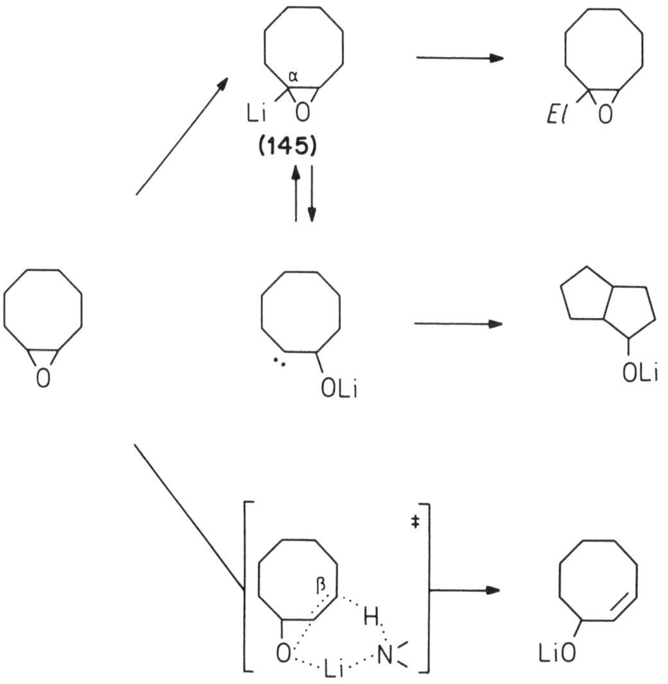

(145)

As we have seen already (pp. 72–73), resonance-impotent alkenes do undergo superbase metallation at olefinic sites, although slowly. Heterosubstituted alkenes are far more reactive. 'Carbenoid' species such as **146** can be smoothly generated by metallation of 1-alkenyl chlorides at very low temperatures [409]. Alkyl vinyl ethers are easily α-deprotonated by butyllithium in the presence of TMEDA [327]. Unbranched 1-alkenyl alkyl ethers require pentylsodium [271] or superbasic mixed metal reagents [343]. All these methods fail when 2-alkyl-2-alkenyl alkyl ethers are the substrates. However, after replacement of the alkoxy group with an acetal moiety, the desired metallation can be easily brought about with *sec*-butyllithium (affording intermediate **147**, for example) [410].

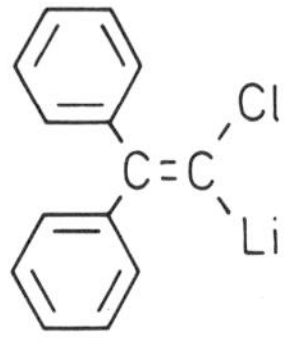

(146) 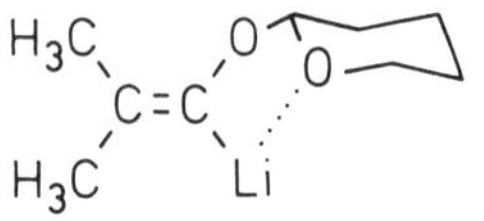 **(147)**

The acidifying effect of a heterosubstituent diminishes with distance but does not vanish. Therefore, 1,1-difluoroethylene [411], α-(methoxymethoxy)styrene [412] and 1-ethoxy-2-bicyclo[2.2.2]octene [413] smoothly undergo a hydrogen–metal exchange reaction with *sec*- or *tert*-butyllithium to produce the 'vicinally counterpolarized' [414] species **148**, **149** and **150**. If a vinyl group is part of a tertiary allyl alcohol, treatment with butyllithium in the presence of TMEDA allows the replacement of even its terminal, γ-distant hydrogen in the *cis* position with lithium (affording **151**, for example) [415].

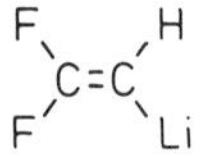

(148) 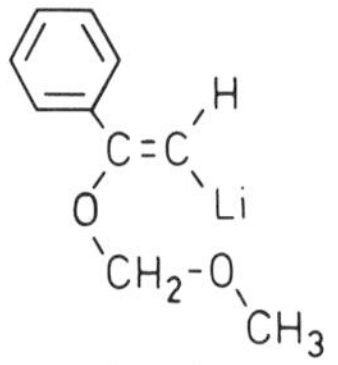 **(149)** 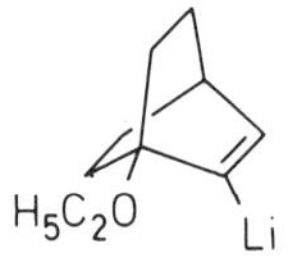 **(150)** 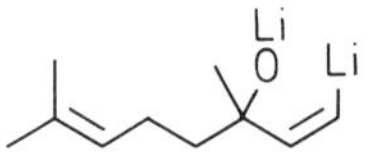 **(151)**

In general, no α-metallation of enamines can be brought about [416]. Noteworthy exceptions are *N*-vinyl- and *N*-1-propenylbenzotriazoles, which readily undergo α-lithiation (to give, e.g., species **152a** and **152b**) [417]. On the other hand, proton abstraction occurs readily at the β-position (to give, e.g., intermediates **153** and **154**) when chelating enamines are submitted to the action of *tert*-butyllithium in hexane (8 h at 25 °C) [418]. Finally, the metallating reagent attacks exclusively at the terminal *cis*-position when *N*-*tert*-butyl- or *N*-(trimethylsilyl)allylamine is treated with two equivalents of butyllithium and TMEDA in refluxing hexane (thus generating the γ-lithio species **155** and **156**) [419].

Furans, thiophenes and pyrroles can be conceived of as concatenate twofold enethers, enethioethers and enamines, respectively. The combined ring strain, heteroatom and

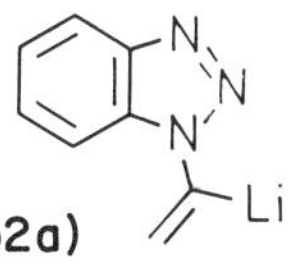

(152a) 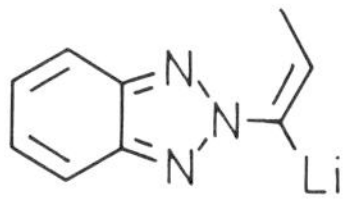 (152b)

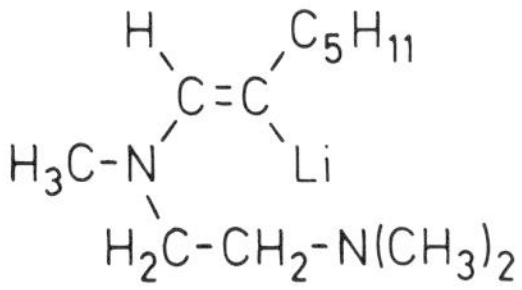

(153) 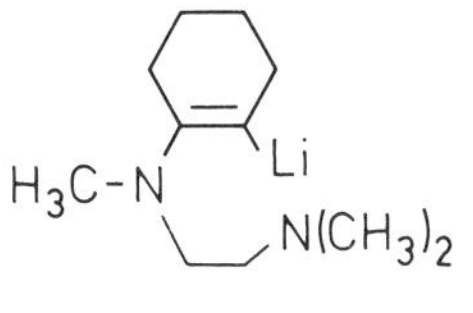 (154)

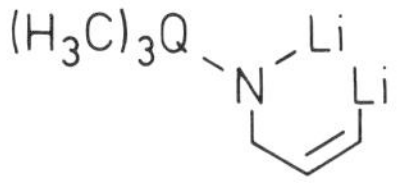

155 : Q = C
156 : Q = Si

polarization effects make the α-position in such heterocycles highly acidic. 2-Furyllithium (**157**) [420], 2-furylsodium [421], 2-thienyllithium [40], 2-thienylsodium (**158**) [421, 423], *N*-trimethylsilyl-2-pyrryllithium (**159**) [266] and numerous derivatives thereof can be very readily prepared.

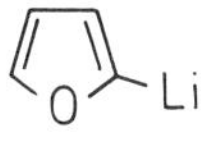

(157) 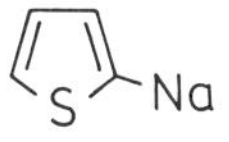 (158) 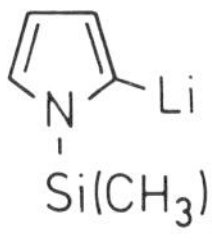 (159)

These heterocyclic compounds are quasi-aromatic. They lead us directly into the area of benzene-type carbocycles.

Functional Arene Substituents as ortho-*Directing Groups* The various neighbouring groups containing halogen, chalcogen or pnictogen heteroatoms will be addressed in order of decreasing electronegativity. In other words, we shall start with halides, turn next to oxygen and sulfur moieties and ultimately deal with nitrogen and phosphorus moieties.

(a) Activating substituent located at a β-position: Fluorobenzene [424], 1-fluoronaphthalene [424] and 9-fluorophenanthrene [425] can be metallated with butyllithium in tetrahydrofuran selectively at positions 2 or 10. Carboxylation products are obtained with yields of *ca* 60%. The organometallic intermediate **160** forms quantitatively if the superbasic mixture of butyllithium and potassium *tert*-butoxide (LIC–KOR) is employed [426]. In the same way, the three isomeric difluorophenylmetals **161** [426], **162** [427] and **163** [426] and also the fluorine-substituted methylphenylmetals **164** [428], **165** [428] and **166** [428] can be effectively generated (formula 160).

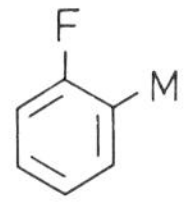

(160)

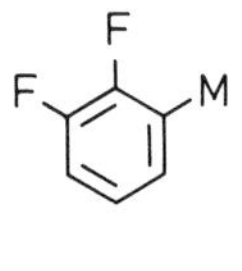

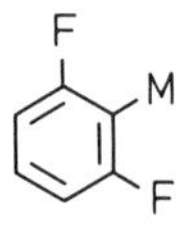

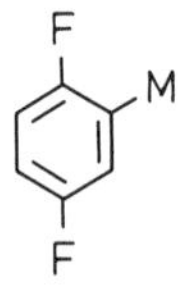

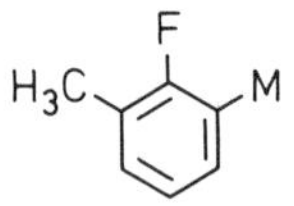

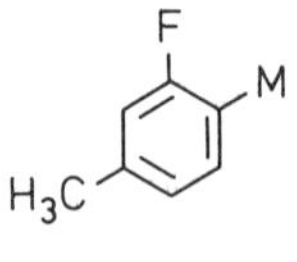

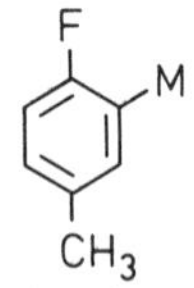

(161) (162) (163) (164) (165) (166)

In general, elimination of lithium (or potassium) fluoride occurs at temperatures around −50 °C, setting free the corresponding 1,2-dehydroarene ('aryne'). *ortho*-Metallated chloroarenes are even more labile. Nevertheless, it is possible to lithiate chlorobenzene [429] and 1,3-dichlorobenzene [430] and to trap the resulting species **167** and **168** with a variety of electrophiles at temperatures around −100 °C.

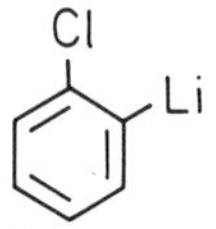

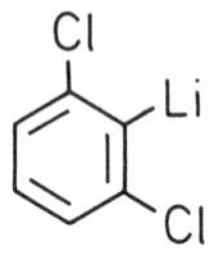

(167) (168)

(Methoxymethoxy)benzene [431] and (2-dimethylaminoethoxy)benzene [432] undergo *ortho*-metallation (affording intermediates **169** and **170**) much more rapidly than the parent compound anisole. The superior neighbouring group assistance provided by an acetal moiety can be also demonstrated by an intramolecular competition experiment. 1-Methoxymethoxy-8-methoxynaphthalene is lithiated at the 2- rather than at the 7-position (intermediate **171**) [433].

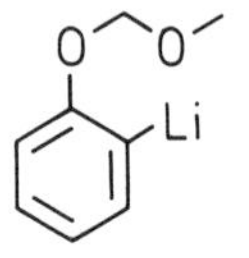

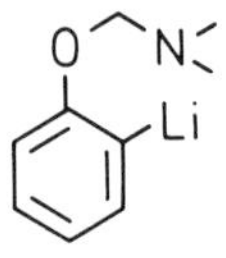

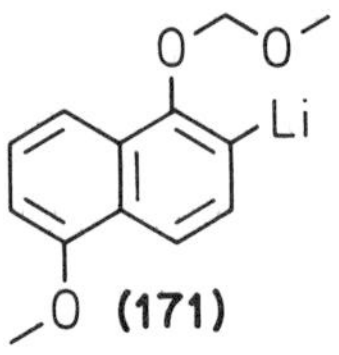

(169) (170) (171)

Phenol can be converted with *tert*-butyllithium in tetrahydrofuran into *o*-lithiooxyphenyllithium [434]. Lithium 1-naphtholate undergoes concomitant metallation at the 2- and 8-positions (products **172** and **173**) while its 2-isomer reacts only at the 3-position (product **174**) [435].

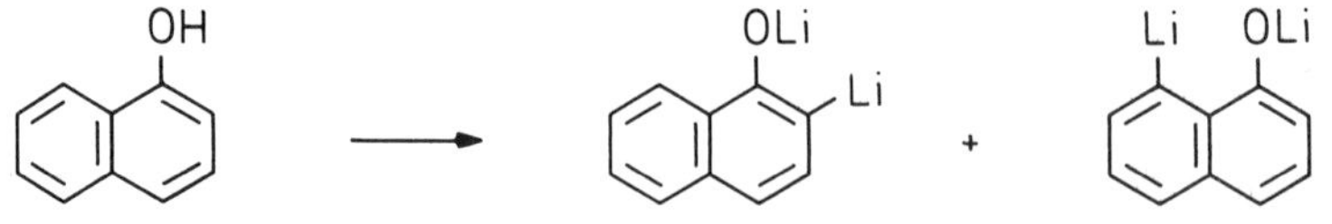

(172) (173)

(174)

Owing to the accumulation of electron excess at the *ortho*- and *para*-positions of the aromatic nucleus, phenolates react more sluggishly than the corresponding alkoxyarenes. This difference in reactivity is well reflected by the outcome of an intramolecular competition. Lithium 4-methoxyphenolate gives the two regioisomeric metallation products **175** and **176** in a 3:97 ratio [434].

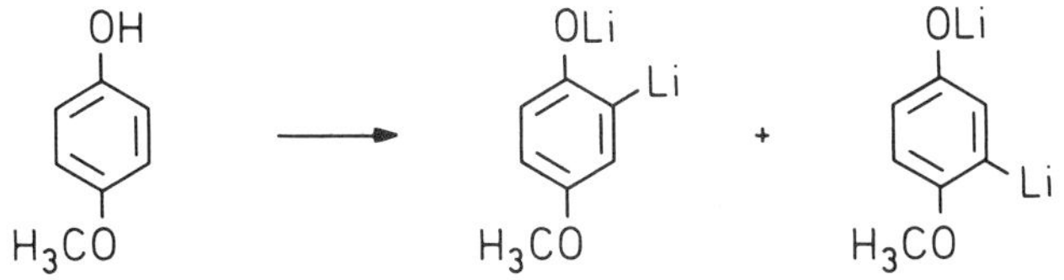

(175) **(176)**

Alkyl- or arylthio substituents do activate the *ortho*-positions although to a considerably smaller extent than do alkoxy groups; 2- and 4-isopropylthioanisole are lithiated at an oxygen rather than a sulphur adjacent position (yielding intermediates **177** and **178**) [436].

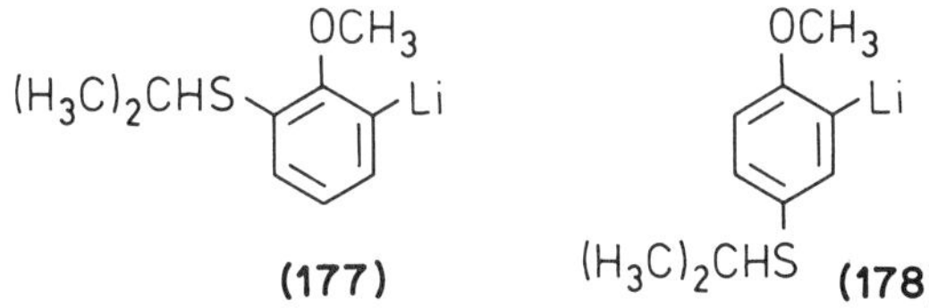

(177) **(178)**

Acyclic or cyclic sulfones can be *ortho*-lithiated provided that they lack 'mobile' hydrogens at an α-carbon atom (e.g. intermediates **179** [437], **180** [438], **181** [439], **182** [440, 441] and **183** [440]. Arenesulfonic esters (e.g. intermediate **184** [61]) and secondary or primary sulfonamides (e.g. intermediates **185** [443], **186** [444], **187** [445] and **188** [444, 446]) are also suitable substrates for selective *ortho*-metallations.

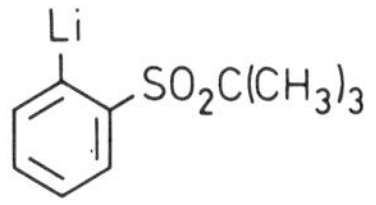

(179) 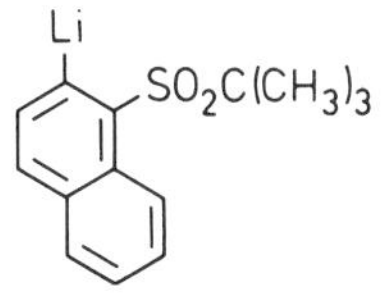 **(180)** 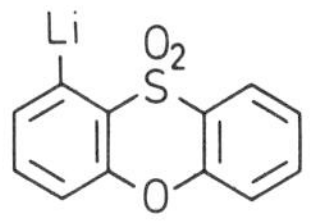 **(181)**

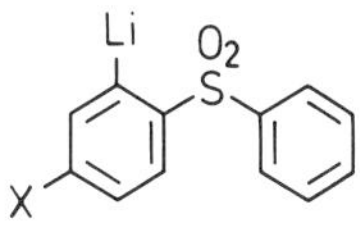

182 : X = H
183 : X = Br

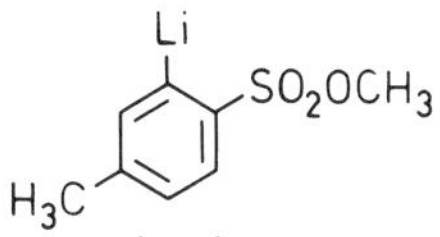

(184) 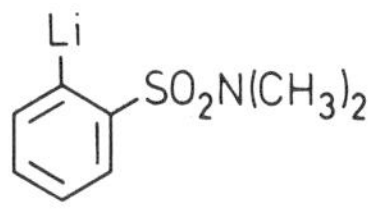 **(185)**

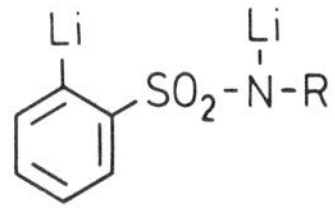

186 : R = CH_3
187 : R = $C(CH_3)_3$
188 : R = C_6H_5

N,*N*-Dimethylaniline [447–449] and alkyl-substituted derivatives [448] readily undergo *ortho*-metallation with butyllithium in diethyl ether or TMEDA-activated butyllithium in hexane. Triphenylamine was claimed to afford the *meta*-isomer of diphenylaminobenzoic acid (29%, via intermediate **189**) when consecutively treated with phenylsodium, carbon dioxide and acid [450]. However, when the LIC–KOR superbase was used as the metallating reagent, only the *ortho*-isomer was obtained (30%, via intermediate **190**) [451].

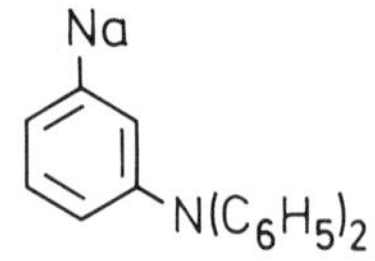

(189) 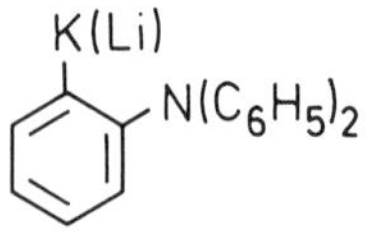 **(190)**

N-Lithiated *N*-alkylanilides are fairly unreactive because of the extensive charge delocalization into the aromatic ring. Lithium diarylamides do undergo *ortho*-metallation as exemplified by the species **191** [452] and **192** [453]. It may still be more advantageous to protect and at the same time activate such a secondary amine by an *N*-pyrrolidinomethyl or lithiooxycarbonyl moiety (see, e.g., intermediates **193** and **194**) [454].

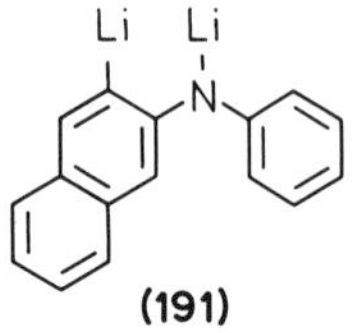

(191) 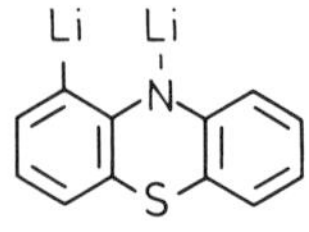 **(192)** 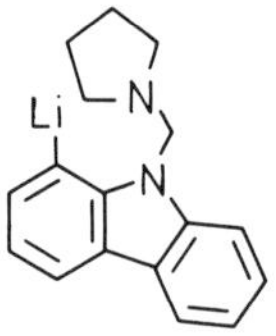 **(193)** 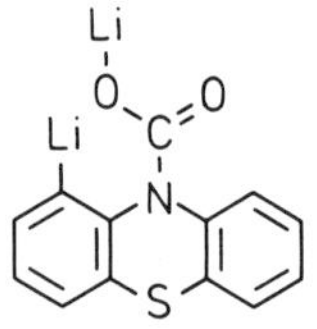 **(194)**

Aniline-derived aldimines or ketimines are prone to undergo nucleophilic addition at the C=N double bond. However, with *N*-pivaloyl- and *N*-*tert*-butoxycarbonyl-(BOC) anilides as the precursors, this addition mode is suppressed since the instantaneously occurring deprotonation of the N—H bond produces an electron-rich imidate or iminocarbonate moiety. Subsequently, smooth *ortho*-metallation can take place generating intermediates such as **195** [455] and **196** [456].

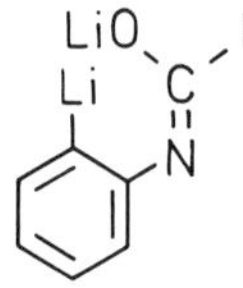

195: R = $C(CH_3)_3$
196: R = $OC(CH_3)_3$

No clean metallation of a triarylphosphine has been reported so far. On the other hand, triphenylphosphine oxide undergoes smooth *ortho*-lithiation when treated with an ethereal solution of phenyllithium [457]. *m*-Anisyldiphenylphosphine oxide is deprotonated by lithium diisopropylamide at −75 °C [458]. The resulting species **197** is a key intermediate on the route to atropisomerically chiral bisphosphines of the biphenyl type [458]. Strangely enough, *tert*-butyllithium promotes a hydrogen–metal exchange with the closely related (2,5-dimethoxyphenyl)diphenylphosphine oxide at the *ortho*-position of one of the unsubstituted nuclei (generating intermediate **198**) rather than at the doubly activated position of the methoxy-bearing ring [459].

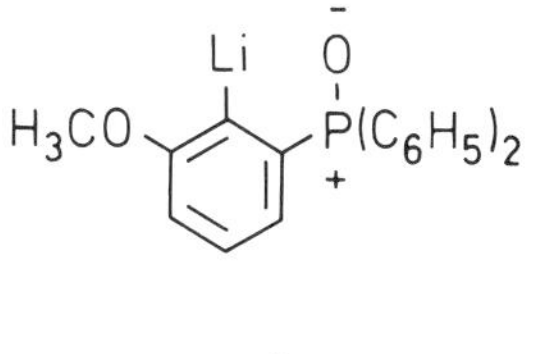

(197) 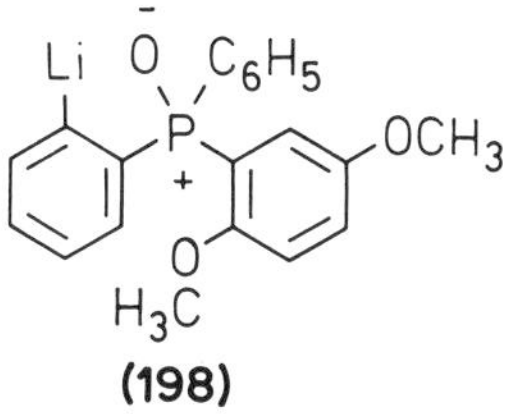 **(198)**

Both phenylphosphonic bis(dimethylamide) and the corresponding thiono compound react with butyllithium under *ortho*-metallation. The lithiated intermediates **199** [460] and **200** [461] can be trapped with standard electrophiles.

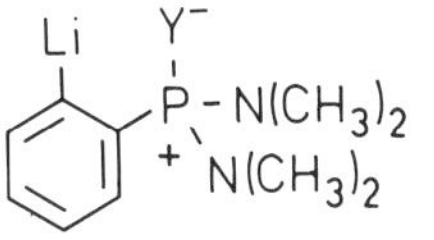

199 : Y = O
200 : Y = S

(b) Activating substituent located at a γ-position: When trifluoromethylbenzene is kept for 6 h in a refluxing ethereal solution of butyllithium, then is poured onto dry-ice and finally is neutralized, a 1:0.4:0.01 mixture of *o*-, *m*- and *p*-trifluoromethylbenzoic acid can be isolated with yields varying in the range 33–48% [462]. In contrast, the LIC–KOR base allows the *ortho*-metallated intermediate **201** to be generated virtually quantitatively and free from regioisomeric contamination [426].

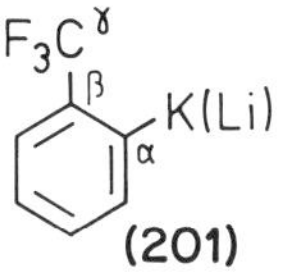

(201)

The accumulation of three halogens does not fully compensate for the relative remoteness of the heteroatoms from the metallation site. A single fluorine atom, if directly attached to the aromatic nucleus, outperforms the trifluoromethyl group by its superior *ortho*-directing power. Thus, the three isomeric fluoro(trifluoromethyl)benzenes give cleanly the metallated species **202**, **203** and **204** [426].

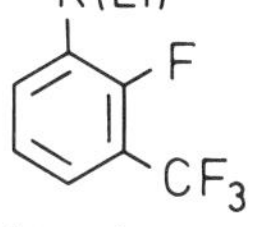

(202) 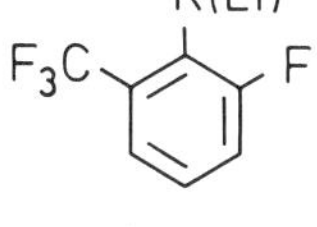 **(203)** 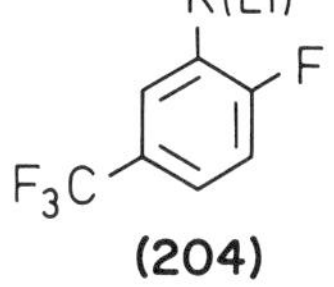 **(204)**

One would expect 1,3-bis(trifluoromethyl)benzene to undergo a hydrogen–metal exchange with particular ease at the position flanked by the two electron-withdrawing substituents. The activating inductive effect, however, appears to be counterbalanced by steric effects to a considerable extent. Butyllithium in diethyl ether attacks the 2- and 4-(6-)positions almost randomly (producing intermediates **205** and **206** in a 2:3 ratio) [463, 464]. TMEDA-activated butyllithium deprotonates selectively the 2-position although the yields are poor (30%) [465]. If the LIC–KOR superbase is used, derivatives of intermediate **205** are formed in 78% yield [426]. On the other hand, the isomer **206** is obtained with satisfactory selectivity if *sec*-butyllithium in the presence of PMDTA is employed as the metallating agent. Finally, the bulky *tert*-butyllithium in tetrahydrofuran (1 h at −20 °C) generates a 45:55 mixture of the organometallic species **206** and **207** [426].

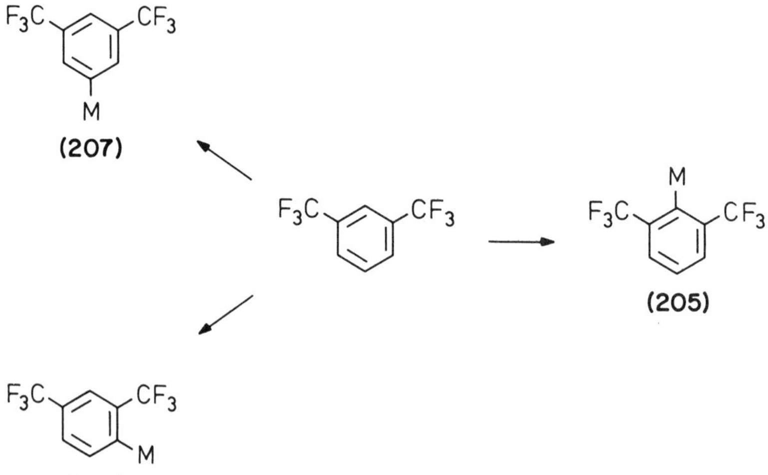

(207) (205) (206)

No such problems are encountered with 1,4-bis(trifluoromethyl)benzene, 1,3,5-tris-(trifluoromethyl)benzene and 5-dimethylamino-1,3-bis(trifluoromethyl)benzene. Treatment with ethereal butyllithium readily converts them into the lithio derivatives **208** [463], **209** [466] and **210** [465].

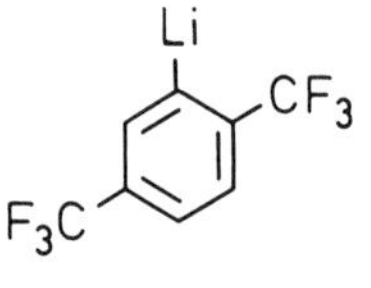

(208) 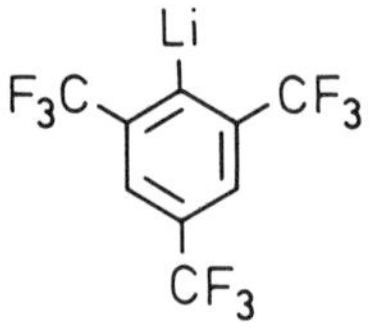 (209) 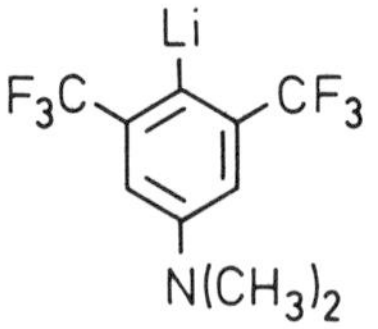 (210)

Benzylic ethers generally undergo a very fast hydrogen–metal exchange at the exocyclic α-position to give intermediates (e.g. **211**) which at temperatures above −50 °C can isomerize by means of a Wittig rearrangement. In view of the high benzylic reactivity, it requires additional neighbouring group participation (e.g. in **212** [467]) if one wishes to divert the metallating agent to an *ortho*-position.

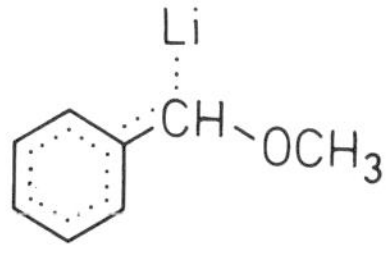

(211) 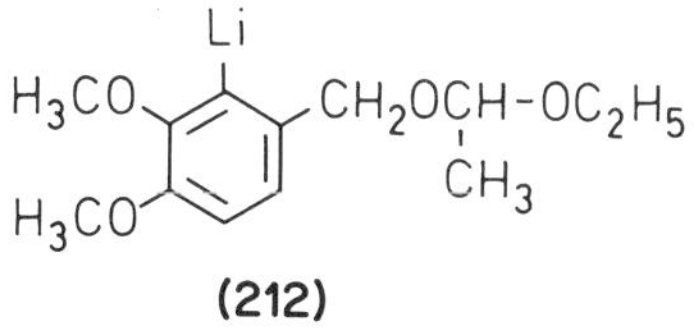 (212)

α-Deprotonation can be successfully prevented by merely replacing the benzylic ether function by a free hydroxy group. On treatment with two equivalents of butyllithium in the presence of TMEDA, a suspension of the lithium alcoholate in hexane will be instantaneously formed. Heating this reaction mixture for several hours under reflux eventually produces *ortho*-metallated species such as **213**, **214** and **215** [468]. Lithium 1-methyl-1-indanolate can still be successfully submitted to *ortho*-lithiation (affording intermediate **216**), despite its rigid structure and the consequently increased distance of the functional group [469].

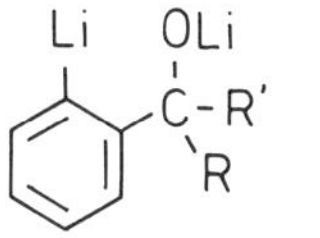

213: R = R' = H_3
214: R = CH_3, R' = H
215: R = R' = CH_3

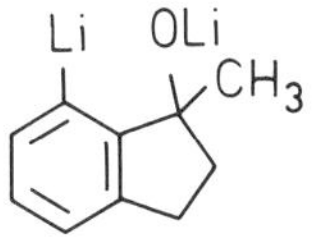

(216)

The nucleophilic addition of a lithium dialkylamide to an aromatic aldehyde generates the *O*-lithio derivative of a hemiaminal, a special type of benzyl alcoholate. Rapid *ortho*-metallation can be performed with those adducts, leading to intermediates such as **217**. After trapping with a suitable electrophile, the formyl group can be liberated again by simple acid hydrolysis [470]. In this manner, aromatic aldehydes can be effectively submitted to electrophilic substitution at the *ortho*-position of the ring.

(217)

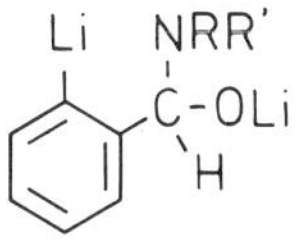

NRR' = $N(CH_3)_2$, morpholino, $N(CH_3)CH_2CH_2N(CH_3)_2$ etc.

The most rigorous way to preclude deprotonation at the benzylic position is to avoid α-hydrogen atoms. Thus, methyl triphenylmethyl ether or 1-phenyl *O*-trimethylsilyl enol ether have no choice other than *ortho*-metallation, generating intermediates **218** [471] and **219** [472]. The same is true for ethyl benzoate except that the corresponding *o*-deprotonated species **220** is unstable and is destroyed by autocondensation [473].

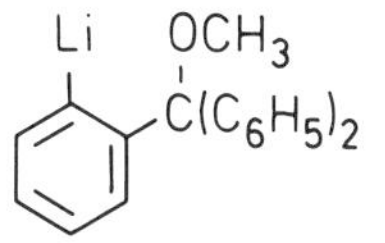

(218) 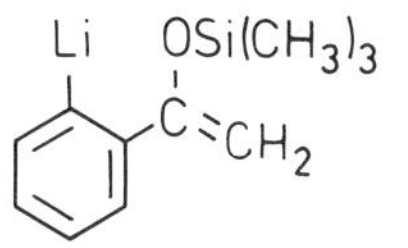 (219) 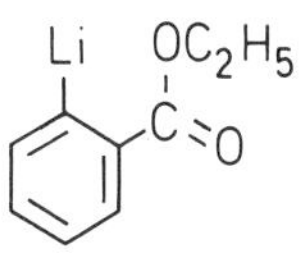 (220)

Practically useful results are obtained with arylcarboxamides. Whereas *N,N*-dimethylbenzamide reacts with *sec*-butyllithium under nucleophilic addition to the carbonyl group, smooth *ortho*-lithiation occurs with the sterically more hindered *N,N*-diethyl- or *N,N*-diisopropylbenzamide (leading to intermediates **221** and **222**) [474]. Even more powerful as an *ortho*-directing group is the nitrogen function of *N*-lithio secondary carboxamides. The corresponding *ortho*-metallated species (e.g. **223** [475], **224** [476] and **225** [475]) are readily accessible regardless of the nature of the variable substituent R.

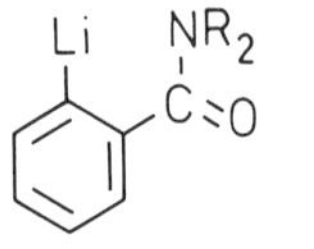

221: R = C_2H_5
222: R = $CH(CH_3)_2$

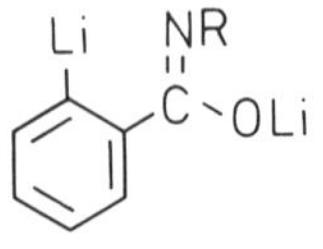

223: R = CH_3
224: R = $C(CH_3)_3$
225: R = C_6H_5

Cyclic iminoethers, in particular 2-aryl-4,4-dimethyl-1,3-oxazolines, belong to the most powerfully activated substrates. The resulting *ortho*-lithiated derivatives (e.g. **226**) are synthetically equivalent to *ortho*-anionized benzoic acids [477]. Simple azomethines (Schiff bases) react more sluggishly (forming, e.g., intermediate **227**). Again, a bulky *N*-alkyl substituent is advisable to avoid nucleophilic addition to the C=N double bond [478].

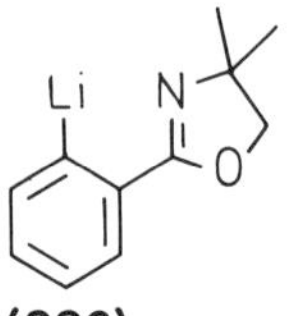

(226) 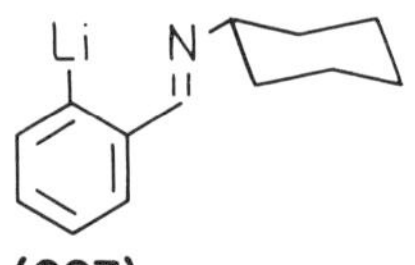 **(227)**

N-Deprotonated *N*-benzyl pivalamides, *O*-*tert*-butyl carbamates or *N′,N′*-dimethylureas tend to undergo α- rather than *ortho*-metallation [479]. However, the latter process can become the favoured reaction mode (intermediates **228**–**230**) [479] if a fluorine atom or another suitable electron-withdrawing substituent provides additional activation of the aromatic position. *N*-Lithio derivatives of secondary benzylamines are attacked by TMEDA-activated butyllithium exclusively at the *ortho*-position (e.g. intermediates **231** and **232**) [480]. Tertiary benzylamines exhibit the same high reactivity and perfect site selectivity (e.g. intermediates **233**–**235**) [481].

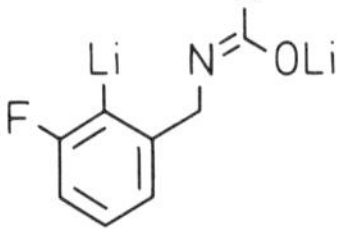

228: R = $C(CH_3)_3$
229: R = $OC(CH_3)_3$
230: R = $N(CH_3)_2$

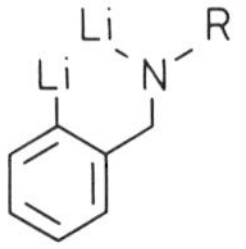

231: R = CH_3
232: R = C_6H_5

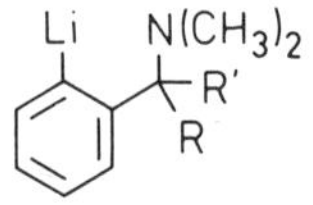

233: R = R′ = H
234: R = CH_3, R′ = H
235: R = R′ = CH_3

ortho-Lithiated benzylamines are particularly versatile modules for organic synthesis. They allow the regioselective introduction of substituents into precursors to a variety of *N*-heterocyclic compounds. In this way, for example, numerous 8-substituted 4-hydroxy-1,2,3,4-tetrahydroisoquinolines (e.g. **236** and **237**) have been prepared [482].

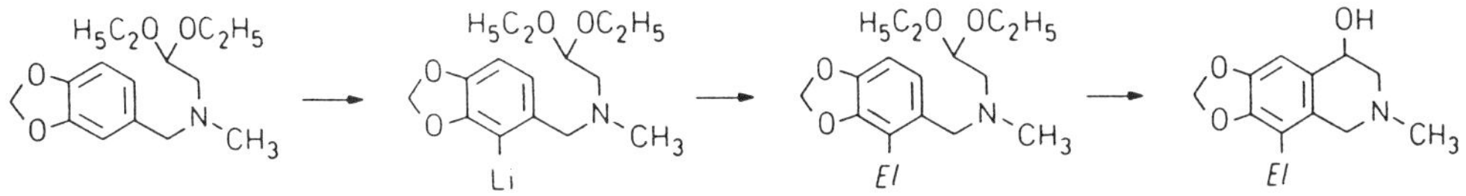

236 : *El* = OCH_3
237 : *El* = I

Tetrahydroisoquinolines can be converted into isoquinolines or *N*-alkyl-3,4-dihydroquinolium salts (e.g. **238** and **239**). Nucleophilic addition of the lithium enolate generated from 6,7-dimethoxy-1-isobenzofuran to such compounds gives *erythro–threo* mixtures of narcotine (**240**, R = CH_3) and its *N*-benzyl analogue, for example. The latter product can be submitted to hydrogenolytic debenzylation to afford nornarcotine (**241**, R = H) [483].

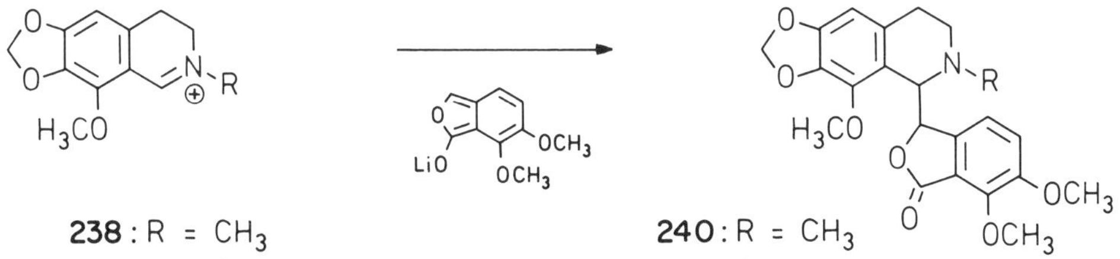

238 : R = CH_3
239 : R = $CH_2C_6H_5$

240 : R = CH_3
241 : R = H

Similarly, *o*-formyl-substituted benzylamines have become readily accessible. They may be employed in the 'off-shore construction' of benzoannelated heterocyclic moieties, in particular for the synthesis of 3-formyl-1,2-dihydroisoquinolines (e.g. **242**) [484].

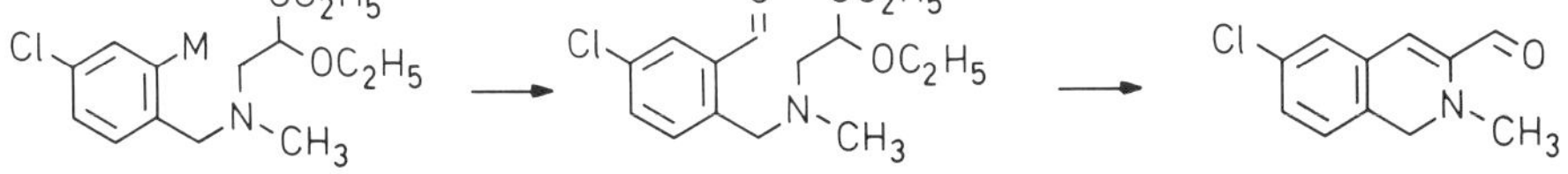

M = H → M = Li

(242)

(c) Activating substituent located at a δ-position: Only a few cases of long-range neighbouring group assistance have been documented. Even then, its manifestation frequently depends on the concurrent action of a second heterosubstituent or functional group.

The adduct of phenyllithium with hexafluoroacetone can rapidly exchange an *ortho*-hydrogen against lithium (intermediate **243**) [485]. The rate-enhancing effect of the six halogen atoms becomes apparent when one compares the ease of this transformation with the sluggish reaction of the halogen-free lithium 2-phenyl-2-propanolate (see p. 91).

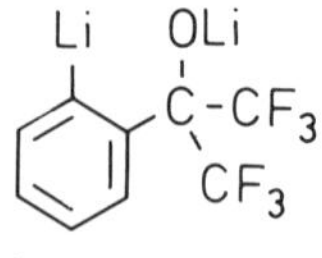

(243)

Alkoxy groups at δ-distance still exert a weak directing effect. 2-(3,4-Dimethoxyphenyl)acetaldehyde dimethyl acetal gives the 2-lithiated species (**244**) although the competing 5-position is sterically less hindered [467].

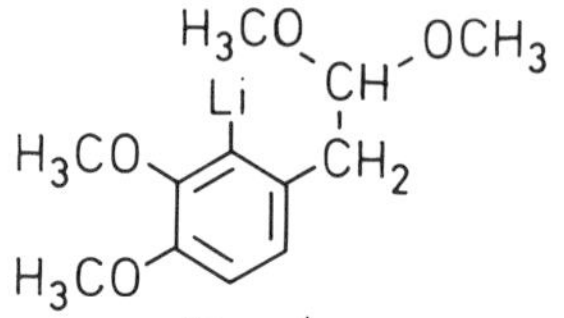

(244)

Only poor yields of *ortho*-metallated derivatives are obtained on treatment of *N*,*N*-dimethyl-2-phenylethylamine or *N*,*N*-dimethyl-2-methyl-2-phenylpropylamine with organolithium reagents [486]. A substantial improvement can be achieved by the introduction of an electronegative substituent into the *meta*-position. The *ortho*-lithiation of 2-(3,4-dimethoxyphenyl)-*N*-(trimethylsilyl)ethylamine (intermediate **245**), followed by formyl transfer from *N*,*N*-dimethylformamide and subsequent acid-catalysed cyclization gives 45% of 7,8-dimethoxy-3,4-dihydroisoquinoline [487].

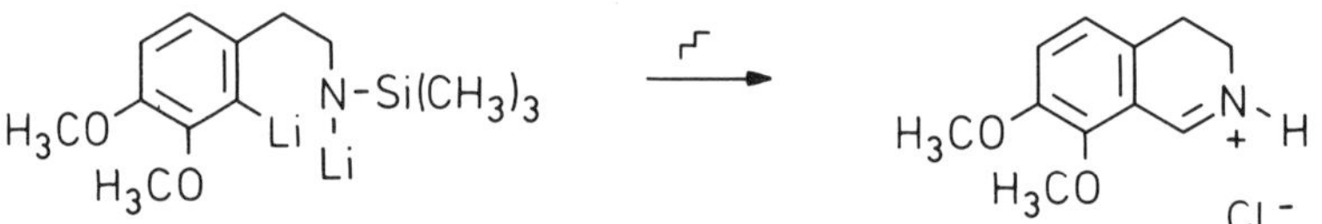

(245)

The same mode of reaction is observed when *N*-pivaloyl-2-(3-methoxyphenyl)ethylamine is treated with two equivalents of butyllithium in diethyl ether at 25 °C (intermediate **246**) [488]. However, with potassium *tert*-butoxide-activated butyllithium in tetrahydrofuran the *para*-position of the ring is attacked (intermediate **247**). Finally, α-deprotonation occurs if *tert*-butyllithium in tetrahydrofuran acts as the base (intermediate **248**) [488].

Interplay of Two or More Activating Groups So far we have dealt with individual *ortho*-directing groups. If two such groups are attached to the same aromatic moiety, they can cooperatively combine their activating powers if they occupy *meta*-positions. In contrast, if located with respect to each other in *ortho*- or *para*-positions, they will have to exert their neighbouring group assistance independently, even competitively.

The evaluation of relative *ortho*-directing aptitudes will provide a guide for predicting the rate and the outcome of metallations involving electronegatively di- or polysubstituted aromatic substrates. At the same time it will raise the question as to the nature of neighbouring group participation in hydrogen–metal interconversion processes.

(a) Relative ortho-*directing aptitudes*: As a brief survey of pertinent literature [391, 392] reveals, the efficiency of *ortho*-directing groups varies within vast limits. If substrates are ranked according to the ease with which they undergo *ortho*-lithiation, a coarse calculation results in which sulfones and carbamates appear at the top and *N*,*N*-dialkylanilines and aryl fluorides at the bottom (Table 1.9).

(b) Intramolecular competition of two ortho-*directing substituents*: The hierarchy given above (Table 1.9) is not without ambiguities, as we shall see later. Nevertheless,

Table 1.9. *ortho*-Directing power of typical heterofunctional substituents [391, 392]

$-OCON(C_2H_5)_2$		$-SO_2C(CH_3)_3$
$-CON(C_2H_5)_2$	4,4-dimethyl-2-oxazolinyl	$-C(=NCH_3)OLi$
$-CH_2N(CH_3)_2$	$-OCH_2OCH_3$	$-N{=}C(OC(CH_3)_3)OLi$
	OCH_3	
$-N(CH_3)_2$	$-F$	$-CF_3$

(arrow upward, left: increasing *ortho*-directing power; arrow downward, right: decreasing *ortho*-directing power) it offers us a first orientation about the approximate rate of a projected *ortho*-hydrogen–metal exchange reaction and, in addition, allows us to anticipate the regioselectivity encountered in the case of an intramolecular competition of two different activating groups. Thus, a whole series of methoxy-substituted *N,N*-dialkylarenecarboxamides were found to be metallated exclusively at a position adjacent to the aminocarbonyl moiety and distant from the methoxy substituent (intermediates **249** [489], **250** [490], **251** [491] and **252** [492]).

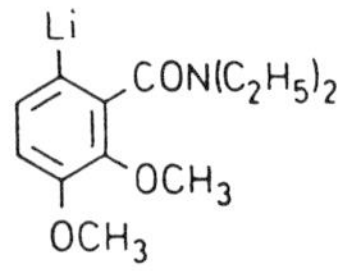

(249) 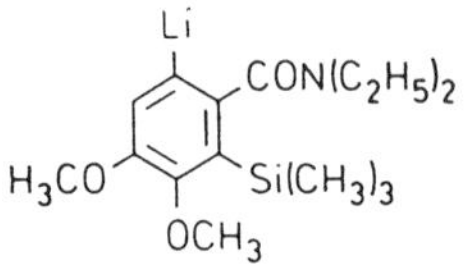 (250) 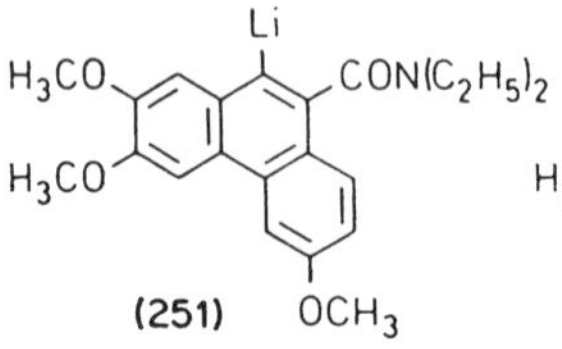 (251) 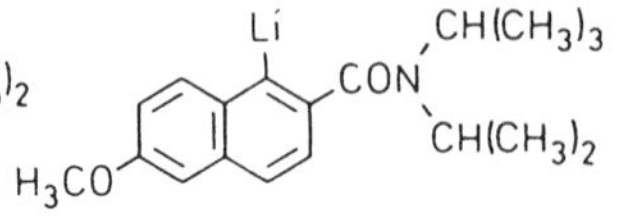 (252)

Whereas two equivalents of *tert*-butyllithium deprotonate *tert*-butyl *N*-*p*-anisylcarbamate at the nitrogen-adjacent *ortho*-position (intermediate **253** [493]), most ether amines, including phenoxazine, give rise to species (e.g. **254** [494] and **255** [495]) which carry the metal in the immediate neighbourhood of the oxygen atom. On the other hand, the 1-diethylamino-6-methoxynaphthalene derivative **256** [496], a benzyl type amine, undergoes metallation in the vicinity of the nitrogen function.

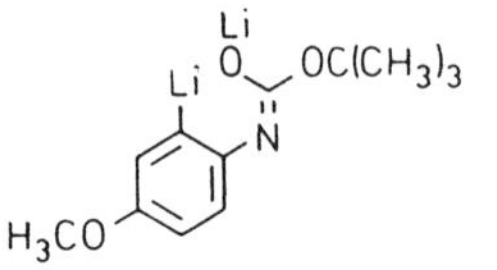

(253) 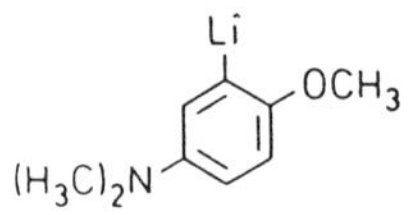 (254) 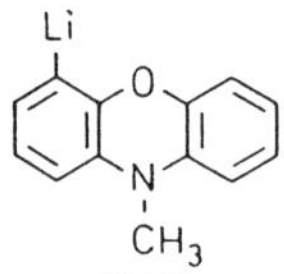 (255) 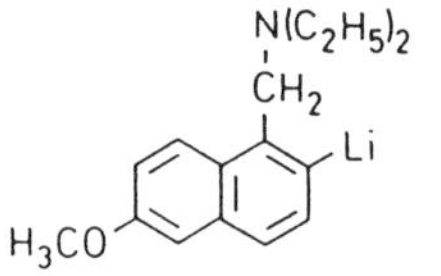 (256)

The three *N*-BOC-trifluoromethylanilines react with two equivalents of *tert*-butyllithium to afford the intermediates **257–259** lithiated at the nitrogen-adjacent position [497]. Although the *meta*-isomer could benefit from additional activation due to the electron-withdrawing effect of the trifluoromethyl moiety, this advantage is overcompensated by steric crowding. Consequently, the metallation occurs at the other *ortho*-position remote from the trifluoromethyl group.

In contrast, the *meta*-isomer of *N*-BOC-fluoroaniline is deprotonated at the doubly activated 2-position (intermediate **261**) [497]. The *ortho*-isomer is attacked at the nitrogen-adjacent position as expected (intermediate **260**) [497].

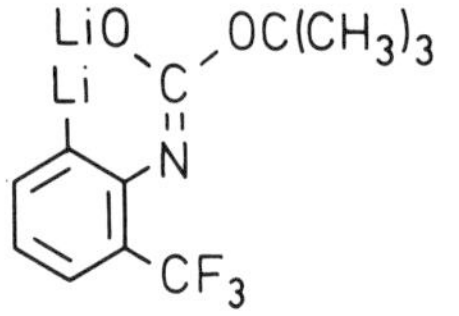

(257) 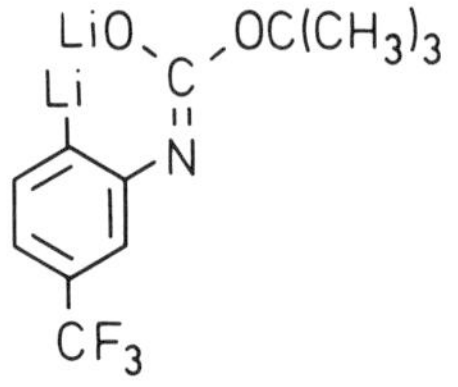 (258) 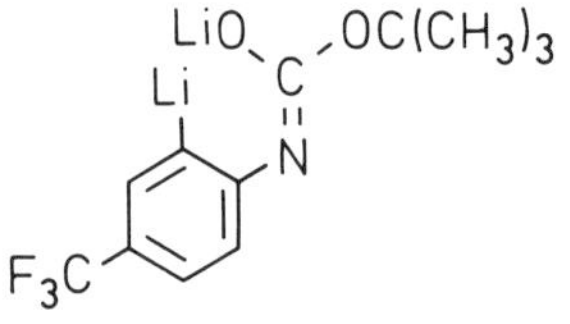 (259)

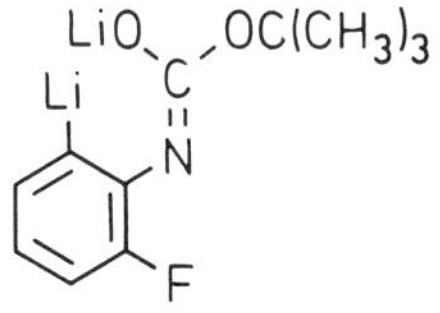

(260) 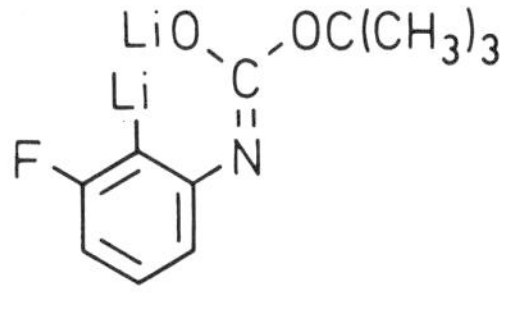 (261)

On the other hand, the hydrogen–metal exchange is reoriented to the halogen-neighbouring position when 1-(*N*,*N*-dimethylamino)fluorobenzene or *N*-(4-fluorophenyl)-*N*′-(*tert*-butoxycarbonyl)piperazine serves as the substrate (intermediates **262** and **263**) [497]. Finally, 3-fluoro-*N*,*N*-bis(trimethylsilyl)aniline is metallated at that *ortho*-position next to the fluoro substituent which is at maximum distance from the amino function (intermediate **264**) [497].

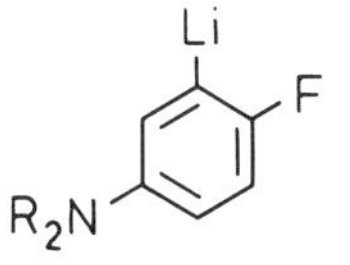

262: R_2N = $(H_3C)_2N$
263: R_2N = H_3COOC-N(piperazinyl)N

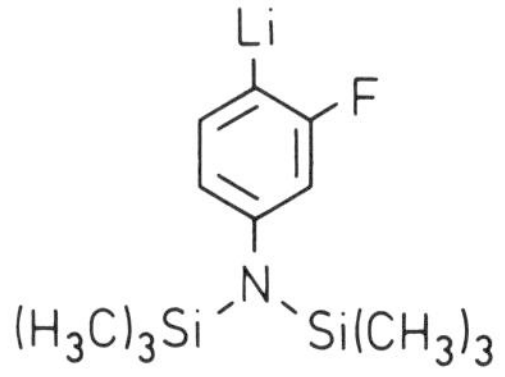

(264)

(c) Optional site selectivities: We have already considered two impressive examples of regioselectivity available 'on command.' Both, 1,3-bis(trifluoromethyl)benzene (p. 90) and *N*-pivaloyl-2-(3-methoxyphenyl)ethylamine (pp. 94–95) can be submitted to selective metallation at either one of three privileged positions merely by variation of the metallating agent and the reaction conditions.

4-Methoxy-*N*,*N*-dimethylbenzylamine represents a similar case of optional site selectivity. With butyllithium alone, deprotonation of the 2-position near to the nitrogen-carrying side-chain occurs (intermediate **265**); with TMEDA-activated butyllithium the oxygen adjacent 3-position is attacked (intermediate **266**) [498].

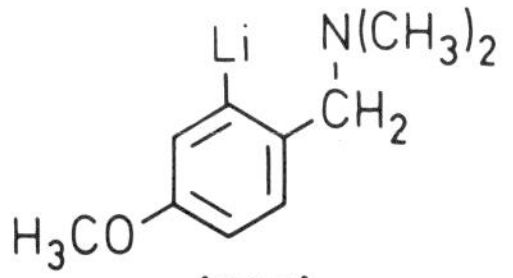

(265) 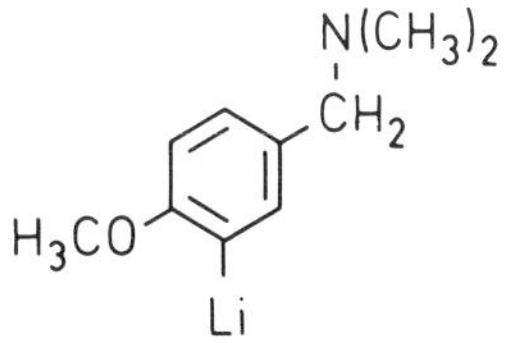 (266)

When treated with *tert*-butyllithium in tetrahydrofuran, *N*-BOC-4-fluoroaniline is almost quantitatively converted into the 2-lithio derivative (intermediate **267**) [497]. However, the 3-position close to the halogen substituent is deprotonated (intermediate **268**) if the superbasic mixture of *tert*-butyllithium and potassium *tert*-butoxide is employed as the reagent [497].

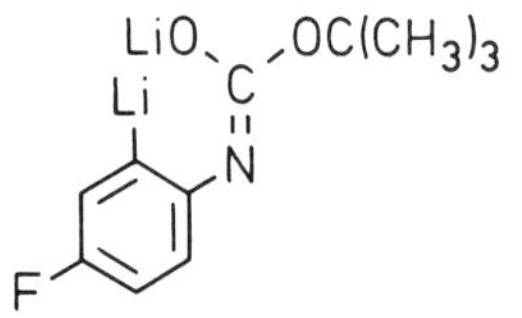

(267) 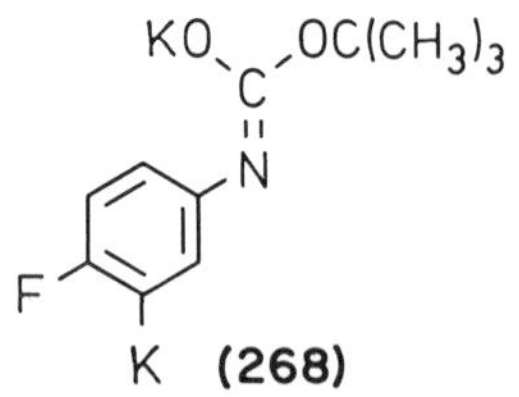 (268)

Butyllithium in tetrahydrofuran metallates the *ortho*- and *para*-isomers of fluoroanisole exclusively at the oxygen-adjacent position (intermediates **269** and **271**). Again the LIC–KOR reagent behaves differently and generates organometallic species (**270** and **272**) showing the opposite regioisomeric pattern [500].

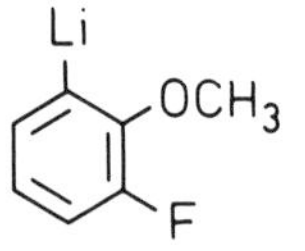

(269) 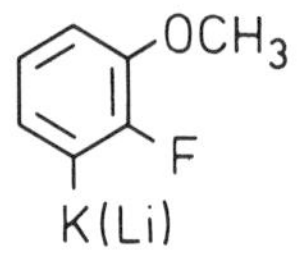 (270) 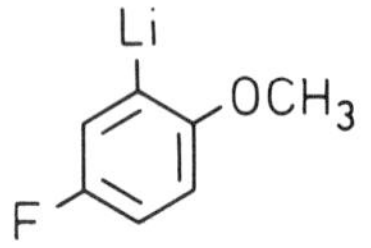 (271) 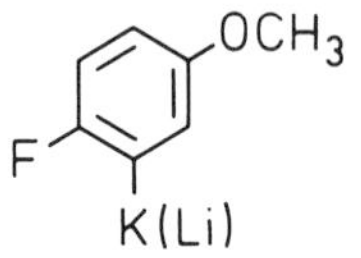 (272)

This compilation of reagent-dependent regioselectivities could be continued. It seems to us more important, however, to understand the phenomena observed. Our explanation [500] argues with the *dual character* of heterosubstituents: electron-withdrawing because of the inductive polarization of σ-bonds and electron-donating because of the availability of lone pairs.

The latter effect is crucial for the complexation of metal compounds. However, for obvious reasons only small-sized metals can interact strongly with solvent molecules or other potential ligands. Therefore, butyllithium has a high affinity for ethers or amines such as anisole or *N,N*-diamethylamine. Once docked at the heteroatom, it can easily reach over to a C—H bond in the vicinity and perform a hydrogen–metal exchange under quasi-intramolecular conditions.

TMEDA- or potassium *tert*-butoxide-complexed butyllithium is coordinatively saturated. There is little tendency left to combine with, say, anisole. Consequently, the 'proximity effect' favouring the metallation of oxygen-adjacent positions is lost. What remains intact is the activation by virtue of the inductive effect. In this respect, the highly electronegative halides are far more powerful than oxygen and nitrogen.

Mechanistic Considerations Inevitably there is an oversimplification if we describe the *ortho*-directing effect of all kinds of heterosubstituents by a set of just two parameters, as was done above. The following discussion will reconfirm the fundamental validity of this first-order approach. At the same time it endeavours to add some colour and some subtlety to the present black-and-white picture.

(a) Intrinsic nature of neighbouring group assistance to metallation reactions: Butyllithium effects the *ortho*-lithiation of 2-methoxyethoxybenzene roughly 15 times faster

than that of anisole [501]. This is an experimental fact; there was some controversy about how to rationalize it. Is it the chelating capacity of the bis(ether) moiety or the inductive effect of the second oxygen atom that causes the acceleration [502]? Rephrased in more general terms, the question reads as follows: how can one differentiate between the manifestation of an acidifying inductive and that of a metal complexing, entropy-saving effect? We suggest applying three different tests.

'Proximity' probe. Inductive effects rely mainly on *through-bond* transmission while the probability for a reactive encounter between a C—H bond and a 'docked-on' organometallic base depends on the *through-space* distance between the crucial centres. 1-Methoxynaphthalene reacts with *tert*-butyllithium in cyclohexane exclusively at the 8-position [503]. Apparently, complexation of the organometallic reagent with the oxygen atom precedes the hydrogen–metal interconversion, which, under such circumstances, can take place more conveniently at the *peri*- rather than *β*-position (transition states **273** and **274**, respectively).

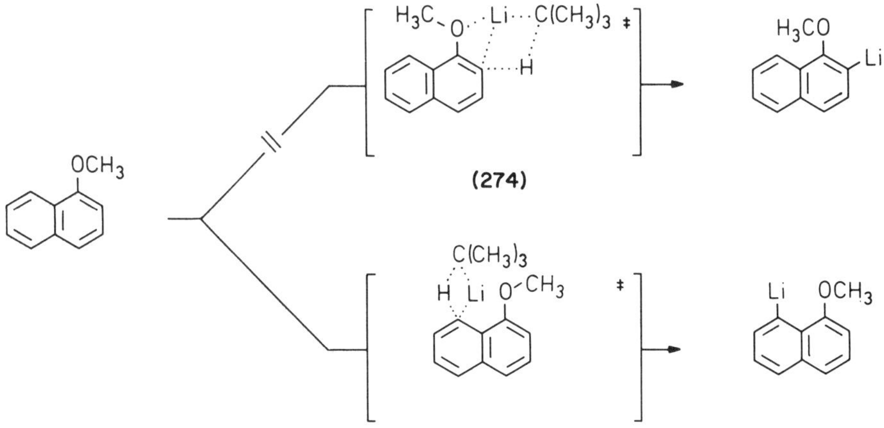

(274)

(273)

TMEDA-activated butyllithium does not always require additional complexation. Consequently, metallation occurs only at the 2-position this time [503]. 1-Dimethylaminomethylnaphthalene undergoes random hydrogen–lithium exchange at both *peri*- (8-) and *β*- (2-) positions [504]. 4-Dimethylaminomethyl-1-methoxynaphthalene gives the 5- and no 2-, 3- or 8-lithiated derivative (**275**) [505]. Excess of butyllithium abstracts instantaneously two protons from the nitrogen centre of 2′-amino-3,4-dimethoxybiphenyl and then slowly another one from the 2-, not the 5-position (intermediate **276**) [506].

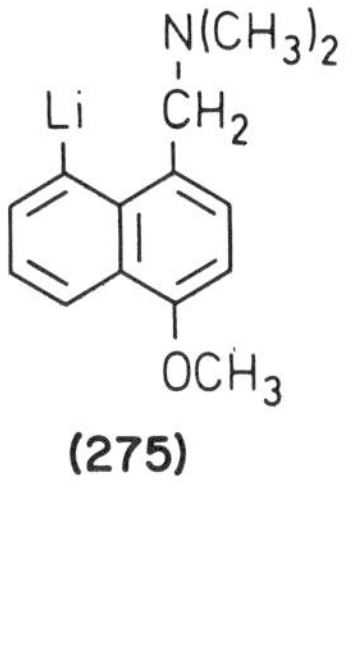

(275)

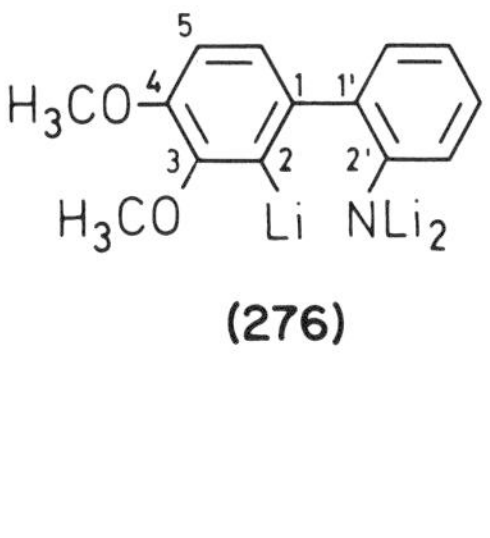

(276)

Methoxymethoxy-, 2-methoxyethoxy- and (2-dimethylaminoethoxy)benzene undergo *ortho*-metallation with comparable ease (intermediates **277–279**). If the heteroatoms were to act only by their inductive electron-withdrawing effect, the reactivity should decrease drastically from left to right.

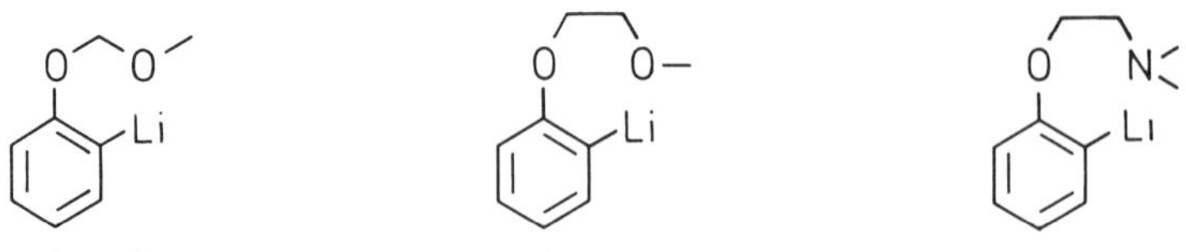

(277) (278) (279)

'*Extreme model*' *probe.* One can devise model substrates that will allow one to identify the nature of the neighbouring group effect without ambiguity. The idea will be illustrated by two paradigms, in one case only inductive, in the other only coordinative forces being operative. The bis(o,o'-biphenyl-ene)ammonium and -phosphonium salts **280** and **281** (M = H), respectively, are devoid of any possibility of binding metals by coordination with an electron-donor atom. Nevertheless, the positively charged heteroatom causes a strong acidification of the *ortho*-positions (generating **280** and **281**, M = : [507]). The primary adduct of butyllithium with diphenylacetylene (**282**, M = H) cannot be intercepted as such since it immediately undergoes a subsequent *ortho* deprotonation to give an o,α-dimetallated species (**282**, M = Li) [508]. No electronegative substituent being present, the reaction must be triggered by the formation of a mixed aggregate between the initial adduct and unconsumed butyllithium.

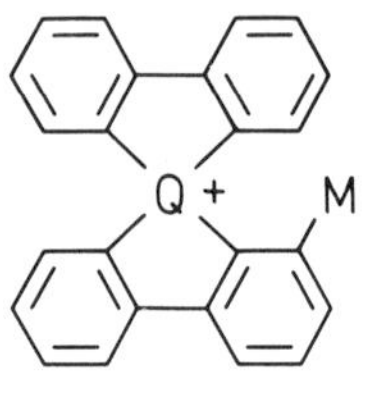

280: Q = N
281: Q = P

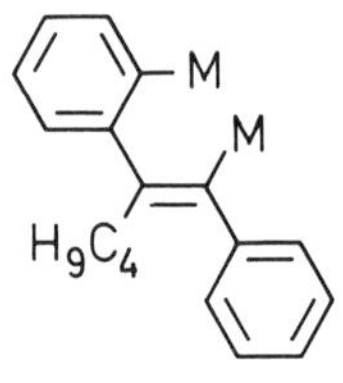

(282)

'*Naked anion*' *probe.* Without metal there will be no metal complexation. If we could employ metal-free anions as bases, the coordinative potential of any heterofunctional substituent would remain unutilized and only the inductive effect would manifest itself. Of course, free carbanions do not exist except in the gas-phase and their investigation is troublesome even under such esoteric conditions. Since information about their relative thermodynamic stabilities is still sparse, one has frequently to resort to a comparison with isoelectronic species. In this sense, pyridines may be considered as metal-free analogues of aryl anions. As their gas-phase proton affinities reveal, only fluorine, trifluoromethyl and cyano substituents have a pronounced electron-withdrawing effect while alkoxy and, in particular, dialkylamino groups are globally donors owing to a dominant electron-releasing resonance effect (Table 1.10) [509].

On the basis of such data, we conclude that halogens facilitate transmetallation

Table 1.10. Gas-phase proton affinities, $\Delta\Delta H^{\circ}_{prot}$ (kcal mol^{-1}), of 2-, 3- and 4-substituted pyridines relative to the parent compound pyridine: effect of dimethylamino, methoxy, methyl, fluoro, trifluoromethyl and cyano substituents [509]

Substituent	at 2-position	at 3-position	at 4-position
$N(CH_3)_2$	+9	+10	+15
OCH_3	+1	+2	+6
CH_3	+3	+2	+3
F	−9	−6	−3
CF_3	−9	−8	−7
CN	−12	−11	−10

reactions chiefly by inductive stabilization of the incipient negative charge whereas nitrogen-containing substituents primarily display their coordinative properties. In the case of alkoxy or aroxy groups it depends on the individual situation which factor becomes dominant.

(b) Primary and secondary metallation sites: In an idealized manner we can differentiate between three archetypes of transition states of the hydrogen–metal exchange process [318]. The carbanion mechanism (transition state **283**), extremely rare anyway, has little chance of realization in the arene field. It would require a strongly acidic substrate (say, like a thiazolium salt) and an ionized base (e.g. fluorenylpotassium in the presence of a macrocyclic polyether such as 18-crown-6). The four-centre mechanism (transition state **284**) is very common. When olefinic or aromatic substrates are involved, organosodium reagents fit particularly well into such a scheme. Finally, there is the multi-centre mechanism with participation neighbouring group X (transition state **285**). Organolithium reagents are most suitable to establish this kind of interaction with a heterofunctional substituent.

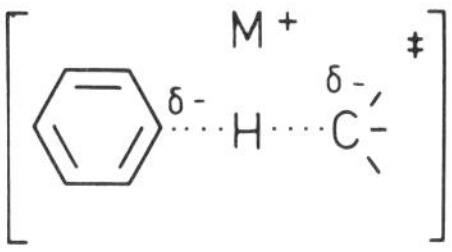

(283) 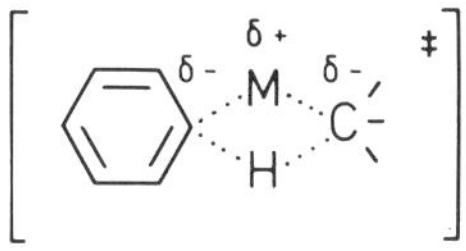 (284) 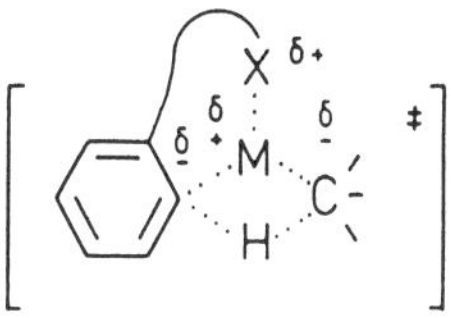 (285)

A given structural feature will stabilize or destabilize both the transition state and the final product. However, the two entities may not experience the substituent effect to the same extent. In plain words, the most favourable transition state may lead to an organometallic intermediate which is not necessarily the most stable among all possible isomers. Under such circumstances, a subsequent transmetallation may take place and convert the primary into a less basic secondary metallation product.

This happens, for example, when 2-(3,4-dimethoxyphenyl)ethyl-*N*,*N*-dimethylamine is treated with butyllithium in the presence of TMEDA. The hydrogen at the 5-position is rapidly replaced with lithium [510]. The resulting intermediate (**286**) exists only temporarily, however. It gradually undergoes isomerization to afford a more stable intermediate (**287**) in which the organometallic bond is flanked by two heterosubstituents [510]. If this second intermediate is thermodynamically more favourable, then why did it need the detour via the isomeric precursor **286** rather than to generate it directly? Apparently, the transition states of metallation reactions are particularly sensitive to steric strain. Therefore, attack of the base at a readily accessible position may be preferred even if it is less acidic than another site.

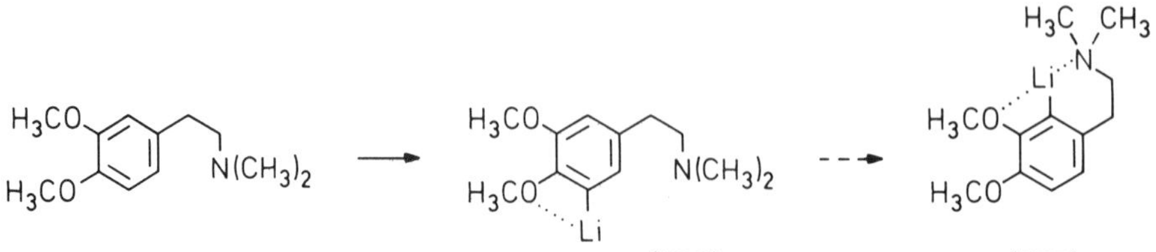

(286) (287)

Not only steric hindrance but also complexation may play a more critical role in the transition state when compared with the ground state of the organometallic compounds. 2-(2-Imidazolin-2-yl)furan and -thiophene react with excess of butyllithium in tetrahydrofuran at -75 °C to afford the 3-lithiated species (**288**, Y = O, and **289**, Y = S) [511]. Obviously the 3-position, owing to the coordination offered by the neighbouring imidazolinyl group, is the most reactive but not the most acidic site. In fact, in (mono)ethylene glycol dimethyl ether (MEGME) as the solvent or with the weaker base lithium diisopropylamide (LIDA), the 5-metallated isomers (**290**, Y = O, and **291**, Y = S) are selectively obtained [511].

288 : Y = O
289 : Y = S

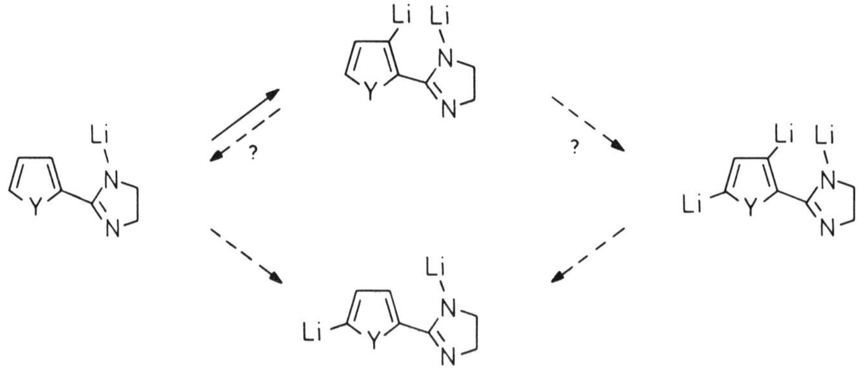

290 : Y = O
291 : Y = S

It is plausible to assume that the latter intermediates are not formed in a straightforward manner but rather via the 3-lithiated isomers (**288** and **289**). Either a *C*-zero- or a *C,C*-dimetallated species then has to act as the turnable for the metal transposition.

Recently, a dilithio species (**293**) was found to be involved in the reaction between *N*-phenylpyrrole and TMEDA-activated butyllithium in hexane (65 °C) or diethyl ether (0 °C). The α-metallation is apparently a relatively slow process. Most of the resulting monolithio species (**292**) rapidly undergoes an aggregate-mediated second hydrogen–metal exchange, which this time occurs at the *ortho*-position of the aromatic ring. The dimetal compound (**293**) thus obtained slowly delivers the aryl-bound second lithium atom to unconsumed *N*-phenylpyrrole and, in this way, selectively reverts to the monometal species (**292**) [512].

N Li Li Li

(292) (293)

(c) Outlook: Scepticism has already been expressed about the possibility of rationalizing all phenomena of neighbouring group participation solely with the dual nature of heterosubstituents which are inductive electron attractors and at the same time metal-complexing electron donors. Nevertheless, this model, simple as it is, has been found to explain amazingly well the experimental findings, at least qualitatively. What has to be considered in addition in order to make the predictions more accurate and more quantitative?

First, we should recognize the interdependence that exists between the acidifying inductive polarization and the entropy-saving coordination. When the oxygen atom of an ether or the nitrogen atom of an amine binds a metal, it has to share electrons with the latter. This imposes a partial positive charge on the heteroatom and, as a corollary, enhances its electron-withdrawing inductive effect.

Another feature that merits attention is the non-additivity of substituent effects. Formaldehyde dimethyl acetal (dimethoxymethane) is virtually inert towards butyllithium, and diphenylmethane reacts only sluggishly. On the other hand, benzyl methyl ether is instantaneously deprotonated even at −75 °C. Evidently the combination of two different substituents is advantageous, since one alkoxy group is enough to ensure complexation and one phenyl group provides sufficient resonance stabilization.

Finally, the subtle, although still efficacious, action of heteroatom-free neighbouring groups should not be overlooked. 1,5-Hexadiene reacts five times faster than 1-hexene with LIC–KOR in tetrahydrofuran, affording the corresponding allyl-type organopotassium compounds (**294** and **295**, respectively) [360]. The acceleration suggests an extra coordination of the metal by the 'passive' second ene moiety. In tetrahydrofuran solution, the reaction of 1,ω-dienes having moderate chain lengths stops cleanly at the monometallation level even if the reagent is employed in large excess. In contrast, the dimetallated species (e.g. **296**) is formed preferentially, if not exclusively, when the reaction mixture is suspended in hexane. This selectivity is strictly obeyed even when under-stoichiometric amounts of the LIC–KOR reagent are used, provided that the chain length does not exceed eleven carbon atoms [310, 360]. Apparently, the monometallation products form with unconsumed LIC–KOR reagent in paraffinic media a mixed aggregate (e.g. **297**) within which the second hydrogen–metal interconversion can be accomplished at a much faster rate than the preceding first one.

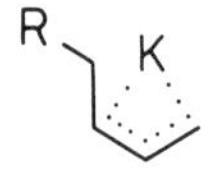

294: R = $H_2C{=}CH-$
295: R = H_3C-CH_2-

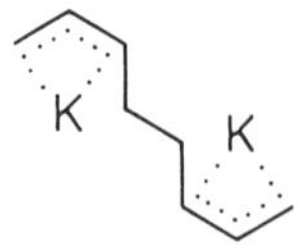

(296)

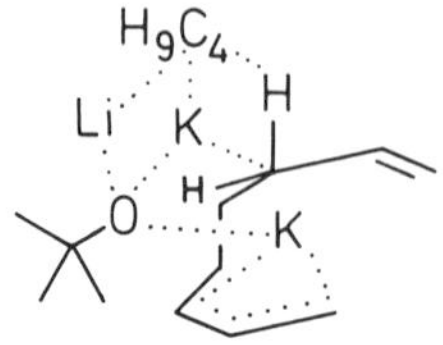

(297)

1.5 HANDLING OF ORGANOMETALLIC REAGENTS

This section covers all practical aspects of the work with organometallic compounds such as safety precautions, special equipment, analytical determinations and shelf-life. First, a collection of detailed procedures for the preparation of organometallic reagents and their application in organic synthesis is given.

1.5.1 STANDARD REAGENTS

The most common organolithium compounds, such as methyllithium, butyllithium ('*n*-butyllithium'), hexyllithium, *sec*-butyllithium, *tert*-butyllithium and phenyllithium, can be purchased from several chemical suppliers. Nevertheless, it may occasionally be advantageous, even mandatory, to prepare such reagents oneself. This is the case, for example, if one needs lithium bromide containing phenyllithium [1]. The salt remains in solution if the phenyllithium is made from bromobenzene in diethyl ether whereas it precipitates from the commercial 3 : 7 ether–benzene solvent mixture.

1.5.1.1 Alkyl-, 1-Alkenyl- and Aryllithiums

All reactions should be carried out under a blanket of ⩾ 99.99% pure nitrogen. Appropriate glassware for work under inert gas protection and convenient methods for titration of organometallic concentrations are described below (pp. 137–138).

Ethereal Methyllithium [2]

Diethyl ether (0.4 l) and lithium shred (69 g, 10 mol) are placed in a 4 l three-necked round-bottomed flask which is hermetically closed with a mercury seal. Under vigorous magnetic stirring, methyl chloride (0.25 kg, 5.0 mol) is bubbled into the mixture at a sufficiently slow rate to prevent most of the gas from escaping through the mercury seal. As soon as the reaction starts, the suspension becomes turbid while the metal surface becomes shiny and silvery white. Now the flask is cooled with an ice-bath and a larger quantity of diethyl ether (0.2 l) is added. After approximately 1 h, all the methyl chloride is absorbed and most of the lithium is consumed. The mixture is heated for 30 min under reflux in order to expel residual methyl chloride from the solution. Finally, it is siphoned through a plug of glass-wool into storage burettes. Roughly 2 l of a 2 M solution are obtained.

Lithium Bromide Containing Ethereal Methyllithium

The same protocol is applied as above but methyl bromide (350 g, 3.7 mol) is used as the starting material besides lithium (56 g, 8.1 mol).

Lithium Bromide Containing Ethereal Butyllithium [3]

At 25 °C and under vigorous stirring, a small quantity of butyl bromide is added to lithium shred (69 g, 10 mol) covered with diethyl ether (0.4 l). As soon as the reaction starts, the flask is placed in a salt–ice-bath and butyl bromide (0.44 l, 0.55 kg, 4.0 mol) in diethyl ether (2.0 l) is added dropwise in the course of *ca* 60 min at such a rate that the interior temperature is kept around −10 °C. Towards the end of the reaction, the temperature is allowed to rise to 0 °C. The solution has to be stored in a freezer (at −20 or −25 °C) to avoid rapid decomposition.

Lithium Bromide Containing Ethereal Cyclopropyllithium [4]

Analogously as described above for butyllithium, cyclopropyllithium and alkyl-substituted derivatives thereof can be prepared as ethereal, lithium bromide-containing solutions.

Ethereal 1,4-Tetramethylenedilithium [5]

Lithium (20 g, 2.9 mol), in the form of powder or dispersion (from which the mineral oil has been removed by thorough washing), is suspended in diethyl ether (0.6 l). Under vigorous stirring, a solution of 1,4-dibromobutane (60 ml, 108 g, 0.5 mol) in diethyl ether (0.40 l) is added dropwise in the course of 90 min. As soon as the reaction starts, the mixture is cooled from 25 to −15 °C. In addition to the expected dilithio compound (*ca* 65%), cyclobutane and other by-products are formed.

Vinyllithium in Tetrahydrofuran [6]

The hexane is removed from the commercial dispersion of 98 : 2 (w/w) lithium–sodium alloy (6.9 g, 1.00 mol) and replaced with tetrahydrofuran (0.1 l). Vinyl chloride (25 g, 0.40 mol) is slowly condensed into this slurry under stirring. As soon as the reaction starts, the flask is placed in an ice-bath and more tetrahydrofuran (0.7 l) is added, while the addition of the halide is continued. When the entire quantity is absorbed, stirring is continued for another 60 min at 0 °C. Owing to side-reactions, only 60–70% of vinyllithium is obtained.

Lithium Bromide Containing Ethereal Phenyllithium [7]

The reaction is started by pouring bromobenzene (10 ml, 15 g, 0.10 mol) into the vigorously stirred slurry of lithium shred or wire (35 g, 5.0 mol) in diethyl ether (0.1 l). As soon as the reaction starts, the flask is placed in an ice-bath and bromobenzene (0.21 l, 0.31 kg, 2.0 mol) in diethyl ether (2.5 l) is added in the course of *ca* 1 h while the temperature of the mixture is kept below its boiling point. The magenta-coloured solution obtained is *ca* 0.75 M in phenyllithium and contains *ca* 6–8% of biphenyl. Pure phenyllithium is colourless [8], (as all alkyl- and aryllithium compounds are). Phenyllithium in tetrahydrofuran can be prepared from bromobenzene [9] or chlorobenzene [10] around −60 °C, but rapidly loses its titre unless conserved below 0 °C.

1.5.1.2 α-Heterosubstituted Alkyllithiums

Methoxymethyllithium [11]

Sodium (0.89%) containing lithium (6.3 g, 0.91 mol) is finely dispersed by heating it in the presence of paraffin oil (0.1 l) to 220 °C and by operating a vibromixer until the temperature has fallen to 120 °C. The paraffin oil is removed by filtration through a frit and the collected metal dispersion is washed with anhydrous diethyl ether. After drying, it is transferred into a three-necked flask filled with formaldehyde dimethyl acetal (0.15 l). At 0 °C, chloromethyl methyl ether (30 ml, 32 g, 0.40 mol) in formaldehyde dimethyl acetal (0.10 l) is added dropwise under vigorous stirring. As soon as a temperature rise indicates the start of the reaction, the mixture is cooled to −25 °C and the remaining chloride is added in the course of 3 h. After an additional 30 min of stirring at −25 °C, the solution is filtered through a glass-wool plug. The organometallic titre generally ranges from 0.32 to 0.35 M.

Methoxymethylpotassium [12]

Dibutylmercury (1.6 g, 5.0 mmol) is slowly added to a vigorously stirred suspension of potassium–sodium alloy (78:22, w/w; 1.0 ml, 0.86 g, 18 mmol) in pentane (20 ml). After 30 min of stirring, the solvent is stripped off and the residue dissolved in precooled (−75 °C) dimethyl ether (5.0 ml, 4.0 g, 86 mmol). The mixture is kept 2 h at 60 °C before being treated with benzaldehyde (1.0 ml, 1.1 g, 10 mmol). The excess of metal is cautiously destroyed by dropwise addition of ethanol (5 ml) followed by water (10 ml). Extraction with hexane (3 × 10 ml) and distillation give 2-methoxy-1-phenylethanol; 75%; b.p. 123–124 °C/15 mmHg.

Dichloromethyllithium [13]

Butyllithium (41 mmol) in hexane (25 ml) is added dropwise over 45 min to dichloromethane (2.6 ml, 3.2 g, 40 mmol) in tetrahydrofuran (90 ml) at −75 °C.

Trichloromethyllithium [14]

Butyllithium (10 mmol) in hexane (15 ml) is added dropwise over 15 min to chloroform (1.6 ml, 2.4 g, 20 mmol) in tetrahydrofuran (32 ml) and diethyl ether (8 ml) at −110 °C.

1.5.1.3 Organosodium and Organopotassium Reagents

Pentylsodium [15]

At −10 °C, pentyl chloride (60 ml, 53 g, 0.50 mol) is added dropwise during 1 h to a sodium dispersion [16] (23 g, 1.0 mol) in pentane (0.2 l) under vigorous high-speed stirring (5000–10 000 rpm) [17]. After continuous stirring for 30 min at −10 °C, a bluish grey suspension is obtained which may be used as such. If only a small quantity is needed or if an aliquot has to be transferred for analytical purposes, the desired volume can be withdrawn from the well stirred suspension by means of a nitrogen-purged pipette.

Phenylsodium [18]

Under vigorous stirring, chlorobenzene (0.10 l, 0.11 kg, 1.0 mol) is added dropwise to a sodium dispersion [16] (50 g, 2.2 mol) in benzene (0.50 l). After one tenth of the chlorobenzene has been introduced, the reaction should have started (as evidenced by a sudden rise in temperature to 50–80 °C). If this is not the case, the sodium has to be activated by the addition of a small amount

(1 ml) of 1-pentanol. Now an ice–salt bath is used to keep the temperature of the reaction mixture around 35 °C. The addition of chlorobenzene is finished after *ca* 30 min; the yield is almost quantitative.

Trimethylsilylmethylpotassium (Suspension in Pentane) [19]

A solution of bis(trimethylsilylmethyl)mercury (5.0 ml, 7.5 g, 20 mol) in pentane (50 ml) is vigorously stirred for 30 min with potassium–sodium alloy [20] (78 : 22, w/w; 3.5 ml, 3.0 g, 64 mmol potassium). The resulting grey amalgam-containing suspension can be used as such.

Trimethylsilylmethylpotassium (Solution in Tetrahydrofuran) [19]

A suspension of trimethylsilylmethylpotassium (40 mmol) in pentane (50 ml) is prepared according to the procedure given above. The solvent is stripped off under reduced pressure and is replaced with precooled (−75 °C) tetrahydrofuran (80 ml). After 15 min of vigorous stirring at −75 °C, one waits until a sediment has deposited and transfers the colourless supernatant liquid by means of a pipette. The organometallic solution is stored at temperatures around or below −50 °C.

1.5.2 WORKING PROCEDURES

Numerous well elaborated protocols have been published elsewhere [21]. The examples given below are intended to illustrate the progress achieved recently, in particular by the use of superbasic reagents. For simplicity, the intermediates resulting from metallation reactions with the LIC–KOR mixture [22] are designated throughout as potassium species, although they invariably contain substantial amounts of organolithium compounds also.

1.5.2.1 Metallation of Simple Hydrocarbons

Cyclopropanes and resonance-inactive alkenes and arenes belong to the least acidic hydrocarbons. Only ordinary alkanes are still more reluctant to undergo deprotonation. Consequently, the most powerful reagents are required in order to accomplish the hydrogen–metal exchange process with synthetically useful efficiency. In this section the metallation and subsequent electrophilic substitution of tricyclo[2.2.1.0^{2,6}]heptane (nortricyclene), 3,3-dimethyl-1-butene (*tert*-butylethylene), bicyclo[2.2.1]hepta-2,5-diene (norbornadiene), spiro[4.4]nona-2.4-diene and bicyclo[2.2.2]oct-2-ene are described in detail. The case of 2,2-dimethyl-3-methylenebicyclo[2.2.1]heptane (fenchene) is included in Section 1.5.2.8 (p. 128).

1-Tricyclo[2.2.1.0^{2,6}]heptylpotassium (Nortricyclylpotassium) [23]

NaC_5H_{11} / $KOC(CH_3)_3$ → K ; $ClSi(CH_3)_3$ → $Si(CH_3)_3$

Tricyclo[2.2.1.$0^{2,6}$]heptane (1.9 g, 20 mmol) is added to a vigorously stirred suspension of pentylsodium (20 mmol) and potassium *tert*-butoxide (2.2 g, 20 mmol) in pentane (40 ml). After 24 h, the mixture is treated with chlorotrimethylsilane (3.2 ml, 2.7 g, 25 mmol). 1-(Trimethylsilyl)-tricyclo[2.2.1.$0^{2,6}$]heptane is isolated by distillation under reduced pressure; 60%; b.p. 52–55 °C/17 mmHg. Starting material (*ca* 35%) is recovered.

(E)-*3,3-Dimethyl-1-butenylsodium [24]*

NaC_5H_9 / NaO–C(CH3)2–C(CH3)2–ONa → Na; $CH_2=O$ → OH

3,3-Dimethyl-1-butene (3.0 ml, 2.0 g, 23 mmol) is added to the suspension of pentylsodium (20 mmol) and disodium pinacolate (20 mmol) in pentane (50 ml). After 50 h of vigorous stirring, the sealed Schlenk tube is opened and the mixture is treated with a 0.8 M solution of formaldehyde [25] (10 mmol) in tetrahydrofuran (25 ml). After neutralization, 4,4-dimethyl-2-penten-1-ol having a (*Z*/*E*) ratio of 0.2:99.8 is isolated; 88%; b.p. 76–77 °C/20 mmHg.

2-Bicyclo[2.2.2]oct-2-enylpotassium [24]

NaC_5H_{11} / $KOC(CH_3)_3$ → K; 1. $FB(OCH_3)_2$ 2. H_2O_2 → O

Bicyclo[2.2.2]oct-2-ene (27 g, 25 mmol) is added to a suspension of pentylsodium (25 mmol) and potassium *tert*-butoxide (2.8 g, 25 mmol) in pentane (25 ml). After 10 h of vigorous stirring, the mixture is consecutively treated with fluorodimethoxyboron etherate (5.0 ml, 4.5 g, 27 mmol) [26] and 35% aqueous hydrogen peroxide (3.0 ml, 3.4 g, 35 mmol) to which a few pellets of sodium hydroxide have been added. 2-Bicyclo[2.2.2]octanone is isolated by bulb-to-bulb (Kugelrohr) distillation; b.p. 130–135 °C/10 mmHg; 70%; m.p. 177–178 °C.

2-Bicyclo[2.2.1]hepta-2,5-dienylsodium [27]

LiC_4H_9 / $NaOC(CH_3)_3$ → Na; $Cl\text{-}Si(CH_3)_3$ → $Si(CH_3)_3$

Butyllithium (50 mmol), from which the commercial hexane solvent has been stripped off, is dissolved in precooled (−75 °C) tetrahydrofuran (0.10 l). Sublimed sodium *tert*-butoxide (5.3 g,

55 mmol) and bicyclo[2.2.1]hepta-2,5-diene (norbornadiene) (5.1 ml, 4.5 g, 50 mmol) are consecutively added. The reaction mixture is stirred until it becomes homogeneous. After 15 h at −50 °C, it is treated with chlorotrimethylsilane (7.6 ml, 6.5 g, 60 mmol). On distillation under reduced pressure, 2-(trimethylsilyl)bicyclo[2.2.1]hepta-2,5-diene is obtained; 92%; b.p. 55–57 °C/10 mmHg.

3-Spiro[4.4]nona-2,4-dienylsodium [24]

NaC_5H_{11} ; 1. $FB(OCH_3)_2$ 2. H_2O_2

Spiro[4.4]nona-2,4-diene (3.0 g, 25 mmol) is added to a suspension of pentylsodium (25 mmol) in pentane (25 ml). After 100 h of vigorous stirring, the reaction mixture is consecutively treated with fluorodimethoxyboron etherate [26] (5.0 ml, 4.5 g, 27 mmol), 35% aqueous hydrogen peroxide (3.0 ml, 3.4 g, 35 mmol) and 5 M aqueous sodium hydroxide (5 ml). Spiro[4.4]non-4-en-3-one is formed in 45% yield; b.p. 92–94 °C/10 mmHg. If the organometallic intermediate is trapped with chlorotrimethylsilane (30 mmol), a 65% yield of the corresponding silane is obtained.

2-Cyclohepta-1,3,5-trienylsodium [27]

LiC_4H_9 / $NaOC(CH_3)_3$; Na ; $ClSi(CH_3)_3$; $Si(CH_3)_3$

Butyllithium (30 mmol), from which the commercial hexane solvent has been stripped off, is taken up in precooled (−90 °C) tetrahydrofuran (50 ml). At −50 °C, cycloheptatriene (3.1 ml, 2.8 g, 30 mmol) and sodium *tert*-butoxide (2.9 g, 30 mmol) are dissolved under stirring in the reaction mixture, which then is kept for 15 h at −50 °C. Chlorotrimethylsilane (4.0 ml, 3.4 g, 32 mmol) is added and the 2-(trimethylsilyl)cyclohepta-1,3,5-triene is isolated by distillation; 57%; b.p. 100–102 °C/10 mmHg.

*3,5-Di-*tert*-butylphenylpotassium [28]*

LiC_4H_9 / $KOC(CH_3)_3$; K ; CO_2 ; COOH

1,3-Di-*tert*-butylbenzene (11 ml, 9.5 g, 50 mmol) and potassium *tert*-butoxide are added to a 1.5 M solution of butyllithium (53 mmol) in hexane (35 ml). After 2 h of stirring at 25 °C, the mixture is poured on to dry-ice covered with tetrahydrofuran (25 ml). After evaporation of the solvents, the residue is dissolved in water (100 ml) and washed with diethyl ether (3 × 20 ml). On acidification of the aqueous layer, 3,5-di-*tert*-butylbenzoic acid precipitates. It is collected, washed and dried; 56%; m.p. 173–174 °C.

5-(1,1,3,3-Tetramethyl-1,3-dihydroisobenzofuryl)potassium [29]

LiC_4H_9 / $KOC(CH_3)_3$; $O{=}CH{-}N(CH_3)_2$

K ; O=CH

Potassium *tert*-butoxide (2.5 g, 22 mmol) is added to a solution of 1,1,3,3-tetramethyl-1,3-dihydroisobenzofuran (3.5 g, 20 mmol) and butyllithium (22 mmol) in hexane (45 ml). The mixture is sonicated for 2 h at 25 °C before being consecutively treated with lithium bromide (25 mmol) in tetrahydrofuran (50 ml) and dimethylformamide (3.0 ml, 2.9 g, 39 mmol). Elution from silica gel with a 1:5 (v/v) mixture of ethyl acetate and hexane affords 5-(1,1,3,3-tetramethyl-1,3-dihydroisobenzofuran)carbaldehyde; 50%; m.p. 83–85 °C.

1.5.2.2 Neighbouring Group-Assisted Metallation of Olefinic Positions

Enechlorides (1-alkenyl chlorides) [30] and enesulfides [31] undergo a hydrogen–metal exchange at the hetero-adjacent position with particular ease. The α-metallation of enethers has been reported previously [32] and has the greatest practical utility since the resulting intermediates are synthetic equivalents of acyl anions [33].

1-Ethoxyvinyllithium [34]

$H_2C{=}CH{-}OC_2H_5 \xrightarrow[TMEDA]{LiC(CH_3)_3} H_2C{=}C(OC_2H_5)Li \xrightarrow{O=CH-C_6H_5} H_2C{=}C(OC_2H_5){-}CH(OH){-}C_6H_5$

At −30 °C, ethyl vinyl ether (9.6 mL, 7.2 g, 0.10 mol) and *N,N,N′,N′*-tetramethylethylenediamine (15 ml, 12 g, 0.10 mol) are added to *tert*-butyllithium (0.10 mol) in pentane (0.10 L). The reaction is complete after 1 h at 25 °C. Addition of benzaldehyde (1.0 ml, 1.1 g, 10 mmol) and neutralization with acetic acid (0.60 mL, 0.63 g, 11 mmol), immediately followed by distillation, afford 2-ethoxy-1-phenyl-2-propen-1-ol; 57%; b.p. 93–95 °C/1 mmHg.

1-(2-Tetrahydropyranyloxy)-1-propenyllithium [35]

Under stirring, potassium *tert*-butoxide (2.8 g, 25 mmol) and *sec*-butyllithium (25 mmol) in hexane (25 mL) are consecutively added to a solution of *cis*-1-propenyl 2-tetrahydropyranyl ether (3.6 g, 25 mmol) in tetrahydrofuran (50 mL). After 1 h at −75 °C, the reaction mixture is treated with methyl iodide (1.8 ml, 4.0 g, 28 mmol). (*Z*)-1-Methyl-1-propenyl 2-tetrahydropyranyl ether is isolated by distillation, 66% (83% formed according to gas chromatographic analysis); b.p. 87–89 °C/15 mmHg.

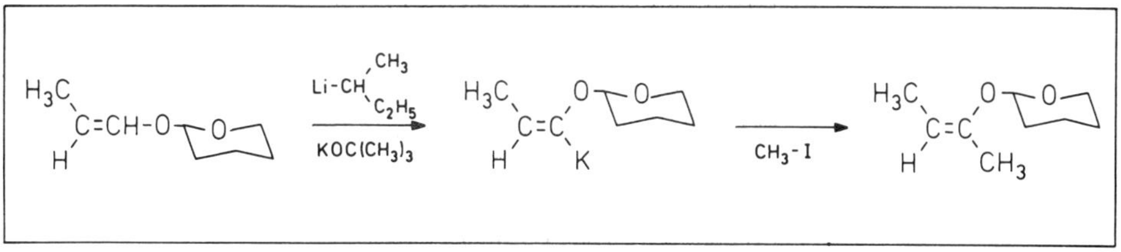

5-Chloro-3,4-dihydro-2H-pyran-6-yllithium [36]

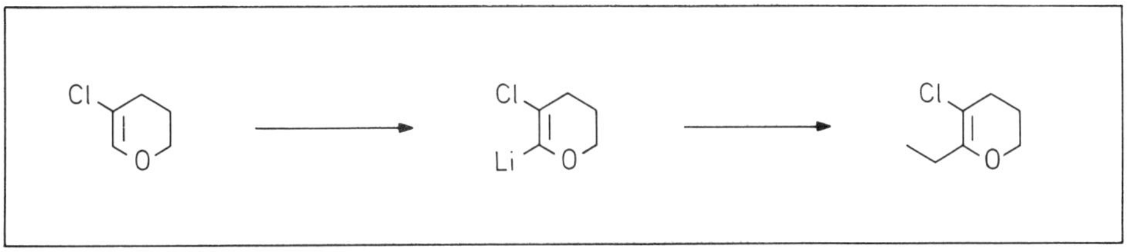

A 1.5 M solution of butyllithium (0.11 mol) in hexane (70 ml) and 5-chloro-3,4-dihydro-2*H*-pyran [37] (12 g, 0.10 mol) in tetrahydrofuran (30 ml) are mixed at −75 °C. After 5 min at 25 °C, the mixture is placed in an ice–salt bath (−20 °C) and dimethyl sulfate (17 g, 0.11 mol) is added in the course of 10 min under stirring. The mixture is washed with brine (3 × 25 ml) and concentrated. Distillation affords 5-chloro-6-ethyl-3,4-dihydro-2*H*-pyran; 79%; b.p. 56–58 °C; n_D^{20} 1.4688. The latter product undergoes clean ring opening when a 2 M solution of it in tetrahydrofuran is treated for 15 min at 25 °C with finely dispersed sodium sand (3.0 molar equiv.). After careful hydrolysis with methanol and water, 4-heptyn-1-ol is isolated in 84% yield; b.p. 93–95 °C/ 14 mmHg.

1.5.2.3 Neighbouring Group-Assisted Metallation of Aromatic Positions

Fluorine or trifluoromethyl substituents, alkoxy or aroxy moieties and nitrogen functional groups in aniline or benzylamine derivatives all activate the neighbouring aromatic positions strongly enough to make them amenable to hydrogen–metal exchange. Such *ortho*-directed metallations play a prominent role as key steps in modern organic synthesis. Optional site selectivity in favour of one among different activated positions can be established when two *ortho*-directing substituents are attached to the same nucleus. Several of the following selected examples emphasize this possibility. Further, it is shown how diphenyl ether can be submitted to either *o*-mono- or *o,o'*-dimetallation.

2-Fluorophenylpotassium [38]

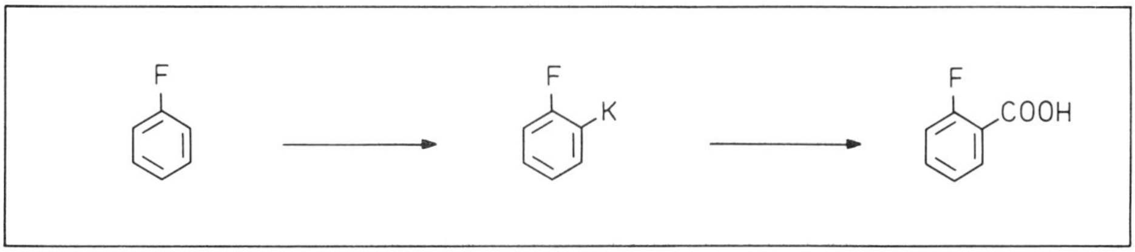

A solution of fluorobenzene (4.7 ml, 4.8 g, 50 mmol) and potassium *tert*-butoxide (5.6 g, 50 mmol) in tetrahydrofuran (0.10 l) is cooled to −75 °C before it is added to butyllithium (50 mmol) from which the commercial hexane solvent has been stripped off. After 3 h at −75 °C, the mixture is poured on to freshly crushed dry-ice. The solvents are evaporated and the residue is treated with ethereal hydrogen chloride (2.5 M; 75 mmol). Extraction with hot hexane, filtration and crystallization afford pure 2-fluorobenzoic acid; 87% (98% by gas chromatographic analysis of the crude mixture); m.p. 123–124 °C.

2-(Trifluoromethyl)phenylpotassium [38]

F_3C F_3C K F_3C COOH

In exactly the same way as described above, trifluoromethylbenzene (benzotrifluoride, α,α,α-trifluorotoluene) (8.2 ml, 9.8 g, 50 mmol) is converted into 2-(trifluoromethyl)benzoic acid; 84% (94% by gas chromatography); m.p. 110–113 °C (from cyclohexane).

2,6-Bis(trifluoromethyl)phenylpotassium [38]

F_3C CF_3 F_3C K CF_3 F_3C COOH CF_3

Under the same conditions as above, 1,3-bis(trifluoromethyl)benzene (hexafluoro-*m*-xylene) (7.8 ml, 10.7 g, 50 mmol) gives 2,6-bis(trifluoromethyl)benzoic acid; 78%; m.p. 133–135 °C (from hexane).

2,4-Bis(trifluoromethyl)phenyllithium [38]

F_3C CF_3 F_3C CF_3 Li F_3C CF_3 COOH

A precooled (−75 °C) solution of 1,3-bis(trifluoromethyl)benzene (7.8 ml, 10.7 g, 50 mmol) and *N,N,N′,N″,N″*-pentamethyldiethylenetriamine (PMDTA) (10.2 ml, 8.7 g, 50 mmol) in tetrahydrofuran (50 ml) is added to *sec*-butyllithium from which the commercial solvent (hexane) has been stripped off. After 10 h at −75 °C, the mixture is poured on to crushed dry-ice. The mixture is worked up as described for the 2-fluorobenzoic acid to give 56% of 2,4-bis(trifluoromethyl)benzoic acid; m.p. 109–110 °C (from hexane).

3-Fluoro-2-methoxyphenyllithium [39]

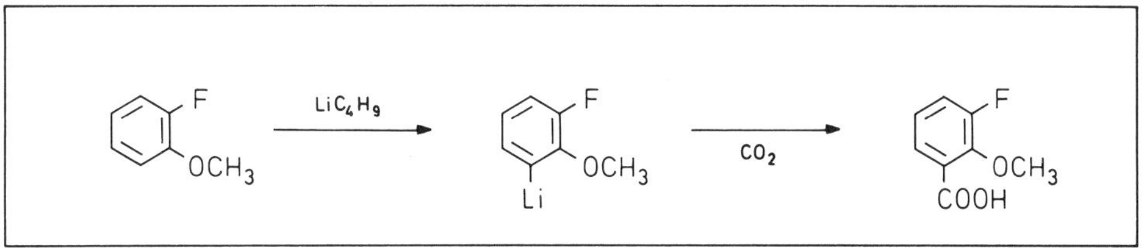

2-Fluoroanisole (5.6 ml, 6.3 g, 50 mmol) is added to a solution of neat butyllithium (50 mmol) in tetrahydrofuran kept at −75 °C. After 50 h, the mixture is poured on to crushed dry-ice. Evaporation of the solvent, neutralization of the residue with ethereal hydrogen chloride (2.5 M, 70 ml) and recrystallization from toluene affords pure 3-fluoro-2-methoxybenzoic acid; 50%; m.p. 154–156 °C.

2-Fluoro-3-methoxyphenyllithium [39]

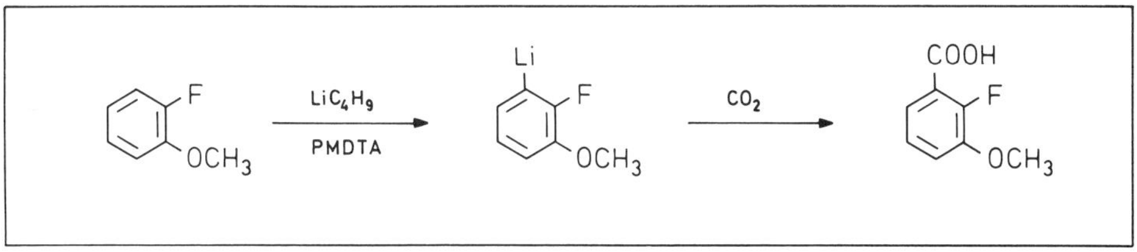

A solution containing 2-fluoroanisole (5.6 ml, 6.3 g, 50 mmol), butyllithium (50 mmol) and *N,N,N′,N″,N″*-pentamethylethylene diamine (PMDTA) (10.2 ml, 8.7 g, 50 mmol) in a mixture of tetrahydrofuran (60 ml) and hexane (40 ml) is kept for 2 h at −75 °C. The 2-fluoro-3-methoxybenzoic acid is isolated as described above; 87%; m.p. 154–156 °C (from water).

5-Fluoro-2-methoxyphenyllithium [39] and 2-Fluoro-5-methoxyphenyllithium [39]

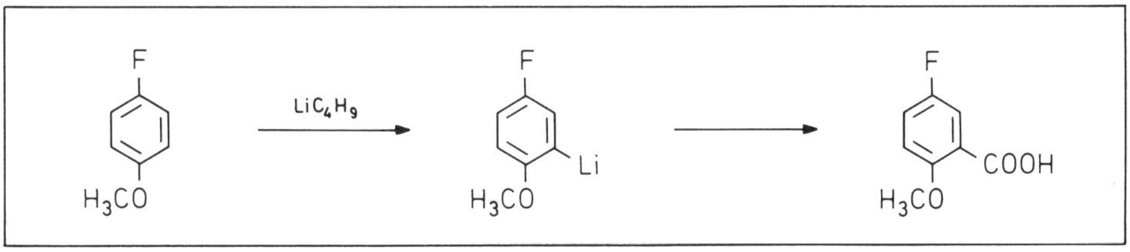

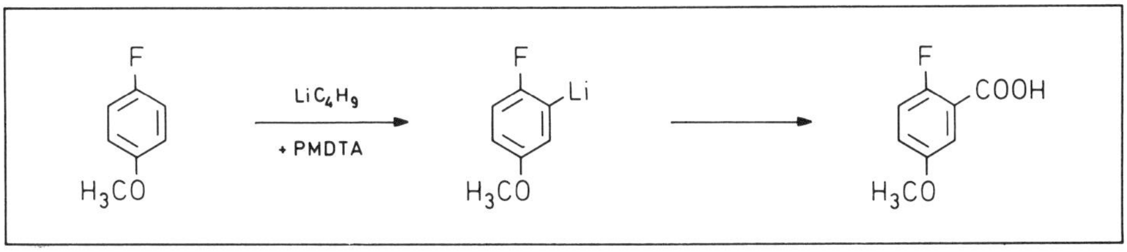

When the protocols given above are applied to 4-fluoroanisole (5.7 ml, 6.3 g, 50 mmol) rather than to its *ortho*-isomer, 5-fluoro-2-methoxybenzoic acid (50%; m.p. 84–85 °C, from hexane) and 2-fluoro-5-methoxybenzoic acid (85%; m.p. 144–145 °C, from toluene) are obtained.

5-[N-(1-Lithiooxy-2,2-dimethylpropylidene)]-1,3-benzodioxolan-4-yllithium (Conversion into the Carboxylic Acid) [40]

LiC_4H_9 (2 equiv.); 1. CO_2 2. H^+/H_2O

At −75 °C, butyllithium (0.24 mol) in hexane (0.15 l) is added to *N*-pivaloylpiperonylamine (24 g, 0.10 mol) in tetrahydrofuran (0.25 l). After the mixture has been allowed to stand for 1 h at 0 °C, a suspension has formed which is poured on to crushed dry ice. Water (0.5 l) and concentrated hydrochloric acid are added until a pH of 2 is reached. 5-Pivaloylamidomethyl-1,3-benzodioxane-4-carboxylic acid precipitates as a white solid, which is collected and recrystallized from ethyl acetate; 70%; m.p. 181–182 °C.

5-[N-1-Lithiooxy-2,2-dimethylpropylidene)]-1,3-benzodioxolan-4-ylithium (Conversion into the Phenol) [40]

LiC_4H_9 (2 equiv.); 1. $FB(OCH_3)_2$ 2. H_2O_2

Alternatively, the suspension formed on metallation of *N*-pivaloylpiperonylamine (0.10 mol) may be treated, at −75 °C, with the fluorodimethoxyborane diethyl etherate (28 ml, 25 g, 0.15 mol). After evaporation of the solvents, the residue is dissolved in methanol (0.15 l) and 35% aqueous hydrogen peroxide (25 ml, 28 g, 0.29 mol) is added. A white pasty precipitate deposits in the course of 30 min. The methanol is evaporated and the remaining mixture is partitioned between saturated aqueous ammonium chloride (0.25 l) and dichloromethane (0.25 l). The aqueous layer is again extracted with dichloromethane (2 × 0.10 l) and the combined organic layers are washed with water (0.10 l) and evaporated. Recrystallization of the residue from ethanol (or 2-propanol) affords 4-hydroxy-5-pivaloylamidomethyl-1,3-benzodioxolane; 68%; m.p. 197–198 °C.

2-Phenoxyphenylpotassium [41]

Potassium *tert*-butoxide (2.8 g, 25 mmol) is dissolved in a solution containing butyllithium (25 mmol) and diphenyl ether (4.3 g, 25 mmol) in tetrahydrofuran (30 ml) at −75 °C. The mixture is allowed to stand 5 h at −75 °C, before sulfur (0.96 g, 30 mmol) is added. At + 25 °C, the mixture is treated with 5 M hydrogen chloride in diethyl ether and is then evaporated to dryness. The residue is extracted with hot hexane (3 × 25 ml). Upon concentration and cooling of the combined organic phases, 2-phenoxybenzenethiol crystallizes as colourless needles; 66%; m.p. 68–69 °C.

Bis(2-lithiophenyl) Ether [41, 42]

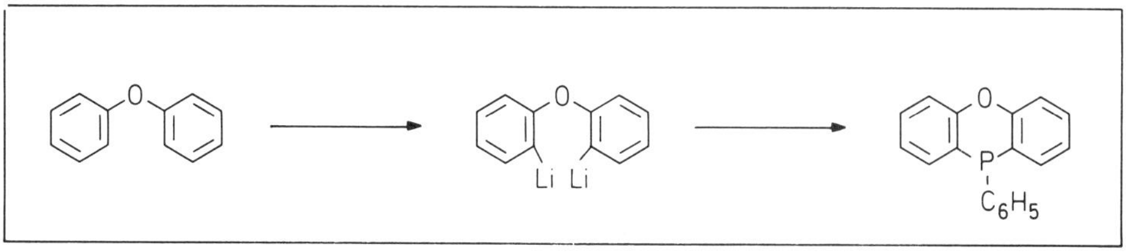

A solution of diphenyl ether (17 g, 0.10 mol), *N*,*N*,*N*′,*N*′-tetramethylethylenediamine (TMEDA) (30 ml, 23 g, 0.20 mol) and butyllithium (0.20 mol) in diethyl ether (0.10 l) and hexane (0.15 l) is allowed to stand for 3 h at 25 °C. A brownish red precipitate settles out. On cooling with an ice-bath, *P*,*P*-dichlorophenylphosphine (benzenephosphonous dichloride) (27 ml, 36 g, 0.20 mol) is added dropwise. The mixture is evaporated to dryness and the residue extracted with hot hexane. On concentration and cooling of the solution, 10-phenylphenoxaphosphine crystallizes as colourless platelets; 59%; m.p. 95–96 °C.

1.5.2.4 Generation of Benzyl-Type Organometallics

The benzylic metallation of diaryl- and triarylmethanes is trivial. It can already be accomplished with ordinary butyllithium [22] in tetrahydrofuran or glycol dimethyl ether and with potassium amide [43] in liquid ammonia. However, it requires butylpotassium [44] or the superbasic butyllithium–potassium *tert*-butoxide mixture [22] for clean deprotonation at the methyl group of toluene. All other reagents concomitantly attack the aromatic positions to a considerable extent (>10%) [45]. The clean α-metallation of isopropylbenzene (cumene) can only be brought about with trimethylsilylmethylpotassium [23] (see p. 107); even butylpotassium [46] fails this test on site selectivity. Toluene-type substrates carrying electron-donating substituents, e.g. *p*-fluorotoluene [47], *p*-cresyl methyl ether [22, 48] and *N*,*N*,*p*-trimethylaniline [49], are particularly inert. The difficulties can again be overcome with superbasic mixed metal reagents. In addition, such species tend to discriminate very efficiently between different benzylic positions if the latter are subject to unequal electronic or steric shielding. Thus, *p*-ethyltoluene and 1-ethyl-4-isobutylbenzene give *p*-ethylbenzylpotassium and 1-(4-isobutylphenyl)ethylpotassium exclusively [50].

4-(Dimethylamino)benzylpotassium [49]

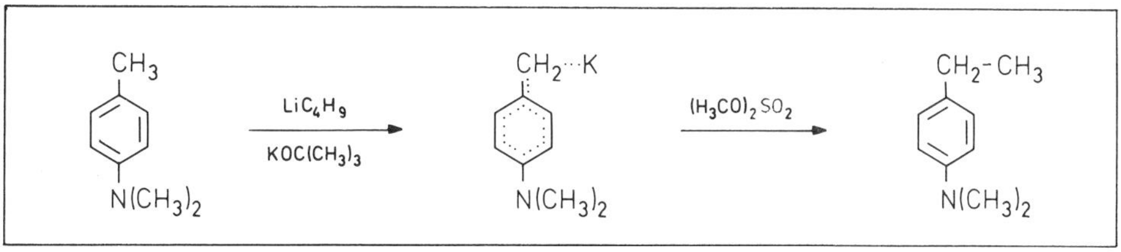

At −75 °C, butyllithium (15 mmol) in hexane (10 ml) and potassium *tert*-butoxide (1.7 g, 15 mmol) are added to a solution of *N*,*N*,*p*-trimethylaniline (*N*,*N*-dimethyl-*p*-toluidine) (2.2 ml, 2.0 g, 15 mmol) in tetrahydrofuran (15 ml). The mixture is vigorously stirred until it becomes homogeneous, then kept for 48 h at −75 °C using a cryogenic unit. After the addition of dimethyl sulfate

(1.4 ml, 1.9 g, 15 mmol), it is allowed to reach 25 °C. Gas chromatographic analysis by means of an internal reference substance ('standard') reveals the presence of 65% of *N,N*-dimethyl-*p*-ethylaniline in addition to some unconsumed starting material. The main product is isolated by distillation; b.p. 96–97 °C/10 mmHg.

4-Ethylbenzylpotassium [50]

CH_3 / C_2H_5 —(LiC_4H_9, $KOC(CH_3)_3$)→ $CH_2 \cdots K$ / C_2H_5 —($Br\text{-}CH(CH_3)_2$)→ $CH_2\text{-}CH(CH_3)_2$ / C_2H_5

A mixture of *p*-ethyltoluene (2.8 ml, 2.4 g, 20 mmol), butyllithium (22 mmol), potassium *tert*-butoxide (2.5 g, 22 mmol) and hexane (15 ml) is vigorously stirred for 2 h at 25 °C. At −25 °C, isopropyl bromide (2.1 ml, 2.7 g, 22 mmol) is dropwise added in the course of 15 min. 1-Ethyl-4-isobutylbenzene is isolated by distillation; 87%; b.p. 87–89 °C/10 mmHg.

1-(4-Isobutylphenyl)ethylpotassium [50]

$CH_2\text{-}CH(CH_3)_2$ / C_2H_5 —(LiC_4H_9, $KOC(CH_3)_3$)→ $CH_2\text{-}CH(CH_3)_2$ / $H_3C\text{-}CH \cdots K$ —(CO_2)→ $CH_2\text{-}CH(CH_3)_2$ / $H_3C\text{-}CH\text{-}COOH$

Under vigorous stirring, a mixture of 1-ethyl-4-isobutylbenzene (3.2 g, 20 mmol), butyllithium (25 mmol), potassium *tert*-butoxide (2.8 g, 25 mmol) and hexane (20 ml) is heated for 3 h to 60 °C before being poured onto crushed dry-ice covered with anhydrous tetrahydrofuran (50 ml). After evaporation of the volatiles, the residue is dissolved in water (0.10 l). The aqueous phase is washed with diethyl ether (3 × 30 ml), acidified to pH 1 and extracted again with diethyl ether (3 × 30 ml). The organic layer is washed with a 1% aqueous solution of sodium hydrogen carbonate (1 × 10 ml, then 1 × 15 ml) and dried. The oily material left behind on evaporation of the solvent is crystallized from hexane to afford pure 2-(4-isobutylphenyl)propanoic acid; 79%; m.p. 70–72 °C.

1.5.2.5 Generation of Allyl-Type Organometallics

The superbasic LIC–KOR (butyllithium–potassium *tert*-butoxide) mixture is particularly suitable for the conversion of simple alkenes into highly reactive 2-alkenylpotassium species. These versatile synthetic intermediates can be generated in a stereocontrolled manner. Depending on the reaction conditions, the original configuration of the olefin is either retained (*stereodefensive* or *stereoconservative* mode [51]) or it undergoes a torsional isomerization to establish to thermodynamically more stable structure (*stereoselective* mode [51]). The last three procedures given below illustrate this option.

2-Phenylallylpotassium [52]

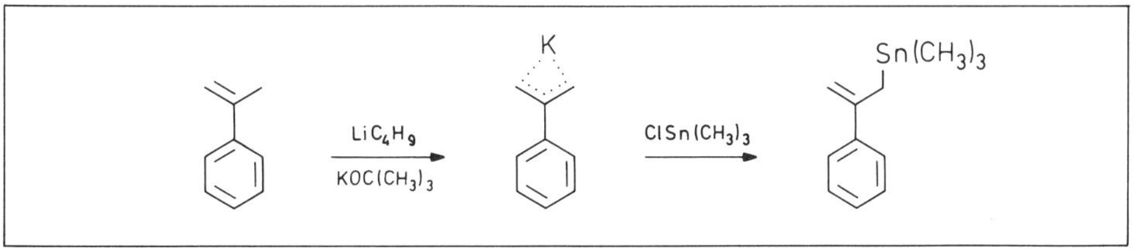

Butyllithium (40 mmol) in neat tetrahydrofuran (25 ml) is added dropwise, over 30 min, to a well stirred solution of α-methylstyrene (5.2 ml, 4.7 g, 40 mmol) and potassium *tert*-butoxide (4.5 g, 40 mmol) in tetrahydrofuran (120 ml) at −75 °C. After 1 h at −50 °C, the mixture is added portionwise to chlorotrimethylstannane (8.0 g, 40 mmol) in tetrahydrofuran (30 ml) at −75 °C. The turbid liquid is poured into water (0.2 l). The aqueous phase is extracted with hexane (3 × 40 ml) and the combined organic layers are washed with brine (2 × 0.10 l) before being dried and evaporated. Distillation through a short (10 cm long) Vigreux column gives colourless trimethyl(2-phenyl-2-propenyl)stannane; 31%; b.p. 68–69 °C/0.1 mmHg; n_D^{20} 1.5542.

2-(Trimethylsilyl)allylpotassium [52]

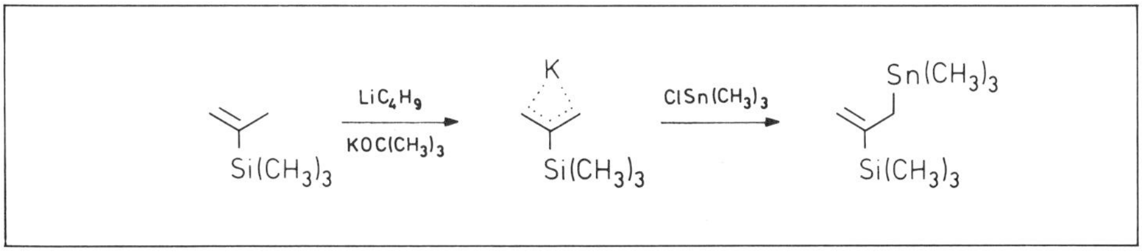

Trimethyl(1-methylethenyl)silane (2.3 g, 20 mmol) and potassium *tert*-butoxide (2.2 g, 20 mmol) are consecutively added to butyllithium (20 mmol) in neat tetrahydrofuran (50 ml) at −75 °C. After 5 h at −50 °C, the mixture is treated with chlorotrimethylstannane and worked up as described in the preceding paragraph. Trimethyl[1-(trimethylstannylmethyl)ethenyl]silanc is obtained as a colourless liquid; 45%; b.p. 33–34 °C/0.7 mmHg; n_D^{20} 1.4816.

Pinenylpotassium [26, 53]

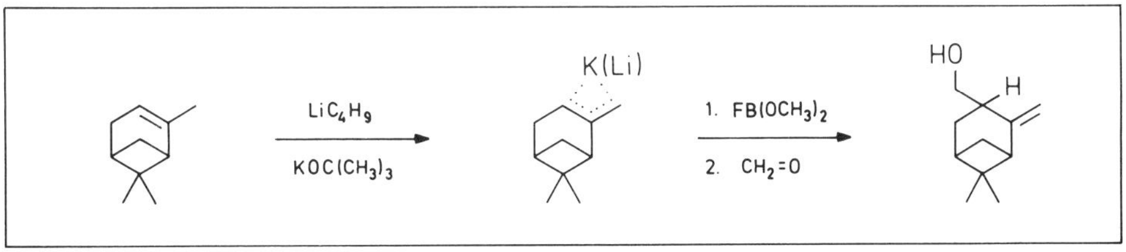

A well stirred mixture of (1*S*)-α-pinene (0.8 ml, 6.8 g, 50 mmol) and hexane (35 ml) is heated for 30 min to 50 °C. At −75 °C, precooled tetrahydrofuran (25 ml) and fluorodimethoxyborane dietherate (11 ml, 10 g, 60 mmol) are added. At 0 °C, the mixture is treated with a 40% aqueous solution of formaldehyde (10 ml, 11 g, 0.14 mol) before being heated for 40 min at 50 °C. After addition of 2 M sulfuric acid (50 ml), the organic layer is washed with 3 M aqueous sodium hydroxide (25 ml) and brine (25 ml), then dried and evaporated. Distillation affords [1*R*-(1α,3α,5α)]-6,6-dimethyl2-methylene-bicyclo[3.11]heptane-3-methanol; 58%; b.p. 82–84 °C/2 mmHg; n_D^{20} 1.5021; $[\alpha]_D^{20}$ + 44° (c = 1, CH_3OH).

endo-*2-Butenylpotassium [54]*

LiC_4H_9 / $KOC(CH_3)_3$; 1. $FB(OCH_3)_2$ 2. H_5C_2-CH=O ; HO

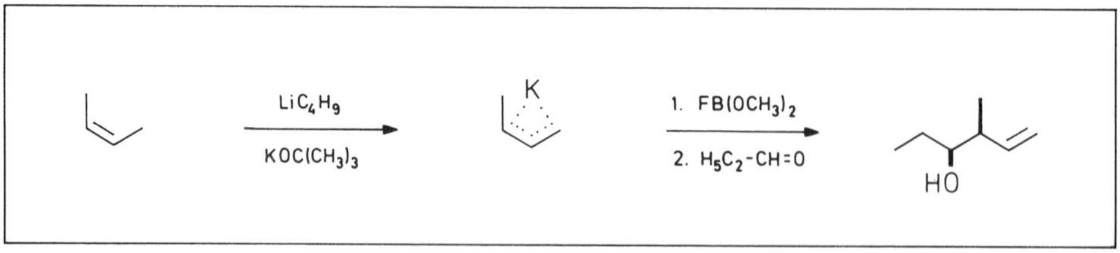

(*Z*)-2-Butene (14 ml, 8.5 g, 0.15 mol) is condensed into a solution of butyllithium (0.10 mol) and potassium *tert*-butoxide (11 g, 0.10 mol) in neat tetrahydrofuran (75 ml) at −75 °C. After 15 min of stirring at −50 °C, the mixture is transferred into a −75 °C dry-ice–methanol bath and treated with fluorodimethoxyborane diethyl etherate (50 ml, 45 g, 0.27 mol). The orange–yellow mixture immediately decolorizes. After 30 min at −75 °C, propanal (8.0 ml, 6.4 g, 0.11 mol) is added dropwise. After 30 min at −75 °C, the mixture is poured into 2 M aqueous sodium hydroxide (0.2 l) and vigorously stirred for 2 h at 25 °C before being extracted with hexane (3 × 25 ml). Distillation affords 4-methyl-5-hexen-3-ol having an *erythro*/*threo* [55] ratio of 96:4; 62%; b.p. 66–68 °C/50 mmHg.

exo-*2-Butenylpotassium [54]*

LiC_4H_9 / $KOC(CH_3)_3$; 1. $FB(OCH_3)_2$ 2. H_5C_2-CH=O ; HO

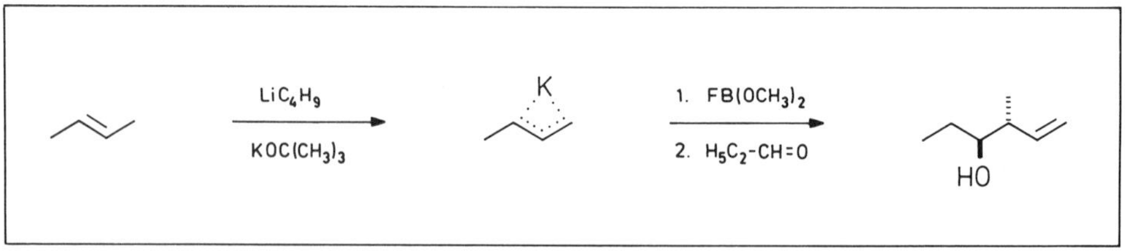

The same procedure applied to (*E*)-2-butene (14 ml, 8.5 g, 0.15 mol) furnishes 4-methyl-5-hexen-3-ol with an *erythro*/*threo* [55] ratio of 3:97; 61%; b.p. 64–67 °C/50 mmHg.

endo-*5-(2,3-Dimethyltricyclo[2.2.1.0^{2,6}]hept-3-yl)-2-methyl-2-pentenylpotassium ("α-santalenylpotassium") [41, 56]*

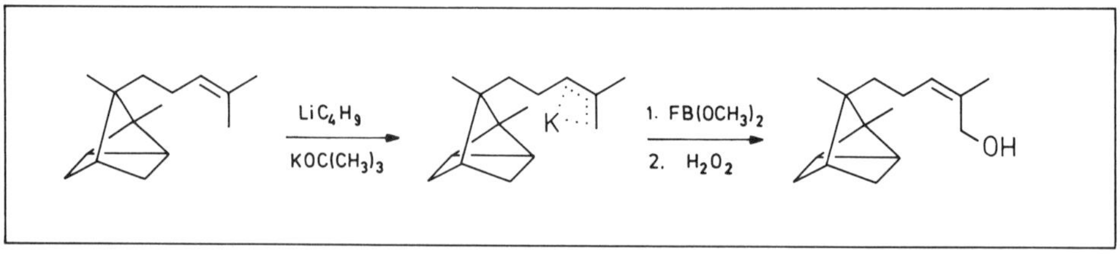

LiC_4H_9 / $KOC(CH_3)_3$; K ; 1. $FB(OCH_3)_2$ 2. H_2O_2 ; OH

Potassium *tert*-butoxide (2.5 g, 22 mmol) is added to a solution of butyllithium (24 mmol) and 5-(2,3-dimethyltricyclo[2.2.1.0^{2,6}]hept-3-yl)-2-methyl-2-pentene (α-santalene) (4.1 g, 20 mmol) in neat tetrahydrofuran (20 ml) at −75 °C. The mixture is stirred until homogeneous, then kept for 2 h at −50 °C. At −75 °C, it is consecutively treated with fluorodimethoxyborane diethyl etherate (7.5 ml, 6.6 g, 40 mmol) and 35% aqueous hydrogen peroxide (4.0 ml, 4.5 g, 47 mmol). After 1 h of stirring at 25 °C, the aqueous phase is saturated with sodium chloride before being extracted with diethyl ether (2 × 25 ml). The combined organic layers are absorbed on silica gel (25 ml). The dried powder is placed on top of a chromatographic column filled with fresh silica gel (100 ml) and hexane. Elution with a 15:85 (v/v) mixture of ethyl acetate and hexane followed by evaporation gives a colourless oil having the characteristic smell of α-santalol; 46%; b.p. 92–93 °C/0.1 mmHg; n_D^{20} 1.5022; $[\alpha]_D^{20}$ + 7° (c = 0.1, $CHCl_3$). No trace of the unnatural (*E*) isomer is detected by gas chromatography. α-Santalene (77%) is readily prepared from 7-iodomethyl-1,7-dimethyltricyclo[2.2.1.$0^{2,6}$]heptane and 3-methyl-2-butenylpotassium (prenylpotassium) [23] in tetrahydrofuran. Without isolation of the hydrocarbon thus obtained, the reaction can be continued according to the procedure given above to afford α-santalol with a 35% overall yield.

1.5.2.6 Generation of Pentadienyl-Type Organometallics

The metallation of 1,3-dienes and, in particular, 1,4-dienes can be accomplished with great ease. Open-chain, i.e. mobile, pentadienylmetal species may adopt different conformations depending on the choice of the metal involved or the solvent (see pp. 79–81). In contrast to cyclohexadienyllithium [57], the corresponding potassium compound (or derivatives thereof [58]) is relatively stable towards hydride elimination and survives long enough to be trapped by electrophiles.

1,3-Cyclohexadienylpotassium [59]

At −75 °C, 1,4-cyclohexadiene (1.9 ml, 1.6 g, 20 mmol) is added dropwise to a solution of trimethylsilylmethylpotassium (20 mmol; see p. 107) in tetrahydrofuran (40 ml). After 15 h at −50 °C, the mixture is poured onto freshly crushed dry-ice. The solid material is neutralized with ethereal hydrogen chloride (10 ml of a 2.3 M solution) before being treated with ethereal diazomethane until the yellow colour persists. The solution is filtered and concentrated. Bulb-to-bulb (Kugelrohr) distillation (bath temperature 130–140 °C/10 mmHg) affords 44% of methyl 2,5-cyclohexadienylcarboxylate in addition to small amounts (*ca* 5%) of methyl benzoate.

(E)-2,4-Pentadienylpotassium, Suspended in Hexane [60, 61]

1,4-Pentadiene (4.1 ml, 2.7 g, 40 mmol) and butyllithium (40 mmol) in hexane (25 ml) are rapidly added to a well stirred slurry of potassium *tert*-butoxide (4.5 g, 40 mmol) in hexane (25 ml) kept in an ice-bath. After 20 min at 0 °C, the mixture is cooled to −75 °C before being treated with fluorodimethoxyboron dietherate (8.0 ml, 7.1 g, 43 mmol) in diethyl ether (12 ml). Aqueous sodium hydroxide (10 ml of a 5 M solution) and 35% aqueous hydrogen peroxide (5 ml, 5.7 g, 58 mmol) are added and the two-phase mixture is vigorously stirred for 2 h. The organic layer is decanted, briefly dried and concentrated. Distillation of the residue gives 2,4-pentadien-1-ol having a *cis/trans* ratio of 3:97; 88%; b.p. 55–57 °C/12 mmHg.

(Z)-2,4-Pentadienylpotassium, Dissolved in Tetrahydrofuran [60, 61]

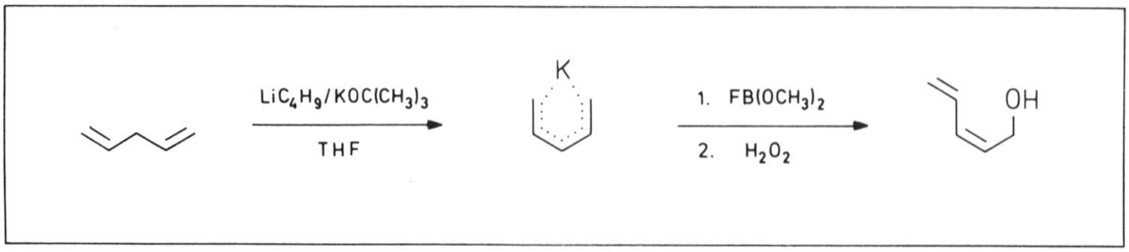

1,4-Pentadiene (40 mmol) is metallated as described above. The hexane is stripped off and the residue is dissolved in precooled (−75 °C) tetrahydrofuran (50 ml). After 2 h at −75 °C, the mixture is treated with fluorodimethoxyboron etherate (43 mmol) and 35% aqueous hydrogen peroxide (5 ml). 2,4-Pentadien-1-ol is isolated having a *cis/trans* ratio of 98:2; 59%; b.p. 62–63 °C/18 mmHg.

1.5.2.7 Ether Cleavage vs Hydrogen–Metal Exchange

All procedures given so far refer to the 'metallation' (hydrogen–metal interconversion) mode. The preference for this method is well justified by its evident advantages. Nevertheless, this method also has its limits, and other ways leading to organometallic intermediates should always be taken into consideration.

As the following comparison reveals, it is not necessary to prepare allyllithium and allylsodium by reductive ether cleavage. Allylpotassium can be very easily obtained by the superbase-promoted metallation of propene. If a less electropositive metal is required, potassium can be replaced with lithium or sodium in a metathetical exchange with lithium bromide [62] and sodium tetraphenylborate [63], respectively. On the other hand, the C_{15} building blocks β-ionylideneethyllithium and -potassium can be more conveniently generated by treatment of methyl vinyl-β-ionyl ether with the corresponding metal rather than by metallation of the unsaturated hydrocarbon, the latter being not readily accessible.

Allyllithium [64]

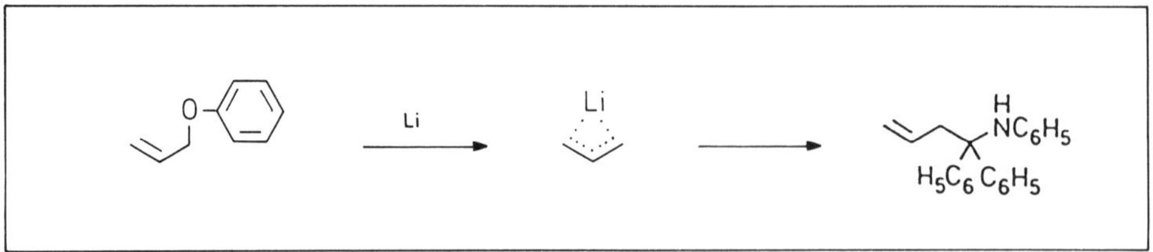

Allyl phenyl ether (6.8 ml, 6.7 g, 50 mmol) in diethyl ether (25 ml) is added over 45 min to a rapidly stirred suspension of freshly cut lithium wire (4.2 g, 0.61 mol) in tetrahydrofuran (0.10 l) kept in a −15 °C cooling bath. Usually after the first few minutes a slightly exothermic reaction sets in, manifesting itself by the appearance of a pale bluish green colour. If this is not the case, the metal has to be activated by introducing a spatula tip amount of biphenyl or a few drops of 1,2-dibromoethane. The reaction mixture is then stirred for 15 min at 25 °C before the solution, which has become dark red in the meantime, is filtered through polyethylene tubing filled with glass-wool into a storage burette. Reaction with equivalent amount of (diphenylmethylene)aniline gives *N*-(1,1-diphenyl-3-butenyl)aniline; 68%; m.p. 78–79 °C (isolated by elution from a chromatography column filled with silica gel and recrystallized from ethanol).

Allylsodium [65]

Na; 1. CO_2 2. H⊕; COOH

A dispersion of sodium (5.8 g, 0.25 mol) is prepared in octane [16, 17], which is then replaced by hexane (0.5 l). Diallyl ether (18 ml, 14 g, 0.15 mol) is added over a period of 1 h under high-speed stirring (5000 rpm) and at an average temperature of 35 °C. After a further 30 min of stirring the reaction mixture is siphoned on to dry-ice. After acidification, the 3-butenoic acid is extracted with diethyl ether and distilled; 77%, b.p. 61–62 °C/3 mmHg.

Allylpotassium [66]

LiC_4H_9, $KOC(CH_3)_3$; K; O, H+; OH

Propene (0.2 l, 11 g, 0.25 mol), potassium *tert*-butoxide (11 g, 0.10 mol) and hexane (25 ml) are placed in a three-necked flask equipped with a dry-ice condenser. Butyllithium (0.10 ml) in hexane (65 ml) is added over 30 min under vigorous stirring. The reaction mixture is kept at reflux for a further 5 h before oxirane (6.0 ml, 5.3 g, 0.13 mol) is added at −75 °C. After evaporation to dryness, the residue is treated with ethereal hydrogen chloride (0.12 mol, 2.4 M). Immediate distillation gives pure 4-penten-1-ol; 63%; b.p. 82–85 °C/60 mmHg.

1-Methyl-1-phenylethylpotassium [67] ('2-Phenylisopropylpotassium,' α,α-Dimethylbenzylpotassium) from Cumyl Methyl Ether

OCH_3 H_3C-C-CH_3; K H_3C-C-CH_3; H_5C_6 CH_3 C=C HOOC $C(CH_3)_2$

Potassium–sodium alloy [20] (78 : 22 w/w; 18 ml, 15 g, 0.32 mol potassium) is transferred by means of a pipette into a solution of cumyl methyl ether [67] (2-methoxy-2-phenylpropane) (15 g, 0.10 mol) in diethyl ether (0.50 l). The mixture is vigorously stirred or, if placed in a sealed tube, shaken. The bright red solution is carefully decanted through polyethylene tubing fitted with a glass-wool plug into a burette. Tolane (11.6 g, 65 mmol) was added in portions to an aliquot (200 ml) of this roughly 3 M solution. An exothermal reaction caused the solvent to boil. The reaction mixture was poured on dry ice. Extraction with water (3 × 25 ml) and evaporation gave a colourless resinous material; 26%.

1-Methyl-1-phenylethylpotassium from Cumene [23]

$CH(CH_3)_2$ $KCH_2Si(CH_3)_3$ H_3C K CH_3 C CH_3I $C(CH_3)_3$

Isopropylbenzene (cumene) (0.50 ml, 0.43 g, 3.6 mmol) is added to a precooled solution of trimethylsilylmethylpotassium (see p. 107) (3.6 mmol) in tetrahydrofuran (5.0 ml). After 24 h at −50 °C, the reaction is quenched with methyl iodide (0.30 ml, 0.68 g, 4.8 mmol). While 74% of *tert*-butylbenzene were identified by gas chromatography, no trace of cymols (isopropyltoluenes) was found.

2-(β-Ionylidene)ethyllithium [3-Methyl-5-(2,6,6-trimethyl-1-cyclohexenyl)-2,4-pentadienyllithium] [61]

OCH_3 Li Li 1. $FB(OCH_3)_2$ 2. H_2O_2 OH

At −75 °C, a 0.5 M solution of the lithium–biphenyl (1 : 1) adduct [68] (20 mmol) in tetrahydrofuran (40 ml) is added to (*E*)-3-methoxy-3-methyl-1-(2,6,6-trimethyl-1-cyclohexenyl)-1,4-pentadiene (methyl vinyl-β-ionyl ether) (4.9 g, 21 mmol) in tetrahydrofuran (10 ml). Immediately a bright red colour develops. Still at −75 °C, the mixture is consecutively treated with fluorodimethoxyboron etherate (3.8 ml, 3.4 g, 20 mmol), 35% aqueous hydrogen peroxide (2.0 ml, 2.3 g, 23 mmol) and 4 M aqueous sodium hydroxide (20 ml). After 30 min of stirring, the aqueous phase is saturated with sodium chloride. The organic layer is decanted and evaporated. Chromatography of the residue on active alumina using a 1 : 1 (v/v) mixture of diethyl ether and hexane as the eluent gives 75% of analytically pure (2*E*,4*E*)-3-methyl-5-(2,6,6-trimethyl-1-cyclohexenyl)-2,4-pentadien-1-ol.

3,7-Dimethyl-2,4,6-octadienyllithium [61]

3,7-Dimethyl-1,3,6-octatriene (ocimene; mixture of *Z*- and *E*-isomers) (1.4 g, 10 mmol) is added to a solution of lithium diisopropylamide (prepared from diisopropylamine and butyllithium) (10 mmol) in hexamethylphosphoric triamide (1.0 ml), tetrahydrofuran (7.0 ml) and hexane (7.0 ml) at −75 °C. After 5 h at −75 °C, the reddish black mixture decolorizes when fluorodimethoxyborane diethyl etherate (2.0 ml, 1.8 g, 11 mmol) and 70% aqueous hydrogen peroxide (10 mmol) are added consecutively. After 15 h of stirring at 25 °C, the organic phase is diluted with diethyl ether (10 ml) and, protected by nitrogen against contact with air, washed with brine (3 × 10 ml). It is dried and evaporated. Distillation affords (2*E*,4*E*)-3,7-dimethyl-2,4,6-octatrien-1-ol (*trans*-dehydrogeraniol); 51%; b.p. 84–85 °C/2 mmHg; m.p. 47–49 °C (from pentane). The yield increases to 68% if the metallation of ocimene is carried out with a 0.5 M solution of trimethylsilylmethylpotassium (see p. 107) for 1 h at −75 °C.

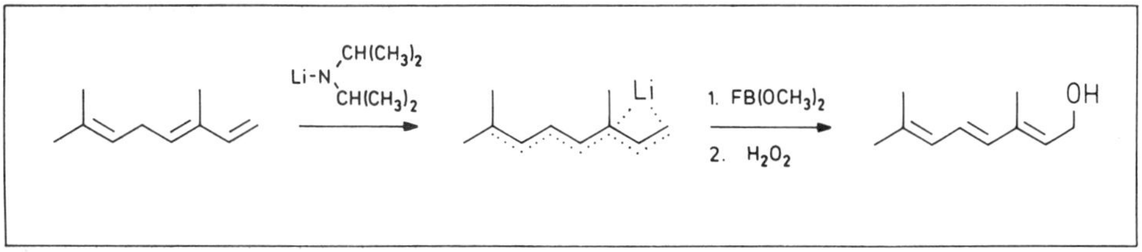

1.5.2.8 Halogen–Metal vs Hydrogen–Metal Exchange

Frequently it is difficult to make a choice between the two most important methods known for the generation of organometallic intermediates. The halogen–metal exchange route offers the advantage of a particularly easy and rapid execution. On the other hand, one has to consider the accessibility of the starting materials: unlike the simple hydrocarbons required for the hydrogen–metal exchange, the halogenated precursors are in general not commercially available.

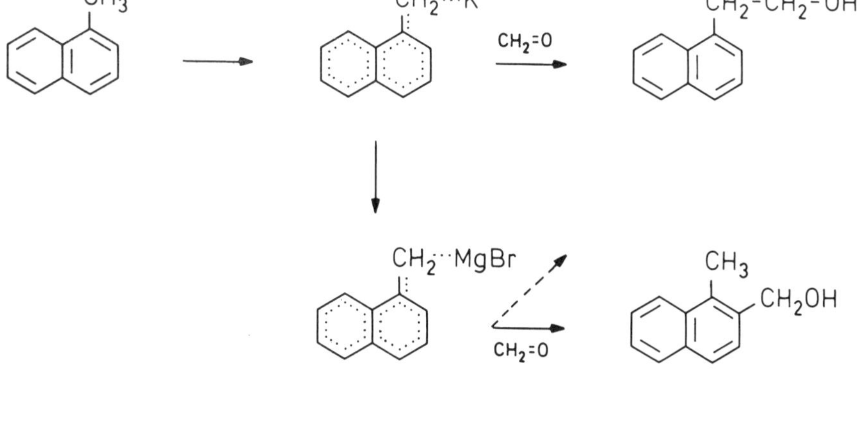

There are, however, still other factors to be taken into account. Let us assume we wish to prepare a series of ω-(1-naphthyl)alkanols. The two lowest members of this family, 1-naphthylmethanol and 2-(1-naphthyl)ethanol, can be easily prepared by treatment of 1-bromonaphthalene with butyllithium and reaction of the resulting 1-naphthyllithium [69] with formaldehyde or oxirane. However, if longer chains have to be attached, it is better to start with 1-methylnaphthalene, submit it to a hydrogen–metal exchange employing the superbasic LIC–KOR reagent [70, 71] and treat the resulting benzyl-type intermediate with oxirane, oxetane or even tetrahydrofuran. Further, the 1-naphthylmethylpotassium can be converted with magnesium dibromide into the corresponding Grignard reagent (see scheme on preceding page). Addition of formaldehyde now produces mainly (53%) 2-(1-methylnaphthyl)methanol in addition to small amounts (8%) of 2-(1-naphthyl)ethanol [70].

(*Z*)-2-Ethoxyvinyllithium is a versatile synthetic equivalent of the acetaldehyde enolate. It can be very easily generated from *cis*-2-bromovinyl ether by halogen–metal exchange [72]. The hydrogen–metal exchange with ethyl vinyl ether as the substrate would lead to a different, α-metallated species [35]. β-Deprotonation can be achieved, however, if the enether is replaced by ethyl ethynyl ether [73]. The ethoxyacetylide may then be added to the carbonyl group of an aldehyde or ketone, giving rise to a propargyl-type alcohol. Partial hydrogenation of the latter leads to the same (*Z*)-γ-ethoxyallyl alcohol which can be obtained more directly with the (*Z*)-2-ethoxyvinyllithium reagent.

$$\text{(H)(Br)C=C(H)(OC}_2\text{H}_5) \rightarrow \text{(H)(Li)C=C(H)(OC}_2\text{H}_5) \rightarrow \text{(H)(RR'C(OH))C=C(H)(OC}_2\text{H}_5) \rightarrow \text{R'RC=CH-CH=O}$$

$$\uparrow \text{H}_2\{\text{Pd}\}$$

$$\text{H-C}\equiv\text{C-OC}_2\text{H}_5 \rightarrow \text{BrMg-C}\equiv\text{C-OC}_2\text{H}_5 \rightarrow \text{RR'C(HO)-C}\equiv\text{C-OC}_2\text{H}_5 \rightarrow \text{R'RC=CH-COOH}$$

Other powerful synthetic equivalents of enolates are α-deprotonated Schiff bases (azomethines) derived from acetaldehyde or other alkanals [74]. Such entities are the key intermediates in the so-called ‘directed aldol condensation,’ a method which has also been successfully applied to the conversion of β-ionone into β-ionylideneacetaldehyde [74]. Finally, the lithium enolate of acetaldehyde (lithium ethenolate) can be generated by the spontaneous fragmentation of α-lithiated tetrahydrofuran [75]. On reaction with carbonyl compounds, however, this species gives only moderate yields, presumably as a consequence of tight aggregate formation and poor solubility.

The halogen–metal exchange is a flexible method having little, if any, structural limitations. 2-, 3- and 4-Bromoanisole can be converted with the same ease into 2-, 3- and 4-anisyllithium [76]. In contrast, the metallation of anisole [77] leads invariably to a 2-anisylmetal species no matter what conditions are applied.

$$\text{2-BrC}_6\text{H}_4\text{OCH}_3 \longrightarrow \text{2-LiC}_6\text{H}_4\text{OCH}_3 \longleftarrow \text{C}_6\text{H}_5\text{OCH}_3$$

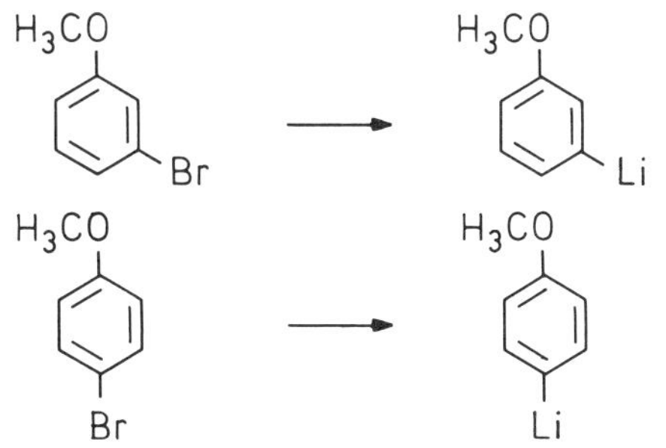

The metallation route is restricted with respect not only to regioisomers but also to stereoisomers. Treatment of camphene with a superbasic mixed metal reagent gives exclusively a species having the (*E*)-configuration [23], whereas the halogen–metal exchange method allows the generation of either (*Z*)- or (*E*)-ω-camphenyllithium selectively, provided that the corresponding bromides or iodides are available [78].

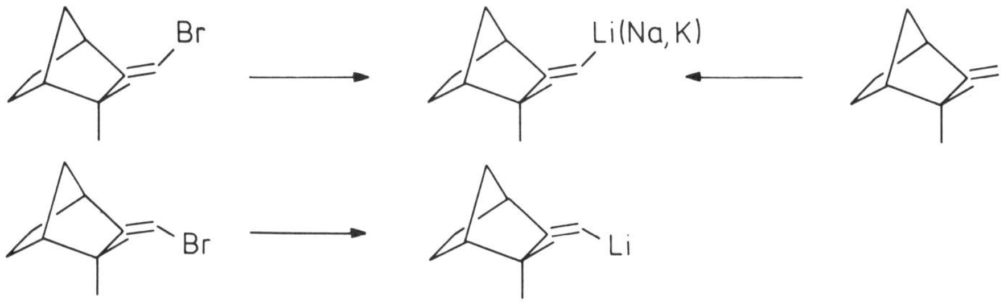

The halogen–metal exchange procedure can be also applied to all kinds of heterocyclic substrates [79]. In contrast, the metallation method works well only with five-membered ring systems, which, in addition, should not carry acidic methyl or methylene substituents. Pyridines, pyridazine, pyrimidines, pyrazines, triazines and other six-membered nitrogen heterocycles undergo preferentially nucleophilic addition of an organometallic reagent followed by rearomatization through elimination of metal hydride [80, 81].

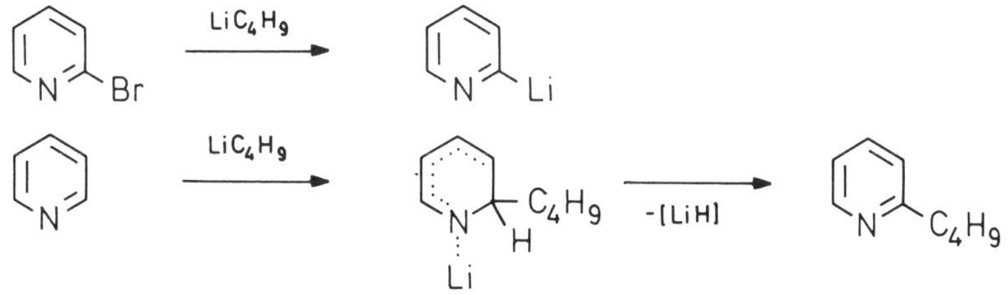

The halogen–metal exchange between 1-bromonaphthalene and butyllithium can be followed visually if executed in a 1 : 1 (v/v) mixture of diethyl ether and hexane. In this solvent, 1-naphthyllithium precipitates. It is not recommended to work in tetrahydrofuran or to use 1-iodonaphthalene as the starting material, since otherwise extensive condensation between 1-naphthyllithium and the butyl halide formed as a by-product will occur.

1-Naphthyllithium [41]

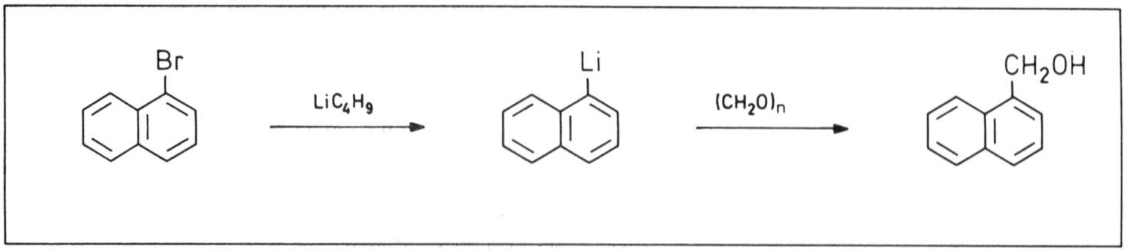

At 0 °C, 1-bromonaphthalene (18 ml, 41 g, 0.20 mol) and dry paraformaldehyde (7.5 g, 0.25 mol) are consecutively added to butyllithium (0.20 mol) in diethyl ether (0.12 l) and hexane (0.13 l). After 1 h of stirring at 25 °C, the mixture is neutralized with ethereal hydrogen chloride (0.20 mol, 3.5 M), the solvents are stripped off and the residue is distilled; b.p. 163–164 °C/12 mmHg. The 1-naphthalenemethanol crystallizes spontaneously; 91%; m.p. 61–62 °C.

1-Naphthylmethylpotassium [41, 71]

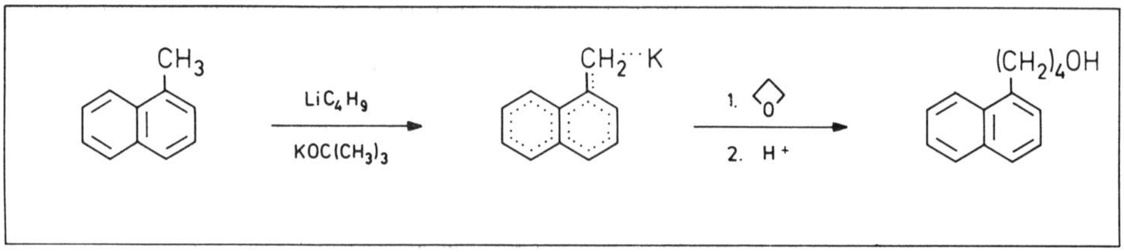

A solution of butyllithium (0.10 mol) in hexane (60 ml) is evaporated to dryness and the residue is taken up in precooled (−75 °C) tetrahydrofuran (0.10 l). 1-Methylnaphthalene (14 ml, 14 g, 0.10 mmol(and potassium *tert*-butoxide (11 g, 0.10 mol) are consecutively added. The mixture is stirred 3 min at −50 °C until the alcoholate has dissolved and is kept 2 h at −50 °C. At −75 °C, oxetane (trimethylenoxide: 7.8 ml, 7.0 g, 0.12 mol) is introduced. The mixture is allowed to reach 25 °C and the solvent is evaporated. Brine (50 ml) is added and the product is extracted with hexane (3 × 25 ml). Distillation affords 4-(1-naphthyl)1-butanol; 81%; b.p. 150–152 °C/3 mmHg.

(Z)-2-Ethoxyvinyllithium [41, 72]

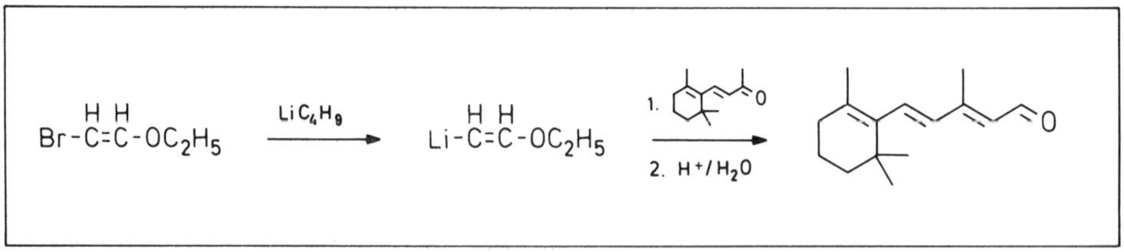

Butyllithium (0.10 mol), from which the commercial solvent hexane has been stripped off, is dissolved in precooled (−75 °C) diethyl ether (50 ml). Always at −75 °C, (*Z*)-2-bromovinyl ethyl ether [72] (11 ml, 15 g, 0.10 mol) and, 5 min later, 4-(1,6,6-trimethyl-1-cyclohexen-1-yl)-3-buten-2-one (*β*-ionone) (20 ml, 19 g, 0.10 mol) are added. When the mixture has reached 25 °C, it is treated with 0.5 M sulphuric acid (50 ml) and vigorously stirred for 15 min. The organic phase is decanted and the aqueous phase is rapidly extracted with diethyl ether (2 × 25 ml). The combined ethereal layers are washed (2 × 25 ml of a saturated aqueous solution of sodium hydrogen-carbonate, 1 × 25 ml of brine) and dried. Distillation affords *β*-ionylideneacetaldehyde as a 1:3 (*Z*/*E*) mixture; 73%; b.p. 103–105 °C/0.1 mmHg; n_D^{20} 1.5765.

2-Ethoxyethynylmagnesium bromide [73]

$HC{\equiv}C{-}OC_2H_5 \xrightarrow{BrMgC_2H_5} BrMg{-}C{\equiv}C{-}OC_2H_5$ — 1. (β-ionone); 2. H^+/H_2O; 3. $H_2\{Pd/Pb\}$; 4. H^+/H_2O →

Under ice cooling, ethoxyacetylene (ethyl ethynyl ether; 28 ml, 21 g, 0.30 mol) and, 10 min later, β-ionone (42 ml, 40 g, 0.21 mol) are dropwise added to a solution of ethylmagnesium bromide (0.31 mol) in diethyl ether (0.25 l). The mixture is poured into a saturated aqueous solution of ammonium chloride (0.15 l) and shaken until the precipitate formed has dissolved. The ethereal layer is decanted, washed with water (3 × 0.10 l) and dried. After evaporation of the solvent, the residue (54 g, 99% of crude products) is absorbed on deactivated alumina and eluted with hexane, benzene and diethyl ether. The last fraction contains relatively pure carbinol (23 g). It is concentrated and taken up in ethyl acetate (0.10 l). Palladium (2.5% on barium sulphate, 0.5 g) is added and the suspension is shaken under a hydrogen atmosphere. The reaction starts slowly, then proceeds rapidly. When the calculated amount of hydrogen (85 mmol) has been absorbed, the catalyst is removed by centrifugation. The ethyl acetate is evaporated and replaced by diethyl ether (0.10 l). The ethereal solution is shaken for 30 min at 25 °C with 0.2 M hydrochloric acid (0.10 l), then washed with aqueous sodium hydrogen carbonate and water, dried and concentrated. Distillation gives β-ionylideneacetaldehyde as a (*Z*/*E*) isomeric mixture; 36%; b.p. 118–122 °C/ 0.01 mmHg.

3-Methoxyphenyllithium [76]

A solution of butyllithium (50 mmol) in hexane (30 ml) is diluted with tetrahydrofuran (70 ml). At −75 °C, 3-bromoanisole (5.4 ml, 8.6 g, 50 mmol) and, 5 min later, while a white precipitate has formed, 6-methyl-5-hepten-2-one (7.4 ml, 6.3 g, 50 mmol) are added. The volatiles are evaporated and the residue is triturated with brine (35 ml). 2-(3-Methoxyphenyl)-6-methyl-5-hepten-2-ol is extracted with hexane (3 × 20 ml) and isolated by distillation; 65%; bp 118–120 °C/ 0.5 mmHg; n^{20}_D 1.4219.

2-Methoxyphenyllithium (o-Anisyllithium) [82]

$\xrightarrow{LiC_4H_9}$; 1. CO_2, 2. H^+

A solution of anisole (11 ml, 11 g, 0.10 mol) and butyllithium (0.10 mol) in diethyl ether (0.15 l) is heated for 21 h under reflux. The mixture is poured onto an excess of crushed dry-ice (20 g) covered with diethyl ether (50 ml). Extraction of the ethereal layer with 2% aqueous sodium hydroxide followed by acidification with hydrochloric acid gives a precipitate which exhibits the properties of 2-methoxybenzoic acid; 65%. Trace amounts of phenol can be detected by gas chromatography.

(Z)-2-(3,3-Dimethylbicyclo[2.2.1]hept-2-ylidene)methyllithium ['Z'-ω-Camphenyllithium] [78]

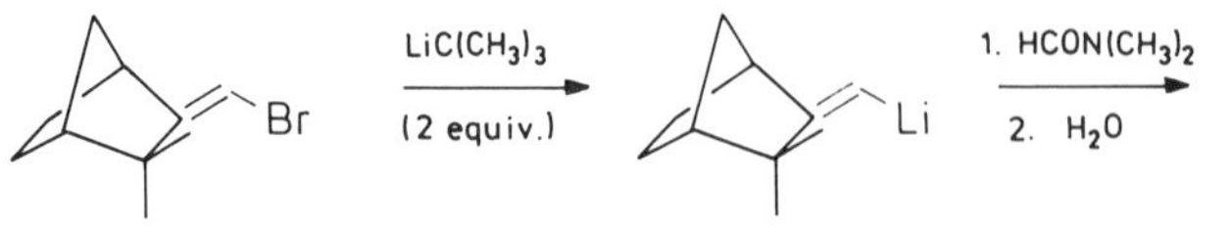

$LiC(CH_3)_3$ (2 equiv.) ; 1. $HCON(CH_3)_2$ 2. H_2O

(*Z*)-2-(3,3-Dimethylbicyclo[2.2.1]hept-2-ylidene)methyl bromide (4.3 g, 20 mmol; by gas chromatographic separation of the stereoisomers [79] and, 5 min. later *N*,*N*-dimethylformamide (2.0 ml, 1.9 g, 26 mmol) were added to a solution of *tert*-butyllithium (40 mmol) in neat tetrahydrofuran (50 ml) at −75 °C. Evaporation of the solvent, addition of water 25 ml, extraction with hexane (3 × 20 ml), washing with brine (20 ml) and distillation afforded the product as a colourless oil; 53%; b.p. 91–94 °C/2 mmHg.

(E)-2-(3,3-Dimethylbicyclo[2.2.1]heptylidene)methylpotassium (['(E)-ω-Camphenylpotassium'] [23, 41]

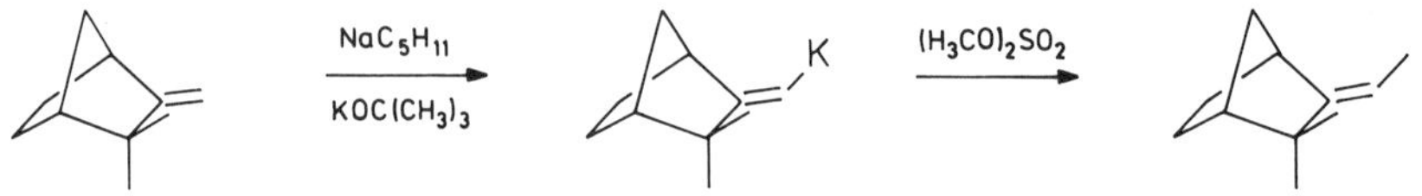

NaC_5H_{11} / $KOC(CH_3)_3$; $(H_3CO)_2SO_2$

A mixture of camphene (8.1 ml, 6.8 g, 50 mmol), pentylsodium (50 mmol) and potassium *tert*-butoxide (8.0 g, 71 mmol) in pentane (50 ml) is submitted during 1 h to high-speed stirring (20 000 rpm) before being allowed to stand overnight (15 h). On dropwise addition of dimethyl sulfate 10.0 ml, 12.6 g, 105 mmol, an exothermic reaction takes place. In addition to some starting material (11%), (*E*)-3-ethylidene-2,2-dimethylbicyclo[2.2.1]heptane is collected on distillation; 60%; b.p. 86–89 °C/12 mmHg.

2-Pyridinyllithium [41, 83]

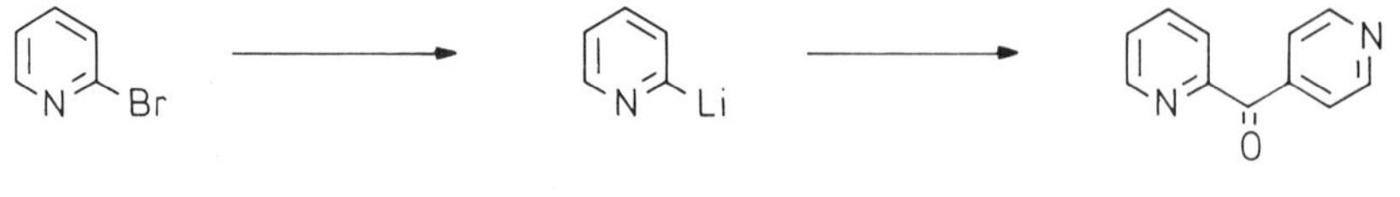

A solution of *tert*-butyllithium, (50 mmol) in hexane is evaporated to dryness and the residue is taken up in precooled (−75 °C) tetrahydrofuran (50 ml). 2-Bromopyridine (4.9 ml, 7.9 g, 50 mmol) and, 5 min later, 4-cyanopyridine (5.2 g, 50 mmol) are added. The solvent is evaporated and the residue triturated with brine (25 ml). The 2-pyridyl 4-pyridyl ketone is extracted with diethyl ether and purified by crystallization from a 1:10 (v/v) mixture of acetone and hexane; 59%; m.p. 122–123 °C.

1-(2-Butyl-1,2-dihydropyridinyl)lithium [84]

$\xrightarrow{LiC_4H_9}$ C_4H_9 H Li $\xrightarrow{-[LiH]}$ C_4H_9

Butyllithium (50 mmol), from which the commercial solvent (hexane) has been stripped off, is dissolved in toluene (25 ml). Pyridine (4.0 ml, 4.0 g, 50 mmol) is added and the mixture is heated for 2 h to 100 °C. The clear solution obtained after filtration [in the presence of diatomite (kieselguhr)] is concentrated and distilled. 2-Butylpyridine is obtained with a yield of 86%; b.p. 64–66 °C/12 mmHg; n_D^{20} 1.4883.

1.5.3 STORABILITY OF ORGANOMETALLIC REAGENTS

As we have seen in the preceding section, organometallic compounds employed as key intermediates for synthetic purposes are used as soon as prepared. Commercial reagents are generally purchased in large quantities and are consumed only over a longer period of time. In such cases, the researcher has to know how to conserve the material.

Two modes of decomposition tend to shorten the shelf-life of organometallic species even if stored under inert atmosphere. They may be destroyed by reaction with the solvent or by spontaneous β-elimination of metal hydride (or other metal-bearing fragments).

Stability Towards Hydride Elimination

The hydride elimination is difficult to control. Fortunately, with most standard reagents it occurs to a significant extent only at elevated temperatures (50–150 °C) [85, 86]. There is, however, one important exception. A commercial solution of *sec*-butyllithium should be stored in the cold (around 0 or −25 °C), otherwise it will lose *ca* 1% of its organometallic titre per week owing to the formation of lithium hydride in addition to 1- and 2-butene. This process will be drastically accelerated if more than trace amounts of a lithium alkoxide are present [87]. Alkoxides are inevitably formed when air enters into contact with an organometallic solution.

Stability Towards Solvent Attack

Hexane and other alkanes are almost inert towards standard organometallics. However, whenever an ethereal reaction medium is employed, one has to worry about competitive reactions between the reagent and the solvent.

Numerous careful studies on the stability of methyl- [88], primary alkyl [88–90], *sec*-alkyl- [90, 91], cyclopropyl- [90], cyclohexyl- [90], *tert*-alkyl- [91], vinyl- [92] and phenyllithium [88] have been performed. Butyllithium has received particular attention [88, 90, 93]. Nevertheless, the accuracy of the reported data should not be overestimated. So far, rate constants and reaction orders have been measured in not more than a single case [93] and a detailed kinetic analysis of solvent-mediated decomposition of organometallic reagents is still lacking. For practical purposes, however, it is amply sufficient to compile half-lifes, $\tau_{1/2}$, of organometallic reagents as a function of the solvent and the temperature in a semi-quantitative way (see Tables 1.11 and 1.12). The relevant trends then become even more clearly visible.

The replacement of a hydrogen atom in methyllithium by an alkyl group diminishes its resistance against ethereal solvents by a factor of 10, if not 100 (see Table 1.11). This

Table 1.11. Stability of typical organolithium compounds towards ethereal solvents: temperature ranges in which half-lifes of *ca* 100 h are observed

[°C]	DEE	THF	MEGME
+ 50	$Li{-}CH_3$		
+ 25	$Li{-}C_6H_5$		
0	$Li{-}CH_2C_3H_7$	$Li{-}CH_3$	
- 25	$Li{-}CH(CH_3)C_2H_5$	$Li{-}C_6H_5$	
- 50	$Li{-}C(CH_3)_3$	$Li{-}CH_2C_3H_7$	$Li{-}CH_3$
- 75		$Li{-}CH(CH_3)C_2H_5$	
-100		$Li{-}C(CH_3)_3$	$Li{-}CH_2C_3H_7$
-125			$Li{-}CH(CH_3)C_2H_5$
-150			$Li{-}C(CH_3)_3$

Table 1.12. Stability of organometallic compounds towards ethereal solvents as expressed by temperature-dependent half-lifes of *ca* 100 h: comparison between alkyllithium, -sodium and -potassium reagents

[°C]	Li-R	Na-R	K-R
+ 50			
+ 25			
0	$Li-CH_3$		
- 25			
- 50	$Li-CH_2C_3H_7$	$Na-CH_3$	
- 75			$K-CH_2Si(CH_3)_3$
-100		$Na-CH_2C_3H_7$	$K-CH_3$
-125			
-150			$K-CH_2C_3H_7$

means in practical terms that one has to lower the temperature by at least 25 °C in order to ensure the same chemical stability as met with the previous reagent. The more polar the solvent, the more it activates the organometallic species and thus provokes its destruction. As a rule of thumb, alkyllithium species react with diethyl ether 100 times more slowly than with tetrahydrofuran and 10 000 times more slowly than with monoethylene glycol dimethyl ether ('glyme', MEGME). Finally, phenyllithium being only loosely aggregated (see pp. 17, 56) proves to be considerably more vulnerable in ethereal solvents than one might have expected on the basis of its much inferior basicity compared with alkyllithium reagents.

Replacement of lithium by a heavier alkali metal makes the organometallic species much more labile. Alkylsodium and -potassium compounds require temperature ranges of -50 to -100 °C and -100 to -150 °C, respectively, to have a chance of survival in tetrahydrofuran (Table 1.12). Trimethylsilylmethylpotassium and the superbasic LIC–KOR (butyllithium–potassium *tert*-butoxide) mixture are less critical to handle. They can be kept in tetrahydrofuran at dry-ice temperature for hours.

So far, we have not taken account of the nature of the reaction that may take place between an organometallic entity and a solvent molecule. In fact, one has to differentiate between three processes, although only one or two of them will manifest themselves with a given solvent.

When treated with strong bases (M—R), diethyl ether simply undergoes a direct, possibly *syn*-periplanar (transition state **1**) [94] β-elimination of metal ethoxide. Unlike benzyl ethyl ether [95–97], diethyl ether is not acidic enough to be susceptible to α-deprotonation. Otherwise an α-metallated species **2** would be generated that would become engaged in either a Wittig rearrangement [98] producing 2-butanol or in an α',β-elimination [95], affording again ethylene and metal ethoxide. The intermediacy of species such as **2** has been ruled out on the basis of deuterium-labelling experiments under some reaction conditions [96], but has been demonstrated to occur under others [97].

$$H_3C\text{-}CH(M)\text{-}O\text{-}CH_2\text{-}CH_2\text{-}H \ \textbf{(2)} \dashrightarrow H_3C\text{-}CH(OM)\text{-}CH_2\text{-}CH_3$$

$$H_3C\text{-}CH_2\text{-}O\text{-}CH_2\text{-}CH_3 \xrightarrow{+\,M\text{-}R} \left[H_3C\text{-}CH_2\text{-}O \cdots (H_2C{=}CH_2) \cdots H \cdots M\text{-}R \right]^{\ddagger} \ \textbf{(1)} \longrightarrow H_3C\text{-}CH_2\text{-}OM + H_2C{=}CH_2$$

(formation of **2** from diethyl ether + M-R is crossed out; **2** $\dashrightarrow$ products of **1** also shown)

β-Elimination is once more the dominant reaction when alkyllithium reagents interact with the chelating solvent ethylene glycol dimethyl ether [99]. In addition, a substitution process has been found to be possible, a methyl group being transferred to the organometallic reagent (M—R) [22].

$$H_3CO\text{-}CH_2\text{-}CH_2\text{-}O\text{-}CH_3 + M\text{-}R \xrightarrow{S_N2} H_3CO\text{-}CH_2\text{-}CH_2\text{-}OM + H_3C\text{-}R$$

$$H_3CO\text{-}CH_2\text{-}CH_2\text{-}O\text{-}CH_3 + M\text{-}R \xrightarrow{E2} H_3CO\text{-}M + H_2C{=}CH\text{-}OCH_3 + H\text{-}R$$

A similar alkylation of the organometallic reagent may occasionally occur if tetrahydrofuran acts as the solvent. This was found to be the case with allyl-type organopotassium species [100], with lithium dimethylphosphide [101] and with organocuprates [102], all of them being excellent nucleophiles. The more basic alkylmetal reagents, however, effect a hydrogen–metal exchange at the α-position. The resulting *gem*-counterpolarized intermediate **3** (M = K) can be trapped at −75 °C with electrophiles [103]. Above −60 °C, species **3** (M = K or Li) performs a fragmentation and decomposes to ethylene and acetaldehyde enolate **4** [75, 104]. The latter component may be intercepted with carbonyl compounds, its nucleophilic addition affording aldol-type products [104, 105].

1.5.4 ANALYSIS

In order to check the stability of a reagent in a given solvent under given conditions, reliable methods for the determination of organometallic concentrations must be available. In general, titration techniques are used that allow one to distinguish between the 'reagent basicity,' i.e. the alkalinity originating from the hydrolysis of the organometallic, and the 'residual basicity,' i.e. the alkalinity due to the presence of metal oxides, alkoxides, hydroxides and carbonates. The latter compounds form when reagent solutions are exposed to contact with air.

The same methods can be used to evaluate the concentration of an organometallic species generated as a key intermediate in a synthesis sequence. Frequently, however, one prefers to probe the presence of the expected intermediate in a more pragmatic way. An aliquot of the reaction mixture is withdrawn and is poured on to dry-ice or is added to an ethereal solution of benzaldehyde. The carboxylation and α-arylhydroxymethylation of organometallics being virtually quantitative transformations, they provide a meaningful estimate of the concentration with which the crucial intermediate has been obtained.

If more detailed information is required, NMR studies may provide the answer. Of course, 'invisible' solvents such as perdeuteriated tetrahydrofuran, dioxane, dimethyl ether, diethyl ether or benzene have to be employed and the sample-containing tubes have to be well stoppered or, better, flame sealed. The findings can be quantified if a known amount of an inert reference compound (e.g. cyclohexane or neopentane) is added as an internal standard.

On the other hand, often only qualitative information is sought. One would like to know whether an organometallic reagent has been essentially consumed rather than to specify the trace amount that possibly has remained unchanged. Colour tests are the most convenient probe if one simply wants to monitor the disappearance of an organometallic compound.

1.5.4.1 Quantitative Determination of Organometallic Concentrations

The most popular method is the so-called double titration according to Gilman and Haubein [106]. It is carried out in two steps. First, an aliquot of the reagent solution is hydrolysed with ice and the aqueous phase is neutralized with 1 M hydrochloric acid (against phenolphthalein). In this way, the 'total basicity' is accounted for. Next, the organometallic compound (MR) is selectively destroyed by the addition of excess of 1,2-dibromoethane from which either bromine or hydrogen bromide is eliminated (producing BrR and ethylene or HR and vinyl bromide) whereas it is not affected by hydroxide or alkoxide bases. Hence the 'residual basicity' found after the treatment with 1,2-dibromoethane can be ascribed to non-organometallic by-products. The difference between total and residual basicity corresponds to the organometallic concentration.

$$\begin{array}{ccccc}
OH^- & \xleftarrow{H_2O} & MR & \xrightarrow[2.\ H_2O]{1.\ BrCH_2CH_2Br} & Br^-\ (!) \\
OH^- & \longleftarrow & M_2O & \longrightarrow & OH^- \\
OH^- & \longleftarrow & MOH & \longrightarrow & OH^- \\
OH^- & \longleftarrow & MOR & \longrightarrow & OH^- \\
HCO_3^- & \longleftarrow & M_2CO_3 & \longrightarrow & HCO_3^- \\
\hline
\text{"total basicity"} & & & & \text{"residual basicity"}
\end{array}$$

The double titration method has the advantage of simplicity and wide applicability. It can even be adapted to the titre evaluation of radical anions such as the naphthalene–lithium (1 : 1) adduct (lithium dihydronaphthylide) [107].

Nevertheless, many attempts have been devoted to the development of so-called direct titration methods. The basic idea is trivial. Benzyl-type organometallics usually exhibit a characteristic and intense orange–red or cherry red colour. Hence they may be taken as self-indicators and be titrated directly with a normalized solution of benzoic acid in an ethereal or aromatic solvent [108].

A possible extension of this approach to colourless organoalkali reagents is obvious. One has only to add an excess of triphenylmethane to their solution in diethyl ether or, better, tetrahydrofuran. If a reactive alkyl- or arylmetal species is involved, an instantaneous hydrogen–metal exchange will occur, generating triphenylmethyllithium [109]. Alternatively, 1,1-diphenylethylene could be allowed to react with the reagent, whereupon an adduct of the 1,1-diphenylethylmetal type will result. Either species being deeply coloured, its concentration can be again determined by direct titration with an anhydrous acid.

Particularly intense colours appear if an organolithium reagent combines with 1,10-phenanthroline. On titration with 2-butanol, sharp end points can be observed [110]. Also, the deprotonation of amines and imines carrying aromatic substituents can lead to intensely coloured species. Thus, the treatment of either *N*-benzylidenebenzylamine [111] or *N*-phenyl-1-naphthylamine [112] produces yellow solutions which can be accurately titrated with anhydrous acids.

In the meantime, profound modifications of the transmetallation protocol have been proposed. The troublesome preparation of standard solutions of benzoic acid or 2-butanol is avoided by using a known amount of a divalent solid proton source such as diphenylacetic acid [113]. When an alkyllithium reagent is added, first the carboxy group is deprotonated to give a colourless carboxylate. Additional alkyllithium will produce a yellow enolate-type species **5** and thus indicate that precisely one equivalent of base has already been consumed [113]. Similar results can be achieved with 4-biphenylmethanol [114], 2,5-dimethoxybenzyl alcohol [115] and 1,3-diphenylacetone *p*-toluenesulphonylhydrazone [116].

$$(C_6H_5)_2CH{-}COOH \longrightarrow (C_6H_5)_2CH{-}COOLi \longrightarrow (C_6H_5)_2C{=}C(O{-}Li)_2 \qquad \textbf{(5)}$$

All these methods suffer from lack of generality since they work well only with strongly basic alkyllithium reagents. Moreover, as the colours of the doubly deprotonated species are relatively pale, changes cannot be recognized unambiguously. The recent introduction of 1-pyreneacetic acid as an indicator appears to obviate these shortcomings [117]. This time the doubly deprotonated enolate species **6** (M = metal) has an intense red colour and its carboxylate precursor is acidic enough to react

$$\text{(1-pyrenyl)}{-}CH_2COOH \longrightarrow \text{(1-pyrenyl)}{-}CH_2{-}COOLi \longrightarrow \text{(1-pyrenyl)}{-}CH{=}C(OLi)_2 \qquad \textbf{(6)}$$

not only with alkyllithiums but also with weakly basic lithium acetylides, lithium diisopropylamide and even with Grignard reagents [117].

There remain, however, doubts about how strictly such a method can discriminate between organometallic and residual basicity. In other words, one has to consider whether the conversion of diphenylacetic acid and 1-pyreneacetic acid into the corresponding carboxylates really requires the action of an organometallic reagent or could also be accomplished with one of the alkaline impurities, say lithium oxide. Under these circumstances an entirely new approach is particularly welcome. A very promising recent suggestion is to monitor the decolouration of brick-red diphenyl ditelluride (**7**) when titrated with an organometallic reagent [118]. The trick is to replace a CH acid giving rise to a coloured 'anion' by a coloured nucleophile that produces colourless reaction products, namely a telluride and a metal tellurolate [118]. Again, the method appears to be applicable to a variety of organometallic intermediates including 1-alkynyllithium and organomagnesium reagents [118].

$$\underset{(\mathbf{7})}{H_5C_6\text{-}Te\text{-}Te\text{-}C_6H_5} + M\text{-}R \longrightarrow H_5C_6\text{-}Te\text{-}R + M\text{-}Te\text{-}C_6H_5$$

1.5.4.2 Qualitative Tests

Inspired by earlier work dealing with the addition of Grignard reagents to diarylketones [119], Gilman has devised colour tests that allow one to probe very easily the presence of organometallic reagents. The most widely used Gilman–Schulze test (Colour Test I) [120] responds to virtually all polar organometallics. A small sample (0.1–0.5 ml) of the dissolved reagent is transferred by means of a pipette into a 1 M solution of 4,4′-bis(dimethylamino)benzophenone (Michler's ketone) in benzene (1–2 ml). On addition of water (1–2 ml) and a few drops of a 0.2% solution of iodine in glacial acetic acid, a very intense malachite green–blue colour appears.

M-R

HO R

R

The characteristic colour is associated with the formation of a di- or triarylmethyl-type cation depending on whether an alkyl- or arylmetal compound was present. In the latter case the ionization of the carbinol precursor does not require acetic acid; it can be brought about in a buffered medium (e.g. with 20% aqueous pyrocatechol [121]) and even in the absence of iodine.

A distinction between aliphatic and other organometallic reagents may be very valuable, for example when primary, secondary or tertiary butyllithium is used in a neighbouring group-assisted metallation of an aromatic position. The Gilman–Swiss test (Colour Test II) [122] is specific to alkyllithium reagents. The key step is a halogen–metal interconversion. A small sample of the presumed organometallic solution

(0.5 ml) is introduced into a 15% solution (1 ml) of 4-bromo-*N*,*N*-dimethylaniline in benzene. Then a 15% solution (1 ml) of benzophenone in benzene and, a few minutes later, concentrated hydrochloric acid (1 ml) are added. A bright red colour results only if an alkyllithium reagent was present.

1.5.5 EQUIPMENT AND HAZARDS

The physical handling of reagents is the 'alpha' and 'omega' of their application to laboratory-scale and industrial-scale syntheses, the beginning and the end. Therefore, this contribution will close with some practical hints.

1.5.5.1 Anaerobic Devices

The manufacturers supply butyllithium in steel containers ranging from small cylinders to 35 000 l tank cars. Detailed instructions concerning the discharge and cleaning of the vessels, the safe handling and the properties of the reagent and emergency routines can be obtained on request [123, 124].

Alkyllithium solutions can be purchased for laboratory purposes in quantities as small as 100 ml in screw-capped or serum sealed glass bottles. If such reagents are used regularly, it is far more economic to order 5–10 l quantities. Siphoning is the most convenient technique for the transfer of the liquid to Schlenk type storage burettes (see Figure 1.10).

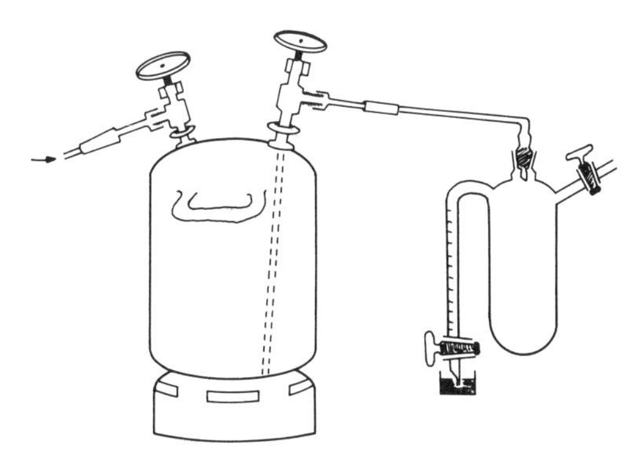

Figure 1.10. Transfer of butyllithium from a steel cylinder to a Schlenk burette under a nitrogen atmosphere

The various possibilities of carrying out reactions in the laboratory under an inert atmosphere have been summarized in several reviews [125]. Synthetically oriented chemists performing small-scale (0.01–0.1 mol) experiments are well advised to work routinely under >99.99% pure nitrogen using a gas/vacuum line apparatus. Suitably designed and reasonably priced Schlenk-type glassware can be obtained from the specialized trade [126].

1.5.5.2 Safety Precautions

Any alkaline material may cause irreversible damage to the eyes. Therefore, it is essential to wear goggles with sideshields before entering any organometallic laboratory. If, nevertheless, eyes have accidentally come into contact with an alkaline substance, they should be immediately rinsed with 3% aqueous boric acid and a physician should be consulted urgently. If organolithium solutions have been spilled over parts of the skin, dilute (e.g. 5%) acetic acid may be used for neutralization.

Some organometallic compounds, notably *tert*-butyllithium, are pyrophoric (self-ignitable), especially if highly concentrated. Potassium–sodium alloy catches fire particularly rapidly. In the case of such an accident, extinguishers operating with watery foam, halocarbons or carbon dioxide (!) must not be used. Dry sand, uncontaminated with mud and butts, is very efficient for fighting such a fire.

1.6 REFERENCES AND NOTES

1 The Organometallic Approach: A Look Back

1. G. Bugge (Ed.), *Das Buch der Grossen Chemiker*, Vol. 2, Verlag Chemie, Weinheim, 1929; J. R. Partington, *A History of Chemistry*, Vol. 4, Macmillan, London, 1964; *Poggendorff: Biographisch-literarisches Handwörterbuch der Exakten Naturwissenschaften*, Vol. 7b, Part 6, Akademie-Verlag, Berlin, 1980.
2. F. Wöhler, *Ann. Phys. (Leipzig)* **12** (1828) 253.
3. Prior to publication, on 22 February 1828 Wöhler sent a letter to his friend and former teacher Berzelius informing him with unconcealed triumph that 'ich Harnstoff machen kann, ohne dazu Nieren oder überhaupt ein Thier, sei es Mensch oder Hund, nöthig zu haben' (I can make urea without needing kidneys or an animal at all, be it man or dog).
4. F. Wöhler, *Ann. Chem. Pharm. (Heidelberg)* **35** (1840) 111; see also J. E. Mallet, *Ann. Chem. Pharm. (Heidelberg)* **79** (1851) 223.
5. R. Bunsen, *Ann. Chem. Pharm. (Heidelberg)* **42** (1843) 14.
6. E. Frankland, *Ann. Chem. Pharm. (Heidelberg)* **71** (1849) 171, 213; *J. Chem. Soc.* **11** (1849) 263, 297.
7. J. A. Wanklyn, *Ann. Chem. Pharm.* **108** (1858) 67; *Proc. R. Soc. London*, **9** (1858) 341.
8. G. Wittig, P. Davis, G. Koenig, *Chem. Ber.* **84** (1951) 627; and later publications, e.g. G. Wittig, G. Closs, F. Mindermann, *Liebigs Ann. Chem.* **594** (1955) 89, spec. 101; see also D. Seebach, *Angew. Chem.* **81** (1969) 690; *Angew. Chem., Int. Ed. Engl.* **8** (1969) 639.
9. M. Schlosser, *Struktur und Reaktivität Polarer Organometalle*, Springer, Berlin, 1973.

2 Structures

1. M. Schlosser, *Struktur und Reaktivität Polarer Organometalle*, Springer, Berlin, 1973.
2. G. Wittig, *Organische Anionochemie*, *Experientia* **14** (1958) 389; E. Buncel, *Carbanions: Mechanistic and Isotope Aspects*, Elsevier, Amsterdam, 1975.
3. R. C. Weast (Ed.), *Handbook of Chemistry and Physics*, 67th edn, pp. E-76–E-77. CRC Press, Boca Raton, FL.
4. P. Coppens, *Angew. Chem.* **89** (1977), 33; *Angew. Chem. Int. Ed. Engl.* **16** (1977), 32; K. Angermund, K. H. Claus, R. Goddard, C. Krüger, *Angew. Chem.* **97** (1985), 241; *Angew. Chem. Int. Ed. Engl.* **24** (1985), 237; P. Coppens, M. B. Hall (Eds), *Electron Distributions and the Chemical Bond*, Plenum Press, New York, 1982; K. Toriumi, Y. Saito, *Adv. Inorg. Chem. Radiochem.* **27** (1983), 27; P. Coppens, *Coord. Chem. Rev.* **65** (1985), 285; F.L. Hirshfield, *Cryst. Rev.* **2** (1991), 169; W. Weyrich (ed.), Proceedings of the 10th Sagamore Conference on Charge, Spin and Momentum Densitites, *Z. Naturforsch.* **48a** (1993), 11-462.
5. R. A. Eades, P. G. Gassman, D. A. Dixon, *J. Am. Chem. Soc.* **103** (1981) 1066; C. R. A. Catlow, A. M. Stoneham, *J. Phys. C.* **16** (1983), 4321; *Chem. Abstr.* **99** (1983), 200'617u; E. N. Maslen, M. A. Spackman, *Austr. J. Phys.* **38** (1985), 273; *Chem. Abstr.* **103** (1985), 147'349z; P. Seiler, J. Dunitz, *Helv. Chim. Acta* **69** (198), 1107; P. Seiler, *Acta Cryst.* **B49** (1993), 223.
6. F. Escudero, M. Yanez, *Mol. Phys.* **45** (1982), 617.
7. The author owes this hint to Professor J. D. Dunitz, Zurich.
8. P. P. Power, *Acc. Chem. Res.* **21** (1988) 147.
9. S. S. Al-Juaid, C. Eaborn, P. B. Hitchcock, C. A. McGeary, J. D. Smith, *J. Chem. Soc., Chem. Commun.* (1989) 273.
10. D. Belluš, B. Klingert, R. W. Lang, A. Rihs, *J. Organomet. Chem.* **339** (1988) 17.
11. A. C. Belch, M. Berkowitz, J. A. McCammon, *J. Am. Chem. Soc.* **108** (1986) 1755, 1762.
12. K. Neupert-Laves, M. Dobler, *Helv. Chim. Acta* **58** (1975) 432.
13. Y. Pocker, R. F. Buchholz, *J. Am. Chem. Soc.* **93** (1981) 2905.
14. R. Schwesinger, K. Piontek, W. Littke, H. Prinzbach, *Tetrahedron Lett.* **26** (1985) 1201.
15. R. Sarma, F. Ramirez, B. McKeever, Y. F. Chaw, J. F. Marecek, D. Nierman, T. M. MacCaffrey, *J. Am. Chem. Soc.* **99** (1977) 5289.
16. J. D. Owen, *J. Chem. Soc., Dalton Trans.* (1980) 1066.
17. P. Veya, C. Floriani, A. Chiesi-Villa, C. Guastini, *Organometallics* **10** (1991) 1652.
18. A. D. Pajerski, G. L. Bergstresser, M. Parvez, H. G. Richey, *J. Am. Chem. Soc.* **110** (1988) 4844.
19. C. Riche, C. Pascard-Billy, *J. Chem. Soc., Chem. Commun.* (1977) 183.
20. M. Dobler, R. P. Phizackerley, *Helv. Chim. Acta* **57** (1974) 664.
21. J. A. T. Norman, G. P. Pez, *J. Chem. Soc., Chem. Commun.* (1991) 971.
22. G. W. Mappes, S. A. Friedman, T. P. Fehlner, *J. Phys. Chem.* **74** (1970) 3307; S. J. Ashcroft, G. Beech, *Inorganic Thermodynamics*, Van Nostrand Reinhold, New York, 1973; see also R. Ahlrichs, *Theor. Chim. Acta* **35** (1974) 59; O. S. Akkerman, G. Schat, E. A. I. M. Evers, F. Bickelhaupt, *Recl. Trav. Chim. Pays-Bas* **102** (1983) 109.
23. J. F. Liebmann, A. Greenberg, R. E. Williams (Eds), *Advances in Boron and the Boranes*, VCH, Weinheim, 1988; P. v. R. Schleyer, M. Bühl, U. Fleischer, W. Koch, *Inorg. Chem.* **29** (1990) 153; G. A. Olah, K. Wade, R. E. Williams (Eds), *Electron Deficient Boron and Carbon Clusters*, Wiley, New York, 1991.
24. R. P. Bell, H. J. Emeleus, *Q. Rev. Chem. Soc.* **2** (1948) 141; B. L. Caroll, L. S. Bartell, *J. Chem. Phys* **42** (1965, 1135, 3076; *Inorg. Chem.* **7** (1968) 219.

25. E. G. Hoffmann, *Liebigs Ann. Chem.* **629** (1960) 104.
26. A. W. Laubengayer, W. F. Gilliam, *J. Am. Chem. Soc.* **63** (1941) 477; C. H. Henrickson, D. P. Eyman, *Inorg. Chem.* **6** (1967) 1461.
27. K. S. Pitzer, H. S. Gutowsky, *J. Am. Chem. Soc.* **68** (1946) 2204.
28. P. H. Lewis, R. E. Rundle, *J. Chem. Phys.* **21** (1953) 986.
29. G. E. Coates, F. Glockling, N.D. Huck, *J. Chem. Soc.* **1952** 4496; J. Goubeau, K. Walter, *Z. Anorg. Allg. Chem.* **322** (1963), 58.
30. A. I. Snow, R. E. Rundle, *Acta Cryst.* **4** (1951), 348.
31. E. Kaufmann, K. Raghavachari, A. E. Reed, P. V. R. Schleyer, *Organometallics* **7** (1988) 1597; for early examples, see: P. Hubberstey, *Coord. Chem. Rev.* **40** (1982), 48–49.
32. T. L. Brown, M. T. Rogers, *J. Am. Chem. Soc.* **79** (1957) 1859; R. West, W. Glaze, *J. Am. Chem. Soc.* **83** (1961) 3580; J. Goubeau, K. Walter, *Z. Anorg. Allg. Chem.* **322** (1963) 58; J. W. Chinn, R. J. Lagow, *Organometallics* **3** (1984) 75; D. Plavsic, D. Srzic, L. Klasinc, *J. Phys. Chem.* **90** (1986) 2075.
33. L. Andrews, *J. Chem. Phys.* **47** (1967) 4834.
34. J. L. Atwood, T. Fjeldberg, M. F. Lappert, N. T. Luong-Thi, R. Shakair, A. J. Thorne, *J. Chem. Soc., Chem. Commun.* (1984) 1163.
35. A. Haaland, *Top. Curr. Chem.* **53** (1975) 1; B. Beagley, in *Molecular Structure by Diffraction Methods*, ed G. A. Sim, L. E. Sutton, Vol. 1, p. 111, London, 1973.
36. H. Dietrich, *Acta Crystallogr.* **16** (1963) 681; *J. Organomet. Chem.* **205** (1981) 291.
37. W. E. Whine, G. Stucky, S. W. Peterson, *J. Am. Chem. Soc.* **97** (1975) 6401.
38. R. P. Hughes, J. Powell, *J. Chem. Soc., Chem. Commun.* (1971) 275.
39. D. R. Armstrong, W. Clegg, H. M. Colquhoun, J. A. Daniels, R. E. Mulvey, I. R. Stephenson, K. Wade, *J. Chem. Soc., Chem. Commun.* (1987) 630.
40. R. E. Mulvey, P. G. Perkins, *Polyhedron* (1988) 2119; J. J. Novoa, M.-H. Whangbo, G. D. Stucky, *J. Org. Chem.* **56** (1991) 3181.
41. B. Teclé, A. F. M. Maqsudur Rahman, J. P. Oliver, *J. Organomet. Chem.* **317** (1986) 267.
42. E. Weiss, T. Lambertsen, B. Schubert, J. K. Cockcroft, A. Wiedemann, *Chem. Ber.* **123** (1990) 79, see also E. Weiss, E. A. C. Lucken, *J. Organomet. Chem.* **2** (1964) 197; E. Weiss, G. Hencken, *J. Organomet. Chem.* **21** (1969) 265.
43. E. Weiss, T. Lambertsen, B. Schubert, J. K. Cockcroft, *J. Organomet. Chem.* **358** (1988) 1; see also E. Weiss, G. Sauermann, *Chem. Ber.* **103** (1979) 265.
44. E. Weiss, H. Köster, *Chem. Ber.* **110** (1977) 717.
45. E. Weiss, S. Corbelin, J. K. Cockcroft, A. N. Fitch, *Chem. Ber.* **123** (1990) 1629; see also E. Weiss, G. Sauermann, G. Thirase, *Chem. Ber.* **116** (1983) 74.
46. G. Stucky, R. E. Rundle, *J. Am. Chem. Soc.* **86** (1964) 4825.
47. A. L. Spek, P. Voorbergen, G. Schaf, C. Blomberg, F. Bickelhaupt, *J. Organomet. Chem.* **77** (1974) 147; see also L. J. Guggenberger, R. E. Rundle, *J. Am. Chem. Soc.* **86** (1964) 5344.
48. W. Uhl, M. Layn, W. Hiller, *J. Organomet. Chem.* **361** (1989) 139.
49. M. A. Beno, H. Hope, M. M. Olmstead, P. P. Power, *Organometallics* **4** (1985) 2117.
50. R. A. Bartlett, H. V. R. Dias, P. P. Power, *J. Organomet. Chem.* **341** (1988) 1.
51. M. Geissler, J. Kopf, B. Schubert, E. Weiss, W. Neugebauer, P. v. R. Schleyer, *Angew. Chem.* **99** (1987) 569; *Angew. Chem., Int. Ed. Engl.* **26** (1987) 587.
52. J. Toney, G. D. Stucky, *J. Chem. Soc., Chem. Commun.* (1967) 1168.
53. U. Schümann, J. Kopf, E. Weiss, *Angew. Chem.* **97** (1985) 222; *Angew. Chem., Int. Ed. Engl.* **24** (1985) 215.
54. R. P. Zerger, G. D. Stucky, *J. Chem. Soc., Chem. Commun.* (1973) 44.

55. F. Hein, H. Schramm, *Z. Phys. Chem. (Leipzig)* **151** (1930) 234.
56. T. L. Brown, M. T. Rogers, *J. Am. Chem. Soc.* **79** (1957) 1859.
57. T. L. Brown, R. L. Gerteis, D. A. Bafus, J. A. Ladd, *J. Am. Chem. Soc.* **86** (1964) 2135.
58. H. L. Lewis, T. L. Brown, *J. Am. Chem. Soc.* **92** (1970) 4664.
59. P. West, R. Waack, *J. Am. Chem. Soc.* **89** (1967) 4395.
60. L. D. McKeever, R. Waack, M. A. Doran, E. B. Baker, *J. Am. Chem. Soc.* **90** (1968) 3244.
61. R. H. Baney, R. J. Krager, *Inorg. Chem.* **3** (1964) 1657; G. E. Hartwell, T. L. Brown, *Inorg. Chem.* **5** (1966) 1257.
62. W. H. Glaze, G. M. Adams, *J. Am. Chem. Soc.* **88** (1966) 4653; S. Bywater, D. J. Worsfold, *J. Organomet. Chem.* **10** (1967) 1.
63. M. Weiner, G. Vogel, R. West, *Inorg. Chem.* **1** (1962) 654.
64. W. Bauer, W. R. Winchester, P. v. R. Schleyer, *Organometallics* **6** (1987) 2371.
65. G. Wittig, F. J. Meyer, G. Lange, *Liebigs Ann. Chem.* **571** (1951) 167; R. Waack, M. A. Doran, *J. Am. Chem. Soc.* **91** (1969) 2456; M. Schlosser, V. Ladenberger, *Chem. Ber.* **100** (1967) 3877.
66. D. Seebach, R. Hässig, J. Gabriel, *Helv. Chim. Acta* **66** (1983) 308.
67. L. M. Jackman, L. M. Scarmoutzos, *J. Am. Chem. Soc.* **106** (1984) 4627; E. Wehman, J. T. B. H. Jastrzebski, J. M. Ernsting, D. M. Grove, G. V. Koten, *J. Organomet. Chem.* **133** (1988) 353; O. Eppers, H. Günther, *Helv. Chim. Acta* **75** (1992) 2553.
68. W. Bauer, D. Seebach, *Helv. Chim. Acta* **67** (1984) 1972; see also G. Fraenkel, P. Pramanik, *J. Chem. Soc., Chem. Commun.* (1983) 1527.
69. M. Morton, L. J. Fetters, *J. Polym. Sci.* **A2** (1964) 3311; *Chem. Abstr.* **61** (1964) 9585d.
70. M. Stähle, M. Schlosser, *J. Organomet. Chem.* **220** (1981) 277.
71. O. Desponds, *Doctoral Thesis* Université de Lausanne, 1991.
72. For a contradictory result, see W. R. Winchester, W. Bauer, P. v. R. Schleyer, *J. Chem. Soc., Chem. Commun.* (1987) 177.
73. G. E. Hartwell, T. L. Brown, *J. Am. Chem. Soc.* **88** (1966) 4625; M. Y. Darensbourg, B. Y. Kimura, G. E. Hartwell, T. L. Brown, *J. Am. Chem. Soc.* **92** (1970) 1236; J. F. McGarrity, C. A. Ogle, Z. Brich, H. R. Loosli, *J. Am. Chem. Soc.* **107** (1985) 1810.
74. M. A. Weiner, R. West, *J. Am. Chem. Soc.* **85** (1963) 485; D. E. Applequist, D. F. O'Brien, *J. Am. Chem. Soc.* **85** (1963) 743; T. L. Brown, *Acc. Chem. Res.* **1** (1968) 23.
75. D. T. Hurd, *J. Org. Chem.* **13** (1948) 711; L. M. Seitz, T. L. Brown, *J. Am. Chem. Soc.* **88** (1966) 4140; N. O. House, R. A. Latham, G. M. Whitesides, *J. Org. Chem.* **32** (1967) 2481.
76. G. Wittig, R. Ludwig, R. Polster, *Chem. Ber.* **88** (1955) 294.

3 Reactivity and Selectivity

1. B. de B. Darwent, *Bond Dissociation Energies in Simple Molecules*, National Bureau of Standards, Washington, DC, 1970.
2. S. G. Lias, J. E. Bartmess, J. F. Liebman, J. L. Holmes, R. D. Levin, W. G. Mallard, *J. Phys. Chem., Ref. Data* **17** Suppl. 1 (1988).
3. C. H. Wu, *J. Chem. Phys.* **65** (1976) 2040; K. F. Zmbov, C. H. Wu, H. R. Ihrle, *J. Chem. Phys.* **67** (1977) 4603; H. Partridge, C. W. Bauschlicher, L. G. M. Pettersson, A. D. McLean, B. Liu, M. Yoshimine, A. Komornicki, *J. Chem. Phys.* **92** (1990), 5'377.
4. On the basis of Pauling's formula and electronegativities (L. Pauling, *The Nature of the Chemical Bond*, 3rd edn, Cornell University Press, Ithaca NY, 1960), the following bond

ionicities can be derived: C—H 4%, C—Hg 10%, C—Zn 18%, C—Mg30%, C—Li 43%, C—Na 47%, C—K 51%; see also T. H. Chan, I. Fleming, *Synthesis* (1979) 761.

5. J. N. Brønstedt, K. J. Pederson, *Z. Phys. Chem. (Leipzig)* **108** (1924) 185.
6. M. DePaz, A. Guidoni-Guardini, L. Friedman, *J. Chem. Phys.* **52** (1970) 687; M. Arshadi, R. Yamdagni, P. Kebarle, *J. Phys. Chem.* **74** (1970) 1475.
7. T. Holm, *J. Organomet. Chem.* **56** (1973) 87.
8. T. Holm, *J. Organomet. Chem.* **77** (1974) 27.
9. E. Moret, O. Desponds, M. Schlosser, unpublished results (1987).
10. M. Hanack, H. J. Schneider, *Angew. Chem.* **79** (1967) 709; *Angew. Chem., Int. Ed. Engl.* **6** (1967) 666; S. Sarel, J. Yovell, M. Sarel-Imber, *Angew. Chem.* **80** (1968) 592; *Angew. Chem., Int. Ed. Engl.* **7** (1968) 577.
11. V. W. Bowry, J. Lusztyk, K. U. Ingold, *J. Am. Chem. Soc.* **113** (1991) 5687.
12. J. D. Roberts, R. H. Mazur, *J. Am. Chem. Soc.* **73** (1951) 2509; D. J. Patel, C. L. Hamilton, J. D. Roberts, *J. Am. Chem. Soc.* **87** (1965) 5144.
13. P. T. Lansbury, V. A. Pattison, W. A. Clement, J. D. Sidler, *J. Am. Chem. Soc.* **86** (1964) 2247.
14. M. E. H. Howden, A. Maercker, J. Burdon, J. D. Roberts, *J. Am. Chem. Soc.* **88** (1966) 1732; A. Maercker, J. D. Roberts, *J. Am. Chem. Soc.* **88** (1966) 1742.
15. E. Grovenstein, *Angew. Chem.* **90** (1978) 317; *Angew. Chem., Int. Ed. Engl.* **17** (1978) 313.
16. H. E. Zimmerman, F. J. Smentowski, *J. Am. Chem. Soc.* **79** (1957) 5455; E. Grovenstein, L. P. Williams, *J. Am. Chem. Soc.* **83** (1961) 412; E. Grovenstein, A. Beres, Y.-M. Cheng, J. A. Pegolotti, *J. Org. Chem.* **37** (1972) 1281.
17. E. Moret, J. Fürrer, M. Schlosser, *Tetrahedron* **44** (1988) 3539.
18. E. Grovenstein, P.-C. Lu, *J. Am. Chem. Soc.* **104** (1982) 6681; *J. Org. Chem.* **47** (1982) 2928.
19. E. Grovenstein, R. E. Williamson, *J. Am. Chem. Soc.* **97** (1975) 646.
20. A. Maercker, M. Passlack, *Chem. Ber.* **116** (1983) 710.
21. G. Fraenkel, J. W. Cooper, *Tetrahedron Lett.* **9** (1968) 599.
22. F. R. Jensen, R. L. Bedard, *J. Org. Chem.* **24** (1959) 874.
23. R. B. Bates, D. W. Gosselink, J. A. Kaczynski, *Tetrahedron Lett.* **8** (1967) 199.
24. R. Paul, S. Tchelitcheff, *C.R. Acad. Sci.* **239** (1954) 1222.
25. R. Lehmann, M. Schlosser, unpublished results (1979); see p. 162 (§ 5).
26. M. Schlosser, H. Bosshardt, A. Walde, M. Stähle, *Angew. Chem.* **92** (1980) 302; *Angew. Chem., Int. Ed. Engl.* **19** (1980) 303.
27. M. Lwow, *Z. Chem. [2]* **7** (1871) 257; C. R. Noller, *J. Am. Chem. Soc.* **51** (1929) 594.
28. C. Moureu, R. Delange, *C.R. Acad. Sci.* **138** (1940) 1339; A. Wohl, B. Mylo, *Ber. Dtsch. Chem. Ges.* **45** (1912) 322; H. Lohaus, *J. Prakt. Chem. [2]* **119** (1928) 235; R. T. Dillon, H. J. Lucas, *J. Am. Chem. Soc.* **50** (1928) 1712; J. P. Wibaut, R. Huls, H. G. P. van der Voort, *Recl. Trav. Chim. Pays-Bas* **71** (1952) 798; 1012; C. A. Dornfeld, G. H. Coleman, *Org. Synth., Coll. Vol.* **3** (1955) 701; H. Stetter, R. Reske, *Chem. Ber.* **103** (1970) 643.
29. E. L. Eliel, F. Nader, *J. Am. Chem. Soc.* **91** (1969) 536; **92** (1970) 584.
30. G. Courtois, M. Harama, L. Miginiac, *J. Organomet. Chem.* **198** (1980) 1; see also H. J. Bestmann, G. Wölfel, K. Mederere, *Synthesis* (1987) 848.
31. H. Wasserman, R. P. Dion, *Tetrahedron Lett.* **23** (1982) 785.
32. M. Bockmühl, G. Ehrhart, *Liebigs Ann. Chem.* **561** (1949) 52.
33. H. W. H. J. Bodewitz, C. Blomberg, F. Bickelhaupt, *Tetrahedron* **29** (1973), 719; J. F. Garst, F. E. Barton *J. Am. Chem. Soc.* **96** (1974), 523; J. F. Garst, J. E. Deutch, G. M. Whitesides, *J. Am. Chem. Soc.* **108** (1986), 2490; H. M. Walborsky, *Acc. Chem. Res.* **23** (1990), 286; J. F. Garst, *Acc. Chem. Res.* **24** (1991), 95; C. Walling, *Acc. Chem. Res.* **24** (1991), 255; H. M.

Walborsky, C. Zimmermann, *J. Am. Chem. Soc.* **114** (1992), 4996; C. Hamdouchi, M. Topolski, V. Goedken, H. M. Walborsky, *J. Org. Chem.* **58** (1993), 3148.

34. W. Schlenk, R. Ochs, *Ber. Dtsch. Chem. Ges.* **49** (1916), 608; K. Ziegler, K. Bähr, *Ber. Dtsch. Chem. Ges.* **61** (1928), 253; E. Müller, T. Töpel, *Ber. Dtsch. Chem. Ges.* **72B** (1939), 273; P. D. Bartlett, C. G. Swain, R. B. Woodward, *J. Am. Chem. Soc.* **63** (1941), 3229; G. A. Russell, E. G. Janzen, E. T. Strom, *J. Am. Chem. Soc.* **86** (1964), 1807; R. C. Lamb, P. W. Ayers, M. K. Toney, J. F. Garst, *J. Am. Chem. Soc.* **88** (1966), 4261; E. J. Panek, G. M. Whitesides, *J. Am. Chem. Soc.* **94** (1972), 8769; Y.-S. Zhang, B. Wenderoth, W.-Y. Su, E. C. Ashby, *J. Organomet. Chem.* **292** (1985), 29; E. C. Ashby, T.N. Pham. *J. Org. Chem.* **87** (1955), 1291; J. Tanaka, M. Nojima, S. Kusabayshi, *J. Am. Chem. Soc.* **109** (1987), 3391; C. Walling, *J. Am. Chem. Soc.* **110** (1988), 6846; R. Okazaki, K. Shibata, N. Tokitoh, *Tetrahedron Lett.* **32** (1991), 6601.
35. R. A. Raphael, *Acetylenic Compounds in Organic Synthesis*, Butterworth, London, 1955; H. G. Viehe (Ed.), *The Chemistry of Acetylenes*, Marcel Dekker, New York, 1969; L. Brandsma, H. D. Verkruijsse, *Synthesis of Acetylenes, Allenes and Cumulenes*, Elsevier, Amsterdam, 1981; E. Winterfeldt, in *Modern Synthetic Methods*, ed. R. Scheffold, Vol. 6, p. 103, 1992.
36. H. Gilman, A. L. Jacoby, H. Ludeman, *J. Am. Chem. Soc.* **60** (1938) 2336; R. G. Jones, H. Gilman, *Chem. Rev.* **54** (1954) 835.
37. M. Stähle, M. Schlosser, unpublished results (1978). For simplicity, we do not discuss the aggregation state of metal acetylides in this context.
38. J. Hartmann, M. Schlosser, *Helv. Chim. Acta* **59** (1976) 453.
39. H. Gilman, R. H. Kirby, *J. Am. Chem. Soc.* **63** (1941) 2046; see also W. I. O'Sullivan, F. W. Swamer, W. J. Humphlett, C. R. Hauser, *J. Org. Chem.* **26** (1961) 2306.
40. W. I. O'Sullivan, F. W. Swamer, W. J. Humphlett, C. R. Hauser, *J. Org. Chem.* **26** (1961) 2306.
41. E. E. Blaise, *C.R. Acad. Sci.* **132** (1901) 38: **133** (1901) 1217; P. L. Pickard, T. L. Tolbert, *J. Org. Chem.* **26** (1961) 4886; R. A. Benkeser, *Synthesis* (1971) 347.
42. J. Fauvarque, J. F. Fauvarque, *C.R. Acad. Sci.* **263** (1966) 488; *Chem. Abstr.* **65** (1966) 19989e.
43. C. A. Brown, *J. Org. Chem.* **39** (1974) 1324, 3913.
44. J. Hartmann, M. Schlosser, *Synthesis* **1975**, 328; C. Margot, *Doctoral Dissertation*, Ecole Polytechnique Fédérale de Lausanne, 1985; K. Koch, *Diploma Work*, Université de Lausanne, 1988.
45. G. A. Russell, *J. Am. Chem. Soc.* **81** (1959) 2017.
46. Y. Guggisberg, F. Faigl, M. Schlosser, *J. Organomet. Chem.* **415** (1991) 1; see also M. Tiffeneau, R. Delange, *C.R. Acad. Sci.* **137** (1903) 573; R. A. Benkeser, D. C. Snyder, *J. Org. Chem.* **47** (1982) 1243.
47. H. Felkin, C. Frajerman, *Tetrahedron Lett.* **11** (1970) 1045; H. Felkin, C. Frajerman, G. Roussi, *Bull. Soc. Chim. Fr.* (1970) 3704.
48. M. Schlosser, M. Stähle, *Angew. Chem.* **92** (1980) 497; *Angew. Chem., Int. Ed. Engl.* **19** (1980) 487; M. Stähle, M. Schlosser, *J. Organomet. Chem.* **220** (1981) 277.
49. S. D. Patterman, I. L. Karle, G. D. Stucky, *J. Am. Chem. Soc.* **92** (1970) 1150; H. Köster, E. Weiss, *J. Organomet. Chem.* **168** (1979) 273; R. A. Bartlett, H. V. R. Dias, P. P. Power, *J. Organomet. Chem.* **341** (1988) 1; M. Bühl, N. R. V. E. Hommes, P. v. R. Schleyer, U. Fleischer, W. Kutzelnigg, *J. Am. Chem. Soc.* **113** (1991) 2459; A. Sygula, P. W. Rabideau, *J. Am. Chem. Soc.* **114** (1992) 821; H. Ahlbrecht, J. Harbach, T. Hauck, H. O. Kalinowski, *Chem. Ber.* **125** (1992) 1753.
50. M. Schlosser, P. Schneider, *Helv. Chim. Acta* **63** (1980) 2404; see also E. Moret, *Doctoral*

Thesis, Université de Lausanne, 1980, pp. 19–20, 64; C. Margot, *Doctoral Thesis*, Ecole Polytechnique Fédérale de Lausanne, 1985, pp. 15–18, 84–85.

51. H. D. Zook, J. A. Miller, *J. Org. Chem.* **36** (1971) 1112; C. Cambillau, P. Sarthou, G. Bram, *Tetrahedron Lett.* **17** (1976) 281; Y. Hara, M. Matsuda, *Bull. Chem. Soc. Jpn.* **49** (1976) 1126; R. Gomper, H. U. Wagner, *Angew. Chem.* **88** (1976) 389; *Angew. Chem., Int. Ed. Engl.* **15** (1976) 321; D. Seebach, *Angew. Chem.* **100** (1988) 1685; *Angew. Chem., Int. Ed. Engl.* **27** (1988) 1624.
52. H. Kolbe, *Ann. Chem. Pharm.* **113** (1860) 125; *J. Prakt. Chem. [2]* **10** (1874) 89; **11** (1875) 24; R. Schmitt, *J. Prakt. Chem. [2]* **31** (1885) 397; *Ber. Dtsch Chem. Ges.* **20** (1887) 2702; see also G. Casnati, G. Casiraghi, A. Pochini, G. Sartori, R. Ungaro, *Pure Appl. Chem.* **55** (1983) 1677.
53. G. Wittig, H. Reiff, *Angew. Chem.* **80** (1968); *Angew. Chem., Int. Ed. Engl.* **7** (1963) 7.
54. G. R. Kieczykowski, R. H. Schlessinger, R. B. Sulsky, *Tetrahedron Lett.* **8** (1967) 597; W. Oppolzer, W. Fröstl, *Helv. Chim. Acta* **58** (1975) 587; K. Takabe, H. Fujiwara, T. Katagiri, J. Tanaka, *Tetrahedron Lett.* **16** (1975) 1237.
55. M. Okubo, S. Ueda, *Bull. Chem. Soc. Jpn.* **53** (1980) 281.
56. G. Ciamician, P. Silber, *Ber. Dtsch. Chem. Ges.* **17** (1884) 1437; A. Pictet, *Ber. Dtsch. Chem. Ges.* **37** (1904) 2797; G. Ciamician, *Ber. Dtsch. Chem. Ges.* **37** (1904) 4225, 4238; B. Oddo, *Gazz. Chim. Ital.* **39** (1909) 649; *Ber. Dtsch. Chem. Ges.* **43** (1910) 1012; P. S. Skell, G. P. Bean, *J. Am. Chem. Soc.* **84** (1962) 4655.
57. M. Gaudemar, *Tetrahedron* **32** (1976) 1689.
58. D. J. Cram, F. A. A. Elhafez, *J. Am. Chem. Soc.* **74** (1952) 5828; D. J. Cram, K. R. Kopecky, *J. Am. Chem. Soc.* **81** (1959) 2748; T. A. Nguyên, *Top. Curr. Chem.* **88** (1980) 151.
59. C. W. Jefford, D. Jaggi, J. Boukouvalas, *Tetrahedron Lett.* **27** (1986) 4011.
60. M. Schlosser, *Pure Appl. Chem.* **60** (1988), 1627.
61. M. Schlosser, O. Desponds, R. Lehmann, E. Moret, G. Rauchschwalbe, *Tetrahderon* **49** (1993), 10175.
62. J. E. Nordlander, J. D. Roberts, *J. Am. Chem. Soc.* **81** (1959) 1769; P. West, J. I. Purmort, S. V. McKinley, *J. Am. Chem. Soc.* **90** (1968) 797; H. E. Zieger, J. D. Roberts, *J. Org. Chem.* **34** (1969) 1976.
63. M. Schlosser, J. Hartmann, V. David, *Helv. Chim. Acta* **57** (1974) 1567; M. Schlosser, J. Hartmann, *J. Am. Chem. Soc.* **98** (1976) 4674.
64. R. B. Bates, D. W. Gosselink, J. A. Kaczynski, *Tetrahedron Lett.* **7** (1967) 205.
65. M. Schlosser, G. Rauchschwalbe, *J. Am. Chem. Soc.* **100** (1978) 3258.
66. G. J. Heiszwolf, H. Kloosterziel, *Recl. Trav. Chim. Pays-Bas* **86** (1967) 807.
67. H. Yasuda, Y. Ohnuma, M. Yamauchi, H. Tani, A. Nakamura, *Bull. Chem. Soc. Jpn.* **52** (1979) 2036; H. Yasuda, A. Nakamura, *J. Organomet. Chem.* **285** (1985) 15; H. Yasuda, A. Nakamura, *Angew. Chem.* **99** (1987) 745, *Angew. Chem., Int. Ed. Engl.* **26** (1987) 723.
68. H. Bosshardt, M. Schlosser, *Helv. Chim. Acta* **63** (1980) 2393.
69. M. Schlosser, *Tetrahedron* **34** (1978) 3; G. Wolf, E. U. Würthwein, *Chem. Ber.* **124** (1991) 889.
70. D. A. Evans, J. V. Nelson, T. R. Taber, *Top. Stereochem.* **13** (1982) 1; M. Braun, *Angew. Chem.* **99** (1987) 24; *Angew. Chem., Int. Ed. Engl.* **26** (1987) 24; C. H. Heathcock, in *Modern Synthetic Methods*, ed R. Scheffold, Vol 6. p. 1, spec. 51–84, 1992.

4 Preparation of Organometallic Reagents and Intermediates

1. M. Schlosser, *Angew. Chem.* **76** (1964) 124; 258; *Angew. Chem., Int. Ed. Engl.* **3** (1964) 287, 362.

2. K. Nützel, in *Houben/Weyl: Methoden der Organischen Chemie*, ed. E. Müller, Vol. 13/2a, pp. 47–527, Georg Thieme, Stuttgart, 1973.
3. K. Nützel, in *Houben/Weyl: Methoden der Organischen Chemie*, ed. E. Müller, Vol. 13/2a, pp. 553–858, Georg Thieme, Stuttgart, 1973.
4. R. D. Rieke, *Acc. Chem. Res.* **10** (1977) 301.
5. K. J. Klabunde, *Chemistry of Free Atoms and Particles*, Academic Press, New York, 1980; D. N. Cox, R. R. Roulet, *Organometallics* **4** (1985) 2001; **5** (1986) 1886; *J. Organomet. Chem.* **342** (1988) 87; s.a.: P. S. Skell, M. J. McGlinchey, *Angew. Chem.* **87** (1975), 215; *Angew. Chem. Int. Ed. Engl.* **14** (1975), 195; W. Reichelt, *Angew. Chem.* **87** (1975), 239; *Angew. Chem. Int. Ed. Engl.* **14** (1975), 218; K. J. Klabunde, *Angew. Chem.* **87** (1975), 309; *Angew. Chem. Int. Ed. Engl.* **14** (1975), 287; E. P. Kündig, M. Moskovits, G. A. Ozin, *Angew. Chem.* **87** (1975), 314; *Angew. Chem. Int. Ed. Engl.* **14** (1975), 292.
6. E. Linnemann, *Ber. Dtsch. Chem. Ges.* **10** (1877) 1111; R. L. Shriner, F. W. Neumann, *Org. Snth, Coll. Vol.* **3** (1955), 73; R. L. Frank, P. V. Smith, *Org. Synth., Coll. Vol.* **3** (1955), 73.
7. R. Wilkinson, *J. Chem. Soc.* (1931) 3057; see also C. R. Noller, *Org. Synth., Coll. Vol.* **2** (1943) 184; G. F. Hennion, J. J. Sheehan, *J. Am. Chem. Soc.* **71** (1949) 1964.
8. K. S. Suslick, in *Modern Synthetic Methods*, ed. R. Scheffold, Vol. 4, p. 1, Springer, Berlin, 1986; K. S. Suslick (Ed.), *Ultrasound: Its Chemical, Physical and Biological Effects*, VCH, New York, 1988.
9. D. E. Pearson, D. Cowan, J. D. Beckler, *J. Org. Chem.* **24** (1959) 504.
10. J. F. Garst, F. Ungváry, R. Batlaw, K. E. Lawrence, *J. Am. Chem. Soc.* **113** (1991) 6697.
11. C. Walling, *Acc. Chem. Res.* **24** (1991) 255; see also J. F. Garst, J. M. Deutch, G. M. Whitesides, *J. Am. Chem. Soc.* **108** (1986) 2490; J. F. Garst, B. L. Swift, D. W. Smith, *J. Am. Chem. Soc.* **111** (1989) 234.
12. H. M. Walborsky, *Acc. Chem. Res.* **23** (1990) 286; see also M. S. Kharasch, O. Reinmuth, *Grignard Reactions of Non-Metallic Substances*, Prentice Hall, Englewood Cliffs, NJ, 1954.
13. D. Bryce-Smith, *J. Chem. Soc.* (1956) 1603; H. R. Ward, R. G. Lawler, R. A. Cooper, *J. Am. Chem. Soc.* **91** (1969) 746; A. R. Lepley, R. L. Landau, *J. Am. Chem. Soc.* **91** (1969) 748; G. A. Russell, D. W. Lamson, *J. Am. Chem. Soc.* **91** (1969) 3967; J. Sauer, W. Braig, *Tetrahedron Lett.* **10** (1969) 4275.
14. K. Ziegler, H. Colonius, *Liebigs Ann. Chem.* **479** (1930) 136.
15. M. Stiles, R. P. Mayer, *J. Am. Chem. Soc.* **81** (1959) 1501, footnote 38b; see also J. B. Wright, E. S. Gutsell, *J. Am. Chem. Soc.* **81** (1959) 5193, footnote 10; C. W. Kamienski, D. L. Esmay, *J. Org. Chem.* **25** (1960) 1807.
16. W. L. Borkowski, to Foote Mineral Co., *AP* 3 293 313 (1966); *Chem. Abstr.* **66** (1967) 46477b.
17. C. Giancaspro, G. Sleiter, *J. Prakt. Chem.* **321** (1979), 876.
18. G. Wittig, *Angew. Chem.* **53** (1940) 242; H. Gilman, J. W. Morton, *Org. React.* **8** (1954) 286.
19. H. Gilman, T. S. Soddy, *J. Org. Chem.* **22** (1957), 565; H. Gilman, B. J. Gau, *J. Org. Chem.* **22** (1957), 1165; D. L. Esmay, *Adv. Chem. Ser.* **23**, 47, *Am. Chem. Soc.*, Washington, 1959; R. Waack, M. A. Doran, *J. Org. Chem.* **32** (1967), 3396.
20. U. Schöllkopf, H. Küppers, H. J. Traencker, W. Pitteroff, *Liebigs Ann. Chem.* **704** (1967) 120.
21. J. P. Dulcère, J. Grimaldi, M. Santelli, *Tetrahedron Lett.* **22** (1981) 3179.
22. C. B. Wooster, *Chem. Rev.* **11** (1932) 37, 48; E. DeBoer, *Adv. Organomet. Chem.* **2** (1964) 115; E. T. Kaiser, L. Kevan (Eds), *Radical Ions*, Wiley-Interscience, New York, 1968.
23. J. Barluenga, J. Flórez, M. Yus, *J. Chem. Soc., Chem. Commun.* (1982) 1153; J. Barluenga,

J. R. Fernández, M. Yus, *J. Chem. Soc., Perkin Trans. 1* (1988) 302; J. Barluenga, F. Foubelo, F. J. Fananás, *Tetrahedron* **45** (1989) 2183.
24. P. K. Freeman, L. L. Hutchinson, *J. Org. Chem.* **45** (1980), 1924; **48** (1983) 4705; R. E. Ireland, *J. Am. Chem. Soc.* **107** (1985) 3285; H. Choi, *J. Chem. Soc., Chem. Commun.* (1987) 225.
25. J. Stapersma, G. W. Klumpp, *Tetrahedron* **37** (1981) 187.
26. K. Ziegler, *Angew. Chem.* **49** (1936) 459.
27. M. Bockmühl G. Ehrhart, to Farbwerke Hoechst, *Ger. Pat.* 622 875 (1935), 633 083 (1936); see also H. Ruschig, R. Fugmann, W. Meixner, *Angew. Chem.* **70** (1958) 71.
28. A. A. Morton, G. M. Richardson, A. T. Hallowell, *J. Am. Chem. Soc.* **63** (1941) 327.
29. A. A. Morton, J. B. Davidson, H. A. Newey, *J. Am. Chem. Soc.* **64** (1942) 2240.
30. A. A. Morton, H. A. Newey, *J. Am. Chem. Soc.* **64** (1942) 2248.
31. A. A. Morton, M. L. Brown, M. E. T. Holden, R. L. Letsinger, E. E. Magat, *J. Am. Chem. Soc.* **67** (1945) 2224.
32. A. A. Morton, J. B. Davidson, R. J. Best, *J. Am. Chem. Soc.* **64** (1942) 2239.
33. R. N. Meals, *J. Org. Chem.* **9** (1944) 211.
34. A. A. Morton, E. J. Lanpher, *J. Org. Chem.* **23** (1958) 1636.
35. R. G. Anderson, M. B. Silverman, D. M. Ritter, *J. Org. Chem.* **23** (1958) 750.
36. C. D. Broaddus, *J. Org. Chem.* **29** (1964) 2689.
37. A. Maercker, *Angew. Chem.* **99** (1987) 1002; *Angew. Chem., Int. Ed. Engl.* **26** (1987) 972.
38. E. Grovenstein, E. P. Blanchard, D. A. Gordon, R. W. Stevenson, *J. Am. Chem. Soc.* **81** (1959) 4842.
39. E. Bartmann, *J. Organomet. Chem.* **284** (1985) 149; see also E. Bartmann, *Angew. Chem.* **98** (1986) 629; *Angew. Chem., Int. Ed. Engl.* **25** (1986) 653.
40. B. Mudryk, T. Cohen, *J. Am. Chem. Soc.* **113** (1991) 1866.
41. B. Mudryk, T. Cohen, *J. Org. Chem.* **56** (1991) 5760.
42. Z. Jedlinsky, M. Kowalczuk, A. Misiołek, *J. Chem. Soc., Chem. Commun.* (1988) 1261; Z. Jedlinsky, A. Misiołek, P. Kurcok, *J. Org. Chem.* **54** (1989) 1500.
43. K. Freudenberg, F. Klinck, E. Flickinger, A. Sobek, *Ber. Dtsch. Chem. Ges.* **72** (1939) 217; K. Freudenberg, W. Lautsch, G. Piazolo, *Ber. Dtsch. Chem. Ges.* **74** (1941) 1879, spec. 1886; see also C. D. Hurd, G. Oliver, *J. Am. Chem. Soc.* **81** (1959) 2795.
44. Another argument could be based on the solvent dependence of the metal–oxygen bond polarity and, as a corollary, the resonance stabilization of the phenolate: M. Schlosser, *Struktur und Reaktivität Polarer Organometalle*, pp. 102–103, Springer, Berlin, 1973.
45. R. R. Dewald, N. J. Conlon, W. M. Song, *J. Org. Chem.* **54** (1989) 261.
46. R. L. Letsinger, J. G. Traynham, *J. Am. Chem. Soc.* **70** (1948) 3342.
47. A. A. Morton, E. E. Magat, R. L. Letsinger, *J. Am. Chem. Soc.* **69** (1947) 950, spec. 959.
48. J. J. Eisch, A. M. Jacobs, *J. Org. Chem.* **28** (1963) 2145.
49. V. Rautenstrauch, *Helv. Chim. Acta* **57** (1974) 496.
50. H. Gilman, H. A. McNinch, D. Wittenberg, *J. Org. Chem.* **23** (1958) 2044.
51. H. Gilman, G. L. Schwebke, *J. Org. Chem.* **27** (1962) 4259.
52. P. T. Lansbury, V. A. Pattison, *J. Org. Chem.* **27** (1962) 1933.
53. K. Ziegler, B. Schnell, *Liebigs Ann. Chem.* **437** (1924) 227.
54. K. Ziegler, H. Dislich, *Chem. Ber.* **90** (1957) 1107.
55. K. Ziegler, F. Thielmann, *Ber. Dtsch. Chem. Ges.* **56** (1923) 1740.
56. G. Wittig, W. Happe, *Liebigs Ann. Chem.* **557** (1947) 205.
57. A. Kliegl, *Ber. Dtsch. Chem. Ges.* **62** (1929) 1327.

58. C. E. Claff, A. A. Morton, *J. Org. Chem.* **20** (1955) 440.
59. M. Laguerre, J. Dunoguès, N. Duffault, R. Calas, *J. Organomet. Chem.* **193** (1980) C17.
60. H. Gilman, J. J. Dietrich, *J. Org. Chem.* **22** (1957) 851; *J. Am. Chem. Soc.* **80** (1958) 380.
61. K. Ziegler, F. Thielmann, *Ber. Dtsch. Chem. Ges.* **56** (1923) 1740; R. Gerdil, E. A. C. Lucken, *J. Chem. Soc.* (1963) 2857, 5444; (1964) 3916; C. Rücker, H. Prinzbach, *Tetrahedron Lett.* **24** (1983) 4099; R. K. Boeckman, E. J. Enholm, D. M. Demko, A. B. Charette, *J. Org. Chem.* **51** (1986) 4743; C. Rücker, *J. Organomet. Chem.* **310** (1986) 135; B.-S. Guo, W. Doubleday, T. Cohen, *J. Am. Chem. Soc.* **109** (1987) 4710; S. D. Rychnovsky, D. E. Mickus, *Tetrahedron Lett.* **30** (1989) 3011; S. D. Rychnovsky, *J. Org. Chem.* **54** (1989) 4982.
62. T. Cohen, M. Bhupathy, *Acc. Chem. Res.* **22** (1989) 152.
63. C. G. Screttas, I. C. Smonou, *J. Organomet. Chem.* **342** (1988) 143.
64. R. W. Hoffmann, B. Kemper, *Tetrahedron Lett.* **22** (1981) 5263.
65. A. Schönberg, E. Petersen, H. Kaltschmitt, *Ber. Dtsch. Chem. Ges.* **66** (1933) 233; T. Cohen, R. B. Weisenfeld, *J. Org. Chem.* **44** (1979) 3601; A. Krief, B. Kenda, P. Barbereau, *Tetrahedron Lett.* **32** (1991) 2509; see also D. Pandy-Szekeres, G. Déléris, J. P. Piccard, J. P. Pillot, R. Calas, *Tetrahedron Lett.* **21** (1980) 4267.
66. C. S. Shiner, T. Tsunoda, B. A. Goodman, S. Ingham, S.-H. Lee, P. E. Vorndam, *J. Am. Chem. Soc.* **111** (1989) 1381.
67. T. Tsunoda, K. Fujiwara, Y. Yamamoto, S. Itô, *Tetrahedron Lett.* **32** (1991) 1975.
68. T. Cohen, T. R. Matz, *J. Am. Chem. Soc.* **102** (1980) 6900; T. Cohen, M. T. Lin, *J. Am. Chem. Soc.* **106** (1984) 1130.
69. T. Cohen, T. R. Matz, *Synth. Commun.* **10** (1980) 311.
70. Corresponding sodium species: S. Bank, M. Platz, *Tetrahedron Lett.* **14** (1973) 2097.
71. H. Gilman, J. J. Dietrich, *J. Am. Chem. Soc.* **80** (1958) 380.
72. D. Wittenberg, H. Gilman, *J. Org. Chem.* **23** (1958) 1063.
73. A. M. Agular, J. Giacin, A. Mills, *J. Org. Chem.* **27** (1962) 674; A. M. Agular, J. Beisler, A. Mills, *J. Org. Chem.* **27** (1962) 1001.
74. M. Berthelot, *Ann. Chim. Phys. (Paris) [4]* **9** (1866) 385; C. Matignon, *C.R. Acad. Sci.* **124** (1897) 775, 1026; **125** (1897) 1033.
75. H. Moissan, *C.R. Acad. Sci.* **127** (1898) 911; P. Lebeau, M. Picon, *C.R. Acad. Sci.* **156** (1913) 1077; T. H. Vaughn, G. F. Hennion, R. R. Vogt, J. A. Nieuwland, *J. Org. Chem.* **2** (1938) 1; T. L. Jacobs, *Org. React.* **5** (1949) 1, spec. 26; J. H. Saunders, *Org. Synth., Coll. Vol.* **3** (1955) 416; A. Fisch, J. M. Coisne, H. P. Figeys, *Synthesis* (1982) 211.
76. K. Hess, H. Munderloh, *Ber. Dtsch. Chem. Ges.* **51** (1918) 377; A. v. Antropoff, J. F. Müller, *Z. Anorg. Allg. Chem.* **204** (1932) 305; see also G. F. Hennion, E. P. Bell, *J. Am. Chem. Soc.* **65** (1943) 1847; T. F. Rutledge, *J. Org. Chem.* **22** (1957) 649; R. Nast, K. Vetter, *Z. Anorg. Allg. Chem.* **279** (1955) 146; E. Weiss, H. Plass, *Ber. Dtsch. Chem. Ges.* **101** (1968) 2947.
77. M. Corbellini, L. Turner, *Chim. Ind. (Milan)* **42** (1960) 251; *Chem. Abstr.* **54** (1960) 19250.
78. C. Glaser, *Liebigs Ann. Chem.* **154** (1870) 137, 161; J. V. Nef, *Liebigs Ann. Chem.* **308** (1899) 264; C. Moureu, H. Desmots, *Bull. Soc. Chim. Fr. [3]* **27** (1902) 360; P. Ivitzky, *Bull. Soc. Chim. Fr.* (1924) 357; H. Gilman, R. V. Young, *J. Org. Chem.* **1** (1936) 315; A. L. Henne, K. W. Greenlee, *J. Am. Chem. Soc.* **65** (1943) 2020; B. B. Elsner, P. F. M. Paul, *J. Chem. Soc.* (1951) 893.
79. J. Thiele, *Ber. Dtsch. Chem. Ges.* **43** (1901) 68.
80. K. Ziegler, H. Froitzheim-Kühlhorn, K. Hafner, *Chem. Ber.* **89** (1956) 434; see also K. Hafner, H. Kaiser, *Liebigs Ann. Chem.* **618** (1958) 140; G. Wilkinson, *Org. Synth., Coll.*

Vol. **4** (1963) 473; R. B. King (Ed.), *Organometallic Syntheses*, Vol. 1, p. 64, Academic Press, New York, 1965.
81. E. O. Fischer, W. Hafner, H. O. Stahl, *Z. Anorg. Allg. Chem.* **282** (1955) 45.
82. R. Weissgerber, *Ber. Dtsch. Chem. Ges.* **42** (1909) 569; R. Riemschneider, W. Grunow, *Monatsh. Chem.* **92** (1961) 1191; A. Bosch, R. K. Brown, *Can. J. Chem.* **42** (1964) 1718.
83. H. Gilman, R. D. Gorsich, *J. Org. Chem.* **23** (1958) 550; J. J. Eisch, W. Kaska, *J. Org. Chem.* **27** (1962) 3745.
84. R. Weissgerber, *Ber. Dtsch. Chem. Ges.* **41** (1908) 2913.
85. R. Meier, *Chem. Ber.* **86** (1953) 1483; G. W. H. Scherf, R. K. Brown, *Can. J. Chem.* **38** (1960) 2450.
86. C. A. Kraus, R. Rosen, *J. Am. Chem. Soc.* **47** (1925) 2739; C. B. Wooster, N. W. Mitchell, *J. Am. Chem. Soc.* **52** (1930) 688.
87. H. O. House, V. Kramar, *J. Org. Chem.* **27** (1962) 4146.
88. J. de Postis, *C.R. Acad. Sci.* **222** (1946) 398; **224** (1947) 579.
89. H. Normant, T. Cuvigny, J. Normant, B. Angelo, *Bull. Soc. Chim. Fr.* (1965) 1561, 3441, 3446.
90. C. E. Claff, A. A. Morton, *J. Org. Chem.* **20** (1955) 981.
91. H. Gilman, F. Breuer, *J. Am. Chem. Soc.* **56** (1934) 1123.
92. K. Ziegler, F. Thielmann, *Ber. Dtsch. Chem. Ges.* **56** (1923) 1740, spec. 1745; J. W. Schick, H. D. Hartough, *J. Am. Chem. Soc.* **70** (1948) 286.
93. A. Schönberg, E. Petersen, H. Kaltschmitt, *Ber. Dtsch. Chem. Ges.* **66** (1933) 233.
94. K. Mochida, K. Kojima, Y. Yoshida, *Bull. Chem. Soc. Jpn.* **60** (1987) 2255.
95. L. Hackspill, *Ann. Chim. Phys. (Paris) [8]* **28** (1913) 653.
96. E. Grovenstein, T. H. Longfield, D. E. Quest, *J. Am. Chem. Soc.* **99** (1977) 2800.
97. J. B. Wilkes, *J. Org. Chem.* **32** (1967) 3231.
98. H. Yasuda, Y. Ohnuma, M. Yamauchi, H. Tani, A. Nakamaura, *Bull. Chem. Soc. Jpn.* **52** (1979) 2036.
99. V. Rautenstrauch, *Angew. Chem.* **87** (1975) 254; *Angew. Chem., Int. Ed. Engl.* **14** (1975) 259; see also J. R. V. E. Hommes, F. Bickelhaupt, G. W. Klumpp, *Angew. Chem.* **100** (1988), 1100; *Angew. Chem. Int. Ed. Engl.* **27** (1988), 1083; A. Maercker, B. Grebe, *J. Organomet. Chem.* **334** (1987), C 21.
100. L. Hackspill, R. Rohmer, *C.R. Acad. Sci. [2]* **217** (1943) 152.
101. B. Bogdanović, A. Cordi, H. Stepowska, P. Locatelli, B. Wermeckes, U. Wilczok, K. Haertel, H. Richtering, *Chem. Ber.* **117** (1984) 42.
102. B. Bogdanović, B. Werneckes, *Angew. Chem.* **93** (1981) 691; *Angew. Chem., Int. Ed. Engl.* **20** (1981), 684
103. A. Maercker, K. D. Klein, *J. Organomet. Chem.* **410** (1991) C35.
104. A. Maercker, W. Berkulin, P. Schiess, *Angew. Chem.* **95** (1983) 248; *Angew. Chem., Int. Ed. Engl.* **22** (1938) 246; *Tetrahedron Lett.* **25** (1984) 1701.
105. K. N. Campbell, L. T. Eby, *J. Am. Chem. Soc.* **63** (1941), 216, 2683; N. A. Dobson, R. A. Raphael, *J. Chem. Soc.* **1955**, 3558.
106. A. Maercker, T. Graule, U. Girreser, *Angew. Chem.* **98** (1986) 174; *Angew. Chem., Int. Ed. Engl.* **25** (1986) 167.
107. W. Schlenk, E. Marcus, *Ber. Dtsch. Chem. Ges.* **47** (1914) 1664.
108. J. G. Stampfli, C. S. Marvel, *J. Am. Chem. Soc.* **53** (1931) 4057.
109. J. B. Conant, B. S. Garvey, *J. Am. Chem. Soc.* **49** (1927) 2599.
110. G. Wittig, E. Stahnecker, *Liebigs Ann. Chem.* **605** (1957) 69.
111. G. Wittig, H. H. Schlör, *Suomen Kem. B* **31** (1958) 2; *Chem. Abstr.* **53** (1959) 12183.

112. G. Wittig, M. Leo, *Ber. Dtsch. Chem. Ges.* **63** (1930) 943.
113. P. L. Salzberg, C. S. Marvel, *J. Am. Chem. Soc.* **50** (1928) 1737.
114. E. Grovenstein, A. M. Bhatti, D. E. Quest, D. Sengupta, D. VanDerveer, *J. Am. Chem. Soc.* **105** (1983), 6290.
115. H. Gilman, R. D. Gorsich, *J. Org. Chem.* **23** (1958) 550; T. D. Walsh, T. L. Megremis, *J. Am. Chem.* **103** (1981), 3897.
116. W. Theilacker, E. Möllhoff, *Angew. Chem.* **74** (1962) 781; *Chem. Abstr.* **58** (1963), 5596a; for the cleavage of strained small-ring compounds see, e.g., A. Maercker, W. Berkulin, P. Schiess, *Angew. Chem.* **95** (1983), 248; *Angew. Chem. Int. Ed. Engl.* **22** (1983), 246; A. Maercker, K. D. Klein, *Angew. Chem.* **101** (1989), 63; *Angew. Chem. Int. Ed. Engl.* **28** (1989), 83; *J. Organomet. Chem.* **401** (1991), C 1; A. Maercker, W. Berkulin, *Chem. Ber.* **123** (1990), 185.
117. J. A. Wanklyn, *Liebigs Ann. Chem.* **107** (1858) 125; **108** (1858) 68; **111** (1859) 234; A. v. Grosse, *Ber. Dtsch. Chem. Ges.* **59** (1926) 2646.
118. G. B. Buckton, *Liebigs Ann. Chem.* **112** (1859) 220; C. F. Acree, *Am. Chem. J.* **29** (1903) 588; *Chem. Zentralbl.* **II** (1903) 195; P. Schorigin, *Ber. Dtsch. Chem. Ges.* **41** (1908) 2711, 2717, 2723.
119. W. Schlenk, J. Holtz, *Ber. Dtsch. Chem. Ges.* **50** (1917) 262.
120. D. Braun, W. Betz, W. Kern, *Makromol. Chem.* **42** (1960) 89.
121. H. E. Zimmerman, A. Zweig, *J. Am. Chem. Soc.* **83** (1961) 1196.
122. G. Wittig, F. J. Meyer, G. Lange, *Liebigs Ann. Chem.* **571** (1951) 200; R. Waack, M. A. Doran, *J. Am. Chem. Soc.* **85** (1963) 1651.
123. D. E. Applequist, D. F. O'Brien, *J. Am. Chem. Soc.* **85** (1963) 743.
124. A. N. Nesmeyanov, A. E. Borisov, I. S. Saveleva, E. I. Golubeva, *Izv. Akad. Nauk SSSR, Ser. Khim.* (1958) 1490; *Chem. Abstr.* **53** (1959) 7973a.
125. G. Wittig, E. Benz. *Ber. Dtsch. Chem. Ges.* **91** (1958) 879; see also H. J. S. Winkler, H. Winkler, *J. Am. Chem. Soc.* **88** (1966) 968.
126. G. Wittig, F. Bickelhaupt, *Chem. Ber.* **91** (1958) 883.
127. G. Wittig, W. Herwig, *Chem. Ber.* **87** (1954) 1511; **88** (1955) 972.
128. W. H. Carothers, D. D. Coffman, *J. Am. Chem. Soc.* **52** (1930) 1254.
129. W. H. Carothers, D. D. Coffman, *J. Am. Chem. Soc.* **51** (1929) 588.
130. F. C. Whitmore, H. D. Zook, *J. Am. Chem. Soc.* **64** (1942) 1783.
131. H. Gilman, R. H. Kirby, *J. Am. Chem. Soc.* **58** (1936) 2074.
132. H. Gilman, J. C. Bailie, *J. Org. Chem.* **2** (1937) 84; see also M. Morton, A. A. Rembaum, J. L. Hall, *J. Polymer Sci.* **A1** (1963) 461; *Chem. Abstr.* **59** (1963), 8876h.
133. E. Weiss, G. Sauermann, *Angew. Chem.* **80** (1968) 123; *Angew. Chem., Int. Ed. Engl.* **7** (1968) 133.
134. A. A. Morton, R. L. Letsinger, *J. Am. Chem. Soc.* **69** (1974) 172.
135. A. J. Hart, D. H. O'Brien, C. R. Russell, *J. Organomet. Chem.* **72** (1974) C19; J. Hartmann, M. Schlosser, *Synthesis* (1975) 328; *Helv. Chim. Acta* **59** (1976) 453.
136. M. Stähle, M. Schlosser, *J. Organomet. Chem.* **220** (1981) 277.
137. H. Gilman, R. V. Young, *J. Org. Chem.* **1** (1936) 315.
138. D. Bryce-Smith, *J. Chem. Soc.* (1954) 1079; (1963) 5983.
139. D. Bryce-Smith, E. E. Turner, *J. Chem. Soc.* (1953) 861.
140. W. Schlenk, J. Holtz, *Ber. Dtsch. Chem. Ges.* **50** (1917) 271; see also F. Hein, E. Petzchner, K. Wagler, A. Segitz, *Z. Anorg. Allg. Chem.* **141** (1924) 161; E. Weiss, *Chem. Ber.* **97** (1964) 3241.
141. L. M. Seitz, T. L. Brown, *J. Am. Chem. Soc.* **88** (1966) 2174.

142. D. Y. Curtin, W. J. Koehl, *J. Am. Chem. Soc.* **84** (1962) 1967.
143. U. Schöllkopf, F. Gerhart, *Angew. Chem.* **79** (1967) 819; *Angew. Chem., Int. Ed. Engl.* **6** (1967) 805.
144. A. Maercker, M. Theis, *Angew. Chem.* **96** (1984), 990; *Angew. Chem. Int. Ed. Engl.* **23** (1984), 995.
145. C. Chung, R. J. Lagow, *J. Chem. Soc., Chem. Commun.* **1972**, 1078.
146. E. C. Juenge, D. Seyferth, *J. Org. Chem.* **26** (1961) 563.
147. D. Seyferth, G. J. Murphy, R. A. Woodruff, *J. Organomet. Chem.* **66** (1974) C29; *J. Am. Chem. Soc.* **96** (1974) 5011.
148. D. Seyferth, H. M. Cohen, *Inorg. Chem.* **2** (1963) 625; see also A. Maercker, T. Graule, W. Demuth, *Angew. Chem.* **99** (1987), 1075; *Angew. Chem. Int. Ed. Engl.* **26** (1987), 1023.
149. D. Seyferth, M. A. Weiner, *J. Am. Chem. Soc.* **83** (19621) 3583.
150. D. Seyferth, L. G. Vaughan, *J. Am. Chem. Soc.* **86** (1964) 883.
151. D. Seyferth, D. E. Welch, G. Raab, *J. Am. Chem. Soc.* **84** (1962) 4266.
152. E. Piers, V. Karunaratne, *Tetrahedron* **45** (1989) 1089.
153. D. Seyferth, M. A. Weiner, *J. Org. Chem.* **24** (1959) 1395; **26** (1961) 4797; D. Seyferth, T. F. Jula, *J. Organomet. Chem.* **8** (1967) P13.
154. R. Waack, M. A. Doran, *J. Org. Chem.* **32** (1967) 3395.
155. H. Gilman, S. D. Rosenberg, *J. Org. Chem.* **24** (1959) 2063.
156. H. J. R. de Boor, O. S. Akkerman, F. Bickelhaupt, *Organometallics* **9** (1990) 2898.
157. O. Desponds, M. Schlosser, *J. Organomet. Chem.* **409** (1991) 93.
158. G. E. Keck, J. H. Byers, A. M. Tafesh, *J. Org. Chem.* **54** (1988), 1127; V. Farina, S. I. Hauck, *J. Org. Chem.* **56** (1991) 4317.
159. A. J. Ashe, L. L. Lohr, S. M. Al-Taweel, *Organometallics* **10** (1991), 2424.
160. U. Schöllkopf, in *Houben-Weyl: Methoden der Organischen Chemie*, ed. E. Müller, Vol. 13/1, p. 133, footnote 8, Georg Thieme, Stuttgart, 1970; D. Peterson, *Organomet. Chem. Rev., Section A* **7** (1972) 295; W. C. Still, *J. Am. Chem. Soc.* **100** (1978) 1481; W. C. Still, A. Mitra, *J. Am. Chem. Soc.* **100** (1978) 1927; W. C. Still, C. Sreekumar, *J. Am. Chem. Soc.* **102** (1980) 1201.
161. U. Schollkopf, *Angew. Chem.* **82** (1970), 795; *Angew. Chem. Int. Ed. Engl.* **9** (1970), 763; T. Nakai, K. Michami, *Chem. Rev.* **86** (1986), 885.
162. N. Meyer, D. Seebach, *Chem. Ber.* **113** (1980) 1290.
163. T. K. Jones, S. E. Denmark, *J. Org. Chem.* **50** (1985) 4037.
164. J. S. Sawyer, T. L. McDonald, G. J. McGarvey, *J. Am. Chem. Soc.* **106** (1984), 3376; C. R. Johnson, J. R. Medich, *J. Org. Chem.* **53** (1988) 4131.
165. S. D. Burke, S. A. Shearouse, D. J. Burch, R. W. Sutton, *Tetrahedron Lett.* **21** (1980) 1285.
166. D. K. Hutchinson, P. L. Fuchs, *J. Am. Chem. Soc.* **109** (1987) 4930.
167. G. J. McGarvey, M. Kimura, *J. Org. Chem.* **50** (1985) 4655.
168. J. S. Sawyer, A. Kucerovy, T. L. McDonald, G. J. Garvey, *J. Am. Chem. Soc.* **110** (1988) 842.
169. O. J. Taylor, J. L. Wardell, *J. Chem. Res. (S)* (1989) 98.
170. Reviews: M. Schlosser, *Angew. Chem.* **76** (1964) 124, 258; *Angew. Chem., Int. Ed. Engl.* **3** (1964) 287, 362.
171. M. Schlosser, J. Hartmann, *Angew. Chem.* **85** (1973) 544; *Angew. Chem., Int. Ed. Engl.* **12** (1973) 508.
172. M. Schlosser, M. Stähle, *Angew. Chem.* **94** (1982) 142; *Angew. Chem., Int. Ed. Engl.* **21** (1982) 145; *Angew. Chem., Suppl.* (1982) 198.
173. L. Lochmann, J. Pospísil, D. Lím, *Tetrahedron Lett.* **7** (1966) 257; L. Lochmann, J. Trekoval, *J. Organomet. Chem.* **99** (1975) 329.

174. M. Schlosser, J. Hartmann, *J. Am. Chem. Soc.* **98** (1976) 4674.
175. Reviews: R. G. Jones, H. Gilman, *Org. React.* **6** (1951) 339; W. F. Bailey, J. J. Patricia, *J. Organomet. Chem.* **352** (1988) 1.
176. D. E. Applequist, D. F. O'Brien, *J. Am. Chem. Soc.* **85** (1963) 743.
177. R. R. Sauers, S. B. Schlosberg, P. E. Pfeffer, *J. Org. Chem.* **33** (1968) 2175.
178. E. D. Bergmann, I. Shahak, Z. Aizenshtat, *Tetrahedron Lett.* **10** (1969) 2007.
179. W. E. Parham, L. D. Jones, Y. A. Sayed, *J. Org. Chem.* **41** (1976) 1184.
180. Review: G. Köbrich, *Angew. Chem.* **79** (1967) 15; *Angew. Chem., Int. Ed. Engl.* **6** (1967) 41.
181. D. Seyferth, G. J. Murphy, R. A. Woodruff, *J. Organomet. Chem.* **66** (1974), C29.
182. F. Tellier, R. Sauvêtre, J. F. Normant, Y. Dromzee, Y. Jeannin, *J. Organomet. Chem.* **331** (1987) 281; T. Dubuffet, R. Sauvêtre, J. F. Normant, *J. Organomet. Chem.* **341** (1988) 11.
183. A. Solladié-Cavallo, S. Quazzotti, *J. Fluorine Chem.* **46** (1990), 221; *J. Organomet. Chem.* **46** (1990) 221.
184. A. A. Morton, J. B. Davidson, B. L. Hakon, *J. Am. Chem. Soc.* **64** (1942) 2242.
185. C. D. Broaddus, T. J. Logan, T. J. Flautt, *J. Org. Chem.* **28** (1963) 1174.
186. A. G. Lindstone, I. A. Morris, *Chem. Ind. (London)* (1958) 560; see also H. Gilman, F. W. Moore, O. Baine, *J. Am. Chem. Soc.* **63** (1941) 2479.
187. G. Wittig, V. Wahl, G. Kolb, unpublished results; see also G. Kolb, *Doctoral Thesis*, University of Heidelberg, 1958.
188. G. Wittig, U. Schöllkopf, *Tetrahedron* **3** (1958) 91.
189. H. J. Reich, N. H. Phillips, I. L. Reich, *J. Am. Chem. Soc.* **107** (1985) 4101.
190. W. B. Farnham, J. C. Calabrese, *J. Am. Chem. Soc.* **108** (1986) 2449.
191. O. R. Pierce, E. T. McBee, G. F. Judd, *J. Am. Chem. Soc.* **76** (1954) 474; T. F. McGrath R. Levine, *J. Am. Chem. Soc.* **77** (1955) 3656; P. G. Gassman, N. J. O'Reilly, *J. Am. Chem. Soc.* **52** (1987) 2481.
192. M. Schlosser, V. Ladenberger, *Chem. Ber.* **100** (1967) 3893.
193. T. Ishihara, K. Hayashi, T. Ando, H. Yamanaka, *J. Org. Chem.* **40** (1975) 3264.
194. A. Schmidt, G. Köbrich, *Tetrahedron Lett.* **15** (1974) 2561; see also K. Kitatani, T. Hiyama, H. Nozaki, *J. Am. Chem. Soc.* **97** (1975), 949; D. Seyfert, R. L. Lambert, M. Massol, *J. Organomet. Chem.* **88** (1975), 255; H. J. J. Loozen, W. A. M. Castenmiller, E. J. M. Buter, H. M. Buck, *J. Org. Chem.* **41** (1976), 2965; H. J. J. Loozen, W. M. M. Robben, H. M. Buck, *Recl. Trav. Chim. Pays-Bas* **95** (1976), 245.
195. R. C. Gadwood, *Tetrahedron Lett.* **25** (1984) 5851.
196. M. Schlosser, K. S. Y. Lau, *J. Org. Chem.* **43** (1978); 1595; see also J. Ficini, J. C. Depezay, *Tetrahedron Lett.* **9** (1968) 937.
197. L. Duhamel, J. Gralak, B. Ngono, *J. Organomet. Chem.* **363** (1989) C4.
198. J. Ficini, J. C. Depezay, *Bull. Soc. Chim. Fr.* (1966) 3878.
199. J. Ficini, J. C. Depezay, *Tetrahedron Lett.* **10** (1969) 4797; J. C. Depezay, Y. LeMerrer, *Tetrahedron Lett.* **15** (1974) 2751;2755; J. C. Depezay, Y. LeMerrer, M. Sanière, *Synthesis* (1985), 766.
200. M. J. Manning, P. W. Raynolds, J. S. Swenton, *J. Am. Chem. Soc.* **98** (1976) 5008.
201. E. J. Corey, G. N. Widiger, *J. Org. Chem.* **40** (1975), 2975.
202. M. Schlosser, E. Hammer, *Helv. Chim. Acta* **57** (1974) 2547.
203. C. J. Kowalski, A. E. Weber, K. W. Fields, *J. Org. Chem.* **47** (1982) 5088.
204. C. Neukom, D. P. Richardson, J. H. Myerson, P. A. Bartlett, *J. Am. Chem. Soc.* **108** (1986) 5559.
205. A. I. Meyers, R. F. Spohn, *J. Org. Chem.* **50** (1985) 4872.
206. A. F. Kluge, K. G. Untch, J. H. Fried, *J. Am. Chem. Soc.* **94** (1972) 7827.

207. J. S. Skotnicki, R. E. Schaub, K. F. Bernady, G. J. Siuta, J. F. Poletto, M. J. Weiss, *J. Med. Chem.* **20** (1977) 1551; see also M. Suzuki, T. Suzuki, T. Kawagishi, Y. Morita, R. Noyori, *Isr. J. Chem.* **24** (1984) 118.
208. B. C. Borer, R. J. K. Taylor, *Synlett* (1990) 601.
209. B. Contreras, L. Duhamel, G. Plé, *Synth. Commun.* **20** (1990) 2983.
210. L. Duhamel, G. Plé, Y. Ramondenc, *Tetrahedron Lett.* **30** (1989) 7377.
211. H. Gilman, S. M. Spatz, *J. Org. Chem.* **16** (1951) 1485.
212. J. P. Wibaut, A. P. de Jonge, H. G. P. van der Voort, P. P. H. L. Otto, *Recl. Trav. Chim. Pays-Bas* **70** (1951), 1054; J. Wibaut, L. G. Heeringa, *Recl. Trav. Chim. Pays-Bas* **74** (1955), 1993; see also H. Gilman, S. M. Spatz, *J. Org. Chem.* **16** (1951), 1485.
213. H. Gilman, D. S. Melstrom, *J. Am. Chem. Soc.* **68** (1946) 103.
214. Z.-j. Zeng, S. C. Zimmerman, *Tetrahedron Lett.* **29** (1988) 5123.
215. H. Gilman, T. S. Soddy, *J. Org. Chem.* **22** (1957) 565; **23** (1958) 1584.
216. D. L. Nguyen, M. Schlosser, *Helv. Chim. Acta* **60** (1977) 2085; see also S. Gronowiz, G. Sorlin, *Ark. Kemi* **19** (1962) 515; Y. Fukuyama, Y. Kawashima, T. Miwa, T. Tokoroyama, *Synthesis* (1974) 443; H. Haarmann, W. Eberbach, *Tetrahedron Lett.* **32** (1991) 903.
217. S. Gronowitz, P. Moses, A. B. Hörnfeldt, *Ark. Kemi* **17** (1961) 237.
218. J. M. Muchowski, R. Naef, *Helv. Chim. Acta* **67** (1984), 1168; see also B. L. Bray, P. H. Mathies, R. Naef, D. R. Solas, T. T. Tidwell, D. R. Artis, J. M. Muchowski, *J. Org. Chem.* **55** (1990), 6317; P. W. Shum, A. P. Kozikowski, *Tetrahedron Lett.* **31** (1990), 6785.
219. J. S. Pudlo, M. R. Nassiri, E. R. Kern, L. L. Wotring, J. C. Drach, L. B. Townsend, *J. Med. Chem.* **33** (1990) 1984.
220. B. H. Lipshutz, W. Hagen, *Tetrahedron Lett.* **33** (1992), 5865.
221. A. Dondoni, A. R. Mastellari, A. Medici, E. Negrini, P. Pedrini, *Synthesis* (1986) 757.
222. S. Gronowitz, T. Frejd, *Acta Chem. Scand., Ser. B* **30** (1976), 485, S. Gronowitz, B. Holm, *Acta Chem. Scand. B* **30** (1976), 505.
223. M. P. Groziak, L.-l. Wei, *J. Org. Chem.* **56** (1990) 4296.
224. H. J. S. Winkler, H. Winkler, *J. Am. Chem. Soc.* **88** (1964) 964, 969.
225. M. Schlosser, V. Ladenberger, *J. Organomet. Chem.* **8** (1967) 193.
226. H. Neumann, D. Seebach, *Tetrahedron Lett.* **17** (1976) 4839.
227. A. I. Meyers, K. A. Babiak, A. L. Campbell, D. L. Comins, M. P. Fleming, R. Henning, M. Heuschmann, J. P. Hudspeth, J. M. Kane, P. J. Reider, D. M. Roland, K. Shimizu, K. Tomioka, R. D. Walkup, *J. Am. Chem. Soc.* **105** (1983) 5015.
228. W.-Y. Wang, S. V. d'Andrea, J. P. Freeman, J. Szmuszkovicz, *J. Org. Chem.* **56** (1991) 2914; see also for comparison, K. Green, *J. Org. Chem.* **56** (1991) 4325.
229. J. Barluenga, J. M. Montserrat, J. Flórez, *Tetrahedron Lett.* **33** (1992) 6183.
230. M. Schlosser, T. Kadibelban, O. Desponds, unpublished results (1969, 1993).
231. H. Gilman, G. E. Brown, *J. Am. Chem. Soc.* **67** (1945) 824; G. Wittig, A. Maercker, *J. Organomet. Chem.* **8** (1967) 491.
232. M. Schlosser, T. Kadibelban, G. Steinhoff, *Liebigs Ann. Chem.* **743** (1971) 25, footnote 39; H. J. Reich, D. P. Green, N. H. Phillips, J. P. Borst, I. L. Reich, *Phosphorus Sulfur* **67** (1992) 83.
233. H. J. Reich, B. Ö. Gudmundsson, R. R. Dykstra, *J. Am. Chem. Soc.* **114** (1992) 7937.
234. H. J. Reich, D. P. Green, N. H. Phillips, *J. Am. Chem. Soc.* **113** (1991) 1414.
235. K. J. Irgolic, In *Houben-Weyl: Methods of Organic Chemistry* (D. Klamann, ed.), Vol. E 12b, Thieme, Stuttgart, 1990; T. Hiro, Y. Morita, T. Inoue, N. Kambe, A. Ogawa, I. Ryu, N. Sonoda, *J. Am. Chem. Soc.* **112** (1990) 455.

236. Reviews and monographs: H. J. Reich, *Acc. Chem. Res.* **12** (1979) 22; M. Braun, *Nachr. Chem. Tech. Labor.* **33** (1985) 964; C. Paulmier, *Selenium Reagents and Intermediates in Organic Synthesis*, Pergamon, Oxford, 1986; D. Liotta, *Organoselenium Chemistry*, Wiley, New York, 1987; A. Krief, L. Hevesi, *Organoselenium Chemistry*, Springer. Berlin, 1988.
237. J. F. Arens, M. Fröling, A. Fröhling, *Recl. Trav. Chim. Pays-Bas* **78** (1959), 663; A. Fröhling, J. F. Arens, *Recl. Trav. Chim. Pays-Bas* **81** (1009); E. J. Corey, D. Seebach, *Angew. Chem.* **77** (1965), 1134, 1135; *Angew. Chem. Int. Ed. Engl.* **4** (1965), 1975, 1077; *J. Org. Chem.* **31** (1966), 4097; T. Mukaiyama, R. Narasaka, M. Furusato, *J. Am. Chem. Soc.* **94** (1972), 8641; G. Schill, C. Merke, *Synthesis* **1975**, 387.
238. D. Seebach, N. Peleties, *Chem. Ber.* **105** (1972) 511; A. Krief, W. Dumont, M. Clarembeau, G. Bernard, E. Badaoui, *Tetrahedron* **45** (1989), 2005; A. Krief, W. Dumont, M. Clarembeau, E. Badaoui, *ZTetrahedron* **45** (1989), 2023.
239. L. Set, D. R. Cheshire, D. L. J. Clive, *J. Chem. Soc., Chem. Commun.* (1985) 1205.
240. K. Ziegler, H. G. Gellert, *Liebigs Ann. Chem.* **567** (1950) 195; A. Maercker, W. Theysohn, *Liebigs Ann. Chem.* **747** (1971), 70.
241. P. D. Bartlett, S. Friedmann, M. Stiles, *J. Am. Chem. Soc.* **75** (1953) 1771; see also P. D. Bartlett, S. J. Tauber, W. P. Weber, *J. Am. Chem. Soc.* **91** (1969) 6362.
242. G. Wittig, E. Hahn, *Angew. Chem.* **72** (1960) 781: G. Wittig, J. Otten, *Tetrahedron Lett.* **4** (1963) 601; J. E. Mulvaney, Z. G. Gardlund, *J. Org. Chem.* **30** (1965) 917.
243. R. M. Magid, J. G. Welch, *J. Am. Chem. Soc.* **88** (1966) 5681.
244. C. M. Hill, L. Haynes, D. E. Simmons, M. E. Hill, *J. Am. Chem. Soc.* **75** (1953) 5408.
245. J. J. Eisch, G. R. Husk, *J. Am. Chem. Soc.* **87** (1965) 4194.
246. J. K. Crandall, A. C. Clark, *Tetrahedron Lett.* **10** (1969) 325; H. Felkin, G. Swierczewski, A. Tambute, *Tetrahedron Lett.* **10** (1969) 707.
247. C. Margot, *Doctoral Thesis*, Ecole Polytechnique Fédérale, Lausanne, 1985, pp. 15–16, 18, 70–71; see also M. Schlosser, *Pure Appl. Chem.* **60** (1988) 1627, spec. 1629.
248. M. Schlosser, P. Schneider, *Helv. Chim. Acta* **63** (1980) 2404.
249. A. V. Tobolsky, C. E. Rogers, *J. Polymer Sci.* **40** (1959), 73; D, J, Worsfold, S. Bywater, *Canad, J. Chem.* **38**, (1960), 1891; S. Bywater, D. J. Warsfold, *Canad, J. Chem.* **42** (1964), 2884; H. L. Hsieh, D. J. Worsfold, *J. Polymeer Sci.* **3A** (1965), 624; W. K. R. Barnikol., G. V. Schulz, *Makromolek. Chem.* **86** (1965), 298; *Z. Phys. Chem.* **47** (1965), 89; B. J. Schmitt, G. V. Schulz, *Makromolek. Chem.* **142** (1971), 325.
250. M. Szwarc, *Carbanions, Living Polymers and Electron Transfer Processes*, Wiley-Interscience, New York, 1968.
251. K. Ziegler, W. Schäfer, *Liebigs Ann. Chem.* **511** (1934) 101; G. R. Knox, P. L. Pauson, *J. Chem. Soc.* (1961) 4610.
252. D. Nicholls, M. Szwarc, *J. Am. Chem. Soc.* **88** (1966), 5757.
253. K. Hafner, C. Bernhard, R. Müller, *Liebigs Ann. Chem.* **650** (1961) 35.
254. J. J. Eisch, W. C. Kaska, *J. Am. Chem. Soc.* **84** (1962) 1501; see also J. E. Mulvaney, L. J. Carr, *J. Org. Chem.* **33** (1968) 3286.
255. P. Schorigin, *Ber. Dtsch. Chem. Ges.* **41** (1908) 2723; **43** (1910) 1938.
256. J. B. Conant, G. W. Wheland, *J. Am. Chem. Soc.* **54** (1932) 1212; G. W. Wheland, *J. Chem. Phys.* **2** (1934) 474; W. K. McEwen, *J. Am. Chem. Soc.* **58** (1936) 1124.
257. W. Schlenk, E. Bergmann, *Liebigs Ann. Chem.* **463** (1928) 98.
258. H. Gilman, H. A. Pacevitz, O. Baine, *J. Am. Chem. Soc.* **62** (1940) 1514.
259. R. A. Finnegan, *Tetrahedron Lett.* **4** (1963) 429.
260. R. J. Gillespie, R. S. Nyholm, *Q. Rev. Chem. Soc.* **11** (1957) 339; R. J. Gillespie, *Angew.*

Chem. **79** (1967) 885; *Angew. Chem., Int. Ed. Engl.* **6** (1967) 819; L. S. Bartell, *J. Chem. Educ.* **45** (1968) 754.
261. M. Schlosser, *Struktur und Reaktivität Polarer Organometalle*, pp. 64–71, Springer, Berlin, 1973.
262. G. L. Closs, L. E. Closs, *J. Am. Chem. Soc.* **85** (1963) 2022; see also G. L. Closs, R. B. Larrabee, *Tetrahedron Lett.* **6** (1965) 287; K. B. Wiberg, G. M. Lampman, R. P. Ciula, D. S. Connor, P. Schertler, J. Lavanish, *Tetrahedron* **21** (1965) 2749.
263. G. L. Closs, L. E. Closs, *J. Am. Chem. Soc.* **83** (1961) 1003, 2015; G. L. Closs, *Proc. R. Soc. London* (1962) 152.
264. L. Pauling, *J. Am. Chem. Soc.* **53** (1931) 1367; L. Pauling, *The Nature of the Chemical Bond and the Structure of Molecules and Crystals*, Cornell University Press, Ithaca, NY, 1960.
265. M. Schlosser, J. Hartmann, M. Stähle, J. Kramar, Aa. Walde, A.a Mordini, *Chimia* **40** (1986) 306.
266. D. J. Chadwick, S. T. Hodgson, *J. Chem. Soc., Perkin Trans. 1* (1982) 1833.
267. A. Mordini, M. Schlosser, *Chimia* **40** (1986) 309.
268. M. Stähle, R. Lehmann, J. Kramar, M. Schlosser, *Chimia* **39** (1985) 229.
269. Partial metallation with butyllithium: G. Wittig, E. Hahn, *Angew. Chem.* **72** (1960) 781.
270. M. Schlosser, P. Schneider, *Angew. Chem.* **91** (1979) 515; *Angew. Chem., Int. Ed. Engl.* **18** (1979) 489.
271. O. Riobé, A. Lebouc, J. Delaunay, *C.R. Séances Acad. Sci.* **284** (1977), 281; *Chem. Abstr.* **86** (1977), 189'646e; R. K. Boeckman, K. J. Bruza, *Tetrahedron Lett.* **18** (1977), 4187; see also R. Paul, S. Tchelitcheff, *C.R. Acad. Sci.* **235** (1952) 1226; *Chem. Abstr.* **48** (1954), 1944b; *Bull. Soc. Chim. Fr., Documentat.* **19** (1952) 808.
272. S. Akiyama, J. Hooz, *Tetrahedron Lett.* **14** (1973), 275.
273. R. B. Bates, W. A. Beavers, *J. Am. Chem. Soc.* **96** (1974), 5001.
274. G. G. Eberhardt, U.S.-Pat. 3'206'519 (to Sun Oil Co., appl. 15 June 1962, filed 14 Sept, 1965); *Chem. Abstr.* **63** (1965), 14'757h; A. W. Langer, *Trans N.Y. Acad. Sci.* **27** (1965), 741; *Chem. Abstr.* **63** (1965), 18'131f; A. J. Chalk, T. J. Hoogeboom, *J. Organomet. Chem.* **11** (1968), 615.
275. R. Paul, S. Tchelitcheff, *Bull. Soc. Chim. Fr.* **1948**, 108, 1199; *C.R. Séances Acad. Sci.* **239** (1954), 1222; *Chem. Abstr.* **49** (1955), 13'915d; R. B. Bates, D. W. Gosselink, J. A. Kaczynski, *Tetrahedron Lett.* **8** (1967), 199; G. Heiszwolf, H. Kloosterziel, *Recl. Trav. Chim. Pays-Bas* **86** (1967), 807.
276. R. B. Bates, W. H. Deines, D. A. McCombs, D. E. Potter, *J. Am. Chem. Soc.* **91** (1969) 4608.
277. M. Schlosser, *J. Organomet. Chem.* **8** (1967) 9.
278. G. Wittig, P. Davis, G. Koenig, *Ber. Dtsch. Chem. Ges.* **84** (1951) 627.
279. N. H. Cromwell, D. B. Capps, *J. Am. Chem. Soc.* **74** (1952) 4448; R. Meier, *Chem. Ber.* **86** (1953) 1483; A. Melera, M. Claesen, H. Vanderhaeghe, *J. Org. Chem.* **29** (1964) 3705.
280. K. Ziegler, L. Jacob, *Liebigs Ann. Chem.* **511** (1934) 45.
281. G. Wittig, M. Wetterling, *Liebigs Ann. Chem.* **557** (1956) 1; G. Wittig, D. Krauss, *Liebigs Ann. Chem.* **679** (1964), 34.
282. U. Schöllkopf, F. Gerhart, *Angew. Chem.* **80** (1968) 842; *Angew. Chem., Int. Ed. Engl.* **7** (1968) 805; D. Hoppe, *Angew. Chem.* **86** (1974), 878; *Angew. Chem. Int. Ed. Engl.* **13** (1974), 789.
283. K. M. Downard, J. C. Sheldon, J. H. Bowie, D. E. Lewis, R. N. Hayes, *J. Am. Chem. Soc.* **111** (1989), 8112.
284. W. V. E. Doering, A. K. Hoffmann, *J. Am. Chem. Soc.* **77** (1955) 521.

285. For a recent account of the electron distribution in phosphorus ylids, see M. Schlosser, T. Jenny, B. Schaub, *Heteroatom. Chem.* **1** (1990) 151.
286. G. Wittig, M. Rieber, *Liebigs Ann. Chem.* **562** (1949) 187; see also P. J. Wheatley, G. Wittig, *Proc. Chem. Soc. London* (1962) 251.
287. D. Hellwinkel, *Chem. Ber.* **98** (1965) 576.
288. A. Rauk, L. C. Allen, K. Mislow, *Angew. Chem.* **82** (1970) 453, spec. 455; *Angew. Chem., Int. Ed. Engl.* **9** (1970), 400 spec. 402; S. Wolfe, L. A. LaJohn. F. Bernardi, A. Mangini, G. Tonachini, *Tetrahedron Lett.* **24** (1983) 3789.
289. D. J. Peterson, *J. Organomet. Chem.* **8** (1971) 199.
290. D. Seyferth, D. E. Welch, J. K. Heeren, *J. Am. Chem. Soc.* **85** (1963) 642; **86** (1964) 1100; see also L. Horner, H. Hoffmann, H. Wippel, *Chem. Ber.* **91** (1958), 61.
291. M. Schlosser, in *Methodicum Chemicum* (F. Kore, ed.) **7**, 529, spec. 543, 547, Thieme, Stuttgart, 1976; **7B**, 506, spec. 519, 523–524, Academic Press, New York, 1978; T. Huynh Ba, doctoral dissertation, Université de Lausanne, 1976; E. Weił, J. Kopf, T. Gardein, S. Corberlin, U. Schümann, M. Kirilov, G. Petrov, *Chem. Ber.* **118** (1985), 3529; S. E. Denmark, R. L. Dorow, *J. Am. Chem. Soc.* **112** (1990), 864; W. Zarges, M. Marsch, K. Harms, F. Haller, G. Frenking, G. Boche, *Chem. Ber.* **124** (1991), 861; S. E. Denmark, P. C. Miller, S. R. Wilson, *J. Am. Chem. Soc.* **113** (1991), 1468.
292. H. Gilman, F. J. Webb, *J. Am. Chem. Soc.* **71** (1949) 4062; D. A. Shirley, B. J. Reeves, *J. Organomet. Chem.* **16** (1969) 1.
293. Reviews: D. Seebach, *Synthesis* (1969) 17; *Angew. Chem.* **81** (1969) 690; *Angew. Chem., Int. Ed. Engl.***8** (1969).
294. D. Seebach, *Angew. Chem.* **79** (1967) 468; *Angew. Chem., Int. Ed. Engl.* **6** (1967) 442; G. A. Wildschut, H. J. T. Bos, L. Brandsma, J. F. Arens, *Monatsh. Chem.* **98** (1967) 1043.
295. D. J. Peterson, *J. Organomet. Chem.* **9** (1967) 373.
296. G. A. Gornowicz, R. West, *J. Am. Chem. Soc.* **90** (1968) 4478.
297. D. F. Hoeg, D. I. Lusk, A. L. Crumbliss, *J. Am. Chem. Soc.* **87** (1965) 4147; G. Köbrich, H. R. Merkle, *Chem. Ber.* **99** (1966) 1782; G. Köbrich, K. Flory, R. H. Fischer, *Chem. Ber.* **99** (1966) 1793.
298. S. Andreades, *J. Am. Chem. Soc.* **86** (1964) 2003; D. K. Bohme, E. Lee-Ruff, L. B. Young, *J. Am. Chem. Soc.* **94** (1972) 5152; K. J. Klabunde, D. J. Burton, *J. Am. Chem. Soc.* **94** (1972) 5985; J. E. Bartmess, J. A. Scott, R. T. McIver, *J. Am. Chem. Soc.* **101** (1979) 6047; D. M. McMillen, D. M. Golden, *Annu. Rev. Phys. Chem.* **33** (1982) 493.
299. G. Köbrich, H. Trapp, *Chem. Ber.* **99** (1966) 670, 680; G. Köbrich, K. Flory, *Chem. Ber.* **99** (1966) 1773; G. Köbrich, W. Drischel, *Tetrahedron* **22** (1966) 2621.
300. K. Oshima, K. Shimoji, H. Takahashi, H. Yamamoto, H. Nozaki, *J. Am. Chem. Soc.* **95** (1973) 2694; R. Muthukrishnan, M. Schlosser, *Helv. Chim. Acta* **59** (1976) 13.
301. J. W. Schick, H. D. Hartough, *J. Am. Chem. Soc.* **70** (1948) 286; H. Gilman, D. A. Shirley, *J. Am. Chem. Soc.* **71** (1949) 1870.
302. W. Theilacker, E. Wegner, *Liebigs Ann. Chem.* **664** (1963), 125.
303. W. V. E. Doering, L. K. Levy, *J. Am. Chem. Soc.* **77** (1955), 509.
304. S. Oae, W. Tagaki, A. Ohno, *J. Am. Chem. Soc.* **83** (1961), 5036.
305. M. D. Rausch, D. J. Ciappenelli, *J. Organomet. Chem.* **10** (1967), 127; F. L. Hedberg, H. Rosenberg, *Tetrahedron Lett.* **10** (1969), 4011; J. J. Bishop, A. Davison, M. L. Katcher, D. W. Lichten etfg, R. E. Merrill, J. C. Smart, *J. Organomet, Chem.* **27** (1971), 241; M. Walczak, K. Walczak, R. Mink, M.D. Rausch, G. Stucky, *J. Am. Chem. Soc.* **100** (1978), 6382; F. Rebière, O. Samuel, H. B. Kagan, *Tetrahedron Lett.* **31** (1990), 3121; U. Mueller-Westerhoff, Y. Zheng, G. Ingram, *J. Organomet. Chem.* **463** (1993), 163.

306. C. Elschenbroich, *J. Organomet. Chem.* **14** (1968), 157; C. Elschenbroich, G. Heikenfeld, M. Wünsch, W. Marsa, G. Baum, *Angew. Chem.* **100** (1988), 397; *Angew. Chem. Int. Ed. Engl.* **27** (1988), 414.
307. D. J. Peterson, *J. Organomet. Chem.* **9** (1967), 373; F. H. Köhler, N. Hertkorn, J. Blümel, *Chem. Ber.* **120** (1987), 2081; M. Schakel, M. P. Aarnts, G. W. Klumpp, *Recl. Trav. Chim. Pays-Bas* **109** (1990), 305.
308. G. Rauchschwalbe, M. Schlosser, unpublished work (1976).
309. M. Schlosser, S. Strunk, *Tetrahedron Lett.* **25** (1984) 741.
310. M. Schlosser, *Pure Appl. Chem.* **60** (1988), 741.
311. A. Mordini, in: *Advances in Carbanion Chemistry*, ed. V. Snieckus, Chap. 1, Jai Press, Greenwich, CT, 1992.
312. M. Schlosser, J. H. Choi, S. Takagishi, *Tetrahedron* **46** (1990), 5633.
313. H. Gilman, J. C. Bailie, *J. Org. Chem.* **2** (1937), 84.
314. R. Lehmann, M. Schlosser, *Tetrahedron Lett.* **25** (1984) 745.
315. R. A. Finnegan, R. S. McNees, *J. Org. Chem.* **29** (1964) 3234.
316. E. J. Lanpher, L. M. Redmen, A. A. Morton, *J. Org. Chem.* **23** (1958) 1370.
317. R. A. Finnegan, R. S. McNees, *J. Org. Chem.* **29** (1964) 3241.
318. J. Hartmann, M. Schlosser, *Helv. Chim. Acta* **59** (1976) 453.
319. M. Stähle, A. Walde, A. Mordini, M. Schlosser, unpublished results (1980–85).
320. J. L. Giner, C. Margot, C. Djerassi, *J. Org. Chem.* **54** (1989) 2117.
321. G.-f. Zhong, M. Schlosser, *Tetrahedron Lett.* **34** (1993), 5441.
322. A. A. Morton, R. A. Finnegan, *J. Polym. Sci.* **38** (1959) 19.
323. C. D. Broaddus, *J. Org. Chem.* **29** (1964) 2689; see also C. D. Broaddus, T. J. Logan, T. J. Flautt, *J. Org. Chem.* **28** (1963) 1174.
324. L. Franzini, M. Schlosser, unpublished results (1991–92).
325. A. Mordini, unpublished results (1992).
326. L. Brandsma, H. D. Verkruijsse, C. Schade, P. v. R. Schleyer, *J. Chem. Soc., Chem. Commun.* (1986) 260.
327. U. Schöllkopf, in *Houben-Weyl, Methoden der Organischen Chemie*, ed. E. Müller, Vol. 13/1, p. 116, Georg Thieme, Stuttgart, 1970; U. Schöllkopf, P. Hänssle, *Liebigs Ann. Chem.* **763** (1972) 208; J. E. Baldwin, G. A. Höfle, O. W. Lever, *J. Am. Chem. Soc.* **96** (1974) 7125.
328. S. Hanessian, M. Martin, R. C. Desai, *J. Chem. Soc., Chem. Commun.* (1986) 926; see also R.R. Schmidt, R. Preuss, R. Betz, *Tetrahedron Lett.* **28** (1987) 6591.
329. R. K. Boeckmann, A. B. Charette, T. Asberom, B. H. Johnston, *J. Am. Chem. Soc.* **113** (1991) 5337.
330. A. A. Morton, E. L. Little, *J. Am. Chem. Soc.* **71** (1949) 487; A. A. Morton, C. E. Claff, F. W. Collins, *J. Org. Chem.* **20** (1955) 428; D. Bryce-Smith, *J. Chem. Soc.* (1954) 1079.
331. H. Hommes, H. D. Verkruijsse, L. Brandsma, *J. Chem. Soc., Chem. Commun.* (1981) 366; P. A. A. Klusener, J. C. Hanekamp, L. Brandsma, P. von. R. Schleyer, *J. Org. Chem.* **55** (1990), 1311; see also K. C. Eberly, H. E. Adams, *J. Organomet. Chem.* **3** (1965) 165; N. R. Pearson, G. Hahn, G. Zweifel, *J. Org. Chem.* **47** (1982) 3364.
332. A. J. Chalk, T. J. Hoogeboom, *J. Organomet. Chem.* **11** (1968) 615; C. D. Broaddus, *J. Org. Chem.* **35** (1970) 10; see also A. W. Langer, *Trans. N.Y. Acad. Sci.* **27** (1965) 741; *Chem. Abstr.* **63** (1965), 18131f.
333. F. Faigl, M. Schlosser, *Tetrahedron Lett..* **32** (1991) 3369.
334. R. D. Bach, R. C. Klix, *Tetrahedron Lett..* **27** (1986) 1983.
335. G. B. Trimitsis, A. Tuncay, R. D. Beyer, K. J. Ketterman, *J. Org. Chem.* **38** (1973), 1491; J. Klein, A. Medlik, A. Y. Meyer, *Tetrahedron* **32** (1976), 51.

336. B. Gordon, J. E. Loftus, *J. Org. Chem.* **51** (1986) 1618.
337. E. Moret, M. Schlosser, unpublished results (1987).
338. R. E. Ludt, G. P. Crowther, C. R. Hauser, *J. Org. Chem.* **35** (1970) 1288.
339. P. Maccaroni, M. Schlosser, unpublished results (1991–92).
340. S. Takagishi, M. Schlosser, *Synlett* (1991) 119.
341. R. B. Bates, T. J. Siahaan, K. Suvannachut, *J. Org. Chem.* **55** (1990), 1328.
342. J. Hübscher, R. Barner, *Helv. Chim. Acta* **73** (1990), 1068, spec. 1080.
343. J. Hartmann, R. Muthukrishnan, M. Schlosser, *Helv. Chim. Acta* **57** (1974) 2261.
344. G.-f. Zhong, M. Schlosser, unpublished results (1991–1993).
345. M. Schlosser, R. Dahan, S. Cottens, *Helv. Chim. Acta* **67** (1984) 284.
346. M. Schlosser, M. Stähle, *Angew. Chem.* **92** (1980) 497; *Angew. Chem., Int. Ed. Engl.* **19** (1980) 487.
347. O. Desponds, M. Schlosser, unpublished results (1991–92).
348. T. Jenny, M. Schlosser, unpublished results (1983); see also E. W. Thomas, *Tetrahedron Lett.* **24** (1983), 2347; E. Sternberg, P. Binger, *Tetrahedron Lett.* **26** (1985), 301.
349. N. Hertkorn, F. H. Köhler, *J. Organomet. Chem.* **355** (1988) 19.
350. G. Rauchschwalbe, M. Schlosser, *Helv. Chim. Acta* **58** (1975) 1094.
351. H. C. Brown, M. Zaidlewicz, K. S. Bhat, *J. Org. Chem.* **54** (1989) 1764.
352. C. Margot, *Doctoral Thesis*, Université de Lausanne.
353. M. Schlosser, H. Bosshardt, R. Lehmann, E. Moret, G. Rauchschwalbe, *Tetrahedron Rep.* **49** (1993), 10'175.
354. M. Schlosser, J. Hartmann, V. David, *Helv. Chim. Acta* **57** (1974) 1567.
355. M. Schlosser, J. Hartmann, *J. Am. Chem. Soc.* **98** (1976) 4674.
356. H. C. Brown, K. S. Bhat, *J. Am. Chem. Soc.* **108** (1986) 293; W. R. Roush, R. L. Halterman, *J. Am. Chem. Soc.* **108** (1986) 294.
357. L. Franzini, M. Schlosser, unpublished results (1991–92).
358. M. Stähle, J. Hartmann, M. Schlosser, *Helv. Chim. Acta* **60** (1977) 1730.
359. E. Moret, *Doctoral Thesis*, Université de Lausanne, 1980.
360. E. Moret, O. Desponds, M. Schlosser, *J. Organomet. Chem.* **409** (1991) 93.
361. K. Fujita, C. Margot, E. Moret, M. Schlosser, unpublished results (1980–85).
362. R. Lehmann, G. Rauchschwalbe, M. Schlosser, unpublished results (1973–81).
363. A. Maercker, W. Berkulin, *Chem. Ber.* **123** (1990), 185.
364. H. Ahlbrecht, G. Rauchschwalbe, *Synthesis* (1973) 417; H. Ahlbrecht, J. Eichler, *Synthesis* (1974) 672; H. Ahlbrecht, *Chimia* **31** (1977) 391.
365. A. Ahlbrecht, J. Eichler, *Synthesis* **1974**, 672; M. Julia, A. Schouteeten, M. M. Baillarge, *Tetrahedron Lett.* **15** (1974), 3433; S. F. Martin, m. DuPriest, *Tetrahedron Lett.* **18** (1977), 3925; R. J. P. Corriu, V. Huynh, J. J. E. Moreau, *J. Organomet. Chem.* **259** (1983), 283; J. J. Eisch, J. H. Shah, *J. Org. Chem.* **56** (1991), 2955; A. Degl'Innocenti, A. Mordini, D. Pinzani, G. Reginato, A. Ricci, *Synlett* **1991**, 712.
366. D. A. Evans, G. C. Andrews, B. Buckwalter, *J. Am. Chem. Soc.* **96** (1974) 5560.
367. W. C. Still, T. L. Macdonald, *J. Am. Chem. Soc.* **96** (1974) 5561.
368. D. Hoppe, *Angew. Chem.* **96** (1984) 930; *Angew. Chem., Int. Ed. Engl.* **23** (1984) 932.
369. E. Moret, M. Schlosser, *Tetrahedron Lett.* **25** (1984) 4491.
370. E. Moret, P. Schneider, C. Margot, M. Stähle, M. Schlosser, *Chimia* **39** (1985) 231.
371. M. Schlosser, S. Strunk, *Tetrahedron* **45** (1989) 2649.
372. R. M. Carlson, L. L. White, *Synth. Commun.* **13** (1983) 237.
373. G. Cardillo, M. Contento, S. Sandri, *Tetrahedron Lett.* **15** (1974) 2215.
374. J. A. Katzenellenbogen, A. L. Crumrine, *J. Am. Chem. Soc.* **96** (1974) 5662; **98** (1976) 4925.

375. K. Koumaglo, T. H. Chan, *Tetrahedron Lett.* **25** (1984) 717.
376. A. Mordini, G. Palio, A. Ricci, M. Taddei, *Tetrahedron Lett.* **29** (1988) 4991.
377. L.-H. Li, D. Wang, T. H. Chan, *Tetrahedron Lett.* **32** (1991) 2879.
378. R. L. P. DeJong, L. Brandsma, *J. Organomet. Chem.* **238** (1982) C17; **316** (1986) C21.
379. R. Mantione, Y. Leroux, *Tetrahedron Lett.* **12** (1971) 593; Y. Leroux, C. Roman, *Tetrahedron Lett.* **14** (1973) 2585.
380. M. Stähle, M. Schlosser, *Angew. Chem.* **91** (1979) 938; *Angew. Chem., Int. Ed. Engl.* **18** (1979) 875.
381. S. Hoff, L. Brandsma, J. F. Arens, *Recl. Trav. Chim. Pays-Bas* **87** (1968) 916, 1179; **88** (1969) 609; see also J. C. Clinet, G. Linstrumelle, *Tetrahedron Lett.* **19** (1978) 1137.
382. L. Rothen, M. Schlosser, *Tetrahedron Lett.* **32** (1991) 2475.
383. G. Rauchschwalbe, H. Bosshardt, A. Walde, J. Homberger, M. Schlosser, unpublished results (1978–82).
384. M. Schlosser, G. Rauchschwalbe, *J. Am. Chem. Soc.* **100** (1978) 3258.
385. M. Schlosser, H. Bosshardt, A. Walde, M. Stähle, *Angew. Chem.* **92** (1980) 302; *Angew. Chem., Int. Ed. Engl.* **19** (1980) 303.
386. E. Moret, M. Schlosser, *Tetrahedron Lett.* **25** (1984) 1449.
387. H. Bosshardt, M. Schlosser, *Helv. Chim. Acta* **63** (1980) 2393.
388. G. Rauchschwalbe, M. Schlosser, unpublished results (1978–79).
389. A. A. Morton, I. Hechenbleikner, *J. Am. Chem. Soc.* **58** (1936) 2599; G. Lüttringhaus, G. v. Sääf, *Liebigs Ann. Chem.* **542** (1939) 241; G. Wittig, E. Benz, *Chem. Ber.* **91** (1958) 874.
390. G. Wittig, U. Pockels, H. Dröge, *Ber. Dtsch. Chem. Ges.* **71** (1938) 1903; H. Gilman, R. L. Bebb, *J. Am. Chem. Soc.* **61** (1939) 109.
391. H. W. Gschwend, H. R. Rodriguez, *Org. React.* **26** (1979) 1.
392. N. S. Narasimhan, R. S. Mali, *Synthesis* (1983) 957; *Top. Curr. Chem.* **138** (1987) 63; V. Snieckus, *Bull. Soc. Chim. Fr.* (1988) 67; *Chem. Rev.* **90** (1990) 879.
393. M. Schlosser, B. Spahić, Le Van Chau, *Helv. Chim. Acta* **58** (1975) 2586.
394. E. J. Corey, T. M. Eckrich, *Tetrahedron Lett.* **24** (1983) 3165.
395. D. J. Peterson, H. R. Hays, *J. Org. Chem.* **30** (1965) 1939.
396. H. Ahlbrecht, H. Dollinger, *Tetrahedron Lett.* **25** (1984) 1353.
397. D. J. Peterson, *J. Organomet. Chem.* **21** (1970) P63; F. H. Köhler, N. Hertkorn, J. Blümel, *Chem. Ber.* **120** (1987) 2081; M. Schakel, M. P. Aarnts, G. W. Klumpp, *Recl. Trav. Chim. Pays-Bas* **109** (1990) 305.
398. S. T. Kerrick, P. Beak, *J. Am. Chem. Soc.* **113** (1991), 9708.
399. A. Katritzky, S. Sengupta, *J. Chem. Soc., Perkin Trans. 1* (1989) 17.
400. A. Hosomi, M. Ando, H. Sakurai, *Chem. Lett.* (1984) 1384.
401. L. Brandsma, H. Verkruijsse, in *Organometallic Synthesis Workshop*, ed. M. Schlosser, p. 141, UNIL, Lausanne, 1987.
402. G. Wittig, L. Löhmann, *Liebigs Ann. Chem.* **550** (1942) 260.
403. See also, however, M. Franciotti, A. Mordini, M. Taddei, *Synlett* (1992) 137.
404. S. G. Lias, J. E. Bartmess, J. F. Liebman, J. L. Holmes, R. D. Levin, W. G. Mallard, *J. Phys. Chem. Ref. Data* **17**, Suppl. 1 (1988).
405. R. C. Gadwood, M. R. Rubino, S. C. Nagarajan, S. T. Michel, *J. Org. Chem.* **50** (1985) 3255.
406. J. J. Eisch, J. E. Galle, *J. Am. Chem. Soc.* **98** (1976) 4646; *J. Organomet. Chem.* **121** (1976) C10; *J. Org. Chem.* **55** (1990) 4835.
407. A. C. Cope, H. H. Lee, H. E. Petree, *J. Am. Chem. Soc.* **80** (1958) 2849; A. C. Cope, M. Brown, H. H. Lee, *J. Am. Chem. Soc.* **80** (1958) 2855; A. C. Cope, M. M. Martin, M. A.

McKervey, *Q. Rev. Chem. Soc* **20** (1966) 143; C. J. M. Sterling, *Chem. Rev.* **78** (1978) 517, spec. 528; J. K. Whitesell, P. D. White, *Synthesis* (1975) 602.
408. L. J. Haynes, I. Heilbron, E. H. R. Jones, F. Sondheimer, *J. Chem. Soc.* (1947) 1583; J. K. Crandall, M. Apparu, *Org. React.* **29** (1983) 345; A. Mordini, E. Ben Rayana, C. Margot, M. Schlosser, *Tetrahedron* **46** (1990) 2401.
409. G. Köbrich, *Angew. Chem.* **79** (1967) 15; *Angew. Chem., Int. Ed. Engl.* **6** (1967) 41.
410. J. Hartmann, M. Stähle, M. Schlosser, *Synthesis* (1974) 888.
411. T. Taguchi, T. Morikawa, O. Kitagawa, T. Mishima, Y. Kobayashi, *Chem. Pharm. Bull.* **33** (1985) 5137.
412. P. G. McDougal, J. C. Rico, D. VanDerveer, *J. Org. Chem.* **51** (1986) 4492.
413. K. J. H. Kruithof, R. F. Schmitz, G. W. Klumpp, *Recl. Trav. Chim. Pays-Bas* **104** (1985) 1.
414. M. Schlosser, V. Ladenberger, *Angew. Chem.* **78** (1966) 677; *Angew. Chem., Int. Ed. Engl.* **5** (1966) 667.
415. T. Cuvigny, M. Julia, C. Rolando, *J. Chem. Soc., Chem. Commun.* (1984) 8; *Tetrahedron Lett.* **28** (1987) 2587.
416. D. Seebach, M. Kolb, *Chem. Ind. (London)* (1974) 687; C. Wiaux-Zamar, J. P. Dejonghe, L. Ghosez, J. F. Normant, J. Villieras, *Angew. Chem.* **88** (1976) 417; *Angew. Chem., Int. Ed. Engl.* **15** (1976) 371; see also R. R. Schmidt, J. Talbiersky, P. Russegger, *Tetrahedron Lett.* **20** (1979) 4273.
417. A. R. Katritzky, J.-q. Li, N. Malhotra, *Liebigs Ann. Chem.* (1992) 843.
418. G. Stork, C. S. Shiner, C.-W. Cheng, R. L. Polt, *J. Am. Chem. Soc.* **108** (1986) 304.
419. D. Hänsggen, E. Odenhausen, *Chem. Ber.* **112** (1979) 2389; J. Schulze, R. Boese, G. Schmidt, *Chem. Ber.* **114** (1981) 1297; S. A. Burns, R. J. P. Corriu, V. Huynh, J. J. E. Moreau, *J. Organomet. Chem.* **333** (1987) 281.
420. R. A. Benkeser, R. B. Currie, *J. Am. Chem. Soc.* **70** (1948) 1780; W. E. Truce, E. Wellisch, *J. Am. Chem. Soc.* **74** (1952) 5177; V. Ramanathan, R. Levine, *J. Org. Chem.* **27** (1962) 1216.
421. H. Gilman, F. Breuer, *J. Am. Chem. Soc.* **56** (1934) 1123.
422. D. W. Adamson, *J. Chem. Soc.* (1950) 885; S. Gronowitz, P. Moses, *Acta Chem. Scand.* **16** (1960) 155.
423. J. W. Schick, H. D. Hartough, *J. Am. Chem. Soc.* **70** (1948) 286.
424. H. Gilman, T. S. Soddy, *J. Org. Chem.* **22** (1957) 1715.
425. G. Wittig, W. Uhlenbrock, P. Weinhold, *Chem. Ber.* **95** (1962) 1692.
426. M. Schlosser, G. Katsoulos, S. Takagishi, *Synlett* (1990) 747.
427. A. M. Roe, R. A. Burton, D. R. Reavill, *J. Chem. Soc., Chem. Commun.* (1965) 582; C. Tamborski, E. J. Soloski, *J. Org. Chem.* **31** (1966) 746; A. M. Roe, R. A. Burton, G. L. Willey, M. W. Baines, A. C. Rasmussen, *J. Med. Chem.* **11** (1968) 814.
428. S. Takagishi, M. Schlosser, *Synlett* (1991) 119.
429. M. Iwao, *J. Org. Chem.* **55** (1990) 3622.
430. T. H. Kress, M. R. Leanna, *Synthesis* (1988) 803.
431. B. M. Dunn, T. C. Bruice, *J. Am. Chem. Soc.* **92** (1979) 2410; R. G. Harvey, C. Cortez, T. P. Ananthanarayan, S. Schmolka, *J. Org. Chem.* **53** (1988) 3936; S. Jeganathan, M. Tsukamoto, M. Schlosser, *Synthesis* (1990) 109; see also W. E. Parham, E. L. Anderson, *J. Am. Chem. Soc.* **70** (1948) 4187; R. Stern, J. English, H. G. Cassidy, *J. Am. Chem. Soc.* **79** (1957) 5797; C. A. Townsend, L. M. Bloom, *Tetrahedron Lett.* **22** (1981) 3923.
432. A. Wada, S. Kanatomo, S. Nagai, *Chem. Pharm. Bull.* **33** (1985) 1016.
433. T. Kamikawa, I. Kubo, *Synthesis* (1986) 431.
434. G. Posner, K. A. Canella, *J. Am. Chem. Soc.* **107** (1985) 2571.

435. G. Coll, J. Morey, A. Costa, J. M. Saá, *J. Org. Chem.* **53** (1988) 5345; G. A. Suner, P. M. Deyá, J. M. Saá, *J. Am. Chem. Soc.* **112** (1990) 1467.
436. S. Cabbidù, S. Melis, P. P. Piras, M. Secci, *J. Organomet. Chem.* **132** (1977) 321.
437. F. M. Stoyanovich, B. P. Fedorov, *Angew. Chem.* **78** (1966) 116; *Angew. Chem., Int. Ed. Engl.* **5** (1966) 127.
438. F. M. Stoyanovich, R. G. Karpenko, Y. L. Goldfarb, *Tetrahedron* **27** (1971) 433.
439. D. A. Shirley, E. A. Lehto, *J. Am. Chem. Soc.* **77** (1955) 1841.
440. W. E. Truce, M. F. Amos, *J. Am. Chem. Soc.* **73** (1951) 3013.
441. H. Gilman, D. L. Esmay, *J. Am. Chem. Soc.* **75** (1953) 278; see also G. Köbrich, *Chem. Ber.* **92** (1959) 2981.
442. J. N. Bonfiglio, *J. Org. Chem.* **51** (1986) 2833.
443. H. Watanabe, R. A. Schwarz, C. R. Hauser, J. Lewis, D. W. Slocum, *Can. J. Chem.* **47** (1969) 1543.
444. H. Watanabe, R. L. Gay, C. R. Hauser, *J. Org. Chem.* **33** (1968) 900; H. Watanabe, C. L. Mao, C. R. Hauser, *J. Org. Chem.* **34** (1969) 1786.
445. J. G. Lombardo, *J. Org. Chem.* **36** (1971) 1843.
446. H. Watanabe, C. L. Mao, I. T. Barnish, C. R. Hauser, *J. Org. Chem.* **34** (1969) 919.
447. A. R. Lepley, W. A. Khan, A. B. Giumanini, A. G. Giumanini, *J. Org. Chem.* **31** (1966) 2047.
448. R. E. Ludt, G. P. Growther, C. R. Hauser, *J. Org. Chem.* **35** (1970) 1288.
449. D. W. Slocum, G. Bock, C. A. Jennings, *Tetrahedron Lett.* **11** (1970) 3443.
450. H. Gilman, G. E. Brown, *J. Am. Chem. Soc.* **62** (1940) 3208.
451. S. Takagishi, M. Schlosser, unpublished work (1990).
452. N. S. Narasimhan, A. C. Ranade, *Indian J. Chem.* **7** (1969) 538; *Chem. Abstr.* **71** (1969) 49743z.
453. G. Cauquil, A. Casadevall, E. Casadevall, *Bull. Soc. Chim. Fr.* (1969) 1049.
454. A. J. Katritzky, L. M. Vazquez de Miguel, G. W. Rewcastle, *Synthesis* (1988) 215; A. J. Katritzky, G. W. Rewcastle, L. M. Vazquez de Miguel, *J. Org. Chem.* **53** (1988) 794.
455. W. Fuhrer, H. W. Gschwend, *J. Org. Chem.* **44** (1979) 1133.
456. J. M. Muchowski, M. C. Venuti, *J. Org. Chem.* **45** (1980) 4798.
457. B. Schaub, T. Jenny, M. Schlosser, *Tetrahedron Lett.* **25** (1984) 4097.
458. R. Schmid, J. Foricher, M. Cereghetti, P. Schönholzer, *Helv. Chim. Acta* **74** (1991) 370.
459. J. M. Brown, S. Woodward, *J. Org. Chem.* **56** (1991) 6803.
460. D.-S. Liu, S. Trippett, *Tetrahedron Lett.* **24** (1983) 2039.
461. D. C. Craig, N. K. Roberts, J. L. Tanswell, *Aust. J. Chem.* **44** (1990) 1487.
462. J. D. Roberts, D. Y. Curtin, *J. Am. Chem. Soc.* **68** (1946) 1658; D. A. Shirley, J. R. Johnson, J. P. Hendrix, *J. Organomet. Chem.* **11** (1968) 209.
463. K. D. Bartle, G. Hallas, J. D. Hepworth, *Org. Magn. Reson.* **5** (1973) 479; see also W. J. Houlihan, to Sandoz-Wander, *US Pat.* 3 751 491, 1973; *Chem. Abstr.* **79** (1973) 91787g.
464. P. Aeberli, W. J. Houlihan, *J. Organomet. Chem.* **67** (1974) 321.
465. D. E. Grocock, T. K. Jones, G. Hallas, J. D. Hepworth, *J. Chem. Soc. C* (1971) 3305.
466. G. E. Carr, R. D. Chambers, T. F. Holmes, D. G. Parker, *J. Organomet. Chem.* **325** (1987) 13; R. Filler, W. K. Gnandt, W. Chen, S. Lin, *J. Fluorine Chem.* **52** (1991) 99.
467. E. Napolitano, E. Giannone, R. Fiaschi, A. Marsili, *J. Org. Chem.* **48** (1983) 3653.
468. N. Meyer, D. Seebach, *Angew. Chem.* **90** (1978) 553; *Angew. Chem., Int. Ed. Engl.* **17** (1978) 521.
470. D. L. Comins, J. D. Brown, *J. Org. Chem.* **49** (1984) 1078; **51** (1986) 3566; see also L. Barsky, H. W. Gschwend, J. McKenna, H. R. Rodriguez, *J. Org. Chem.* **41** (1976) 3651.

471. H. Gilman, W. J. Meikle, J. W. Morton, *J. Am. Chem. Soc.* **74** (1952) 6282.
472. J. Klein, A. Medlik-Balan, *J. Org. Chem.* **41** (1976) 3307.
473. C. J. Upton, P. Beak, *J. Org. Chem.* **40** (1975) 1094.
474. P. Beak, R. A. Brown, *J. Org. Chem.* **47** (1982) 34.
475. C. L. Mao, I. T. Barnish, C. R. Hauser, *J. Heterocycl. Chem.* **6** (1969) 475; see also W. H. Putersbaugh, C. R. Hauser, *J. Org. Chem.* **29** (1964) 853; N. S. Narasimhan, B. H. Bide, *Tetrahedron* **27** (1971) 6171; D. W. Slocum, C. A. Jennings, *J. Org. Chem.* **41** (1976) 3653.
476. W. Fuhrer, H. W. Gschwend, *J. Org. Chem.* **44** (1979) 1133.
477. P. Beak, B. Siegel, *J. Am. Chem. Soc.* **96** (1974) 6803; M. Reuman, A. I. Meyers, *Tetrahedron* **41** (1985) 837.
478. F. E. Ziegler, K. W. Fowler, *J. Org. Chem.* **41** (1976) 1564.
479. G. Simig, M. Schlosser, *Tetrahedron Lett.* **29** (1988) 4277; G. Simig, M. Schlosser, unpublished results (1988–90); G. Katsoulos, M. Schlosser, unpublished results (1988–91).
480. R. E. Ludt, C. R. Hauser, *J. Org. Chem.* **36** (1971) 1607.
481. F. N. Jones, R. L. Vaulx, C. R. Hauser, *J. Org. Chem.* **28** (1963) 3461; see also K. P. Klein, C. R. Hauser, *J. Org. Chem.* **32** (1967) 1479; R. L. Gay, C. R. Hauser, *J. Am. Chem. Soc.* **89** (1967) 2297.
482. G. Simig, M. Schlosser, *Tetrahedron Lett.* **31** (1990) 3125.
483. M. Schlosser, G. Simig, *J. Chem. Soc., Perkin Trans. 1* (1992) 1613.
484. M. Schlosser, G. Simig, *J. Chem. Soc., Perkin Trans. 1* (1992) 1613; (1993) 163.
485. E. F. Perozzi, R. S. Michalak, G. D. Figuly, W. H. Stevenson, D. B. Dess, M. R. Ross, J. C. Martin, *J. Org. Chem.* **46** (1981) 1049.
486. D. W. Slocum, T. R. Engelmann, C. A. Jennings, *Aust. J. Chem.* **21** (1968) 2319; see also R. L. Vaulx, F. N. Jones, C. R. Hauser, *J. Org. Chem.* **30** (1965) 58; N. S. Narasimhan, A. C. Ranade, *Tetrahedron Lett.* **7** (1966) 603.
487. C. Lamas, L. Castedo, D. Domínguez, *Tetrahedron Lett.* **29** (1988) 3865.
488. M. Schlosser, G. Simig, *Tetrahedron Lett.* **32** (1991) 1965.
489. S. O. de Silva, J. N. Reed, V. Snieckus, *Tetrahedron Lett.* **19** (1978) 5099; R. J. Mills, N. J. Taylor, V. Snieckus, *J. Org. Chem.* **54** (1989) 4372.
490. R. J. Mills, V. Snieckus, *Tetrahedron Lett.* **25** (1984) 479, 483.
491. M. Iwao, M. Watanabe, S. O. de Silva, V. Snieckus, *Tetrahedron Lett.* **22** (1981) 2349.
492. R. D. Bindal, J. A. Katzenellenbogen, *J. Org. Chem.* **52** (1987) 3181.
493. J. N. Reed, J. Rotchford, D. Strickland, *Tetrahedron Lett.* **29** (1988) 5725.
494. D. W. Slocum, C. A. Jennings, *J. Org. Chem.* **41** (1976) 3563.
495. R. H. B. Galt, J. Horbury, Z. S. Matusiak, R. J. Pearce, J. S. Shaw, *J. Med. Chem.* **32** (1989) 2357.
496. R. G. Harvey, C. Cortez, T. Sugiyama, Y. Ito, T. W. Sawyer, J. DiGiovanni, *J. Med. Chem.* **31** (1988) 154.
497. S. Takagishi, G. Katsoulos, M. Schlosser, *Synlett* (1992) 360.
498. K. P. Klein, C. R. Hauser, *J. Org. Chem.* **32** (1967) 1479.
499. D. W. Slocum, G. Book, C. A. Jennings, *Tetrahedron Lett.* **11** (1970) 3443.
500. G. Katsoulos, S. Takagishi, M. Schlosser, *Synlett* (1991) 731.
501. R. A. Ellison, F. N. Kotsonis, *J. Org. Chem.* **38** (1973) 4192.
502. S. D. Young, K. E. Coblens, B. Ganem, *Tetrahedron Lett.* **22** (1981) 4887.
503. D. A. Shirley, C. F. Cheng, *J. Organomet. Chem.* **20** (1969) 251; see also R. A. Barnes, L. J. Nehmsmann, *J. Org. Chem.* **27** (1962) 1939; N. S. Narasimhan, R. S. Mali, *Tetrahedron* **31** (1975) 1005.

504. J. T. B. H. Jastrzebski, G. Van Koten, K. Goubitz, C. Arlen, M. Pfeffer, *J. Organomet. Chem.* **246** (1983) C75.
505. G. van Koten, A. J. Leusink, J. G. Noltes, *J. Organomet. Chem.* **84** (1975) 117.
506. N. S. Narasimhan, P. S. Chandrachood, N. R. Shete, *Tetrahedron* **37** (1981) 825.
507. M. Schlosser, Z.-P. Liu, G. Katsoulos, unpublished work (1992).
508. J. E. Mulvaney, Z. G. Garlund, S. L. Garlund, D. J. Newton, *J. Am. Chem. Soc.* **88** (1966) 476; J. E. Mulvaney, L. H. Carr, *J. Org. Chem.* **33** (1968) 3286; D. Y. Curtin, R. P. Quirk, *Tetrahedron* **24** (1968) 5791.
509. D. H. Aue, H. M. Webb, W. R. Davidson, P. Toure, H. P. Hopkins, S. P. Moulik, D. V. Jahagirdar, *J. Am. Chem. Soc.* **113** (1991) 1770.
510. C. D. Liang, *Tetrahedron Lett.* **27** (1986) 1971.
511. D. J. Chadwick, D. S. Ennis, *Tetrahedron* **47** (1991) 9901.
512. F. Faigl, M. Schlosser, *Tetrahedron* **49** (1993) 10271.

5 Handling of Organometallic Reagents

1. M. Schlosser, K. F. Christmann, *Angew. Chem.* **77** (1965) 682; *Angew. Chem., Int. Ed. Engl.* **4** (1965) 689; M. Schlosser, Huynh Ba Tuong, B. Schaub, *Tetrahedron Lett.* **26** (1985) 311.
2. Adapted from K. Ziegler, K. Nagel, M. Patheiger, *Z. Anorg. Allg. Chem.* **282** (1955) 345.
3. Adapted from H. Gilman, J. A. Geel, C. G. Brannen, M. W. Bullock, G. E. Dunn, L. S. Miller, *J. Am. Chem. Soc.* **71** (1949) 1499.
4. Adapted from D. Seyferth, H. M. Cohen, *J. Organomet. Chem.* **1** (1963) 15.
5. R. West, E. G. Rochow, *J. Org. Chem.* **18** (1953) 1739.
6. Adapted from R. West, W. H. Glaze, *J. Org. Chem.* **26** (1961) 2096.
7. Adapted from G. Wittig, *Angew. Chem.* **53** (1940) 242; H. Gilman, J. W. Morton, *Org. React.* **8** (1954) 286.
8. M. Schlosser, V. Ladenberger, *J. Organomet. Chem.* **8** (1967) 193.
9. H. Gilman, B. J. Gaj, *J. Org. Chem.* **22** (1957) 1165.
10. H. Glman, T. S. Soddy, *J. Org. Chem.* **22** (1957(, 565; H. Filman, B. J. Gaj,*J. Org. Chem.* **22** (1957), 1165; D. L. Esmay, *Adv. Chem. Ser.* **23** 47, *Am. Chem. Soc.*, Washington, 1959; R. Waack, M. A. Doran, *J. Org. Chem.* **32** (1967), 3396.
11. U. Schöllkopf, H. Küppers, H. J. Traenckner, W. Pitteroff, *Liebigs Ann. Chem.* **704** (1967) 120.
12. R. Lehmann, M. Schlosser, *Tetrahedron Lett.* **25** (1984) 745.
13. G. Köbrich, H. R. Merkle, *Chem. Ber.* **99** (1966) 1782.
14. G. Köbrich, K. Flory, R. H. Fischer, *Chem. Ber.* **99** (1966) 1793.
15. Adapted from A. A. Morton, I. Hechenbleikner, *J. Am. Chem. Soc.* **58** (1936) 1697; A. A. Morton, F. D. Marsh, R. D. Coombs, A. L. Lyons, S. E. Penner, H. E. Ramsden, V. B. Baker, E. L. Little, R. L. Letsinger, *J. Am. Chem. Soc.* **72** (1950) 3785; R. Y. Mixer, W. G. Young, *J. Am. Chem. Soc.* **78** (1956) 3379.
16. The dispersion can be prepared by heating sodium in octane at 110 °C under high speed stirring. More conveniently, the toluene is removed from a commercial dispersion by filtration over a frit and replaced with pentane.
17. Equipment: Ultraturrax line of IKA Labortechnik, D-79219 Staufen, Germany.
18. Adapted from H. Gilman, H. A. Pacevitz, O. Baine, *J. Am. Chem. Soc.* **62** (1940) 1514; spec. 1517; J. F. Nobis, L. F. Moormeier, *Ind. Eng. Chem.* **46** (1954) 539; H. Ruschig, R. Fugmann, W. Meixner, *Angew. Chem.* **70** (1958), 71.

19. J. Hartmann, M. Schlosser, *Synthesis* (1975) 328; M. Stähle, M. Schlosser, *J. Organomet. Chem.* **220** (1981) 277.
20. K. Ziegler,F. Crössmann, H. Kleiner, O. Schäfer, *Liebigs Ann. Chem.* **473** (1929) 1, spec. 19; H. Gilman, R. V. Young, *J. Org. Chem.* **1** (1936) 315; H. Gilman, T. C. Wu, *J. Org. Chem.* **18** (1953) 753.
21. U. Schöllkopf, H. F. Ebel, A. Lüttringhaus, in *Houben-Weyl: Methoden der Organischen Chemie*, Vol. 13/1, pp. 97–226; 263–665, Georg Thieme, Stuttgart, 1970; H. W. Gschwend, H. R. Rodrigez, *Org. React.* **26** (1979) 1, spec. 96–103; B. J. Wakefield, *Organolithium Methods*, Academic Press, London, 1988.
22. M. Schlosser, *J. Organomet. Chem.* **8** (1967) 9; M. Schlosser, S. Strunk, *Tetrahedron Lett.* **25** (1984) 741; M. Schlosser, *Pure Appl. Chem.* **60** (1988) 1627.
23. J. Hartmann, M. Schlosser, *Helv. Chim. Acta* **59** (1976) 453.
24. M. Schlosser, J. Hartmann, M. Stähle, J. Kramař, A. Walde, A. Mordini, *Chimia* **40** (1986) 306.
25. M. Schlosser, T. Jenny, Y. Guggisberg, *Synlett* (1990) 704.
26. G. Rauchschwalbe, M. Schlosser, *Helv. Chim. Acta* **58** (1975) 1094.
27. M. Stähle, R. Lehmann, J. Kramař, M. Schlosser, *Chimia* **39** (1985) 229.
28. M. Schlosser, J. H. Choi, S. Takagishi, *Tetrahedron* **46** (1990) 5633.
29. G.-q. Shi, S. Cottens, S. A. Shiba, M. Schlosser, *Tetrahedron* **48** (1992) 10569.
30. G. Köbrich, *Angew. Chem.* **84** (1972), 557; *Angew. Chem. Int. Ed. Engl.* **11** (1972), 473.
31. K. Oshima, K. Simoji, H. Rakahashi, H. Yamamoto, H. Nozaki, *J. Am. Chem. Soc.* **95** (1973), 2694; R. Muthukrishnan, M. Schlosser, *Helv. Chim. Acta* **59** (1976), 13.
32. O. Riobé, A. Lebouc, J. Delaunay, *C.R. Séances Acad. Sci.* **284** (1977), 281; *Chem. Abstr.* **86** (1977), 189'646e; R. K. Boeckman, K. J. Bruza, *Tetrahedron Lett.* **18** (1977), 4187; see also R. Paul, S. Tchelitcheff, *C.R. Séances Acad. Sci.* **235** (1952), 1226; *Chem. Abstr.* **48** (1954), 1944b; *Bull. Soc. Chim. Fr.* **1952**, 808.
33. O. W. LEver, *Tetrahedron* **32** (1976), 1943.
34. U. Schöllkopf, in *Houben-Weyl: Methoden der organischen Chemie* (E. Müller, ed.), **13/1**, 116, Thieme, Stuttgart, 1970; U. Schöllkopf, P. Hänßle, *Leibis Ann. Chem.* **763** (1974), 208.
35. J. Hartmann, M. Stähle, M. Schlosser, *Synthesis* (1974) 888.
36. M. A. Hassan, D. L. Nguyen, Y. Bessard, M. Schlosser, unpublished results (1979–87); see also O. Riobé, A. Lebouc, J. Delaunay, *C.R. Acad. Sci.* **284** (1977) 281; *Chem. Abstr.* **86** (1977), 189'646e; M. A. Hassan, *J. Chem Soc. Pak.* **5** (1983), 103; *Chem. Abstr.* **99** (1983), 212'107j.
37. R. Paul, *C.R. Acad. Sci.* **218** (1944) 122; O. Riobé, *Bull. Soc. Chim. Fr* **1951**. 829.
38. M. Schlosser, G. Katsoulos, S. Takagishi, *Synlett* (1990) 747.
39. G. Katsoulos, S. Takagishi, M. Schlosser, *Synlett* (1991) 731.
40. G. Simig, M. Schlosser, *Tetrahedron Lett.* **29** (1988) 4277; **31** (1990) 5213.
41. G.-f. Zhong, X.-h. Wang, L. Garamszegi, M. Schlosser, unpublished results (1992–1993).
42. See also: K. Oita, H. Gilman, *J. Am. Chem. Soc.* **79** (1957), 339; H. Gilman, W. J. Trepka, *J. Org. Chem.* **26** (1961), 5202; I. Granoth, J. B. Levy, C. Symmes, *J. Chem. Soc., Perkin Trans. II* **1972**, 697.
43. C. A. Kraus, R. Rosen, *J. Am. Chem. Soc.* **47** (1925) 2739; R. Levine, E. Baumgarten, C. R. Hauser, *J. Am. Chem. Soc.* **66** (1944) 1230; C. R. Hauser, D. S. Hoffenberg, W. H. Puterbaugh, F. C. Frostik, *J. Org. Chem.* **20** (1955) 1531.
44. C. D. Broaddus, *J. Am. Chem. Soc.* **88** (1966) 4174.
45. A. J. Chalk, T. J. Hoogeboom, *J. Organomet. Chem.* **11** (1968) 615; C. D. Broaddus, *J. Org. Chem.* **35** (1970) 10.

46. R. A. Benkeser, T. V. Liston, *J. Am. Chem. Soc.* **82** (1960) 3221; see also A. A. Morton, I. Hechenbleikner, *J. Am. Chem. Soc.* **58** (1936) 2599.
47. S. Takagishi, M. Schlosser, *Synlett* (1991) 119.
48. P. Maccaroni, M. Schlosser, unpublished results (1992).
49. E. Moret, M. Schlosser, unpublished results (1987).
50. F. Faigl, M. Schlosser, *Tetrahedron Lett.* **32** (1991) 3369.
51. H. Bosshardt, M. Schlosser, *Helv. Chim. Acta* **63** (1980) 2393.
52. O. Desponds, M. Schlosser, *J. Organomet. Chem.* **409** (1991) 93; O. Desponds, *Doctoral Thesis*, Université de Lausanne (1991).
53. E. Moret, L. Franzini, M. Schlosser, unpublished results (1987–92); see also M. Zaidlewicz, *J. Organomet. Chem.* **293** (1985) 139.
54. K. Fujita, M. Schlosser, *Helv. Chim. Acta* **65** (1982) 1258; C. Margot, M. Schlosser, unpublished results (1985).
55. The stereolabels *erythro* and *threo* are used as defined by R. Noyori, I. Nishida, J. Sakata, *J. Am. Chem. Soc.* **103** (1981) 2106.
56. M. Schlosser, G.-f. Zhong, *Tetrahedron Lett.* **34** (1993), 5441.
57. R. B. Bates, D. W. Gosselink, J. A. Kaczynski, *Tetrahedron Lett.* **8** (1967), 199; see also R. Paul, S. Tchelitcheff, *C.R. Séances Acad. Sci.* **239** (1954), 1222; *Chem. Abstr.* **49** (1955), 13'915d.
58. M. Schlosser, H. Bosshardt, A. Walde, M. Stähle, *Angew. Chem.* **92** (1980) 302; *Angew. Chem., Int. Ed. Engl.* **19** (1980) 303.
59. R. Lehmann, M. Schlosser, unpublished results (1979).
60. M. Schlosser, G. Rauchschwalbe, *J. Am. Chem. Soc.* **100** (1978) 3258.
61. G. Rauchschwalbe, A. Zellner, M. Schlosser, unpublished results (19977; 93).
62. G. Wittig, R. Ludwig, R. Polster, *Chem. Ber.* **88** (1955) 294; G. Wittig, F. Bickelhaupt, *Chem. Ber.* **91** (1958) 865; W. H. Puterbaugh, C. R. Hauser, *J. Org. Chem.* **28** (1963) 3465.
63. A. Maercker, J. D. Roberts, *J. Am. Chem. Soc.* **88** (1966) 1742.
64. J. J. Eisch, A. M. Jacobs, *J. Org. Chem.* **28** (1963) 2145.
65. R. L. Letsinger, J. G. Traynham, *J. Am. Chem. Soc.* **70** (1948) 3342.
66. J. Hartmann, R. Muthukrishnan, M. Schlosser, *Helv. Chim. Acta* **57** (1974) 2261.
67. K. Ziegler, H. Dislich, *Ber. Dtsch. Chem. Ges.* **90** (1957), 1107, spec. 1115.
68. J. J. Eisch, W. C. Kaska, *J. Org. Chem.* **27** (1962), 3745.
69. H. Gilman, S. M. Spatz, *J. Org. Chem.* **16** (1951) 1485.
70. Y. Guggisberg, F. Faigl, M. Schlosser, *J. Organomet. Chem.* **415** (1991) 1.
71. UM. Schlosser, J. Hartmann, *Angew. Chem.* **85** (1973), 544; *Angew. Chem. Int. Ed. Engl.* **12** (1973) 508.
72. K. S. Y. Lau, M. Schlosser, *J. Org. Chem.* **43** (1978) 1595.
73. J. F. Arens, D. A. van Dorp, *Recl. Trav. Chim. Pays-Bas* **67** (1948) 973.
74. G. Wittig, H. Reif, *Angew. Chem.* **80** (1968), 8; *Angew. Chem. Int. Ed. Engl.* **7** (1968), 7.
75. A. Rembaum, S. P. Siao, N. Indictor, *J. Polymer Sci.* **56** (1962), 517; *Chem. Abstr.* **56** (1962), 13'083a; R. B. Bates, L. M. Kroposki, D. E. Potter, *J. Org. Chem.* **37** (1972), 560; A. Maercker, W. Theysohn, *Liebigs Ann. Chem.* **747** (1971), 70.
76. T. R. Hoye, S. J. Martin, D. R. Peck, *J. Org. Chem.* **47** (1982), 331; see also H. Gilman, E. A. Zoellner, W. M. Selby, *J. Am. Chem. Soc.* **54** (1932) 1957.
77. G. Wittig, U. Pockels, H. Dröge, *Ber. Dtsch. Chem. Ges.* **71** (1938) 1903; H. Gilman, R. L. Bebb, *J. Am. Chem. Soc.* **61** (1939) 109; see also A. A. Morton, I. Hechenbleikner, *J. Am. Chem. Soc.* **58** (1936) 2599.

78. X.-P. Zhang, L. Garamszegi, M. Schlosser, unpublished results (1993).
79. U. Schöllkopf, in *Houben-Weyl: Methoden der Organischen Chemie*, ed. E. Müller, Vol. 13/1, pp. 157–158, Georg Thieme, Stuttgart, 1970.
80. K. Ziegler, H. Zeiser, *Ber. Dtsch. Chem. Ges.* **63** (1930) 1847; *Liebigs Ann. Chem.* **485** (1931) 174.
81. L. A. Walters, S. M. McElvain, *J. Am. Chem. Soc.* **55** (1933) 4625; H. Gilman, G. C. Gainer, *J. Am. Chem. Soc.* **69** (1947) 877; H. Gilman, R. D. Nelson, *J. Am. Chem. Soc.* **70** (1948) 3316; H. Gilman, J. Eisch, *J. Am. Chem. Soc.* **79** (1957) 4423; H. Bredereck, R. Gompper, H. Herlinger, *Chem. Ber.* **91** (1958) 2832, 2844; R. Grashey, R. Husigen, *Chem. Ber.* **92** (1959) 2641; R. A. Abramovitch, J. G. Saha, *Adv. Heterocycl. Chem.* **6** (1966) 229; G. Wittig, U. Thiele, *Liebigs Ann. Chem.* **726** (1969); A. E. Hauck, C. S. Giam, *J. Chem. Soc., Perkin Trans. 1* (1979) 2393.
82. D. A. Shirley, J. R. Johnson, J. P. Hendrix, *J. Organomet. Chem.* **11** (1968) 209.
83. J. P. Wibaut, A. P. de Jonge, H. G. P. van der Voort, P. P. H. L. Otto, *Recl. Trav. Chim. Pays-Bas* **70** (1951) 1054; see also H. Gilman, S. M. Spatz, *J. Org. Chem.* **16** (1951) 1485; H. E. French, K. Sears, *J. Am. Chem. Soc.* **73** (1951) 469; J. P. Wibaut, L. G. Heeringa, *Recl. Trav. Chim. Pays-Bas* **74** (1955) 1003; D. W. Boykin, A. R. Patel, R. E. Lutz, A. Burger, *J. Heterocycl. Chem.* **4** (1967) 459.
84. Adapted from K. Ziegler, H. Zeiser, *Ber. Dtsch. Chem. Ges.* **63** (1930) 1847.
85. K. Ziegler, H. Gellert, *Liebigs Ann. Chem.* **567** (1950) 179, 185; K. Ziegler, K. Nagel, M. Patheiger, *Z. Anorg. Allg. Chem.* **285** (1955) 345.
86. W. H. Carothers, D. D. Coffman, *J. Am. Chem. Soc.* **51** (1929) 588; **52** (1930) 1254; A. A. Morton, H. A. Newey, *J. Am. Chem. Soc.* **64** (1942) 2247; A. A. Morton, E. J. Lanpher, *J. Org. Chem.* **21** (1956) 93.
87. R. A. Finnegan, H. W. Kutte, *J. Org. Chem.* **30** (1965) 4139; W. H. Glaze, J. Lin, E. G. Felton, *J. Org. Chem.* **31** (1966) 2643.
88. H. Gilman, A. H. Haubein, H. Hartzfeld, *J. Org. Chem.* **19** (1954) 1034; H. Gilman, B. J. Gaj, *J. Org. Chem.* **22** (1957) 1165.
89. H. Gilman, G. L. Schwebke, *J. Organomet. Chem.* **4** (1965) 483.
90. D. Seyferth, H. M. Cohen, *J. Organomet. Chem.* **1** (1963) 15.
91. P. D. Bartlett, S. Friedman, M. Stiles, *J. Am. Chem. Soc.* **75** (1953) 1771.
92. D. Seyferth, M. A. Weiner, *J. Am. Chem. Soc.* **83** (1961) 3583.
93. S. C. Honeycutt, *J. Organomet. Chem.* **29** (1971) 1.
94. R. L. Letsinger, E. Bobko, *J. Am. Chem. Soc.* **75** (1953) 2649.
95. G. Wittig, R. Polster, *Liebigs Ann. Chem.* **599** (1956) 13.
96. R. L. Letsinger, D. F. Pollart, *J. Am. Chem. Soc.* **78** (1956) 6179; R. L. Letsinger, *Angew. Chem.* **70** (1958), 153.
97. A. Maercker, W. Demuth, *Liebigs Ann. Chem.* **1977**, 1909.
98. G. Wittig, L. Löhmann, *Liebigs Ann. Chem.* **599** (1956) 13.
99. J. J. Fitt, H. Gschwend, *J. Org. Chem.* **49** (1984) 209.
100. E. Moret, O. Desponds, M. Schlosser, *J. Organomet. Chem.* **409** (1991) 83.
101. R. E. Goldberg, D. E. Lewis, K. Cohn, *J. Organomet. Chem.* **15** (1968) 491.
102. J. Millon, G. Linstrumelle, *Tetrahedron Lett.* **17** (1976), 1095; C. Huynh, F. Derguinni-Boumechal, G. Linstrumell, *Tetrahedron Lett.* **20** (1979), 1503.
103. R. Lehmann, M. Schlosser, *Tetrahedron Lett.* **25** (1984), 745.
104. P. Tomboulian, D. Anick, S. Beare, K. Dumke, D. Hart, R. Hites, A. Metzger, R. Nowak, *J. Org. Chem.* **38** (1973); see also R. B. Bates, L. M. Kroposki, D. E. Potter, *J. Org. Chem.* **37** (1972) 560.

105. M. E. Jung, R. B. Blum, *Tetrahedron Lett.* **18** (1977), 3791; K. Kamata, M. Terashima, *Heterocycles* **14** (1980) 205.
106. H. Gilman, A. H. Haubcin, *J. Am. Chem. Soc.* **66** (1944) 1515; see also H. Gilman, F. K. Cartledge, *J. Organomet. Chem.* **2** (1964) 447; K. C. Eberly, *J. Org. Chem.* **26** (1961) 1309.
107. D. J. Ager, *J. Organomet. Chem.* **241** (1982) 139.
108. J. B. Conant, G. W. Wheland, *J. Am. Chem. Soc.* **54** (1932) 1212.
109.
110. S. C. Watson, J. F. Eastham, *J. Organomet. Chem.* **9** (1967) 165.
111. L. Duhamel, J.-C. Plaquevent, *J. Org. Chem.* **44** (1979) 3404.
112. D. E. Bergbreiter, E. Pendergras, *J. Org. Chem.* **46** (1981) 219.
113. W. G. Kofron, L. M. Baclawski, *J. Org. Chem.* **41** (1976) 1879.
114. E. Juaristi, A. Martinez-Richa, A. Garcia-Rivera, J. S. Cruz-Sanchez, *J. Org. Chem.* **48** (1983) 2603.
115. M. R. Winkle, J. M. Lansinger, R. C. Ronald, *J. Chem. Soc., Chem. Commun.* (1980) 87.
116. M. F. Lipton, C. M. Sorensen, A. C. Sadler, R. H. Shapiro, *J. Organomet. Chem.* **186** (1980) 155.
117. H. Kiljunen, T. A. Hase, *J. Org. Chem.* **56** (1991) 6950.
118. Y. Aso, H. Yamashita, T. Otsubo, F. Ogura, *J. Org. Chem.* **54** (1989) 5627.
119. A. v. Baeyer, V. Villiger, *Ber. Dtsch. Chem. Ges.* **36** (1903) 2775; P. Ehrlich, F. Sachs, *Ber. Dtsch. Chem. Ges.* **36** (1903) 4296; F. Sachs, L. Sachs, *Ber. Dtsch. Chem. Ges.* **37** (1904) 3088.
120. H. Gilman, F. Schulze, *J. Am. Chem. Soc.* **47** (1925) 2002.
121. J. M. Gaidis, *J. Organomet. Chem.* **8** (1967) 385.
122. H. Gilman, J. Swiss, *J. Am. Chem. Soc.* **62** (1940) 1847.
123. FMC Corporation, Lithium Division, 449 North Cox Road, Gastonia, NC 28054, USA.
124. Chemetall, Sparte Lithium, Postfach 101501, D-60323 Frankfurt, Germany; see also J. Deberitz, *Lithium: Production and Application of a Fascinating and Versatile Element*, Verl. Mod. Ind., Landsberg, 1993.
125. U. Schöllkopf, in *Houben-Weyl: Methoden der Organischen Chemie*, ed. E. Müller, Vol. 13/1, pp. 16–18, Georg Thieme, Stuttgart, 1970; D. F. Shriver, M. A. Drezdzon, *The Manipulation of Air-Sensitive Compounds*, Wiley, New York, 1986; G. B. Gill, D. A. Whiting, *Aldrichim. Acta* **19** (1986) 31; B. J. Wakefield, *Organolithium Methods*, pp. 4–15, Academic Press, London, 1988.
126. Peco Laborbedarf, Sensfelder Weg 8, D-64293 Aldrich Chem. Co., P.O. Box 355, Milwaukee, Wisconsin 53201, USA.

2

Organolithium Compounds—Industrial Applications and Handling

FRANZ TOTTER and PETER RITTMEYER
Chemetall GmbH, Frankfurt am Main, Germany

Organometallics in Synthesis—A Manual. Edited by M. Schlosser

2.1 INTRODUCTION

Although known for almost 80 years [1] and readily available even in large quantities since the thirties [2], organolithium compounds have been used in the chemical industry for only three decades. For a long time the only organolithium compound manufactured on an industrial scale was butyllithium. Its main use was as a catalyst in the anionic polymerization of alkenes to produce synthetic rubbers [3].

A search in the WPAT database shows a substantial increase in the number of patents dealing with organolithium compounds during the past 20 years:

- 46 patents/year in 1970–80;
- 62 patents/year in 1981–85;
- 89 patents/year in 1986–90.

In the same manner, the demand for organolithium compounds has increased. Today the commercial production of butyllithium has reached an order of approximately 1000 tons/year of neat butyllithium. Other organolithium compounds, e.g. methyllithium and phenyllithium, have meanwhile also become commercially available, although they are manufactured in considerable smaller quantities.

The most important application of butyllithium is still its use as a polymerization catalyst in the synthetic rubber industry. However, butyllithium and other organolithium compounds are more and more recognized as valuable tools for industrial organic synthesis, e.g. for the production of pharmaceuticals and agrochemicals. Owing to their strong basicity, which can be modified by various additives such as metal alkoxides and amines, they can be used as versatile reagents for the metallation of organic substrates. This opens the field to more sophisticated chemistry in industry too.

Organometallic chemistry and its application in organic synthesis play an important role in the professional training of chemists. Every student has experience of using butyllithium during his or her practical work on preparative chemistry. Although the advantages of organolithium compounds are well known to chemists in production facilities, they still try to avoid them as organolithium compounds are regarded as dangerous and not to be handled on an industrial scale. For the synthesis of a target molecule they often prefer a more complicated and often troublesome way.

The aim of this chapter is to reduce this psychological barrier. It will be shown by means of important examples that organolithium compounds are already used in industrial organic synthesis.

With the knowledge of the properties and the equipment for handling air- and moisture-sensitive substances, it is possible to take advantage of organolithium compounds in organic synthesis even on a larger scale. If attention is paid to some safety precautions, their handling is feasible without problems.

It is intended to give an introduction to the handling of organolithium compounds on an industrial scale. Detailed information for potential users is available from the producers.

2.2 SYNTHESIS AND PROPERTIES

2.2.1 INDUSTRIAL SYNTHESIS OF ORGANOLITHIUM COMPOUNDS

Although many methods are known for preparing organolithium compounds in laboratory [4], the commercially available derivatives are manufactured by reacting an organic halide, mainly chlorides, with lithium metal:

$$\mathrm{RCl} + 2\,\mathrm{Li} \longrightarrow \mathrm{RLi} + \mathrm{LiCl} \qquad (2.1)$$

To realize this reaction on an industrial scale, the following points are of special interest.

Organolithium compounds are usually prepared as solutions in hydrocarbon solvents. The reaction is highly exothermic, and therefore the temperature has to be controlled carefully. The rate of the reaction is determined by the rate of addition of the halide and is limited by the capacity of the reflux condenser.

An important point with regard to the economy of the process is the quality of the lithium metal used. A satisfactory yield and a high rate constant are obtained if the lithium metal, which is applied as a fine dispersion, contains 0.5–2% of sodium. The sodium content supports the initiation and accelerates the reaction. After completion of the reaction, excess of lithium metal and lithium chloride are removed.

Concerning butyllithium and *sec*-butyllithium, which are the commercially most important organolithium compounds, the hydrocarbon solution obtained from synthesis is normally concentrated to $>90\%$. The concentrate is stored and can be diluted with different solvents to different concentrations according to the customers' needs.

The manufacture of *tert*-butyllithium is somewhat problematic because *tert*-butyllithium and unreacted *tert*-butyl chloride easily react to give hydrocarbon by-products. Even under optimum conditions the yields are considerably lower than for butyllithium and *sec*-butyllithium [5].

Because of its insolubility in hydrocarbon solvents, methyllithium is produced normally in diethyl ether as solvent. The preparation in aromatic hydrocarbon–tetrahydrofuran (THF) mixtures containing some magnesium has been described [6].

Phenyllithium, which up to now has not been produced in substantial quantities, is prepared in a mixture of 70% cyclohexane and 30% diethyl ether.

2.2.2 AVAILABILITY

Butyllithium and *sec*-butyllithium are available in different solvents and different concentrations:

butyllithium	15%, 23%, 50% and 90% in hexane
	20% in cyclohexane
	20% in toluene
sec-butyllithium	10% in isopentane
	12% in cyclohexane
tert-butyllithium	15% in pentane
	15% in hexane

The highest concentration in which butyllithium is shipped is 90%. This solution is recommended if the user's process requires a solvent other than hexane, cyclohexane or toluene. It also helps to avoid additional transportation costs.

It should be stressed that hexyllithium is now available on an industrial scale. The common concentration is 33%. On a molar basis this concentration corresponds to the common concentration of 23% butyllithium. This reagent could help to overcome the barrier in using organolithium compounds in organic synthesis because a disadvantage of butyllithium, the formation of the gaseous by-product butane (see Section 2.4.1) is eliminated. Instead of butane the commonly used solvent *n*-hexane is formed.

The following organolithium compounds are also available on an industrial scale; other may be manufactured on request.

methyllithium	5% in diethyl ether
	6.4% in diethyl ether containing *ca* 10% LiBr
	2.6% in cumene–THF containing 0.5% methylmagnesium
ethyllithium	2% in hexane
phenyllithium	20% in cyclohexane–diethyl ether
lithium acetylide	solid complex with ethylenediamine

2.2.3 PROPERTIES OF ORGANOLITHIUM COMPOUNDS

Organolithium compounds in the eyes of a production chemist show properties which a university chemist does not consider so important. Although some monographs have been published on their preparation and constitution [7], little information about the technical and economic properties of organolithium compounds is available. In the opinion of a process development chemist, the kind of aggregates of organolithium compounds are not as important as e.g. the stability, the reactivity and, of course, the price. The following sections deal with these aspects and give some examples.

2.2.3.1 Stability and Solvents

Organolithium compounds are not thermally stable. Ziegler and Gellert [8] have studied the thermal decomposition of concentrated butyllithium. The main reaction results in the formation of lithium hydride and 1-butene in the course of a β-elimination:

$$C_4H_9Li \longrightarrow CH_3CH_2CH{=}CH_2 + LiH \qquad (2.2)$$

In an open system the evolved gas consists almost exclusively of 1-butene. In a closed system, however, butane and some polymeric material (from butadiene produced by elimination of lithium hydride from crotyllithium?) are also formed to a certain extent owing to secondary reactions:

$$\begin{aligned} CH_3CH_2CH{=}CH_2 + LiC_4H_9 &\longrightarrow CH_3CH(Li)CH{=}CH_2 + C_4H_{10} \\ CH_3CH(Li)CH{=}CH_2 &\longrightarrow CH_2{=}CHCH{=}CH_2 + LiH \end{aligned} \qquad (2.3)$$

Since most commercially available organolithium compounds are supplied as solutions in hydrocarbon solvents, their thermal stability in these solvents is of special interest. It is reported that the thermal decomposition of butyllithium in hydrocarbon solution is a first-order reaction catalysed by the corresponding lithium *n*-butoxide, which may be formed in the presence of traces of oxygen [9]. The decompositions of *sec*-butyllithium [10] and *tert*-butyllithium [11] are best described by a 0.5-order kinetic law. Nevertheless, in a closed system, e.g. a transport container, there is no appreciable pressure build-up as the olefin formed in the decomposition process dissolves well in the applied solvent, which is also a hydrocarbon.

In general, the decomposition rate of an organolithium compound depends strongly on the concentration of the solution, the solvent, the temperature and the content of free base, i.e. lithium alkoxide. Table 2.1 gives the decomposition rates of commercially important butyllithium solutions as a function of temperature and concentration [12].

The isomers of butyllithium differ remarkably in their stability. *tert*-Butyllithium is more stable than butyllithium. *sec*-Butyllithium is by far the least stable isomer and therefore has to be stored and shipped in isolated containers below 0 °C. The stability of *n*-hexyllithium seems to be higher than that of butyllithium.

The solvation of organolithium compounds in hydrocarbons normally does not

Table 2.1. Decomposition rates (% material lost per day)

Storage temperature (°C)	C_4H_9Li 15–20% in hexane	C_4H_9Li 90% in hexane	*sec*-C_4H_9Li 10–12% in isopentane
0	0.00001	0.0005	0.003
5	0.0002	0.0011	0.006
10	0.0004	0.0025	0.012
20	0.0018	0.013	0.047
35	0.017	0.11	0.32

Table 2.2. Decomposition of butyllithium in ethers [15]

Ether	Temperature (°C)	$t_{1/2}$
Diethyl ether	25	6 d
	35	31 h
Diisopropyl ether	25	18 d
Glycol dimethyl ether	25	10 min
THF	0	23.5 h
	−30	5 d

produce significant heat. The heat of solvation for, e.g., butyllithium in light petroleum is *ca* $-0.7\ \text{kcal mol}^{-1}$ [13]. This confirms the experience that the dilution of concentrated solutions can be done without problems.

Common solvents for reactions with organolithium compounds are ethers. The addition of donor solvents to hydrocarbon solutions results in a break-up of the existing clusters due to solvation of the lithium ion:

$$(RLi)_n \rightleftharpoons (RLi)_6 \rightleftharpoons (RLi)_4 \rightleftharpoons (RLi)_2 \quad (2.4)$$

The enhanced nucleophilicity of the residual alkyl increases the reactivity, resulting in cleavage of the ether C—O bonds [14]. To slow this undesired side-reaction, low temperatures are necessary using ethers as co-solvents (Table 2.2).

The cleavage of diethyl ether results in the formation of ethylene and lithium methoxide. The decomposition products of THF are ethylene and the lithium enolate of acetaldehyde. It is remarkable that the temperature coefficient of the ether cleavage becomes negative if equimolar amounts of symmetrical ethers are added to heptane solutions of organolithium compounds [15].

2.2.3.2 Behaviour Towards Air and Water

Organolithium compounds react with oxygen of the air to form lithium alkoxides. Reaction with moisture in air results in the formation of lithium hydroxide and the corresponding hydrocarbons, butane gas in the case of butyllithium. These reactions, of course, decrease the content of active material.

The risk of self-ignition in air due to the pyrophoric properties of organolithium compounds depends on several factors:

- nature of the organolithium compound;
- concentration of the solution;
- humidity of the air;
- surface on which the solution is spilled;
- age of the solution.

Various experiments summarized below show the pyrophoric properties of butyllithium solutions [16]. A video tape which illustrates the experiments and gives some practical hints in handling organolithium compounds is available from the authors.

If a 15% solution of butyllithium is poured into a porcelain cup, self-ignition normally does not occur. Nevertheless, a reaction with air takes place. When a filter-paper is dipped in the solution and then exposed to air it begins to char but does not ignite. In the case of a 50% solution, again no self-ignition of the solution is observed but the filter-paper immediately catches fire. This demonstrates that the risk of self-ignition increases with the concentration of the solution. If the surface is large enough that evaporation of the solvent is sufficiently fast, self-ignition is possible. Therefore, special care has to be taken not to spill butyllithium or other organolithium compounds on to materials with large specific surface areas, e.g. clothing, mineral wool or filter-papers.

The risk of self-ignition is also dependent on the age of the butyllithium solution, as during storage lithium hydride is formed. If the amount of finely dispersed lithium hydride is large enough, spilled butyllithium solutions normally ignite.

The addition of water to solutions of organolithium compounds leads to immediate hydrolysis with the evolution of heat and the formation of the respective hydrocarbon. In the presence of oxygen spontaneous ignition can occur. The heat of hydrolysis for, e.g., butyllithium (15% in light petroleum) is $-54.7\ \mathrm{kcal\ mol^{-1}}$ [13].

Generally, although butyllithium solutions at the usual concentrations do not catch fire when spilled, self-ignition cannot be ruled out. Therefore, they should always be kept under an inert gas (see Section 2.4; pp. 181–183).

2.2.3.3 Toxicology

The toxicological properties of organolithium compounds refer to the properties of the hydrolysis products. The caustic effect of lithium hydroxide leads to severe irritation of the skin, mucous membranes, eyes and respiratory tract. The lithium ion is considerably more toxic than the sodium ion. On the other hand, certain forms of manic-depressive diseases are treated with dosages of lithium carbonate [17].

The properties of the organic hydrolysis products can be found in the literature [18].

2.3 APPLICATION

2.3.1 POLYMER CHEMISTRY

After the identification of natural rubber as *cis*-1,4-polyisoprene at the beginning of this century, attempts were made to produce rubbers synthetically [19]. The first industrial processes started with conjugated dienes and alkali metals as catalysts. In the 1920s the former IG Farbenindustrie manufactured a sodium-processed polybutadiene called BUNA [BUtadiene-NAtrium (sodium)].

The first styrene–butadiene copolymers and butadiene–acrylonitrile rubbers were made in the late 1930s with persulfates and peroxides as radical catalysts. The styrene–butadiene rubbers are still the most important synthetic rubbers.

In the 1950s a new development in the field of synthetic rubbers was started. Organometallic complex catalysts of the Ziegler–Natta type [20, 21] and also lithium [22] and lithium alkyls [23] catalyse the polymerization of isoprene to *cis*-1,4-polyisoprene, the structure of natural rubber.

2.3.1.1 Polybutadiene

After the BUNA production in the 1930s, the polymerization of butadiene was considerably improved when the Ziegler–Natta catalysts based on trialkylaluminium–titanium(IV) chloride were discovered in 1956. They polymerize butadiene mainly in the 1,4-*cis*-position [24]. BUNA, on the other hand, contains 10% 1,4-*cis*-polybutadiene, 20% 1,4-*trans*- and about 70% 1,2-linked butadiene with vinyl groups.

The initial step in the polymerization reaction is the addition of butyllithium to the butadiene monomer to form a butylbutenyllithium [25]:

$$H_9C_4{-}Li + H_2C{=}CHCH{=}CH_2 \longrightarrow H_9C_4CH_2{-}CH{=}CHCH_2Li \quad (2.5)$$

The next step is the growth of the chain [26], which continues until all monomer is consumed. Under normal reaction conditions no termination of the polymerization reaction occurs by elimination of lithium hydride, the polymer staying active ('living polymers' [27]). The resulting polymers have a narrow molecular weight distribution which can be adjusted by varying the concentration of the butyllithium catalyst. The microstructure can be modified over a wide range by varying the counter ion [28] (Li, Na, K), the solvent and/or the temperature [29].

Owing to its good availability and the resulting lower price, rubbers based on butadiene (butadiene rubbers, BR) are much more important than those of isoprene (isoprene rubbers, IR; see below).

Owing to their excellent abrasion resistance and high resilience, the main use for butadiene rubbers is in the field of car and truck tyres. They are also used for shoe soles and mechanical rubber goods, e.g. conveyor belts.

2.3.1.2 Polyisoprene

The polymerization of isoprene is carried out in solution with pentane or hexane as solvent. Lithium alkyls are more advantageous than Ziegler–Natta catalysts as they are less sensitive to impurities. The lithium alkyls are added to the isoprene at a concentration of *ca* 0.1–0.2 mmol per 100 g.

The polymerization reaction is again divided into two steps [30]. A slow initial step is followed by a fast chain growth, again forming 'living polymers':

$$\begin{aligned} LiC_4H_9 + C_5H_8 &\longrightarrow LiC_5H_8C_4H_9 \\ LiC_5H_8C_4H_9 + nC_5H_8 &\longrightarrow Li(C_5H_8)_{n+1}C_4H_9 \end{aligned} \quad (2.6)$$

Natural rubber is nearly exclusively built up from *cis*-1,4-polyisoprene units. Synthetic isoprene rubber contains, depending on the kind of catalyst, between 90–92% (Li type) and 98% (Ziegler–Natta type) *cis*-1,4-polyisoprene. In the case of the lithium-type catalyst the remainder consists of 2–3% *trans*-1,4-polyisoprene and 6–7% 3,4-polyisoprene [31].

Synthetic isoprene rubber types are used in the same fields as natural rubber, e.g. as vehicle tyres. Ziegler–Natta-type rubber is an adequate replacement for natural rubber whereas lithium-type rubber can only be used in combination with other rubber types.

2.3.1.3 Styrene–Butadiene Rubber

The production of styrene–butadiene rubbers (SBR) can be carried out both in emulsion and in solution. Emulsion-polymerized styrene–butadiene rubbers consists of *ca* 25% styrene and *ca* 75% butadiene in a random distribution. The catalysts for this copolymerization are radical-forming compounds such as peroxides or persulfates.

Solution styrene–butadiene rubber is generally made with lithium-based catalysts such as butyllithium. The resulting copolymers are either random copolymers such as the emulsion-polymerized styrene–butadiene rubber or block copolymers. The block copolymers are made by polymerization of styrene or butadiene and continuation of the polymerization by addition of the counterpart (butadiene or styrene monomers).

With dilithium compounds, three-block polymers which are of economic interest are in principle accessible [32]. They could be synthesized by a discontinuous polymerization of butadiene and styrene.

As dilithium compounds are still not readily accessible, polymerization is carried out in practice with butyllithium. Polymers made with lithium alkyls are 'living polymers,' i.e. still reactive. Terminating reactions, such as elimination of LiH, do not occur under the reaction conditions normally used. The 'living polymers' are finally coupled by reagents such as 1,2-dibromoethane or silicon tetrachloride to give the SBS three-block polymer. These polymers are deformable at higher temperatures (thermoplastic elastomers).

	Polymer	Catalyst
Block copolymers:	–SSSSSSS–BBBBBBBBBB–	Butyllithium
	–SSS–BBBBBBB–SSSSSS–	Lithium alkyls and dilithium compounds
Random copolymers:	–S–BBB–SS–BB–SS–BBB–	Butyllithium

2.3.2 *ORGANIC SYNTHESIS*

As a result of its widespread application as an initiator for anionic polymerization, butyllithium is today available in large quantities at a reasonable price (*ca* 75DM kg^{-1} butyllithium content for bulk quantities). On the other hand, the equipment and the know-how for safe handling of air- and moisture-sensitive chemicals also exists. Therefore, chemists and engineers began to transfer the long experience with organolithium compounds as useful reagents in organic synthesis from the laboratory to the large-scale production of organic intermediates. Today an increasing percentage of butyllithium production is used for the synthesis of organic derivatives, especially in the pharmaceutical industry, but also for the manufacture of agrochemicals, flavours and fragrances [33]. The use of organolithium compounds in industrial organic synthesis is still a rapidly growing field.

In many cases the advantages of using organolithium compounds in chemical synthesis outweigh the necessary investment in appropriate technical equipment. Reactions with organolithium compounds are normally very economical because of the mild reaction conditions, high yields, few by-products and easy separation of the products from the resulting lithium salts. In some cases it is also possible to recycle the lithium residues (see Section 2.4.5), thus avoiding additional costs for their disposal.

Most frequently butyllithium (or hexyllithium) is used to generate lithiated intermediates by substitution reactions via metal–hydrogen or metal–halogen exchange followed by further reaction of the resulting carbanions with electrophiles. Special organolithium compounds such as methyllithium and ethyllithium are mainly used to perform direct addition of the alkyl group to multiple bonds, e.g. carbonyl groups.

In the following, some selected reactions for the use of organolithium compounds in organic synthesis on an industrial scale are given as examples. Although arbitrary to some extent, the choice made is meant to illustrate the wide range of chemistry and applications where organolithium compounds are useful.

2.3.2.1 Deprotonation Reactions

Some reactions involving a carbanion generated via proton abstraction by the strong base butyllithium are industrially important:

$$\mathrm{RLi + R'H \longrightarrow RH + R'Li} \qquad (2.7)$$

Hetero-substituted unsaturated, heteroaromatic and aromatic systems are important structural fragments in many pharmaceuticals and agrochemicals. Since these compounds can easily be metallated with organolithium bases, the corresponding carbanions are useful for numerous derivatization reactions with different electrophiles.

Direct metallation of heteroaromatic systems normally takes place at the α-position adjacent to the heteroatom. Since, e.g., thiophenes and furans are readily available and the 2-lithio derivatives are easy to obtain and relatively stable, their derivatization reactions have been studied extensively [4a, 34]. In the following some industrially important examples are discussed.

The first example is the synthesis of ticlopidine [35]. Although several methods of preparation have been described, the most convenient is with butyllithium. Thiophene is lithiated in the 2-position with butyllithium in THF–hexane. Then the organolithium species formed is reacted with oxirane and after hydrolysis 2-(2-thienyl)ethanol, a key intermediate in the synthesis of this drug, is obtained (reaction 2.8).

$\mathrm{LiC_4H_9}$ Ticlopidine (2.8)

The pharmacological activity of ticlopidine has been thoroughly examined and described in many patents. It is used as a blood platelet aggregation inhibitor. In combination with aspirine the activity is potentiated [36]. It is also considered to be an anticancer drug which prevents metastasis and, unlike most other anticancer drugs, it has few side effects [37].

Another example is the preparation of 2-(2,2′-dicyclohexylethyl)piperidine, also known as perhexiline [38]. This drug is used for the prevention of angina pectoris in case of coronary–arterial diseases. The synthesis starts with α-picoline, which is metallated with butyllithium at the methyl group. The organometallic intermediate is allowed to react with either benzophenone or dicyclohexylketone. After elimination of water from the resulting tertiary alcohols, the residual double bonds are catalytically hydrogenated.

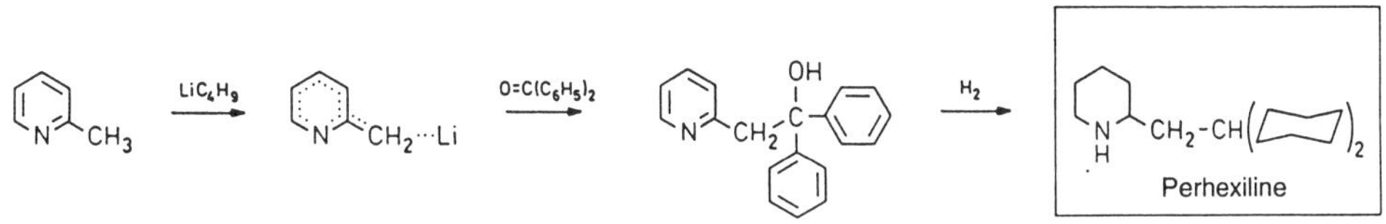

LiC_4H_9 $O{=}C(C_6H_5)_2$ H_2 Perhexiline (2.9)

Deprotonation of hetero-substituted arenes at the *ortho* position with organolithium bases and subsequent reaction with an electrophile is a versatile tool in the chemistry of aromatics.

2,6-Dimethoxybenzoic acid is an intermediate in the synthesis of the antibiotic methicillin [39]. A useful method of preparation is the lithiation of 1,3-dimethoxybenzene and subsequent reaction of the organometallic with carbon dioxide.

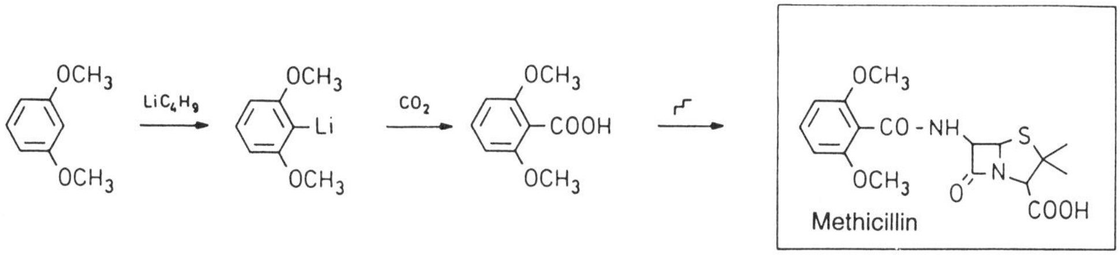

LiC_4H_9 CO_2 Methicillin (2.10)

Another application of the so-called *ortho*-directed lithiation is the preparation of suitable intermediates in the synthesis of liquid crystal compounds. Regioselective deprotonation of 2,3-difluorobenzene derivatives in the 4-position and subsequent reaction with electrophiles yields 1,4-disubstituted-2,3-difluorobenzenes which are valuable as intermediates in the synthesis of liquid crystal compounds or as components which increase, e.g., the dielectric properties of liquid crystal phases [40].

The lithiated intermediates are unstable above −50 °C. Therefore, it is necessary to work at relatively low temperatures. In this reaction it is advantageous to increase the basicity of butyllithium by addition of alkoxides or complexing agents, e.g. amines or amides.

The addition of tetramethylethylenediamine (TMEDA) to increase the basicity of butyllithium is applied in the synthesis of 3-(2,2-difluoro-1,3-benzodioxol-4-yl)-4-cyanopyrrole, a new pyrrole fungicide. 2,2-Difluoro-1,3-benzodioxole is lithiated at the 4-position using 20% *n*-butyllithium in toluene and TMEDA as metallating agent. The reaction temperature is −10 °C. Subsequent reaction with electrophiles yields the desired intermediates (reaction 2.11) [41]:

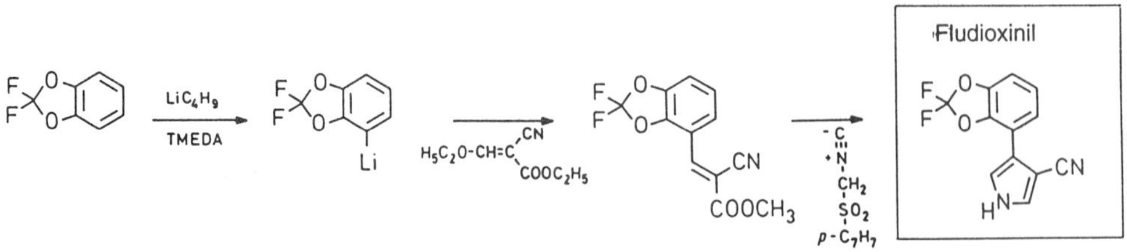

LiC_4H_9 / TMEDA; $H_5C_2O\text{-}CH{=}C(CN)COOC_2H_5$; $\text{-}C{\equiv}N^+\text{-}CH_2\text{-}SO_2\text{-}p\text{-}C_7H_7$ → Fludioxinil (2.11)

In this context it should be pointed out that another variation of metallation conditions is the use of lithium amides, especially lithium diisopropylamide (LDA), as a base. LDA has the advantage over butyllithium that its (thermodynamic) basicity is lower, so it lacks reactivity towards electrophilic substituents present in the substrate to be metallated. LDA can be prepared *in situ* by reaction of butyllithium with diisopropylamine. The use of hexyllithium instead of butyllithium avoids the formation of gaseous butane.

Lithiation of 1,3-bis(trifluoromethyl)benzene with butyllithium and subsequent reaction with an electrophile results in the formation of a mixture of 2,4- and 2,6-isomers. A dramatic shift of the isomer distribution towards the 2,4-isomer is observed when the lithiation is performed with lithium 2,2′,6,6′-tetramethylpiperide, which is readily obtained by treatment of the corresponding amine with butyllithium [42]. Reaction of the organometallic intermediate with *N*,*N*-dimethylformamide (DMF) yields 2,4-bis(trifluoromethyl)benzaldehyde with more than 98% selectivity. Bis-2,4-(trifluoromethyl) benzene derivatives are valuable intermediates for the production of pharmaceuticals containing 2,4-bis(trifluoro)benzene groups, e.g. antimalarial drugs.

In the recent literature, advances concerning enantioselective deprotonation reactions using butyllithium as a base and, e.g., optically active amines as chiral auxiliaries have been described [43]. Since the market for optically pure fine chemicals in the field of pharmaceuticals, aromas and agrochemicals is expected to grow from US$900 million in 1988/89 to US$2000 million in 2000 [44], the producers of organolithium compounds are quite confident about the future evolution. Without doubt, many more laboratory processes will be developed to the industrial scale.

2.3.2.2 Lithium–Halogen Exchange Reactions

Like lithium–hydrogen exchange, lithium–halogen exchange reactions are also very useful in the industrially important field of aromatic and heteroaromatic chemistry. In general, the equilibrium of the reversible reaction is shifted far the side of the less basic organolithium component.

$$RLi + R'Hal \rightleftharpoons RHal + R'Li \qquad (2.12)$$

Lithium–halogen exchange reactions can be accompanied by alkylation since butyl bromide or iodide are invariably formed as a by-product. However, the high rate of the reaction even at very low temperatures always makes it possible to avoid such undesirable side reactions. Actually, already several plants have been built that allow reactions to be performed even at −90 °C [45].

The following examples from the field of drug and agrochemical synthesis demonstrate that lithium–halogen exchange reactions are often used to prepare lithiated isomers which are not accessible by deprotonation.

In the manufacture of the vasodilator cetiedil, 3-thienyllithium is used for further reaction with lithium cyclohexylglyoxylate to afford the tertiary alcohol. The deprotonation of thiophene with butyllithium yields the 2-thienyl derivative (see synthesis of ticlopidine, reaction 2.8) and the conversion of 3-bromothiophene to the Grignard compound is not possible. However, the 3-isomer can be prepared by lithium–bromine exchange from readily accessible 3-bromothiophene and butyllithium (reaction 2.13). The reaction conditions, diethyl ether as solvent and reaction temperature −70 °C, are typical for lithium–halogen exchange reactions. At higher temperatures, 3-metallated thiophenes may undergo ring opening reactions [4a].

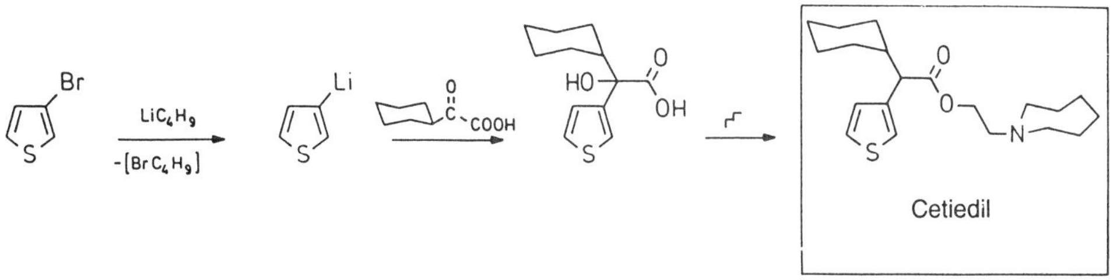

LiC_4H_9, $-[BrC_4H_9]$

Cetiedil

(2.13)

In the field of agrochemicals, 3-pyridinemethanol derivatives are useful as fungicides [46] or as growth inhibitors for aquatic weeds [47]. They can be synthesized by reaction of 3-lithiopyridine with a suitable ketone (reaction 2.14).

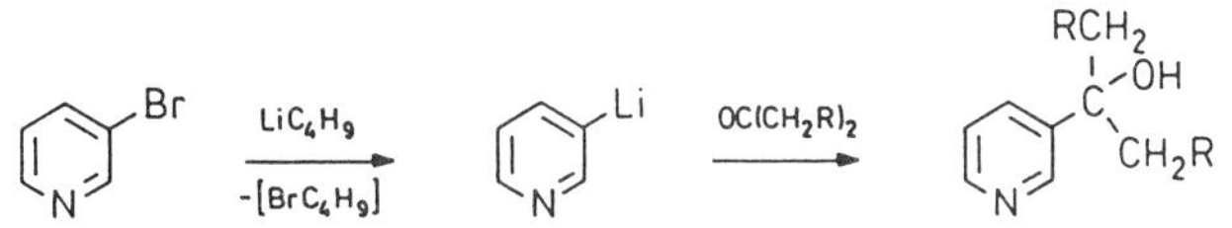

LiC_4H_9, $-[BrC_4H_9]$; $OC(CH_2R)_2$

(2.14)

A 5-pyrimidinemethanol derivative is the well known fungicide fenarimol. Its synthesis implies the generation of 5-pyrimidyllithium by treatment of 5-bromopyrimidine with butyllithium.

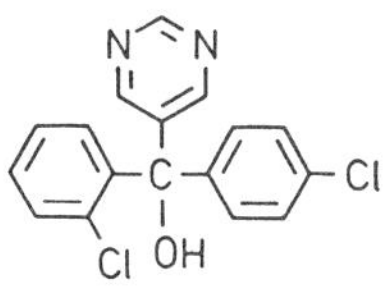

Fenarimol

2.3.2.3 Wittig Reactions

Organolithium compounds are widely used as bases for the conversion of phosphonium salts into ylids, which are useful in Wittig reactions. Lithium halide, which is a by-product when the ylid is generated by means of an organolithium compound, favours the formation of *trans*-alkenes.

11-[3-(Dimethylamino)propylidine]-6,11-dihydrodibenz[*b*,*e*]oxepin, the generic name of which is doxepin, is as an antidepressant drug. Some carboxylic acid derivatives of doxepin and carbocyclic analogues thereof surprisingly possess potent antihistaminic and antiasthmatic properties [48]. They are prepared industrially by means of a Wittig reaction, the required phosphorus ylid being generated by treatment of the corresponding phosphonium salt with butyllithium (reaction 2.15)

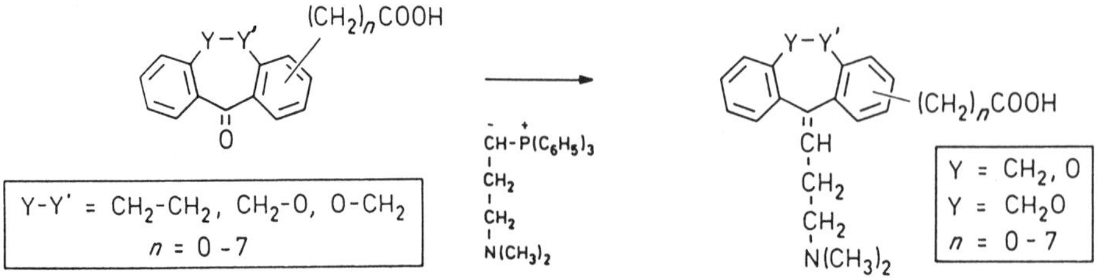

Y-Y' = CH_2-CH_2, CH_2-O, O-CH_2
n = 0 - 7

$\overset{-}{C}H$-$\overset{+}{P}(C_6H_5)_3$ / CH_2 / CH_2 / $N(CH_3)_2$

Y = CH_2, O
Y = CH_2O
n = 0 - 7

(2.15)

2.3.2.4 Direct Addition Reactions

For economic reasons, only the butyllithium isomer and hexyllithium are commercially used as strong bases whereas the other organolithium compounds serve as sources for the direct introduction of the corresponding alkyl group.

A typical field for the use of methyllithium in industrial organic synthesis is steroid chemistry. An important step in the synthesis of the androgen mesterolone is the introduction of a methyl-group at C-1. The most economic way to perform this transformation is a conjugate addition of lithium dimethylcuprate to the $\Delta^{1,4}$-3-oxo-steroid structure (reaction 2.16) [49, 50].

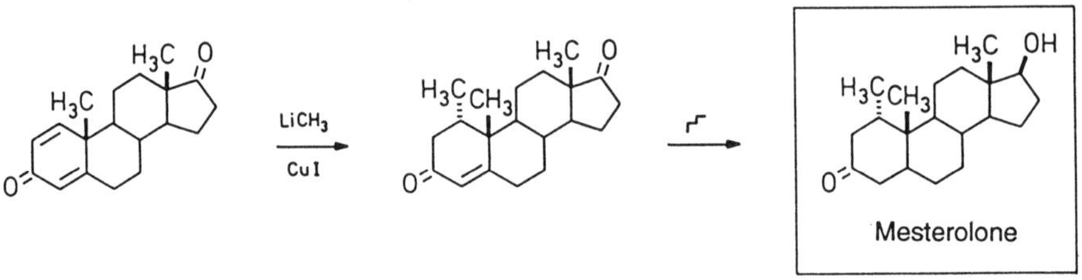

$LiCH_3$ / CuI

Mesterolone

(2.16)

The organocopper reagent is prepared *in situ* from methyllithium solution in diethyl ether and copper(I) iodide. Referring to the methyl group attached to C-10, the *trans*-product is selectively formed. This one-step reaction replaces a multi-step synthesis including a halogenation, reductive dehalogenation, addition of a Grignard reagent and acid-induced rearrangement.

Another example from the field of steroid chemistry is the synthesis of hydroxyprogesterone. The addition of methyllithium to a nitrile group is used to construct the acetyl side chain. Beforehand, the hydroxy group has to be protected. The primarily formed imine is hydrolysed to the desired ketone. This reaction provides a general access to the oxo function.

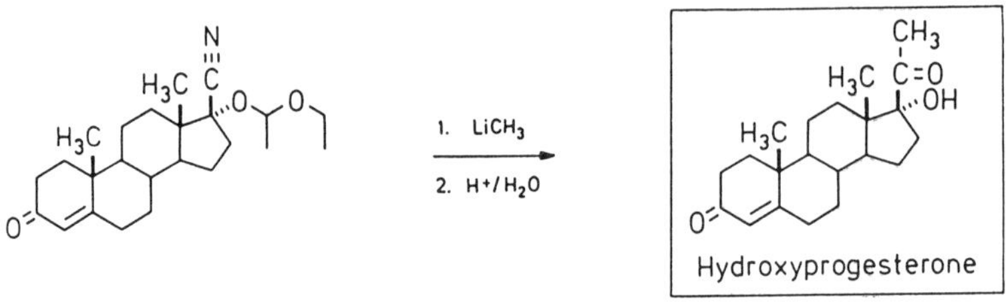

1. $LiCH_3$
2. H^+/H_2O

Hydroxyprogesterone

(2.17)

C-8 Alkyl-substituted 1,5-dimethylbicyclo[3.2.1]octan-8-ols possess exceptional odoriferous properties and are useful as components for perfumes and perfumed products. Since their odour is close to the original woody odour of Patchouli oil, they can be used as substitutes for this natural product. They are prepared by addition of organolithium compounds to the corresponding ketones (reaction 2.18) [51].

(2.18)

If a Grignard reagent is used instead of organolithium compounds, considerable amounts of by-products, resulting from a simple reduction of the keto group, are formed.

For addition reactions of this type, the work-up can be performed in a very simple way. After the reaction, a 2–5-fold excess of water is added to the reaction mixture. After decomposition of excess of organolithium reagent (*caution*: formation of methane or ethane), the resulting lithium hydroxide precipitates in the form of a pasty mass which sometimes sticks on the wall of the reactor. The organic phase is easily removed by decantation.

2.3.2.5 Coupling Reactions

An industrial application of a coupling reaction between an organolithium compound and a halide is the manufacture of *tert*-butyldimethylchlorosilane. It is accomplished by condensng dichlorodimethylsilane with *tert*-butyllithium.

$$(H_3C)_2SiCl_2 \xrightarrow{LiC(CH_3)_3} Cl\text{-}Si(CH_3)_2\text{-}C(CH_3)_3 \tag{2.19}$$

This reaction is the most important technical application of *tert*-butyllithium. The reaction product is known as an excellent protective agent for hydroxy and amino groups. Increasing amounts of this protective agent are expected to be required for the synthesis of a new generation of β-lactam antibiotics of the penem and carbapeneme type. Several synthetic routes have been developed [52]. Most of them rely on a temporary protection of the hydroxyethyl side-chain by a *tert*-butyldimethyllsilyl moiety.

Penem   Carbapenem derivative

Methyllithium is also useful for the preparation of high-purity trimethylgallium and trimethylindium.

$$MCl_3 + 3CH_3Li \longrightarrow M(CH_3)_3 + 3LiCl \qquad M = Ga, In \tag{2.20}$$

These compounds are used as Group III precursors for the preparation of III–V epitaxial layers by metal–organic chemical vapour deposition (MOCVD):

$$Ga(CH_3)_3 + AsH_3 \longrightarrow GaAs + 3CH_4 \qquad (2.21)$$

These are used in the manufacture of microelectronic and optoelectronic devices.

2.4 HANDLING

2.4.1 TECHNICAL EQUIPMENT

2.4.1.1 General Remarks

As mentioned in the Introduction, preparative chemists in industry are well informed about the advantages of using organolithium compounds in organic synthesis. However, when they have to scale up a special synthesis from the laboratory to production, they often try to avoid these reagents and develop other, often more complicated ways to obtain the target molecule. In the following it is shown that it is possible to work with organolithium compounds on an industrial scale without any problems provided that certain safety aspects of plant design and personal safety measures are borne in mind [53].

Organolithium compounds are normally available in organic solvents which are flammable. Therefore, the entire electrical installation of the plant has to fulfil all regulations which are necessary to prevent electrical discharges corresponding to the explosive and temperature classes of the solvent. This means that among other things all apparatus and machines must be earthed (grounded).

Normally there should be no solvent vapour–air mixtures as the whole production unit is equipped with an inert gas safety system (see below). However, owing to unforeseen circumstances, e.g. leakages, such solvent vapour–air mixtures can occur. Hence the whole production area must be ventilated to prevent the accumulation of such mixtures, taking into consideration that solvent vapour–air mixtures normally are heavier than air.

In addition to these general safety precautions, the sensitivity of organolithium compounds towards air and water requires their handling under an inert gas and the avoidance of accidental contact with water.

2.4.1.2 Inert Gas System

The primary protective measure for eliminating atmospheric air and moisture from the production process and the storage tanks of organolithium compounds is the inert gas respiratory and safety system (Figure 2.1).

Suitable inert gases for the handling of organolithium compounds are nitrogen and argon. All parts of the plant are connected to the closed circular respiratory main line via an individual respiratory line. The respiratory collecting main leads into a gas buffer

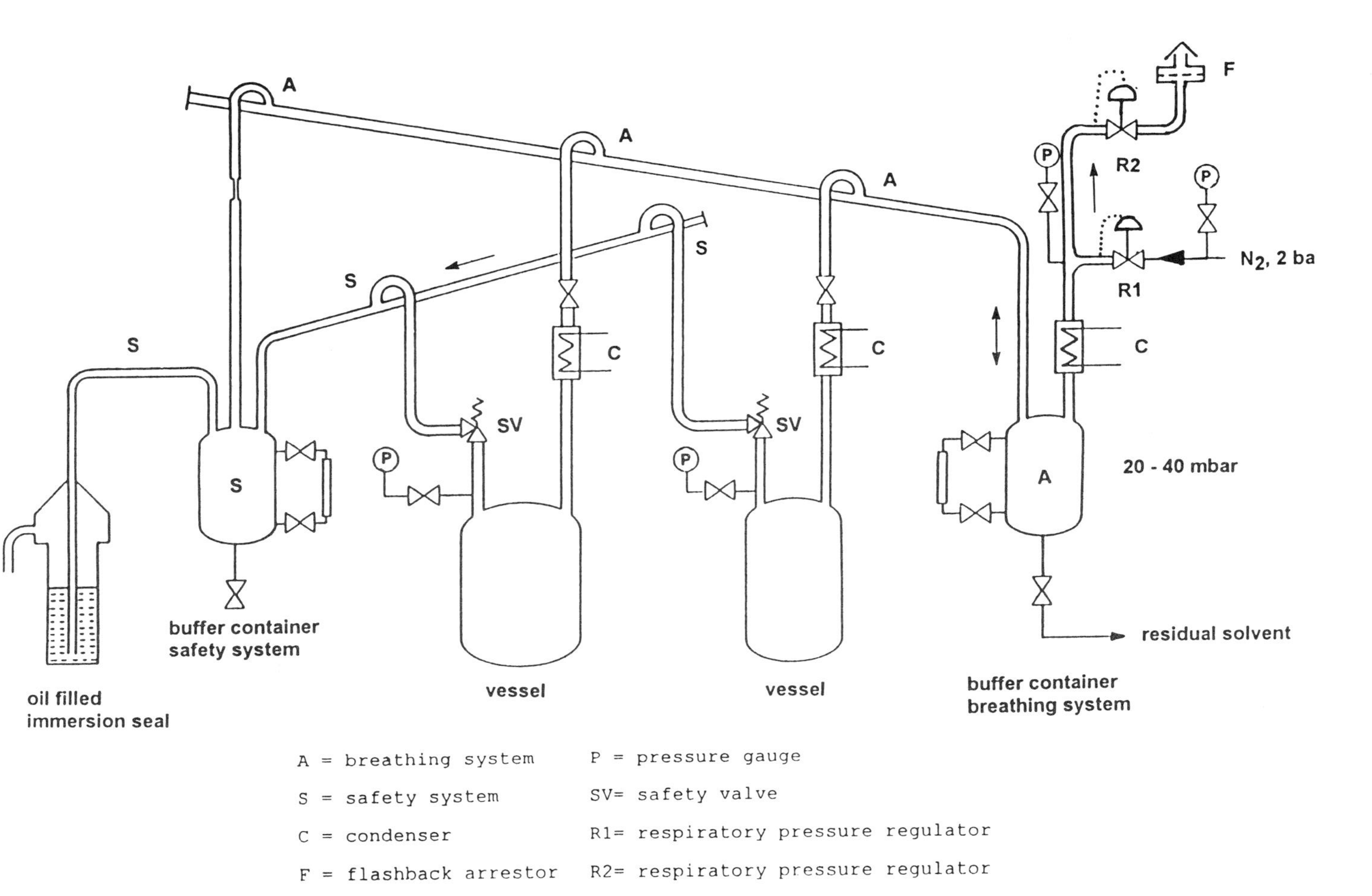

A = breathing system P = pressure gauge

S = safety system SV= safety valve

C = condenser R1= respiratory pressure regulator

F = flashback arrestor R2= respiratory pressure regulator

Figure 2.1. Schematic gas-flow and ventilation system

vessel which simultaneously serves as entrainment separator (A). This gas buffer vessel is connected via a second line to the respiratory pressure regulator which is used to maintain an inert gas overpressure of 20–40 mbar. The pressure regulator R2 allows the inert gas to flow into the equipment and the spill valve R1 allows gases to escape from the equipment when the pressure rises above a threshold value. The inert gas overpressure ensures that in case of leakages in the respiratory system or the connected vessels air and moisture cannot enter. Because all vessels of a plant are connected to one respiratory system via the respiratory main line, inert gas losses during product transfer processes are avoided and of course emissions are minimized. The condenser between gas buffer vessel and pressure regulator should be operated at temperatures as low as possible (e.g. -40 °C) in order to recover also solvents with high vapour pressure.

In addition to the breathing system, the vessels should also be equipped with a safety valve system leading into a safety buffer container (SV, S). The system is closed by an oil-filled immersion seal which closes the safety valves against the environment and keeps them under inert gas atmosphere.

All breathing and safety pipelines enter the main line from above to prevent solvent carryover or possible product transfer by foaming. The main lines should be constructed without pockets and should have a slight inclination towards the buffer containers.

2.4.1.3 Reactors

In general, organolithium compounds do not corrode ordinary steel. Owing to the special conditions of reactions at low temperatures it is recommended to use low-carbon stainless steel. Using this material it is also possible to work with liquid lithium metal which normally, owing to the small size of the lithium atom, attacks carbon at the grain boundary.

Another possibility is the use of glass enamelled vessels. They are resistant to most of the media used and are available in heat- and pressure-resistant modifications. The main drawback is that they are not completely resistant to alkaline aqueous solutions, e.g. lithium hydroxide solutions produced during work-up. The normally smooth surface becomes slightly dull after a certain time. The inferior heat-exchange properties of enamelled vessels are compensated for by means of an effective stirring system and the installation of an internally mounted heat exchanger.

An effective stirring system is also advantageous to achieve good heat dissipation if the reaction mass becomes viscous during the reaction or the work-up procedure.

Examples are known in which lithiation reactions are performed continuously in flow reactors on a pilot-plant scale. This is advantageous if the reaction requires very low temperatures and/or if short residence times are necessary because the lithiated intermediate is unstable.

Special aspects such as pumps, seals and fittings are not discussed here, but the producers of organolithium compounds are of course willing to share their relevant experience with potential users of organolithium compounds.

2.4.1.4 Heating and Cooling

As already mentioned, organolithium compounds react very vigorously with water. Therefore, any accidental contact with water must be strictly avoided. The best method is to use inert liquid heat transfer media for cooling and heating the reaction vessels and for the reflux condensers.

If water has to be used, the following safety precautions are strictly recommended:

- installation of the cooling/heating pipes outside the vessel;
- installation of very fast self-closing valves;
- use of X-ray-examined and pressure-tested reflux condensers together with automatic water alarm systems;
- reduction of the corrosive properties of the cooling water by adjusting the pH to *ca* 8.2 by adding e.g. sodium hydroxide solution;
- avoiding the formation of condensation water in the installation by insulating the respective pipes;
- no water-taps in the plant.

It is, of course, advantageous to have the possibility of switching very quickly from heating to cooling in case of unforeseen exothermic reactions.

2.4.1.5 Treatment of the By-Product Butane

In all deprotonation reactions butyllithium yields butane as a by-product. Depending on the reaction temperature, it escapes from the reaction mixture in the form of a gas, either immediately during the reaction or during following work-up. As butane is a flammable gas (lower explosion limit 1.5 vol.%, upper explosion limit 8.5 vol.%), precautions have to be taken. Some possibilities are direct burning in a plant flame, burning after trapping the gas in chilled hydrocarbon solutions or dilution with air by a strong ventilator.

All operations have to be carried out in accordance with governmental regulations. In order to avoid the formation of butane gas, butyllithium can be replaced by *n*-hexyllithium (see Section 2.2.2), which forms liquid hexane as a by-product.

2.4.1.6 Personal Safety

The most important part of ensuring safe handling of organolithium compounds is the proper training of the operating personnel in the plant. This includes primarily providing the necessary information about the properties and potential hazards and also the correct action to be taken in the case of spillage and/or fire.

All personnel working with organolithium compounds on an industrial scale should be protected by appropriate measures, i.e.

- face and eye protection;
- leather gloves;
- protective clothing against fire;
- protective shoes.

The staff in production plants wear leather suits, which are recommended because of their high thermal insulation capacity. The clothing is made from impregnated leather as it soaks up much less liquid than unprotected leather. The leather suits must not have any pockets outside in order to avoid any trapping of liquids. As leather shrinks somewhat in a fire or it may be necessary to remove this clothing quickly, it has rows of snap fasteners along the arms and legs. Another suitable material for protective clothing is Nomex, an aluminium-coated fibre. Compared with leather it has a lower thermal insulation capacity and it is more sensitive to alkali.

The face and eyes are protected by goggles and face shields fixed on the worker's helmet. Gloves should be made of leather. To prevent electrostatic discharges, the personnel should wear protective shoes with conductive soles.

2.4.2 CONTAINERS AND SHIPPING REGULATIONS

Organolithium compounds are normally delivered as solutions in a hydrocarbon solvent. Depending on the desired quantities, different types of transport containers are used. Small amounts for laboratory use are delivered in glass bottles with PTFE-sealed screw-caps. For use on an industrial scale various types of containers up to rail tanks with a maximum content of 35 m^3 are available. All containers, except the glass bottles are equipped with a dip tube for safe filling and emptying and a connection for inert gas. Normally the valve of the dip tube is red and the inert gas connection is green.

For reasons of safety and compliance with governmental regulations, the containers may be filled to no more than 90% of their capacity. A slight inert gas overpressure is maintained during shipment in the containers. After emptying, this inert gas overpressure should continue to be maintained.

Containers for temperature-sensitive organolithium compounds, e.g. *sec*-butyllithium, are insulated. Solutions of this are filled into the containers at a temperature below −15 °C and reach the customer under normal transport conditions still below 0 °C.

Organolithium compounds have to be shipped in officially approved containers. The following regulations apply (example butyllithium). Emptied containers are still dangerous goods according to the following regulations.

Postal shipment: not permitted.

Rail: GGVE/RID: Class 4.2 (No. 31a).
Warning label: 4.2/4.3.
Consignment note designation: Spontaneously combustible, butyllithium.

Road: GGVS/ADR: Class 4.2 (No. 31a).
Warning label: 4.2/4.3.
Note in accompanying documents: spontaneously combustible, butyllithium.

Sea: GGVSee/IMDG: Class 4.2/UN-No. 2445.
Label: 'Spontaneous combustible.'

Air: not permitted.

The containers should be labelled as follows:

F	'Highly flammable.'
C	'Caustic.'
Umbrella	'Moisture sensitive.'
R 14/15	Reacts violently with water liberating highly flammable gases.
R 17	Spontaneously flammable in air.
R 20	Harmful by inhalation.
R 34	Causes burns.
R 48	Danger of serious damage to health by prolonged exposure.
S 6	Keep under nitrogen or argon.
S 7/9	Keep container tightly closed in a well ventilated place.
S 16	Keep away from sources of ignition. No smoking.
S 26	In case of contact with eyes, rinse immediately with plenty of water and seek medical advice.
S 33	Take precautionary measures against static discharges.
S 36/37/39	Wear suitable protective clothing, gloves and eye/face protection.
S 43	In case of fire use extinguishing powder on base of sodium chloride. Never use water.

2.4.3 STORAGE

Storage tanks (see Figure 2.2) for solutions of organolithium compounds usually are of a normal steel construction and can be positioned either vertically or horizontally. If the storage temperature is to be kept below -10 °C because of the thermal sensitivity of the product (e.g. *sec*-butyllithium), then storage tanks made of fine-grain steel must be used.

In case the tanks should be emptied via a dip-tube with the aid of inert gas pressure or a self-priming centrifugal pump, the tanks are designed for pressure use. The tanks can have a bottom outlet to empty the tanks completely for cleaning operations or for changes to other products, e.g. butyllithium to hexyllithium or in case a solvent is changed.

Two pressure gauges are recommended on the tank for measuring and reading both the minimum respiration pressure range and the pressure range up to the maximum permissible container pressure. All equipment included in the tank farm must meet governmental regulations for storing flammable liquids, e.g. earthing of the components according to the respective regulation.

For the measurement of the level in the tank, a level indicator has to be installed. This level indicator should be fitted with an alarm device to indicate when the product level reaches 90% of the container capacity.

In addition to filling and emptying lines, the tank has to be equipped with additional connectors for nitrogen and reserve flanges for tank cleaning, i.e. for steam, water and waste gas.

It is recommended to equip the tank with a stirrer for product homogenization after

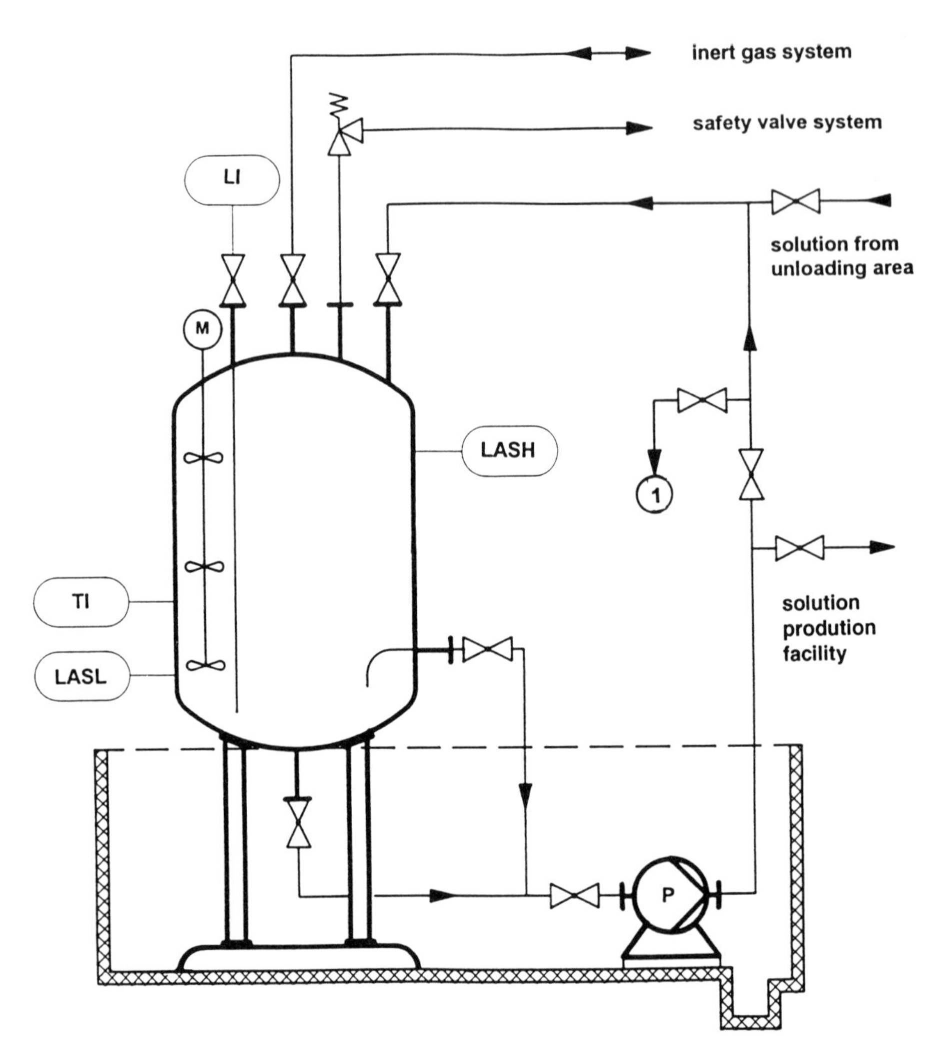

P = magnetic coupled pump
1 = sampling point
LASH = level alarm switch high
LASL = level alarm switch low
TI = temperature indicator
LI = level indicator

Figure 2.2. Storage tank

the addition of solvent and for the cleaning of the tank. From time to time it is necessary to clean the storage tanks and the pipes. A detailed description of the procedure and technical advice are available from the producers of organolithium compounds.

2.4.4 SAMPLING AND ANALYSIS

Many methods are known for the determination of active organolithium reagents. Numerous colorimetric single-titration methods have been developed recently [54]. They can be classified into three categories:

- Coordination of an organolithium compound to polycyclic aromatic bases forms coloured charge-transfer complexes, which are decolorized by titration with an alcohol.
- Single deprotonation of various compounds forms coloured anions, which are also titrated with an alcohol.
- Double deprotonation of various compounds forms a coloured dianion. This method requires no standard acidic solution. The reagent is titrated with the organolithium reagent by using a syringe.

For quality reasons, the producers of organolithium compounds are not only interested in the determination of the active reagent. They also have to know the content of the free base (mainly lithium alkoxide). Therefore, a modified procedure of the old double titration method of Gilman and Haubein [55] is frequently preferred for an exact analysis. First the total base is determined by acidimetric titration of the lithium hydroxide formed after hydrolysis of the sample, then the active lithium organic compound is reacted with an organic halide leaving the 'free' base (lithium alkoxide and oxide) unaffected. The assay of the lithium organic compound (active base) is the difference between the total base and free base.

In the original method of Gilman and Haubein [55], diethyl ether was used as a solvent. Because of possible ether-cleavage reactions this method leads to inaccurate results. On the other hand, it is known that the reaction of organic halides with lithium organic compounds in the absence of ethers is slow and not quantitative. It was found that the reason for the incomplete reaction is the large amount of solvent, which absorbs the resulting heat of reaction. When the halide and the organolithium compounds are mixed using only the amount of hydrocarbon solvent present in the organolithium solution, the reaction suddenly starts after an induction period of a few minutes, causing evolution of heat, gas (!) and precipitation of lithium halide. It is complete after a few seconds. For the analysis of butyllithium and *sec*-butyllithium, benzyl chloride is recommended as the organic halide; for the determination of *tert*-butyllithium and methyllithium (solution in diethyl ether), allyl bromide is preferred. Detailed descriptions of the procedures are available from the authors.

In the production plant samples of lithium organic compounds are usually analysed when the solution is transferred from the transport containers to the storage tanks, or from there to the reaction vessel. For this purpose special equipment is available which is mounted in the transfer lines and allows the sample to be taken under inert conditions. To obtain a representative sample first several litres of organolithium solution should be removed from the container. This is to avoid incorrect results if finely dispersed lithium hydride is present and has accumulated near the end of the dip tube.

2.4.5 RECYCLING AND WASTE DISPOSAL

The batches of a reaction with organolithium reagents are usually worked up with water or acid. Lithium, being present, e.g. as alkoxide, sulfide or halide, goes into the aqueous phase in the form of lithium hydroxide or lithium halide, in most cases lithium chloride. The desired product remains in the organic phase. The question arises of what should be done with the lithium in the aqueous phase.

For economic and ecological reasons, it is of course desirable to recycle the lithium content. A major problem is the unavoidable organic contamination from the solvent and the organic reaction products. The content of organic carbon in 30% lithium chloride solutions from the synthesis of phamaceuticals ranges typically from 1000 up to 20 000 mg l^{-1} TOC (total organic carbon). If such amounts of impurities are present in the aqueous phase it is nearly impossible to recover pure lithium hydroxide or lithium chloride which meets the quality demands for other applications.

Because of the special properties of highly concentrated lithium chloride solutions, it is necessary to develop specific techniques for the treatment of such solutions. Because every organic reaction produces another series of impurities, the user of organolithium compounds should contact the supplier to discuss the recycling of contaminated lithium-containing solutions in detail. The general protocol (Figure 2.3) can be adapted to any given problem.

For the disposal of residual organolithium compounds or spilled solution, the following procedures are recommended.

Smaller quantities can be decomposed by slow, controlled addition of or to water with vigorous stirring and cooling under an inert gas. The resulting aqueous solution of lithium hydroxide can be removed from the solvent by phase separation. Larger amounts can usually be returned to the supplier.

Spilled solutions should be covered with ground limestone or sodium hydrogencarbonate for absorption and then placed in an open area where the solvent can evaporate and the organolithium is oxidized and hydrolysed by air. Subsequently the remainder can be flushed away in the water treatment system and treated according the applicable regulations and laws.

2.4.6 FIRE FIGHTING

When working with organolithium compounds one should always keep in mind that there is the possibility of spontaneous ignition. Therefore, all necessary equipment for fire fighting should be kept available, e.g. portable extinguishers for fighting small fires and mobile, larger extinguishers for fighting larger fires. In addition, boxes with ground limestone powder and shovels should be kept on hand for covering spilled solutions

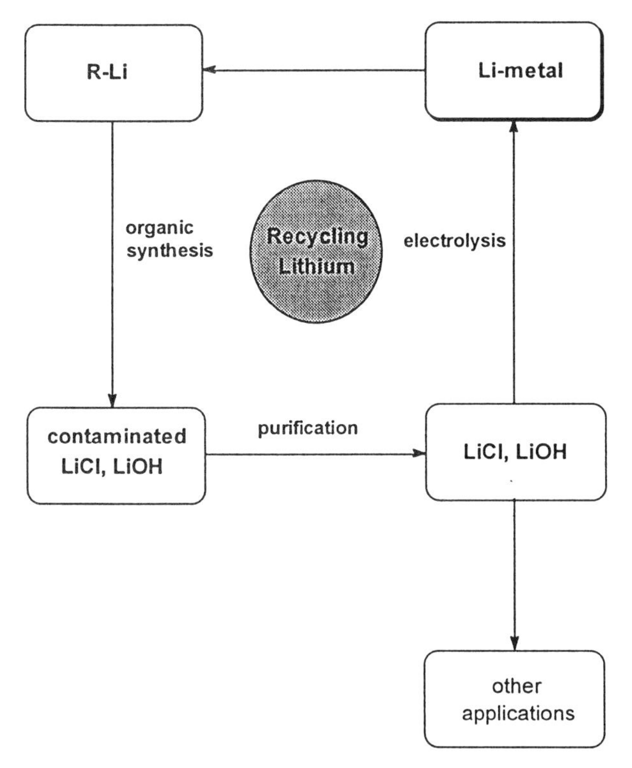

Figure 2.3. Recycling lithium

and extinguishing small fires. *Fires must not be fought with water, carbon dioxide or halogenated hydrocarbons (halons) as they react vigorously with organolithium compounds.*

Suitable media for fire fighting are ground limestone powder (for small fires), sodium hydrogen carbonate-based extinguishing agents (not for concentrated organolithium compounds) and sodium chloride-based extinguishing agents (e.g. Totalit M).

For personnel, protection blankets and self-rescue showers should be readily accessible. Self-rescue showers are devices that release a shower of extinguishing powder when activated by stepping on a floor plate. It makes sense to install water showers or tempered water basins outside a production plant to extinguish the fire of a person's clothing or to wash off alkalinity from the skin.

2.4.7 SOURCES OF SUPPLY

Ventilation pressure regulators:
Regel- und Messtechnik GmbH,
Osterholzstrasse 45,
D-34123 Kassel-Bettenhausen, Germany.

Protective leather suits:
August Schwan GmbH & Co.,
Postfach 120268,
D-41721 Viersen, Germany.

Extinguishing powders based on sodium chloride:
Total Walther GmbH,
Postfach 1127,
D-68526 Ladenburg, Germany.

Samplers:
Dovianus BV,
P.O. Box 47,
Kompasstraat 40,
NL-2900AA Capelle an der Issel, The Netherlands.

Absorption of butane in chilled hydrocarbons:
Citex Maschinen- und Apparatebau GmbH,
Im Hagen 3,
D-22113 Oststeinbeck, Germany.

2.5 REFERENCES

1. W. Schlenk, J. Holtz, *Ber. Dtsch. Chem. Ges.* **50** (1917) 262.
2. K. Ziegler, H. Colonius, *Liebigs Ann. Chem.* **479** (1930) 135.
3. M. Morton, *Anionic Polymerization: Principles and Practice*, Academic Press, New York, 1983; S. Bywater, in *Encyclopedia of Polymer Science and Engineering*, Vol. 2, p. 2, Wiley, New York, 1985; M. Fontanille, in *Comprehensive Polymer Science*, Ed. G. Allen, J. C. Bevington, Vol. 3, p. 356, Pergamon Press, Oxford, 1989.
4. For example: (a) L. Brandsma, H. D. Verkruijsse, *Preparative Polar Organometallic Chemistry*, Vol. 1, Springer, Berlin, 1987; (b) J. J. Zuckerman (Ed.), *Inorganic Reactions and Methods*, Vol. 14, VCH Publishers, New York, 1988; (c) B. J. Wakefield, *Organolithium Methods*, Academic Press, New York, 1988.
5. W. N. Smith, *J. Organomet. Chem.* **82** (1974) 1.
6. T. L. Rathman, R. C. Morrison, *Eur. Pat.* EP 0 285 374 (to Lithium Corporation of America, filed 29 March 1988); *Chem. Abstr.* **110** (1989) 213071.
7. M. Schlosser, *Struktur und Reaktivität Polarer Organometalle*, Springer, Berlin, 1973.
8. K. Ziegler, H. G. Gellert, *Liebigs Ann. Chem.* **567** (1950) 179.
9. R. A. Finnegan, H. W. Kutta, *J. Org. Chem.* **30** (1965) 4138.
10. W. H. Glaze, G. M. Adams, *J. Am. Chem. Soc.* **88** (1966) 4653.
11. R. L. Eppley, J. A. Dixon, *J. Organomet. Chem.* **11** (1968) 174.
12. *Handling of Butyllithium*, Chemetall, Frankfurt, 1993; *Organometallics in Organic Synthesis*, FMC Corporation, Lithium Division, Gastonia, NC, 1987.
13. P. A. Fowell, C. T. Mortimer, *J. Chem. Soc.* (1961) 3793.
14. A. Maercker, *Angew. Chem.* **99** (1987) 1002; *Angew. Chem. Int. Ed. Engl.* **26** (1987) 972.

15. R. Bach, J. R. Wasson, in *Kirk-Othmer Encyclopedia of Chemical Technology*, 3rd edn., Vol. 14, p. 467, Wiley, New York, 1981.
16. K. J. Niehues, in *Proceedings of the IVth Hydride Symposium*, p. 169, Chemetall, Frankfurt, 1987.
17. J. W. Jefferson, J. H. Greist, M. Baudhuin, in *Lithium: Current Applications in Science, Medicine, and Technology*, ed. R. O. Bach, p. 345, Wiley, New York, 1985; R. R. Fieve, E. D. Peselow, in *Lithium: Current Applications in Science, Medicine, and Technology*, ed. R. O. Bach, p. 353, Wiley, New York, 1985.
18. E.g. N. I. Sax and R. J. Lewis, *Dangerous Properties of Industrial Materials*, 7th edn., Van Nostrand Reinhold, New York, 1989.
19. C. D. Harries, *Fr. Pat.* 434 989 (to Farbenfabriken vorm. Friedr. Bayer, filed 7 Oct. 1911, patented 17 Jan. 1912); F. E. Mathews, E. H. Strange, *Br. Pat.* 24 790 (filed 25 Oct. 1910, patented 25 Oct. 1911).
20. K. Ziegler, *Angew. Chem.* **68** (1956) 581.
21. S. E. Horne, *US Pat.* 3 114 743 (to Goodrich-Gulf Chemicals Inc., filed 2 Dec. 1954, patented 17 Dec. 1963).
22. L. E. Forman, *US Pat.* 3 285 901 (to the Firestone Tire & Rubber Company, filed 11 Oct. 1962, patented 15 Nov. 1966); *Chem. Abstr.* **66** (1967) 116486.
23. L. E. Forman, R. W. Kibler, F. A. Bozzacco, *US Pat.* 3 208 988 (to The Firestone Tire & Rubber Company, filed 24 Oct. 1961, patented 28 Sept. 1965); *Chem. Abstr.* **63** (1965) 18424.
24. *Br. Pat.* 848 065 (to Phillips Petroleum Company, filed 17 Oct. 1956, patented); *Chem. Abstr.* **55** (1961) 15982.
25. H. L. Hsieh, *J. Polym. Sci.* **A-3** (1965) 153.
26. K. Ziegler, F. Dersch, W. Wollthan, *Liebigs Ann. Chem.* **511** (1934) 13; K. Ziegler, L. Jakob, *Liebigs Ann. Chem.* **511** (1934) 45; K. Ziegler, L. Jakob, H. Wollthan, A. Wenz, *Liebigs Ann. Chem.* **511** (1934) 65.
27. M. Szwarc, *Nature (London)* **178** (1956) 1168; M. Szwarc, M. Levy, R. Milkovich, *J. Am. Chem. Soc.* **78** (1956) 2656.
28. S. Bywater, *Fortschr. Hochpolymer. Forsch.* **4** (1965) 66; *Chem. Abstr.* **62** (1965) 14827.
29. A. E. Oberster, T. C. Bouton, J. K. Valitis, *Angew. Makromol. Chem.* **29/30** (1973) 291; *Chem. Abstr.* **79** (1973) 32639.
30. F. Bandermann, H. Sinn, *Makromol. Chem.* **96** (1966) 150; *Chem. Abstr.* **65** (1966) 12287.
31. R. S. Stearns, L. E. Forman, *J. Polym. Sci.* **41** (1959) 381; *Chem. Abstr.* **54** (1960) 17934; M. Bruzzone, G. Corradini, F. Amato, *Rubber Plast. Age* **46** (1965) 278; *Chem. Abstr.* **62** (1965) 13355.
32. G. Holden, R. Milkovich, *US Pat.* 3 265 765 (to Shell Oil Company, filed 29 Jan. 1962, patented 9 Aug. 1966).
33. W. Krauss, in *Proceedings of the IVth Hydrid Symposium*, p. 53, Chemetall, Frankfurt, 1987.
34. H. W. Gschwend, H. R. Rodriguez, *Org. React.* **26** (1976) 1.
35. *Ger. Pat.* DE 24 04 308 (to Centre d'Etudes pour l'Industrie Pharmaceutique, filed 30 Jan. 1974); *Ger. Pat.* DE 26 04 248 (to Parcor, filed 4 Feb. 1976, patented 13 Feb. 1986); *Chem. Abstr.* **90** (1979) 87431.
36. *Ger. Pat.* DE 26 12 491 (to Centre d'Etudes pour l'Industrie Pharmaceutique, filed 14 April 1976, patented 6 Sept. 1984); *Chem. Abstr.* **91** (1979) 186862.
37. T. Suzuki, *Ger. Pat.* DE 29 00 203 (to Sopharma SA, filed 4 Jan. 1979); *Chem. Abstr.* **91** (1991) 186862.

38. S. W. Horgan, F. P. Palopoli, E. J. Schwoegler, *Ger. Pat.* DE 27 13500 (to Richardson-Merrell Inc., filed 26 March 1977, patented 2 Jan. 1987; corresp. *US Pat.* 4 069 222); *Chem. Abstr.* **88** (1978) 22658.
39. F. P. Doyle, J. H. C. Nayler, G. N. Rolinson, *US Pat.* 2 951 839 (filed 2 May 1960, patented 6 Sept. 1960).
40. V. Reifenrath, J. Krause, *World Pat.* WO 89/08629 (to Merck Patent Gessellschaft, filed 27 Feb. 1989, corresp. *Ger. Pat.* DE 38 97 910); *Chem. Abstr.* **112** (1990) 169747.
41. P. Ackerman, H.-R. Känel, B. Schaub, *Eur. Pat.* EP 0 333 661 (to Ciba Geigy AG, filed 9 March 1989); *Chem. Abstr.* **112** (1990) 139019; B. Schaub, H.-R. Känel, P. Ackerman, *Eur. Pat.* EP 0 333 658 (to Ciba Geigy AG, filed 9 March 1989); *Chem. Abstr.* **112** (1990) 158229.
42. R. Masciardi, *Eur. Pat.* EP 0 442 340 (to Hoffmann-La Roche AG, filed 4 Feb. 1991), *Chem. Abstr.* **115** (1991) 182797.
43. H. Waldmann, *Nachr. Chem. Tech. Lab.* **39** (1991) 413; S. T. Kerrick, P. Beak, *J. Am. Chem. Soc.* **113** (1991) 9708.
44. E. Polastro, *Speciality Chem.* No. 3–4 (1991) 136.
45. J. Vit, *Speciality Chem.* No. 10 (1990) 380.
46. E. V. Krumkalns, *US Pat.* 4 043 791 (to Eli Lilly and Company, filed 2 April 1976, patented 23 Aug. 1977); *Chem. Abstr.* **87** (1977) 162796.
47. E. V. Krumkalns, *Ger. Pat.* DE 1 670 647 (to Eli Lilly and Company, filed 10 Jan 1968, patented 28 Oct. 1982).
48. O. W. Lever, H. J. Jefferson, *Eur. Pat.* EP 0 351 887 (to Wellcome Foundation Ltd., filed 15 Aug. 1986).
49. R. Philipson, E. Kaspar, *Ger. Pat.* DE 2 046 640 (to Schering AG, filed 14 Sept. 1970); *Chem. Abstr.* **76** (1972) 154025.
50. G. H. Posner, *An Introduction to Synthesis Using Organanocopper Reagents*, Wiley, New York, 1980.
51. E. J. Brunke, H. Struwe, *Ger. Pat.* DE 3 128 790 (to Dragoco Gerbering & Co GmbH, filed 21 July 1981); *Chem. Abstr.* **98** (1983) 160298.
52. G. Sedelmeier, *Nachr. Chem. Tech. Lab.* **38** (1990) 61; *Chem. Abstr.* **114** (1991) 61718.
53. B. Schneider, in *Proceedings of the IVth Hydrid Symposium*, p. 109, Chemetall, Frankfurt, 1987; W. Huter, in *Proceedings of the IVth Hydrid Symposium*, pp. 135, Chemetall, Frankfurt, 1987.
54. Y. Aso, H. Yamashita, T. Otsubo, F. Ogura, *J. Org. Chem.* **54** (1989) 5627.
55. H. Gilman, A. H. Haubein, *J. Am. Chem. Soc.* **66** (1944), 1515.

3

Titanium in Organic Synthesis

MANFRED T. REETZ
Max-Planck-Institut für Kohlenforschung, Mülheim/Ruhr, Germany

Organometallics in Synthesis—A Manual. Edited by M. Schlosser

3.1 INTRODUCTION

Titanium is the seventh most abundant metal on earth and has been used in a multitude of reactions in organic, inorganic and polymer chemistry. Several important processes will not be discussed in this chapter, including the Ziegler–Natta polymerization [1], Friedel–Crafts [2] and Diels–Alder reactions induced by stoichiometric amounts of $TiCl_4$ [3], $Ti(OR)_4$-mediated transesterifications [4], Ti-based olefin metathesis [5] and most of the inorganic chemistry associated with titanium-based sandwich compounds [6, 7].

The emphasis is on the use of titanium(IV) in C—C bond formation, especially on the titanation of classical carbanions which generates selective reagents for Grignard, aldol and Michael additions and substitution reactions. Another important aspect to be treated is the $TiCl_4$-mediated carbonyl addition and alkylation reactions of allysilanes, enolsilanes and enolizable C—H acidic compounds, processes that may or may not involve organotitanium intermediates, depending on the conditions and substrates. A variety of titanium-promoted olefin-forming reactions are also described, as are enantioselective C—C bond forming reactions catalyzed by chiral titanium compounds. The latter include Grignard-type and aldol additions, Diels–Alder and [2 + 2]-cycloadditions and ene reactions. The Sharpless epoxidation has been presented in detail elsewhere, and only the essential aspects are reiterated in this chapter.

Since the appearance of the first book on organotitanium reagents in organic synthesis in 1986 [8], the amount of information on this topic has more than doubled. Space limitations do not allow for an exhaustive treatment. Rather, basic principles illustrated by typical examples are emphasized.

A number of titanium compounds such as CH_3TiX_3 have been known for decades, but were not tested in organic synthesis [6, 7]. Such physical organic aspects as bond energies, structure, spectroscopy, MO treatments and kinetics have been delineated elsewhere [8]. Only a few central points are repeated here.

The Ti—C bond is not particularly weak in a thermodynamic sense, as was assumed for a long time. However, kinetically low-energy decomposition pathways such as β-hydride elimination may limit the use of certain organotitanium compounds [8]. The Ti—O bond is fairly strong ($\sim$115 kcal mol^{-1}), which means that any reaction in which a Ti—C bond converts to a Ti—O entity is expected to have a strong driving force. The Ti—C bond is typically 2.1 Å long, similarly to Li—C and Mg—C bonds, but shorter than Zr—C (2.2 Å). Importantly, the Ti—O bond is short (~1.7 Å) relative to Li—O (~2.0 Å), Mg—O (~2.1 Å) and Zr—O (~2.2 Å) bonds. This means that transition states in such reactions as Grignard or aldol additions are expected to be most compact in the case of titanium. The ligands in Ti(IV) compounds are tetrahedrally arranged around the metal.

Reagents of the type $RTiCl_3$ and $RTi(NR_2)_3$ are monomeric in solution. Compounds such as $CH_3Ti(OR)_3$ are monomeric or show little aggregation if the alkoxy ligand is as bulky such as isopropoxy (or bulkier), whereas analogs having smaller ligands (ethoxy) are dimeric. The electronic nature of titanium changes greatly in the series $TiCl_4$, $Cl_3TiOiPr$, $Cl_2Ti(OiPr)_2$, $ClTi(OiPr)_3$ and $Ti(OiPr)_4$ in that Lewis acidity decreases as electron-donating alkoxy ligands are introduced. The same effect operates in reagents such as CH_3TiCl_3 vs $CH_3Ti(OiPr)_3$. Amino ligands also decrease Lewis acidity dramatically, as do pentahaptocyclopentadienyl groups [8].

3.2 SYNTHESIS, STABILITY AND CHEMOSELECTIVE GRIGNARD AND ALDOL ADDITIONS OF ORGANOTITANIUM REAGENTS

3.2.1 TITANATION OF CLASSICAL CARBANIONS

3.2.1.1 Titanating Agents

Grignard and alkyllithium compounds and a host of heteroatom-substituted and resonance-stabilized 'carbanions' generated by halogen–metal exchange, reductive cleavage of sulfides or deprotonation of CH-acidic organic compounds constitute a large class of reagents traditionally used in carbonyl addition reactions [9]. The reagents are generally basic and reactive, which means that the degree of chemo- and stereo-selectivity in reactions with carbonyl compounds may be limited. For example, almost five decades ago it was noted that phenylmagnesium bromide hardly discriminates between aldehyde and ketone functionalities [10]. A similar lack of chemoselectivity has been observed for alkyllithium reagents and for lithium ester enolates, deprotonated nitriles, sulfones and similar reactive species [11, 12]. This means that reactions of these carbanions with ketoaldehydes or diketones have not been part of rational synthetic design. Some of the classical reagents also lead to undesired enolization and to indiscriminate attack at polyfunctional molecules [8, 11–16].

In 1979–80, it was discovered that certain organotitanium reagents prepared by transmetallation of RMgX, RLi or R_2Zn undergo chemo- and stereoselective C—C bond-forming reactions with carbonyl compounds and certain alkyl halides [17–21]. These observations led to the working hypothesis that titanation of classical 'carbanions' increases chemo- and stereoselectivity (Scheme 3.1) [22]. It became clear that steric and electronic properties of the reagents can be adjusted in a predictable way by proper choice of the ligands X at titanium [8, 11, 14].

If the ligand X is Cl, the organyltitanium trichlorides $RTiCl_3$ are fairly Lewis acidic, forming octahedral bisetherates with diethyl ether or tetrahydrofuran (THF). This

$$\boxed{\text{ORGANYL}}\!-\!M \xrightarrow{ClTiX_3} \boxed{\text{ORGANYL}}\!-\!TiX_3 \rightarrow$$

M = Li, Mg, Zn X = Cl, OR, NR_2 etc.

Scheme 3.1.

property can be modulated by introducing alkoxy or amino ligands which increase the electron density at the metal. Thus, reagents of the type $RTi(OR)_3$ or $RTi(NR_2)_3$ do not form bisetherates. This electronic difference is important in chelation- or non-chelation-controlled additions to alkoxy carbonyl compounds (Section 3.3) and in certain S_N1 substitution reactions of alkyl halides (Section 3.6). By varying the size of the alkoxy or amino ligands, steric properties are also easily tuned. Finally, chiral modification is possible (Sections 3.3.3 and 3.7.1). For the purpose of controlling chemoselectivity, the nature of the ligand is not crucial in most cases, whereas stereoselectivity depends critically on the type of ligand (Section 3.3). An important parameter not apparent in Scheme 3.1 is the possibility of increasing the number of ligands in the reagents. For example, if $Ti(OR)_4$ is used as the titanating agent for RLi, ate complexes $RTi(OR)_4Li$ evolve which are bulkier than the neutral $RTi(OR)_3$ counterparts.

Common titanating agents are $TiCl_4$, $ClTi(OiPr)_3$ [11, 12], $ClTi[N(CH_3)_2]_3$ [23], $ClTi[N(C_2H_5)_2]_3$ [24] and $CpTiCl_3$ [25]. Most of the compounds are easily accessible (and commercially available):

$$TiCl_4 + 3Ti(OiPr)_4 \longrightarrow 4ClTi(OiPr)_3$$

$$TiCl_4 + 3LiNR_2 \longrightarrow ClTi(NR_2)_3$$

$$R = CH_3,\ C_2H_5$$

$$TiCl_4 \xrightarrow[CpSiMe_3]{CpNa\ or} CpTiCl_3$$

One of the dramatic results of titanation is the pronounced increase in chemo- and stereoselectivity of Grignard and aldol additions. For example, most reagents $RTiX_3$ react with ketoaldehydes solely at the aldehyde function. Mechanistically, this can also be studied by reacting a 1 : 1 mixture of an aldehyde and a ketone with one part of the titanium reagent (reaction 3.1) [11–15].

$$R'CHO + R''C(=O)R''' \longrightarrow R'CH(OH)R + R''R'''C(R)OH \qquad (3.1)$$

RLi	~ 50	:	50
$RTiX_3$	> 99	:	< 1

Titanium reagents are less basic and less reactive than the lithium and magnesium precursors. Steric factors are primarily responsible for aldehyde selectivity [26]. The terms 'metal tuning' and 'ligand tuning' have become popular in describing the methodology summarized in Scheme 3.1. Indeed, it has stimulated the search for other metals in controlling carbanion selectivity. Analogs involving zirconium [27–30], cerium [31], chromium [32], hafnium [33], manganese [34], iron [35] and ytterbium [36] are prominent examples. The influence of ligands still needs to be studied before a final comparison with titanium can be made. The important virtues of titanium reagents include low cost, high degrees of chemo- and stereoselectivity and non-toxic waste on work-up (ultimately TiO_2).

3.2.1.2 Chlorotitanium Reagents

A few general rules should be kept in mind when attempting to titanate carbanions [8]. Approaches that do *not* work well include titanation of primary alkyllithium reagents with $TiCl_4$, because this leads to the undesired reduction of titanium [14, 37], probably via β-hydride elimination. However, by choosing dialkylzinc reagents R_2Zn as the precursor, $TiCl_4$-mediated addition reactions with carbonyl compounds are possible [38]. These may not involve titanium trichlorides of the type $RTiCl_3$, but rather activation of the carbonyl compound by $TiCl_4$ complexation followed by reaction of the zinc reagents.

The parent compound CH_3TiCl_3 is easily accessible in >95% yield either by the titanation of methyllithium in diethyl ether [13] or by Zn–Ti exchange using $(CH_3)_2Zn$ in CH_2Cl_2 [6, 17, 39]. In the former case the reagent exists in equilibrium with the mono- and bisetherate [13, 40]. Although ether-free CH_3TiCl_3 can be distilled prior to reactions with electrophiles, *in situ* reaction modes are preferred.

$$CH_3Li(\text{etherate}) \xrightarrow[\text{diethyl ether}]{TiCl_4} CH_3TiCl_3(\text{etherate})$$

$$(CH_3)_2Zn \text{ or } (CH_3)_3Al \xrightarrow[CH_2Cl_2]{TiCl_4} CH_3TiCl_3$$

Since dimethylzinc is pyrophoric [41], the first reaction is the method of choice. However, in certain chelation-controlled carbonyl additions [42] and substitution reactions [17, 18], ether-free solutions are required which makes the use of dimethylzinc or trimethylaluminum mandatory. Solutions of $(CH_3)_2Zn$ in CH_2Cl_2 are much less hazardous and can be handled like *n*-butyllithium [39, 43].

Currently it is not known how general the problem of undesired reduction in reactions of organolithium reagents with $TiCl_4$ actually is [37]. Cases in which *no* difficulties are encountered include the titanation of cyclopropyllithium **1** → **2** [44] and of resonance-stabilized carbanions of the type **3** and **5** to form the reagents **4** [45] and **6** [46], respectively. All of them are stable in diethyl ether solutions at −78 to −10 °C and undergo smooth aldehyde additions.

$$\text{c-C}_3H_5\text{–Li} \; (1) \xrightarrow{TiCl_4} \text{c-C}_3H_5\text{–TiCl}_3 \; (2)$$

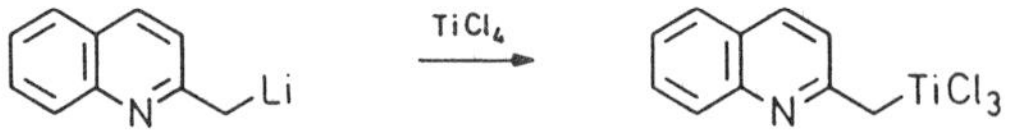

(3) (4)

Occasionally, trichlorides $RTiCl_3$ are not completely chemoselective, specifically when the precursor is extremely reactive, e.g. the lithium reagent **3**. The trichloride **4** is formed quantitatively, but the aldehyde selectivity is only 73% [45, 47]. Problems of this kind are solved by using alkoxy ligands as delineated below.

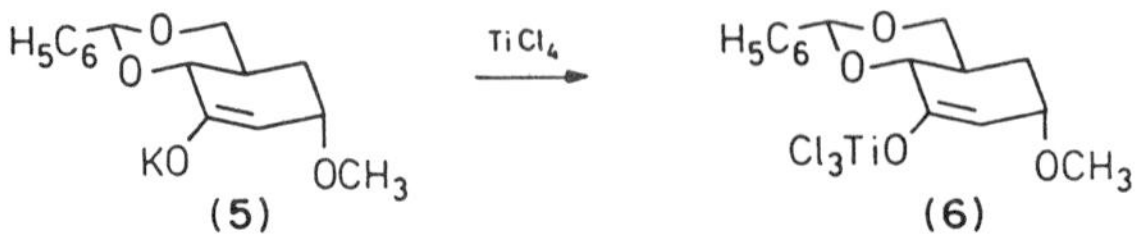

(5) (6)

3.2.1.3 Alkoxytitanium Reagents

The majority of titanation reactions of organolithium and magnesium reagents has been performed with $ClTi(OiPr)_3$. This generates compounds of the type $RTi(OiPr)_3$, usually in >95% yield [11]. Typical examples are reagents **7** [11, 12, 48], **8** [11], **9** [11, 12], **10** [44], **11** [11], **12** [11], **13** [14, 49, 50, 51], **14** [52], **15** [53, 54, 55], **16** [56, 57], **17** [57], **18** [11, 12, 58], **19** [58], **20** [59], **21** [44], **22** [60], **23** [11], **24** [61], **25** [11], **26a** [44], **26b** [62], **27** [63], **28** [64], **29** [8, 44], **30** [8, 44], **31** [8, 44], **32** [45], **33** [65] and **34** [66]. The precise structures have not been elucidated. However, in the case of several ambide nucleophiles, ^{13}C NMR spectroscopy has been helpful in determining the position of titanium. For example, α-lithio nitriles and hydrazones are titanated at nitrogen (cf. **25**, **26** and **33**), whereas α-lithio sulfones are titanated at carbon (**23**) [47].

$H_3CTi(O^iPr)_3$ (7) $H_5C_2Ti(O^iPr)_3$ (8) $H_9C_4Ti(O^iPr)_3$ (9) $H_5C_6CH_2Ti(O^iPr)_3$ (10)

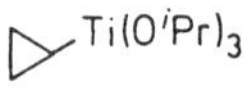

(11) 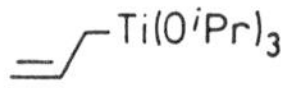 (12) 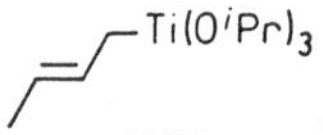 (13) 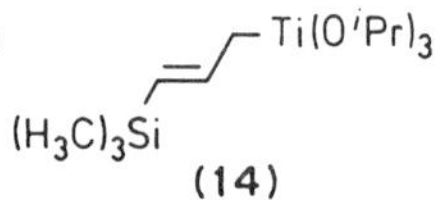 (14)

$RC{\equiv}CTi(O^iPr)_3$ (15) (16) 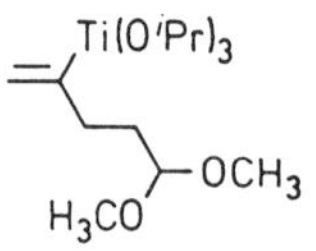 (17) 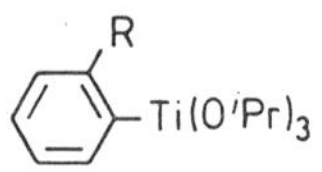 (18a) R = H (18b) R = F

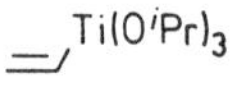

$H_5C_6SCH_2Ti(O^iPr)_3$ (19) $(H_5C_6)_2SbCH_2Ti(O^iPr)_3$ (20) $(H_3C)_3SiCH_2Ti(O^iPr)_3$ (21) $(H_5C_6)_2P(O)CH_2Ti(O^iPr)_3$ (22)

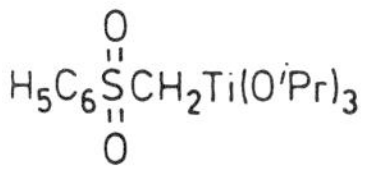

$H_5C_6S(O_2)CH_2Ti(O^iPr)_3$ (23)　$Cl_2CHTi(O^iPr)_3$ (24)　$CH_2{=}C{=}N{-}Ti(O^iPr)_3$ (25)　$R{-}CH{=}C{=}N{-}Ti(O^iPr)_3$ (26a) R = C_6H_5; (26b) R = CN

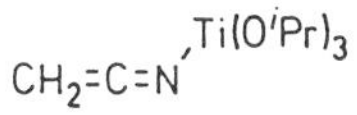

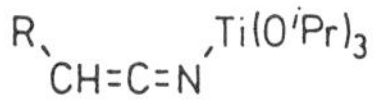

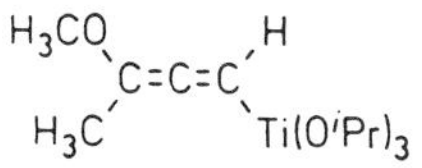

(27) 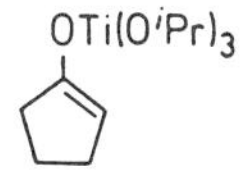　(28) 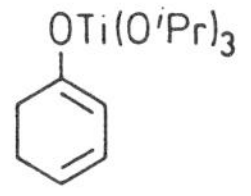　(29) 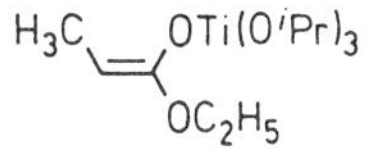　(30)

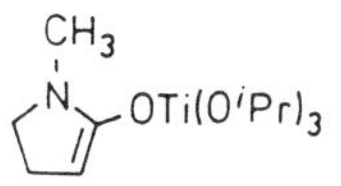

(31) 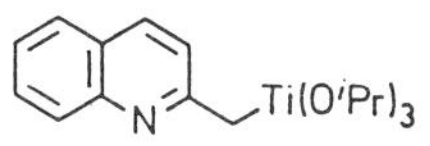　(32) 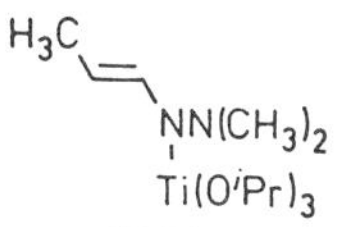　(33) 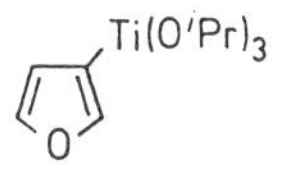　(34)

Further reagents are described in Section 3.3. $CH_3Ti(OiPr)_3$ (**7**) is so stable that it can be distilled [11, 48], but not the alkyl homologs such as **8** and **9**. For synthetic purposes an *in situ* reaction mode is sufficient. This means that lithium salts are present in the reaction mixture. Pronounced differences in the selectivity of carbonyl addition reactions in the presence or absence of lithium or magnesium salts have not been observed to date, although this needs to be studied.

All of the reagents add smoothly to most aldehydes. The rate of addition to ketones is much lower, which means high degrees of aldehyde selectivity. Typical examples concern the reaction of the ketoaldehyde **35** with reagents **7** and **23** leading solely to products **36** and **37**, respectively. In contrast, the lithium precursors do not discriminate between the two carbonyl functionalities and lead to complex mixtures [11].

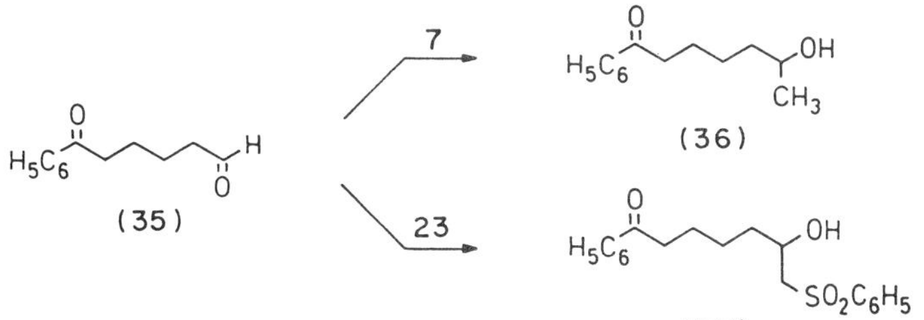

(35) —7→ (36); (35) —23→ (37)

When performing reactions with polyfunctional compounds, it is generally important to add the organotitanium reagent to the organic substrate in order to obtain the maximum degree of chemoselectivity. The other reaction mode, i.e. addition of the substrate to the reagent, means a local excess of the latter, which may lead to reactions at two or more sites [11]. This could be the reason why the yield of the reaction of the allyltitanium reagent **12** with the trifunctional substrate **38** to form adduct **39** turned out to be only 58% [67]. Other titanium reagents react completely chemoselectively with compounds of the type **38** [67].

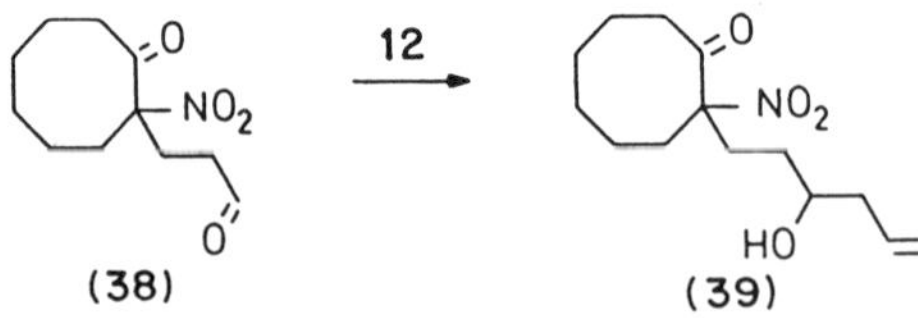

(38) —12→ (39)

A number of other points should be remembered when performing carbonyl addition reactions with $RTi(OiPr)_3$ reajents [8]. The reactivity varies widely, depending on the nature of the R group. Reagents derived from 'resonance-stabilized' carbanions, e.g. **12**, **13**, **14**, **23**, **25**, **26**, **27**, **28**, **29**, **30**, **31**, **32** and **33** are most reactive, rapidly adding to aldehydes at low temperatures (−20 to −78 °C/1 h); addition to ketones is considerably slower, but generally fast enough for yields to be good under reasonable conditions (−10 to −78 °C, 3–4 h). The less reactive compound, $CH_3Ti(OiPr)_3$, adds smoothly to aldehydes at low temperatures, but ketones require room temperature and long reaction times (6–72 h). Working in the absence of solvents speeds up the addition to bulky ketones dramatically [68]. Saturated analogs such as reagents **8**, **9**, **10**, **11**, **19**, **20** and **21** are even less reactive and generally fail to add to ketones and bulky aldehydes. When reacting the somewhat sluggish alkynyl compounds **15** with aldehydes, an excess of reagent is required [54, 55].

Certain vinyltitanium reagents have been reported to decompose via reductive dimerization at −60 °C in THF, rendering them useless for synthetic purposes [27]. However, the rate of reductive elimination depends on the solvent, such undesired decomposition being slowest in diethyl ether [57]. Thus, besides the parent reagent **16** [56, 57], a fairly wide variety of vinyltitanium reagents undergo Grignard-type addition to aldehydes in diethyl ether at −78 °C (4–8 h) [57]. The reactivity of the phenyltitanium reagent **18a** [11] is similar to that of the methyl analog **7**, but *ortho*-substituted derivatives such as **18b** require room temperature for effective aldehyde additions [12, 58].

In most cases an equivalent amount or a slight excess (10–20%) of titanating agent suffices [11, 12]. However, it has recently been observed that in the titanation of the extremely reactive quinoline reagent **3** a double or even triple excess of $ClTi(OiPr)_3$ is necessary to ensure maximum chemoselectivity [45]. The reason for this is currently not clear, but may involve ate complexes **40**. Collapse of this intermediate with formation of the reagent **32** may be promoted by additional $ClTi(OiPr)_3$ acting as a Lewis acid [45]. These observations suggest that whenever similar problems arise in other situations, an excess of titanating agent should be tried. In fact, cases in which stereoselectivity is influenced favorably by an excess of $ClTi(OiPr)_3$ have been reported [69] (see Section 3.3.3).

Secondary and tertiary alkyllithium and -magnesium reagents cannot be used in the

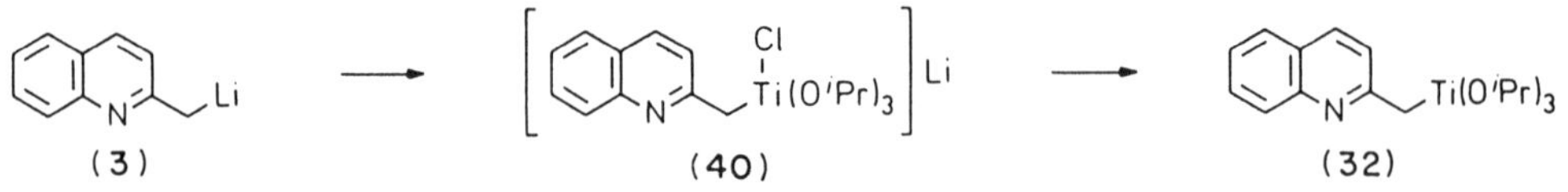

(3) → (40) → (32)

titanation/carbonyl addition sequence owing to undesired β-hydride elimination, reduction and/or rearrangement [8]. This applies to titanating agents of the type $ClTi(OiPr)_3$, $Cl_2Ti(OiPr)_2$ and $TiCl_4$. However, sometimes these transition metal properties can be exploited in a useful manner, as in the titanium mediated hydromagnesiation of olefins and acetylenes [70–72] (see also Section 3.5.3):

$$(H_3C)_2CH{-}MgCl \xrightarrow[(cat.)]{TiCl_4} (H_3C)(H_2C{\cdots}H)CH{-}TiCl_3 \longrightarrow H_3C{-}CH{=}CH_2 + HTiCl_3 \longrightarrow CH_3CH_2CH_2TiCl_3 \longrightarrow CH_3CH_2CH_2MgCl$$

$$\xrightarrow[2.\ Br_2]{1.\ TiCl_4/LiAlH_4} Br$$

Although zinc reagents of the type R_2Zn and RZnX generally do not add smoothly to carbonyl compounds, amino alcohols [73] or Lewis acids such as $TiCl_4$ [38] lead to rapid Grignard-type additions. A synthetically important variation concerns generation of zinc reagents RZnBr followed by addition of $ClTi(OiPr)_3$ [74]. Presumably, this generates the corresponding titanium reagents which add smoothly to aldehydes. Another potentially important development concerns the addition of Et_2Zn to aldehydes mediated by $Ti(OiPr)_4$ [38b]; chiral versions occur enantioselectively (Section 3.7.1).

$$RZnBr \xrightarrow[2.\ R'CHO]{1.\ ClTi(O^iPr)_3} R'CH(OH)R$$

The reactivity of reagents of the type $CH_3Ti(OR)_3$ varies according to the nature of the alkoxy group [8, 15, 26]. For example, the reaction of $CH_3Ti(OiPr)_3$ with heptanal at -30 °C is about 40–50 times faster than that of the triethoxy analog $CH_3Ti(OC_2H_5)_3$ [26]. This is because the latter is largely dimeric via Ti—O bridging and therefore less reactive, whereas the bulkier $CH_3Ti(OiPr)_3$ is only partially aggregated [8, 26]. Such phenomena should be useful in designing catalytic processes. The triphenoxy reagent $CH_3Ti(OPh)_3$ is considerably less reactive than $CH_3Ti(OiPr)_3$, probably owing to steric reasons [50]. Further variation of alkoxy groups needs to be examined, especially with respect to stereoselectivity (Section 3.3). Acyloxy ligands have yet to be tested systematically [75].

Dialkyl- and dialkynyltitanium dialkoxides, e.g. **42** [11], **43** [76a] and **44** [76b] are easily prepared via the dichloride **41**, which is accessible by mixing equivalent amounts of $TiCl_4$ and $Ti(OiPr)_4$ [11]. An X-ray structural analysis of **43** has been reported [76a].

$$TiCl_4 + Ti(O^iPr)_4 \longrightarrow Cl_2Ti(O^iPr)_2 \quad \mathbf{(41)}$$

Reactivity toward carbonyl compounds increases dramatically in the series $CH_3Ti(OiPr)_3 < (CH_3)_2Ti(OiPr)_2 < Ti(CH_3)_4$ [11, 28]. Thus, reagent **42** reacts smoothly with sterically hindered ketones such as **45** (22 °C/4 h; 98% yield of adduct **46**), whereas $CH_3Ti(OiPr)_3$ requires 2 days for 90–94% conversion [11]. Similarly, dialkynyltitanium reagents **44** provide much higher yields in aldehyde additions than the alkynyltrialkoxytitanium counterparts [76b].

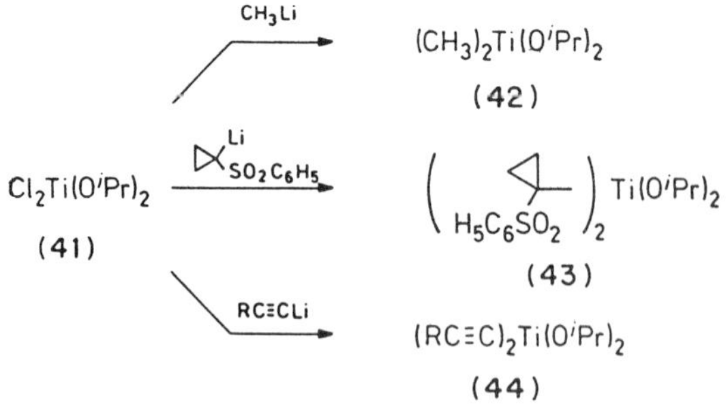

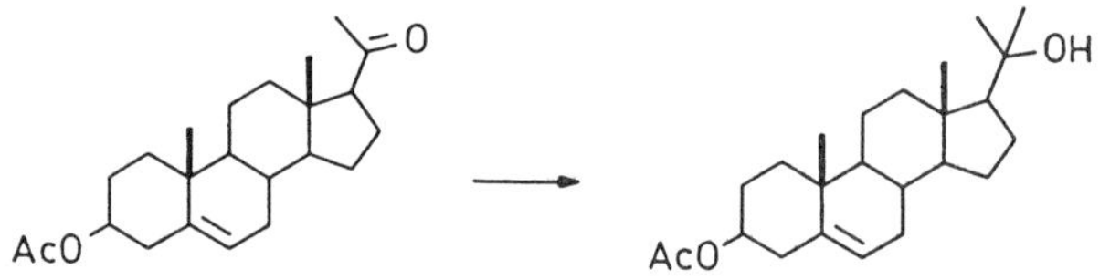

A powerful method for the generation of functionalized carbon nucleophiles concerns the reaction of the corresponding iodides (e.g. **47**) with zinc (activated by 4% 1,2-dibromoethane and 3% chlorotrimethylsilane) to produce zinc reagents (e.g. **48**), which can be transmetallated with CuCN/2LiCl [77] or $Cl_2Ti(OiPr)_2$ [78]. In the latter case reagents of the type **49** are formed, which readily add to aldehydes in high yield [78]. A disadvantage of this method is the fact that an excess of reagent (1.5 equiv.) is required, in contrast to an alternative reagent involving three metals (1.3 equiv. of RCu(Cn)ZnI and 2 equiv. of BF_3/OEt_2) [77].

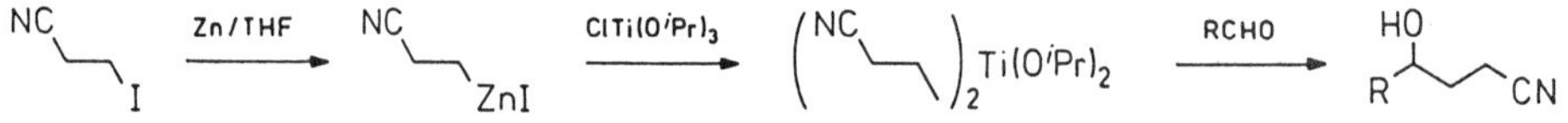

Although compounds of the type $(CH_3)_3TiOiPr$ [6] have not been utilized in synthetic organic chemistry, the extremely reactive and non-basic tetramethyltitanium $(CH_3)_4Ti$ adds to sterically hindered enolizable ketones which normally fail to undergo Grignard reactions with CH_3Li, CH_3MgX and $(CH_3)_2Ti(OiPr)_2$ [11]. Organocerium reagents are ideal reagents for easily enolizable ketones [31], but highly sterically hindered substrates have not been tested. Tetramethylzirconium surpasses even $(CH_3)_4Ti$ as a supermethylating agent in such extreme situations [11].

3.2.1.4 Titanium ate Complexes

A different type of organotitanium reagent containing alkoxy ligands concerns ate complexes [8, 11, 14, 50, 64]. For example, when adding methyllithium to $Ti(OiPr)_4$ at low temperatures, the expected $CH_3Ti(OiPr)_3$ is not formed, but rather an ate complex of yet undefined structure, formally $CH_3Ti(OiPr)_4Li$ (**50**) [8, 14, 44]. Reagent **50** adds Grignard-like to aromatic aldehydes (70–80% yields), but the conversion in the case of aliphatic aldehydes is poor owing to the basic nature of the reagent (probably competing aldol condensation) [8, 14, 44]. Nevertheless, the cheap $Ti(OiPr)_4$ can be used as a titanating agent if the carbon nucleophile is particularly reactive. This requirement is fulfilled in the case of resonance-stabilized precursors such as allylic lithium or magnesium reagents, lithium enolates and other heteroallylic species. Typical examples are **51** [11, 52], **52** [11, 52], **53a** [11, 50, 52], **53b** [79] and **54** [64] (see also Section 3.3). In some cases the ate complex may be converted into the neutral reagent $RTi(OiPr)_3$, especially if an excess of $Ti(OiPr)_4$ is used [45, 69].

$H_3CTi(O^iPr)_4Li$ (50) — $Ti(O^iPr)_4Li$ (51) — $Ti(O^iPr)_4MgCl$ (52) — $Ti(O^iPr)_4Li$, R (53a) R = $Si(CH_3)_3$; (53b) R = SC_2H_5 — $OTi(O^iPr)_4Li$ (54)

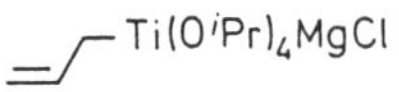

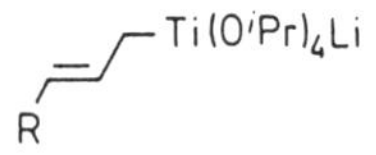

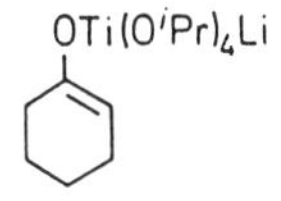

Many of the reagents have been shown to be aldehyde selective and/or stereoselective (Section 3.3). For example, adding $Ti(OiPr)_4$ to a solution of the very reactive allylmagnesium chloride results in the ate complex **52**, which reacts with the ketoaldehyde **35** solely at the aldehyde function [11]. In the absence of $Ti(OiPr)_4$ a complex mixture of products is obtained. Hence this titanation is a simple way to tame the very reactive allylmagnesium chloride.

3.2.1.5 Aminotitanium Reagents

A wide variety of alkyl-, aryl- and vinyllithium (or -magnesium) reagents can be titanated with $ClTi(NR_2)_3$ to form distillable compounds of the type $R'Ti(NR_2)_3$ [80]. Surprisingly, even the *tert*- butyltitanium analogs are thermally stable. Unfortunately, most of the trisamino reagents do not undergo Grignard-type additions, because aminoalkylation occurs [11, 14, 81]. Nevertheless, these ligands can be used if the carbon nucleophile is very reactive. This requirement is fulfilled in the case of 'resonance-stabilized' carbanions, e.g. **55–57** [11, 52, 64, 65]. All of them are accessible by titanation of the Li or Mg precursor with $ClTi(NEt_2)_3$ and transfer the carbon nucleophile in preference to undesired amino addition. Since chemoselectivity is achieved using $TiCl_4$, $Cl_2Ti(OiPr)_2$ or $ClTi(OiPr)_3$, the more expensive titanating agents $ClTi(NR_2)_3$ should be employed only if stereoselectivity needs to be controlled (Section 3.3).

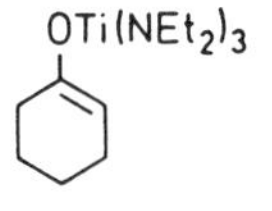

$OTi(NEt_2)_3$ (55) — $Ti(NEt_2)_3$ (56) — $Ti(NEt_2)_3$, $NN(CH_3)_2$, H_3C (57)

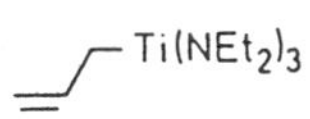

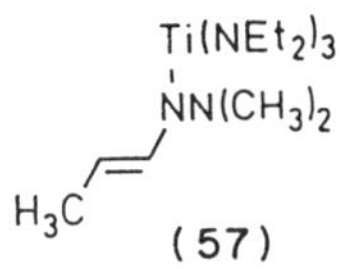

3.2.1.6 Further Types of Chemoselectivity

Organotitanium reagents are also chemoselective in reactions with diketones, dialdehydes [11, 26] and carbonyl compounds containing additional functional groups [11, 13, 15]. In the case of aldehyde/aldehyde and ketone/ketone discrimination, competition experiments point to surprisingly high degrees of molecular recognition [11, 14, 26, 45]. Small differences in steric environment can tip the balance in favor of the less shielded reaction site. This type of molecular recognition is related to Brown's work on amine–borane equilibria [82a] and to Yamamoto's report on Lewis acid–Lewis base interactions involving bulky aluminum compounds and carbonyl compounds [82b].

The mechanism of all of the Ti-based reactions most likely involves reversible complexation of the carbonyl group by the metal followed by irreversible C—C bond formation [26]. This may lead to enhanced selectivities through molecular recognition.

R'CHO

$L_3TiO-C(R')(H)(R)$

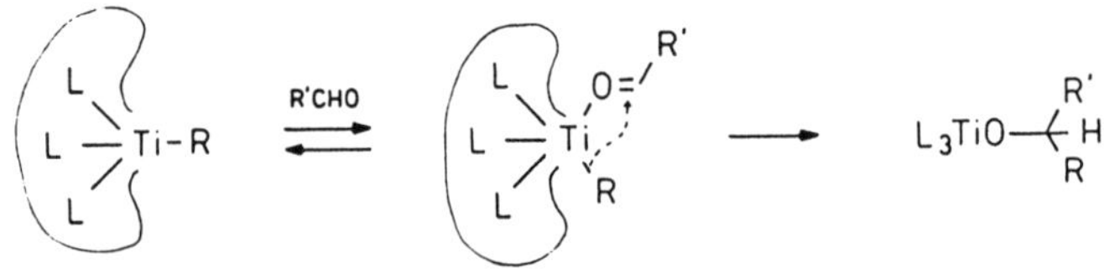

Scheme 3.2.

Other than steric factors, chelation effects in the case of α- and β-alkoxy and amino carbonyl compounds also influence the sense and degree of molecular recognition [14, 26, 33, 42]. In fact, detailed kinetic experiments [26] (which are more precise than simple 1 : 1 : 1 competition experiments) show that steric effects are completely overriden in the case of α-alkoxy and amino ketones.

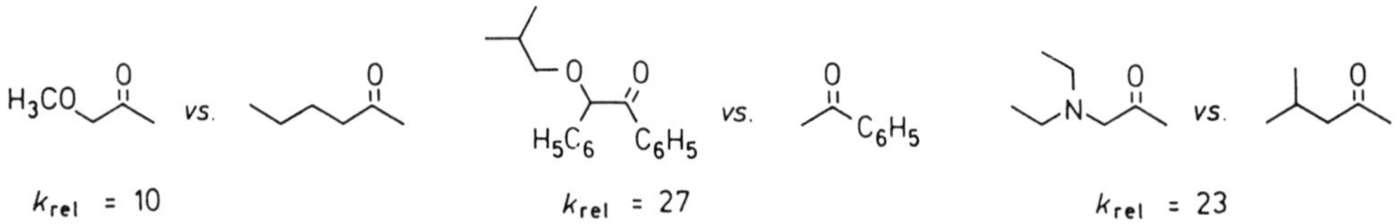

k_{rel} = 10 k_{rel} = 27 k_{rel} = 23

Scheme 3.3.

In stereochemically relevant cases a single, chelation-controlled diastereomer is formed [26] (see also Section 3.3), corroborating the hypothesis that chelation enhances the rate of addition [26]. α-Alkoxy-*aldehydes* do *not* react faster, and in fact lead to non-chelation-controlled adducts [26] (Section 3.3.1.3). Similar observations have been made in related cases involving titanium and other metal reagents [62b, 83, 84]. Other activating factors such as field effects have been invoked in the aldol addition of lithium enolates [85]. Whatever all of the phenomena behind site selectivity may be, they form the basis of a potentially powerful synthetic tool in reactions of polyfunctional molecules.

Typical examples of chemoselective additions to di- or polyfunctional compounds are shown below. They are all cases in which the lithium or magnesium precursors alone are unsuitable, conversion to the desired products being less than 45%. For example, the addition of CH_3MgX or CH_3Li to the nitro-aromatic ketone **58** results in attack at the aromatic ring, whereas prior titanation with $TiCl_4$ leads to smooth carbonyl addition, the yield of addition product **59** being 80% [13]. α,β-Unsaturated aldehydes and ketones react in a 1,2-manner [1, 12]. An exception is the reaction of tetrabenzyltitanium with certain α,β-unsaturated ketones [86]. Aldehyde **60**, having a sensitive acetoxy moiety, reacts with $CH_3Ti(OiPr)_3$ to form the adduct **61** in 81% yield [11], whereas CH_3MgCl affords less than half of this amount. Other notable examples are the reactions of the difunctional carbonyl compounds **62** and **64**, which react with the titanium reagents shown to form the adducts **63** (82% yield) [12, 21] and **65** (76%) [13], respectively.

(58) $\xrightarrow{CH_3Li/TiCl_4}$ (59)

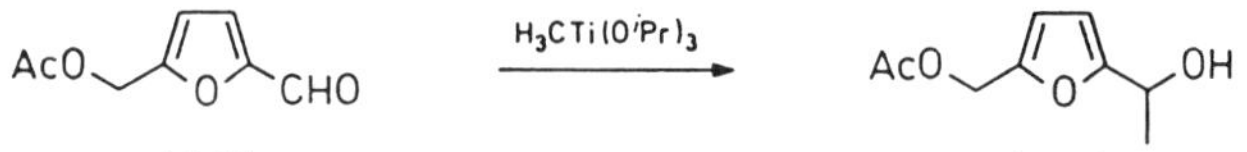

(60) $\xrightarrow{H_3CTi(O^iPr)_3}$ (61)

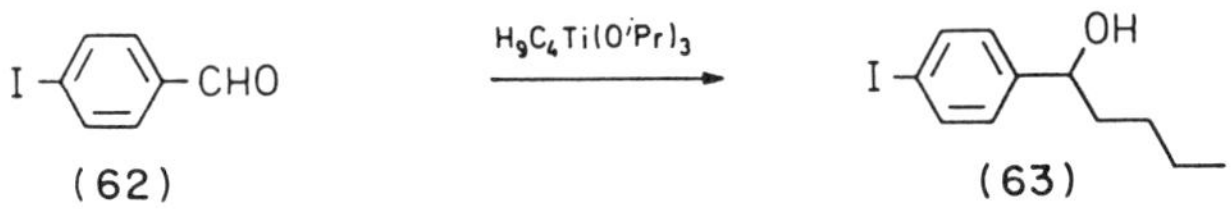

(62) $\xrightarrow{H_9C_4Ti(O^iPr)_3}$ (63)

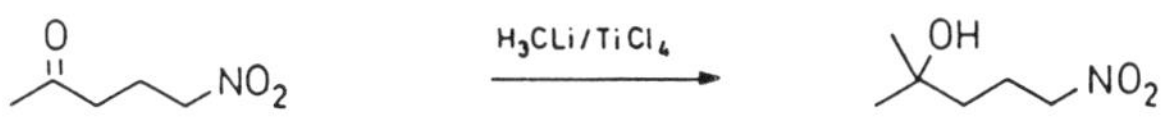

(64) $\xrightarrow{H_3CLi/TiCl_4}$ (65)

Saturated reagents of the type $RTiCl_3$ and $RTi(OiPr)_3$ generally do not add to aldimines. Exceptions are more reactive reagents such as allylic compounds [87], enolates [87] or substrates with reactive C—N double bonds such as pyrimidones [53], yields being 80–82%:

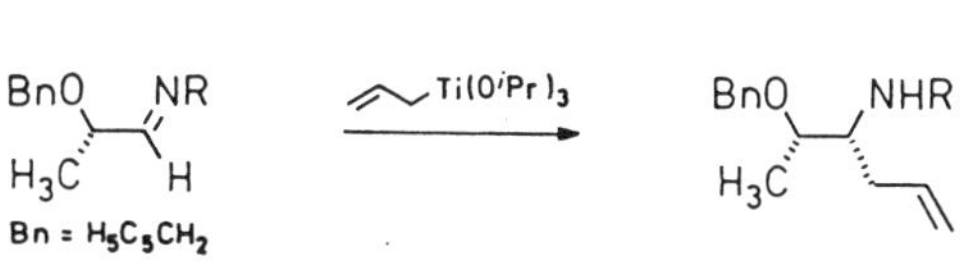

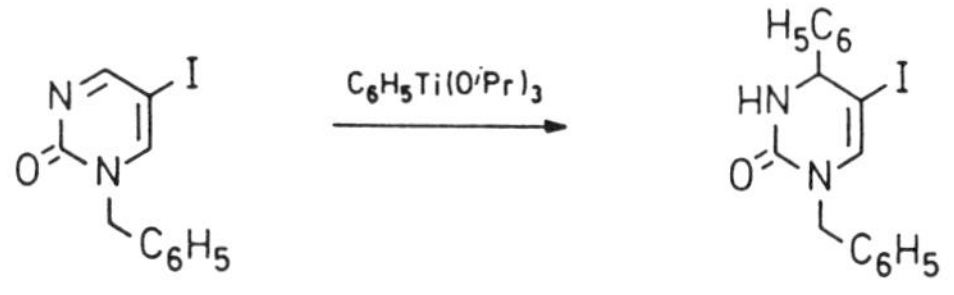

$\xrightarrow{\text{allyl}Ti(O^iPr)_3}$

Bn = $H_5C_6CH_2$

$\xrightarrow{C_6H_5Ti(O^iPr)_3}$

3.2.2 ALTERNATIVE SYNTHESES OF ORGANOTITANIUM REAGENTS

In addition to the titanation of classical 'carbanions' and zinc reagents, several other routes to organotitanium reagents are available. Some involve metal–metal exchange reactions of silicon, tin or lead compounds, especially with the reactive $TiCl_4$ to form trichlorotitanium reagents, e.g. **66** [88], **67** [89], **68** [90] and **69** [91a]. Adding $Ti(OiPr)_4$ or $Ti(NR_2)_4$ to such reagents has not been tried. This would lead to the ligand exchange, thereby making ligand tuning possible. All of the trichlorotitanium reagents add smoothly to aldehydes. $Et_4Pb/TiCl_4$ ethylates chiral aldehydes stereoselectively [91b–e]. In an interesting extension, chiral α-alkoxy lead compounds have been shown to react stereoselectively with steroidal aldehydes in the presence of $TiCl_4$ [91c]. Ketones were not tested, but are likely to react much less readily.

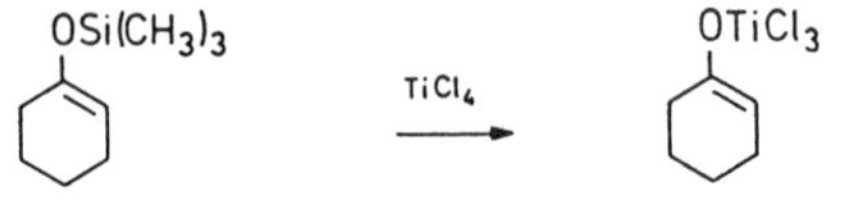

OSi(CH3)3 —TiCl4→ OTiCl3 (66)

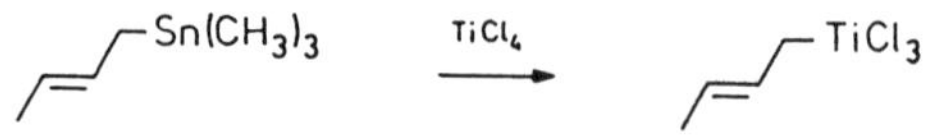

—Sn(CH3)3 —TiCl4→ —TiCl3 (67)

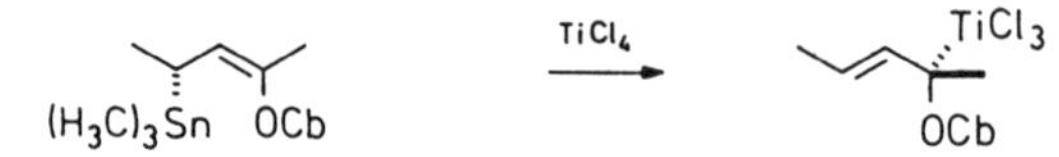

(H3C)3Sn OCb —TiCl4→ TiCl3 OCb (68)

OCb = OCON(ⁱPr)2

(H3C)3Sn N C6H5 O C6H5 —TiCl4→ Cl3Ti N C6H5 O C6H5 (69)

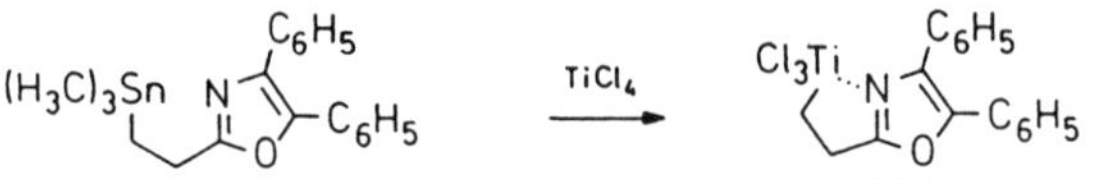

In the case of enolsilanes, rapid Si–Ti exchange using $TiCl_4$ is possible only with *Z*-configured compounds [88]. The trichlorotitanium enolate derived from cyclopentanone decomposes as it is formed. An exception is the enolsilane from cyclohexanone which undergoes fairly fast Si–Ti exchange [88]. Reaction with aldehydes is related to the Mukaiyama aldol addition [92] (Section 3.2.3), in which enolsilanes are treated with aldehydes in the presence of $TiCl_4$, a process, which does *not* involve titanium enolates [92d].

Sn–Ti exchange reactions have been exploited to generate such interesting reagents as **67**, **68** and **69**. It is likely that the stannylated form of many other resonance-stabilized carbanions will react with $TiCl_4$ in the desired way. Whereas tetramethyltin $(CH_3)_4Sn$ does not add to aldehydes in the presence of $TiCl_4$, the benzyl compound **70** affords the desired adducts **71** in yields of 60–75% [44].

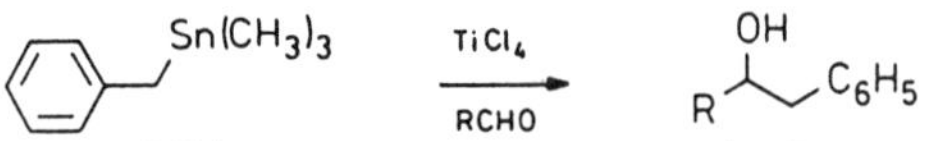

Sn(CH3)3 (70) —TiCl4, RCHO→ OH R C6H5 (71)

Titanium homoenolates such as **73** are accessible by $TiCl_4$-mediated ring cleavage of the silicon compounds **72** [93]. The reagent is dimeric and adds smoothly to aldehydes to form adducts **74**, but reactions of ketones are sluggish. Treatment of the titanium homoenolate **73** with $Ti(OiPr)_4$ results in the dichloroisopropoxy analog, which is monomeric and therefore more reactive, adding to ketones and sterically hindered aldehydes [94]. In a synthetically important extension, siloxycyclopropanes containing further functionalities have been shown to undergo Ti-mediated homo-aldol additions [95a, b]. β-Stannylated ketones react similarly [95c].

(72) $\xrightarrow{TiCl_4}$ (73) $\xrightarrow{R'CHO}$ (74)

Along a different line, isonitriles were reported to react with $TiCl_4$ to form the α,α-adducts **75** [96], which add chemoselectively to aldehydes [97]. Following aqueous work-up, the products are *N*-methyl-α-hydroxycarboxamides **76**.

$RN\equiv C$ $\xrightarrow{TiCl_4}$ (75) $\xrightarrow{1.\ R'CHO;\ 2.\ H_2O}$ (76)

In a completely different and powerful approach, CH-acidic compounds are treated with $TiCl_4$ and a tertiary amine, thereby inducing titanation. An early example pertains to the Knoevenagel condensation of malonates with aldehydes in the presence of $TiCl_4$ and triethylamine (Sections 3.3.2.2 and 3.5.1) [98], a process which probably involves intermediate chlorotitanium enolates. In more recent times a stepwise procedure for the formation of chlorotitanium enolates has been devised [100] e.g. **77** → **78** [99]. This important development allows for S_N1 substitution reactions (Section 3.6) and aldol (Section 3.3) and Michael additions (Section 3.4).

(77) $\xrightarrow{TiCl_4,\ H_5C_2N(^iPr)_2}$ (78)

Bn = $CH_2C_6H_5$

Claisen condensations utilizing $Cl_2Ti(OTf)_2$ are also based on titanation of carbonyl compounds [101]. For example, the diester **79** readily affords the Claisen condensation product **80** (83%) regioselectively.

(79) $\xrightarrow{Cl_2Ti(OTf)_2,\ N\text{-}C_2H_5\text{-piperidine},\ -23^\circ C,\ 3h}$ (80)

Interestingly, even $Ti(OR)_4$ will induce enolization of aldehydes and ketones, albeit at higher temperatures (20–140 °C). A convenient procedure for aldol condensations has been devised [102]. For example, heating a mixture of acetone and benzaldehyde in the presence of $Ti(OiPr)_4$ affords 75% of the aldol condensation product **81** [102]. In order to avoid undesired Meerwein–Ponndorf side-reactions, $Ti(OtBu)_4$ may be used in certain cases.

$$CH_3COCH_3 + C_6H_5CHO \xrightarrow{Ti(O^iPr)_4} CH_3COCH{=}CHC_6H_5 \quad (81)$$

Finally, ketene reacts with $Ti(OiPr)_4$ to form reagents **82** (or more likely the enolates) which undergo smooth aldol additions [103].

$$CH_2{=}C{=}O \xrightarrow{Ti(O^iPr)_4} ({}^iPrO)_3Ti{-}CH_2COO^iPr \ (82) \xrightarrow{RCHO} RCH(OH)CH_2COO^iPr$$

3.2.3 $TiCl_4$-MEDIATED ADDITIONS OF ENOL- AND ALLYLSILANES TO CARBONYL COMPOUNDS

One of the synthetically reliable methods for performing crossed aldol reactions is the Mukaiyama aldol addition [92]. Accordingly, enolsilanes are reacted with aldehydes (at −78 °C) or ketones (at 0 °C) in the presence of stoichiometric amounts of $TiCl_4$. Generally, it is believed that $TiCl_4$ activates the carbonyl component by complexation and that the enolsilane then undergoes C—C bond formation. In recent mechanistic studies it has been demonstrated that the silyl groups do not become bonded to the aldehyde oxygen atoms, i.e. open transition states are involved [92b, d]. However, in rare cases prior Si–Ti exchange may occur (Section 3.2.2). Various aspects of stereoselectivity are discussed in Section 3.3.2.2. Perhaps one of the disadvantages is the fact that stoichiometric amounts of $TiCl_4$ are needed. Catalytic amounts of other Lewis acids have been shown to the effective [92b, 104]. *O*-Silyl ketene ketals undergo $TiCl_4$-promoted additions to imines, making *β*-lactams accessible [92b, c].

$$RCHO + R'R''C{=}C(OSi(CH_3)_3)R''' \xrightarrow{TiCl_4} RCH(OH){-}CR'R''{-}C(O)R'''$$

Closely related are allysilane additions [105]. They too have been exploited widely, especially in diastereofacial additions to chiral aldehydes (Section 3.3.1.2). Although in the original procedure ambient temperature was chosen, addition to aldehydes is in fact rapid at −78 °C, whereas ketones require higher temperatures (−10 to 0 °C). This means that ketoaldehydes react completely aldehyde selectively [67]. A detailed NMR study of the mechanism of the reaction has recently appeared [105d]. Si–Ti metathesis plays no role, and the initial products are titanium alkoxides. A similar conclusion has been reached in a rapid injection NMR study of the $TiCl_4$-mediated chelation controlled addition of allysilanes to chiral α-alkoxyaldehydes [92d]. Intramolecular allylsilane additions constitute a useful way to synthesize carbocycles [106].

3.2.4 *REVERSAL OF CHEMOSELECTIVITY VIA SELECTIVE* IN SITU *PROTECTION OF CARBONYL FUNCTIONS*

Although titanium reagents allow for high levels of site selectivity in Grignard and aldol additions to ketoaldehydes, dialdehydes and diketones as delineated in Sections 3.2.1–3.2.3, selective C—C bond formation at the sterically more hindered functionality may well be required in other situations. Classically, this problem can be solved to some extent by a sequence of reactions involving protection, reaction and deprotection. Unfortunately, protection is not always completely site selective, even if modern reagents are chosen, as in the protective ketalization of the less hindered ketone functionality in steroidal diketones using silicon reagents of the type $R_3SiOCH_2CH_2OSiR_3/R_3SiOTf$ [107]. Titanium chemistry provides a partial solution to such problems. It was discovered that the amides $Ti(NR_2)_4$ ($R = CH_3, C_2H_5$) readily add to aldehydes at -78 °C to form adducts **83**, whereas ketones react only with the less bulky reagent $Ti[N(CH_3)_2]_4$ at -40 °C to form the analogs **84** [108].

(83a) $R = CH_3$
(83b) $R = C_2H_5$

(84a) $R = CH_3$
(84b) $R = C_2H_5$

This difference in reactivity can be exploited to protect aldehydes selectively in the presence of ketones. Following *in situ* protection, the addition of carbanions such as reactive alkyllithium reagents or enolates forces C—C bond formation to occur at the less reactive ketone site. Aqueous work-up regenerates the aldehyde function, e.g. **35** → **85**. Ketone selectivity in this one-pot sequence is >99%, the isolated yield of the aldol adduct being 88% [108].

(35) (85)

Differentiation between two ketone sites is also possible, as is addition to ester functions in the presence of aldehyde moieties [108]. On the basis of this methodology selective olefination (Section 3.5.1) of ketoaldehydes at the ketone function has been accomplished [109].

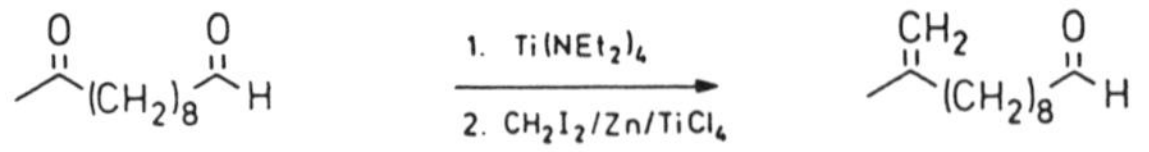

1. $Ti(NEt_2)_4$
2. $CH_2I_2/Zn/TiCl_4$

This methodology is powerful but limitations have become apparent. Only very reactive carbanions can be used because rapid addition needs to occur at temperatures below −30 °C. This is because decomposition of the amino adducts **83** and **84** to form enamines occurs at temperatures above −25 °C [108, 110]. A related method based on *in situ* protection by $TiCl_4/PPh_3$ is very simple to perform, but its synthetic scope remains to be explored [111].

3.3 STEREOSELECTIVE GRIGNARD AND ALDOL ADDITIONS

3.3.1 DIASTEREOFACIAL SELECTIVITY

Since the two π-faces of a chiral aldehyde **86** (or ketone) are diastereotopic, the addition of Grignard-like reagents or enolates may lead to unequal amounts of diastereomeric products **87** and **88**. If R^S is hydrogen and R^M and R^L constitute medium and large alkyl or aryl groups, the problem of Cram/anti-Cram selectivity pertains, for which various models have been proposed [112, 113]. If one of the substituents involves a heteroatom (alkoxy, amino groups), the problem of chelation vs non-chelation is relevant [42]. Although a great deal of progress has been made in these areas, including reactions based on organotitanium chemistry [8, 42], a number of problems persist. If the stereogenic center is positioned further away from the carbonyl function, 1,*n*-asymmetric induction becomes even more difficult.

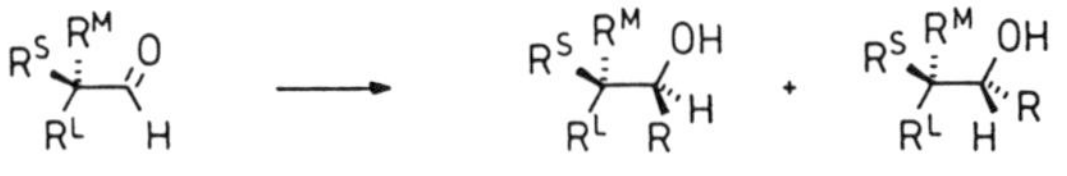

(86) (87) (88)

3.3.1.1 Cram Selectivity

Although Cram selectivity is generally acceptable in addition reactions of bulky Grignard reagents (e.g. *tert*-butyl, isopropyl), mediocre diastereoselectivities are observed in the case of smaller analogs such as methyl, *n*-alkyl, allyl and phenyl reagents [112]. In a number of cases the power of metal and ligand tuning has become apparent as shown below, e.g. [19, 22, 50]:

Reagent			
CH_3MgBr	66	:	34
$CH_3Ti(OiPr)_3$	90	:	10
$CH_3Li/TiCl_4$	90	:	10
$CH_3Ti(OPh)_3$	93	:	7
$CH_2{=}CHCH_2MgCl$	65	:	35
$CH_2{=}CHCH_2Ti[N(CH_3)_2]_3$	88	:	12
$CH_2{=}CHCH_2Ti(NEt_2)_3$	94	:	6
$PhSO_2CH_2Ti(OiPr)_3$	65	:	35
$PhSO_2CH_2Li$	64	:	36

The results show that increasing the size of the ligands generally increases diastereofacial selectivity. The reaction of the titanated sulfone is a painful reminder that the problem of Cram selectivity has not been solved in a general way and that more work with other ligands and metals is necessary. It should be mentioned that in a recent study the reaction of methyl- and *n*-butyllithium with 2-phenylpropanal was carefully re-examined [44b]. In contrast to earlier reports in the literature, Cram selectivities of 89–94% were observed at $-78\ ^\circ C$. Hence 2-phenylpropanal is not a good probe to test Cram-type stereoselectivity, at least in reactions involving methyl- and *n*-butylmetal reagents.

An impressive application of titanium chemistry concerns the regio- and stereoselective reaction of 2-phenylpropanal with the homoenolate **89** [114]. Of the eight possible diastereomers of γ-attack, only two are formed in a ratio of **90**:**91** = 93:7. This means high Cram preference and complete simple diastereoselectivity (see Section 3.3.2), in addition to exclusive formation of the *cis*-form of the enol moiety. The lithium precursor affords a complex mixture of products.

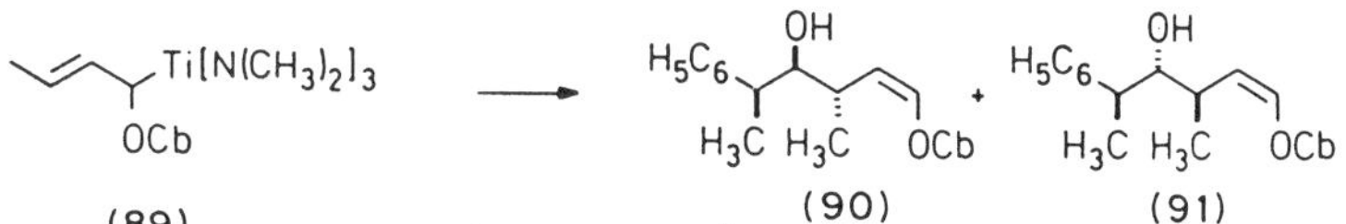

(89) (90) (91)

OCb = $OCON(^iPr)_2$

Side-chain extending reactions of steroidal aldehydes and ketones are often fairly selective using classical reagents [115], but titanation may increase asymmetric induction markedly [50]. A clear limitation concerns *n*-alkyltitanium reagents $RCH_2Ti(OiPr)_3$, which are too sluggish to add to ketones [8]. Alkynyltin reagents add stereoselectively to steroidal aldehydes in the presence of $TiCl_4$ [116]. Several interesting cases of Cram-selective additions of alkynyltitanium reagents in prostaglandin intermediates have been reported [55], although simple chiral aldehydes react non-selectively [54]. Sometimes vinyllithium reagents are more selective than the titanium analogs [56b].

3.3.1.2 Chelation-Controlled Grignard and Aldol Reactions

In contrast to the Cram case in which the carbonyl compounds may react in a variety of conformations, alkoxy carbonyl compounds are potentially capable of chelation. In such cases the number of degrees of freedom is greatly reduced, making asymmetric induction much easier. The principles on which depend this approach are illustrated by the examples given below [42]. The problem is to find the right metal for each particular situation.

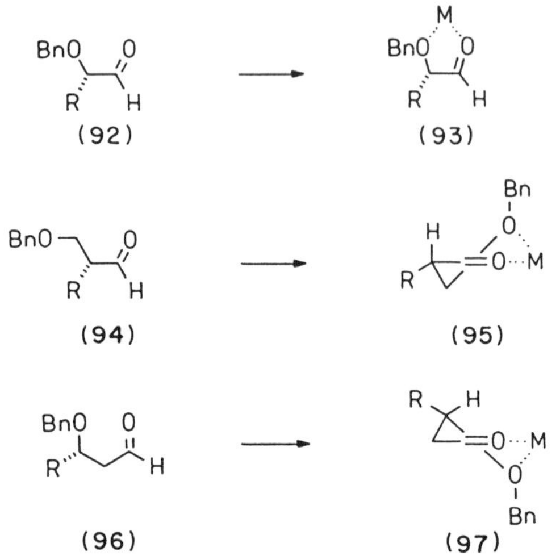

Scheme 3.4.

Efforts up to 1980 provided solutions to only two general problems. According to Cram's classical work on α-alkoxy ketones, simple Grignard reagents RMgX react with high levels of chelation control [117]. Secondly, α-chiral β-alkoxyaldehydes **94** react with cuprates R_2CuLi to form the chelation-controlled Grignard-type adducts [118]. However, these methods do not generally extend to such aldehydes as **92** or **96**. Further, aldol additions of lithium enolates usually afford mixtures of diastereomers in reactions with aldehydes **92**, **94** and **96** [42]. Finally, the classical reagents cannot be used to induce non-chelation control. In many cases metal and ligand tuning based on titanium chemistry solve such problems [42].

Since CH_3TiCl_3 forms bisetherates (Section 3.2.1.2), it was speculated that such aldehydes as **92** should form intermediate Cram-like chelates **98** [119]. Indeed, the ratio of chelation- to non-chelation-controlled adducts turned out to be **99**:**100** = 93:7 [119]. Later the intermediacy of octahedral complexes of the type **98** was proved by ^{13}C NMR spectroscopy [120]. They are the only cases of direct spectroscopic evidence of Cram-type chelates. Chelation-controlled reactions of $RTiCl_3$ need to be performed in an ether-free medium (e.g. CH_2Cl_2) since diethyl ether (or THF) competes for the Lewis acidic titanium, thereby breaking up the chelate. A number of chelates involving $TiCl_4$ or CH_3TiCl_3 have been calculated usng *ab initio* methods in combination with pseudopotentials [42c].

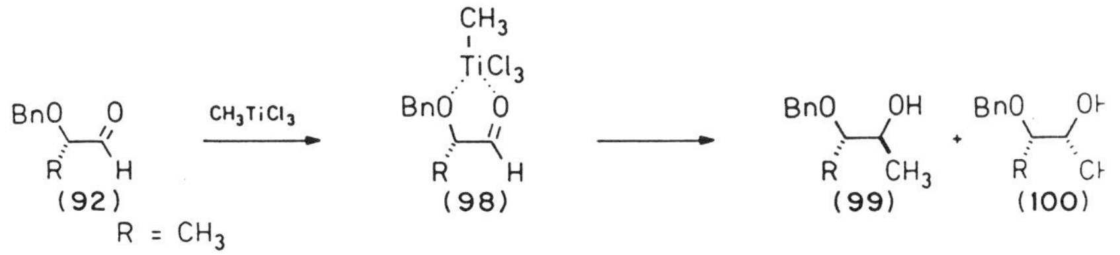

Although other highly Lewis acidic titanium reagents also show excellent levels of chelation control, the range of reagents of the type $RTiCl_3$ is limited (Section 3.2.1.2), especially since ether-free solvents such as CH_2Cl_2 are mandatory [42]. Therefore, a different methodology had to be developed. Accordingly, the alkoxyaldehydes **92** are 'tied up' by Lewis acids such as $TiCl_4$ to form chelates of the type **101**, which are then reacted with allyl- and enolsilanes [119], dialkylzinc reagents [119], tetraalkyllead compounds [91b, e] or $(CH_3)_3SiCN$ [121]. In most cases conversion is >90% with chelation control being >95%.

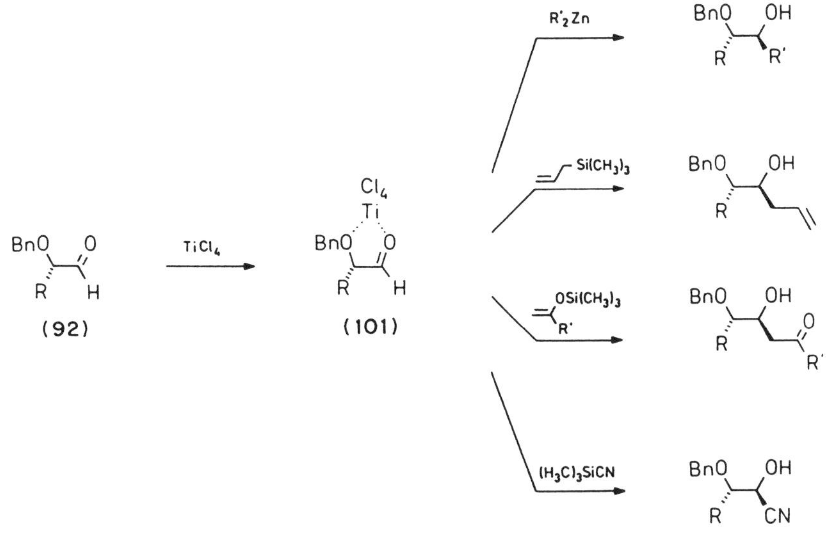

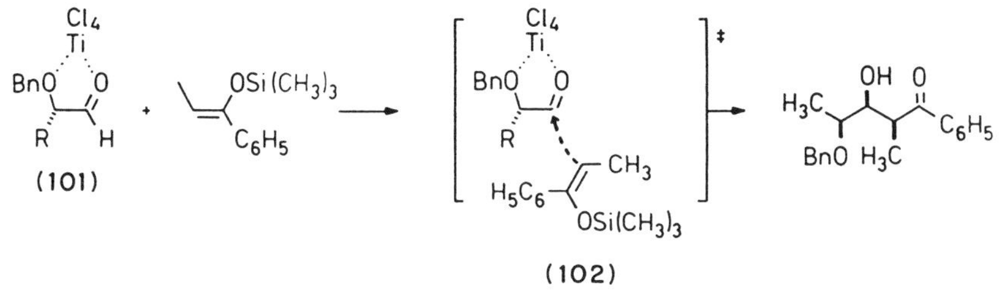

If the carbon nucleophile is prochiral, the problem of simple diastereoselectivity (Section 3.3.2) is also relevant, i.e. four diastereomeric products are possible. In some cases complete control is possible, simple diastereoselectivity being *syn* [122]. Open transition states of the type **102** have been proposed [122]. The mechanistic details of the $TiCl_4$-promoted chelation-controlled Mukaiyama aldol addition had been illuminated by a rapid injection NMR study [92d].

It should be mentioned that in some cases Lewis acids other than $TiCl_4$ should be used (e.g. $SnCl_4$, MgX_2) [42, 122, 123, 124]. Perhaps a disadvantage is the fact that stoichiometric amounts of Lewis acid have to be employed. The fact that catalytic amounts of $LiClO_4$ induce chelation-controlled additions in some cases is a promising development [92d, 92e]. Nevertheless, the enormous generality observed in the case of $TiCl_4$-induced chelation controlled additions remains to be shown for the $LiClO_4$-mediated process. The concept of using Lewis acids capable of chelation in combination with carbon nucleophiles has turned out to be a general principle in stereoselective C—C bond formation [42, 125]. It has been extended to α-chiral β-alkoxyaldehydes (**94**) [122, 126], β-chiral β-alkoxyaldehydes (**96**) [127] and α-chiral α,β-dialkoxyaldehydes [128]. This means that in Scheme 3.4 the metal M does not participate in the form of an organometallic reagent such as RMgX or $RTiCl_3$, but rather as a normal Lewis acid. NMR [8, 119, 129] and X-ray structural data [119b] have been presented, e.g. in the case of $TiCl_4$ and $SnCl_4$ chelates. Along related lines, α- and β-alkoxy acid nitriles react analogously, forming tertiary cyanohydrins stereoselectively [121].

A typical example is the reaction of β-benzyloxyaldehydes **96** with allylsilane in the presence of $TiCl_4$ in CH_2Cl_2 at -78 °C. A chelate **103** is involved which is attacked at the less hindered π-face, 1,3-asymmetric induction being >95% [127]. Intramolecular versions lead to a reversal of diastereoselectivity (**104** vs **107**) [130a]. In these cases the aldehydes **105** become chelated in the form of intermediates **106**, which spontaneously react to form the products **107** with diastereoselectivities of 90%. Chelation-controlled intramolecular addition of allylsilanes in the presence of $TiCl_4$ affords carbocycles in other cases [130b].

BnO O R H (96) Bn = $H_5C_6CH_2$ —$TiCl_4$→ R H Si(CH$_3$)$_3$ O···$TiCl_4$ O Bn (103) → BnO OH R (104)

O Si O R H (105) —$TiCl_4$→ R H Cl O···Ti Cl Cl O Cl Si (106) → OH OH R (107)

The types of prochiral *C*-nucleophiles capable of participating in chelation-controlled reactions vary widely, e.g. **108** [131], **109** [122], **110** [132], **111a** [133a], **111b** [133b], **112** [134], **113** [124] and **114** [135]. In most cases chelation control is >95% and

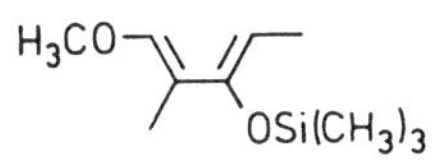

(108) 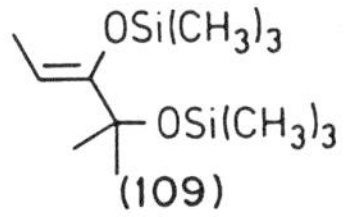 (109)

simple diastereoselectivity is *syn*. Transition states such as **102** (or synclinal analogs) have been proposed [92b, 122]. In related work, chelation-controlled Mukaiyama aldol additions of chiral β-formyl esters have been shown to proceed with >90% diastereofacial selectivity [133c]. The products are γ-lactones having additional keto-functionality.

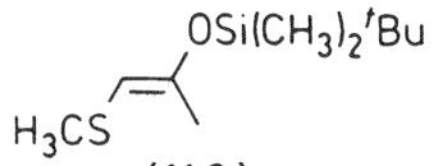

(110) 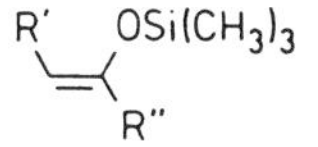 (111) 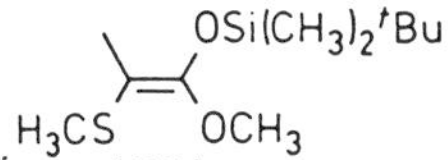 (112)

R' = CH_3; R'' = S^tBu
R' = $Si(CH_3)_3$; R'' = OC_2H_5

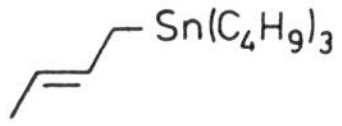

(113) 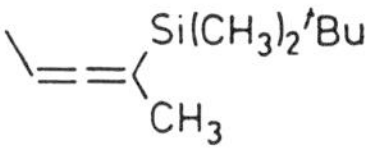 (114)

N,N-Dibenzylaminoaldehydes (**115**) undergo $TiCl_4$-mediated chelation-controlled allylsilane and $(CH_3)_2Zn$ additions, although the diastereoselectivity is not consistently acceptable (the **116**:**117** ratio ranges between 65:35 and 95:5) [136]. This is due to steric inhibition of chelation. If the nitrogen has only one protective group, as in **118**, chelation-controlled allyl- and enolsilane additions are highly effective [137], as are $TiCl_4$-mediated ene reactions [138].

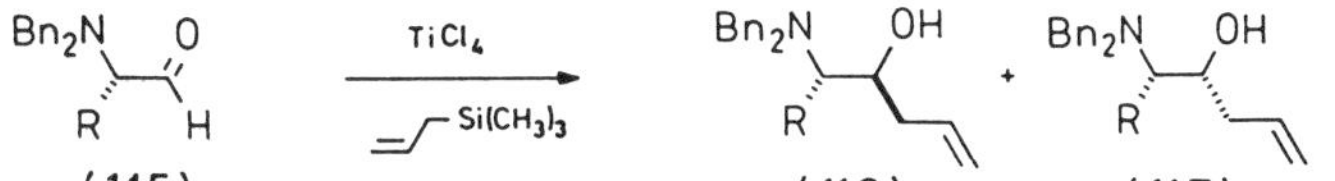

(115) (116) (117)

Bn = $H_5C_6CH_2$

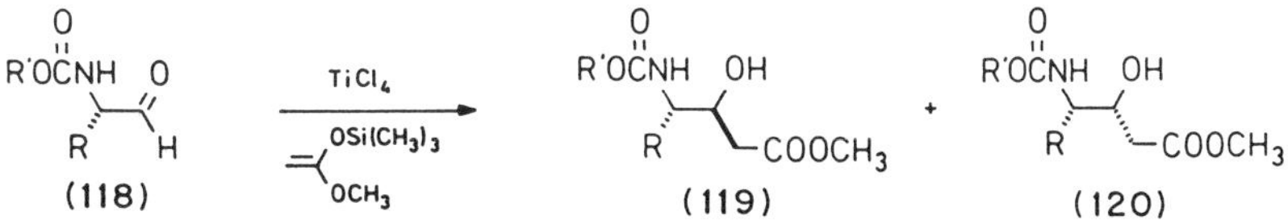

(118) (119) (120)

R'O = $(H_3C)_2CHO$, $(H_3C)_3CO$

Finally, titanium enolates from thioesters prepared by the Evans method ($TiCl_4/NR_3$) add to chiral α-alkoxyaldimines to produce the chelation-controlled adducts [139]. This recent development is of considerable interest in the stereoselective synthesis of β-lactams. Chelation-controlled additions of allylstannanes and Grignard reagents to α,β-epoxyaldehydes mediated by $TiCl_4$ and $MgBr_2$ have also been reported (diastereoselectivity $ds > 95\%$) [140].

3.3.1.3 Non-Chelation-Controlled Grignard and Aldol Additions

Non-chelation control is more difficult to achieve because there are no general ways to reduce the number of rotamers of non-complexed alkoxycarbonyl compounds [42]. Reagents incapable of chelation must be used. Electronic and/or steric factors, notably those defined by the Felkin–Anh model [113], must then be relied on. Since Lewis acidity of organotitanium reagents decreases drastically on going from $RTiCl_3$ to $RTi(OR')_3$ (Section 3.2), the latter could be expected to undergo non-chelation-controlled reactions. Indeed the parent compound triisopropoxymethyltitanium, $CH_3Ti(OiPr)_3$, adds to a variety of α-alkoxyaldehydes in exactly this manner, as shown below [119].

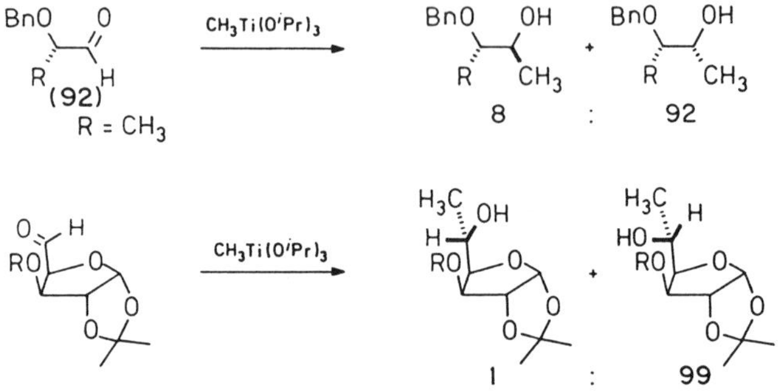

BnO O / R H (92) R = CH_3 —$CH_3Ti(O^iPr)_3$→ BnO OH / R CH_3 (8) : BnO OH / R CH_3 (92)

O H / RO … —$CH_3Ti(O^iPr)_3$→ H_3C OH / H / RO … (1) : H_3C H / HO / RO … (99)

n-Butyltriisopropoxytitanium (**9**) is too sluggish to react with α-alkoxyaldehydes [44]. However, it adds to α,β-dialkoxyaldehydes with non-chelation control [56].

Titanium enolates derived from acetic acid esters do not add diastereoselectively to α-alkoxyaldehydes [141]. The method also fails in the case of crotyltitanium reagents [142]. In contrast, prochiral titanium enolates such as **121** having alkoxy or amino ligands show good degrees of non-chelation control [141], as do certain homoenolate equivalents based on titanium chemistry [114]. Simple diastereoselectivity in the aldol additions is *syn* (see Section 3.3.2 for this type of diastereoselectivity).

BnO O / (92) R H + $OTi(O^iPr)_3$ / C_6H_5 (121) → OH O / C_6H_5 / BnO CH_3 (13) : OH O / C_6H_5 / BnO CH_3 (85)

R = CH_3; Bn = $H_5C_6CH_2$

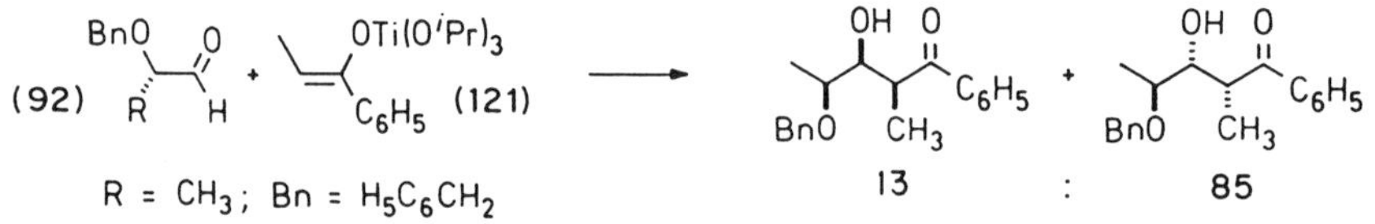

Since α-alkoxy ketones have a pronounced tendency to undergo efficient chelation control even with simple Grignard reagents, reversing the diastereoselectivity is expected to be difficult. Indeed, even $CH_3Ti(OiPr)_3$ reacts with 99% chelation control [68]. However, the combination of metal, ligand and protective group tuning [143] can tip the balance completely. By using bulky silyl protective groups, addition of methyl- and allyltitanium reagents or titanium enolates of low Lewis acidity results in >99% non-chelation control, as in the reaction of the enolate **122** with α-siloxy ketones [68]. The use of bulky silyl protective groups in combination with classical reagents RMgX results in low degrees of selectivity. Similarly, the trimethylsilyl protective group is generally inefficient. Thus, steric factors are more important than possible electronic effects originating from silicon [144]. Clearly, it is the combination of metal tuning, ligand tuning and protective group tuning which ensures success! A clear limitation of this method concerns the less reactive *n*-alkyltitanium reagents which do not add to ketones.

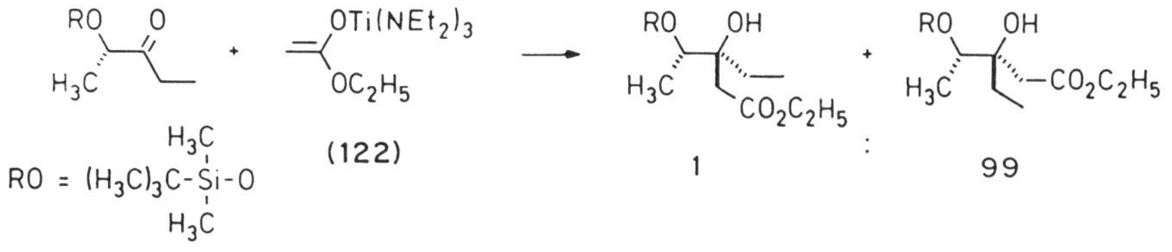

N,*N*-Dibenzylaminoaldehydes (**115**) react with a variety of classical reagents RMgX and RLi with surprisingly high degrees of non-chelation control (*ds* > 90%) [136]. An exception is the very reactive allylmagnesium chloride (3:1 diastereomer mixtures). In this case prior addition of $ClTi(NEt_2)_3$ to the Grignard reagent results in >90% non-chelation control [136]. Classically protected α-aminoaldehydes (e.g. those having Boc groups) generally react with Grignard or organolithium reagents to form mixtures [145], and titanation in one case did not improve the diastereoselectivity greatly [146]. In such cases the interplay of metal, ligand *and* protective group tuning is necessary [147].

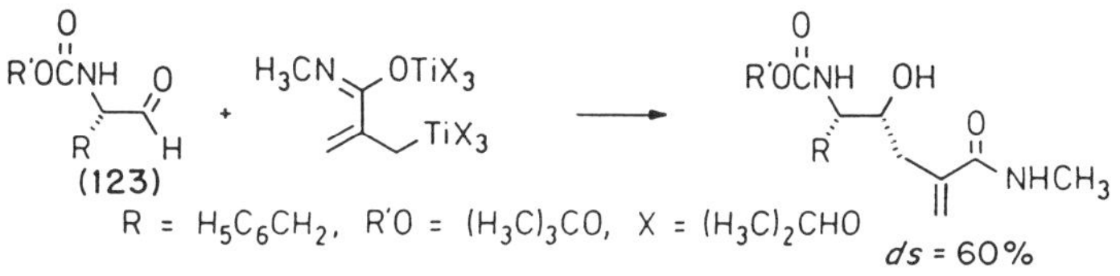

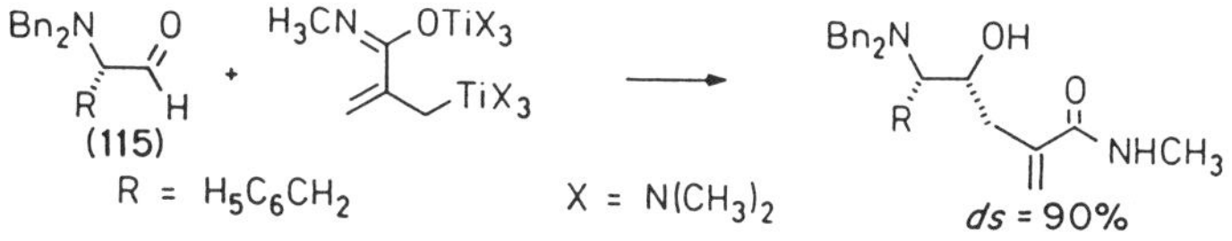

In summary, non-chelation control in reactions of *N*,*N*-dibenzylaminoaldehydes is easily achieved using classical Grignard reagents and lithium enolates, titanation being necessary only in difficult cases [148]. The possible reasons for the general behavior of these aldehydes have been delineated elsewhere [148]. In complete contrast, there are no general ways to obtain consistently high levels of non-chelation control in reactions of α-alkoxyaldehydes or -ketones, in spite of progress achieved with a number of titanium reagents. Perhaps the combination of metal and protective group tuning will provide a solution to this problem. Reagent control [149], albeit more expensive, has been successful in the case of boron [150] and titanium [151–153] reagents (Section 3.3.3).

A number of chelation- and non-chelation-controlled titanium mediated reactions of chiral lactols [154–156], 2-alkoxy-1-(1,3-dithian-2-yl)-1-propanones [157a], α-keto-amides [157b], arylsulfinylacetophenones [158] and α-alkoxyimines [139d] have been reported. In another interesting development, titanium phenolates were shown to arylate glyceraldehyde acetonide and α-aminoaldehydes selectively with non-chelation control [159a, d]:

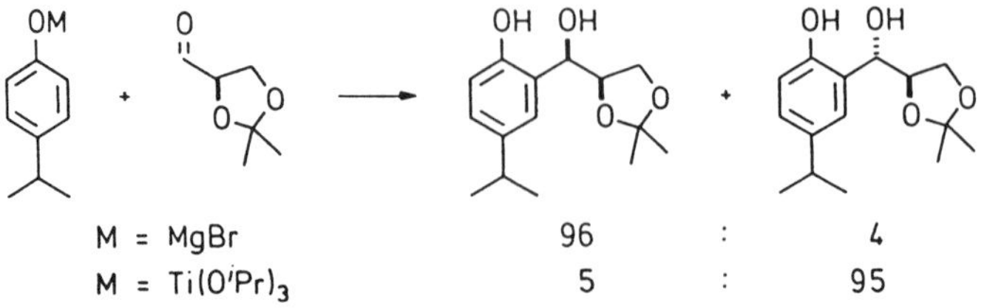

M = MgBr 96 : 4
M = $Ti(O^iPr)_3$ 5 : 95

Along similar lines, titanated pyrroles reacted stereoselectively with glyceraldehyde acetonide and with arabinofuranose and glucopyranose derivatives to form pyrrole *C*-glycoconjugates [159c]. The method constitutes an elegant route to heteroarylate sugars in a highly stereoselective manner (*ds* > 98%). In certain cases the cerium analog affords diastereomers having the opposite relative configuration (chelation control). The bromomagnesium salt of pyrrole produces mixtures.

3.3.2 REACTIONS INVOLVING SIMPLE DIASTEREOSELECTIVITY

3.3.2.1 Prochiral Allylic Titanium Reagents

The reaction of an aldehyde with a prochiral reagent such as a *Z*- or *E*-configured crotyl metal reagent affords *syn*- and *anti*-adducts, respectively [160], each in racemic form (only one enantiomer shown in Scheme 3.5). Perhaps the most efficient reagents are crotylboron compounds, the *E*-configured form resulting in >95% anti-selectivity and the *Z*-analog showing opposite simple diastereoselectivity (>95% *syn* selectivity) [160]. Compact chair-like transition states are the source of stereoselectivity (Scheme 3.5). Nevertheless, other metal systems have been tested, some of which are easily accessible and fairly stereoselective [160]. One of the drawbacks of most of these reagents has to do with the fact that addition to ketones fails chemically or occurs non-stereoselectively.

The titanation of crotylmagnesium chloride (which is a mixture of stereo- and regioisomers) provides stereoconvergently *E*-configured reagents, crotyltriisopropoxy-

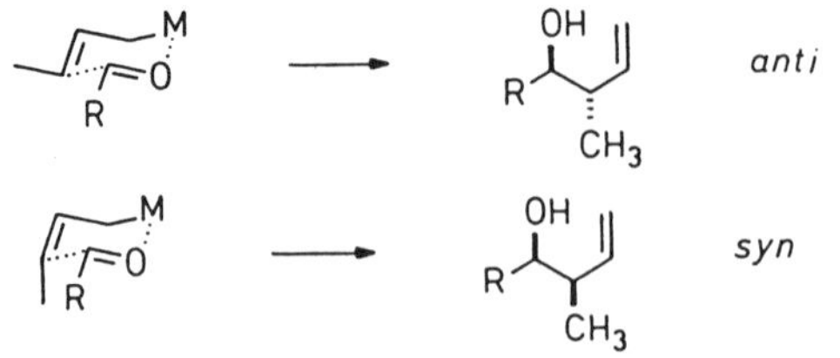

anti

syn

Scheme 3.5.

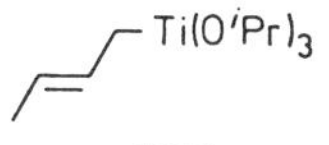

(13) 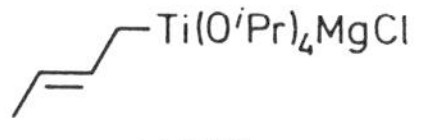 (123) 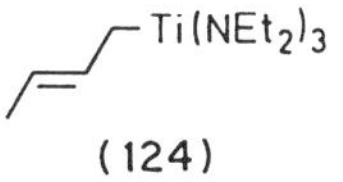 (124)

titanium (**13**), the crotyltitanium ate complex **123** or crotyltris(diethylamino)titanium (**124**) [14, 50].

The addition of aldehydes occurs with moderate to good degrees of *anti* selectivity [14, 15, 50]. Real benefits of the crotyltitanium reagents become apparent in reactions with prochiral ketones, diastereoselectivity generally being >90% [14, 15, 50]. In the case of acetophenone, the ate complex **123** is best suited, but purely aliphatic ketones require the trisamide **124** for the best results. Crotyltriphenoxytitanium also adds stereoselectively to ketones, but is less easily accessible [161]. In all cases chair-like transition states have been proposed.

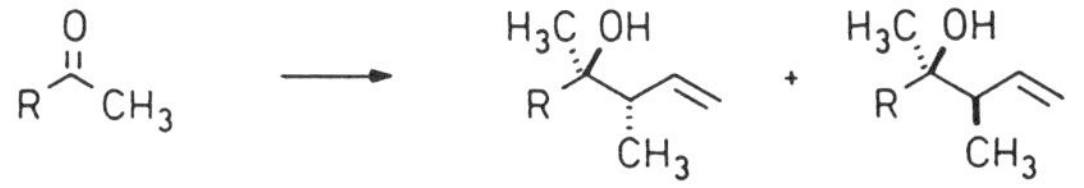

>90 : <10

Bis(cyclopentadienyl)titanium(IV) reagents **125** [162] and the related Ti(III) compounds **126** [163] add to aldehydes highly *anti* selectively. Reversal of diastereoselectivity in reactions of **125** occurs if BF_3-etherate is used as an additive [164]. In this case the BF_3 adduct of the aldehyde reacts with the titanium reagent via an open transition state. This is of synthetic importance because so far it has not been possible to generate *Z*-configured crotyltitanium reagents which are required for *syn* selectivity. Nevertheless, the simplicity and low cost involved in the production of such reagents as **13** and **123** does not pertain to the Cp_2-titanium compounds.

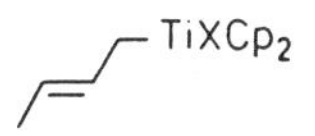

(125, X = Cl, Br, I) 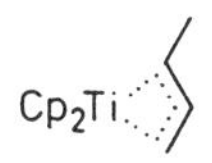 (126)

A large number of heteroatom-substituted allyltitanium reagents have been generated and reacted with aldehydes to form the corresponding *anti* adducts, generally with more than 98% diastereoselectivity [8]. In most cases a very simple experimental protocol is used: generation of the allylic lithium precursor by deprotonation of the neutral compound followed by titanation with $Ti(OiPr)_4$ and *in situ* aldehyde addition. Reductive cleavage of thiothers followed by titanation is another elegant procedure [165]. Titanium ate complexes (Section 3.2.1.4) have been postulated, although structural ambiguities remain. An early example is the silyl-substituted titanium ate complex **53a** which adds regio- and stereoselectively to aldehydes with exclusive formation of the adducts **127** [50, 52]. This one-pot procedure is simpler than the multi-step process based on boron chemistry [166]. Titanation of the Li precursor using $ClTi(OiPr)_3$ [50],

$ClTi(NEt_2)_3$ [50] or Cp_2TiCl [167] also results in species which react *anti* selectively, but the cheap $Ti(OiPr)_4$ is clearly the titanating agent of choice. Reactions with ketones are also stereoselective. The products are synthetically useful because stereospecific conversion into dienes via the Peterson elimination under basic or acidic conditions is possible [50]. Prolonged reaction times in the actual addition reactions of **53a** lead directly to the dienes.

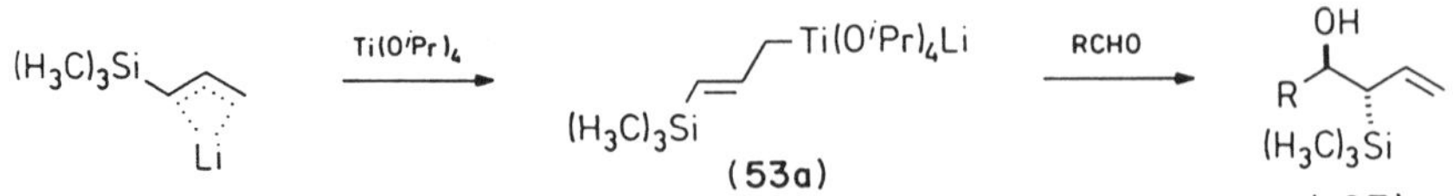

(53a)

(127)

Using the above techniques, reagents **128** [79], **129** [79], **130** [169], **131** [170], **132** [171], **133** [172], **134** [172] and **135** [8, 44] were prepared and reacted with aldehydes. Again, the simplicity and high *anti* selectivity of this procedure make it more efficient than alternative methods based on other metals. For example, reagent **131** has been used in a highly efficient synthesis of the diterpene (±)-aplysin-20 [170]. It is important to point out that some of the reagents are chiral, but were used in racemic form, e.g. **130, 133** and **134**. Therefore, enantiomerically pure forms should provide diastereomerically *and* enantiomerically pure adducts. This interesting aspect has been studied recently [168] (Section 3.3.3).

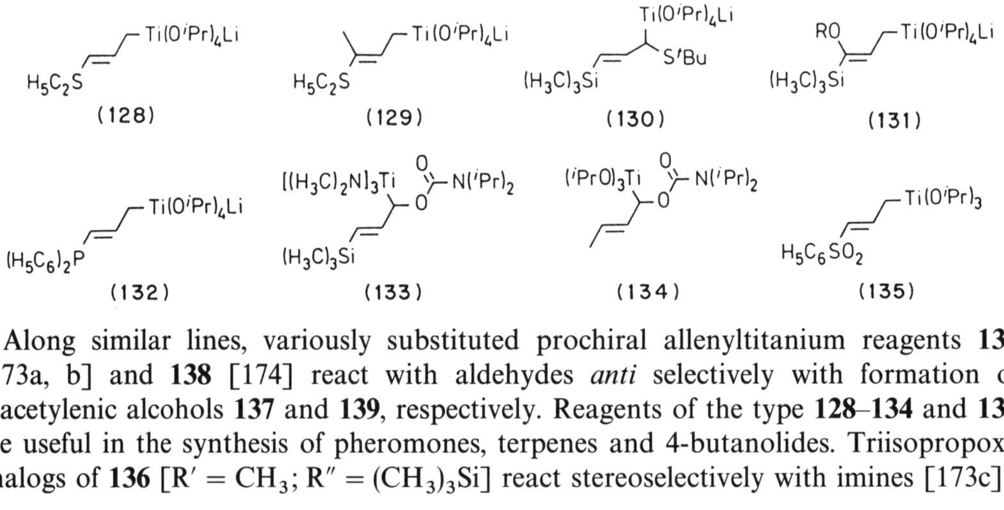

(128) (129) (130) (131)

(132) (133) (134) (135)

Along similar lines, variously substituted prochiral allenyltitanium reagents **136** [173a, b] and **138** [174] react with aldehydes *anti* selectively with formation of β-acetylenic alcohols **137** and **139**, respectively. Reagents of the type **128–134** and **136** are useful in the synthesis of pheromones, terpenes and 4-butanolides. Triisopropoxy analogs of **136** [$R' = CH_3$; $R'' = (CH_3)_3Si$] react stereoselectively with imines [173c].

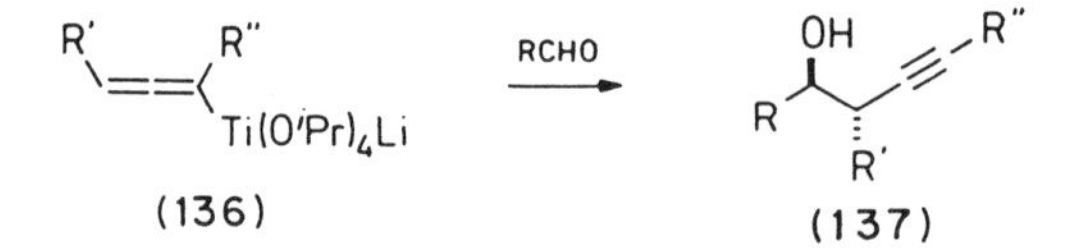

(136) (137)

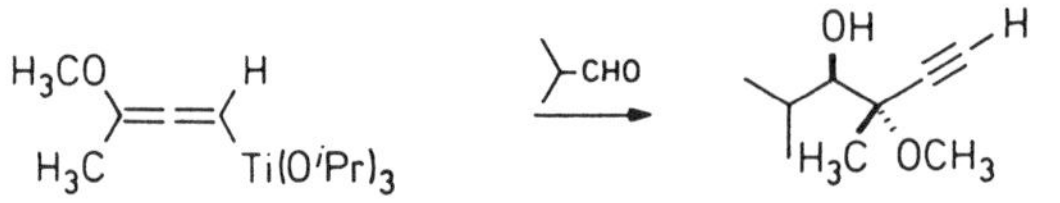

(138) (139)

3.3.2.2 Titanium-Based Aldol Additions

Aldol additions using prochiral titanium enolates (and heteroatom analogs) are useful reagents in certain situations, particularly when they are complementary to existing methodologies based on other metals [175]. The titanation of lithium enolates with $ClTi(OiPr)_3$ or $ClTi(NR_2)_3$ affords titanium enolates which add *syn* selectively to aldehydes, irrespective of the geometry of the enolate [64]. Since diastereoselectivity is not consistently above 90%, the method is limited in scope. However, it is useful in the case of cyclic ketones [64], because it is complementary to boron enolates which provide the opposite diastereomeric products (*anti* selectivity >95%) [175]. A prime example is the enolate derived from cyclohexanone, a case which also illustrates the power of ligand tuning [64]. Certain tin enolates also react *syn* selectively [176a]. The triisopropoxytitanium enolate derived from isobutyric acid methyl ester reacts stereoselectively with chirally modified imines to produce (4*R*)-β-lactams [176b]. The use of the lithium enolate results in the formation of the (4*S*)-β-lactams. Chirally modified triisopropoxytitanium enolates react highly selectively (Section 3.3.3). In fact, this is their primary area of application.

OTiX3 —H5C6CHO→ (O, H, OH, C6H5) + (O, H, OH, C6H5)

X = O^{i}Pr 86 : 14
X = N(CH3)2 92 : 8
X = N(Et)2 97 : 3

The Mukaiyama crossed aldol addition, i.e. the reaction of an aldehyde and an enolsilane in the presence of $TiCl_4$, does not involve titanium enolates (Section 3.2.3) and generally affords mixtures of *syn* and *anti* adducts, although progress has been made [92].

RCHO + R', R'', OSi(CH3)3 —TiCl4→ (OH O, R, R'', R') + (OH O, R, R'', R')

syn *anti*

The proper choice of substrates and conditions may result in high levels of simple diastereoselectivity irrespective of the geometry of the double bond [92b]. If R′ is small (methyl), good to excellent *anti* selectivity results, provided R″ is a bulky group as in sterically demanding ketones [177, 178]. Some ketene ketals (R′ = CH_3; R″ = OEt) also react *anti* selectively [179]. In the case of bulky R′ groups (silyl, *tert*-butyl), *syn* selectivity results [179]. Open transition states in which the enolsilane attacks the activated aldehyde $RCHO/TiCl_4$ were proposed to explain the stereochemical results. Sometimes other Lewis acids are more effective [178, 180]. One of the important applications of the Mukaiyama reaction is the chelation-controlled addition of enolsilanes to $TiCl_4$-chelated forms of chiral α- and β-alkoxyaldehydes [42] (Section 3.3.1.2).

As mentioned in Section 3.2.2, some but not all enolsilanes derived from ketones react smoothly with $TiCl_4$ to form the corresponding titanium trichlorides [88]. These undergo aldol additions with aldehydes, but simple diastereoselectivity is moderate (65–89% *syn* selectivity). Certain chiral enolsilanes undergo the Si–Ti exchange reaction, leading to titanium trichlorides which react diastereo- and enantioselectively (Section 3.3.3) [92b].

Significant progress was made with the discovery that ketones and carboxylic acid derivatives can be titanated by treatment with $TiCl_4$ and a tertiary amine [99, 181]. Formally, this generates the trichlorotitanium enolate and the amine hydrochloride, which may actually interact to form the tetrachlorotitanium ate complex having the ammonium salt as the counter ion, e.g. **140**. It is also not clear whether monomers or aggregates are involved. In any event, the reagents add to aldehydes with high degrees of *syn* selectivity [181]. For example, diastereoselectivity in the formation of adduct **141** is 92%. This methodology has been applied successfully in the aldol addition of thioesters to aldehydes and aldimines [139].

TiCl4 NR3 (140) OTiCl4 NHR3 CHO (141) OH O CH3

The sense and degree of stereoselectivity are comparable to those of the well known boron-mediated processes [175], but the yields are often higher. Zimmerman–Traxler transition states accommodate the results [181]. Chirally modified enolates react diastereo- and enantioselectively [181] (Section 3.3.3). Hence this methodology is expected to be the process of the future, although the enolates derived from cyclic ketones still need to be studied.

Titanium enolates derived from lactams show moderate to good levels of simple diastereoselectivity [8, 44]. Ligand tuning needs to be looked at. No method for the diastereoselective addition of aldehyde-enolates to aldehydes is known. Although aldehyde titanium enolates also provide *syn–anti* mixtures, titanated hydrazones **142** derived from aldehydes add to aldehydes *syn* selectively (**143**:**144** $\leqslant 10 : \geqslant 90$) [65]. The reagents **142** are *E*-configured, which means that *syn* selectivity is difficult to explain (boat transition state?) [65]. Titanated chiral bislactims react enantio- and diastereoselectively (Section 3.3.3).

RCH2CH NN(CH3)2 1. LDA 2. ClTi(OⁱPr)3 Ti(OⁱPr)3 NN(CH3)2 R (142) R'CHO OH NN(CH3)2 R H R (143) OH NN(CH3)2 R H R (144)

$TiCl_4/NEt_3$-mediated Knoevenagel condensations using active methylene carbonyl compounds probably proceed via chlorotitanium enolates [98]. In the case of the phosphonate **145**, the thermodynamically more stable *E*-configured condensation products **146** are formed preferentially [98]. In contrast, the sodium salt of **145** reacts with $ClTi(OiPr)_3$ to form an isolable mixture of *E*/*Z*-enolates [182] which condense stereoconvergently with aldehydes to form selectively the *Z*-configured olefins **147** with concomitant transesterification [183]. Triethylamine as the base is also efficient [183]. Mechanistically, a kinetically controlled *syn*-selective aldol addition is involved followed by stereospecific *O*-titanate elimination [182]. Malonates and other CH-acidic compounds also undergo this type of Knoevenagel condensation [183, 98]. The method is mild and works in cases in which traditional Knoevenagel conditions fail [184].

$(H_5C_2O)_2P(O)CH_2COOC_2H_5$ (145) → [$TiCl_4/N(C_2H_5)_3$, RCHO] → RCH=C($P(O)(OC_2H_5)_2$)($COOC_2H_5$) (146)

(145) → [$NaH/ClTi(O^iPr)_3$, RCHO] → RCH=C($P(O)(OC_2H_5)_2$)(COO^iPr) (147)

3.3.3 ENANTIOSELECTIVE ALDOL AND GRIGNARD ADDITIONS

Chiral modification of titanium reagents is possible in one of several ways [8]:

1. Using reagents $RTiX_3$ in which chirality occurs in the organyl moiety R; the chiral information may or may not be removed after C—C bond formation.
2. Using reagents having chiral ligands X.
3. Using reagents having a center of chirality at titanium.
4. Using reagents which incorporate two or more of the above features.

The first known example of type 1 reactions involves aldol addition of the enolate **148**, which reacts with benzaldehyde to form essentially one of four possible adducts [28]. The original chiral information remains in the product **149**. Related reactions have been reported for titanium enolates derived from *β*-lactams [185] and titanated terpenes [186].

H_3CO–steroid–$OTi[N(C_2H_5)_2]_3$ (148) → [C_6H_5CHO] → H_3CO–steroid ketone with $CH(OH)C_6H_5$ (149)

This methodology has been extended to the enolate **151**, which adds to aldehydes to form adducts **152** preferentially (*ds* $\geqslant$ 98%) [69]. This means essentially complete simple diastereoselectivity (*syn* type) and complete diastereofacial selectivity. Since the original chiral information can be cleaved off oxidatively, enolate **151** is a 'chiral propionate.' It competes well with the boron analog [187]. As a result of a systematic study it became clear that an excess of titanating agent $ClTi(OiPr)_3$ enhances stereoselectivity dramatically. It is possible that less selective ate complexes are formed in the initial titanation step which rapidly lose chloride ions in the presence of additional $ClTi(OiPr)_3$. Based on earlier work on the titanation of lithium enolates using $Ti(OiPr)_4$ to form titanium ate complexes [64], the lithium enolate derived from **150** was reacted with $Ti(OiPr)_4$ [69]. The use of an excess of this titanating agent resulted in an enolate which showed the same selectivity as **151**. The influence of excess of $ClTi(OiPr)_3$ or $Ti(OiPr)_4$ [45, 188] needs to be studied in other systems also.

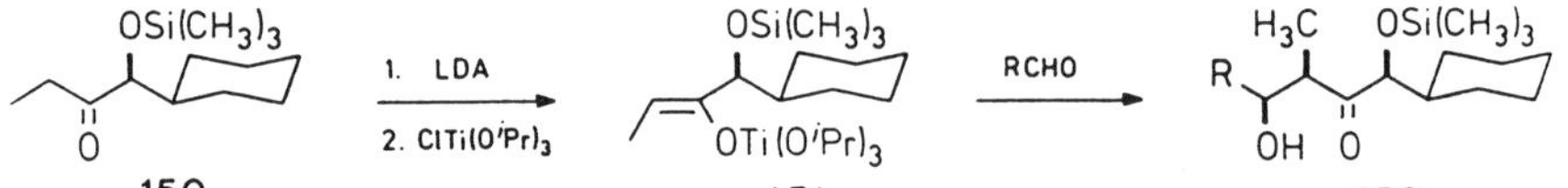

OSi(CH3)3 / 150 → 1. LDA, 2. ClTi(O^{i}Pr)3 → OSi(CH3)3, OTi(O^{i}Pr)3 / 151 → RCHO → R, H3C, OSi(CH3)3, OH, O / 152

Along similar lines, the enolate **154** was prepared and reacted with benzaldehyde [188]. Using a threefold excess of $ClTi(OiPr)_3$ in diethyl ether, a 92:5:3:0 ratio of diastereomers was observed, adduct **155** being the major product. The result is synthetically significant because the analogous boron enolate derived from **153** shows opposite diastereofacial selectivity [175]. This is due to the fact that the titanium reagent undergoes internal chelation, in contrast to the boron analog [188].

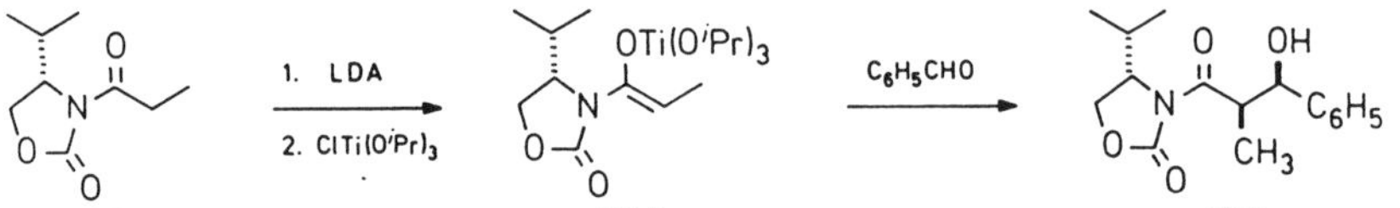

153 → 1. LDA, 2. ClTi(O^{i}Pr)3 → OTi(O^{i}Pr)3 / 154 → C6H5CHO → O, OH, C6H5, CH3 / 155

The previously mentioned titanation of carbonyl compounds using $TiCl_4/EtN(iPr)_2$ (Section 3.3.2.2) is particularly effective in the case of chiral reagents [181]. For example, ketone **150** undergoes the Ti-mediated aldol addition to benzaldehyde with 99% *syn* selectivity. Carbonyl compounds **156–158** show similar levels of stereoselectivity. The yields (70–95%) and occasionally even the selectivities are higher than in the case of the *syn*-selective boron analogs [181, 189].

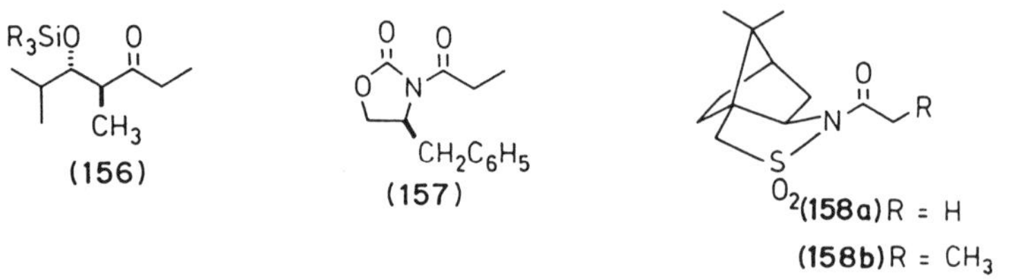

R3SiO, O, CH3 (156) · O, O, N, CH2C6H5 (157) · O, N, R, S, O2 (158a) R = H, (158b) R = CH3

Enantioselective versions of the Mukaiyama aldol additions based on chiral *O*-silyl ketene ketals have also been described. Reagents **159a** [190], **161** [190] and **163a** [191] add to aldehydes to form aldol adducts **160** or **162** on an optional basis. Although the diastereoselectivity (92–99%) is comparable to that of other leading 'chiral acetate synthons' [192], the chemical yields are moderate (47–69%). The Oppolzer-sultam **158a** provides an alternative solution to this long-standing problem [193]. Conversion of **158a** into the corresponding enolsilane affords a 'chiral acetate' which undergoes $TiCl_4$-mediated aldol additions. The products are solids and can be isolated by recrystallization in yields of 54–73%; such work-up provides stereochemically pure products [193].

$SO_2C_6H_5$ $OSi(CH_3)_2{}^tBu$ $\xrightarrow[TiCl_4]{RCHO}$ R*O C(=O) ... OH (160)

(159a) R = H
(159b) R = CH_3

$SO_2C_6H_5$ $OSi(CH_3)_2{}^tBu$ (161) $\xrightarrow[TiCl_4]{RCHO}$ R*O C(=O) ... OH (162)

SO_2 $OSi(CH_3)_2{}^tBu$ $\xrightarrow[TiCl_4]{RCHO}$ R*O C(=O) ... OH (160)

(163a) R = H
(163b) R = CH_3

In the case of the chiral propionates **159b** [190], **163b** [191] and **164** [194], enantioselectivity is generally >90%, but simple diastereoselectivity is not uniformly good, *anti/syn* ratios of 5 : 1 being common. Nevertheless, in a number of cases excellent results are obtained, depending on the type of aldehyde. For example, reagent **164** in combination with $TiCl_4/PPh_3$ adds to *aromatic* aldehydes with essentially exclusive formation of adducts of the type **165** ($ds = 97\%$) [194]. The silicon reagent **164** undergoes an exchange reaction with $TiCl_4$ to form the trichlorotitanium enolate in which the titanium is complexed intramolecularly by the amino group [194]. The method has been applied in a highly efficient synthesis of *trans-β*-lactones possessing interesting biological properties [194c]. Such reagents have also been added stereoselectively to imines [195]. A number of other chiral enolsilanes have been shown to undergo $TiCl_4$-mediated aldol additions [92b]. In most of the above reactions, the *O*-silyl reagents react via open transition states, although under certain conditions Si–Ti exchange prior to aldol additions may occur [92b]. A number of stereoselective additions to imines as a means to produce β-lactams has been reported [92b, c].

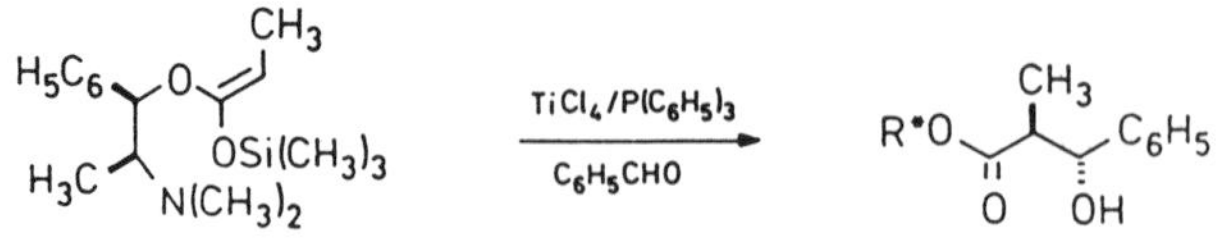

Lithiated bis-lactim ethers **166** are alkylated stereoselectivity *trans* to the isopropyl group, hydrolysis then affording amino acids in >98% enantiomeric purity (Schöllkopf amino acid synthesis) [196]. This powerful method is inefficient if aldehydes are used as electrophiles because simple diastereoselectivity in the aldol-type process is low. This problem was nicely solved by titanating the lithium reagent with $ClTiN[(CH_3)_2]_3$ (or the *N,N*-diethyl analog). The titanated form **167** reacts with aldehydes to form essentially only one of the four possible diastereomers **168** [197]. This means that simple diastereoselectivity is *anti*, which was explained in terms of a chair transition state [197]. The adducts **168** can be hydrolyzed, cleaving off the recyclable chiral auxiliary (valine). Reagents of the type **167** have also been added to chiral aldehydes with complete reagent control [198]. The titanium reagent **167** also reacts stereoselectively with nitroolefins and α,β-unsaturated carbonyl compounds (Section 3.4).

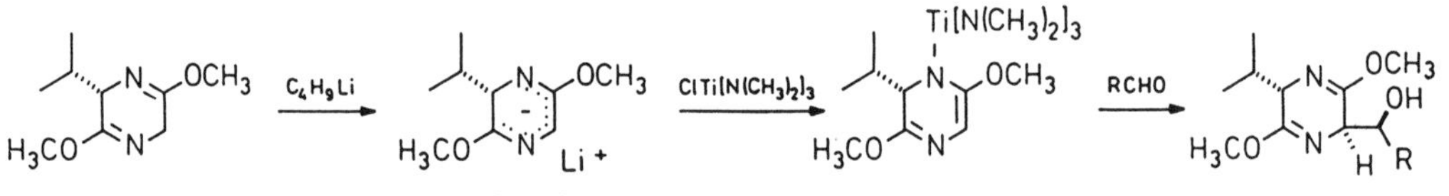

One of the most efficient methods for enantioselective homo-aldol addition concerns the reaction of the titanium reagent **169**, which adds to aldehydes to produce essentially a single adduct (**170**) [199]. Cleavage of the chiral auxiliary affords furans (**171**), which are readily oxidized to the enantiomerically pure lactones (**172**).

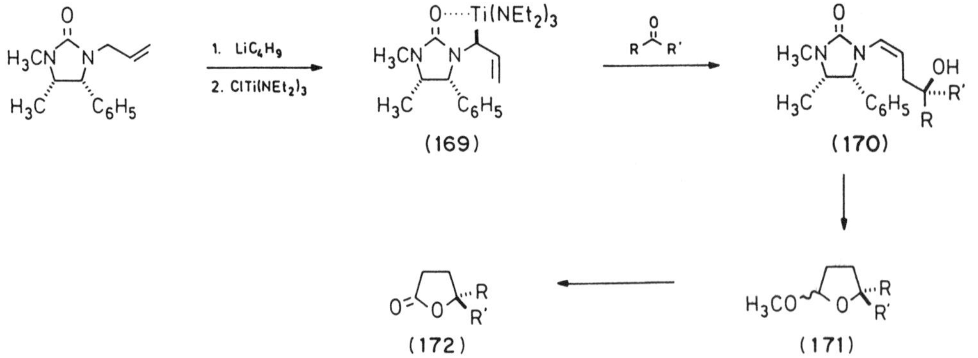

Methyl-substituted analogs of **169** would involve not only enantioselectivity, but also simple diastereoselectivity. Although such reagents have not been developed, a simple approach based on chiral 1-oxyallyltitanium compounds has been reported [200] (Section 3.3.2 describes racemic versions). For example, lithiation of the enantiomerically pure alkenyl carbamate **173** results in the (unselective) lithium reagent **174** which can be titanated by $Ti(OiPr)_4$ with retention of configuration to form the ate complex **175**. The latter reacts with 2-methylpropanal to form essentially a single diastereomer (**176**), having an enantiomeric excess (*ee*) of 86% [200]. Interestingly, titanation with $ClTi(OiPr)_3$ leads to racemization, whereas $ClTi(NEt_2)_3$ results in inversion of configuration. The reasons for the dramatic differences are not fully understood [201]. The stannylated form of **174** undergo stereoselective $TiCl_4$-mediated aldehyde additions which involve prior Si–Ti exchange [90].

(173) $\xrightarrow{C_4H_9Li}$ (174) $\xrightarrow{Ti(O^iPr)_4}$ (175) $\xrightarrow{\text{(CH}_3)_2\text{CHCHO}}$ (176)

OCb = $OCON(^iPr)_2$

Chiral α-alkoxy alkyllead compounds undergo $TiCl_4$-mediated additions to aldehydes via an S_E2 retention pathway to provide the corresponding *syn* adducts ($ds >$ 96%) [91c].

Particularly exciting is the observation that the *achiral* carbamate **177** can be deprotonated enantioselectively by *sec*-butyllithium in the presence of (−)-sparteine to form a lithium intermediate, which after titanation with $Ti(OiPr)_4$ (cf. **178**) undergoes diastereo- and enantioselective homo-aldol additions to form adducts **179** ($ee =$ 80–95%) [202]. The titanium ate complex **178** is configurationally stable in solution below −30 °C. The lithium precursor is configurationally labile in solution, but stable in the solid state [202]. Since only one form of the lithium reagent precipitates, second-order asymmetric induction is involved. Originally it was assumed that the lithium precursor having the *R*-configuration reacts with the transmetallating agent $Ti(OiPr)_4$ with retention of configuration to form the titanium ate complex **178** [202a]. Later it was reported that in fact the *S*-configured lithium precursor is transmetallated by $Ti(OiPr)_4$ with inversion of configuration to produce reagent **178** [202b]. The methodology has been applied in the synthesis of the insect pheromone (+)-eldanolide ($ee = 92\%$) [202c]. It is likely that chiral diamines which are even more efficient than sparteine will be developed and utilized soon.

(177) $\xrightarrow[\text{2. Ti(O}^i\text{Pr)}_4]{\text{1. LiC}_4\text{H}_9\text{/Sparteine}}$ (178) $\xrightarrow{RCHO}$ (179)

The alkenyl carbamate chemistry has been exploited efficiently in the synthesis of natural products [201–203]. It has also been instrumental in the development of an elegant method for the determination of configurational stability of chiral organometallic compounds [174].

Early attempts at devising organotitanium compounds $RTiX_3$ having chiral ligands X were not very successful [8, 15], but paved the way to more efficient reagents. Thus, compounds of the type **180** [12, 14, 44, 204], **181** [12, 204] and **182** [205] add enantioselectively to aromatic aldehydes, the *ee* values ranging between 70% and 95%. Unfortunately, enantioselectivity in the addition to aliphatic aldehydes turned out to be meager. It is important to stress that the precise structure and aggregation state were not determined. In fact, evidence accumulated which showed that monomeric cyclic structures are not always involved, but rather ring-opened oligomers [8, 205].

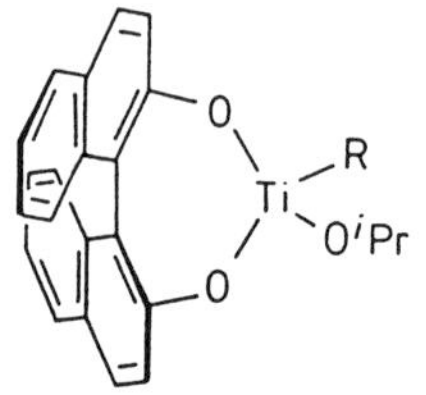

(180a) R = C_6H_5
(180b) R = CH_3

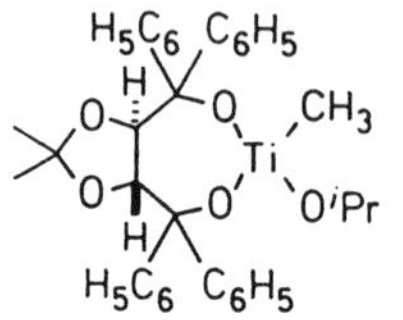

(181)

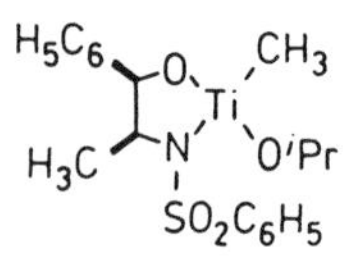

(182)

The power of metal and ligand tuning (Section 3.2) in terms of enantioselectivity really became evident with the introduction of the chiral titanating agent **184** [206]. It contains a bulky, electron-donating Cp group and two sugar ligands based on the cheap diacetoneglucose **183**. Allyl Grignard-type additions and aldol additions are highly enantioselective (*ee* > 90%) [206]. Structural and mechanistic features of this powerful methodology have been delineated elsewhere [207].

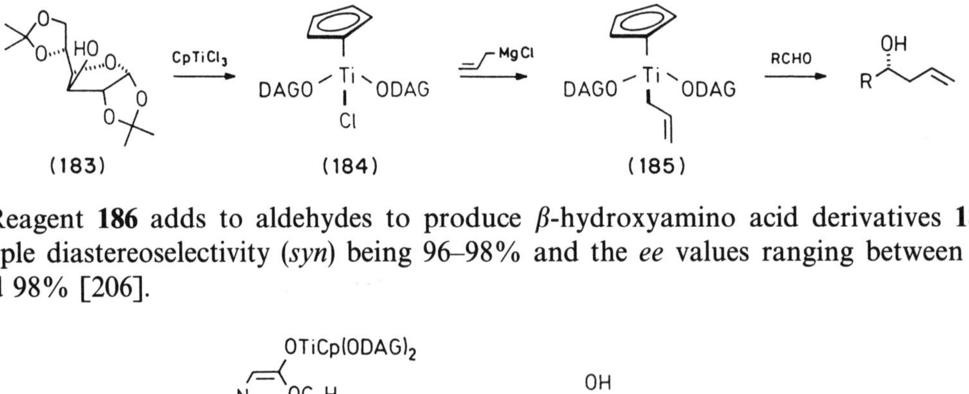

(183) (184) (185)

Reagent **186** adds to aldehydes to produce *β*-hydroxyamino acid derivatives **187**, simple diastereoselectivity (*syn*) being 96–98% and the *ee* values ranging between 87 and 98% [206].

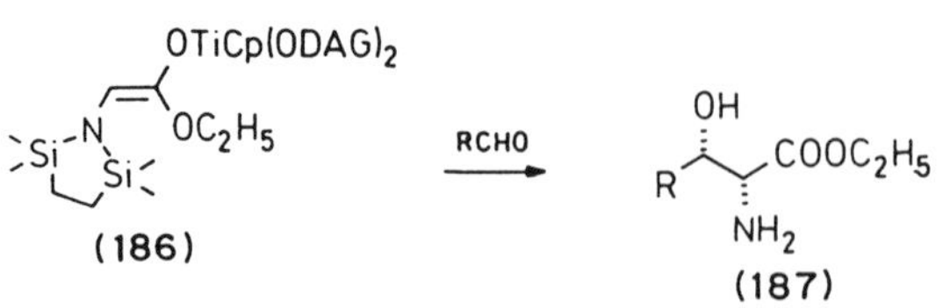

(186) (187)

A stereochemically complementary allyl transfer reagent is **190** [206c]. The C_2-symmetric titanating agent **189**, prepared from the Seebach-diol **188**, does not undergo ring opening to form oligomers. Allyl addition to aldehydes occurs with *ee* values of >95%. In the case of analogous crotyltitanium reagents (and other substituted allylic and enolate compounds), enantioselectivity is also essentially complete. The reagents are so powerful that complete reagent control is observed in reactions with chiral aldehydes. Diols of the type **188** have been used to prepare chiral cyanotitanium reagents which add enantioselectively to aldehydes (*ee* = 68–96%) [206d].

(188) (189) (190)

A limitation of using Cp ligands is related to the fact that alkyl groups are not transferred chemically to aldehydes very well [8], owing to the strongly electron releasing effect of such ligands. In such cases zirconium and hafnium analogs are better suited because they are more Lewis acidic [207]. Titanium reagents having two Cp groups are even less reactive. However, allyl transfer reactions are possible. The first case of such a titanium reagent having two different Cp groups and the stereogenic center at the metal has been described [208]. Since the two Cp groups which were used are too 'similar,' the *ee* values in aldehyde additions turned out to be poor. Nevertheless, reagents with a stereogenic center at titanium (or zirconium) constitute an intriguing goal for the future (see also reagent **182** [205]).

3.4 MICHAEL ADDITIONS

Two outstanding Michael-type reactions are $TiCl_4$-mediated additions of allylsilanes [209] and enolsilanes [210]. Neither of them involve Si–Ti exchange prior to C—C bond formation [105d, 210]. Rather, complexation of the carbonyl group by $TiCl_4$ initiates the reaction.

Conjugate additions of allylsilanes to enones are fairly general, whereas α,β-unsaturated esters react poorly [211]. If the enone is chiral, diastereofacial selectivity is possible, provided one of the π-faces is sterically shielded in the ground state. An example is the reaction of the enone **191** which affords the adduct **192** in 85% yield, diastereoselectivity being >99% [209]. Substrates capable of chelation such as **193**

H
(191)
$\Longrightarrow$ Si(CH$_3$)$_3$ / TiCl$_4$
H
(192)

react via $TiCl_4$ intermediates to afford good yields of adduct **194** (*ds* = 91%) [212]. In an interesting recent development, crotylsilanes were shown to undergo $TiCl_4$-promoted stereoselective conjugate additions to chiral α,β-unsaturated keto esters and to 2-(*S*)-tolylsulfinyl-2-cyclopentenone [213]. Stereoselective $TiCl_4$-induced additions of allyltrimethylsilane to chiral α,β-unsaturated *N*-acyloxazolidinones and *N*-enoylsultams constitute another example of conjugate C—C bond formation [214].

BnO, H_3C (193) → $TiCl_4$ → BnO···$TiCl_4$ H_3C → Si(CH$_3$)$_3$ → BnO, H_3C (194)

Conjugate additions to α,β-unsaturated acid nitriles are also possible [215], as are selective 1,6-additions to dienones [216]. A wide variety of intramolecular allylsilane additions to form polycyclic and/or spiro compounds have been reported [217], as in the cyclization of enone **195** [217c]. Sometimes undesired protodesilylation occurs (proton source?), in which case Lewis acids such as $EtAlCl_2$ (proton sponge!) or fluoride ions are the promoters of choice [217]. Little is known concerning high levels of simple diastereoselectivity in intermolecular conjugate additions of crotylsilanes to achiral enones [209].

(195) Si(CH$_3$)$_3$ → products (H_3C, H_3C) + (H_3C, H_3C)

$TiCl_4$	80	:	20
F -	14	:	86

Related are allenylsilane reactions which involve 1,4-addition, silyl group migration and ring closure to form good yields of cyclopentenes, e.g. **196** [218]. Although a variety of enones react similarly, the method cannot be extended to the unsubstituted parent allenyl reagent [218]. Allenylstannanes substituted in the 3-position provide acetylenic products such as **197** [219].

Enolsilanes derived from ketones or esters undergo conjugate additions to enones and acrylic acid esters, the yields ranging between 55 and 95% [210]. If the enolsilane and/or the carbonyl component are sensitive to $TiCl_4$, mixtures of $TiCl_4$ and $Ti(OiPr)_4$ [which form the milder Lewis acids $Cl_3TiOiPr$ or $Cl_2Ti(OiPr)_2$] should be used [220a]. An interesting development concerns the use of compounds of the type $(RO)_2Ti{=}O$

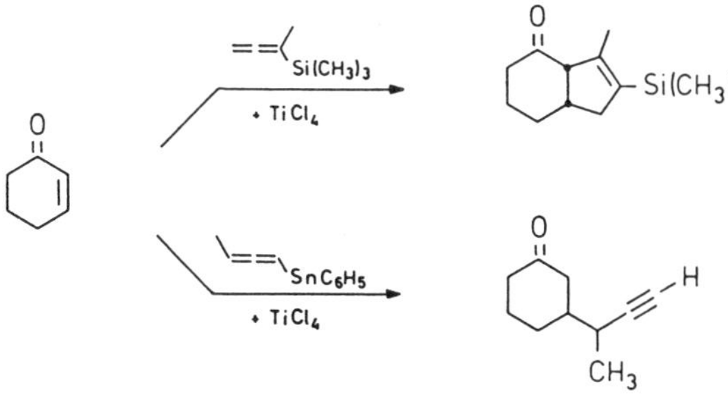

(196)

(197)

as catalysts for the Mukaiyama–Michael addition [220b]. In the $TiCl_4$- and $SnCl_4$-mediated conjugate addition of *O*-silyl ketene ketals, a mechanistic study has shown that electron transfer is involved [221].

Although the problem of simple diastereoselectivity was not addressed in the early work [210], mixtures of diastereomers are probably formed in most of the relevant cases. Later work using similar substrates and $TiCl_4$ or $SnCl_4$ showed that *syn–anti* mixtures are indeed formed, but that in some cases the reaction can be manipulated in a stereoselective sense [222]. Enolsilanes **198** derived from ketones result in moderate to high *anti* selectivity, (cf. **199**), regardless of the geometry of the enolsilane. The (*Z*)-enolsilanes from propiophenone and related aromatic ketones show excellent *anti* selectivity if $SnCl_4$ is used as the Lewis acid ($ds = 90–98\%$) [222]. *Z*- and *E*-configured *O*-silyl ketene ketals react highly *syn* selectively with *tert*-butyl enones to form adducts **200** preferentially ($ds = 96–99\%$), but stereorandomly with most other enones [222].

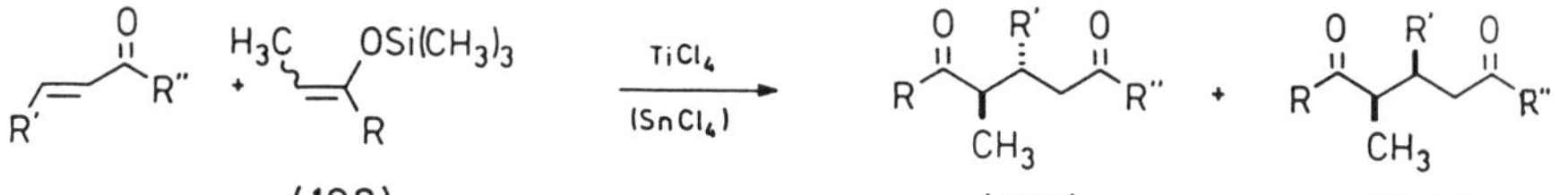

(198) (199) (200)

Chiral enones have been shown to react with remarkable degrees of simple diastereoselectivity *and* diastereofacial selectivity [223]. Finally, chiral *O*-silyl ketene ketals undergo stereoselective $TiCl_4$-mediated conjugate additions ($e = 72–75\%$) [224]. Other metal-mediated Michael additions have been reviewed [225]. Acyclic transition states have been proposed for all of these reactions, but the details of the mechanism remain obscure. Additions to nitroolefins are also mediated by $TiCl_4$, the products being 1,4-diketones following Nef-type work-up [226]:

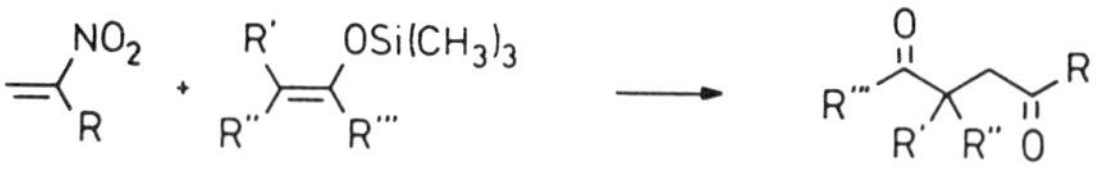

Bona fide organotitanium reagents [8, 15] including triisopropoxytitanium enolates derived from esters [14] generally add to enones in a 1,2-manner (Section 3.2). Exceptions are tetrabenzyltitanium [86] and certain titanium ate complexes of enethiolates (e.g. **201**) [227]. In the case of unsaturated esters (e.g. methyl methacrylate)

the triisopropoxytitanium enolate derived from isobutyric acid ester induces group transfer polymerization involving repetitive Michael additions [125b]. The Evans procedure for Ti-enolate generation (Section 3.2.2) provided reagents capable of undergoing stereoselective Michael additions.

STi(OⁱPr)$_4$Li N(CH$_3$)$_2$ (201) + H$_5$C$_6$–CH=CH–CO–C$_6$H$_5$ → (H$_3$C)$_2$N–C(=S)–…(H$_5$C$_6$)…–CO–C$_6$H$_5$

Allyltitanium reagents add to enones cleanly in a 1,2-manner [44]. However, by increasing the Michael acceptor propensity and by making the 1,2-addition mode less favorable (esters vs ketones), 1,4-additions prevail, as in the reaction of the diester **202** with the crotyltitanium reagent **13** to produce a 90:10 mixture of adducts **203** and **204** [228]. The same stereoselectivity is observed using boron reagents, but crotyltriisopropoxytitanium (**13**) is much easier to prepare.

(202) + Ti(OⁱPr)$_3$ (13) → (203) + (204)

The question of regioselective addition of carbon nucleophiles to 1-acylpyridinium salts has been addressed fairly often. For example, lithium enolates react with the pyridinium salt **205** to produce 1:1 mixtures of 1,2- and 1,4-adducts [229]. Simply treating the enolates with Ti(OiPr)$_4$ results in high degrees of 1,4-preference. For example, reaction of **205** with the titanium ate complex **206** results in a 92:8 mixture of **207** and **208** [229].

(205) COOC$_6$H$_5$ + OTi(OⁱPr)$_4$Li (206) → (207) H$_5$C$_6$OOC + (208) H$_5$C$_6$OOC

Of the large number of other types of Michael acceptors known in the literature, few have been reacted with organotitanium reagents. One of the early examples involves chiral sulfoxides, e.g. **209**, which undergoes chelation controlled Michael additions with methyl- and ethyltitanium reagents [230]. Reductive cleavage of the sulfur moiety results in enantiomerically pure (*ee* > 98%) ketones of the type **210**. Zinc reagents are less selective (*ee* = 42%), but work well for the cyclopentenone analog [230].

Tol (209) → 1. CH$_3$Ti(OⁱPr)$_3$ 2. Al/Hg → H$_3$C (210)

The Schöllkopf bis-lactim methodology, which is successful in the alkylation of the lithium reagents and in aldol-type additions of the titanium analogs [196] (Section 3.3.3), can be used in Michael additions to prochiral nitroolefins [231a, b]. Thus, the titanium reagent **167** adds to nitroolefins such as **211** to produce essentially one of four possible diastereomers, e.g. **212**. The lithium precursor is more reactive and considerably less selective [231a]. The titanium reagent also undergoes highly selective conjugate additions to α,β-unsaturated esters [231c].

H_5C_6–CH=CH–NO_2 (211) + $Ti[N(CH_3)_2]_3$ (167) ⟶ (212)

In summary, Michael additions involving organotitanium reagents have not been studied in such detail as titanium-mediated Grignard-type and aldol additions. More work is necessary in this interesting area.

In a recent study, lithium ketone enolates were treated with $Ti(OiPr)_4$ and the resulting Ti-ate complexes reacted with (*E*)-enones to form Michael adducts having the *anti* configuration [232]. Ti-ester enolates also provide *anti* adducts, in contrast to the Li enolates [232]. Tetrakis(dialkylamino)titanium reagents, $Ti(NR_2)_4$, have been added to α,β-unsaturated ketones and esters in a conjugate manner; the intermediate trisaminotitanium enolates were then reacted with aldehydes [233]. The process amounts to an interesting tandem conjugate addition–aldol reaction. Imines of α-amino esters undergo cycloaddition reactions with acrylates in the presence of triethylamine and $ClTi(OiPr)_3$ or $Cl_2Ti(OiPr)_2$ [234]. Titanium enolates are involved which undergo Michael-type addition/cyclization. Finally, reagents of the type $RTi(OiPr)_3$ and $RTi(OiPr)_4Li$ undergo Cu(I)-catalyzed 1,4-additions [235].

3.5 OLEFIN-FORMING REACTIONS

3.5.1 *WITTIG-TYPE AND KNOEVENAGEL OLEFINATIONS*

The Wittig olefination is one of the most widely used synthetic reactions [236]. However, in the case of enolizable and/or highly functionalized ketones, the yields are often poor. Epimerization at stereogenic centers may also occur owing to the basic nature of the ylides. In the case of methylenation, titanium chemistry provides a general solution to these problems [237].

The addition of $TiCl_4$ (0.7 parts) in dichloromethane to a mixture of CH_2Br_2 (1 part) and zinc dust (3 parts) in THF at room temperature leads within 15 min to a reagent which smoothly olefinates a variety of ketones, e.g. **213** [237a] and **215** [238]. The yield of methylenecycloheptane (**214**) is 83%, compared with the 10% yield obtained by the reaction of the Wittig reagent $Ph_3P{=}CH_2$. The sensitive ketone **215** yields no olefination product using classical methodology, whereas the titanium-based procedure affords 60% of the desired olefin **216**. The use of CH_2I_2 results in a more reactive and in some cases more selective reagent [237b]. Another reagent based on $CH_2I_2/Zn/Ti(OiPr)_4$ reacts aldehyde selectively with ketoaldehydes [109].

(213) $\xrightarrow{CH_2Br_2/Zn/TiCl_4}$ (214)

(215) $\xrightarrow{CH_2Br_2/Zn/TiCl_4}$ (216)

On attempting to apply the titanium-mediated reaction to the synthesis of gibberellins, difficulties were encountered [239]. However, letting the reagent age in THF at 5 °C for 3 days resulted in dramatic improvements [239]. For example, reaction of the non-protected polyfunctional ketone **217** affords a 93% yield of the olefin **218**. In fact, it was claimed that this version is the method of choice in other cases also, including the conversion of (+)-isomenthone (**219**) into (+)-2-methylene-*cis*-*p*-menthane (**220**), which proceeds without any epimerization [240]. Several other impressive applications of the $CH_2X_2/Zn/TiCl_4$ reagents have been described [241].

(217) $\xrightarrow[\text{'aged'}]{CH_2Br_2/Zn/TiCl_4}$ (218)

(219) $\xrightarrow[\text{'aged'}]{CH_2Br_2/Zn/TiCl_4}$ (220)

In an important extension of these reactions, alkylidination of carboxylic acid esters **221** by means of $RCHBr_2/Zn/TiCl_4/TMEDA$ to produce predominantly (*Z*)-alkenyl esters (**222**) has been reported [242]. The method has unprecedented generality and can be extended to the synthesis of *Z*-configured enolsilanes from trimethylsilyl esters [243], enamines from amides and alkenyl sulfides from thioesters [244]. Classical Wittig reactions fail in all of these transformations. A limitation of the titanium-based method is the poor yields in the case of olefination reactions of aldehydes; in these cases low-valent chromium reagents are better suited [245].

(221) RCO_2R' $\xrightarrow[Zn/TiCl_4/TMEDA]{R''CHBr_2}$ (222)

The titanation of Wittig-Horner compounds results in reagents which undergo stereoselective Knoevenagel condensations [182, 183] (not Wittig-type) (Section 3.3.2.2), a process which is general for active methylene compounds [98]:

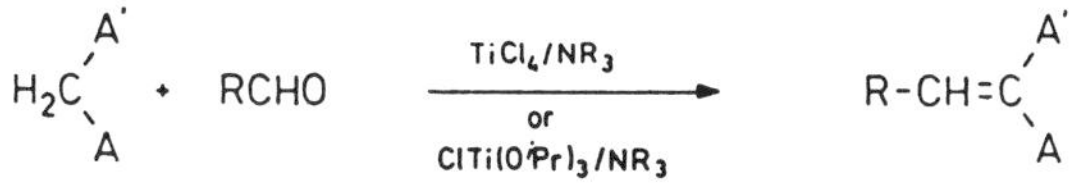

$$H_2C(A)(A') + RCHO \xrightarrow[\text{or } ClTi(O^iPr)_3/NR_3]{TiCl_4/NR_3} R\text{-}CH{=}C(A)(A')$$

A, A' = COR, COOR, CN, $PO(OR)_2$, NO_2

The Tebbe reagent **223** is an isolable and well characterized compound which olefinates carbonyl compounds, including enolizable ketones [246]. The reactive species is actually the carbene complex $Cp_2Ti{=}CH_2$. Thus, titanacyclobutanes can also be used as a source for $Cp_2Ti{=}CH_2$. Its prime virtue is the ability to olefinate esters (e.g. **224** → **225** in 70% yield) [247] and sensitive carbonyl compounds [248]. An alternative (and perhaps cheaper) method for olefination is based on the thermolysis of $Cp_2Ti(CH_3)_2$ at 60–65 °C [249]. Under such conditions ketones and esters are readily olefinated.

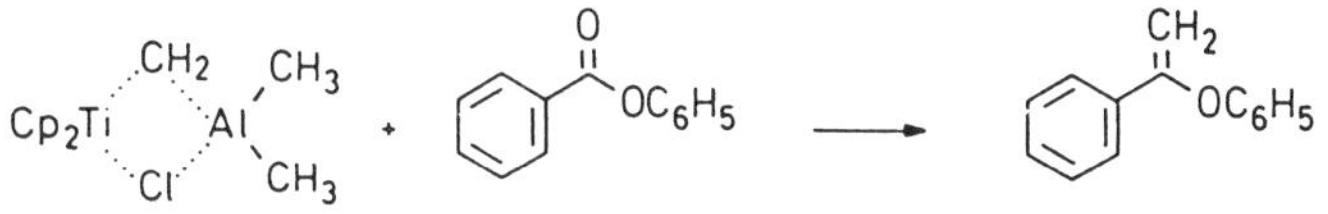

(223) (224) (225)

Related methylenating agents such as **226** [250] and **227** [246, 251] have been described. Reagent **227** is not as Lewis acidic as the Tebbe reagent itself and is thus better suited for the olefination of acid-sensitive ketones and lactones. It remains to be seen which reagents will be used most often. The main advantages of the dibromoalkane/Zn/$TiCl_4$ systems are ready availability, ease of performance and high olefination yields.

Cp_2TiCH_2ZnX (226)

(227)

The chemistry of titanacyclobutanes is fascinating, particularly in their role in olefin metathesis as applied to ring-opening polymerization [5, 252]. They also react with acid chlorides to form titanium enolates [253]. Such reagents as **228** and **229** react with carbonyl compounds to form allenes [254].

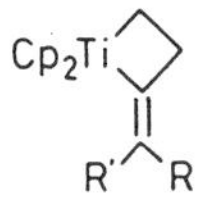

(228) 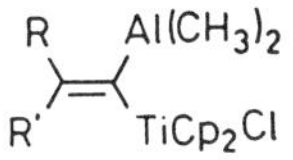 (229)

3.5.2 DEOXYGENATIVE COUPLING OF CARBONYL COMPOUNDS

In 1973–74, three different groups independently discovered that low-valent titanium induces the deoxygenative coupling of ketones and aldehydes to form olefins. The methods differ in the source of low-valent titanium: $TiCl_3$/3THF/Mg [255], $TiCl_3$/$LiAlH_4$ [256] and Zn/$TiCl_4$ [257]. Subsequently, the 'McMurry-reagent' ($TiCl_3$/$LiAlH_4$) was used most often. The substrates can be saturated or unsaturated, and the reaction works in an intermolecular sense to yield acyclic alkenes and intramolecularly on dicarbonyl compounds to yield cycloalkenes. Mechanistically, zerovalent titanium in a pinacol-type coupling followed by reductive elimination has been postulated. However, it has recently been demonstrated that Ti(II) species (specifically polymeric HTiCl) actually induce the McMurry-type couplings [258a]. In the case of titanium on graphite, zero valency may pertain; this reagent undergoes a number of interesting reactions including olefin-forming coupling of ketones and cyclizations with formation of furans and indoles [258b, c]. Several extensive reviews on McMurray-type couplings and other reductive processes have appeared [259].

$$RC(=O)R' \xrightarrow{\text{'Ti'}} \underset{R\ R'\ \ R'\ R}{L_nTiO\quad OTiL_n} \xrightarrow{\text{'Ti'}} RR'C{=}CRR'$$

Unfortunately, in the early version the reproducibility depended on the age, history and source of trichlorotitanium ($TiCl_3$). Although improvements based on a number of other low-valent titanium sources were reported [259], some confusion as to the best procedure persisted. This uncertainty ended with a definitive study of an optimized procedure based on $TiCl_3$/Zn–Cu in dimethoxyethane (DME) [260]. Accordingly, intermolecular couplings and intramolecular processes leading to small-ring cycloalkenes are best performed by rapid addition of the carbonyl component to a reagent prepared by using three equivalents of $TiCl_3(DME)_{1.5}$ per equivalent of carbonyl compound. More difficult intramolecular couplings require four or more equivalents of titanium reagent per carbonyl moiety and also very slow addition to achieve high dilution [260]. The value of this new version is illustrated by the smooth coupling of diisopropyl ketone to form tetraisopropylethylene in 87% yield [260], compared with 12% by the original $TiCl_3$/$LiAlH_4$ method and 37% by the $TiCl_4$/Zn–Cu procedure in the absence of DME.

$$(i\text{-}Pr)_2C{=}O \xrightarrow[\text{Zn/Cu}]{TiCl_3(DME)_{1.5}} (i\text{-}Pr)_2C{=}C(i\text{-}Pr)_2$$

In special cases it may be advantageous to use titanium as a fine dispersion on graphite [258d].

3.5.3 HYDROMETALLATION AND CARBOMETALLATION OF ACETYLENES

Hydrometallation of acetylenes using metals such as boron, aluminum and zirconium are part of standard synthetic organic methodology. In the hydromagnesiation of acetylenes using isobutylmagnesium bromide as the hydride source, Cp_2TiCl_2 has been found to be an efficient catalyst [71, 261]. Disubstituted acetylenes react at room temperature to produce (*E*)-alkenyl Grignard reagents in excellent yields. After hydrolysis, (*Z*)-olefins of high purity are obtained. The reaction occurs with low regioselectivity for unsymmetrical dialkylacetylenes (which is of no consequence in the hydrolysis) [72, 262].

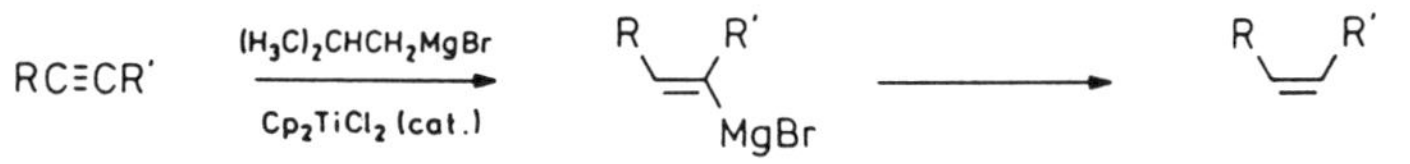

RC≡CR' —$(H_3C)_2CHCH_2MgBr$ / Cp_2TiCl_2 (cat.)→ R, R', MgBr (alkenyl Grignard) ——→ R, R' (olefin)

Regioselectivity is high in the case of alkylarylacetylenes, which means that quenching with electrophiles other than protons becomes meaningful [72, 262]:

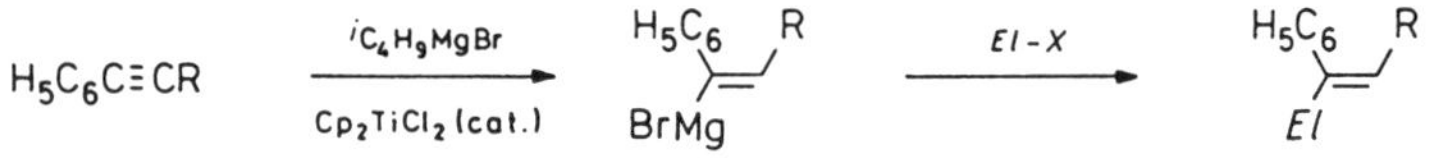

$H_5C_6C≡CR$ —$^iC_4H_9MgBr$ / Cp_2TiCl_2 (cat.)→ H_5C_6, R, BrMg —*El*-*X*→ H_5C_6, R, *El*

The hydromagnesiation of 1-trimethylsilyl-1-alkylenes proceeds with complete regioselectivity if the reaction is carried out at 25 °C for 6 h. The intermediate vinylmagnesium reagents react with a variety of electrophiles, e.g. protons, RI/CuI, I_2, RCHO. This constitutes a powerful method for the synthesis of prochirally pure vinylsilanes [262, 263]:

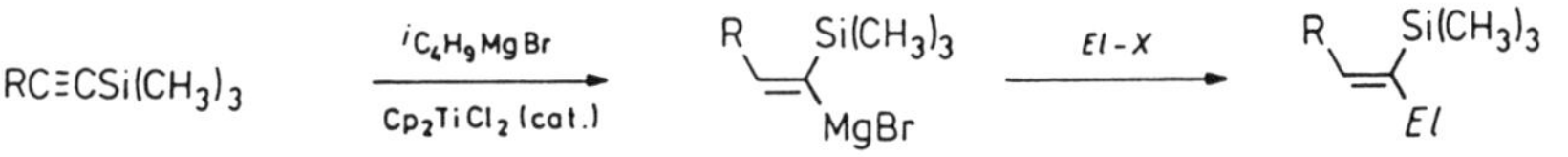

$RC≡CSi(CH_3)_3$ —$^iC_4H_9MgBr$ / Cp_2TiCl_2 (cat.)→ R, $Si(CH_3)_3$, MgBr —*El*-*X*→ R, $Si(CH_3)_3$, *El*

Titanium has also played a role in the carbometallation of acetylenes [264], although zirconium seems to be more versatile, e.g. in Cp_2ZrCl_2-catalyzed additions of R_3Al [29, 265]. However, titanium-based reactions appear to be superior in the carbometallation of homo-propargylic alcohols in which the hydroxy group dictates regioselectivity [266]. For example, the one-pot reaction shown below proceeds fully regioselectively with 70% yield [266].

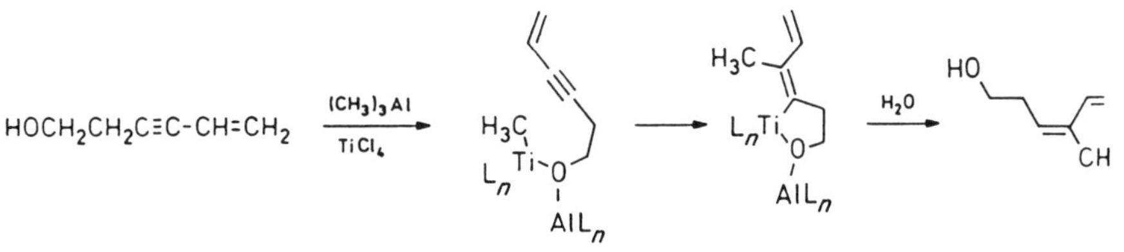

$HOCH_2CH_2C≡C\text{-}CH{=}CH_2$ —$(CH_3)_3Al$ / $TiCl_4$→ H_3C, L_nTi–O–AlL_n ——→ H_3C, L_nTi, O, AlL_n —H_2O→ HO, CH

3.6 SUBSTITUTION REACTIONS

Substitution reactions involving carbon nucleophiles are synthetically important processes. In carbanion chemistry deprotonated carbonyl compounds (and their nitrogen analogs), sulfones, sulfoxides, thioesters, nitriles, etc., undergo smooth reactions with S_N2- active alkyl halides [9]. A serious synthetic gap becomes apparent when attempting to perform these classical reactions with tertiary alkyl halides and other base-sensitive alkylating agents not amenable to S_N2 substitution. Similarly, cuprates R_2CuLi and higher order analogs do not undergo substitution reactions with tertiary alkyl halides. Many of these problems can be solved using titanium reagents of high Lewis acidity which induce S_N1 reactions. Some of these reagents also allow for a combination of two C—C bond-forming reactions in a one-pot sequence, namely addition to carbonyl compounds followed by S_N1-type substitution of the oxygen function. Conversely, titanium reagents, regardless of the type of ligands, are generally not nucleophilic enough to undergo S_N2 reactions with primary alkyl halides. For example, compounds of the type CH_3TiX_3 (X = Cl, OR) do not react with RCH_2X (X = Cl, Br, I), nor do titanium enolates or α-titanated sulfones [8, 14, 44].

The Lewis acidic character of trichloromethyltitanium CH_3TiCl_3 (Section 3.2) is so pronounced that the reagent will ionize S_N1-active alkyl halides to form carbocations such as **231**, which are spontaneously captured by the non-basic carbon nucleophile [17, 18]. Essentially all tertiary **230** and aryl-activated secondary alkyl halides **233** are methylated at -78 to 0 °C to form high yields of products **232** and **234**, respectively.

$$\underset{(230)}{R_3C\text{-}Cl} + H_3CTiCl_3 \longrightarrow \underset{(231)}{\left(R_3\overset{+}{C}\ H_3C\text{-}\overset{-}{Ti}Cl_4\right)} \longrightarrow \underset{(232)}{R_3C\text{-}CH_3}$$

$$\underset{(233)}{H_5C_6CH(Cl)R} \xrightarrow{H_3CTiCl_3} \underset{(234)}{H_5C_6CH(CH_3)R}$$

Ether-free solvents such as CH_2Cl_2 are mandatory, since diethyl ether or THF complexes CH_3TiCl_3, making it incapable of inducing S_N1 ionization. This means that precursors such as $(CH_3)_2Zn$ or $(CH_3)_3Al$ must be used (solutions of the more readily accessible CH_3Li and CH_3MgX contain diethyl ether or THF!). There are several ways to perform these reactions:

1. Solutions of $(CH_3)_2Zn$ in CH_2Cl_2 are treated with two equivalents of $TiCl_4$ to form CH_3TiCl_3, followed by addition of an S_N1-active alkyl halide.
2. Solutions of $(CH_3)_2Zn$ are treated with one equivalent of $TiCl_4$ to form $(CH_3)_2TiCl_2$, followed by addition of an S_N1-active alkyl halide:

$$(CH_3)_2Zn + TiCl_4 \longrightarrow (CH_3)_2TiCl_2$$

3. Only catalytic amounts of $TiCl_4$ are used to produce $(CH_3)_2TiCl_2$, which reacts *in situ* with the S_N1-active substrate.

All of these methods result in excellent yields of alkylated products, the last version being the mildest [18]. The fact that in some cases (e.g. **235**) rearrangements occur is in line with the proposed S_N1 mechanism [18]. Functional groups such as primary and

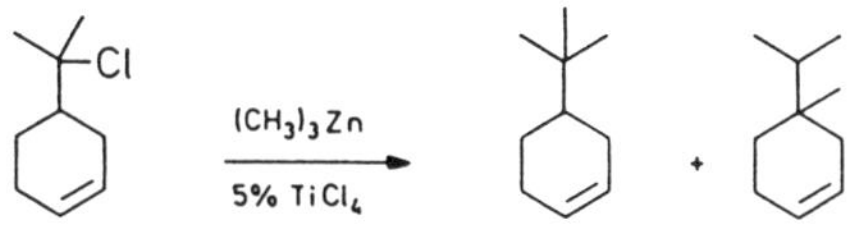

(235)

$(CH_3)_2Zn$ / 5% $TiCl_4$

-30°C	50	50
-78°C	90	10

secondary alkyl halides and esters are tolerated. $(CH_3)_3Al$ in the presence of catalytic amounts of $TiCl_4$ is also effective [20].

Tertiary ethers and even S_N1-active alcohols are methylated by excess $(CH_3)_2TiCl_2$ [267]. The conversion of the tertiary alcohol **236** into the geminal dimethyl compound **237** (74% yield) is a typical example.

(236) 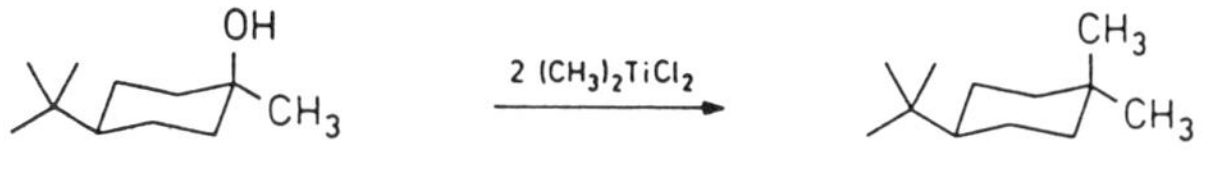 OH, CH_3 — 2 $(CH_3)_2TiCl_2$ → CH_3, CH_3 (237)

Homologs (e.g. n-$BuTiCl_3$) do not induce efficient coupling because reduction of the carbocations via β-hydride abstraction prevails [20]. In this case dialkylzinc reagents in the presence of $ZnCl_2$ is the method of choice [20]. Allylsilanes react smoothly with S_N1-active alkylating agents in the presence of $TiCl_4$ [268]. Chromium-complexed benzylic alcohols undergo stereoselective substitution reactions with $(CH_3)_3Al/TiCl_4$ [269], as do β-chlorosulfides [270]. Both processes involve neighboring group participation, CH_3TiCl_3 or $(CH_3)_2TiCl_2$ being the reactive species. In the case of β-chlorosulfides only catalytic amounts of $TiCl_4$ are needed in a process which involves intermediate episulfonium ions. Substitution (60–75% yields) therefore occurs with formal retention of configuration [270]. Chiral alkenes show pronounced degrees of diastereofacial selectivity. These two-step reactions provide a simple means to carbosulfenylate olefins regio- and stereoselectively [270].

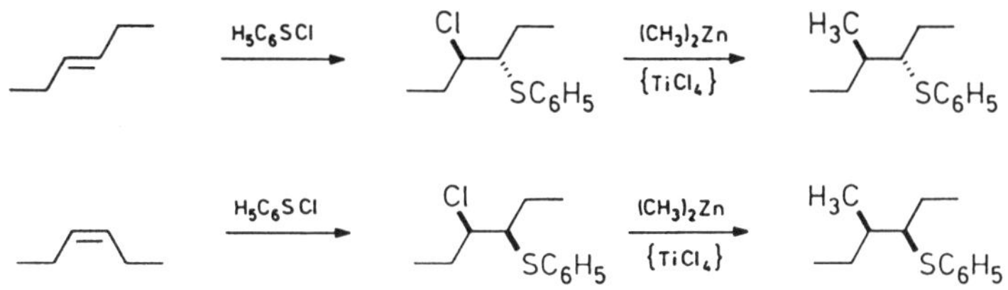

H_5C_6SCl → Cl, SC_6H_5 — $(CH_3)_2Zn$ {$TiCl_4$} → H_3C, SC_6H_5

H_5C_6SCl → Cl, SC_6H_5 — $(CH_3)_2Zn$ {$TiCl_4$} → H_3C, SC_6H_5

Direct geminal dimethylation of ketones using an excess of $(CH_3)_2TiCl_2$ works well for most ketones [39, 267] and aromatic aldehydes [43], a process which involves Grignard-type addition with formation of intermediates **238** followed by S_N1 ionization and further C—C bond formation. The yields generally range between 60 and 90%, but conversion may be lower in the case of extremely sterically shielded ketones.

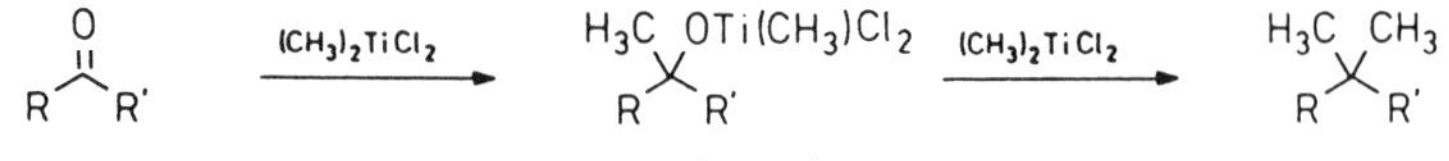

$R(C{=}O)R'$ — $(CH_3)_2TiCl_2$ → H_3C $OTi(CH_3)Cl_2$, R R' (238) — $(CH_3)_2TiCl_2$ → H_3C CH_3, R R'

The limitation of this interesting method has to do with the fact that only a few types of additional functionalities are tolerated and that rearrangement may occur [271]. Ncvcrthclcss, in thc synthcsis of ccrtain molecules with quaternary carbon atoms [272], it is the method of choice [267]. $(CH_3)_2Zn$ may be replaced by $(CH_3)_3Al$ [39].

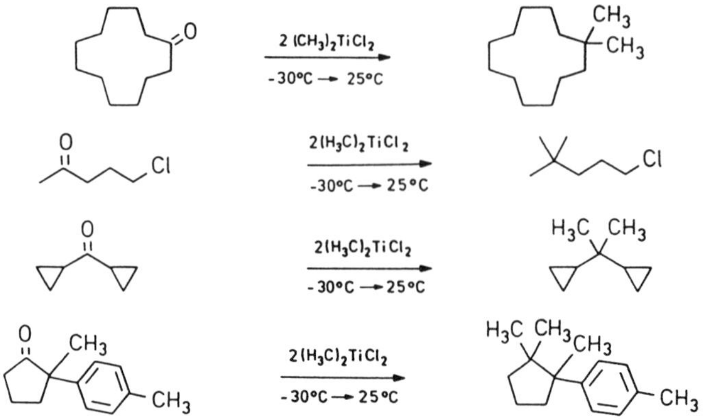

Direct geminal dimethylation of ketones in which one of the newly introduced groups is an alkyl moiety and the other a methyl group is possible in some cases by a one-pot procedure in which an alkyllithium reagent (in ether-free solvent) is added to a ketone and the resulting tertiary alkoxide is treated with an excess of dichlorodimethyltitanium in CH_2Cl_2, as in the preparation of synthetic tetrahydrocannabinoids **239** [273]. The one-pot geminal dialkylation proceeds typically with yields of 70–90%.

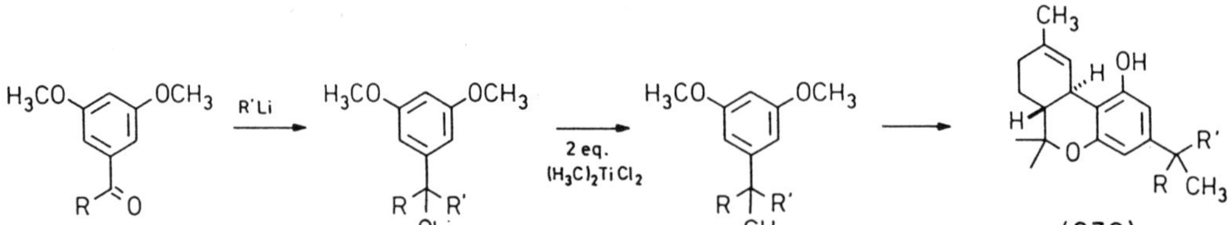

Acetals, ketals and geminal dichlorides also undergo methylation reactions [8, 274]. Certain acetals react with Grignard reagents in the presence of $TiCl_4$ to form the corresponding substitution products, e.g. **240** (82% yield) [275]. An interesting version pertains to chiral acetals of the type **241** [276]. The reaction with $CH_3MgCl/TiCl_4$ affords a 96:4 mixture of substitution products **242** and **243**, respectively.

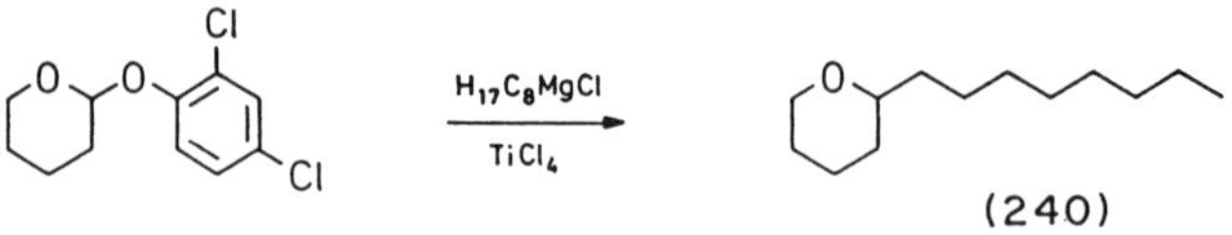

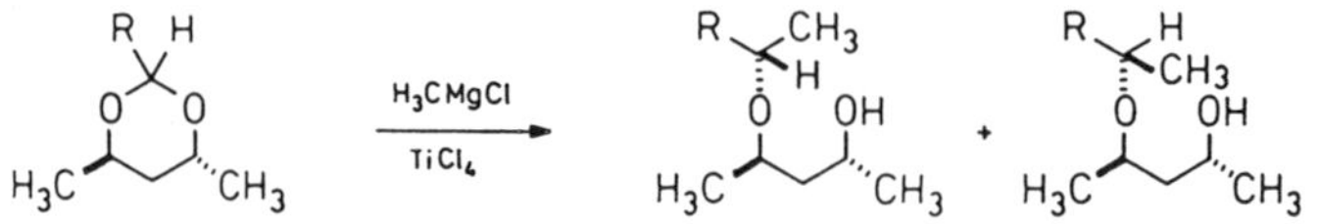

S_N2-like processes have been postulated. Transition state **244** appears to be more favorable than transition state **245** [277]. However, tight ion pairs in which 1,3-alylic strain is operating have not been excluded [278].

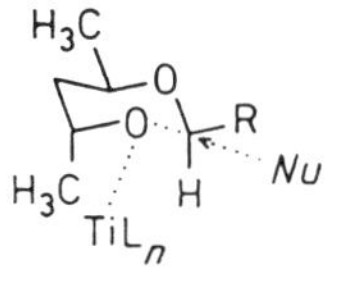

(244) 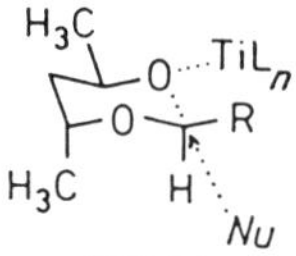 (245)

Not only Grignard and alkyllithium reagents, but also allylsilanes [279], cyano-trimethylsilane [277] and silylated acetylenes [280] function as efficient *C*-nucleophiles. Sometimes mixtures of $TiCl_4$ and $Ti(OiPr)_4$ are better suited. Many of these reactions may not involve the titanium intermediates $RTiCl_3$. However, CH_3TiCl_3, generated from $(CH_3)_3Zn/TiCl_4$, is in fact capable of undergoing such stereoselective substitution reactions [281]. The whole area of stereoselective substitution reactions of chiral non-racemic acetals using titanium reagents and other organometallics has been reviewed [282].

The $TiCl_4$-mediated allylation of the chiral substrate **246** affords a single diastereomer **248** in 68% yield [283a]. This interesting transformation proceeds via the acyliminium ion **247** with complete retention of configuration. Intramolecular allylsilane additions to *N*-acyliminium ions are very useful in heterocyclic chemistry [283b].

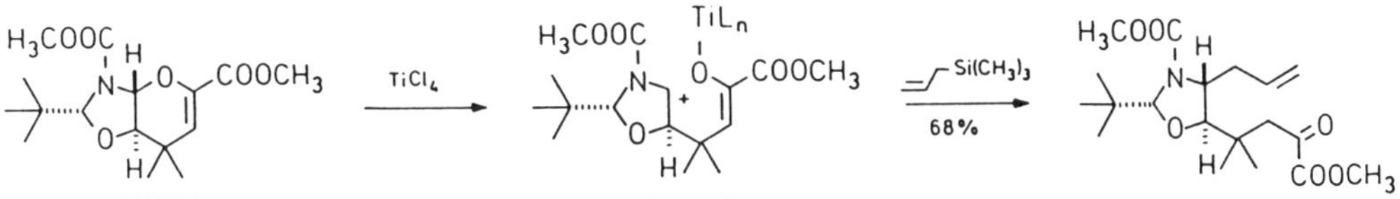

(246) (247) (248)

The reactions of metallated *N*,*O*-hemiacetals bear some resemblance to the above processes [284].

$$RCHO + R'_2NTiCl_3 \longrightarrow RCH(OTiCl_3)(NR'_2) \xrightarrow{R''Li} RCH(R'')(NR'_2)$$

The classical alkylation of lithium enolates using S_N2-active alkyl halides is one of the important C—C bond-forming reactions [9]. However, tertiary and many base-labile secondary alkyl halides fail to react. This long-standing problem can be solved by reacting the enolsilanes derived from carbonyl compounds with tertiary and other S_N1-active alkyl halides in the presence of Lewis acids [285]. In the case of ketones, $TiCl_4$ is the Lewis acid of choice. The mechanism involves S_N1 ionization followed by addition of the carbocation to the enolsilane (Scheme 3.6). The alternative mechanism,

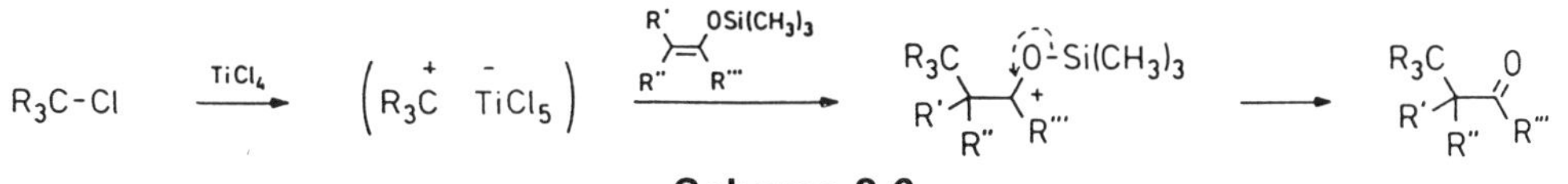

$$R_3C\text{-}Cl \xrightarrow{TiCl_4} \left(R_3\overset{+}{C}\ \ \overset{-}{TiCl_5}\right) \longrightarrow \ldots$$

Scheme 3.6.

i.e. Si–Ti exchange followed by ionization of R_3CCl and alkylation of the tetrachlorotitanium enolate, may occur in rare systems in which alkylation of the enolsilane is slow relative to Si–Ti exchange [88b, 286].

Typical examples illustrating the scope of the reaction are shown below [287–289]. The yields of isolated products range between 60% and 95%. Regioselective *tert*-alkylation [288] is possible since the corresponding enolsilanes [290] are readily accessible. Undesired polyalkylation has never been observed. The reaction is best performed by treating the mixture of an enolsilane and a tertiary alkyl halide in CH_2Cl_2 with the equivalent amount of $TiCl_4$ at -40 to -50 °C. Alternative reaction modes, e.g. adding the enolsilane to a mixture of $TiCl_4$ and tertiary alkyl halide, result in poor yields (30–54%) [291]. In the case of enolsilanes which are sensitive to $TiCl_4$ (e.g. those derived from aldehydes or esters), milder Lewis acids such as $ZnCl_2$ or $BiCl_3$ need to be employed [288].

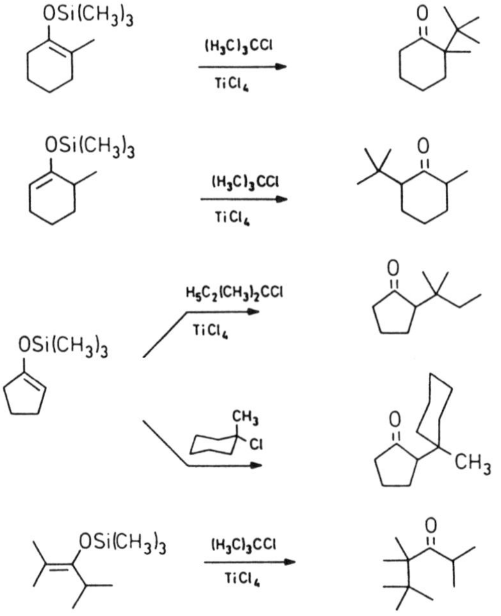

These and other Lewis acid-induced S_N1 alkylations are synthetically significant because they are complementary to the classical alkylations of lithium enolates using S_N2-active alkyl halides [285]. Any alkylating agent which has a higher S_N1 activity than isopropyl halides is likely to be suitable. Neighboring group participation is an important factor in relevant cases [292].

Iodo-carbocyclization of malonates containing olefin functions occurs in the presence of $Ti(OtBu)_4/I_2$ [293]. This interesting synthetic development probably involves titanium enolates which attack intermediate iodonium ions intramolecularly.

Acetals [294], ketals [294] and chloro ethers [295] can also be used in $TiCl_4$-mediated alkylations of enolsilanes. Intramolecular versions result in carbocycles [296]. Again, chlorotitanium enolates are probably not involved. However, the previously mentioned enolization method based on $TiCl_4/NR_3$ (Sections 3.2.2, 3.3.2.2 and 3.3.3) can be used to induce α-alkoxyalkylation of carbonyl compounds [99]. In the case of chiral chlorotitanium enolates, stereoselectivity often approaches 100%, as in the conversion of the amide **249** into the α-alkylated product **250** [99].

(249) Bn = Benzyl; 1. $TiCl_4$ 2. $EtN(^iPr)_2$ 3. $BnOCH_2Cl$ → (250)

Chirality can also be incorporated in the electrophile, as in the case of the chiral formyl cation equivalent derived from the ephedrine derivative **251**, which reacts stereoselectively with enolsilanes. Diastereoselectivity in the case of the substitution product **252** is 88% [297]. The reaction can also be induced by other Lewis acids such as $SnCl_4$, BF_3/OEt_2 and $(CH_3)_3SiOTf$ [297]. Related approaches have been used in other enol- and allylsilane reactions [298].

(251) + H_3C $OSi(CH_3)_2{}^tBu$ —$TiCl_4$→ (252)

Reports of epoxides as alkylating agents are rare. Undesired ring opening followed by a hydride shift to form aldehydes occurs if the medium is too Lewis acidic ($TiCl_4$). For example, treatment of a mixture of the enolsilane **253** and the epoxide **254** does not afford the alkylation product **255**, but rather the aldol adduct **256** [44]. Perhaps trichlorotitanium enolates are better suited. However, such reactions have not been reported to date.

(253) + (254) —$TiCl_4$→ (255) / (256)

Less Lewis acidic reagents such as $CH_3Ti(OiPr)_3$ are not reactive enough to undergo smooth C—C bond formation with most epoxides [8, 44]. However, more reactive species such as allyltitanium reagents react with styrene oxide **257** [8, 44].

Regioselectivity in favor of the ring-opened product **258** (rather than **259**, see overleaf) contrasts well with the non-selective behavior of the Grignard reagent. Conversion amounts to >80% in all cases. Other expoxides react less selectively. However, on going from isopropoxy to phenoxy ligands, a dramatic improvement results, ring opening of epoxides **260** (R = alkyl) occurring consistently at the more substituted carbon atom (80–85% yields of alcohols **261**) [299]. This is of practical importance because cuprates generally react at the less substituted position.

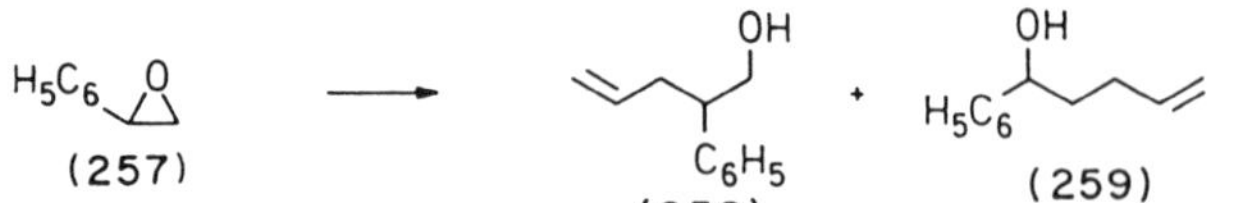

(257) (258) (259)

Reagent	(258)	(259)
$CH_2{=}CHCH_2MgCl$	70	30
$CH_2{=}CHCH_2Ti(OiPr)_3$	>99	<1
$CH_2{=}CHCH_2Ti(OiPr)_4MgCl$	>99	<1

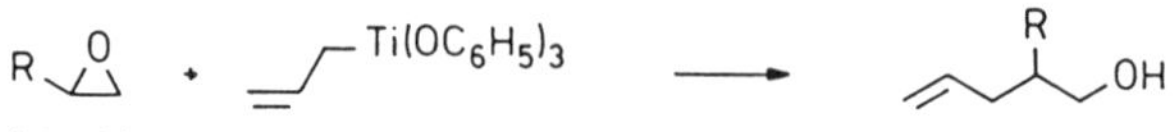

(260) (261)

Epoxides have also been reacted with titanium reagents in completely different ways. One of them is based on the Cp_2TiCl-induced cyclization of epoxyolefins, a process which involves radical intermediates, e.g., **262**/**263** [300a]. Radicals **263** are reduced to the organotitanium compound **264**, which can be intercepted with iodine to form the final product **265** (63% overall yield).

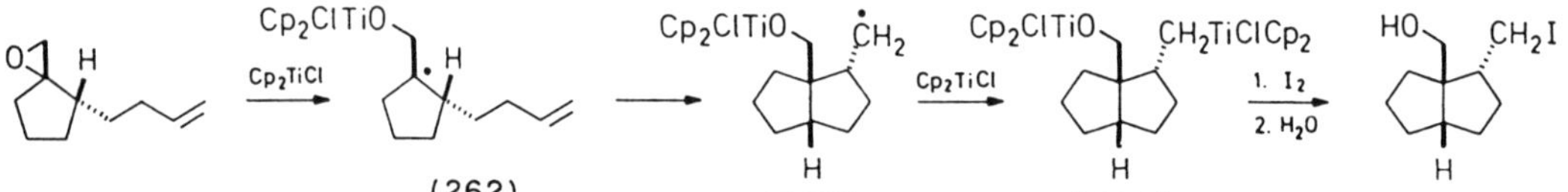

(262) (263) (264) (265)

A synthetically attractive access to bicyclic cyclopentenones employs catalytic amounts of titanocene equivalents ('Cp_2Ti') [300b].

Organotitanium reagents have also been used in Pd- or Ni-catalyzed C—C bond formation [301]. An interesting example is the phenylation of the acetates **266** using triisopropoxyphenyltitanium, $PhTi(OiPr)_3$, which occurs regioselectively to form the products **267** and **268** in ratios of greater than 91:9 [302]. In a highly interesting development, titanium reagents $RTi(OiPr)_3$ and $RTi(OiPr)_4Li$ were shown to undergo selective Cu(I)-catalyzed S_N2' allylation with allylic chlorides and phosphates [235].

A few reports of Heck-type substitution reactions of vinyl and aryl hydrogen atoms are known, e.g. **269** → **270** (65%) [303] and **271** → **272** (32%) [304]. The mechanism of these unusual reactions appears to be related to the Ziegler–Natta polymerization and/or to Cp_2TiCl_2- or $TiCl_4$-mediated carbo- and hydrometallation of olefins and acetylenes (Section 3.5.3).

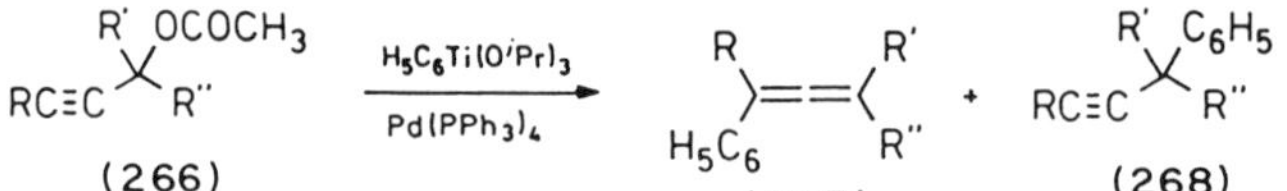

(266) (267) (268)

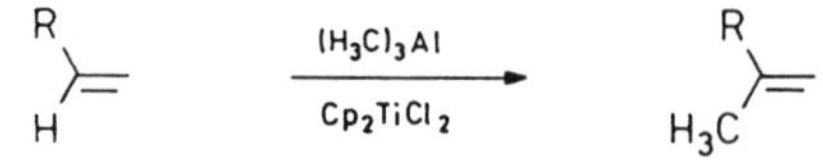

(269) (270)

(271) $\xrightarrow[TiCl_4]{(H_3C)_3Al}$ (272)

The latter substitution reaction has been optimized and is general for homoallylic alcohols. In some cases trimethylaluminum, $(CH_3)_3Al$, can be replaced by CH_3Li. Homoallylic alcohols having a terminal olefinic double bond are converted stereoselectively into (*E*)-3-alken-1-ols, whereas 3-alken-1-ols such as **273** and **275** with internal double bonds afford branched 4-methyl products **274** and **276**, respectively (55–60% yields). In both cases an intriguing 'configurational inversion' of the double bond occurs [305].

(273) $\xrightarrow[TiCl_4]{(H_3C)_3Al}$ (274)

(275) $\xrightarrow[TiCl_4]{(H_3C)_3Al}$ (276)

Acylation reactions of organotitanium reagents have not been studied extensively. $CH_3Ti(OiPr)_3$ reacts with acid chlorides to form the isopropyl esters, not the ketones [8]. In contrast, imidazolides **277** react with allylic titanium reagents to form the *β,γ*-unsaturated ketones **278** in 46–89% yield [306], a process that fails using allylmagnesium chloride [306, 207]. A few cases of Ti-promoted Claisen condensations have been reported [101] (Section 3.2.2).

(277) + R'–CH=CH–CH$_2$–$Ti(NEt_2)_3$ ⟶ (278)

3.7 CATALYTIC ASYMMETRIC REACTIONS

3.7.1 CARBON–CARBON BOND-FORMING REACTIONS

The development of chiral catalysts for asymmetric C—C bond formation is an active area of current research [73, 308]. Titanium has had some part in this exciting race. The requirements for such a process to be synthetically useful are stringent, because the products are enantiomers, not separable diastereomers. Since purification of enantiomeric products in most cases does not result in enantiomeric enrichment, *ee* values of >95% in product formation are really required.

Whereas a variety of titanium-mediated Diels–Alder reactions employing stoichiometric amounts of chiral information have been highly successful [3], only a few efficient titanium-promoted catalytic versions are known. The currently most successful one employs the Narasaka catalyst **280**, which is based on the diol **279**. The latter interacts with dichlorodiisopropoxytitanium, $Cl_2Ti(OiPr)_2$, in an equilibrium reaction in favor of the dichloride **280** [309]. In the presence of molecular sieves (MS 4 Å), the equilibrium is shifted from 87:13 to 94:6. Although $Cl_2Ti(OiPr)_2$ should be more reactive than the bulky analog **280**, the isopropanol which is liberated is believed to cause deactivation of $Cl_2Ti(OiPr)_2$ by inducing aggregation [309]. Compound **280** catalyzes the Diels–Alder reaction of acycloxazolidinone derivatives of α,β-unsaturated carboxylic acids with various dienes [310]. Maximum enantioselectivity is achieved by using mesitylene, $CFCl_3$ or a mixture of toluene and light petroleum as solvents in the presence of molecular sieves.

(279) + $Cl_2Ti(O^iPr)_2$ ⇌ (280) + iPrOH (41)

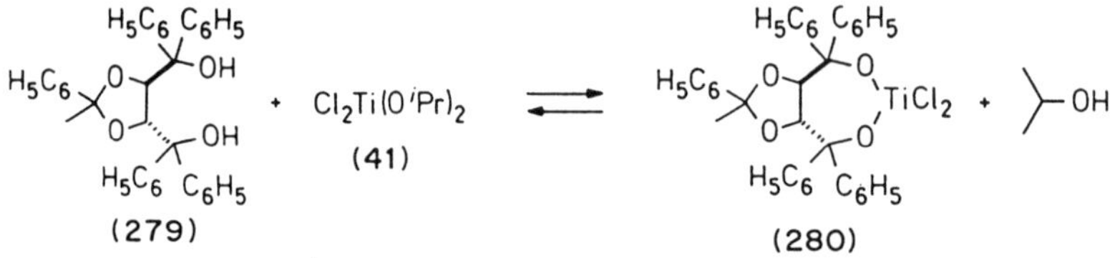

For example, using 10 mol% of **280**, *ee* values of >90% and *endo* selectivities of 90% can be achieved in reactions of dienophiles **281** and **282** [310a]. In other cases enantioselectivity is lower (64–90%), which means that this catalytic system is less general than the boron-based CAB catalyst or other catalysts [311]. However, the catalyst **280** is highly efficient in Diels–Alder reactions of 1,3-oxazolidin-2-one derivatives of 3-borylpropenoic acids [310b]. This development is synthetically significant because the dienophiles are β-hydroxyacrylic acid equivalents. The titanium catalyst is believed to form a rigid chelate between the two amide-carbonyl oxygen atoms.

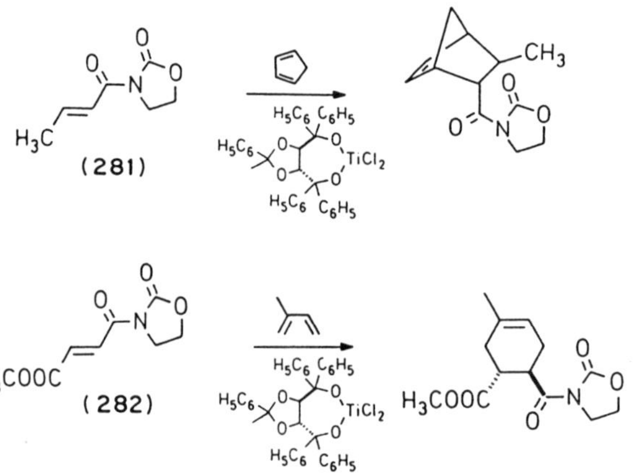

(281) (282)

In a series of intriguing [2 + 2] cycloaddition reactions between dienophiles of the type **281** and **282** and alkenyl or alkynyl sulfides catalyzed by the titanium species **280**

(10 mol%), high levels of enantioselectivity were observed (*ee* = 98%) [312]. Here again molecular sieves and non-donor solvents (toluene/light petroleum) are essential. Related reactions of substituted 1,4-benzoquinones with substituted styrenes using the diol **279** under slightly different conditions provide cyclobutane derivatives having *ee* values of 86–92% [313]. The importance of such reactions is related to the fact that four contiguous asymmetric centers are formed in a single reaction.

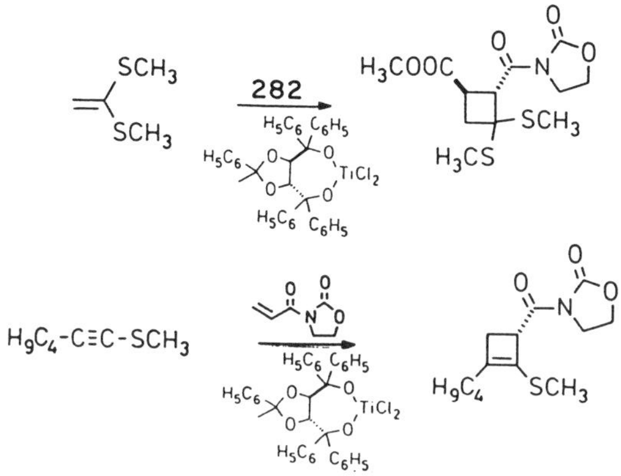

Enantioselective hetero-Diels–Alder reactions of the glyoxylate **283** are catalyzed efficiently by 10 mol% of a different chiral titanium compound, dichloro-1,1′-binaphthalene-2,2′-dioxytitanium, prepared from dinaphthol [314a]. Depending on the reaction conditions, the *cis/trans* ratio **284**/**285** varies between 3:1 and 4:1, the *ee* values of the major product **284** being 92–96%. Substituted derivatives of the diene react similarly.

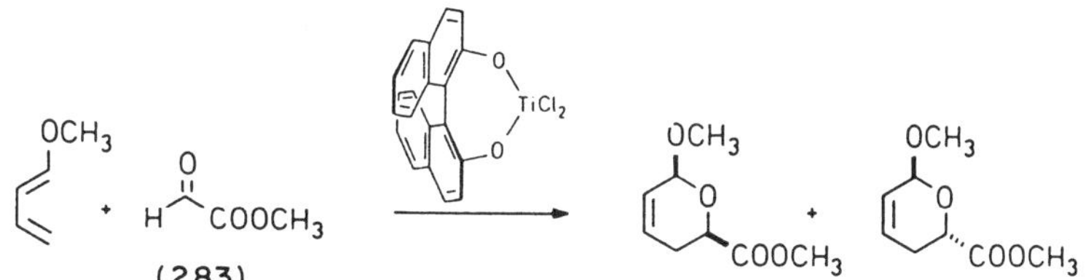

It is believed that of the two transition states leading to the *cis*-product **284**, the *syn-endo* transition state **287** should be less favored owing to steric repulsion in the bulky titanium complex, and that the catalyst should complex in an *anti* fashion, the Diels–Alder reaction then proceeding through the *anti-endo* orientation **286** [314a].

Dichloro-1,1′-binaphthalene- 2,2′-dioxytitanium catalyzes the Diels–Alder reaction of cyclopentadiene and α-methylacryloin, but the *ee* value is poor (16%) [315]. However, *ee* values of up to 86% are obtained in the reaction of α-methylacryloin with electron-rich butadienes in the presence of molecular sieves [314b]. The same catalyst is highly effective in the glyoxylate–ene reaction, e.g. **288** → **289** [316], provided that molecular sieves are used. Accordingly, *ee* values of 95–97% for the adducts **289** (R = CH_3, Ph) are observed.

H_5C_6 (288) + $HC(O)COOCH_3$ (283) → [binaphthol-$TiCl_2$] → $H_2C{=}C(R)CH_2CH(OH)COOCH_3$ (289)

The role of the molecular sieves is not to trap any moisture (as has been claimed in the Sharpless epoxidation [317]), but to accelerate ligand exchange in the formation of the catalyst [316]. Indeed, if an alternative preparation of the catalyst is used based on the reaction of the dilithium salt of dinaphthol with $TiCl_4$ [315], molecular sieves are not necessary. The glyoxylate–ene reaction is fairly general for a variety of 1,1-disubstituted olefins, including methylenecyclohexane and methylenecyclopentane [316]. Sometimes the dibromo analog is more efficient, and as little as 1 mol% of catalyst can be used. In a recent mechanistic study, remarkable positive non-linear effects were discovered which are of great synthetic value [316b]. For example, using 1 mol% of a catalyst derived from dinaphthol having an *ee* value of only 33%, an ene reaction product having an *ee* value of 91% can be obtained!

Other chiral titanium dichlorides have been used in Diels–Alder reactions, but their scope and efficiency remain to be established [318].

The unusual chiral Lewis acid **291** catalyzes the Mukaiyama aldol addition of enolsilane **290**, the *ee* values ranging between 36% and 85%, depending on the nature of the substituents in the enolsilane [319]. Although an optimization would be desirable, the results are already significant since most other efficient catalytic processes involve substituted prochiral enolates [320].

H_5C_6CHO + $H_2C{=}C(OSiR_3)SR'$ (290) → [binaphthol-Ti=O (291)] → $H_5C_6CH(OSiR_3)CH_2C(O)SR'$

Dialkylzinc reagents add highly enantioselectively to aldehydes in the presence of catalytic amounts of chiral β-amino alcohols [73]. A highly interesting alternative has been devised based on catalytic titanium chemistry [38b, 321]. Accordingly, the combination of the bis-sulfonamide **292** and $Ti(OiPr)_4$ catalyzes the enantioselective addition of Et_2Zn and Bu_2Zn to benzaldehyde [38b, 321]. With as little as 0.0005 equiv. of chiral ligand, *ee* values of 98% and nearly quantitative conversions are obtained. The method tolerates the presence of functional groups in the organozinc reagents [321b].

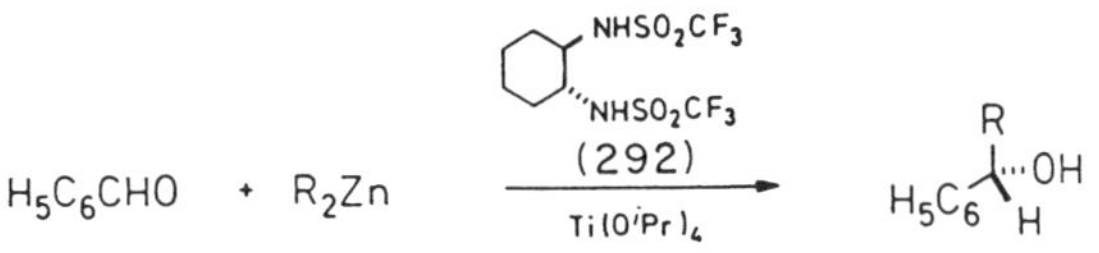

$H_5C_6CHO + R_2Zn \xrightarrow[Ti(O^iPr)_4]{(292)} H_5C_6CH(R)OH$

Although concise mechanistic studies still need to be carried out, alkyltitanium intermediates formed by ligand exchange are believed to be intermediates, Scheme 3.7 [386, 321].

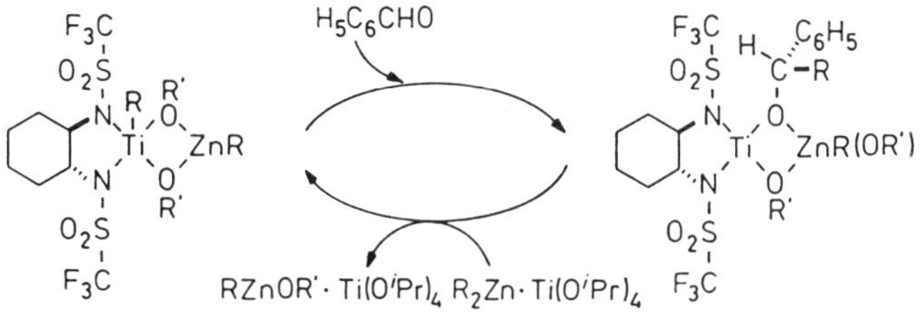

Scheme 3.7.

The idea of using chiral titanium compounds to catalyze dialkylzinc additions to aldehydes has been extended to alkoxytitanium catalysts based on the Seebach diol **188** [322a]. These reactions are highly efficient (*ee* > 90%) and have been studied in great detail with respect to scope and limitation [322a–c]. A third version involving chiral *N*-sufonylamino alcohols in combination with $Ti(OiPr)_4$ also shows great promise (*ee* = 90–97%) [323].

Tartrate-modified titanium compounds catalyze the addition of $(CH_3)_3SiCN$ to aldehydes, the enantioselectivity being mediocre [322d]. Chirally modified $Cp_2Ti(IV)$ complexes have been found to be very promising as catalysts for the enantioselective hydrogenation of non-functionalized olefins [322f] or imines [322g] and for the epoxidation of non-functionalized olefins [322h9].

In summary, several notable titanium-based chiral catalysts have been developed for enantioselective C—C bond formation. Most of the reactions require judicious choices of ligands, solvent, temperature and/or additives.

3.7.2 SHARPLESS EPOXIDATION

A fundamentally important development in organic synthesis is the Sharpless enantioselective epoxidation of allylic alcohols using *tert*-butyl hydroperoxide, tetraisopropoxytitanium and tartaric acid esters [324] (Scheme 3.8). In the original version [324], stoichiometric amounts of tartaric acid esters had to be used. However, by

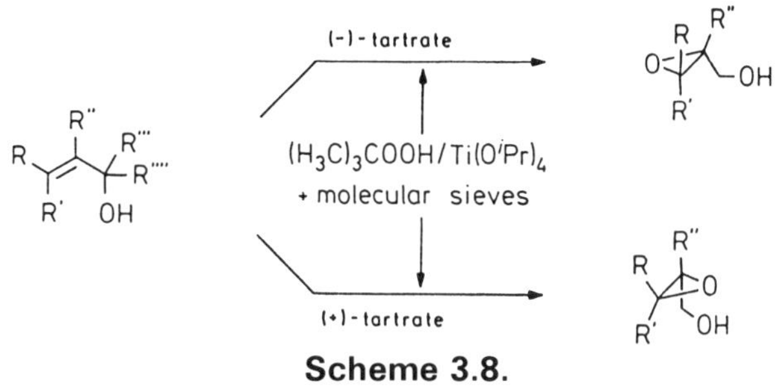

Scheme 3.8.

working in the presence of molecular sieves which bind water, only catalytic amounts are necessary [317].

Dozens of successful examples have been reported by Sharpless and by other groups. Since the epoxides can be used for further ring-opening reactions using cuprates, amines, etc., they constitute useful chiral non-racemic synthetic vehicles. Several extensive reviews have appeared [325]. Table 3.1 contains a number of typical examples which illustrate substrate generality. The Sharpless group has recently published a definitive mechanistic study [326].

Table 3.1. Typical allylic alcohols successfully epoxidated by the Sharpless procedure (Scheme 3.8)

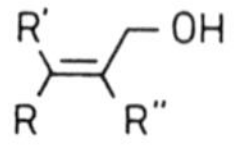

R′, R, R″, OH

Substrate	*ee* (%)	Yield (%)
Unsubstituted:		
R = R′ = R″ = H	95	15
trans-Substituted (R′ = R″ = H):		
R = CH_3	>95	45
R = $C_{10}H_{21}$	>95	79
R = $(CH_2)_3CH{=}CH_2)$	>95	80
R = Me_3Si	>95	60
R = *tert*-Bu	>95	65
R = Ar	95	0–90
R = CH_2OCH_2Ph	98	85
R 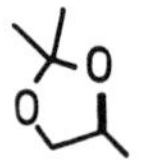 = (O, O)	>95	78–95
R 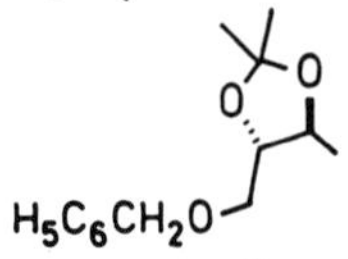 = (O, O) $H_5C_6CH_2O$	>95	70
R 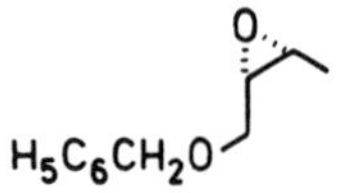 = (O) $H_5C_6CH_2O$	>99	76
R 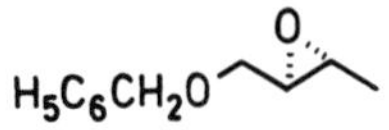 = $H_5C_6CH_2O$ (O)	>99	70
cis-Substituted (R = R″ = H):		
R′ = $C_{10}H_{21}$	90	82
R′ = CH_2Ph	91	83
R′ = CH_2OCH_2Ph	92	84
1,1-Substituted (R = R′ = H):		
R″ = cyclohexyl	>95	81
R″ = $C_{14}H_{29}$	>95	51
R″ = *tert*-Bu	85	60

Table 3.1. (*continued*)

Substrate	*ee* (%)	Yield (%)
trans-1,1,2-Substituted (R′ = H):		
R = R″ = C_6H_5	>95	87
R = C_2H_5, R″ = CH_3	>95	79
R = H_3COO [structure], R″ = CH_3	>95	70
R = [structure], R″ = CH_3	>95	92
cis = 1,1,2-Substituted (R = H):		
R′ = $H_5C_6CH_2$, R″ = CH_3	91	90
1,2,2-Substituted (R″ = H):		
R = $(H_3C)_2C{=}CHCH_2CH_2$, R′ = H_3C	>95	77
R = H_3C, R′ = $(H_3C)_2C{=}CHCH_2CH_2$	94	79
1,1,2,2-Substituted:		
R = C_6H_5, R′ = $C_6H_5CH_2$, R″ = CH_3	94	90

There are some limitations to the method, i.e. certain types of allylic alcohols cannot be successfully epoxidated, as shown in Table 3.2 [325]. First, some substrates react very slowly with poor to moderate enantioselectivity, such as certain *Z*-configured allylic alcohols and some sterically hindered compounds of other substitution patterns. Second, some substrates lead to epoxy alcohols stereoselectively, but the products are unstable to the reaction conditions. Such problems can often be solved by *in situ* derivatization prior to final work-up [317]. Third, stereoselectivity is sensitive to stereogenic centers already present in the substrate. In fact, kinetic resolution is possible in many instances [325, 327]. In terms of reagent control, this means that the mismatched cases may pose problems.

In summary the titanium–tartrate asymmetric epoxidation constitutes a significant synthetic procedure. In spite of several limitations, it is surprisingly versatile and has been applied in a multitude of natural product syntheses and other preparations.

3.8 EXPERIMENTAL PROCEDURES

All manipulations should be carried out in dry flasks under an atmosphere of an inert gas (nitrogen or argon). Generally, $TiCl_4$ taken from fresh bottles can be used without prior purification. However, if moisture has entered the bottle, the $TiCl_4$ should be distilled (b.p. 135–136 °C) and kept under nitrogen or argon. For manipulation, syringe techniques are best employed. The preceding sections contain information concerning the stability of organotitanium reagents and hints on what to avoid. Concerning work-up, simple quenching with water usually poses no problems. If problems with TiO_2-containing emulsions arise, saturated solutions of NH_4F or KF should be used, extraction times being short [8].

Table 3.2. Poor substrates for the Sharpless epoxidation

Substrate	Result
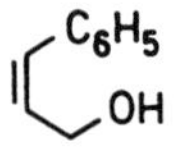 C_6H_5, OH	Slow epoxidation, 65% *ee*
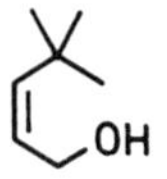 OH	Slow epoxidation, 25% *ee*
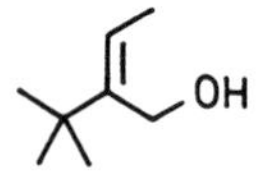 OH	Slow epoxidation, 60% *ee*
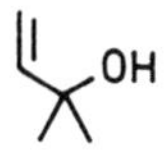 OH	Slow epoxidation, no epoxy alcohol isolated
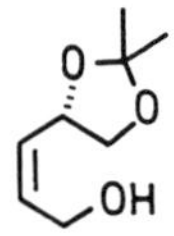 O, O, OH	Slow epoxidation using (−)-tartrate, 23% *ee* (2*R*-enantiomer)
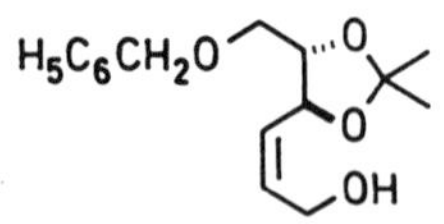 $H_5C_6CH_2O$, O, O, OH	No reaction using (−)-tartrate
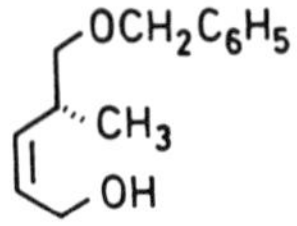 $OCH_2C_6H_5$, CH_3, OH	Slow epoxidation, 67% *ee* using (+)-tartrate; 0% *ee* using (−)-tartrate
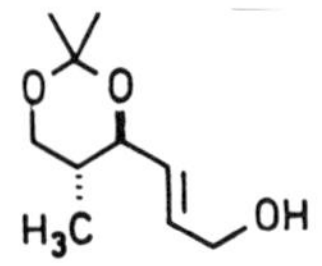 O, O, H_3C, OH	No reaction using (+)-DET
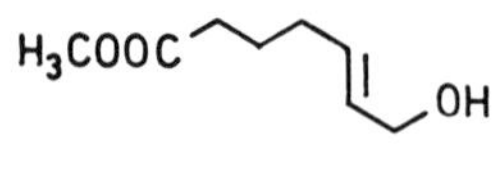 H_3COOC, OH	95% *ee*, 58% yield. Difficult to reproduce owing to lactone diol formation
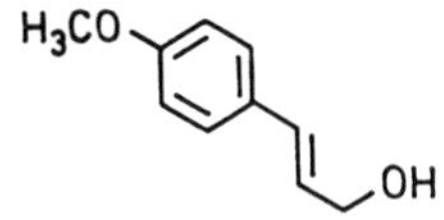 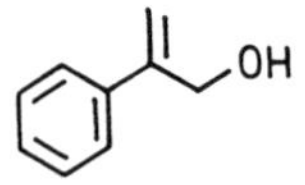 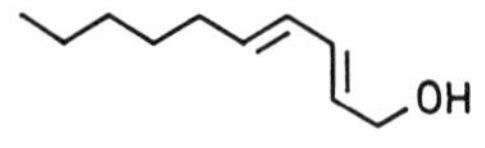 H_3CO, OH	Product epoxy alcohol unstable to reaction conditions; either no product or only very low yields obtained under standard conditions using stoichiometric Ti tartrate
OH	
OH	

3.8.1 SYNTHESIS OF CHLOROTRIISOPROPOXYTITANIUM, ClTi(OiPr)$_3$ [11, 12]

A 1 l three-necked flask equipped with a dropping funnel, magnetic stirrer and nitrogen inlet is charged with 213 g (0.75 mol) of tetraisopropoxytitanium. $TiCl_4$ (47.5 g, 0.25 mol) is then slowly added at about 0 °C. After reaching room temperature, the mixture is distilled (61–65 °C/0.1 Torr) to afford 247 g (95%) of product. The syrupy liquid is >98% pure and slowly solidifies at room temperature. Gentle warming with a heat gun results in liquid formation. This may be necessary during distillation to prevent clogging. Manipulation with a syringe and serum cap presents no problems. Alternatively, the product can be mixed with the appropriate solvent (e.g. pentane, toluene, diethyl ether, THF, CH_2Cl_2) to provide stock solutions. The reagent is hygroscopic, but can be stored in pure form or in solution under nitrogen for months. The actual synthesis can also be performed in solvents. The reagent is commercially available.

3.8.2 SYNTHESIS OF CHLOROTRIPHENOXYTITANIUM, ClTi(OC$_6$H$_5$)$_3$ [12]

$$\mathrm{ClTi(OiPr)_3 + 3C_6H_5OH \longrightarrow ClTi(OC_6H_5)_3 + 3HOiPr}$$

To a solution of chlorotriisopropoxytitanium (71.9 g, 276 mmol) in 500 ml of toluene are added 77.9 g (828 mmol) of phenol. The deep red solution is concentrated by distillation through a 10 cm Vigreux column, which removes most of the isopropanol azeotropically. The residue is distilled in a Kugelrohr at 250 °C/0.001 Torr to give 95.8 g (96%) of product, which crystallizes on cooling. A 0.26 M stock solution in THF can be kept under exclusion of air.

This method of exchanging OR groups is fairly general, including the *in situ* preparation of derivatives with chiral OR groups.

3.8.3 SYNTHESIS OF TETRAKIS(DIETHYLAMINO)TITANIUM, Ti(NEt$_2$)$_4$ [24b]

To a stirred solution of ethylmagnesium bromide (1.5 mmol) in diethyl ether (500 ml) [prepared from ethyl bromide (168.9 g, 1.55 mol) and Mg (37.7 g, 1.55 mol), Et_2NH (109.7 g, 1.5 mol) dissolved in diethyl ether (200 ml) is added dropwise at 0 °C. When no more ethane is liberated, the mixture is stirred for 2 h at 20 °C and filtered through glass-wool. The solution is added dropwise at 0–5 °C to a vigorously stirred suspension of $TiCl_4 \cdot 2THF$ (83.0 g, 0.25 mol) in benzene (250 ml). After the addition, the stirring is continued for 1 h at 0 °C, 2 h at room temperature and 2 h at reflux. The precipitated magnesium salts are filtered and washed with diethyl ether (100 ml). To the combined organic solutions dioxane (150 ml) is added slowly to precipitate the magnesium halides as sparingly soluble dioxane addition complexes. The solvents are removed from the filtrate by distillation *in vacuo*. The residue is distilled at low pressure through a 20 cm Vigreux column, affording 33–42 g (40–50%) of product, b.p. 112 °C/0.3 mbar. ^{13}C NMR (benzene-d_6): $\delta = 45.45$ (CH_2), 15.67 ppm (CH_3).

3.8.4 SYNTHESIS OF CHLOROTRIS(DIETHYLAMINO)TITANIUM, ClTi(NEt$_2$)$_3$ [24a]

A mixture of diethylamine (73.1 g, 1.0 mol) and lithium (7.7 g, 1.1 mol) in dry diethyl ether (400 ml) is heated at reflux temperature under an atmosphere of nitrogen. A solution of styrene (52.1 g, 0.5 mol) in diethyl ether (150 ml) is slowly added over a period of 2 h. The mixture is refluxed for an additional 30 min, and then cooled to 0 °C. At 0 °C, a solution of $TiCl_4$ (63.3 g, 0.33 mol) in toluene (150 ml) is added, the mixture refluxed for 1 h and the solvent stripped off. Vacuum distillation affords the desired reagent; yield, 72–76 g (73–77%); b.p. 94–96 °C/0.02 Torr.

Lithium diethylamide can also be conveniently prepared by treatment of diethylamine with *n*-butyllithium. Treatment with $TiCl_4$ as above then affords $ClTi(NEt_2)_3$ in similar yield. An analogous procedure using dimethylamine affords chlorotris(dimethylamino)titanium, $ClTi[N(CH_3)_2]_3$; b.p. 82–83 °C/0.01 Torr [23].

3.8.5 SYNTHESIS OF DICHLORODIISOPROPOXYTITANIUM, $Cl_2Ti(OiPr)_2$ (41) in solution [11, 328]

The solution of 71 g (0.25 mol) of tetraisopropoxytitanium in 350 ml of diethyl ether is treated with 47.5 g (0.25 mol) of $TiCl_4$ at 0 °C. After stirring for 1 h, diethyl ether is added to a total volume of 0.5 l. This stock solution is 1 M and can be kept in the refrigerator under nitrogen. The same reaction occurs in CH_2Cl_2. The compound can also be obtained as a solid [316] (see procedure for catalytic asymmetric glyoxylate–ene rection below).

3.8.6 SYNTHESIS AND CARBONYL ADDITION REACTIONS OF TRICHLOROMETHYLTITANIUM ETHERATE, CH_3TiCl_3-ETHERATE [13]

$TiCl_4$ (1.9 g, 10 mmol) is added by a syringe to about 50 ml of cooled (−78 °C) diethyl ether, resulting in partial precipitation of the yellow $TiCl_4$-etherate. The equivalent amount of methyllithium or methylmagnesium chloride in diethyl ether is slowly added, causing the color change to black–purple. This mixture of CH_3TiCl_3-etherate is allowed to warm to about −30 °C. Then an aldehyde (9 mmol) is added and the mixture is stirred for 2 h. In the case of ketones, the mixture is stirred for 2–5 h, during which the temperature is allowed to reach 0 °C. In the case of sensitive polyfunctional ketones, it may be better to add the reagent to a cooled solution of substrate. In all cases the cold reaction mixture is poured on to water. In rare cases a TiO_2-containing emulsion may form, in which case a saturated solution of NH_4F should be used. Following the usual work-up (diethyl ether extraction, washing with water, drying over $MgSO_4$), the products are isolated by standard techniques, generally chromatography or Kugelrohr distillation.

3.8.7 SYNTHESIS OF ETHER-FREE TRICHLOROMETHYLTITANIUM, CH_3TiCl_3 [17]

In a dry 100 ml flask equipped with a nitrogen inlet, 8.35 g (4.8 ml, 44 mmol) of clean $TiCl_4$ are mixed with 80 ml of dry CH_2Cl_2. After cooling to −30 °C, 22 mmol of $(CH_3)_2Zn$ (e.g. 5.5 ml of a 4 M CH_2Cl_2 solution [39]) are slowly added via a syringe with stirring. CH_3TiCl_3 is formed quantitatively along with precipitated $ZnCl_2$. This solution is used for *in situ* reactions, although CH_3TiCl_3 can be distilled [6]. Neat dimethylzinc is pyrophoric, although solutions in dichloromethane can be handled safely [39].

3.8.8 SYNTHESIS OF ETHER-FREE DICHLORODIMETHYLTITANIUM, $(CH_3)_2TiCl_2$ [39]

The procedure as above is used except that 44 mmol of $(CH_3)_2Zn$ are employed. Since $(CH_3)_3TiCl_2$ is less stable than CH_3TiCl_3, the solution should be used in reactions as soon as possible. The reaction can also be carried out in pentane [6c].

3.8.9 SYNTHESIS AND CARBONYL ADDITION REACTIONS OF TRIISOPROPOXYMETHYLTITANIUM, $CH_3Ti(OiPr)_3$ (7) [11, 12, 48]

Method A A 2 l three- necked flask equipped with a dropping funnel and magnetic stirrer is charged with 250 ml of diethyl ether and 130.3 g (0.50 mol) of chlorotriisopropoxytitanium (see above) and cooled to −40 °C. The equivalent amount of methyllithium (e.g. 312.5 ml of a 1.6 M diethyl ether solution) is slowly added and the solution is allowed to warm to room temperature within 1.5 h. The solvent is removed *in vacuo* and the yellow product distilled directly from the precipitated lithium chloride at 48–53 °C/0.01 Torr; yield 113 g (94%). ^1H NMR (CCl_4): δ (ppm) 0.5 (s, 3H), 1.3 (d, 18H), 4.5 (m, 3H). The compound is air sensitive, but can be kept under nitrogen in a refrigerator for weeks or months. On standing, slow crystallization begins, which can be reversed by gentle warming with a heat gun. Stock solutions can be prepared by mixing with the desired solvent.

Method B The above procedure is applied on a smaller scale (10–100 mmol, as needed for immediate use), but the diethyl ether solution of $CH_3Ti(OiPr)_3$ containing precipitated lithium chloride is used for carbonyl additions without any further treatment or purification (see below).

*Addition of $CH_3Ti(OiPr)_3$ to 6-Oxo-6-phenylhexanal (**35**) [11]*

The solution of 1.9 g (10 mmol) of 6-oxo-6-phenylhexanal (**35**) in 40 ml of THF is treated with 2.4 g (10 mmol) of distilled $CH_3Ti(OiPr)_3$ at −78 °C. The cooling bath is removed and stirring is continued for 6 h. The reaction mixture is poured on to dilute HCl, diethyl ether is added and the aqueous phase is extracted with diethyl ether. The combined organic phases are washed with water and dried over $MgSO_4$. The solvent is removed and the residue distilled in a Kugelrohr (220 °C/0.02 Torr); yield 1.65 (80%) of the adduct (**36**). An *in situ* reaction mode leads to the same result.

Addition of $CH_3Ti(OiPr)_3$ to 3-Nitrobenzaldehyde [16]

A dry, 500 ml three-necked flask equipped with a pressure-equalizing 100 ml dropping funnel, argon inlet and magnetic stirrer is evacuated and flushed with argon (three cycles). The flask is charged with 16.0 ml (57.7 mmol) of tetraisopropoxytitanium via a plastic syringe and hypodermic needle and 2.1 ml (19.2 mmol) of $TiCl_4$ is added over 5 min, with gentle cooling of the flask in an ice–water bath, to give a viscous oil. After the addition of 70 ml of THF, the clear solution is stirred at room temperature for 30 min. The dropping funnel is charged with 62 ml (77 mmol, 1.24 M in hexane) of methyllithium, which is added to the cooled (ice bath) THF solution over a period of 25–30 min. During the addition the resulting suspension changes from orange to bright yellow. After stirring at ice-bath temperature for 1 h, a solution of 10.6 g (70 mmol) of 3-nitrobenzaldehyde in 60 ml of THF is added from the dropping funnel within 20–25 min at the same temperature. The mixture is stirred at 0–5 °C for 1 h and then 60 ml of 2 M hydrochloric acid are added. The organic phase is separated in a separating funnel and the aqueous phase is extracted with three 150 ml portions of diethyl ether. The combined organic phases are washed with 100 ml of saturated $NaHCO_3$ solution and 100 ml of saturated NaCl solution and then dried over anhydrous $MgSO_4$. After filtration the solution is concentrated on a rotary evaporator and dried at 0.1 mmHg for 1 h. The residue, 11.0–11.1 g (94–95%) of an orange–brown viscous oil, sometimes solidifies on standing (m.p. 55–60 °C); the purity of the crude product is at least 95% (estimated by ^1H NMR). The product (3′-nitro-1-phenylethanol) can be purified by short-path distillation at 120–125 °C (0.15 mmHg) to give 9.9–10.4 g (85–89%) of a yellow oil, which solidifies on standing at room temperature or at −30 °C in a freezer; m.p. 60.5–62.0 °C.

3.8.10 GENERAL PROCEDURE FOR THE FORMATION AND ALDEHYDE ADDITION OF VINYLTITANIUM REAGENTS [57]

A solution (1.5 M) of *tert*-butyllithium (5 mmol) in hexane is cooled to −78 °C under N_2 and vinyl bromide (2.5 mmol) in 3 ml of freshly distilled diethyl ether is added dropwise over 5 min. An immediate precipitate of LiBr forms and the mixture is stirred at −78 °C for 20 min. A 1 M solution of $ClTi(OiPr)_3$ (2.5 ml) in hexane is then added and the solution, which turns brown immediately, is stirred for 20 min. The aldehyde (1 mmol) is added and the reaction mixture stirred at −78 °C until the disappearance of the aldehyde is apparent by TLC (8 h maximum). After quenching with saturated NH_4Cl–Et_2O (33 ml, 1 : 10), the ether layer is decanted. Two fresh portions of diethyl ether are similarly employed to complete extraction of the product(s). The combined ethereal solutions are dried, concentrated and the residue purified by chromatography (SiO_2), to afford the expected allylic alcohol.

3.8.11 GENERAL PROCEDURE FOR FORMATION AND CARBONYL ADDITION OF THE ALLYLTITANIUM ATE COMPLEX $CH_2{=}CHCH_2Ti(OiPr)_4MgCl$ (52) [11] AND THE CROTYLTITANIUM ATE COMPLEX $CH_3CH{=}CHCH_2Ti(OiPr)_4MgCl$ (123) [75]

A solution of 18 mmol of allylmagnesium chloride in about 40 ml of THF is treated with 5.68 g (20 mmol) of tetraisopropoxytitanium at −78 °C. The formation of the orange ate complex **52** is complete after 30 min. Addition of aldehydes (−30 °C/1 h) or ketones (−10 °C/1–2 h) results in 85–95% conversion. The reaction mixtures are poured on to dilute HCl, diethyl ether is added and the aqueous phase is extracted three times with diethyl ether. The combined organic phases are washed with water and dried over $MgSO_4$. In the case of di- or polyfunctional molecules, the solution containing **52** should be added to the substrate (reversed reaction mode). Using crotylmagnesium chloride as a mixture of regio- and stereoisomers, the same protocol is followed to synthesize and react the ate complex **123**.

3.8.12 GENERAL PROCEDURE FOR THE TITANATION OF CARBANIONS AND *IN SITU* CARBONYL ADDITION REACTIONS [11]

Standard techniques employing *n*-butyllithium or lithium diisopropylamide (LDA) are used to lithiate CH-acidic compounds. For example, the solution of 0.82 g (20 mmol) of acetronitrile in 50 ml of THF is treated with 11 ml of 1.8 M *n*-butyllithium solution at −78 °C. After 1 h, 0.52 g (20 mmol) of chlorotriisopropoxytitanium or 0.66 g (22 mmol) of chlorotris(diethylamino)titanium are added and the mixture is stirred for about 45 min at the same temperature, producing reagent **25** in >90% yield. An aldehyde (19 mmol) is added and the mixture is stirred for 2 h and workedup up in the usual way.

Methyl phenyl sulfone is lithiated by slow addition of *n*-butyllithium at 0 °C. Stirring is continued for 30 min and the solution cooled to −78 °C, followed by treatment with $ClTi(OiPr)_3$; the temperature is then allowed to reach −40 °C, which leads to reagent **23**. After cooling to −78 °C, an aldehyde or a ketone is added and the stirred solution is allowed to thaw to room temperature over a period of 3 h. The usual work-up affords the carbonyl addition product in 80–95% yield.

Ethyl acetate is lithiated in THF using LDA at −78 °C; titanation by $ClTi(OiPr)_3$ occurs (0.5 h) at the same temperature to produce the triisopropoxytitanium enolate of the ethyl ester. Aldol addition to aldehydes generally sets in at −78 to −30 °C within 2 h; ketones require 3–4 h (at −78 to 0 °C).

A 10% excess of organometallic reagent is used in all cases. Sometimes a two- or threefold excess of titanating agent, $Ti(OiPr)_4$ or $ClTi(OiPr)_3$, is necessary to maximize chemo- or stereoselectivity [45, 69].

3.8.13 GENERAL PROCEDURE FOR THE FORMATION AND KETONE ADDITION REACTIONS OF CROTYLTRIS(DIETHYLAMINO)TITANIUM (124) [11]

A solution of 3.29 g (11 mmol) of $ClTi(NEt_2)_3$ (see above) in 50 ml of THF is treated with 10 mmol of crotylmagnesium chloride as a mixture of regio- and stereoisomers in THF at 0 °C. The solution is allowed to reach room temperature and is then cooled to −78 °C. A ketone (9 mmol) is added, the mixture is stirred for 2–3 h and then poured on to dilute HCl. Extraction with diethyl ether followed by the usual work-up affords the desired alcohols in 75–85% yield.

3.8.14 GENERAL PROCEDURE FOR THE FORMATION AND ALDEHYDE ADDITION REACTION OF THE (ETHYLTHIO)ALLYTITANIUM ATE COMPLEX $EtSCH{=}CHCH_2Ti(OiPr)_4Li$ (128) [79]

To a solution of allyl ethyl sulfide (0.245 g, 2.4 mmol) in dry THF (8 ml) is added *tert*-butyllithium (2.4 mmol) dropwise at −78 °C and the orange mixture is stirred at 0°C for 30 min. Tetraisopropoxytitanium (0.71 ml, 2.4 mmol) is added at −78 °C. After stirring for 10 min, an aldehyde (2 mmol) is added over a period of 5 min at −78 °C and the mixture is stirred at −78 °C for 10 min and then at 0 °C for 1 h. After the usual work-up, the product is purified by column chromatography to give the *anti*-configured adducts with 99% diastereoselectivity and 87–98% yield.

3.8.15 GENERAL PROCEDURE FOR THE DIASTEREOSELECTIVE HOMO-ALDOL ADDITION BASED ON TITANATED (E)-3-BUTENYL-N,N-DIISOPROPYLCARBAMATES (133 and 134) [114]

To a solution of (*E*)-2-butenyl *N*,*N*-diisopropylcarbamate (21.6 g, 0.108 mol) and tetramethylethylenediamine (TMEDA) (13.9 g, 0.12 mol) in dry ethyl ether (300 ml) in a dry nitrogen atmosphere kept in a dry-ice–acetone bath below −65 °C, a solution of 1.62 M *n*-BuLi in hexane (73.5 ml, 0.119 mol) is added dropwise. Stirring of the light yellow suspension is continued for 30 min before the titanium reagent is added below −60 °C: (Method A) chlorotris(diethylamino)titanium (43.1 g, 0.132 mmol) or (Method B) tetraisopropoxytitanium (169.9 g, 0.60 mol). The solution is stirred for 50 min, and an aldehyde (0.1 mol) is introduced. Stirring is continued for 15 h below −70 °C; the reaction mixture is allowed to warm to −50 °C, and is poured into 5 M HCl (300 ml) and diethyl ether (500 ml). The aqueous layer is extracted with diethyl ether (3 × 200 ml) and the combined ethereal solutions are washed with 2 M HCl (100 ml), saturated aqueous $NaHCO_3$ solution (200 ml) and brine (200 ml). After drying (Na_2SO_4) and evaporation of the solvent in vacuum, the residue is purified by chromatography on silica gel (1000 g) with diethyl ether–pentane (1 : 4 to 1 : 1) as eluent.

3.8.16 GENERAL PROCEDURE FOR THE $TiCl_4$-MEDIATED CHELATION-CONTROLLED GRIGNARD-TYPE ADDITION TO ALKOXYALDEHYDES [119]

A solution of 5.0 mmol of an α- or β-alkoxyaldehyde (**92**, **94** or **96**) in 50 ml of dry CH_2Cl_2 is slowly treated with $TiCl_4$ (0.95 g, 5.0 mmol) at −78 °C. After 5 min, dialkylzinc (5.0 mmol), neat or as a CH_2Cl_2 solution, is added and the mixture is stirred for 2–3 h. In most cases work-up can then begin. However, if the $TiCl_4$–aldehyde complex partially precipitates, the final reaction mixture should be allowed to warm slowly to −35 °C. The mixture is poured onto 50 ml of water and extracted several times with diethyl ether. The combined organic phases are washed with $NaHCO_3$ and NaCl solutions and dried over $MgSO_4$. Following removal of solvents, the residue is purified by standard techniques (chromatography or Kugelrohr distillation). The conversion is >90%.

3.8.17 GENERAL PROCEDURE FOR THE $TiCl_4$-MEDIATED CHELATION-CONTROLLED ALDOL ADDITION TO ALKOXYALDEHYDES [119, 122]

As above, 5.0 mmol of the alkoxyaldehyde in 50 ml of dry CH_2Cl_2 are treated with 5 mmol of $TiCl_4$ at −78 °C. After 5–10 min, 5 mmol of an enolsilane are added using a syringe. In the case of the more reactive ketene ketals, cooled (−78 °C) CH_2Cl_2 solutions are added to the aldehyde–Lewis acid complex. After addition is complete, the mixture is stirred for an additional 1–2 h. Sometimes the initially formed aldehyde–Lewis acid complex precipitates. In these cases final stirring (after the enolsilane has been added) is prolonged (up to 5 h) and/or the temperature is raised to −50 or even −20 °C. The reaction is always complete within 30 min after the solution has become homogeneous (very often within 1–2 min!). The mixture is then poured on to water, extracted twice with diethyl ether and the combined organic phases are neutralized with $NaHCO_3$

solution. In the case of acid-sensitive products, saturated $NaHCO_3$ solution is poured on to the reaction mixture. After drying over $MgSO_4$ the solvent is stripped off and the crude product worked up in the usual way. The conversion is >85% in all cases. The same applies to reactions on a 10–20 mmol scale.

3.8.18 NON-CHELATION-CONTROLLED ADDITION OF $CH_3Ti(OiPr)_3$ (7) TO α-ALKOXYALDEHYDES [119]

A solution of α-alkoxyaldehyde **92** (5.0 mmol) in 25 ml of THF, diethyl ether, CH_2Cl_2 or *n*-hexane is treated with distilled $CH_3Ti(OiPr)_3$ (1.32 g, 5.5 mmol) at −30 °C for 3 d. The mixture is poured on to 50 ml of saturated NH_4F solution, extracted several times with diethyl ether, washed with NaCl solution and dried over $MgSO_4$. The solvents are removed and the residue is Kugelrohr distilled, delivering 80–90% of product. Non-chelation control is highest if hexane is the solvent.

3.8.19 GENERAL PROCEDURE FOR HOMO-ALDOL ADDITIONS BASED ON ALKOXYTITANIUM HOMOENOLATES DERIVED FROM SILOXYCYCLOPROPANES [94]

*Formation of alkyl-3-(trichlorotitanio)propionates (**73**)*

To a water-cooled solution of $TiCl_4$ (110 μl, 1.0 mmol) in 2.0 ml of hexane is added a 1-alkoxy-1-siloxycyclopropane, e.g. **72** (R = iPr) (1.0 mmol) at 21 °C during 20 s. The initially formed milky white mixture turns brown in 10 s, and finally deep purple microcrystals precipitate. Heat evolution continues for several minutes. The mixture is allowed to stand for 30 min; NMR analysis of the supernatant reveals the quantitative formation of chlorotrimethylsilane (with 1,1,2,2-tetrachloroethane as an internal standard). The supernatant is removed with a syringe and the crystals are washed three times with hexane. The homoenolate **73** (R = iPr) weighs 223 mg (83%). This procedure can be scaled up to the 10 g level without modification. Recrystallization from CH_2Cl_2–hexane gives an analytical sample as thin, deep purple needles. The titanium alkyl melts at 90–95 °C with a color change to reddish brown, and sublimes with some decomposition at 90–110 °C/0.005 Torr.

Reaction of the alkoxide-modified homoenolate

The purified homoenolate **73** (1.5 mmol) is dissolved in 3 ml of CH_2Cl_2 at 0 °C and $Ti(OiPr)_4$ (0.75 mmol) is added. After 5 min, 1 mmol of a carbonyl compound is added. The mixture is stirred for 1 h (up to 18 h in the case of bulky ketones). The mixture is poured into a stirred mixture of diethyl ether and water. After 10 min, the ethereal layer is separated and the aqueous layer is extracted three times with diethyl ether. The combined extract is washed with water, aqeous $NaHCO_3$, and saturated NaCl. After drying and concentration, the product is purified to obtain the desired 4-hydroxy ester or lactone. For unreactive substrates, 2 equiv. of **73** and 1 equiv. of $Ti(OiPr)_4$ are used. For unreactive ketones, $Ti(OtBu)_4$ should be employed.

3.8.20 TYPICAL PROCEDURE FOR THE MUKAIYAMA ALDOL ADDITION [92a]

A solution of 1-trimethylsiloxycyclohexene (0.426 g, 2.5 mmol) in 10 ml of dry CH_2Cl_2 is added dropwise to a solution of benzaldehyde (0.292 g, 2.75 mmol) and $TiCl_4$ (0.55 g, 2.75 mmol) in CH_2Cl_2 (20 ml) under an argon atmosphere at −78 °C. The mixture is stirred for 1 h and hydrolyzed at that temperature. Following extraction with diethyl ether, washing with water and drying over Na_2SO_4, the solution is concentrated and the residue is purified by column chromatography on SiO_2 using CH_2Cl_2 to afford 115 mg (23%) of *erythro*-2-(hydroxyphenylmethyl)cyclohexanone (m.p. 103.5–105.5 °C from 2-propanol) and 346 mg (69%) of the *threo* isomer (m.p. 75% from *n*-hexane–diethyl ether).

3.8.21 TYPICAL PROCEDURE FOR THE SYN-SELECTIVE $TiCl_4/NR_3$-MEDIATED ALDOL ADDITION OF CARBONYL COMPOUNDS TO ALDEHYDES [181]

$TiCl_4$ (1.1 equiv.) is added dropwise to a 0.2 M solution of 1.0 equiv. of the ketone or carboxylic acid derivative in CH_2Cl_2 at −78 °C under N_2, giving a yellow slurry. After 2 min, 1.2 equiv. of either Et_3N or $EtN(iPr)_2$ are added dropwise, and the resulting deep red solution is stirred at −78 °C under N_2 for 1.5 h. After the dropwise addition of isobutyraldehyde (1.2 equiv.), stirring is continued at −78 °C for 1.5 h. The reaction is terminated by addition of 1 : 1 v/v of saturated aqueous NH_4Cl, the mixture is warmed to ambient temperature and the product is isolated by a conventional extraction. The *syn* selectivity is 90–99%.

3.8.22 TYPICAL PROCEDURE FOR THE $TiCl_4$-MEDIATED KNOEVENAGEL CONDENSATION [98a]

To a stirred solution of 200 ml of dry THF is added $TiCl_4$ (11 ml, 100 mmol) in 25 ml of dry CCl_4 at 0 °C, which leads to a flaky yellow precipitate. An aldehyde (50 mmol) and a CH-acidic methylene compound such as a malonic acid ester, acetoacetic ester or nitroacetic acid ester (50 mmol) are added in pure form or as a THF solution. Dry pyridine (16 ml, 200 mmol) or *N*-methylmorpholine (22 ml) in 30 ml of THF is slowly added to the stirred mixture within 1–2 h. The mixture is stirred at 0 °C (or 23 °C) for 16–24 h and diluted with 50 ml of diethyl ether. The usual work-up and purification (distillation or recrystallization) afford the Knoevenagel products in yields of 40–95%. Ketones react similarly at 23 °C/1 d [98b].

3.8.23 TYPICAL PROCEDURE FOR THE $ClTi(OiPr)_3$-MEDIATED KNOEVENAGEL CONDENSATION [183]

Procedure A A suspension of 0.25 g (10.5 mmol) of freshly washed (THF) NaH containing no NaOH or Na_2O in 30 ml of THF is treated with 2.24 g (10.0 mmol) of ethyl (diethoxyphosphoryl)acetate (**145**) at room temperature and the mixture stirred until an almost clear solution forms. It is refluxed for 1 h, cooled to −78 °C and treated with 2.73 g (10.5 mmol) of chlorotriisopropoxytitanium. The cooling bath is removed and the mixture stirred for 1.5 h at room temperature. After the addition of an aldehyde (9.5 mmol), the mixture is stirred at room temperature for 4 h (overnight in the case of 4-methoxybenzaldehyde) and poured on to dilute hydrochloric acid. The aqueous phase is extracted twice with diethyl ether and the combined organic phase is washed with water and dried over $MgSO_4$. After stripping off the solvent, the crude product is distilled using a Kugelrohr. *Z*-configured vinyl phosphonates **147** are obtained in 57–62% yield. Transesterification occurs under the reaction conditions.

Procedure B A mixture of 5.2 g (20 mmol) of chlorotriisopropoxytitanium, 2.24 (10.0 mmol) of ethyl (diethoxyphosphoryl)acetate (**145**) and 10.0 mmol of an aldehyde in 30–40 ml of THF is treated with 2.02 g (20 mmol) of triethylamine at 0 °C. The amine hydrochloride precipitates and the mixture is stirred at 0 °C for 2–3 h, followed by work-up according to Procedure A. Procedure B can be applied to the Knoevenagel condensation of malonic acid esters with aldehydes.

3.8.24 GENERAL PROCEDURE FOR $TiCl_4$-MEDIATED ALLYLSILANE ADDITIONS TO ALDEHYDES AND KETONES [105]

To a solution of a carbonyl compound (2 mmol) and CH_2Cl_2 (3 ml) is added $TiCl_4$ (1 mmol) dropwise with a syringe at room temperature under nitrogen. An allylsilane (2 mmol) is then added rapidly and the mixture is stirred for 1 min (aldehyde additions) or 3 min (ketone additions).

The usual aqueous work-up affords 50–91% of the homo-allylic alcohols. The reaction can also be carried out at low temperatures [105, 127].

3.8.25 TYPICAL PROCEDURE FOR THE $TiCl_4$-MEDIATED ADDITION OF METHYL ISONITRILE TO CARBONYL COMPOUNDS [97]

A solution of $TiCl_4$ (1.1 ml, 10 mmol) in 40 ml of dry CH_2Cl_2 is treated with methyl isonitrile (0.57 ml, 10 mmol) at −5 to 0 °C. A white solid precipitates, and the mixture is stirred for 2 h at 0 °C, forming reagent **75** (R = CH_3). After cooling to −60 °C, heptanal (1.14 g, 9.5 mmol) is added. The temperature is allowed to reach 0 °C, which results in a homogeneous solution. It is treated with 25 ml of 2 M HCl and stirred for 1.5 h. The organic phase is separated and the NaCl-saturated aqueous phase is extracted with CH_2Cl_2. The combined organic phases are washed with saturated NaCl solution and dried over $MgSO_4$. On removal of the solvent, 1.45 g (96%) of *N*-methyl-2-hydroxyheptanoic acid amide (**76**); R = CH_3, R′ = *n*-C_6H_{13}) are obtained (m.p. 109.5–110 °C).

3.8.26 PREPARATION OF THE CHIRAL TITANATING AGENT [(4R,5R)-2,2-DIMETHYL-1,3-DIOXOLAN-4,5-BIS(DIPHENYLMETHOXY)]CYCLOPENTADIENYL-CHLOROTITANATE (189) [206, 207]

A solution/suspension of freshly sublimed cyclopentadienyltrichlorotitanium, 11 g (50 mmol) in 400 ml of diethyl ether (distilled over Na–benzophenone), is treated with 23.3 g (50 mmol) of (4*R*,5*R*)-2,2-dimethyl-4,5- [bis(diphenylhydroxymethyl)]-1,3-dioxolane (**188**) under argon in the absence of moisture. After 2 min at room temperature, a solution of NEt_3 (12.65 g, 110 mmol) in 125 ml of diethyl ether is added dropwise to the stirred mixture within 1 h (efficient stirring is essential). After further stirring for 12 h, $Et_3N{\cdot}HCl$ (13.8 g) is filtered off under argon and washed three times with *ca* 50 ml of diethyl ether. This stock solution (610 ml) is assumed to be 82 mM and can be used directly if desired.

For isolation of the complex **189**, the stock solution is concentrated under reduced pressure to *ca* 75 ml and 300 ml of hexane are added. After stirring for 30 min, the suspension is filtered and the residue is washed three times with 10 ml of hexane, yielding 26.8 g (87%) of analytically pure product **189**.

3.8.27 TYPICAL PROCEDURE FOR THE ENANTIOSELECTIVE ALLYLATION OF ALDEHYDES USING [4R,5R)-2,2-DIMETHYL-1,3-DIOXOLAN-4,5-BIS(DIPHENYLMETHOXY)]CYCLOPENTADIENYL CHLOROTITANATE (189) [206]

A 5.3 ml volume of a 0.8 M solution of allylmagnesium chloride in THF (4.25 mmol) is added dropwise within 10 min at 0 °C under argon to a solution of [(4*R*,5*R*)-2,2- dimethyl-1,3-dioxolane-4,5-bis(diphenylmethoxy)]cyclopentadienyl chlorotitanate (**189**) (3.06 g, 5 mmol) in 60 ml of diethyl ether. After stirring for 1.5 h at 0 °C the slightly orange suspension is cooled to −74 °C and treated within 2 min with benzaldehyde (403 mg, 3.8 mmol, dissolved in 5 ml of diethyl ether). The mixture is stirred for 3 h at −74 °C. After hydrolysis with 20 ml of aqueous 45% NH_4F solution and stirring for 12 h at room temperature, the reaction mixture is filtered over Celite and extracted twice with diethyl ether (50 ml). The combined extracts are washed with brine, dried with $MgSO_4$ and concentrated. The solid residue is stirred with 50 ml of pentane. Subsequent filtration furnishes 1.54 g of crude alcohol and 1.68 g of white crystalline material (ligand **188**). Chromatography [200 g of silica gel, CH_2Cl_2–hexane–diethyl ether (4::4:1)] affords 521 mg (93%) of (*S*)-1-phenyl-3-buten-1-ol having an *ee* value of 95%.

3.8.28 TYPICAL PROCEDURE FOR THE $TiCl_4$-MEDIATED CONJUGATE ADDITION OF ENOLSILANES [210]

A 500 ml three-necked flask is fitted with a mechanical stirrer, rubber septum and a two-way stopcock which is equipped with a balloon of argon gas. To the flask is added 100 ml of dry CH_2Cl_2, and the flask is cooled in a dry-ice–acetone bath. $TiCl_4$ (7.7 ml) is added by syringe through the septum. The septum is removed and replaced by a 100 ml pressure-equalizing dropping funnel containing a solution of 11.2 g of isopropylideneacetophenone in 30 ml of CH_2Cl_2. This solution is added over 3 min and the mixture is stirred for 4 min. A solution of 13.5 g of the enolsilane of acetophenone in 40 ml of CH_2Cl_2 is added dropwise with vigorous stirring over 4 min and the mixture is stirred for 7 min. The reaction mixture is poured into a solution of 22 g of Na_2CO_3 in 160 ml of water with vigorous magnetic stirring. The resulting white precipitate is removed by filtration through a Celite pad and the precipitate is washed with CH_2Cl_2. The organic layer of the filtrate is separated and the aqueous layer is extracted with two 40 ml portions of CH_2Cl_2. The combined organic extracts are washed with 60 ml of brine and dried over sodium sulfate. The CH_2Cl_2 solution is concentrated in a rotary evaporator and the residue is passed through a short column of silica gel (Baker 200 mesh, 400 ml) using 1.5 l of a 9:1 (v/v) mixture of hexane and ethyl acetate. The eluent is condensed and distilled; the first fraction (b.p. 81–85 °C/0.6 mmHg, 2.04 g) is a mixture of isopropylideneacetophenone and acetophenone; the second fraction (b.p. 85–172 °C/0.6 mmHg, 0.42 g) is a mixture of the above substances and the desired product; the third fraction (b.p. 172–178 °C/0.6 mmHg) gives 14.0–15.2 g (72–78%) of 3,3-dimethyl-1,5-diphenylpentane-1,5-dione.

3.8.29 TYPICAL PROCEDURE FOR THE $TiCl_4$-MEDIATED CONJUGATE ADDITION OF ALLYLSILANES [209]

A 2 l three-necked, round-bottomed flask is fitted with the dropping funnel, mechanical stirrer and reflux condenser attached to a nitrogen inlet. In the flask are placed 29.2 g (0.20 mol) of benzalacetone and 300 ml of CH_2Cl_2. The flask is immersed in a dry-ice–methanol bath (−40 °C) and 22 ml (0.20 mol) of $TiCl_4$ are slowly added by syringe to the stirred mixture. After 5 min, a solution of 30.2 g (0.26 mol) of allyltrimethylsilane in 300 ml of CH_2Cl_2 is added dropwise with stirring over 30 min. The resulting red–violet reaction mixture is stirred for 30 min at −40 °C, hydrolyzed by addition of 400 ml of water and, after the addition of 500 ml of diethyl ether with stirring, allowed to warm to room temperature. The nearly colorless organic layer is separated and the aqueous layer is extracted with three 500 ml portions of diethyl ether. The organic layer and ether extracts are combined and washed successively with 500 ml of saturated sodium hydrogencarbonate and 500 ml of saturated NaCl, dried over anhydrous Na_2SO_4 and evaporated at reduced pressure. The residue is distilled under reduced pressure through a 6 in Vigreux column to give 29.2–30.0 g (78–80%) of 4-phenyl-6-hepten-2-one; b.p. 69–71 °C (0.2 mmHg).

3.8.30 TYPICAL PROCEDURE FOR THE METHYLENATION OF OLEFINS USING $CH_2Br_2/TiCl_4/Zn$ [239, 240]

Into a 1 l round-bottomed flask fitted with a magnetic stirrer and a pressure-equalizing dropping funnel connected to a nitrogen line are placed 28.75 g (0.44 mol) of activated zinc powder (L. F. Fieser, M. Fieser, *Reagents for Organic Synthesis*, Vol. I, p. 1276, Wiley, New York, 1967), 250 ml of dry THF and 10.1 ml (0.144 mol) of dibromomethane. The mixture is stirred and cooled in a dry-ice–acetone bath at −40 °C. To the stirred mixture is added dropwise 11.5 ml (0.103 mol) of $TiCl_4$ over 15 min. The cooling bath is removed and the mixture is stirred at 5 °C (cold room) for 3 days under a nitrogen atmosphere. A reasonable rate of stirring must be maintained as the mixture thickens, but too fast a rate causes splashing up to the neck of the flask. The dark grey slurry is cooled with an ice–water bath and 50 ml of dry CH_2Cl_2 are added. To the stirred mixture are added 15.4 g (0.1 mol) of (+)-isomenthone (**219**) in 50 ml of dry CH_2Cl_2 over 10 min. The cooling bath is removed and the mixture is stirred at room temperature (20 °C) for 1.5 h. The mixture is diluted with 300 ml of pentane and a slurry of 150 g of $NaHCO_3$ in 80 ml of water is added cautiously over 1 h. It is necessary to add the slurry dropwise at the beginning, allowing the effervescence to subside after each drop. After the initial vigorous effervescence, larger portions can be added. During this part of the addition, the stirrer becomes ineffective and gentle shaking by hand is continued until effervescence ceases. The residue is washed three times with 50 ml portions of pentane. The combined organic solutions are dried over a mixture of 100 g of Na_2SO_4 and 20 g of $NaHCO_3$, filtered through a sintered-glass funnel and the solid desiccant is thoroughly washed with pentane. The solvent is removed at atmospheric pressure by flash distillation through a column (40 cm × 2.5 cm i.d.) packed with glass helices. The liquid residue is distilled to give (+)-3-methylene-*cis*-*p*-menthane (**220**) as a clear, colorless liquid, b.p. 105–107 °C/90 mmHg, 13.6 g, 89% yield.

The reagent must be kept cold at all times because at room temperature the active reagent slowly decomposes and the mixture darkens considerably. Once prepared, the reagent can be stored at −20 °C (freezer) in a well sealed flask without a significant loss of activity. A sample stored in this way for 1 year showed only a slight (5–10%) loss of activity. The molar activity of the active reagent is equivalent to the $TiCl_4$ molarity (determined by reaction with excess of ketone followed by GLC analysis); however, an increase in the proportion of $TiCl_4$ makes no difference to the molar activity.

3.8.31 TYPICAL PROCEDURE FOR THE ALKYLIDENATION OF CARBOXYLIC ACID ESTERS USING $RCHBr_2/Zn/TiCl_4/TMEDA$ [242]

A solution of $TiCl_4$ (1.0 M, 4.0 mmol) in CH_2Cl_2 is added at 0 °C to THF (10 ml) under an argon atmosphere. To the yellow solution at 25 °C is added tetramethylethylenediamine (TMEDA) (1.2 ml, 8.0 mmol) and the mixture is stirred at 25 °C for 10 min. Zinc dust (0.59 g, 9.0 mmol) is added to the mixture. The color of the suspension turns from brownish yellow to dark greenish blue in a slightly exothermic process. After stirring at 25 °C for 30 min, a solution of methyl pentanoate (0.12 g, 1.0 mmol) and 1,1-dibromohexane (0.54 g, 2.2 mmol) in THF (2 ml) is added to the mixture. The color of the resulting mixture gradually turns dark brown while stirring at 25 °C for 2 h. Saturated K_2CO_3 solution (1.3 ml) is added at 0 °C to the mixture. After it has been stirred at 0 °C for another 15 min, the mixture is diluted with diethyl ether (20 ml) and then passed rapidly through a short column of basic alumina (activity III) using diethyl ether–triethylamine (200:1, 100 ml). The resulting clear solution is concentrated and the residue is purified by column chromatography on basic alumina (activity III) with pentane to give the desired 5-methoxy-5-undecene (**222**; $R = n\text{-}C_4H_9$, $R' = CH_3$, $R'' = n\text{-}C_5H_{11}$) (0.18 g, $Z/E = 91:9$) in 96% yield.

3.8.32 TYPICAL PROCEDURE FOR THE METHYLENATION OF CARBOXYLIC ACID ESTERS USING THE TEBBE REAGENT (223) [247]

To a 250 ml round-bottom flask equipped with a magnetic stirring bar are added 5.0 g (20.0 mmol) of bis(cyclopentadienyl)dichlorotitanium. The flask is fitted with a rubber septum through which a large-gauge needle is passed to flush the system with dry nitrogen. After the vessel has been thoroughly purged, the nitrogen line flowing to the needle is opened to a mineral oil bubbler and 20 ml of trimethylaluminum solution (2.0 M in toluene, 40 mmol) are added by a nitrogen-purged syringe (**caution**: pyrophoric material). Methane gas evolved by the reaction is allowed to vent as the resulting red solution is stirred at room temperature for 3 days.

The Tebbe reagent thus formed is used *in situ* by cooling the mixture in an ice–water bath, then adding 4.0 g (20 mmol) of phenyl benzoate dissolved in 20 ml of dry THF by syringe or cannula to the cooled stirring solution over 5–10 min. After the addition, the reaction mixture is allowed to warm to room temperature and is stirred for about 30 min. The septum is removed and 50 ml of anhydrous diethyl ether are added. To the stirred reaction mixture are gradually added 50 drops of 1 M sodium hydroxide solution over 10–20 min. Stirring is continued until gas evolution essentially ceases; then to the resulting orange slurry are added a few grams of anhydrous Na_2SO_4 to remove excess of water. The mixture is filtered through a Celite pad on a large coarse frit using suction and liberal amounts of diethyl ether to transfer the product and rinse the filter pad. Concentration of the filtrate with a rotary evaporator to 5–8 ml provides crude product, which is purified by column chromatography on basic alumina (150 g) eluting with 10% diethyl ether in pentane. Fractions which contain the product are combined and evaporated to give 2.69–2.79 (68–70%) of 1-phenoxy-1-phenylethene (**225**) as a pale yellow oil.

The acid lability of enol ether products requires rigorous treatment of all glassware used for the reaction in order to avoid migration of the double bond in susceptible cases. Satisfactory results are obtained by treating the glassware sequentially with ethanolic 0.5 M solutions of hydrogen chloride and potassium hydroxide for *ca* 1 h, thoroughly rinsing with distilled water after each treatment and finally oven drying. This protocol is also effective for removing stubborn deposits on the glassware after the reaction. Methylenation must be carried out at −78 °C in the case of extremely sensitive substrates or products.

3.8.33 OPTIMIZED MCMURRY COUPLING OF KETONES [260]

Preparation of $TiCl_3(DME)_{1.5}$

$TiCl_3$ (25.0 g, 0.162 mol) is suspended in 350 ml of dry dimethoxyethane (DME) and the mixture is refluxed for 2 days under argon. After the mixture has cooled to room temperature, filtration under argon, washing with pentane and drying under vacuum give the fluffly, blue crystalline $TiCl_3(DME)_{1.5}$ (32.0 g, 80%) that is used in the coupling reaction. The solvate is air sensitive but can be stored indefinitely under argon at room temperature.

Preparation of Zinc–Copper couple

The zinc–copper couple is prepared by adding zinc dust (9.8 g, 150 mmol) to 40 ml of nitrogen-purged water, purging the slurry with nitrogen for 15 min and then adding $CuSO_4$ (0.75 g, 4.7 mmol). The black slurry is filtered under nitrogen, washed with deoxygenated (nitrogen-purged) water, acetone and diethyl ether, and then dried under vacuum. The couple can be stored for months in a Schlenk tube under nitrogen.

Typical Procedure for Ketone Coupling

$TiCl_3(DME)_{1.5}$ (5.2 g, 17.9 mmol) and Zn–Cu (4.9 g, 69 mmol) are transferred under argon to a flask containing 100 ml of DME, and the resulting mixture is refluxed for 2 h to yield a black suspension. Cyclohexanone (0.44 g, 4.5 mmol) in 10 ml of DME is added and the mixture is refluxed for 8 h. After being cooled to room temperature, the reaction mixture is diluted with pentane (100 ml), filtered through a pad of Florisil, and concentrated in a rotary evaporator to yield cyclohexylidenecyclohexane (0.36 g, 97%) as white crystals, m.p. 52.5–53.5 °C. If a 3:1 ratio of $TiCl_3(DME)_{1.5}$ to carbonyl compound instead of 4:1 is used, the yield decreases from 97% to 94%; if a 2:1 ratio is used, the yield is 75%.

3.8.34 TYPICAL PROCEDURE FOR THE METHYLATION OF TERTIARY ALKYL CHLORIDES [17, 20]

A solution of 6 mmol $(CH_3)_2Zn$ (e.g. 1.5 ml of a 4 M CH_2Cl_2 solution) [39] in 30 ml of dry CH_2Cl_2 is treated with $TiCl_4$ (180 mg) and subsequently with *trans*-9-chlorodecalin (1.72 g, 10 mmol) at −30 °C. After 15 min the mixture is poured onto ice–water. Following extraction with diethyl ether, washing with $NaHCO_3$ solution and drying over $MgSO_4$, the solvent is removed and the residue is distilled (90 °C/12 Torr) to provide 1.31 g (82%) of 9-methyldecalin as a 1:1 *cis–trans* mixture.

3.8.35 GENERAL PROCEDURE FOR THE GEMINAL DIMETHYLATION OF KETONES USING $(CH_3)_2TiCl_2$ [39]

To a stirred solution of 40 mmol of $(CH_3)_2TiCl_2$ in dichloromethane (see above) is added a ketone (20 mmol) at −30 °C. The mixture is slowly allowed to come to room temperature during a period of about 2 h and is then poured onto ice water. The aqueous phase is extracted with ether and the combined organic phases are washed with H_2O and $NaHCO_3$. After drying over $MgSO_4$, the solvent is removed and the product distilled (e.g., using a Kugelrohr) or crystallized. In the case of aryl or α,β-unsaturated ketones, addition is best performed at −40 °C and the mixture allowed to come to 0 °C prior to the usual workup.

3.8.36 TYPICAL PROCEDURE FOR THE α-TERT-ALKYLATION OF KETONES VIA THEIR ENOLSILANES [288, 289]

A dry 250 ml three-necked, round-bottomed flask is fitted with an argon inlet, gas bubbler, rubber septum and magnetic stirrer. The apparatus is flushed with dry nitrogen or argon and charged with 120 ml of dry CH_2Cl_2, 15.6 g (0.10 mol) of 1-trimethylsilyloxycyclopentene and 11.7 g (0.11 mol) of 2-chloro-2-methylbutane. The mixture is cooled to −50 °C and a cold (−50 °C) solution of 11 ml (0.10 mol) of $TiCl_4$ in 20 ml of CH_2Cl_2 is added within 2 min through the rubber septum with the aid of a syringe. During this operation rapid stirring and cooling are maintained. Direct sunlight should be avoided. The reddish brown mixture is stirred at the given temperature for an additional 2.5 h and then rapidly poured on to 1 l of ice–water. After the addition of 400 ml of CH_2Cl_2, the mixture is vigorously shaken in a separating funnel; the organic phase is separated and washed twice with 400 ml portions of water. The aqueous phase of the latter two washings is extracted with 200 ml of CH_2Cl_2, the organic phases are combined and dried over anhydrous Na_2SO_4. The mixture is concentrated using a rotary evaporator and the residue is distilled at 80 °C (12 mmHg) to yield 9.2–9.5 (60–62%) of 2-*tert*-pentylcyclopentanone as a colorless oil.

3.8.37 TYPICAL PROCEDURE FOR THE ENANTIOSELECTIVE DIELS–ALDER REACTION MEDIATED BY THE NARASAKA CATALYST (280) [310]

*Preparation of the Catalyst (**280**)*

Under an argon atmosphere, to a toluene solution (5 ml) of dichloroiisopropoxytitanium (**41**) (140 mg, 0.59 mmol) is added a toluene solution (5 ml) of the chiral diol **279** (354 mg, 0.67 mmol) at room temperature, and the mixture is stirred for 1 h.

Diels–Alder Reaction

To a toluene suspension (3 ml) of molecular sieves 4 Å (150 mg) is added a toluene solution of the catalyst **280** (about 0.07 mmol) and the mixture is cooled to 0 °C. A toluene solution (4 ml) of the fumaric acid derivative **282** (140 mg, 0.7 mmol) is added to the mixture and then hexane (5 ml) and isoprene (1 ml) are added. The mixture is stirred overnight at 0 °C, then pH 7 phosphate buffer is added, the organic materials are extracted with ethyl acetate and the combined extracts are dried over anhydrous $MgSO_4$. After evaporation of the solvent, the crude product is purified by thin-layer chromatography [ethyl acetate–hexane (1 : 2)] to give the pure Diels–Alder product in 92% yield (*ee* = 94%).

3.8.38 GENERAL PROCEDURE FOR THE ASYMMETRIC GLYOXYLATE–ENE REACTION CATALYZED BY DICHLORO-1,1′-BI-NAPHTHALENE-2,2′- DIOXYTITANIUM [316]

*Isolation of Dichlorodiisopropoxytitanium (**41**)*

To a solution of tetraiisopropoxytitanium (2.98 ml, 10 mmol) in hexane (10 ml) is added $TiCl_4$ (1.10 ml, 10 mmol) slowly at room temperature. On addition of $TiCl_4$, heat evolves. After stirring for 10 min, the solution is allowed to stand for 6 h at room temperature and the precipitate is then collected. The precipitate is washed with hexane (2 × 5 ml) and recrystallized from hexane (3 ml). The crystalline material is dried under reduced pressure and then dissolved in toluene to give a 0.3 M toluene solution.

Catalytic Ene Reaction

To a suspension of activated powdered molecular sieves 4 Å (500 mg) in CH_2Cl_2 (5 ml) is added a 0.3 M toluene solution of the above dichlorodiisopropoxytitanium (0.33 ml, 0.10 mmol) and (*R*)-(+)- or (*S*)-(−)-binaphthol (28.6 mg, 0.10 mmol) at room temperature under an argon atmosphere. After stirring for 1 h at room temperature, the mixture is cooled to −70 °C. An excess of the isobutylene is bubbled through the mixture **288** (R = CH_3) (*ca* 2 equiv.), and freshly distilled methyl glyoxylate (**283**) (88 mg, 1.0 mmol) is added. The mixture is then warmed to −30 °C and stirred for 8 h. The solution is poured on to saturated $NaHCO_3$ solution (10 ml). Molecular sieves 4 Å are filtered off through a pad of Celite, and the filtrate is extracted with ethyl acetate. The combined organic layer is washed with brine. The extract is then dried over $MgSO_4$ and evaporated under reduced pressure. Separation by silica gel chromatography [hexane–ethyl acetate (20:1)] gives a 72% yield of methyl 2-hydroxy-4-methyl-4- pentenoate (**289**, R = CH_3) having an *ee* value of 95%.

3.8.39 TYPICAL PROCEDURE FOR THE ENANTIOSELECTIVE ADDITION OF $ZnEt_2$ TO ALDEHYDES CATALYZED BY CHIRAL SULFONAMIDES IN THE PRESENCE OF $Ti(OiPr)_4$ [38b, 321a]

In a flame-dried round-bottomed flask is placed (1*R*,2*R*)-1,2-*N*,*N*′-bis(trifluoromethylsulfonyl-amino)cyclohexane (189 mg, 0.5 mmol) under an argon atmosphere. To this are added degassed toluene (10 ml) and $Ti(OiPr)_4$ (8.53 g, 30 mmol) and the mixture is stirred at 40 °C for 20 min. After being cooled to −78 °C, Et_2Zn (1.0 M hexane solution, 30 ml, 30 mmol) is added to the solution. The solution rapidly turns orange. To the resulting solution is added benzaldehyde (2.12 g, 25 mmol) in toluene (2 ml) and the mixture is warmed to −20 °C and stirred at that temperature for 2 h. The reaction is quenched by adding 2 M HCl and the product is extracted with diethyl ether. The organic phase is washed with saturated NaCl, dried over anhydrous Na_2SO_4 and concentrated. The residue is chromatographed on a silica gel column (ethyl acetate 2% in hexane as eluent) to obtain crude 1-phenylpropanol. Distillation gives pure 1-phenylpropanol (3.27 g, 98%); b.p. ~103 °C/15 Torr; [α] − 48.6 ° (*c* 5.13, $CHCl_3$). The product has the *S*-configuration and an *ee* value of >99%.

3.8.40 TYPICAL PROCEDURE FOR THE CATALYTIC SHARPLESS EPOXIDATION OF ALLYLIC ALCOHOLS [317]

Caution: Owing to possible explosions, *tert*-butyl hydroperoxide should not be used in pure form. Also, strong acids and transition metal salts known to be good autoxidation catalysts (Mn, Fe, Ru, Co, etc.) should never be added to high-strength *tert*-butyl hydroperoxide.

An oven-dried 1 l three-necked, round-bottomed flask equipped with a magnetic stirbar, pressure equalizing addition funnel, thermometer, nitrogen inlet and bubbler is charged with 3.0 g of powdered, activated molecular sieves 4 Å and 350 ml of dry CH_2Cl_2. The flask is cooled to −20 °C. L-(+)-Diethyl tartrate (1.24 g, 6.0 mmol) and $Ti(O\text{-}iPr)_4$ (1.49 ml, 1.42 g, 5.0 mmol, via a syringe) are added sequentially with stirring. The reaction mixture is stirred at −20 °C as *tert*-butyl hydroperoxide (39 ml, 200 mmol, 5.17 M in isooctane) is added through the addition funnel at a moderate rate (over *ca* 5 min). The resulting mixture is stirred at −20 °C for 30 min. (*E*)-2-Octenol (12.82 g, 100 mmol, freshly distilled), dissolved in 50 ml of CH_2Cl_2, is then added dropwise through the same addition funnel over a period of 20 min, being careful to maintain the reaction temperature between −20 and −15 °C. The mixture is stirred for an additional 3.5 h at −20 to −15 °C.

Work-up [317] is as follows. A freshly prepared solution of 33 g (0.12 mol) of iron(II) sulfate heptahydrate and 10 g (0.06 mol) of tartaric acid [or 11 g (0.06 mol) of citric acid monohydrate instead of tartaric acid] in a total volume of 100 ml of deionized water is cooled to *ca* 0 °C by means of an ice–water bath. The epoxidation reaction mixture is allowed to warm to *ca* 0 °C and is then slowly poured into a beaker containing the precooled stirring iron(II) sulfate solution (external cooling is not essential during or after this addition). The two-phase mixture is stirred for 5–10 min and then transferred into a separating funnel. The phases are separated and the aqueous phase is extracted with two 30 ml portions of diethyl ether. The combined organic layers are treated with 10 ml of a precooled (0 °C) solution of 30% w/v NaOH in saturated brine. The two-phase mixture is stirred vigorously for 1 h at 0 °C. Following transfer to a separating funnel and dilution with 50 ml of water, the phases are separated and the aqueous layer is extracted with diethyl ether (2 × 50 ml). The combined organic layers are dried over sodium sulfate, filtered and concentrated, yielding a white solid (12.6 g, 88% crude yield, 92.3% *ee* by GC analysis of the Mosher ester). After two recrystallizations from light petroleum (b.p. 40–60 °C) at −20 °C, a white solid, (2*S*)-*trans*-3-pentyloxiranemethanol, is obtained (10.5 g, 73% yield, >98% *ee* by GC analysis of the Mosher ester); m.p. 38–39.5 °C.

For alternative workup procedures in other cases, see ref. 317. The best work-up procedure depends on the type of allylic alcohol. The epoxidation of low molecular weight allylic alcohols is facilitated by *in situ* derivatization [317].

3.9 ACKNOWLEDGEMENTS

I wholeheartedly thank my co-workers whose names are listed in the references. Financial support by the Deutsche Forschungsgemeinschaft (Sonderforschungsbereich 260 at the University of Marburg and Leibniz-Program), the Max-Planck-Institut für Kohlenforschung and the Fonds der Chemischen Industrie is gratefully acknowledged. Thanks are due to Mrs E. Schlosser (Lausanne) for drawing the formulae.

3.10 REFERENCES AND NOTES

1. (a) H. Sinn, W. Kaminsky, *Adv. Organomet. Chem.* **18** (1980) 99; (b) P. Pino, R. Mülhaupt, *Angew. Chem.* **92** (1980) 869; *Angew. Chem., Int. Ed. Engl.* **19** (1980) 857.
2. (a) A. Rieche, H. Gross, E. Höft, *Chem. Ber.* **93** (1960) 88; (b) N. M. Cullinane, D. M. Leyshon, *J. Chem. Soc.* (1954) 2942.
3. (a) H. M. Walborsky, L. Barash, T. C. Davis, *Tetrahedron* **19** (1963) 2333; (b) W. Oppolzer, *Angew. Chem.* **96** (1984) 840; *Angew. Chem., Int. Ed. Engl.* **23** (1984) 876; (c) T. Poll, J. O. Metter, G. Helmchen, *Angew. Chem.* **97** (1985) 116; *Angew. Chem., Int. Ed. Engl.* **24** (1985), 112; (d) T. K. Hollis, N. P. Robinson, B. Bosnich, *J. Am. Chem. Soc.* **114** (1992), 5464.

4. (a) D. Seebach, E. Hungerbühler, R. Naef, P. Schnurrenberger, B. Weidmann, M. Züger, *Synthesis* (1982) 138; (b) H. Rehwinkel, W. Steglich, *Synthesis* (1982), 826; (c) R. Imwinkelried, M. Schiess, D. Seebach, *Org. Synth.* **65** (1987) 230.
5. R. H. Grubbs, W. Tumas, *Science* **243** (1989), 907.
6. (a) *Gmelin Handbuch, Titan-Organische Verbindungen*, Part 1 (1977), Part 2 (1980), Part 3 (1984), Part 4 (1984), Springer, Berlin; (b) M. Bottrill, P. D. Gavens, J. W. Kelland, J. McMeeking, in *Comprehensive Organometallic Chemistry*, ed. G. Wilkinson, F. G. A. Stone, E. W. Abel, Chap. 22, Pergamon Press, Oxford, 1982; (c) K. H. Thiele, *Pure Appl. Chem.* **30** (1972), 575.
7. A. Segnitz, in *Houben–Weyl–Müller, Methoden der Organischen Chemie*, Vol. 13/7, p. 261, Thieme, Stuttgart, 1975.
8. M. T. Reetz, *Organotitanium Reagents in Organic Synthesis*, Springer, Berlin, 1986.
9. (a) J. C. Stowell, *Carbanions in Organic Synthesis*, Wiley, New York, 1979; (b) G. Boche, *Angew. Chem.* **101** (1989) 286; *Angew. Chem., Int. Ed. Engl.* **28** (1989) 277; (c) L. Brandsma, H. Verkruijsse, *Preparative Polar Organometallic Chemistry 1*, Springer, Berlin, 1987; L. Brandsma, *Preparative Polar Organometallic Chemistry 2*, Springer, Berlin, 1990.
10. M. S. Kharasch, J. H. Cooper, *J. Org. Chem.* **10** (1945) 46.
11. M. T. Reetz, J. Westermann, R. Steinbach, B. Wenderoth, R. Peter, R. Ostarek, S. Maus, *Chem. Ber.* **118** (1985) 1421.
12. D. Seebach, B. Weidmann, L. Widler, in *Modern Synthetic Methods 1983*, ed. R. Scheffold, p. 217, Salle/Sauerländer, Aarau, 1983.
13. M. T. Reetz, S. H. Kyung, M. Hüllmann, *Tetrahedron* **42** (1986) 2931.
14. M. T. Reetz, *Top. Curr. Chem.* **106** (1982) 1.
15. B. Weidmann, D. Seebach, *Angew. Chem.* **95** (1983) 12; *Angew. Chem., Int. Ed. Engl.* **22** (1983) 31.
16. R. Imwinkelried, D. Seebach, *Org. Synth.* **67** (1988) 180.
17. M. T. Reetz, J. Westermann, R. Steinbach, *Angew. Chem.* **92** (1980) 931; *Angew. Chem., Int. Ed. Engl.* **19** (1980) 900.
18. M. T. Reetz, J. Westermann, R. Steinbach, *Angew. Chem.* **92** (1980) 933; *Angew. Chem., Int. Ed. Engl.* **19** (1980) 901.
19. M. T. Reetz, R. Steinbach, J. Westermann, R. Peter, *Angew. Chem.* **92** (1980) 1044; *Angew. Chem., Int. Ed. Engl.* **19** (1980) 1011.
20. M. T. Reetz, B. Wenderoth, R. Peter, R. Steinbach, J. Westermann, *J. Chem. Soc., Chem. Commun.* (1980) 1202.
21. (a) B. Weidmann, D. Seebach, *Helv. Chim. Acta* **63** (1980) 2451; (b) B. Weidmann, L. Widler, A. G. Olivero, C. D. Maycock, D. Seebach, *Helv. Chim. Acta* **64** (1981) 357.
22. M. T. Reetz, R. Steinbach, B. Wenderoth, J. Westermann, *Chem. Ind. (London)* (1981) 541.
23. E. Benzing, W. Kornicker, *Chem. Ber.* **94** (1961) 2263.
24. (a) M. T. Reetz, R. Urz, T. Schuster, *Synthesis* (1983) 540; (b) D. Steinborn, I. Wagner, R. Taube, *Synthesis* (1989) 304.
25. A. M. Cardoso, R. J. H. Clark, S. Moorhouse, *J. Chem. Soc., Dalton Trans.* (1980) 1156.
26. (a) M. T. Reetz, S. Maus, *Tetrahedron* **43** (1987) 101; (b) M. T. Reetz, H. Hugel, K. Dresely, *Tetrahedron* **43** (1987) 109.
27. B. Weidmann, C. D. Maycock, D. Seebach, *Helv. Chim. Acta* **64** (1981) 1552.
28. M. T. Reetz, R. Steinbach, J. Westermann, R. Urz, B. Wenderoth, R. Peter, *Angew. Chem.* **94** (1982) 133; *Angew. Chem., Int. Ed. Engl.* **21** (1982) 135; *Angew. Chem., Suppl.* (1982) 257.
29. Review of Zr reagents: E. Negishi, T. Takahashi, *Synthesis* (1988) 1.

30. T. Kauffmann, C. Pahde, D. Wingbermühle, *Tetrahedron Lett.* **26** (1985) 4059.
31. (a) T. Imamoto, N. Takiyama, K. Nakamura, T. Hatajima, Y. Kamiya, *J. Am. Chem. Soc.* **111** (1989) 4392; (b) T. Kauffmann, C. Pahde, A. Tannert, D. Wingbermühle, *Tetrahedron Lett.* **26** (1985) 4063; (c) T. Imamoto, in *Comprehensive Organic Synthesis*, ed. B. M. Trost, I. Fleming, Vol. I, p. 231, Pergamon Press, Oxford, 1991.
32. (a) Y. Okude, S. Hirano, T. Hiyama, H. Nozaki, *J. Am. Chem. Soc.* **99** (1977) 3179; (b) T. Kauffmann, R. Abeln, D. Wingbermühle, *Angew. Chem.* **96** (1984) 724; *Angew. Chem., Int. Ed. Engl.* **23** (1984) 729.
33. For an informative comparison of Ti, Zr, Hf, Cr, Sm, Ce, Mn, Nb, Mo and other organometallics, see T. Kauffmann, in *Organometallics in Organic Synthesis 2*, Springer, Berlin, 1989; see also T. Kauffmann, H. Kieper, H. Pieper, *Chem. Ber.* **125** (1992) 899.
34. (a) G. Cahiez, B. Laboue, *Tetrahedron Lett.* **30** (1989) 3545; (b) J. F. Normant, G. Cahiez, *Modern Synthetic Methods* (R. Scheffold, ed.), vol. 3 (1983), p. 173, Salle & Sauerländer, Francfort and Aarau.
35. (a) T. Kauffmann, B. Laarmann, D. Menges, K. U. Voss, D. Wingbermühle, *Tetrahedron Lett.* **31** (1990) 507; (b) T. Kauffmann, B. Laarmann, D. Menges, G. Neiteler, *Chem. Ber.* **125** (1992) 163.
36. (a) G. A. Molander, E. R. Burkhardt, P. Weinig, *J. Org. Chem.* **55** (1990) 4990; (b) K. Utimoto, A. Nakamura, S. Matsubara, *J. Am. Chem. Soc.* **112** (1990) 8189; (c) review of lanthanide reagents: H. Kagan, J. L. Namy, *Tetrahedron* **42** (1986) 6573.
37. H. G. Raubenheimer, D. Seebach, *Chimia* **40** (1986) 12.
38. (a) M. T. Reetz, R. Steinbach, B. Wenderoth, *Synth. Commun.* **11** (1981) 261; (b) M. Yoshioka, T. Kawakita, M. Ohno, *Tetrahedron Lett.* **30** (1989) 1657.
39. M. T. Reetz, J. Westermann, S. H. Kyung, *Chem. Ber.* **118** (1985) 1050.
40. M. T. Reetz, J. Westermann, *Synth. Commun.* **11** (1981) 647.
41. K. Nützel, in *Houben–Weyl–Müller*, *Methoden der Organischen Chemie*, 4th edn, Vol. 13/2, p. 573, Thieme, Stuttgart, 1973.
42. Reviews on chelation and non-chelation controlled additions to chiral alkoxy aldehydes and ketones: (a) M. T. Reetz, *Angew. Chem.* **96** (1984) 542; *Angew. Chem., Int. Ed. Engl.* **23** (1984) 556; (b) M. T. Reetz, *Acc. Chem. Res.*, in press; (c) V. Jonas, G. Frenking, M. T. Reetz, *Organometallics* **12** (1993), 2110.
43. M. T. Reetz, S. H. Kyung, *Chem. Ber.* **120** (1987) 123.
44. (a) M. T. Reetz *et al.*, unpublished results; (b) M. T. Reetz, S. Stanchev, H. Haning, *Tetrahedron* **48** (1992), 6813.
45. M. T. Reetz, T. Wünsch, *J. Chem. Soc., Chem. Commun.* (1990) 1562.
46. K. L. Yu, S. Handa, R. Tsang, B. Fraser-Reid, *Tetrahedron* **47** (1991) 189.
47. M. T. Reetz, *S. Afr. J. Chem.* **42** (1989) 49.
48. (a) D. F. Herman, W. K. Nelson, *J. Am. Chem. Soc.* **75** (1953) 3882; (b) M. D. Rausch, H. B. Gordon, *J. Organomet. Chem.* **74** (1974) 85.
49. L. Widler, D. Seebach, *Helv. Chim. Acta* **65** (1982) 1085.
50. M. T. Reetz, R. Steinbach, J. Westermann, R. Peter, B. Wenderoth, *Chem. Ber.* **118** (1985) 1441.
51. R. E. Dolle, K. C. Nicolaou, *J. Am. Chem. Soc.* **107** (1985) 1691.
52. M. T. Reetz, B. Wenderoth, *Tetrahedron Lett.* **23** (1982) 5259.
53. (a) F. Rise, K. Undheim, *J. Organomet. Chem.* **291** (1985) 139; (b) K. Undheim, T. Benneche, *Heterocycles* **30** (1990) 1155; (c) F. Rise, K. Undheim, *J. Chem. Soc., Perkin Trans. 1* (1985) 1997; (d) L. L. Gundersen, F. Rise, K. Undheim, *Tetrahedron* **48** (1992) 5647.

54. N. Krause, D. Seebach, *Chem. Ber.* **120** (1987) 1845.
55. R. Mahrwald, H. Schick, L. L. Vasileva, K. K. Pivnitsky, G. Weber, S. Schwarz, *J. Prakt. Chem.* **332** (1990) 169.
56. (a) K. Mead, T. L. Macdonald, *J. Org. Chem.* **50** (1985) 422; (b) see also H. Schick, J. Spanig, R. Mahrwald, M. Bohle, T. Reiher, K. K. Pivnitsky, *Tetrahedron* **48** (1992) 5579.
57. R. K. Boeckman, K. J. O'Connor, *Tetrahedron Lett.* **30** (1989) 3271.
58. B. Weidmann, L. Widler, A. G. Olivero, C. D. Maycock, D. Seebach, *Helv. Chim. Acta* **64** (1981) 357.
59. T. Kauffmann, E. Antfang, B. Ennen, N. Klas, *Tetrahedron Lett.* **23** (1982) 2301.
60. T. Kauffmann, P. Schwartze, *Chem. Ber.* **119** (1986) 2150.
61. T. Kauffmann, R. Fobker, M. Wensing, *Angew. Chem.* **100** (1988) 1005; *Angew. Chem., Int. Ed. Engl.* **27** (1988) 943.
62. (a) A. N. Kasatkin, R. K. Biktimirov, G. A. Tolstikov, L. M. Khalilov, *Zh. Org. Khim.* **26** (1990) 1191; (b) see also T. Kauffmann, H. Kieper, *Chem. Ber.* **125** (1992) 907.
63. R. Metternich, *Dissertation*, Universität Marburg, 1985.
64. M. T. Reetz, R. Peter, *Tetrahedron Lett.* **22** (1981) 4691.
65. M. T. Reetz, R. Steinbach, K. Keßeler, *Angew. Chem.* **94** (1982) 872; *Angew. Chem., Int. Ed. Engl.* **21** (1982) 864; *Angew. Chem., Suppl.* (1982) 1899.
66. H. Haarmann, W. Eberbach, *Tetrahedron Lett.* **32** (1991) 903.
67. (a) T. Aono, M. Hesse, *Helv. Chim. Acta* **67** (1984) 1448; (b) K. Kostova, M. Hesse, *Helv. Chim. Acta* **67** (1984) 1713.
68. M. T. Reetz, M. Hüllmann, *J. Chem. Soc., Chem. Commun.* (1986) 1600.
69. (a) C. Siegel, E. R. Thornton, *J. Am. Chem. Soc.* **111** (1989) 5722; (b) reactions of a chiral α-benzoyl titanium enolate: A. Choudhury, E. R. Thornton, *Tetrahedron* **48** (1992) 5701.
70. (a) G. D. Cooper, H. L. Finkbeiner, *J. Org. Chem.* **27** (1962) 1493; (b) B. Fell, F. Asinger, R. A. Sulzbach, *Chem. Ber.* **103** (1970) 3830.
71. Review of titanium-catalyzed hydromagnesiation of olefins and acetylenes: F. Sato, *J. Organomet. Chem.* **285** (1985) 53.
72. F. Sato, H. Ishikawa, M. Sato, *Tetrahedron Lett.* **22** (1981) 85.
73. Review of enantioselective additions of dialkylzinc reagents to aldehydes catalyzed by chiral *β*-amino alcohols: R. Noyori, M. Kitamura, *Angew. Chem.* **103** (1991) 34; *Angew. Chem., Int. Ed. Engl.* **30** (1991) 49.
74. P. Knochel, J. F. Normant, *Tetrahedron Lett.* **27** (1986) 4431.
75. Preliminary results using CH_3CO_2 ligands appear promising [44].
76. (a) H. J. Gais, J. Vollhardt, H. J. Lindner, H. Paulus, *Angew. Chem.* **100** (1988) 1598; *Angew. Chem., Int. Ed. Engl.* **27** (1988) 1540; (b) T. Mukaiyama, K. Suzuki, T. Yamada, F. Tabusa, *Tetrahedron* **46** (1990) 265.
77. M. C. P. Yeh, P. Knochel, L. E. Santa, *Tetrahedron Lett.* **29** (1988) 3887.
78. M. C. P. Yeh, P. Knochel, *Tetrahedron Lett.* **29** (1988) 2395.
79. (a) Y. Ikeda, K. Furuta, N. Meguriya, N. Ikeda, H. Yamamoto, *J. Am. Chem. Soc.* **104** (1982) 7663; (b) K. Furuta, Y. Ikeda, N. Meguriya, N. Ikeda, H. Yamamoto, *Bull. Chem. Soc. Jpn.* **57** (1984) 2781.
80. (a) H. Bürger, H. J. Neese, *Chimia* **24** (1970) 209; (b) H. Bürger, H. J. Neese, *J. Organomet. Chem.* **36** (1976) 101.
81. M. Schiess, D. Seebach, *Helv. Chim. Acta* **65** (1982) 2598.
82. (a) H. C. Brown, R. B. Johannesen, *J. Am. Chem. Soc.* **75** (1953) 16; (b) K. Maruoka, S. Nagahara, H. Yamamoto, *Tetrahedron Lett.* **31** (1990) 5475.

83. (a) X. Chen, E. R. Hortelano, E. Eliel, *J. Am. Chem. Soc.* **112** (1990) 6130; (b) X. Chen, E. R. Hortelano, E. L. Eliel, S. V. Frye, *J. Am. Chem. Soc.* **114** (1992) 1778.
84. T. Kauffmann, T. Möller, H. Rennefeld, S. Welke, R. Wieschollek, *Angew. Chem.* **97** (1985) 351; *Angew. Chem., Int. Ed. Engl.* **24** (1985) 348.
85. G. Das, E. R. Thornton, *J. Am. Chem. Soc.* **112** (1990) 5360.
86. D. Roulet, J. Capéros, A. Jacot- Guillarmod, *Helv. Chim. Acta* **67** (1984) 1475.
87. (a) Y. Yamamoto, T. Komatsu, K. Maruyama, *J. Chem. Soc., Chem. Commun.* (1985) 814; (b) T. Fujisawa, . Hayakawa, M. Shimizu, *Tetrahedron Lett.* **33** (1992) 7903.
88. (a) E. Nakamura, I. Kuwajima, *Tetrahedron Lett.* **24** (1983) 3343; (b) H. Heimbach, *Diplomarbeit* Universität Bonn (1980).
89. G. E. Keck, D. E. Abbott, E. P. Boden, E. J. Enholm, *Tetrahedron Lett.* **25** (1984) 3927.
90. T. Krämer, J. R. Schwark, D. Hoppe, *Tetrahedron Lett.* **30** (1989) 7037.
91. (a) W. R. Baker, *J. Org. Chem.* **50** (1985) 3942; (b) Y. Yamamoto, J. Yamada, *J. Am. Chem. Soc.* **109** (1987) 4395; (c) J. Yamada, H. Abe, Y. Yamamoto, *J. Am. Chem. Soc.* **112** (1990) 6118; (d) T. Furuta, Y. Yamamoto, *J. Org. Chem.* **57** (1992) 2981; (e) Y. Yamamoto, J. Yamada, T. Asano, *Tetrahedron* **48** (1992) 5587.
92. Reviews of $TiCl_4$-mediated aldol additions of enolsilanes: (a) T. Mukaiyama, *Org. React.* **28** (1982) 203; (b) C. Gennari, in *Comprehensive Organic Synthesis*, ed. B. M. Trost, Pergamon Press, Oxford, 1990; (c) review of enolate additions to imines: D. J. Hart, D. C. Ha, *Chem. Rev.* **89** (1989) 1447; (d) mechanism of chelation-controlled Mukaiyama aldol addition: M. T. Reetz, B. Raguse, C. F. Marth, H. M. Hügel, T. Bach, D. N. A. Fox, *Tetrahedron* **48** (1992) 5731. Mukaiyama type aldol additions under catalytic conditions: Reetz, D. N. A. Fox, *Tetrahedron Lett.* **34** (1993), 1119, and references cited therein.
93. E. Nakamura, I. Kuwajima, *J. Am. Chem. Soc.* **105** (1983) 651.
94. E. Nakamura, H. Oshino, I. Kuwajima, *J. Am. Chem. Soc.* **108** (1986) 3745.
95. (a) H. U. Reissig, H. Holzinger, G. Glomsda, *Tetrahedron* **45** (1989) 3139; (b) S. Kano, T. Yokomatsu, S. Shibuya, *Tetrahedron Lett.* **32** (1991) 233; (c) T. Sato, M. Watanabe, E. Murayama, *Tetrahedron Lett.* **27** (1986) 1621.
96. B. Crociani, M. Nicolini, R. L. Richards, *J. Organomet. Chem.* **101** (1975) C1.
97. D. Seebach, G. Adam, T. Gees, M. Schiess, W. Weigand, *Chem. Ber.* **121** (1988) 507.
98. (a) W. Lehnert, *Tetrahedron* **28** (1972) 663; (b) W. Lehnert, *Tetrahedron* **29** (1973) 635; (c) W. Lehnert, *Tetrahedron* **30** (1974) 301; (d) T. Mukaiyama, *Pure Appl. Chem.* **54** (1982) 2455.
99. (a) D. A. Evans, F. Urpi, T. C. Somers, J. S. Clark, M. T. Bilodeau, *J. Am. Chem. Soc.* **112** (1990) 8215; (b) D. A. Evans, M. T. Bilodeau, T. C. Somers, J. Clardy, D. Cherry, Y. Kato, *J. Org. Chem.* **56** (1991), 5750.
100. (a) C. R. Harrison, *Tetrahedron Lett.* **28** (1987) 4135; (b) S. J. Brocchini, M. Eberle, R. G. Lawton, *J. Am. Chem. Soc.* **110** (1988) 5211.
101. Y. Tanabe, T. Mukaiyama, *Chem. Lett.* (1986) 1813.
102. R. Mahrwald, H. Schick, *Synthesis* (1990) 592.
103. L. Vuitel, A. Jacot-Guillarmod, *Synthesis* (1972) 608.
104. S. Kobayashi, S. Matsui, T. Mukaiyama, *Chem. Lett.* (1988) 1491.
105. (a) A. Hosomi, H. Sakurai, *Tetrahedron Lett.* **17** (1976) 1295; (b) H. Sakurai, *Pure Appl. Chem.* **54** (1982) 1; (c) T. Hayashi, K. Kabeta, I. Hamachi, M. Kumada, *Tetrahedron Lett.* **24** (1983) 2865; (d) mechanistic study of allylsilane additions: S. E. Denmark, N. G. Almstead, *Tetrahedron* **48** (1992) 5565.
106. (a) S. R. Wilson, M. F. Price, *Tetrahedron Lett.* **24** (1983) 569; (b) S. E. Denmark, E. J. Weber, *J. Am. Chem. Soc.* **106** (1984) 7970.

107. J. R. Hwu, J. M. Wetzel, *J. Org. Chem.* **50** (1985) 3946.
108. M. T. Reetz, B. Wenderoth, R. Peter, *J. Chem. Soc., Chem. Commun.* (1983) 407.
109. T. Okazoe, J. Hibino, K. Takai, H. Nozaki, *Tetrahedron Lett.* **26** (1985) 5581.
110. H. Weingarten, W. A. White, *J. Org. Chem.* **31** (1966) 4041.
111. T. Kauffmann, T. Abel, M. Schreer, *Angew. Chem.* **100** (1988) 1006; *Angew. Chem., Int. Ed. Engl.* **27** (1988) 944.
112. (a) D. J. Cram, F. A. Abd Elhafez, *J. Am. Chem. Soc.* **74** (1952) 5828; (b) J. D. Morrison, H. S. Mosher, *Asymmetric Organic Reactions*, Prentice-Hall, Englewood Cliffs, NJ, 1971.
113. N. T. Anh, *Top. Curr. Chem.* **88** (1980) 145.
114. Review of homo-aldol additions: D. Hoppe, *Angew. Chem.* **96** (1984) 930; *Angew. Chem., Int. Ed. Engl.* **23** (1984) 932.
115. D. M. Piatak, J. Wicha, *Chem. Rev.* **78** (1978) 199.
116. Y. Yamamoto, S. Nishii, K. Maruyama, *J. Chem. Soc., Chem. Commun.* (1986) 102.
117. (a) E. L. Eliel, in *Asymmetric Synthesis*, ed. J. D. Morrison, Vol. 2, Academic Press, New York, 1983; (b) W. C. Still, J. A. Schneider, *Tetrahedron Lett.* **21** (1980) 1035.
118. W. C. Still, J. H. McDonald, *Tetrahedron Lett.* **21** (1980) 1031.
119. (a) M. T. Reetz, K. Keßeler, S. Schmidtberger, B. Wenderoth, R. Steinbach, *Angew. Chem.* **95** (1983) 1007; *Angew. Chem., Int. Ed. Engl.* **22** (1983) 989; *Angew. Chem., Suppl.* (1983) 1511; (b) M. T. Reetz, K. Harms, W. Reif, *Tetrahedron Lett.* **29** (1988) 5881.
120. (a) CH_3TiCl_3 chelate of a chiral α-alkoxy ketone: M. T. Reetz, M. Hüllmann, T. Seitz, *Angew. Chem.* **99** (1987) 478; *Angew. Chem., Int. Ed. Engl.* **26** (1987) 477; (b) CH_3TiCl_3 chelate of aldehyde **92** (R = CH_3): M. T. Reetz, B. Raguse, T. Seitz, *Tetrahedron* **49** (1993), 8561.
121. M. T. Reetz, K. Keßeler, A. Jung, *Angew. Chem.* **97** (1985) 989; *Angew. Chem., Int. Ed. Engl.* **24** (1985) 989.
122. M. T. Reetz, K. Keßeler, A. Jung, *Tetrahedron* **40** (1984) 4327.
123. M. T. Reetz, K. Keßeler, A. Jung, *Tetrahedron Lett.* **25** (1984) 729.
124. G. E. Keck, D. E. Abbott, *Tetrahedron Lett.* **25** (1984) 1883.
125. (a) M. T. Reetz, in *Selectivities in Lewis Acid Promoted Reactions*, ed. D. Schinzer, NATO ASI Series C, vol. 289, p. 107, Kluwer, Dordrecht, 1989; (b) M. T. Reetz, *Pure Appl. Chem.* **57** (1985) 1781.
126. (a) S. Kiyooka, C. H. Heathcock, *Tetrahedron Lett.* **24** (1983) 4765; (b) concerning cautionary remarks on this paper, see ref. 123.
127. M. T. Reetz, A. Jung, *J. Am. Chem. Soc.* **105** (1983) 4833.
128. M. T. Reetz, K. Kesseler, *J. Org. Chem.* **50** (1985) 5434.
129. G. E. Keck, S. Castellino, *J. Am. Chem. Soc.* **108** (1986) 3847.
130. (a) M. T. Reetz, A. Jung, C. Bolm, *Tetrahedron* **44** (1988) 3889; (b) G. A. Molander, S. W. Andrews, *Tetrahedron* **44** (1988) 3869.
131. S. J. Danishefsky, W. H. Pearson, D. F. Harvey, C. J. Maring, J. P. Springer, *J. Am. Chem. Soc.* **107** (1985) 1256.
132. J. Uenishi, H. Tomozane, M. Yamato, *Tetrahedron Lett.* **26** (1985) 3467.
133. (a) C. Gennari, A. Bernardi, G. Poli, C. Scolastico, *Tetrahedron Lett.* **26** (1985) 2373; (b) F. Shirai, T. Nakai, *Chem. Lett.* (1989) 445; (c) H. Angert, T. Kunz, H.-U. Reissig, *Tetrahedron* **48** (1992) 5681.
134. A. Bernardi, S. Cardani, C. Gennari, G. Poli, C. Scolastico, *Tetrahedron Lett.* **26** (1985) 6509.
135. R. L. Danheiser, D. J. Carini, D. M. Fink, A. Basak, *Tetrahedron* **39** (1983) 935.

136. M. T. Reetz, M. W. Drewes, A. Schmitz, *Angew. Chem.* **99** (1987) 1186; *Angew. Chem., Int. Ed. Engl.* **26** (1987) 1141.
137. (a) J. V. N. Vara Prasad, D. H. Rich, *Tetrahedron Lett.* **31** (1990) 1803; (b) Y. Takemoto, T. Matsumoto, Y. Ito, S. Terashima, *Tetrahedron Lett.* **31** (1990) 217.
138. K. Mikami, M. Kaneko, T. P. Loh, M. Terada, T. Nakai, *Tetrahedron Lett.* **31** (1990) 3909.
139. (a) R. Annunziata, M. Cinquini, F. Cozzi, P. G. Cozzi, *Tetrahedron Lett.* **33** (1992) 1113; (b) R. Annunziata, M. Cinquini, F. Cozzi, P. G. Cozzi, E. Consolandi, *Tetrahedron* **47** (1991) 7897; (c) M. Cinquini, F. Cozzi, P. G. Cozzi, E. Consolandi, *Tetrahedron* **47** (1991) 8767; (d) R. Annunziata, M. Cinquini, F. Cozzi, P. G. Cozzi, *J. Org. Chem.* **57** (1992) 4155.
140. S. Wang, G. P. Howe, R. S. Mahal, G. Procter, *Tetrahedron Lett.* **33** (1992) 3351.
141. M. T. Reetz, K. Kesseler, *J. Chem. Soc., Chem. Commun.* (1984) 1079.
142. S. F. Martin, W. Li, *J. Org. Chem.* **54** (1989) 6129.
143. For a definition of protective 'group tuning,' see M. T. Reetz, J. Binder, *Tetrahedron Lett.* **30** (1989) 5425.
144. (a) L. E. Overman, R. J. McCready, *Tetrahedron Lett.* **23** (1982) 2355; (b) T. Nakata, T. Tanaka, T. Oishi, *Tetrahedron Lett.* **24** (1983) 2653.
145. J. Jurczak, A. Golebiowski, *Chem. Rev.* **89** (1989) 149.
146. D. J. Kempf, *J. Org. Chem.* **51** (1986) 3921.
147. M. W. Drewes, *Dissertation*, Universität Marburg, 1988.
148. (a) M. T. Reetz, *Pure Appl. Chem.* **60** (1988) 1607; (b) Review of reactions of *N,N*-dibenzylaminoaldehydes, -aldimines and α,β-unsaturated esters: M. T. Reetz, *Angew. Chem.* **103** (1991) 1559; *Angew. Chem., Int. Ed. Engl.* **30** (1991) 1531.
149. S. Masamune, W. Choy, J. S. Petersen, L. R. Sita, *Angew. Chem.* **97** (1985) 1; *Angew. Chem., Int. Ed. Engl.* **24** (1985) 1.
150. M. T. Reetz, E. Rivadeneira, C. Niemeyer, *Tetrahedron Lett.* **31** (1990) 3863.
151. M. Grauert, U. Schöllkopf, *Liebigs Ann. Chem.* (1985) 1817.
152. D. Hoppe, T. Krämer, J. R. Schwark, O. Zschage, *Pure Appl. Chem.* **62** (1990) 1999.
153. R. O. Duthaler, A. Hafner, M. Riediker, *Pure Appl. Chem.* **62** (1990) 631.
154. R. Bloch, L. Gilbert, *Tetrahedron Lett.* **28** (1987) 423.
155. K. Tomooka, K. Matsuzawa, K. Suzuki, G. Tsuchihashi, *Tetrahedron Lett.* **28** (1987) 6339.
156. M. T. Reetz, A. Schmitz, X. Holdgrün, *Tetrahedron Lett.* **30** (1989) 5421.
157. (a) T. Sato, R. Kato, K. Gokyu, T. Fujisawa, *Tetrahedron Lett.* **29** (1988) 3955; (b) T. Fujisawa, Y. Ukaji, M. Funabora, M. Yamashita, T. Sato, *Bull. Chem. Soc. Jpn.* **63** (1990) 1894.
158. T. Fujisawa, A. Fujimura, Y. Ukaji, *Chem. Lett.* (1988) 1541.
159. (a) G. Casiraghi, M. Cornia, G. Casnati, G. Gasparri Fava, M. Ferrari Belicchi, L. Zetta, *J. Chem. Soc., Chem. Commun.* (1987) 794; (b) F. Bigi, G. Casnati, G. Sartori, G. Araldi, G. Bocelli, *Tetrahedron Lett.* **30** (1989) 1121; (c) G. Casiraghi, M. Cornia, G. Rassu, C. Del Sante, P. Spanu, *Tetrahedron* **48** (1992) 5619.
160. (a) R. W. Hoffmann, *Angew. Chem.* **94** (1982) 569; *Angew. Chem., Int. Ed. Engl.* **21** (1982) 555; (b) M. Schlosser, *Pure Appl. Chem.* **60** (1988) 1627.
161. D. Seebach, L. Widler, *Helv. Chim. Acta* **65** (1982) 1972.
162. F. Sato, K. Iida, S. Iijima, H. Moriya, M. Sato, *J. Chem. Soc., Chem. Commun.* (1981) 1140.
163. Y. Kobayashi, K. Umeyama, F. Sato, *J. Chem. Soc., Chem. Commun.* (1984) 621.
164. M. T. Reetz, M. Sauerwald, *J. Org. Chem.* **49** (1984) 2292.
165. T. Cohen, B. S. Guo, *Tetrahedron* **42** (1986) 2803.
166. D. J. S. Tsai, D. S. Matteson, *Tetrahedron Lett.* **22** (1981) 2751.

167. F. Sato, Y. Suzuki, M. Sato, *Tetrahedron Lett.* **23** (1982) 4589.
168. T. Krämer, D. Hoppe, *Tetrahedron Lett.* **28** (1987) 5149.
169. (a) J. Ukai, Y. Ikeda, N. Ikeda, H. Yamamoto, *Tetrahedron Lett.* **25** (1984) 5173; (b) Y. Ikeda, J. Ukai, N. Ikeda, H. Yamamoto, *Tetrahedron Lett.* **25** (1984) 5177.
170. (a) A. Murai, A. Abiko, N. Shimada, T. Masamune, *Tetrahedron Lett.* **25** (1984) 4951; (b) A. Murai, A. Abiko, T. Masamune, *Tetrahedron Lett.* **25** (1984) 4955.
171. J. Ukai, Y. Ikeda, N. Ikeda, H. Yamamoto, *Tetrahedron Lett.* **24** (1983) 4029.
172. (a) E. van Hülsen, D. Hoppe, *Tetrahedron Lett.* **26** (1985) 411; (b) R. Hanko, D. Hoppe, *Angew. Chem.* **94** (1982) 378; *Angew. Chem., Int. Ed. Engl.* **21** (1982) 372; *Angew. Chem., Suppl.* (1982) 961.
173. (a) K. Furuta, M. Ishiguro, R. Haruta, N. Ikeda, H. Yamamoto, *Bull. Chem. Soc. Jpn.* **57** (1984) 2768; (b) H. Hiraoka, K. Furuta, N. Ikeda, H. Yamamoto, *Bull. Chem. Soc. Jpn.* **57** (1984) 2777; (c) Y. Yamamoto, W. Ito, K. Maruyama, *J. Chem. Soc., Chem. Commun.* (1984) 1004.
174. R. W. Hoffmann, J. Lanz, R. Metternich, G. Tarara, D. Hoppe, *Angew. Chem.* **99** (1987) 1196; *Angew. Chem., Int. Ed. Engl.* **26** (1987) 1145.
175. (a) D. A. Evans, J. V. Nelson, T. R. Taber, *Top. Stereochem.* **13** (1982) 1; (b) C. H. Heathcock, in *Asymmetric Synthesis*, ed. J. D. Morrison, Vol. 3, p. 111, Academic Press, Orlando, 1984.
176. (a) T. Harada, T. Mukaiyama, *Chem. Lett.* (1982) 467; (b) T. Fujisawa, Y. Ukaji, T. Noro, K. Date, M. Shimizu, *Tetrahedron* **48** (1992) 5629.
177. K. Banno, T. Mukaiyama, *Chem. Lett.* (1976) 279.
178. C. H. Heathcock, S. K. Davidsen, K. T. Hug, L. A. Flippin, *J. Org. Chem.* **51** (1986) 3027.
179. (a) C. Gennari, F. Molinari, P. Cozzi, A. Oliva, *Tetrahedron Lett.* **30** (1989) 5163; (b) T. H. Chan, T. Aida, P. W. K. Lau, V. Gorys, D. N. Harpp, *Tetrahedron Lett.* **20** (1979) 4029; (c) I. Matsuda, Y. Izumi, *Tetrahedron Lett.* **22** (1981) 1805.
180. C. Gennari, M. G. Beretta, A. Bernardi, G. Moro, C. Scolastico, R. Todeschini, *Tetrahedron* **42** (1986) 893.
181. D. A. Evans, D. L. Rieger, M. T. Bilodeau, F. Urpi, *J. Am. Chem. Soc.* **113** (1991) 1047; see also Y. Xiang, E. Olivier, N. Ouimet, *Tetrahedron Lett.* **33** (1992) 457.
182. M. T. Reetz, M. v. Itzstein, *J. Organomet. Chem.* **334** (1987) 85.
183. M. T. Reetz, R. Peter, M. v. Itzstein, *Chem. Ber.* **120** (1987) 121.
184. M. T. Reetz, D. Röhrig, *Angew. Chem.* **101** (1989) 1732; *Angew. Chem., Int. Ed. Engl.* **28** (1989) 1706.
185. J. d'Angelo, F. Pecquet-Dumas, *Tetrahedron Lett.* **24** (1983) 1403.
186. D. Hoppe, in *Enzymes as Catalysts in Organic Synthesis*, ed. M. P. Schneider, p. 177, Reidel, Dordrecht, 1986.
187. S. Masamune, M. Hirama, S. Mori, S. A. Ali, D. S. Garvey, *J. Am. Chem. Soc.* **103** (1981) 1568.
188. (a) M. Nerz-Stormes, E. R. Thornton, *Tetrahedron Lett.* **27** (1986) 897; (b) Camphor-derived Ti enolate: M. P. Bonner, E. R. Thornton, *J. Am. Chem. Soc.* **113** (1991) 1299.
189. W. Oppolzer, J. Blagg, I. Rodriguez, E. Walter, *J. Am. Chem. Soc.* **112** (1990) 2767.
190. G. Helmchen, U. Leikauf, I. Taufer-Knöpfel, *Angew. Chem.* **97** (1985) 874; *Angew. Chem., Int. Ed. Engl.* **24** (1985) 874 (erratum 4057).
191. W. Oppolzer, *Tetrahedron* **43** (1987) 1969.
192. Review of chiral 'acetate synthons:' M. Braun, *Angew. Chem.* **99** (1987) 24; *Angew. Chem., Int. Ed. Engl.* **26** (1987) 24.
193. W. Oppolzer, C. Starkemann, *Tetrahedron Lett.* **33** (1992) 2439.

194. (a) C. Gennari, A. Bernardi, L. Colombo, C. Scolastico, *J. Am. Chem. Soc.* **107** (1985) 5812; (b) C. Gennari, L. Colombo, G. Bertolini, G. Schimperna, *J. Org. Chem.* **52** (1987) 2754; (c) S. Cardani, C. De Toma, C. Gennari, C. Scolastico, *Tetrahedron* **48** (1992) 5557.
195. C. Gennari, G. Schimperna, I. Venturini, *Tetrahedron* **44** (1988) 4221.
196. U. Schöllkopf, *Pure Appl. Chem.* **55** (1983) 1799.
197. (a) U. Schöllkopf, J. Nozulak, M. Grauert, *Synthesis* (1985) 55; (b) G. Bold, T. Allmendinger, P. Herold, L. Moesch, J. P. Schär, R. O. Duthaler, *Helv. Chim. Acta* **75** (1991), 865.
198. (a) U. Schöllkopf, T. Beulshausen, *Liebigs Ann. Chem.* (1989) 223; (b) T. Beulshausen, U. Groth, U. Schöllkopf, *Liebigs Ann. Chem.* (1991) 1207.
199. H. Roder, G. Helmchem, E. M. Peters, K. Peters, H. G. v. Schnering, *Angew. Chem.* **96** (1984) 895; *Angew. Chem., Int. Ed. Engl.* **23** (1984) 898.
200. T. Krämer, D. Hoppe, *Tetrahedron Lett.* **28** (1987) 5149.
201. D. Hoppe, O. Zschage, in *Organic Synthesis via Organometallics*, ed. K. H. Dötz, R. W. Hoffmann, p. 267, Vieweg, Braunschweig, 1991.
202. (a) D. Hoppe, O. Zschage, *Angew. Chem.* **101** (1989) 67; *Angew. Chem., Int. Ed. Engl.* **28** (1989) 65; (b) O. Zschage, D. Hoppe, *Tetrahedron* **48** (1992) 5657; (c) H. Paulsen, D. Hoppe, *Tetrahedron* **48** (1992) 5667.
203. D. Hoppe, T. Krämer, C. Freire Erdbrügger, E. Egert, *Tetrahedron Lett.* **30** (1989) 1233, and references cited therein.
204. (a) D. Seebach, A. K. Beck, M. Schiess, L. Widler, A. Wonnacott, *Pure Appl. Chem.* **55** (1983) 1807; (b) D. Seebach, G. Adam, T. Gees, M. Schiess, W. Weigand, *Chem. Ber.* **121** (1988) 507; (c) J. T. Wang, X. Fan, X. Feng, Y. M. Qian, *Synthesis* (1989) 291.
205. M. T. Reetz, T. Kükenhöhner, P. Weinig, *Tetrahedron Lett.* **27** (1986) 5711.
206. (a) M. Riediker, R. O. Duthaler, *Angew. Chem.* **101** (1989) 488; *Angew. Chem., Int. Ed. Engl.* **28** (1989) 494; (b) K. Oertle, H. Beyeler, R. O. Duthaler, W. Lottenbach, M. Riediker, E. Steiner, *Helv. Chim. Acta* **73** (1990) 353; (c) R. O. Duthaler, P. Herold, S. Wyler-Helfer, M. Riediker, *Helv. Chim. Acta* **73** (1990) 659; (d) H. Minamikawa, S. Hayakawa, T. Yamada, N. Iwasawa, K. Narasaka, *Bull. Chem. Soc. Jpn.* **61** (1988) 4379; (e) A. Hafner, R. O. Duthaler, R. Marti, G. Rihs, P. Rothe-Streit, F. Schwarzenbach, *J. Am. Chem. Soc.* **114** (1992) 2321.
207. R. O. Duthaler, A. Hafner, M. Riediker, in *Chem. Rev.* **92** (1992), 807.
208. M. T. Reetz, S. H. Kyung, J. Westermann, *Organometallics* **3** (1984) 1716.
209. (a) A. Hosomi, H. Sakurai, *J. Am. Chem. Soc.* **99** (1977) 1673; (b) H. Sakurai, A. Hosomi, J. Hayashi, *Org. Synth.* **62** (1984) 86; (c) T. Tokoroyama, L.-R. Pan, *Tetrahedron Lett.* **30** (1989) 197.
210. (a) K. Narasaka, K. Soai, Y. Aikawa, T. Mukaiyama, *Bull. Chem. Soc. Jpn.* **49** (1976) 779; (b) K. Narasaka, *Org. Synth.* **65** (1987) 12.
211. (a) G. Majetich, A. Casares, D. Chapman, M. Behnke, *J. Org. Chem.* **51** (1986) 1745.
212. C. H. Heathcock, S. Kiyooka, T. A. Blumenkopf, *J. Org. Chem.* **49** (1984) 4214; correction: *J. Org. Chem.* **51** (1986) 3252.
213. L.-R. Pan, T. Tokoroyama, *Tetrahedron Lett.* **33** (1992) 1469.
214. M. J. Wu, C.-C. Wu, P.-C. Lee, *Tetrahedron Lett.* **33** (1992) 2547.
215. D. El-Abed, A. Jellal, M. Santelli, *Tetrahedron Lett.* **25** (1984) 1463.
216. K. Nickisch, H. Laurent, *Tetrahedron Lett.* **29** (1988) 1533.
217. (a) Review of intramolecular conjugate additions of allylsilanes: D. Schinzer, *Synthesis* (1988) 263; (b) G. Majetich, M. Behnke, K. Hull, *J. Org. Chem.* **50** (1985) 3615; (c) D.

Schinzer, C. Allagiannis, S. Wichmann, *Tetrahedron* **44** (1988) 3851; (d) G. Majetich, J. Defauw, *Tetrahedron* **44** (1988) 3833.
218. (a) R. L. Danheiser, D. J. Carini, D. M. Fink, A. Basak, *Tetrahedron* **39** (1983) 935; (b) R. L. Danheiser, D. M. Fink, Y. M. Tsai, *Org. Synth.* **66** (1987) 8.
219. J. Haruta, K. Nishi, S. Matsuda, Y. Tamura, Y. Kita, *J. Chem. Soc., Chem. Commun.* (1989) 1065.
220. (a) K. Narasaka, K. Soai, Y. Aikawa, T. Mukaiyama, *Bull. Soc. Chem. Jpn.* **49** (1976) 779; (b) T. Mukaiyama, R. Hara, *Chem. Lett.* (1989) 1171.
221. T. Sato, Y. Wakahara, J. Otera, H. Nozaki, S. Fukuzumi, *J. Am. Chem. Soc.* **113** (1991) 4028.
222. C. H. Heathcock, M. H. Norman, D. E. Uehling, *J. Am. Chem. Soc.* **107** (1985) 2797.
223. C. H. Heathcock, D. E. Uehling, *J. Org. Chem.* **51** (1986) 279.
224. C. Gennari, L. Colombo, G. Bertolini, G. Schimperna, *J. Org. Chem.* **52** (1987) 2754.
225. D. A. Oare, C. H. Heathcock, *Top. Stereochem.* **19** (1989) 227.
226. M. Miyashita, T. Yanami, T. Kumazawa, A. Yoshikoshi, *J. Am. Chem. Soc.* **106** (1984) 2149.
227. C. Goasdoue, N. Goasdoue, M. Gaudemar, *Tetrahedron Lett.* **25** (1984) 537.
228. Y. Yamamoto, S. Nishii, K. Maruyama, *J. Chem. Soc., Chem. Commun.* (1985) 386.
229. D. L. Comins, J. D. Brown, *Tetrahedron Lett.* **25** (1984) 3297.
230. G. H. Posner, L. L. Frye, M. Hulce, *Tetrahedron* **40** (1984) 1401.
231. (a) U. Schöllkopf, W. Kühnle, E. Egert, M. Dyrbusch, *Angew Chem.* **99** (1987) 480; *Angew. Chem., Int. Ed. Engl.* **26** (1987) 480; (b) K. Busch, U. M. Groth, W. Kühnle, U. Schöllkopf, *Tetrahedron* **48** (1992) 5607; (c) U. Schöllkopf, D. Pettig, U. Busse, *Synthesis* (1986), 737.
232. A. Bernardi, P. Dotti, G. Poli, C. Scolastico, *Tetrahedron* **48** (1992) 5597.
233. A. Hosomi, T. Yanagi, M. Hojo, *Tetrahedron Lett.* **32** (1991) 2371.
234. D. A. Barr, R. Grigg, V. Sridharan, *Tetrahedron Lett.* **30** (1989) 4727.
235. M. Arai, B. H. Lipshutz, E. Nakamura, *Tetrahedron* **48** (1992) 5709.
236. (a) A. Maercker, *Org. React.* **14** (1965) 270; (b) M. Schlosser, B. Schaub, J. de Oliveira-Neto, S. Jeganathan, *Chimia* **40** (1986) 244.
237. (a) K. Takai, Y. Hotta, K. Oshima, H. Nozaki, *Bull. Chem. Soc. Jpn.* **53** (1980) 1698; (b) J. Hibino, T. Okazoe, K. Takai, H. Nozaki, *Tetrahedron Lett.* **26** (1985) 5579.
238. E. Minicone, A. J. Pearson, P. Bovicelli, M. Chandler, G. C. Heywood, *Tetrahedron Lett.* **22** (1981) 2929.
239. L. Lombardo, *Tetrahedron Lett.* **23** (1982) 4293.
240. L. Lombardo, *Org. Synth.* **65** (1987) 81.
241. (a) Y. Ogawa, M. Shibasaki, *Tetrahedron Lett.* **25** (1984) 1067; (b) R. L. Snowden, P. Sonnay, G. Ohloff, *Helv. Chim. Acta* **64** (1981) 25; (c) A. Kramer, H. Pfander, *Helv. Chim. Acta* **65** (1982) 293.
242. T. Okazoe, K. Takai, K. Oshima, K. Utimoto, *J. Org. Chem.* **52** (1987) 4410.
243. K. Takai, Y. Kataoka, T. Okazoe, K. Utimoto, *Tetrahedron Lett.* **29** (1988) 1065.
244. K. Takai, O. Fujimura, Y. Kataoka, K. Utimoto, *Tetrahedron Lett.* **30** (1989) 211.
245. T. Okazoe, K. Takai, K. Utimoto, *J. Am. Chem. Soc.* **109** (1987) 951.
246. (a) F. N. Tebbe, G. W. Parshall, G. S. Reddy, *J. Am. Chem. Soc.* **100** (1978) 3611; (b) K. A. Brown-Wensley, S. L. Buchwald, L. Cannizzo, L. Clawson, S. Ho, D. Meinhardt, J. R. Stille, D. Straus, R. H. Grubbs, *Pure Appl. Chem.* **55** (1983) 1733; (c) J. D. Meinhart, E. V. Anslyn, R. H. Grubbs, *Organometallics* **8** (1989) 583.
247. S. H. Pine, G. Kim, V. Lee, *Org. Synth.* **69** (1990) 72.

248. (a) R. E. Ireland, M. D. Varney, *J. Org. Chem.* **48** (1983) 1829; (b) C. S. Wilcox, G. W. Long, H. Suh, *Tetrahedron Lett.* **25** (1984) 395; (c) W. A. Kinney, M. J. Coghlan, L. A. Paquette, *J. Am. Chem. Soc.* **106** (1984) 6868.
249. N. A. Petasis, E. I. Bzowej, *J. Am. Chem. Soc.* **112** (1990) 6392.
250. J. J. Eisch, A. Piotrowski, *Tetrahedron Lett.* **24** (1983) 2043.
251. J. W. F. L. Seetz, G. Schat, O. S. Akkerman, F. Bickelhaupt, *Angew. Chem.* **95** (1983) 242; *Angew. Chem., Int. Ed. Engl.* **22** (1983) 248; *Angew. Chem., Suppl.* (1983) 234.
252. E. J. Ginsburg, C. B. Gorman, S. R. Marder, R. H. Grubbs, *J. Am. Chem. Soc.* **111** (1989) 7621.
253. J. R. Stille, R. H. Grubbs, *J. Am. Chem. Soc.* **105** (1983) 1664.
254. (a) S. L. Buchwald, R. H. Grubbs, *J. Am. Chem. Soc.* **105** (1983) 5490; (b) T. Yoshida, E. Negishi, *J. Am. Chem. Soc.* **103** (1981) 1276.
255. S. Tyrlik, I. Wolochowicz, *Bull. Soc. Chim. Fr.* (1973) 2147.
256. J. E. McMurry, M. P. Fleming, *J. Am. Chem. Soc.* **96** (1974) 4708.
257. T. Mukaiyama, T. Sato, J. Hanna, *Chem. Lett.* (1973) 1041.
258. (a) L. Aleandrei, S. Becke, B. B. Bogdanović, D. Jones, J. Rozière, *J. Organomet. Chem.*, in press; (b) A. Fürstner, H. Weidmann, *Synthesis* (1987) 1071; (c) A. Fürstner, D. N. Jumbam, H. Weidmann, *Tetrahedron Lett.* **32** (1991) 6695; (d) A. Fürstner, *Angew. Chem.* **105** (1993), 171, *Angew. Chem. Int. Ed. Engl.* **32** (1993), 164.
259. Reviews of the application of low-valent titanium reagents: (a) D. Lenoir, *Synthesis* (1989) 883; (b) J. E. McMurry, *Chem. Rev.* **89** (1989) 1513; (c) H. N. C. Wong, *Acc. Chem. Res.* **22** (1989) 145.
260. J. E. McMurry, T. Lectka, J. G. Rico, *J. Org. Chem.* **54** (1989) 3748.
261. New applications of hydromagnesiation: T. Ito, I. Yamakawa, S. Okamoto, Y. Kobayshi, F. Sato, *Tetrahedron Lett.* **32** (1991) 371.
262. (a) T. Hirao, N. Yamada, Y. Oshshiro, T. Agawa, *Chem. Lett.* (1982) 1997; (b) F. Sato, Y. Tanaka, M. Sato, *J. Chem. Soc., Chem. Commun.* (1983) 165; (c) K. Yamamoto, T. Kimura, Y. Tomo, *Tetrahedron Lett.* **25** (1984) 2155; (d) F. Sato, Y. Kobayashi, *Org. Synth.* **69** (1990) 106.
263. T. Ito, I. Yamakawa, S. Okamoto, Y. Kobayashi, F. Sato, *Tetrahedron Lett.* **32** (1991) 371.
264. E.-i. Negishi, *Acc. Chem. Res.* **20** (1987) 65.
265. C. L. Rand, D. E. Van Horn, M. W. Moore, E. Negishi, *J. Org. Chem.* **46** (1981) 4093.
266. (a) H. E. Tweedy, R. A. Coleman, D. W. Thompson, *J. Organomet. Chem.* **129** (1977) 69; (b) T. J. Zitzelberger, M. D. Schiavelli, D. W. Thompson, *J. Org. Chem.* **48** (1983) 4781.
267. M. T. Reetz, J. Westermann, R. Steinbach, *J. Chem. Soc., Chem. Commun.* (1981) 237.
268. (a) T. Sasaki, A. Usuki, M. Ohno, *J. Org. Chem.* **45** (1980) 3559; (b) I. Fleming, I. Paterson, *Synthesis* (1979) 446.
269. M. Uemura, K. Isobe, Y. Hayashi, *Tetrahedron Lett.* **26** (1985) 767.
270. M. T. Reetz, T. Seitz, *Angew. Chem.* **99** (1987) 1081; *Angew. Chem., Int. Ed. Engl.* **26** (1987) 1028.
271. G. H. Posner, T. P. Kogan, *J. Chem. Soc., Chem. Commun.* (1983) 1481.
272. Reviews of syntheses of compounds having quaternary carbon atoms: (a) S. F. Martin, *Tetrahedron* **36** (1980) 419; (b) C. Rüchardt, H. D. Beckhaus, *Angew. Chem.* **92** (1980) 417; *Angew. Chem., Int. Ed. Engl.* **19** (1980) 429.
273. M. T. Reetz, J. Westermann, *J. Org. Chem.* **48** (1983) 254.
274. M. T. Reetz, R. Steinbach, B. Wenderoth, *Synth. Commun.* **11** (1981) 261.
275. H. Ishikawa, T. Mukaiyama, S. Ikeda, *Bull. Chem. Soc. Jpn.* **54** (1981) 776.

276. S. D. Lindell, J. D. Elliott, W. S. Johnson, *Tetrahedron Lett.* **25** (1984) 3947.
277. W. S. Johnson, P. H. Crackett, J. D. Elliott, J. J. Jagodzinski, S. D. Lindell, S. Natarajan, *Tetrahedron Lett.* **25** (1984) 3951.
278. R. W. Hoffmann, *Chem. Rev.* **89** (1989) 1841 (specifically p. 1855).
279. P. A. Bartlett, W. S. Johnson, J. D. Elliott, *J. Am. Chem. Soc.* **105** (1983) 2088.
280. W. S. Johnson, R. Elliott, J. D. Elliott, *J. Am. Chem. Soc.* **105** (1983) 2904.
281. A. Mori, K. Maruoka, H. Yamamoto, *Tetrahedron Lett.* **25** (1984) 4421.
282. Reviews of Lewis acid-induced substitution reactions of acetals: (a) A. Alexakis, P. Mangeney, *Tetrahedron: Asymmetry* **1** (1990) 477; (b) T. Mukaiyama, M. Murakami, *Synthesis* (1987) 1043; (c) see also refs 285 and 294b.
283. (a) D. Seebach, G. Stucky, E. Pfammatter, *Chem. Ber.* **122** (1989) 2377; (b) H. H. Mooiweer, H. Hiemstra, H. P. Fortgens, W. N. Speckamp, *Tetrahedron Lett.* **28** (1987) 3285.
284. (a) D. Seebach, C. Betschart, M. Schiess, *Helv. Chim. Acta* **67** (1984) 1593; improvements: (b) H. Takahashi, T. Tsubuki, K. Higashiyama, *Synthesis* (1988) 238.
285. Review of Lewis acid-mediated α-alkylation of carbonyl compounds: M. T. Reetz, *Angew. Chem.* **94** (1982) 97; *Angew. Chem., Int. Ed. Engl.* **21** (1982) 96.
286. M. T. Reetz, *Nachr. Chem. Tech. Lab.* **29** (1981) 165.
287. M. T. Reetz, W. F. Maier, *Angew. Chem.* **90** (1978) 50; *Angew. Chem., Int. Ed. Engl.* **17** (1978) 48.
288. (a) M. T. Reetz, W. F. Maier, H. Heimbach, A. Giannis, G. Anastassious, *Chem. Ber.* **113** (1980) 3734; (b) M. T. Reetz, W. F. Maier, J. Chatziiosifidis, A. Giannis, H. Heimbach, U. Löwe, *Chem. Ber.* **113** (1980) 3741.
289. M. T. Reetz, I. Chatziiosifidis, F. Hübner, H. Heimbach, *Org. Synth.* **62** (1984) 95.
290. Review of enolsilane chemistry: P. Brownbridge, *Synthesis* (1983) 1 and 85.
291. T. H. Chan, I. Paterson, J. Pinsonnault, *Tetrahedron Lett.* **18** (1977) 4183.
292. M. T. Reetz, M. Sauerwald, P. Walz, *Tetrahedron Lett.* **22** (1981) 1101.
293. O. Kitagawa, T. Inoue, T. Taguchi, *Tetrahedron Lett.* **33** (1992) 2167.
294. Review: T. Mukaiyama, *Angew. Chem.* **89** (1977) 858; *Angew. Chem., Int. Ed. Engl.* **16** (1977) 817.
295. I. Fleming, T. V. Lee, *Tetrahedron Lett.* **22** (1981) 705, and references cited therein.
296. G. S. Cockerill, P. Kocienski, R. Treadgold, *J. Chem. Soc., Perkin Trans. 1* (1985) 2101.
297. (a) A. Bernardi, S. Cardani, O. Carugo, L. Colombo, C. Scolastico, R. Villa, *Tetrahedron Lett.* **31** (1990) 2779; (b) C. Palazzi, G. Poli, C. Scolastico, R. Villa, *Tetrahedron Lett.* **31** (1990) 4223; (c) A. Pasquarello, G. Poli, D. Potenza, C. Scolastico, *Tetrahedron: Asymmetry* **1** (1990) 429.
298. (a) K. Conde-Friebold, D. Hoppe, *Synlett.* (1990) 99; (b) T. Basile, L. Longobardo, E. Tagliavini, C. Trombini, A. Umani-Ronchi, *J. Chem. Soc., Chem. Commun.* (1990), 759.
299. T. Tanaka, T. Inoue, K. Kamei, K. Murakami, C. Iwata, *J. Chem. Soc., Chem. Commun.* (1990) 906.
300. (a) W. A. Nugent, T. V. RajanBabu, *J. Am. Chem. Soc.* **110** (1988) 8561; (b) S. C. Berk, R. B. Grossman, S. L. Buchwald, *J. Am. Chem. Soc.* **115** (1993), 4912.
301. Certain Ti enolates undergo Pd-catalyzed allylation using allylacetate: refs 8, 44.
302. A. N. Kasatkin, A. N. Kulak, G. A. Tolstikov, *Bull. Acad. Sci. USSR (Div. Chem.)* **35** (1986) 871.
303. J. J. Barber, C. Willis, G. M. Whitesides, *J. Org. Chem.* **44** (1979) 3603.
304. M. Schlosser, K. Fujita, *Angew. Chem.* **94** (1982) 320; *Angew. Chem. Int. Ed. Engl.* **21** 1982), 309; *Angew. Chem., Suppl.* (1982) 646.

305. E. Moret, M. Schlosser, *Tetrahedron Lett.* **26** (1985) 4423.
306. M. T. Reetz, B. Wenderoth, R. Urz, *Chem. Ber.* **118** (1985) 348.
307. H. A. Staab, M. Lüking, F. H. Dürr, *Chem. Ber.* **95** (1962) 1275.
308. (a) T. Hayashi, M. Kumada, in *Asymmetric Synthesis*, ed. J. D. Morrison, Vol. 5, p. 147, Academic Press, Orlando, 1985; (b) K. Narasaka, *Synthesis* (1991) 1.
309. N. Iwasawa, Y. Hayashi, H. Sakurai, K. Narasaka, *Chem. Lett.* (1989) 1581.
310. (a) K. Narasaka, N. Iwasawa, M. Inoue, T. Yamada, M. Nakashima, J. Sugimori, *J. Am. Chem. Soc.* **111** (1989) 5340; (b) K. Narasaka, I. Yamamoto, *Tetrahedron* **48** (1992) 5743.
311. (a) K. Furuta, S. Shimizu, Y. Miwa, H. Yamamoto, *J. Org. Chem.* **54** (1989) 1481; (b) E. J. Corey, T.-P. Loh, *J. Am. Chem. Soc.* **113** (1991) 8966.
312. (a) Y. Hayashi, K. Narasaka, *Chem. Lett.* (1989) 793; (b) Y. Hayashi, K. Narasaka, *Chem. Lett.* (1990) 1295; (c) Y. Ichikawa, A. Narita, A. Shiozawa, Y. Hayashi, K. Narasaka, *J. Chem. Soc., Chem. Commun.* (1989) 1919.
313. T. A. Engler, M. A. Letavic, J. P. Reddy, *J. Am. Chem. Soc.* **113** (1991) 5068.
314. (a) M. Terada, K. Mikami, T. Nakai, *Tetrahedron Lett.* **32** (1991) 935; (b) K. Mikami, M. Terada, Y. Motoyama, T. Nakai, *Tetrahedron: Asymmetry* **2** (1991) 643.
315. M. T. Reetz, S.-h. Kyung, C. Bolm, T. Zierke, *Chem. Ind. (London)* (1986), 824.
316. (a) K. Mikami, M. Terada, T. Nakai, *J. Am. Chem. Soc.* **112** (1990) 3949; (b) K. Mikami, M. Terada, *Tetrahedron* **48** (1992) 5671; (c) F. T. van der Meer, B. L. Feringa, *Tetrahedron Lett.* **33** (1992), 6695.
317. Y. Gao, R. M. Hanson, J. M. Klunder, S. Y. Ko, H. Masamune, K. B. Sharpless, *J. Am. Chem. Soc.* **109** (1987) 5765.
318. P. N. Devine, T. Oh, *Tetrahedron Lett.* **32** (1991) 883.
319. T. Mukaiyama, A. Inubushi, S. Suda, R. Hara, S. Kobayashi, *Chem. Lett.* (1990) 1015.
320. (a) K. Furuta, T. Maruyama, H. Yamamoto, *J. Am. Chem. Soc.* **113** (1991) 1041; (b) S. Kobayashi, Y. Fujishita, T. Mukaiyama, *Chem. Lett.* (1990) 1455.
321. (a) H. Takahashi, T. Kawakita, M. Ohno, M. Yoshioka, S. Kobayashi, *Tetrahedron* **48** (1992) 5691; (b) M. J. Rozema, C. Eisenberg, H. Lütjens, R. Ostwald, K. Belyk, P. Knochel, *Tetrahedron Lett.* **34** (1993), 3115.
322. (a) B. Schmidt, D. Seebach, *Angew. Chem.* **103** (1991) 100; *Angew. Chem., Int. Ed. Engl.* **30** (1991) 99; (b) B. Schmidt, D. Seebach, *Angew. Chem.* **103** (1991) 1383; *Angew. Chem., Int. Ed. Engl.* **30** (1991) 1321; (c) J. L. v.d. Bussche-Hünnefeld, D. Seebach, *Tetrahedron* **48** (1992) 5719; (d) M. Hayashi, T. Matsuda, N. Oguni, *J. Chem. Soc., Chem. Commun.* (1990) 1364; (e) D. Seebach, D. A. Plattner, A. K. Beck, Y. U. Wang, D. Hunzicker, *Helv. Chim. Acta* **75** (1992), 2171; (f) R. L. Halterman, K. P. C. Vollhardt, M. E. Welker, *J. Am. Chem. Soc.* **109**)1987), 8105; (g) C. A. Willoughby, S. L. Buchwald, *J. Am. Chem. Soc.* **114** (1992), 7562; (h) S. L. Colletti, R. L. Halterman, *Tetrahedron Lett.* **33** (1992), 1005.
323. K. Ito, Y. Kimura, H. Okamura, T. Katsuki, *Synlett* (1992) 573.
324. T. Katsuki, K. B. Sharpless, *J. Am. Chem. Soc.* **102** (1980) 5974.
325. Reviews of the Sharpless epoxidation: (a) B. E. Rossiter, in *Asymmetric Synthesis*, ed. J. D. Morrison, Vol. 5, p. 193, Academic Press, Orlando, 1985; (b) M. G. Finn, K. B. Sharpless, in *Asymmetric Synthesis*, ed. J. D. Morrison, Vol. 5. p. 247, Academic Press, Orlando, 1985; (c) A. Pfenniger, *Synthesis* (1986) 89.
326. (a) S. S. Woodard, M. G. Finn, K. B. Sharpless, *J. Am. Chem. Soc.* **113** (1991) 106; (b) M. G. Finn, K. B. Sharpless, *J. Am. Chem. Soc.* **113** (1991) 113.
327. Review of kinetic resolution: H. B. Kagan, J. C. Fiaud, *Top. Stereochem.* **18** (1988) 249.
328. C. Dijkgraaf, J. P. G. Rousseau, *Spectrochim. Acta*, **24A** (1968) 1213.

4

Synthetic Procedures Involving Organocopper Reagents†

BRUCE H. LIPSHUTZ
University of California, Santa Barbara, CA, USA

† Dedicated to Professor Harry H. Wasserman on the occasion of his 70th birthday.

Organometallics in Synthesis—A Manual. Edited by M. Schlosser

4.1 INTRODUCTION

For many organic chemists, carrying out an organocopper reagent-mediated transformation presents special problems. Aside from the more obvious experimental rigors implicit in any sensitive organometallic scheme (i.e. care to exclude air and moisture), there are several issues which must be confronted, including:

- Which copper(I) precursor should be chosen and what level of purity is required?
- Although an ethereal solvent is the norm, which is best for a particular reaction type?
- With so many variations of copper reagents from which to choose, is there one that is likely to be best suited for the goal in mind?
- Under what circumstances should additives be relied upon to assist with a selected coupling reaction?
- When and how can the number of equivalents of potentially valuable organolithiums be minimized without sacrificing efficiency?

These and a host of other questions often come to the fore when contemplating the use of organocopper reagents. Hence, it might well be expected that such a need for decision making may encourage those unfamiliar with this area to seek other avenues for solutions to synthetic problems. It is perhaps a great tribute to copper, therefore, that it continues to serve as the most relied upon transition metal for effecting carbon–carbon bonds [1].

The explanations behind the intense usage of these reagents are not cryptic. Copper reagents, although utilized for several unrelated types of couplings, tend to react under very mild conditions and generally afford synthetically useful yields [2]. The reagents themselves, formally at least, are soft nucleophiles, but in most cases are actually electrophilically driven. That is, without the presence of gegenions capable of Lewis acidic character (e.g. Li^+, MgX^+), most couplings do not take place [3]. Such fundamental properties impart highly valued chemoselectivity patterns not witnessed with other organometallic species, especially those composed of far more ionic metals (e.g. RLi, RMgX, R_3Al, RZnX). Thus, few would consider introducing, in any direct way, an alkyl, vinylic or aromatic moiety in a Michael sense by means other than copper chemistry. The remarkable penchant for copper reagents to add to conjugated carbonyl groups in a 1,4-manner [2c], together with their relative inertness towards carbonyl 1,2-additions [4], combine to enhance their worth as selective agents. In substitution reactions, their low basicity encourages displacement over competing elimination [2d]. Although these two processes (i.e. 1,4-addition and subtraction) make up the lion's share of organocopper reagent usage, other characteristic reactions include carbocupration and metallocupration of an acetylene. Although 1,2-additions to enones are rare [4], non-conjugated ketones and especially aldehydes can also be appropriate reaction partners.

In the light of the comments above, this chapter strives to address many of the concerns which researchers less experienced in organocopper chemistry may harbor. With assistance from the information provided herein, it is hoped that the inertial barrier toward employing an organocopper reagent will be substantially lowered. Indeed, it could be argued that manipulation of copper reagents may well be less technically demanding than the successful preparation and use of a Grignard reagent, thus involving skills normally taken for granted beyond the very early stages of one's career in organic chemistry. It should be mentioned, however, that skill in handling such powerful synthetic weapons by no means implies an appreciation as to the mechanistic details surrounding their use, or an understanding of their composition and structure. It is only over the past 5–10 years of their 40-year history [5] that insight has begun to accrue concerning these otherwise black box phenomena. Therefore, although the emphasis here is mainly on how to utilize copper reagents to form carbon-carbon bonds, it is important for the synthetic practitioner to bear in mind that unanswered, complex physical organic questions envelop virtually every reaction in this expansive field.

4.2 ORGANOCOPPER COMPLEXES DERIVED FROM ORGANOLITHIUM REAGENTS

4.2.1 LOWER ORDER, GILMAN CUPRATES R_2CuLi

Historically, Gilman's orginal recipe for combining two equivalents of an organolithium with either CuI or CuSCN was envisioned to arrive at cuprate R_2CuLi (**1**), proceeding via an intermediate (presumed polymeric) organocopper $(RCu)_n$ [5]. Although it is now known that for X = I, Br and Cl equation 4.1 does hold, thiocyanate ligand loss does not occur but leads to a completely different reagent ($R_2Cu(SCN)Li_2$; see below [6]. Species **1** may consist of virtually any sp^3-, sp^2- or sp-based ligand originating from the corresponding organolithium precursor

$$2RLi + CuX \longrightarrow [(RCu)_n + RLi + LiX] \longrightarrow \underset{(\mathbf{1})}{R_2CuLi} + LiX \quad (4.1)$$

Generation of a homocuprate **1** (i.e. where both R groups are the same) is a very straightforward operation, as long as good-quality CuI, CuBr or $CuBr \cdot SMe_2$ is used. There are a number of methods for purifying these salts [7] and, of course, all copper(I) halides are readily available commercially. When using CuI, care must be exercised to protect it from light and excessive moisture. When pure, it is a readily flowing white powder. Tinges of yellow–orange to pink suggest traces of I_2 are present, although it is the accompanying product of disproportionation, Cu(II), that is the more serious impurity. Hence, whereas Soxhlet extraction will remove halogen, only recrystallization frees the $(CuI)_n$ from other materials. One procedure which works well is as follows.

Purification of $(CuI)_n$ [8]

A dark brown solution of CuI (Fisher, 13.167 g, 69.1 mmol), KI (Fisher, 135 g, 813 mmol) and 100 ml of water were treated with charcoal and filtered through a Celite pad. The yellow filtrate was then diluted with 300 ml of water, causing a fine gray–white precipitate to form. The resulting mixture was cooled in an ice–water bath and additional water was added portionwise to ensure complete precipitation. The precipitate was then suspended and filtered on a medium sintered-glass frit under a cone of purging dry N_2. The filtrant was triturated under purging dry N_2 successively with 4 × 100 ml portions of water, 4 × 80 ml of acetone and 4 × 80 ml of distilled diethyl ether. The remaining off-white solid was allowed to dry with suction on the frit under a purge of dry N_2 and transferred into a 50 ml round-bottomed flask, wrapped in foil and dried under vacuum ($\leqslant$0.20 mmHg) overnight. The evacuated flask was then heated at 90 °C for 4 h. Yield: 9.33 g (49.0 mmol, 70%) of grayish white powdery CuI.

As for CuBr, its use in the uncomplexed state is rare, and can be problematic. As the dimethyl sulfide complex, however, it is an excellent cuprate precursor [9]. Unfortunately, however, aged material tends to lose percentages of volatile Me_2S, and can easily lead to improper stoichiometries. That is, with losses of Me_2S, excess of CuBr will be present and the two equivalents of RLi added cannot fully form R_2CuLi [10]. Hence, relatively fresh bottles of $CuBr \cdot SMe_2$ should be employed and, once opened, they should be well wrapped with Parafilm and stored under a blanket of Ar. Alternatively, the $CuBr \cdot SMe_2$ complex can be prepared fresh following the standard procedure given below.

Purification of the $CuBr \cdot SMe_2$ Complex [9]

To 40.0 g (279 mmol) of pulverized CuBr (Fisher) were added 50 ml (42.4 g, 682 mmol) of Me_2S (Eastman, b.p. 36–38 °C). The resulting mixture, which warmed during dissolution, was stirred vigorously and then filtered through a glass-wool plug. The residual solid was stirred with an additional 30 ml (25 g, 409 mmol) of Me_2S to dissolve the bulk of the remaining solid and this mixture was filtered. The combined red solutions were diluted with 299 ml of hexane. The white crystals that separated were filtered with suction and washed with hexane until the washings were colorless. The residual solid was dried under N_2 to leave 51.6 g (90%) of the complex as white prisms that dissoved in an Et_2O–Me_2S mixture to give a colorless solution. For recrystallization, a solution of 1.02 g of the complex in 5 ml of Me_2S was slowly diluted with 20 ml of hexane to give 0.96 g of the pure complex as colorless prisms, mp 124–129 °C (decomp.). The complex is essentially insoluble in hexane, Et_2O, acetone, $CHCl_3$, CCl_4, MeOH, EtOH and H_2O. Although the complex does dissolve in DMF and in DMSO, the observations that heat is evolved and that the resulting solutions are green suggest that the complex has dissociated and that some oxidation (or disproportionation) to give Cu(II) species has occurred.

Copper(I) triflate (Fluka, $CF_3SO_3Cu \cdot 0.5C_6H_6$) is also a good source from which to generate lower order cuprates, the subsequent chemistry from which works well [11]. It is, however, a far more expensive and more sensitive precursor than the halides, and as such has not found widespread acceptance.

The other major component required in forming R_2CuLi is the organolithium. One of the surest pitfalls in all of cuprate chemistry, irrespective of reagent type, is a lack of knowledge concerning the quality and quantity of RLi being used to form the copper reagent. When relying on commercially available RLi, it is *never* good practice to assume that a molarity as given on a bottle is correct. Usually they are sufficiently off-specification, irrespective of direction (i.e. a high or low titre), to be detrimental. Thus, with a high titre, insufficient RLi will be introduced, leading to mixtures of copper species. On the other hand, with a low titre, excess of RLi (i.e. above the 2.0 equiv. needed) will inadvertently be present, which can have disastrous consequences.

There are several titration procedures which can be followed to insure that correct amounts of RLi are being combined with CuX [12]. One which is frequently called upon relies on complexation between lithium and 1,10-phenanthroline, which is brown–red. Neutralization with *s*-BuOH leads to a lemon-yellow end-point, from which the molarity of an RLi can be determined, as outlined for the titration of *n*-BuLi.

Titration of Organolithium Reagents: Normal Addition [13]

An over-dried 50 ml Erlenmeyer flask and stirring bar were cooled under an Ar stream. A few crystals of 1,10-phenanthroline (Fisher) were placed in the flask and dissolved in 15–20 ml of diethyl ether. The flask was then capped with a rubber septum, maintained under an Ar atmosphere and stirred at 0 °C. The organolithium was added dropwise to this colorless solution until a colored end-point was observed (e.g. for *n*-butyllithium a reddish brown color was noted). Exactly 1 ml of organolithium was then added. To this solution, *s*-BuOH was added dropwise until the colored end-point was obtained; for *n*-butyllithium red–brown → pale yellow). This method was then repeated, again adding 1 ml of organolithium (first drop again indicating end-point) followed by titration with *s*-BuOH. The molarity was determined from the following equations:

$$\text{Molarity (RLi)} = \frac{\text{mmol (RLi)}}{\text{ml (RLi) used}}$$

$$\text{mmol (RLi)} = \text{mmol } (s\text{-BuOH}) = \frac{\text{density } (s\text{-BuOH})}{\text{MW } (s\text{-BuOH})} \times \text{ml } (s\text{-BuOH})$$

$$= \frac{806 \text{ mg ml}^{-1}}{74.12 \text{ mg mmol}^{-1}} \times \text{ml } (s\text{-BuOH})$$

At least three consistent measurements should be made to insure accuracy. (For titration of *t*-BuLi, benzene was used in place of diethyl ether at room temperature.)

Still more accurate is the procedure above, but in reverse, i.e. addition of the RLi to the *s*-BuOH. The control realized by adding a *solution* of RLi to the alcohol, as opposed to introducing *neat* *s*-BuOH to a solution of RLi, accounts for the improvement.

Titration of Organolithium Reagents: Inverse Addition

Methyllithium and *n*-butyllithium were purchased as solutions, 1.4 M in Et_2O and 2.5 M in hexane, respectively, from Aldrich and were titrated as follows. An oven-dried 25 ml Erlenmeyer flask equipped with a magnetic stirring bar was cooled under Ar and charged with 10 ml of freshly distilled Et_2O and one crystal of 1,10-phenanthroline. At ambient temperature, 50 μl of 2-pentanol (distilled from CaH_2) were added via a 100 μl syringe to provide a clear, colorless solution. The solution of organolithium reagent in question was then added dropwise via a 0.5 ml syringe to the vigorously stirred alcoholic–Et_2O solution until an end-point was realized. This was evidenced by a persistent tan–light maroon solution. Beware that if too much base has been added a brown–maroon solution will result, indicating that the proper end-point has been surpassed. This protocol was repeated in the same Erlenmeyer flask until three trials were within 95% agreement of organometallic solution added. The molarity of organolithium reagent may be calculated following the equation given in the alternative procedure above.

With high-quality CuI or $CuBr \cdot SMe_2$ available and accurate control of the molarity of an RLi of interest, obtaining the corresponding Gilman cuprate **1** is relatively trivial. Before preparing an organocopper reagent of any type, however, there are a few points which must be considered: (1) *All* reactions involving organocopper complexes should be run under either an argon or nitrogen atmosphere with a slightly positive pressure applied, being vented to a bubbler. Argon, although more expensive than nitrogen, is strongly recommended for two major reasons: (a) it is heavier than air, and hence blankets any reactions which may have residual amounts of other lighter gases present; (b) its use permits (although not recommended) removal of stoppers for purposes of introducing solids to the reaction vessel; (2) Since essentially every organocopper reaction is sensitive to moisture to some extent, the educt to be added to the reagent must be dry. It is therefore good technique routinely to dry a substrate azeotropically with toluene at room temperature under a high vacuum prior to use; (3) Along similar lines of thought, and as is normally done prior to other organometallic chemistry, the glassware should either be stored in an oven at temperatures >120 °C or flame-dried just before use.

To prepare a lower order cuprate (R_2CuLi), then, the copper salt is slurried in an ethereal solvent and cooled to -78 °C, where two equivalents of the RLi are added. Warming to dissolution completes the process. Depending on the type of reaction to be performed and the specific R_2CuLi being utilized, the cuprate may be recooled to -78 °C prior to addition of the substrate. The cornerstone of modern cuprate chemistry, Gilman's Me_2CuLi, is prepared in this manner, the subsequent chemoselective 1,4-addition of which to, e.g. a bromo enone, is described below.

Preparation of Me_2CuLi; Chemoselective 1,4 Addition to a Primary Bromo Enone [14]

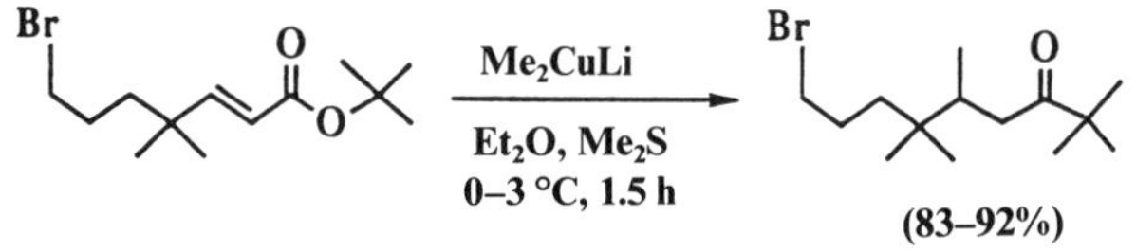

Me_2CuLi

Et_2O, Me_2S
0–3 °C, 1.5 h

(83–92%)

To a cold (0 °C) solution of Me_2CuLi, prepared from 365 mg (1.78 mmol) of $CuBr{\cdot}SMe_2$ in 8 ml of Et_2O and 5 ml of Me_2S at 0 °C to which had been added 3.56 mmol of MeLi (halide free) was added a solution of 295 mg (1.07 mmol) of the bromo enone in 5 ml of Et_2O. The resulting mixture, from which an orange precipitate separated, was stirred at 0–3 °C for 1.5 h and then siphoned into a cold aqueous solution (pH 8) of NH_3 and NH_4Cl. The ethereal extract of this mixture was dried and concentrated and the residual crude product (0.35 g of yellow liquid) was chromatographed on silica gel with Et_2O–hexane (1 : 39 v/v) as eluent. The bromo ketone was collected as 0.28 g (92%) of colorless liquid: n_D^{25} 1.4687; IR (CCl_4), 1710 cm^{-1} (C=O); ^1H NMR (CCl_4), δ 3.2–3.5 (m, 2H, CH_2Br), 2.2–2.4 (m, 2H, CH_2CO), 1.3–2.1 (m, 5H, aliphatic CH), 1.12 (s, 9H, *t*-Bu) and 0.7–1.0 (m, 9H, CH_3 including a CH_3 singlet at 0.85); mass spectrum, *m*/*z* (relative intensity, %) 292 (M^+, 4), 290 (M^+, 4), 276 (16), 274 (16), 234 (100), 232 (100), 127 (16), 83 (18), 69 (33), 57 (57), 55 (22), 43 (20), and 41 (29); ^{13}C NMR ($CDCl_3$, multiplicity in off-resonance decoupling), δ 213.7 (s), 44.0 (s), 38.8 (t), 38.5 (t), 35.2 (t), 34.5 (s), 27.2 (t), 26.2 (q, 5 C atoms), 24.4 (d) and 14.7 (q). Analysis: calculated for $C_{14}H_{27}BrO$, C 57.73, H 9.34, Br 27.43; found, C 57.97, H 9.39, Br 27.21%. From a comparable reaction in Et_2O at 0–5 °C for 2 h, the yield of the bromo ketone was 83%.

For lithium reagents which require prior generation by metal–halide (usually Li–I or Li–Br) or metal–metal (Li–Sn) exchange, warming the 2RLi + CuX mixture to effect dissolution may present complications owing to ligand isomerization and/or cuprate decomposition. To avoid this potential problem, the Cu(I) salt can be solubilized by admixture with any of several possible additives, e.g. $(MeO)_3P$ [15], LiX [16], Me_2S. This trivial modification is exemplified by the formation and ultimate addition of a (*Z*)-2-ethoxyethenyl group to cyclohexenone.

Conjugate Addition of a Divinylic Cuprate to Cyclohexenone [17]

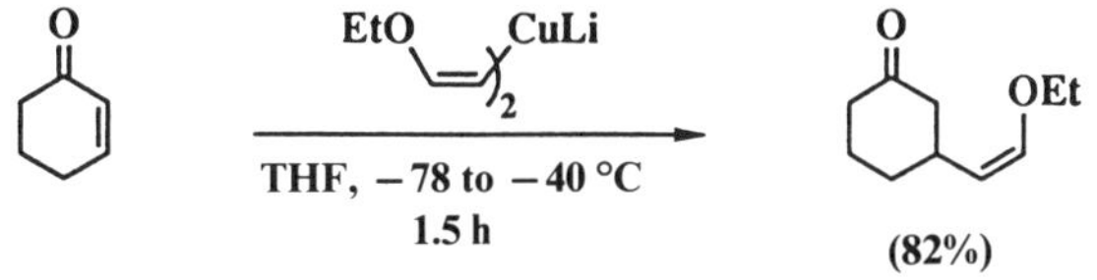

EtO CuLi

THF, −78 to −40 °C
1.5 h

(82%)

A solution of *cis*-2-ethoxyvinyllithium was prepared from 2.18 g (6.04 mmol) of *cis*-1-ethoxy-2-tri-*n*-butylstannylethylene and *n*-butyllithium (1.1 equiv.) in 15 ml of THF at −78 °C over 1 h. A solution of *cis*-2-ethoxyvinyllithium was prepared from 2.18 g (6.04 mmol) of *cis*-1-ethoxy-2-tri-*n*-butylstannylethylene and *n*-butyllithium (1.1 equiv.) in 15 ml of THF at −78 °C over 1 h. A solution of 0.577 g (3.03 mmol) of purified copper(I) iodide and 0.89 ml (12.1 mmol) of Me_2S in 5 ml of THF was then added over 5 min. After stirring for 1 h at −78 °C, 0.264 g (2.75 mmol) of cyclohexenone in 5 ml of THF were added over 10 min. After stirring for 1 h, the mixture was warmed to −40 °C over 30 min, quenched with aqueous 20% NH_4Cl solution and extracted (d, 1H, $J = 6$ Hz), 4.28 (dd, 1H, $J = 6, 9$ Hz), 3.78 (q, 2H, $J = 7$ Hz), 1.42–3.30 (br m, 9H), 1.22 (t, 3H, $J = 7$ Hz). Analysis: calculated for $C_{10}H_{16}O_2$, C 71.39, H 9.59; found, C 71.78, H 9.79%.

Conversion of the lithium enolate of acetone dimethylhydrazone (presumably) to the corresponding azaallyl cuprate occurs on addition of predissolved CuI in diisopropyl sulfide at −78 °C. Thus, 1,4-rather than 1,2-additon of the equivalent of acetone enolate to, e.g. methyl vinyl ketone, occurs in excellent yield.

Conjugate Addition of a Cuprate Derived from a Lithiated N,N-Dimethylhydrazone to Methyl Vinyl Ketone [18]

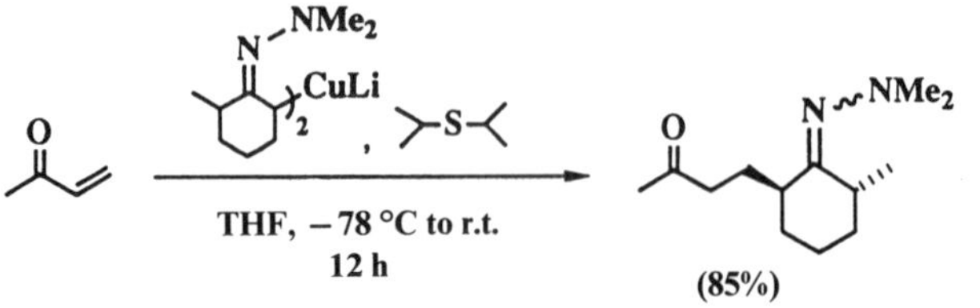

NMe$_2$ CuLi, THF, −78 °C to r.t. 12 h (85%)

A precooled (*ca* −30 °C) clear solution of 0.96 g (5 mmol) of copper(I) iodide in 2.88 ml (20 mmol) of diisopropyl sulfide and 10 ml of THF was added dropwise with stirring at −78 °C to a suspension of 6 lithio-2-methylcyclohexanone dimethylhydrazone [5 mmol, generated from 1.54 g (10 mmol) of 2-methylcyclohexanone dimethylhydrazone and lithium diisopropylamide (10 mmol)] in 40 ml of THF. The lithium compound dissolved during warming of the orange reaction mixture from −78 to −20 °C over 30 min and from −20 to 0 °C over 10 min, resulting in a clear, golden yellow solution. It was cooled again to −78 °C and 0.41 ml (5 mmol) of methyl vinyl ketone were added dropwise. After 2 h, the reaction mixture was slowly warmed to room temperature over a period of 12 h. The black–brown reaction mixture was poured into a solution of saturated NH_4Cl containing ammonia solution (pH 8) and repeatedly extracted with CH_2Cl_2. The organic phase was shaken several times with NH_4Cl–ammonia solution until the aqueous phase was no longer blue. The combined aqueous phase was again extracted with CH_2Cl_2 and the combined organic phases were then dried over sodium sulfate. After removal of the solvent by rotary evaporation *in vacuo*, the crude product (1.19 g, spectroscopic yield 100%) was purified by distillation to give 0.42 g (85%) of a light yellow oil, b.p. 100 °C (0.05 mmHg).

Displacement reactions with R_2CuLi are also extremely valuable, yet equally as straightforward to conduct. The most cooperative electrophiles are primary centers bearing iodide, bromide or tosylate. Epoxides of varying substitution patterns likewise couple fairly well, especially less hindered examples (e.g. monosubstituted oxiranes). Tosylates (in Et_2O) and iodides (in THF) are rated roughly comparable in terms of leaving group ability toward R_2CuLi [19] and relatively unreactive cuprates such as Ph_2CuLi can nonetheless be used to prepare functionalized aromatic systems, as illustrated below [20]. Differences in leaving group ability can also be used to advantage, as in the case of displacement in a chlorobromide by a dicyclopropyl cuprate [21].

Double Displacement of a Ditosylate with Ph_2CuLi [20]

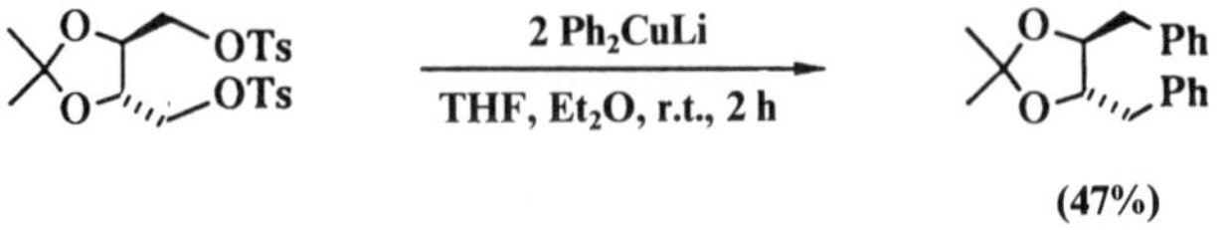

OTs, OTs; 2 Ph_2CuLi, THF, Et_2O, r.t., 2 h; Ph, Ph (47%)

To a solution of 3.0 g of copper(I) iodide in 10 ml of dry diethyl ether, stirred at 0 °C under dry Ar, was added dropwise 20 ml of 2.1 M phenyllithium as a solution in 75% benzene–25% hexane.

A solution of 1.93 g of 2,3-*O*-isopropylidene-L-threitol ditosylate in 12 ml of diethyl ether and 3 ml of THF was added dropwise to the resulting green solution and the mixture was stirred at 25 °C for 2 h. Saturated aqueous NH_4Cl was added and the volatile solvents were removed under reduced pressure. The aqueous residue was extracted with several portions of diethyl ether and the extracts were washed with saturated brine solution, dried and concentrated *in vacuo*. The yellow oily residue was chromatographed on 20 g of silica gel, eluting first with hexane to remove biphenyl, then with hexane–ethyl acetate (3:1) to elute the product. Distillation at 140 °C (0.1 mmHg) yielded 650 mg (47%) of the colorless product; IR (neat) 3080, 3060, 3010, 2940, 2880, 1620, 1500, 1460, 1380, 1370, 1240, 1215, 1160, 1075, 1050, 750, 695 cm^{-1}; ^1H NMR ($CDCl_3$), δ 1.4 (s, 6H), 2.8 (m, 4H), 4.0 (m, 2H), 7.25 (s, 10H).

Selective Displacement of a Bromochloride by a Cyclopropylcuprate [21]

Br–(CH₂)₄–Cl → (c-C₃H₅)₂CuLi, THF, −35 °C, 1.5 h → cyclopropyl–(CH₂)₄–Cl (90%)

A solution of 1.1 M cyclopropyllithium in diethyl ether (660 ml) was added over 45 min at −35 °C to a slurry of 73 g (0.38 mol) of copper(I) iodide in 660 ml of THF. After a Gilman test [22] (see below) was negative, 1-bromo-4-chlorobutane (54 g, 0.32 mol) was rapidly added to the mixture, which was held at −35 °C for 1.5 h. Aqueous saturated ammonium sulfate was then added and the mixture was filtered. The product was extracted with 2 l of diethyl ether-pentane (1:1). The organic layer was washed several times with water, then with brine. After drying over calcium sulfate, the extract was distilled through a 45 cm Vigreux column to remove solvents. The pot residue was then short-path distilled to yield 37.4 g (90%) of the product, b.p. 58–59 °C (17 mmHg); IR (neat) 3084, 3008, 2941, 2864, 1024 cm^{-1}; ^1H NMR (CCl_4), δ 3.48 (t, 2H, $J = 6$ Hz); mass spectrum (70 eV), m/z 55 (base).

Gilman Test for Free RLi (or RMgX) [22]

Michler's ketone (Me_2N–C₆H₄–CO–C₆H₄–NMe_2) → 1. RM; 2. H_2O; 3. I_2/HOAc → blue aqueous layer?

To an ethereal solution of the cuprate (*ca* 2.5 ml sample) is added an equal volume of a 1% solution of Michler's ketone in toluene and the mixture is allowed to warm to room temperature. Water (1.5 ml) is then added and the mixture stirred vigorously for 5 min. A 0.2% solution of I_2 in glacial acetic acid is then added dropwise. If a true blue solution results, this is indicative of a positive test (i.e. there is free RLi/RMgX in the cuprate solution). A solution which develops a yellow, tan or greenish coloration implies a negative test.

Ring opening of an epoxide occurs under conditions reflecting both the nature of the substrate and the relative reactivity of the cuprate. In general, although THF is an acceptable solvent, Et_2O is preferred as the Lewis acidity of lithium cations associated with R_2CuLi is maximized in this medium [23]. Mixed solvent systems, when necessary, are certainly acceptable. The compatibility of functional groups with soft R_2CuLi is often used to advantage.

Opening of a Cyclohexene Epoxide with Me_2CuLi [24]

Me_2CuLi; Et_2O, 20 °C, 18 h (85%)

To a solution of lithium dimethylcuprate [from 6.4 ml of 0.75 M methyllithium (4.8 mmol) and 490 mg (2.58 mmol) of copper(I) iodide] in diethyl ether under N_2 at 0 °C was added methyl *cis*-6-benzyloxy-*trans*-2,3-epoxy-1-methylcyclohexane-*rac*-1-carboxylate in diethyl ether and the mixture was stirred at 20 °C for 18 h. Addition of saturated aqueous NH_4Cl and extraction with diethyl ether gave the product as a colorless oil (125 mg, 85%); IR (film), 3560, 1720, 1270, 1060 cm^{-1}; ^1H NMR ($CDCl_3$), δ 1.05 (d, 3H, $J = 6$ Hz), 1.20 (s, 3H), 1.46–1.89 (m, 5H), 2.99 (d, 1H, $J = 2.9$ Hz), 3.64 (s, 3H), 3.87 (m, 1H), 3.98 (dd, 1H, $J = 2.9$), 10.7 Hz), 4.28 (d, 1H, $J = 11.7$ Hz), 4.52 (d, 1H, $J = 11.7$ Hz), 7.28 (m, 5H).

Another attractive feature of lower order cuprates R_2CuLi is their ability to induce coupling at an sp^2 carbon center bearing an appropriate leaving group. Although procedures exist which augment the action of R_2CuLi alone in this regard (see below and Section 4.2.1.3), there are many circumstances where additives are not required, such as in the cases of a vinyl bromide and vinyl triflate.

Substitution of a Vinylic Bromide Using Ph_2CuLi [25]

Ph_2CuLi; Et_2O, 0 °C, 5 h

(*E*:*Z* 4:1) (60%) (*Z*:*E* 4:1)

Lithium diphenylcuprate was prepared at 0 °C by slowly adding 25 ml of 1.86 M (46.5 mmol) phenyllithium solution to a suspension of 5.03 g (24.4 mmol) of copper(I) bromide–dimethyl sulfide complex in 20 ml of dry diethyl ether. A yellow precipitate formed initially, which changed to a homogeneous green solution after complete addition. After 40 min at 0 °C, a solution of 1.36 g (5.81 mmol) of 1-(1-bromo-2-deuterioethenyl)naphthalene ($E:Z = 4:1$) in 3 ml of dry diethyl ether were then added. After 4.5 h at 0 °C, the reaction mixture was poured into aqueous saturated NH_4Cl solution (pH 9 by addition of ammonia solution), and this was stirred for 1.5 h. The ether layer was separated, washed twice with brine and then dried. Removal of solvent afforded a light yellow oil, which was purified by short-path distillation, collecting the fraction with b.p. 124–134 °C (1 mmHg). The yield was 0.80 g (60%) of the product, which was crystallized from methanol, m.p. 57.5–58.5 °C; ^1H NMR ($CDCl_3$), δ 5.36 (s, 0.2H), 5.93 (s, 0.8H), 6.8–7.9) (m, 12H); the *Z*-isomer predominated (4:1).

Substitution of a Vinylic Trifluoromethanesulfonate (Triflate) with Me_2CuLi [26]

OSO_2CF_3

Me_2CuLi; THF, −15 °C, 12 h

(75%)

A solution of 2.0 M methyllithium in hexane (5.5 ml, 10.8 mmol) was added to a stirred slurry of copper(I) iodide (1.43 g, 7.5 mmol) in 15 ml of THF at 0 °C. A solution of 1-trifluoromethanesulfonyloxy-4-*tert*-butylcyclohexene in 5 ml of THF was added and the reaction mixture was stirred at −15 °C for 12 h. It was then diluted with hexane, filtered through a pad of Florisil and concentrated on a rotary evaporator *in vacuo*. Chromatography of the residue on silica gel provided the product (250 mg, 75%); ^{1}H NMR ($CDCl_3$), δ 5.38 (m, 1H), 1.87 (m, 4H), 1.63 (s, 3H), 1.25 (m, 3H), 0.84 (s, 9H).

Carbocupration of acetylenes with R_2CuLi (and also with magnesiocuprates; see below) is an excellent route to vinylic cuprates [2j, 27]. A 2:1 stoichiometry of the acetylene to R_2CuLi is required so as to arrive at the divinylic species. Once the intermediate reagent has been formed, subsequent introduction of a variety of electrophiles is possible and couplings follow the normal modes of reaction.

Carbocupration/Protio Quench of a 1-Alkyne Using n-Bu_2CuLi [2j]

OEt, OEt → 1. *n*-Bu_2CuLi, Et_2O, −55 °C; 2. NH_4Cl–H_2O → *n*-Bu, OEt, OEt (91%)

n-Butyllithium (50 mmol) was added to a suspension of copper(I) iodide (5.3 g, 28 mmol) in diethyl ether (50 ml) at −40 °C. The mixture was stirred at −35 to −25 °C for 30 min and to the resulting solution was added at −55 °C 2-propynal diethyl acetal (6.4 g, 50 mmol) in diethyl ether (30 ml). After stirring for 30 min at −40 °C, the vinylcuprate (ready for further use with $E^+ \neq H^+$, if desired), was hydrolyzed with saturated aqueous NH_4Cl solution (80 ml) admixed with a concentrated ammonia solution (20 ml). After filtration on Celite and separation of the layers, the organic phase was dried with potassium carbonate, the solvents were evaporated *in vacuo* and the crude product was distilled; yield, 8.4 g (91%); b.p. 96–97 °C (15 mmHg).

Some of the most useful carbocuprations involve acetylene gas (*ca* 2 equiv.) and give rise to bis-(*Z*)-alkenylcuprates, which can then go on to transfer both vinyl groups in many reactions. A generalized procedure is as follows.

Carbocupration of Acetylene Gas: Preparation of a Lithium bis-(Z)-alkenylcuprate [2j]

2 HC≡CH → R_2CuLi, Et_2O, −50 to −20 °C, 15 min → R, R, CuLi → E^+ → products

The organolithium reagent (50 mmol) prepared in diethyl ether (from the corresponding bromide and lithium) was added to a suspension of copper(I) iodide (5.35 g, 28 mmol) or copper(I) bromide–dimethyl sulfide (1:1 complex; 5.75 g, 28 mmol) in diethyl ether (100 ml) at −35 °C. The mixture was stirred for 20 min at −35 °C to effect dissolution. Acetylene, cleared from acetone (through a −78 °C trap) was measured in a water gasometer (50 mmol; 1.2 l) and bubbled into the reaction mixture after being dried over a column packed with calcium chloride. The temperature was allowed to rise from −50 °C (at the start) to −25 °C. Stirring of the pale green solution was maintained for 20 min at −25 °C, at which point the cuprate was ready for use.

Whereas acetylene carbocupration with R_2CuLi leads to (*Z*)-alkenylcuprates (see above), it is also possible in the presence of excess acetylene to effect a double carbocupration to (Z,Z)-dienylcuprates [28]. The first equivalent of acetylene reacts at -50 °C, conversion to the dienylcuprate requiring warming to 0 °C but no higher as decomposition ensues. A range of electrophiles (e.g. enones, aldehydes, CO_2, X_2, activated halides) can then be introduced affording the expected dienes in fair to good yields. Preparation of the navel orangeworm pheromone [hexadeca-(11Z,13Z)-dienal] is illustrative of the method.

Double Carbocupration of Acetylene: Synthesis of a Pheromone [29]

Et_2CuLi —(1. 2 H—≡—H, −50 °C; 2. 4 H—≡—H, −10 °C)→ $(Et\text{-}CH{=}CH\text{-}CH{=}CH)_2CuLi$ —(I(CH₂)₉CH(OEt)₂)→ Et-dienyl-(CH₂)₉CH(OEt)₂ —(H_3O^+)→ $(Et\text{-}CH{=}CH\text{-}CH{=}CH)(CH_2)_9CHO$ (33%)

A stock solution of ethyllithium was prepared by addition of bromoethane (2.61 ml, 25 mmol) to a suspension of finely cut lithium wire (1 g, 143 mmol) in hexane (30 ml) at -20 °C. After being stirred at this temperature for 1 h, the mixture was allowed to warm to room temperature during an additional 2 h. Titration gave the molarity as 0.85 M (73%). A portion of this ethyllithium solution (8 ml, 6.8 mmol) was added dropwise to a stirred suspension of $CuBr{\cdot}SMe_2$ (0.689 g, 3.36 mmol) in diethyl ether (25 ml) at -40 °C. A homogeneous blue–black solution of lithium diethylcuprate formed. After being stirred at -35 °C for 30 min, the solution was cooled to -50 °C and treated with acetylene (165 ml, 7.4 mmol). The mixture was allowed to warm to -25 °C for 30 min and then to -10 °C. More acetylene (300 ml, 13.4 mmol) was added during *ca* 10 min, while the temperature was maintained at -10 °C. Once addition was complete, the solution was cooled to -40°C and treated with the iodoacetal (1.0 g, 2.79 mmol) and HMPA (0.5 ml, 2.8 mmol). After being stirred at -40 to 0 °C for 3 h, the reaction was quenched with water (30 ml), diluted with brine (50 ml), and the product was extracted with diethyl ether (2 × 30 ml). The combined extracts were washed with brine (50 ml) and dried over $MgSO_4$. After chromatography on silica (CH_2Cl_2), the product acetal was treated with oxalic acid (2 g) in water (20 ml). THF was added until the mixture became homogeneous and the solution was stirred for 2 h at 60 °C. The product was extracted with light petroleum (3 × 100 ml) and the extract was dried over $MgSO_4$ and concentrated under reduced pressure. Rapid distillation from a Kugelrohr oven (125 °C, 0.3 mmHg) gave hexadeca-(11Z,13Z)-dienal (0.22 g, 33%); IR, 3040, 3000, 2720, 1730, 1600, 725 cm^{-1}; ^1H NMR (60 MHz), δ 1.00 (t, 3H, $J = 7.2$ Hz), 1.32 (m, 14H), 1.90–2.60 (m, 6H), 5.45 (m, 2H), 6.20 (m, 2H), 0.68 (br t, 1H); ^{13}C NMR, δ 14.21, 20.78, 22.08, 27.48, 29.36 (5 × C), 29.65, 43.86, 123.06, 123.53, 131.92, 133.45, 202.43; mass spectrum, m/z 236 (M^+). The ^1H NMR data were consistent with published values [30].

Addition of R_2CuLi across simple olefins, as opposed to acetylenes, is an unknown reaction. However, with cyclopropenone ketal 2, homo- and mixed Gilman cuprates react at -70 °C in 1 min to afford a product of *cis*-addition [31]. Both activated and unactivated electrophiles (in the presence of HMPA) can be employed to trap the intermediate vinylic cuprate, which occurs with retention of stereochemistry. The ketal can be stored (in an ampoule) at -20 °C, although it is handled at ambient temperatures. A slight excess of R_2CuLi is usually used; greater quantities of cuprate do not improve yields. Given the rapidity of the carbocupration step, the ketal should be added as a solution in Et_2O or THF, rather than as a neat liquid. If vinylic cuprates are used, the vinylcyclopropane products can be rearranged to cyclopentenone derivatives. Likewise, trapping the intermediate of such an addition with a vinylic electrophile to afford a divinylcyclopropane can ultimately lead to seven-membered ring formation. Should these latter two types of transformations be of interest, it should be noted that vinylcyclopropanone ketals are sensitive to acid, opening via a cyclopropylcarbinyl cation to the corresponding β,γ-unsaturated ester. A further application utilizing chiral, non-racemic cyclopropenone ketals to afford regio- and stereoselectively substituted cyclopropyl cuprates has also been achieved [32a].

Carbocupration of a Cyclopropenone Ketal with R_2CuLi [32b]

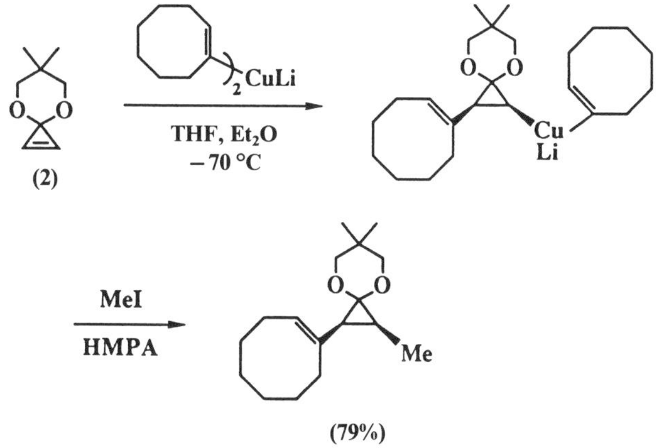

To a THF (13 ml) solution of 1-bromocyclooctene (0.63 ml, 4.4 mmol) was added at −70 °C *t*-BuLi (5.8 ml of a 1.53 M solution in pentane, 8.8 mmol) over 30 s. After stirring for 3 min, the vinyllithium solution was added via a cannula to a suspension of $CuBr{\cdot}SMe_2$ (0.451 g, 2.2 mmol) in Et_2O (3 ml). The mixture was stirred at −40 °C for 20 min, then cooled to −70 °C. The cyclopropenone ketal **2** (0.28 ml, 2 mmol) [33] in Et_2O (1.5 ml) was added to the cuprate solution over 1 min and stirred for 5 min. An Et_2O (2 ml) solution of iodomethane (0.62 ml, 10 mmol) and HMPA (0.38 ml, 2.2 mmol) was added and the mixture was warmed slowly to 0 °C over 3 h, and then stirred 1 h at 0 °C. The solution was poured into saturated NH_4Cl (15 ml) and the water layer was extracted with Et_2O. The combined organic layer was filtered through a short column of silica gel and concentrated *in vacuo*. The residual oil was chromatographed on silica gel (2% ethyl acetate in hexane) to obtain the 2,3-disubstituted cyclopropanone ketal as a colorless oil (0.423 g, 79%); IR (neat), 2950, 2920, 2850, 1470, 1450, 1395, 1260, 1155, 1070, 1025, 915 cm^{-1}; ^1H NMR (200 MHz, $CDCl_3$) δ 0.94 (s, 3H), 0.95–1.75 (m involving s, at 1.29), 2.05–2.49) (m), 3.45–3.65 (m, 4H), 5.63 (dt, $J = 1.9$, 9.0 Hz). Elemental analysis corresponded to $C_{17}H_{28}O_2$.

4.2.1.1 Reactions in the Presence of $BF_3{\cdot}Et_2O$

Exposure of the Gilman dimeric cuprate $(Me_2CuLi)_2$ in THF to $BF_3{\cdot}Et_2O$ at −78 °C was recently shown to give rise to Me_3Cu_2Li, along with $MeLi{\cdot}BF_3$, the former being the species responsible (along with $BF_3 \cdot Et_2O$) for the chemistry of this cuprate–Lewis acid pair (equation 4.2) [34].

$$[R_2CuLi]_2 + 2BF_3 \rightleftharpoons R_3Cu_2Li + RLi \cdot BF_3 + BF_3 \quad (4.2)$$

Regardless of the events which occur rapidly on mixing R_2CuLi with $BF_3 \cdot Et_2O$, of necessity done and maintained at low temperatures (−78 °C up to *ca* −50 °C), this combination is responsible for remarkable accelerations in numerous situations where R_2CuLi alone is either too unreactive or incompatible with increasing temperatures needed for a reaction to ensue [35]. Little or no visible changes in solutions of R_2CuLi occur on addition of $BF_3 \cdot Et_2O$ at −78 °C, and there is no induction period prior to introduction of a substrate. Thus, as is true for Me_3SiCl (see below), the cuprates are prepared exactly as described earlier, as these additives are involved only after the initial preparation. At least one equivalent of $BF_3 \cdot Et_2O$ (or TMS-Cl) is required for the full benefits to be realized, in part owing to the likely formation of boron ate complexes, rather than lithium salts, as the initial products [36]. In each of the examples which follow, it is noteworthy that $BF_3 \cdot Et_2O$ is essential for the desired chemistry to take place. The reactions include the conjugate addition to a highly hindered propellane skeleton [37], displacement of an aziridine to afford a secondary amine [38] and a highly diastereoselective bond formation involving a chiral, non-racemic acetal [39].

$BF_3 \cdot Et_2O$-Assisted Conjugate Addition of Me_2CuLi to a Hindered α,β-Unsaturated Ketone [37]

Me_2CuLi, $BF_3 \cdot Et_2O$

Et_2O, −78 °C, 1.25 h

O O

(71%)

To a cold (−30 °C) slurry of 870 mg (4.6 mmol) of purified copper(I) iodide and 8.0 ml of diethyl ether were added under Ar 6.0 ml (9.6 mmol) of methyllithium (1.2 M). The clear solution of lithium dimethylcuprate was stirred for 5 min and cooled to −78 °C, and 0.19 ml (1.54 mmol) of freshly distilled $BF_3 \cdot Et_2O$ was added. After the mixture had been stirred 5 min, a solution of 310 mg (1.6 mmol) of isomeric enones in 2 ml of diethyl ether was added dropwise; an immediate precipitation of methylcopper was observed. The mixture was stirred at −78 °C for 15 min, an additional 0.08 ml (0.75 mmol) of $BF_3 \cdot Et_2O$ was added and the mixture was stirred at −78 °C for 1 h. After the mixture had slowly warmed to room temperature, the organic material was extracted into diethyl ether and washed with saturated NH_4Cl, water and brine, and then dried. Evaporation of the solvent *in vacuo* afforded 277 mg of crude product, which was shown by IR spectroscopy to consist of approximately a 60:40 mixture of saturated ketone and enone, respectively. Without prior isolation of the ketone, the crude produce mixture was recycled in the same manner as described above. Kugelrohr distillation (b.p. 90–100 °C, 0.4 mmHg), afforded 233 mg (71%) of product as a 2:1 mixture of *anti* and *syn* isomers, respectively; vapour-phase chromatography (VPC) (195 °C) cleanly separated the mixture. The first component (*anti* isomer) had the following spectral data: IR (CCl_4), 2880–3000 (s, br), 1730 (s), 1460 (m) cm^{-1}; NMR (250 MHz, $CDCl_3$), δ 1.01, 1.05, 1.08 (d, $J = 6.5$ Hz, and 2s, 9H), 1.24–2.16 (m, 11H), AB q centered at 2.25 ($J_{AB} = 15.7$ Hz, $\Delta\nu_{AB} = 40$ Hz, δ_A 2.16, δ_B 2.33, 2H, H_A and H_B); mass spectrum, *m/z* 206.1679 (M^+; calculated for $C_{14}H_{22}O$, 206.1675). The second component (*syn* isomer) displayed the following spectral data: IR (CCl_4), 2870–3000 (s, br), 1730 (s), 1450 (m) cm^{-1}; NMR (250 MHz,

$CDCl_3$), δ 0.95, 0.99, 1.08 (s, d, $J = 6.5$ Hz, and s, 9H), 1.16–2.12 (complex m, 12H, containing H_A of an AB pattern at 1.97, $J_{AB} = 15$ Hz), 2.67 (d, 1H, H_B of an AB pattern centered at 2.32, $\Delta\upsilon_{AB} = 175$ Hz, $J_{AB} = 15$ Hz); mass spectrum, *m/z* 206.1673 (M^+; calculated for $C_{14}H_{22}O$, 206.1675).

$BF_3 \cdot Et_2O$-Promoted Alkylation of a Protected Aziridine [38]

1. Ph_2CuLi, THF, $BF_3 \cdot Et_2O$, −78 °C to r.t.
2. HCOOH, 80–85 °C, 90 min

MeO, OMe, N, CH, NH_2 (80%)

A 50 ml round-bottomed flask charged with CuI (1.5 mmol) and THF (6 ml) was cooled under Ar to −40 °C and treated with phenyllithium [3 mmol in cyclohexane–diethyl ether (7:3)]. The resulting black mixture was stirred 15 min, then cooled to −78 °C. To it was rapidly added the 4,4-dimethoxybenzhydryl (DMB)-protected aziridine (0.5 mmol) in THF (0.5 ml) followed by $BF_3 \cdot Et_2O$ (1.5 mmol). After warming the mixture to room temperature, 14% ammonia solution (15 ml) was added along with diethyl ether (10 ml) and solid NH_4Cl (1 g). The resulting dark blue aqueous layer was extracted three times with hexane–diethyl ether (1:1). The combined extracts were dried (K_2CO_3), filtered and concentrated to afford the *N*-DMB derivative of *β*-phenethylamine in 95% yield after flash chromatography [hexane–ethyl acetate (4:1)]. This sample was deprotected by stirring in 88% formic acid (5 ml) at 80–85 °C for 90 min. After removing the solvent *in vacuo* at 15 mmHg and then at 0.5 mmHg, the amine was partitioned between 5% aqueous HCl and diethyl ether to furnish pure *β*-phenethylamine (44 mg, 80%).

$BF \cdot Et_2O$-Assisted Cuprate Alkylation of a Chiral, Non-Racemic Acetal [39]

Me, O, O, Me, H, Me — $(n\text{-}C_6H_{13})_2CuLi$; Et_2O, $BF_3 \cdot Et_2O$, −78 to −55 °C, 15 min — Me, HO, O, Me, $n\text{-}H_{13}C_6$, H, Me (89%)

To a slurry of CuI (2.86 g, 15 mmol) in Et_2O (70 ml) was slowly added, at −40 °C, an ethereal solution of *n*-hexyllithium·LiBr (30 ml of a 1 M solution, 30 mmol; prepared from *n*-hexyl bromide and Li metal). The blue solution of the cuprate was ready after complete dissolution of CuI (30–60 min). After cooling to −78 °C, the chiral acetal (0.58 g, 5 mmol) dissolved in 10 ml of Et_2O was slowly added. If the addition is too rapid it will result in the formation of a yellow–orange precipitate which redissolves on warming. Under vigorous stirring, a solution of $BF_3 \cdot Et_2O$ (1.9 ml, 15 mmol) in Et_2O (10 ml) was added. The reaction was exothermic and the mixture was allowed to warm to −55 °C (internal temperature) for 15 min, whereupon no starting material remained (GC analysis). The orange–red mixture was hydrolyzed by addition of 30 ml of aqueous NH_4Cl and 20 ml of aqueous ammonia. The salts were filtered off and the aqueous layer extracted twice (2 × 50 ml of Et_2O). The combined organic phases were dried over Na_2SO_4 and the solvents were removed *in vacuo*. The residue was chromatographed on silica gel [cyclohexane–EtOAc (80:20)] to afford 897 mg of a pale yellow oil (89%). Only one set of signals was seen by NMR. IR, 3300, 1100 cm^{-1}; ^1H NMR, δ 3.52 (m, 2H), 3.14 (m, 1H), 1.45–1.30 (m, 10H), 1.16 (d, 6H), 1.10 (d, 3H), 0.90 (t, 3H); ^{13}C NMR, δ 77.50 (—CHO—), 73.06 (—CHO—), 71.06 (—CHOH), 37.66, 31.94, 29.44, 25.77, 22.67 (CH_2), 20.00, 18.53, 16.15, 14.09 (CH_3). The diastereomer at C-2 may be prepared in the same manner from Me_2CuLi and 2-hexyl-(*R*,*R*)-4,5-dimethyldioxolane. The corresponding NMR data are as follows: ^1H NMR, δ 3.58 (m, 1H), 3.50 (m, 1H), 3.28 (m, 1H), 1.40–1.30 (m, 10H), 1.14 (d, 3H), 1.12 (d, 3H), 1.08 (d, 3H), 0.90 (t, 3H); ^{13}C NMR, δ 78.18 (—CHO—), 74.55 (—CHO—), 70.97 (—CHOH), 36.92, 31.97, 29.61, 25.71, 22.73 (CH_2), 21.27, 18.62, 16.86, 14.12 (CH_3).

4.2.1.2 Reactions in the Presence of Me_3SiCl

Although Gilman reagents are modified under the influence of $BF_3 \cdot Et_2O$ [34], just how TMS-Cl alters the reaction pathway of lower order cuprates is currently a hot topic of debate. Arguments favoring both trapping of intermediate Cu(III) adducts [40] and trace quantities of carbonyl–TMS-Cl Lewis acid–Lewis base activation [41] have been advanced. Whatever the role, this simple synthetic maneuver usually pays off handsomely, and should be considered whenever hindered unsaturated carbonyl-containing molecules are involved. An illustrative procedure is cited below, the process (aside from the introduction of Me_3SiCl to the cuprate) being otherwise essentially identical with the standard mode of reagent use. Other work also suggests that both TMS-I and TMS-Br [42], in some cases, may provide even a greater boost to reagent activity.

Me_3SiCl-Accelerated 1,4-Addition of Bu_2CuLi to Acrolein [43]

CHO —(Bu_2CuLi, THF / HMPA, Me_3SiCl / −70 °C, 2.5 h)→ *n*-Bu OTMS (80%) (*E*:*Z* 98:2)

To a stirred suspension of $CuBr \cdot SMe_2$ (61.7 mg, 0.3 mmol) in 0.8 ml of THF at −70 °C was added dropwise butyllithium in hexane (0.60 mmol). The mixture was stirred at −40 °C for 30 min and then cooled to −70 °C, after which HMPA (174 μl, 1.0 mmol) was added. After several minutes, a mixture of acrolein (33.4 μl, 0.50 mmol), chlorotrimethylsilane (120 μl, 1.0 mmol) and decane (internal standard) in 0.3 ml of THF was added dropwise. After 2.5 h at −70 °C, 80 μl of triethylamine and pH 7.4 phosphate buffer were added. GLC analysis of the resulting mixture indicated an 80% yield and an *E*:*Z* ratio of 98:2.

4.2.1.3 Reactions in the Presence of Other Additives

Cuprate couplings effected in the presence of highly polar solvents, such as DMF or HMPA [14], or modified by inclusion of other metals can lead to changes in chemoselectivity or product formation that would not otherwise occur. For example, 1,4-diene formation by attachment of a vinylic cuprate (in particular, one derived via carbocupration) to an sp^2 center is best done in the presence of a catalytic amount of $Pd(Ph_3P)_4$ and $ZnBr_2$ (1 equiv.) [44]. Thus, following the preparation of a (*Z*)-vinylic cuprate (see above), a typical procedure is exemplified by the preparation of 7(*E*),9(*Z*)-dodecadien-l-yl acetate.

Coupling of a Vinylic Iodide with a Vinylic Cuprate; 1,3-Diene Synthesis [44]

$AcO(CH_2)_6$–CH=CH–I → [Et–CH=CH]$_2$CuLi, THF, $ZnBr_2$, $Pd(PPh_3)_4$, −20 to 10 °C, 30 min → $AcO(CH_2)_6$–CH=CH–CH=CH–Et (78%)

To a solution of (*Z*)-dibutenyl cuprate (prepared from 6 mmol of EtLi in 30 ml of diethyl ether and acetylene via carbocupration; see above) were added at −40 °C 20 ml of THF and then a solution of 700 mg of $ZnBr_2$ in 10 ml of THF. After stirring for 30 min at −20 °C, a mixture of 3 mmol of 1-iodo-1(*E*)-octen-8-yl acetate (94% *E* purity) and 0.15 mmol of $Pd(PPh_3)_4$ in 10 ml of THF was added. The reaction mixture was slowly warmed to 10 °C and, after 30 min at this temperature, 40 ml of saturated NH_4Cl solution were added. The organic phase was concentrated *in vacuo* and 50 ml of pentane were added to precipitate the inorganic salts, which were filtered off. The organic solution was washed twice with 20 ml of saturated NH_4Cl solution, dried over $MgSO_4$ and the solvent removed *in vacuo*. The crude residue was purified by preparative TLC to afford a 78% yield of the pheromone (97% *E*/*Z* purity).

4.2.1.4 Reactions in Non-Ethereal Media

Although not routine, cuprate reactions can be run in solvents other than Et_2O or THF [45]. The impetus may lie, e.g., in changes expected where stereochemical issues are present, or where increased interactions between substrate and cuprate are desirable. Solvents such as hexane, benzene, toluene or (as in the case below) CH_2Cl_2 may be used to replace Et_2O in which the cuprates are initially formed. NMR analyses confirm, at least in the case of CH_2Cl_2, that all of the Et_2O *is not* removed, the residual solvent serving as ligands on Li^+ [46]. As long as the cuprate is used immediately after solvent exchange (*in vacuo*), decomposition does not appear to be a problem. Occasionally, manipulations of this type can have multiple benefits, such as increasing both the rate and stereoselectivity of 1,4-additions.

Conjugate Addition of Me_2CuLi to an α,β-Unsaturated Ester in CH_2Cl_2 [46]

Aryl enoate (CO_2Me; ortho *CH(Me)NMe_2) → 1. Me_2CuLi, Et_2O; 2. −Et_2O; 3. +CH_2Cl_2, 0 °C, 1 h → CH(Me)CH$_2CO_2Me$ product

results in:				
	Et_2O:	(40%),	*de* :	<35
		(89%),	*de* :	0 (+TMS-I)
	THF:	(<1%),	*de* :	-
	CH_2Cl_2:	(92%),	*de* :	71

Methyllithium (0.5 ml, 1.6 M) was added to a slurry of CuI (167 mg, 0.88 mmol) in 3 ml of diethyl ether at 0 °C. After stirring for 10 min the ether was evaporated *in vacuo* at 0 °C for 30–60 min. CH_2Cl_2 (2 ml) was added and evaporated at 0 °C for 30 min to remove all the ether except the equivalent coordinating to the cuprate. Finally, 8 ml of CH_2Cl_2 were added before the enoate (0.4 mmol) dissolved in 2 ml of CH_2Cl_2 was added and the mixture stirred for 60 min at 0 °C. The reaction was then quenched at 0 °C with concentrated aqueous ammonia–aqueous NH_4Cl and extracted with CH_2Cl_2 (3 × 10 ml). The organic layer was washed with brine, dried (Na_2SO_4) and the solvent evaporated *in vacuo*; yield 92%; *de*, 71%. The same procedure was used for lithium diphenylcuprate (phenyllithium, 0.8 ml, 2 M).

4.2.2 LOWER ORDER, MIXED CUPRATES RR′CuLi

In a reaction of a Gilman homocuprate, transfer of only one R group usually takes place leading to a by-product organocopper $(RCu)_n$ which is lost on work-up (as RH and copper salts). Thus, based on RLi invested, for every equivalent of cuprate (which requires 2RLi to form), the maximum realizable yield with one equivalent of substrate is 50%. While tolerable for most commercially obtained RLi, those cuprates whose precursors must be synthetically prepared and then lithiated are too costly to sacrifice. To conserve valued RLi, mixed cuprates have been developed which derive from two different organolithiums; one is the organometallic of interest, R_tLi; the other, R_rLi, consists of a ligand R_r which is less prone to release by copper. When R_tLi and R_rLi combine with CuX (X = I, Br), R_tR_rCuLi is formed. Subsequent reaction with a suitable substrate leads to transfer of the R_t group in preference to R_r, with loss of the by-product $(R_rCu)_n$ being of no consequence.

Many different residual ligands (R_r) have been developed over the years [47]. Perhaps the most widely used are acetylenic derivatives, first introduced in 1972 [48]. Examples include lithiated *tert*-butylacetylene [49], 1-pentyne [48] and 3-methyl-3-methoxy-1-butyne [50]. Procedures in which each of the above is involved are given below. All of these acetylenes are easily obtained either in a single operation (see procedure below for 3-methyl-3-methoxy-1-butyne) or from commercial sources. Metallation is readily performed with, e.g., MeLi or *n*-BuLi, and the resulting lithiated acetylene is ready for use. When added to CuI, the transmetalated product, a copper(I) acetylide, tends to have reasonable solubility in Et_2O and is readily dissolved in THF as solutions bearing a red–orange color. Owing to their relative volatility, reformation of the neutral acetylenes on work-up does not complicate this methodology.

Michael Addition of a Mixed Acetylenic Cuprate to Isophorone [49]

O
—≡—Cu(Me)Li
Et_2O, 5–7 °C
20 min
O
(76%)

A solution of *t*-BuC≡CLi, prepared from 269 mg (3.28 mmol) of *tert*-butylacetylene and 3.06 mmol of MeLi in 2.4 ml of Et_2O, was added with stirring to a cold (10–13 °C) slurry of 573 mg (3.02 mmol) of purified CuI in 2.0 ml of Et_2O. To the resulting cold (5–7 °C) red–orange solution of the acetylide was added 1.7 ml of an Et_2O solution containing 2.74 mmol of MeLi. This addition resulted in a progressive color change from red–orange to yellow to green. To the resulting cold (5–7 °C) solution of the cuprate was added 2 ml of an Et_2O solution containing

2.08 mmol of isophorone. The color of the reaction mixture changed progressively from green to yellow (1–2 min) to red–orange (20 min), after which the mixture was partitioned between Et_2O and an aqueous solution (pH 8) of NH_4Cl and NH_3. The resulting orange Et_2O solution was washed with 3 × 25 ml of 28% ammonia solution to complete hydrolysis and the remaining colorless Et_2O solution was washed with water, dried and concentrated *in vacuo*. After the residual yellow liquid (288 mg) had been mixed with a known weight of n-$C_{14}H_{30}$ (as an internal standard), analysis (GLC, silicone fluid QF_1 on Chromosorb P, apparatus calibrated with known mixtures) indicated the presence of the tetramethyl ketone (76% yield) and the unchanged enone (12% recovery). None of the reduced ketone was detected. Collected samples of the ketones were identified with authentic samples by comparison of GLC retention times and IR and NMR spectra.

Conjugate Addition of a Lower Order Acetylenic Butyl Cuprate [48]

n-H_7C_3—≡—Cu(n-Bu)Li

Et_2O, HMPA, −78 °C, 15 min

Bu-n

(97%)

A slurry of 0.64 g of dry n-propylethynylcopper (4.90 mmol) in 10 ml of anhydrous diethyl ether was treated with 1.80 ml of dry hexamethylphosphoric triamide (9.80 mmol) and the mixture was stirred at room temperature under Ar until a clear solution was obtained (5–10 min). To the cooled (−78°C) solution were added 3.10 ml of a 1.49 M solution of n-butyllithium (4.62 mmol) in hexane, and the resulting yellow solution was stirred for 15 min at −78 °C. The solution of mixed cuprate so formed was then treated with 2.50 ml of a 1.80 M solution of 2-cyclohexenone (4.50 mmol) in anhydrous diethyl ether, stirred for 15 min at −78 °C, quenched by pouring into ice-cold aqueous ammonium sulfate solution and extracted with diethyl ether. The ethereal layers were extracted with ice-cold 2% v/v sulfuric acid, then filtered through Celite and washed with aqueous sodium hydrogen carbonate (5%). The dried (Na_2SO_4) extracts afforded almost pure 3-n-butylcyclohexanone (0.675 g, 97%), homogeneous by TLC analysis [R_f 0.40; diethyl ether–benzene (1:10)] and >99% pure by GLC analysis (10 ft, 10% SE-30 column, 170 °C, retention time, 5.2 min). The IR and NMR spectra were satisfactory.

Preparation of 3-Methyl-3-methoxy-1-butyne [50]

HO—≡—H

1. NaH, DMF, 1 h

2. Me_2SO_4, 0 °C to r.t. 1 h

MeO—≡—H

(84%)

A slurry of sodium hydride (7.2 g, 150 mmol; 50% in mineral oil) in 150 ml of DMF was cooled to 0 °C and 8.4 g (100 mmol) of 2-methyl-butyn-2-ol dissolved in 100 ml of DMF were added dropwise over 30 min. The reaction mixture was stirred for an additional 30 min and dimethyl sulfate (19 g, 14.3 ml, 150 mmol) was slowly added over a 20 min period. After stirring for an additional 5 min at 0 °C, the flask was allowed to warm to room temperature and stirring was continued for 45 min. Excess of sodium hydride was then destroyed by the dropwise addition of glacial acetic acid to the cooled (0 °C) reaction mixture. Direct distillation through a 30 cm Vigreux column afforded 8.2 g (84%) of pure material, b.p. 77–80 °C; IR (liquid film), 3290, 1080 cm^{-1}; NMR ($CDCl_3$), δ 3.35 (s, 3H), 2.38 (s, 1H), 1.46 (s, 6H).

Displacement of a Benzylic Halide with a Mixed Lower Order Acetylenic Cuprate [50]

MeO OMe I — MeO(C≡C)CuLi THPO — THF, −78 to −20 °C → MeO OMe OTHP (92%)

A 1 M solution of 3-methoxy-3-methyl-1-butyne in THF was treated at 0 °C with 1 equiv. of *n*-butyllithium. The clear, colorless solution was stirred for 5–10 min and transferred into a slurry of copper(I) iodide in THF (1 mmol ml^{-1}), precooled to 0 °C. The resulting red–orange solution was then stirred at this temperature for 30 min and subsequently transferred either by syringe or cannula to a −78 °C solution of the lithio reagent (0.5–1 M). Under the above conditions, a virtually instantaneous reaction occurred, yielding a pale yellow to colorless solution of the mixed cuprate. The use of more concentrated conditions led to the appearance of a white precipitate during cuprate formation (presumably lithium iodide), which readily dissolved at *ca* −30 °C to give a homogeneous solution. Addition of the substrate (0.985 equiv.) as a solution in THF at −78 °C was followed by warming to −20 °C and stirring for several hours. Standard extractive work-up employing pH 8 aqueous ammonia in saturated NH_4Cl gave the product (92%), which was virtually pure (by NMR). The formation of these mixed Gilman reagents can also be accomplished satisfactorily by the addition of an R_tLi reagent to a solution of the copper(I) acetylide (i.e. 'normal addition').

Other non-acetylenic groups have also found favor as residual ligands (R_r) in lower order cuprate reactions. The strong associations of copper with sulfur, and also with phosphorus, have led to the development of the thiophenoxide (PhS—) [51], 2-thienyl (2-Th—) [52] and dicyclohexylphosphido [$(C_6H_{11})_2P$—] [16, 53] ligands in this capacity. Each in lithiated form is readily prepared from commercial materials and has proved to serve in a manner similar to that of the acetylenic unit (see above).

Acylation of a Mixed Phenylthio Cuprate [51a]

PhC(O)Cl — PhS(t-Bu)CuLi, THF, −78 to −20 °C → PhC(O)C(CH$_3$)$_3$ (87%)

A stirred suspension of 4.19 g (22.0 mmol) of copper(I) iodide in 45 ml of THF was treated at 25 °C with 18.3 ml of 1.20 M (22.0 mmol) lithium thiophenoxide in THF–hexane (1:1). A clear, yellow solution formed within 5 min but became a cloudy suspension on cooling to −78 °C. Dropwise addition of 10.6 ml of 2.06 M (21.8 mmol) *tert*-butyllithium in pentane to the cold (−78 °C) suspension gave a fine, nearly white precipitate. Into this cold (−78 °C) suspension was injected after 5 min 15.0 ml of a precooled (−78 °C) solution containing 2.81 g (20.0 mmol) of benzoyl chloride in THF. Addition of the substrate regenerated the cloudy, yellow suspension, and the reaction mixture was stirred for 20 min before quenching was effected by injection of

5.0 ml (125 mmol) of absolute methanol. The reaction mixture was allowed to warm to room temperature, then poured into 200 ml of saturated, aqueous NH_4Cl, and the yellow precipitate thus formed was removed by suction filtration. The aqueous phase was extracted with 3 × 100 ml of diethyl ether and the combined ether phases were washed twice with 50 ml of 1 M NaOH and dried with $MgSO_4$. The solvent was removed *in vacuo* to afford 3.21 g (99%) of a slightly yellow oil with spectral properties essentially identical with those of pure pivalophenone. Short-path distillation gave 2.82 g (87% yield based on benzoyl chloride) of colorless pivalophenone, b.p. 105–106 °C (15 mmHg); n_D^{20} 1.5092; IR (CCl_4), 1680 cm^{-1} (C=O); NMR (CCl_4), δ 7.6–7.8 (m, 2H), 7.2–7.5 (m, 3H), 1.33 (s, 9H).

In the case of mixed thienylcuprates [R_t(2-Th)CuLi], the thiophene obtained from various vendors is usually not of satisfactory quality to be used as received. At least one distillation (b.p. 84 °C) is strongly recommended, and clean thiophene should appear as a close to water-white liquid. Metallation in ethereal media should afford solutions pale yellow (when done in THF at *ca* −25 °C) to yellow (at 0 °C to ambient temperature). When R_t = Me [i.e. Me(2-Th)CuLi], a particularly stable reagent is formed, thereby allowing commercialization of this methyl thienyl cuprate by the Lithium Division of FMC Corporation [52a].

Michael Addition of a Mixed 2-Thienyl Cuprate to an α,β-Unsaturated Ester Assisted by Me_3SiCl [52b]

Methyl cinnamate (CO_2Me) → [Me(2-Th)CuLi; Et_2O, TMS-Cl; −50 to 0 °C] → methyl 3-phenylbutanoate (Me, CO_2Me) (75%)

n-Butyllithium (2.5 mmol) was added to a solution of thiophene (3 mmol) in diethyl ether (5 ml) at 0 °C and the solution was stirred at room temperature for at least 40 min. Then another 2.5 ml of diethyl ether were added, the mixture was cooled in an ice-bath and finally powdered copper iodide (2.5 mmol) was introduced. 2-Thienylcopper formed immediately as a yellowish suspension. The mixture was stirred for about 5 min and then methyllithium (4.85–4.95 mmol) was added. The mixture was stirred until the Gilman test (see Section 4.2.1) for free alkyllithium was negative (about 5 min). The color of the cuprate solution was yellow to light green. The reaction mixture was then cooled to about −50 °C and methyl cinnamate (2 mmol) in diethyl ether (2.5 ml) was added. The addition resulted in a shiny yellow color. Within 1 min after substrate addition, trimethylchlorosilane (5 mmol) was added. The temperature was allowed to rise to 0 °C and the reaction was followed by GLC. After work-up the crude product, dissolved in pentane, was chromatographed through silica gel to separate trimethylsilylthiophene from the conjugate adduct. The silica gel was then eluted with diethyl ether. After filtration, drying with sodium sulfate and evaporation, the yield was 0.268 g (75%) of methyl (3-phenyl)butanoate.

The use of dicyclohexylphosphine, *en route* to the mixed phosphido lower order cuprates $R_t(C_6H_{11})_2PCuLi$, requires some care in terms of handling (i.e. minimized exposure to air). Once formed, however, the cuprates are particularly stable reagents, although they participate readily in substitution and Michael additions.

Conjugate Addition of a Lower Order Phosphido Cuprate to Cyclohexenone [53]

$(C_6H_{11})_2P(n\text{-Bu})CuLi$, Et_2O, −78 °C, 45 min

(80%)

A 10.7 g (54.0 mmol) quantity of dicyclohexylphosphine (K & K Labs or Organometallics) dissolved in 30 ml of dry, oxygen-free diethyl ether in a septum-sealed 100 ml pear-shaped flask was cooled to 0 °C (ice bath) and 37.5 ml (54.0 mmol) of 1.44 M (0.21 M residual base) butyllithium (Aldrich, in hexane) were added. The resulting suspension was stirred at 0 °C for 1 h and then transferred by cannula into a suspension of 10.9 g (53.0 mmol) of $CuBr{\cdot}SMe_2$ (Aldrich) in 60 ml of diethyl ether in a septum-sealed 500 ml round-bottomed flask, which was also at 0 °C. (The flask containing the phosphide was rinsed with 10 ml of diethyl ether.) The homogeneous brown solution was stirred for 15 min at 0 °C and then cooled to −50 °C for 15 min. A 36.8 ml (53.0 mmol) quantity of 1.44 M butyllithium was added and the homogeneous brown solution was stirred for 15 min at −50 °C. It was then cooled to −75 °C for 15 min and 4.80 g (49.9 mmol) of 2-cyclohexen-1-one (Aldrich, distilled and refrigerated) in 20 ml of diethyl ether in a septum-sealed 50 ml pear-shaped flask cooled to −75 °C were added by cannula. (A further 5 ml of diethyl ether were used to rise the 2-cyclohexen-1-one flask.) After 45 min at −75 °C, a 2.0 ml aliquot (out of a total of 200 ml) was withdrawn by syringe and added to 1 ml of 3 M aqueous NH_4Cl in a septum-selaed 2 dram vial, which also contained 29.7 mg of tetradecane (internal standard). Calibrated GLC analysis indicated a 91% yield of product; no starting material remained. After a total of 1 h, 200 ml of 3 M aqueous NH_4Cl (deoxygenated with N_2) were added to the reaction mixture, which was allowed to warm to ambient temperature. The final pH of the aqueous layer was 8. The mixture was filtered through Celite 545 and the filter deposit was washed with 200 ml of diethyl ether. The organic layer was separated and back-extracted with 250 ml of 0.2 M aqueous sodium thiosulfate, 200 ml of NH_4Cl solution, 250 ml of 0.4 M aqueous sodium thiosulfate and finally 100 ml of NH_4Cl. The aqueous layers were sequentially extracted with 50 ml of diethyl ether, which were added to the original organic layer. Drying over anhydrous sodium sulfate (Baker, granular) and evaporation under reduced pressure (<30 °C, 30 Torr) gave 9.31 g of a yellow oil, which was purified by flash chromatography on 180 g of Florisil (Fisher) slurry-packed in a 40 × 3.5 cm i.d. column and eluted with 3.5 l of 5% diethyl ether–hexane followed by 0.5 L of 10% diethyl ether–hexane. All fractions collected were of 50 ml (the column volume was 350 ml), and 5.61 g (98% pure by GLC) of product were recovered from fractions 11–76. An additional 0.53 g (98% pure) was obtained by stripping the column with 20% diethyl ether–hexane (*ca* 750 ml). The total yield of pure product was 6.14 g (80%). Analysis: calculated for $C_{10}H_{18}O$, C 77.87, H 11.76; found, C 77.60, H 11.59%.

All of the above 'dummy' ligands R_r have one feature in common: they are introduced as lithiated species (R_rLi) either to react with CuX to form initially R_rCu + LiX and thence with R_tLi, R_tR_rCuLi, or with R_tCu to form R_tR_rCuLi directly. Another alternative does exist, however, in which the R_r is already bound to copper as the Cu(I) salt. That is, CuCN is special in that treatment with R_tLi affords the lower order cyanocuprate $R_tCu(CN)Li$, rather than the products of metathesis, R_tCu + LiCN (Scheme 4.1) [54].

The strength of the copper-cyanide bond is presumably responsible for this behavior, but what is gained in simplicity of preparation and stability is paid for in reactivity.

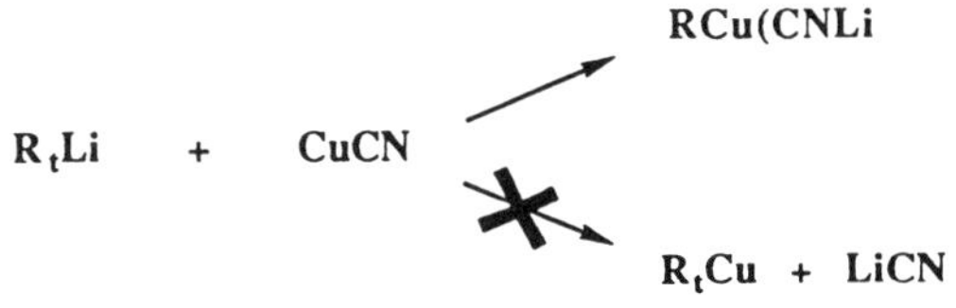

RCu(CNLi

R_tLi + CuCN

R_tCu + LiCN

Scheme 4.1

Hence, while lower order cyanocuprates are easily formed by this simple 1:1 correspondence (Scheme 4.1), the cyano ligand serving as the non-transferable R_r, they are best used in reactions with activated electrophiles, e.g. allylic epoxides.

Ring Opening of an Allylic Epoxide with a Lower Order Cyanocuprate [54a]

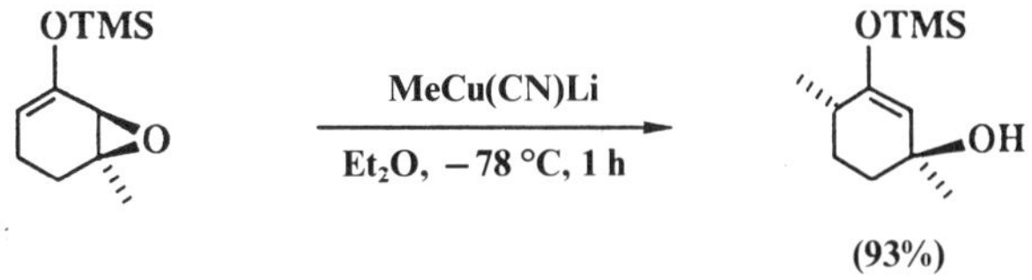

OTMS

MeCu(CN)Li

Et_2O, −78 °C, 1 h

OTMS

OH

(93%)

Caution: All cuprate reactions using CuCN should never be quenched with acidic aqueous solutions, so as to prevent generation of HCN.

CuCN (720 mg, 8 mmol) was placed into a flame-dried round-bottomed flask, which was then filled with *ca* 40 ml of dry diethyl ether and cooled to −40 °C under N_2. A 4.67 ml volume of methyllithium in diethyl ether (1.71 M, 8 mmol) was added and the yellowish suspension stirred for *ca* 30 min at −40 °C, until no CuCN was visible at the bottom of the flask. After cooling to −78 °C, a solution of 400 mg (2.0 mmol) of the epoxy enol ether in 5 ml of dry diethyl ether was added dropwise, with an intensification of the yellow color. The mixture was allowed to warm to room temperature over 5 h and then quenched with 30 ml of saturated NH_4Cl solution. After filtration through a Celite pad and washing of the ether layer with brine solution, the organic phase was dried over sodium sulfate and concentrated *in vacuo* to yield 390 mg (93%) of adduct. The crude reaction product was of high purity, as determined by 360 MHz ^1H NMR and ^{13}C NMR. Only one product was detectable by the aforementioned spectroscopic techniques. IR (neat), 840, 890, 1200, 1255, 1655, 3400 cm^{-1}; ^1H NMR ($CDCl_3$, 90 MHz), δ 0.07 (s, 9H), 0.89 (d, 3H, $J = 7$ Hz), 1.16 (s, 3H), 1.20–2.39 (m, 5H), 4.65 (br s, 1H); ^{13}C NMR ($CDCl_3$), δ 0.027, 17.31, 27.06, 29.93, 32.96, 34.32, 69.48, 110.59, 156.32; mass spectrum, m/z 214 (M), 199, 196, 181, 171, 165, 156, 144, 141, 127, 75 (100%), 73.

4.2.3 HIGHER ORDER CYANOCUPRATES $R_2Cu(CN)Li_2$

Retention of the cyano group on copper when CuCN is exposed to R_tLi (Scheme 4.1) presumably reflects an element of π basicity between the filled *d* orbitals on Cu(I) and π acidity of the vacant π^* orbitals in the nitrile. These interactions, together with the polarization of the C≡N ligand, may be responsible for the lower order species [RCu(CN)Li] being receptive toward a second equivalent of RLi, thereby forming $R_2Cu(CN)Li_2$ (equation 4.3).

These Cu(I) dianions are far more reactive than the corresponding lower order cyanocuprates, and also compare favorably with Gilman reagents in this regard [2a, e–g]. However, while more robust toward, e.g., primary halides and epoxides, they are at the same time more stable than lower order species R_2CuLi [55], another fringe benefit of the cyano ligand.

$$RLi + CuCN \longrightarrow RCu(CN)Li \xrightarrow{RLi} R_2Cu(CN)Li_2 \qquad (4.3)$$

The precursor to both RCu(CN)Li and $R_2Cu(CN)Li_2$ is CuCN, which offers several advantages over CuI or $CuBr{\cdot}SMe_2$. CuCN is far less costly, and requires no special handling (such as recrystallization or protection from light), although it is good practice to store it in an Abderhalden at 56 °C (refluxing acetone) over KOH. It comes in several different forms (according to the Merck Index) which imparts different colors to ethereal solutions of these cuprates. They range from an apple juice light brown (at *ca* 0.3 M) with CuCN in the tan form (Mallinckrodt) to yellowish green (green form, Fluka), to almost water-white solutions (off-white powder, Aldrich). Fortunately, there is no distinction between them in terms of the subsequent chemistry of the derived $R_2Cu(CN)Li_2$ [56].

With both CuCN and the RLi of interest available admixture in THF or Et_2O at −78 °C followed by slight warming effects dissolution to the higher order cuprate. Care should be exercised at this point not to slosh the mixture, placing the CuCN too high up on the walls of the flask, which could engender slight reagent decomposition (observed as a black ring around the flask). Once homogeneity has been reached in THF (or cloudy solutions on occasion in Et_2O), recooling to −78 °C sets the stage for substrate addition. Unsaturated enones tend to react at these low temperatures fairly quickly [57], as do unhindered epoxides [58] and primary bromides [59]. Et_2O is the preferred solvent for most conjugate additions and oxirane couplings, mainly because of the enhanced Lewis acidity of the Li^+ in solution. Halide displacements, on the other hand, are not 'push–pull' events, and therefore THF is the solvent of choice [56]. As with the case above involving a lower order cyanocuprate, work-up procedures should always avoid the use of highly acidic solutions to prevent generation of noxious HCN.

1,4-Addition of a Higher Order Cyanocuprate to Mesityl Oxide [57]

$Ph_2Cu(CN)Li_2$ / Et_2O, −78 °C, 1 h

Ph O

(83%)

$Ph_2Cu(CN)Li_2$ was prepared by the addition of PhLi (0.65 ml, 1.44 mmol) to CuCN (66 mg, 0.74 mmol) in 0.95 ml of Et_2O at −78 °C. Warming this mixture to 0 °C produces a yellowish but not completely homogeneous solution. The temperature was returned to −78 °C, at which point mesityl oxide (57 μl, 0.5 mmol) was added neat via a syringe. Stirring was continued at −78 °C. After 45 min the solution became viscous and further stirring was difficult. Quenching after 1 h and work-up in the usual manner were followed by column chromatography on SiO_2 with pentane–Et_2O (3 : 1), yielding 72.6 mg (83%) of product: R_f [pentane–Et_2O (3 : 1)] 0.43; IR (neat), 3060, 3025, 1722, 1705, 763, 698 cm^{-1}; ^1H NMR, δ 1.45 (s, 6H), 1.80 (s, 3H), 2.70 (s, 2H), 7.10–7.45 (m, 5H); mass spectrum, *m/z* (relative intensity, %) 176 (M^+, 11.8), 119 (100), 91 (45.5), 43 (52.7); HRMS, calculated for $C_{12}H_{16}O$, 176.1201; found, 176.1178.

Coupling of a Divinyl Higher Order Cyanocuprate with a 1,1-Disubstituted Oxirane [58]

$\xrightarrow[\text{THF, 0 °C, 5 h}]{(\text{vinyl})_2Cu(CN)Li_2}$ OH

(94%)

To $(vinyl)_2Cu(CN)Li_2$ at 0 °C, formed via addition of vinyllithium (0.96 ml, 2.0 mmol) to CuCN (89 mg, 1.0 mmol) in 1.5 ml of THF, was added 92 μl (0.77 mmol) of the epoxide. The solution was stirred at 0 °C for 5 h, then quenched and worked up in the usual fashion. Chromatographic purification on SiO_2 with 10% Et_2O–pentane yielded 92.3 mg (94%) as a clear oil: TLC, R_f [Et_2O–pentane (1:1)] 0.73; IR (neat), 3400, 3080, 1640, 1150, 960, 910 cm^{-1}; ^1H NMR, δ 0.85 (t, 6H, $J = 9$ Hz), 1.35 (s, 1H), 1.47 (q, 4H, $J = 9$ Hz), 2.20 (d, 2H, $J = 7.5$ Hz), 5.10 (m, 2H), 5.80 (m, 1H); mass spectrum, m/z (relative intensity, %) 113 (3.0), 99 (5.1), 87 (52.8), 81 (27.6), 67 (38.7), 57 (100), 41 (52.8); HRMS, calculated for $C_8H_{16}OCH_3$, 113.0938; found, 113.0952. Analysis: calculated for $C_8H_{16}OCH_3$, C 74.27, H 13.34; found, C 73.98, H 13.51%.

Displacement of a Primary Bromide by a Higher Order Cyanocuprate [59]

$$Br(CH_2)_4CN \xrightarrow[\text{THF, } -50\ °C,\ 2.5\ h]{n\text{-}Bu_2Cu(CN)Li_2} CH_3(CH_2)_7CN$$

(92%)

Copper cyanide (89.6 mg, 1.0 mmol) was placed in a 25 ml two-necked, round-bottomed flask, evacuated with a vacuum pump and purged with Ar, and the procedure was repeated three times. THF (1.0 ml) was injected via a syringe and the resulting slurry cooled to −75 °C, then *n*-butyllithium (0.8 ml, 2.0 mmol) was added dropwise. Subsequent warming to 0 °C produced a tan solution, which was immediately recooled to −50 °C, followed by dropwise addition, via a syringe, of 5-bromovaleronitrile (0.089 ml, 0.77 mmol). The reaction mixture was stirred at this same temperature for 2.5 h, followed by quenching with 5 ml of saturated NH_4Cl–concentrated ammonia solution (9:1). Extraction with Et_2O (3 × 5 ml), drying over Na_2SO_4 and evaporation of the solvent *in vacuo* resulted in a light yellow oil, which was chromatographed on silica gel with 15% Et_2O–pentane to yield 0.099 g (92%) of a clear liquid: TLC, R_f [Et_2O–pentane (1:4)] 0.5; NMR and IR data identical with those of pelargononitrile (Aldrich).

4.2.3.1 Reactions in the Presence of $BF_3 \cdot Et_2O$

When higher order cyanocuprates are exposed to 1 equiv. or more of $BF_3 \cdot Et_2O$ in THF, two significant events occur immediately, in spite of no visible change in appearance of the solution: (1) an equilibrium is established with the lower order cyanocuprate (equation 4.4), and (2) the BF_3 situates itself to a significant degree on the nitrile ligand in $R_2Cu(CU)Li_2$ [60].

$$R_2Cu(CN)Li_2 + BF_3 \rightleftharpoons R_2Cu(CN{-}BF_3)Li_2 \rightleftharpoons RCu(CN)Li + RLi \cdot BF_3 \quad (4.4)$$

As control experiments demonstrated that the reactive species is surely the higher order reagent, it now seems likely that the boost in cuprate reactivity may be ascribed, at least in part, to the rapid inclusion of this potent Lewis acid into the cuprate cluster.

Thus, as the enone is added, it sees a species bearing a far stronger carbonyl activating moiety than would otherwise be the case (i.e. BF_3 rather than, or in addition to, Li^+). Although the ^1H and ^{11}B NMR data unequivocally attest to the equilibrium as shown, whether the BF_3-complexed cuprate species is the actual reagent responsible for the chemistry is open to debate. What is not a matter for conjecture, however, is the often extraordinary difference this combination of reagents can make in otherwise challenging substrates [35]. The argument is particularly convincing, for example, with isophorone, where only with $BF_3 \cdot Et_2O$ present could a phenyl group be delivered to the β-site in high yield [36].

$BF_3 \cdot Et_2O$-Assisted 1,4-Addition of a Higher Order Cyanocuprate to a Hindered Enone [61]

$Ph_2Cu(CN)Li_2$; THF, $BF_3 \cdot Et_2O$; −78 to −50 °C, 1 h; −15 °C, 0.75 h

Ph

(>95%)

$Ph_2Cu(CN)Li_2$ was prepared as a yellow solution in THF (0.6 ml)–Et_2O (0.6 ml) using PhLi (2.0 mmol, 0.90 ml) and CuCN (1 mmol, 89.6 mg). $BF_3 \cdot Et_2O$ (1.0 mmol, 0.13 ml) was added to the cuprate at −78 °C with no visible change seen. Isophorone (0.50 mmol, 0.074 ml) was added, neat, at −78 °C, followed by stirring at −50 °C for 1 h and then at −15 °C for 0.75 h. Quenching followed by quantitative VPC analysis indicated that phenyl transfer had occurred to the extend of >95%. Filtration through SiO_2 afforded pure material; IR (neat), 1712 cm^{-1}; mass spectrum, m/z (relative intensity, %) 216 (M^+, 52), 201 (85), 159 (43), 145 (64), 118 (100); HRMS, calculated for $C_{15}H_{20}O$, 216.1490; found 216.1501.

Usually this additive, if successful in turning an otherwise sluggish coupling into a facile process, is 'compatible' with $R_2Cu(CN)Li_2$ up to *ca* −50 °C. Temperatures much above this limit start to erode seriously the percentage of educt consumption, presumably owing to side-reactions between 'RLi' and BF_3 (cf equation 4.4), in addition to BF_3-mediated opening of THF by the $RLi \cdot BF_3$ present [62]. Fortunately, S_N2 displacements of epoxides with the $R_2Cu(CN)Li_2$–BF_3 mixture also tend to occur at low temperatures and at greater rates relative to reactions in the absence of this additive [63]. Such is also true for openings of oxetanes.

$BF_3 \cdot Et_2O$-Mediated Opening of an Oxetane by a Higher Order Cyanocuprate [63]

O

1. $Ph_2Cu(CN)Li_2$; Et_2O, $BF_3 \cdot Et_2O$; −78 to −50 °C, 1 h; 2. Ac_2O, pyr

OAc

(87%)

An ethereal solution of PhLi·LiBr (49.2 ml of a 1.3 M solution in Et_2O, 64 mmol) was rapidly added to a suspension of CuCN (3 g, 33 mmol) in Et_2O (100 ml) at −30 °C. Stirring was continued for 10 min at -15 °C until a gray solution was obtained. This solution was cooled to -40°C and trimethylene oxide (1.74 g, 30 mmol) in Et_2O was added. The yellowish solution was again cooled to -78 °C, whereupon $BF_3 \cdot Et_2O$ (4.2 ml, 32 mmol) in Et_2O (20 ml) was slowly added. The reaction was complete after 1 h at -50 °C. The yellow turbid solution was then hydrolyzed by addition of 60 ml of aqueous NH_4Cl and 40 ml of aqueous ammonia. The salts were filtered off and the aqueous layer extracted twice (2×100 ml of Et_2O). The combined organic phases were concentrated *in vacuo*. The residue was directly acetylated by dissolving in pyridine (50 ml) and addition, at 0 °C, of Ac_2O (8.55 ml, 90 mmol). After stirring overnight at room temperature, MeOH (5 ml) was added to destroy excess of Ac_2O. After 1 h, Et_2O (250 ml) was added and this solution was washed once with aqueous $NaHCO_3$ solution (100 ml), then four times with aqueous NH_4Cl (4×100 ml) to remove most of the pyridine, then once with 1 M HCl (100 ml). The organic phase was dried over $MgSO_4$ and concentrated *in vacuo*. The residue was distilled through a 15 cm Vigreux column affording the desired product (4.65 g, 87% yield), b.p. 74 °C (0.05 mmHg); IR, 1740, 748, 701 cm^{-1}; ^1H NMR, δ 7.28 (m, 5H), 4.12 (t, 2H), 2.70 (t, 2H), 2.01 (s, 3H), 2.0 (m, 2H); ^{13}C NMR, δ 170.8 (—COO—), 141.2, 128.4, 126.0 (arom.), 68.7 (—CH_2O—), 32.2, 30.2, 20.8.

4.2.3.2 Reactions in the Presence of Me_3SiCl

Although the extent of impact of Me_3SiCl on lower order homocuprates R_2CuLi has yet to be fully delineated [40–43], admixture of this silyl halide with $R_2Cu(CN)Li_2$ even at -100 °C has a dramatic effect on these species [64]. Remarkably, it is the cyano ligand which is sequestered, giving rise to Me_3SiCN along with the lower order cuprate (equation 4.5).

$$R_2Cu(CN)Li_2 + 2Me_3SiCl \xrightarrow{\text{THF}, < -78\,°\text{C}} Me_3SiCN + R_2CuLi + Me_3SiCl + LiCl \tag{4.5}$$

The spectroscopically established presence of TMS–CN raises questions as to its role in this chemistry; nonetheless, the use of TMSCl (in excess) appears to offer benefits similar to those noted for reactions of lower order cuprates with α,β-unsaturated carbonyl systems (see above) [40–43]. Improvements in rates and diastereoselectivities of 1,2-additions observed due to the *in situ* formation of TMS–CN, together with residual TMSCl, have also been noted [64].

Conjugate Addition of an α-Alkoxy Higher Order Cyanocuprate [65]

OMOM
1. $(\ldots)_2Cu(CN)Li_2$
THF, TMSCl, -78 °C, 1 h
0 °C. 2.5 h
2. 1 M HCl, 10 min
OMOM
(96%)

A solution of the α-alkoxystannane (1.0 mmol) in 5 ml of THF was cooled to -78 °C (CO_2–acetone). A 0.50 ml sample of a 2.6 M solution of *n*-butyllithium in hexane (1.3 mmol) was then added and the solution was stirred for 5 min at -78 °C. A 25 ml round-bottomed flask containing 0.045 g (0.5 mmol) copper(I) cyanide suspended in 2 ml of THF was then cooled to -78 °C (CO_2–acetone). The α-alkoxylithio species was transferred via a cannula to the suspension of copper cyanide at -78 °C. The cuprate mixture was gradually allowed to warm to -60 °C (bath temperature) over a period of 30 min. A clear, homogenous solution was obtained. A third 25 ml round-bottomed flask containing a solution of the enone (0.5 mmol) in 3 ml of THF was cooled to −78 °C (CO_2–acetone). Trimethylsilyl chloride (0.32 ml, 2.5 mmol) was then added to the enone solution. The enone–TMSCl mixture was then added to the cuprate solution (at −78 °C) via a cannula. The resulting mixture was stirred for 1 h at −78 °C, and then gradually warmed to 0 °C (ice-bath) over an additional 2.5 h period. The reaction mixture was quenched by the addition of 1 ml of 1.0 M HCl, stirred for 10 min and then diluted with 100 ml of diethyl ether. The mixture was washed sequentially with a 1:1 mixture of aqueous NH_4Cl and 1.0 M HCl (40 ml), saturated aqueous NaCl (40 ml) and saturated Na_2CO_3 (40 ml). The layers were separated and the organic phase was dried over anhydrous $MgSO_4$. After removal of the solvent under reduced pressure, the crude reaction product was purified by flash chromatography on silica gel using 15–20% ethyl acetate–light petroleum as eluent (yield 96%): IR (neat), 2900, 1710, 1145, 1090 cm^{-1}; ^1H NMR ($CDCl_3$), δ 4.56 (s, 2H), 3.32 (s, 3H), 2.90 (dd, 1H, $J = 4.6$, 6.0 Hz), 2.43–1.36 (m, 10H), 0.85 (d, 6H, $J = 6.7$ Hz); ^{13}C NMR ($CDCl_3$) δ 211.7, 98.7, 88.3, 55.9, 45.7, 42.7, 41.2, 30.1, 29.1, 25.8, 25.0, 19.7, 17.9. Analysis: calculated for $C_{12}H_{22}O_3$, C 67.26, H 10.35; found, C 67.29, H 10.38%.

The above procedure, which effectively permits the introduction of a protected α-hydroxyalkyl appendatge in a 1,4-sense into an enone, is best carried out with fresh, purified stannane (by column chromatography). Clear solutions of the higher order cuprate should be used; cloudy mixtures imply impure stannane and lead to inferior results. The use of the MOM derivative in this case is not essential, as the MEM analog works equally as well. However, the benzyloxy methyl (BOM) derivative is not an acceptable choice.

4.2.4 *HIGHER ORDER MIXED CYANOCUPRATES* $R_tR_rCu(CN)Li_2$

4.2.4.1 Formation Using $R_tLi + R_rLi + CuCN$

The impetus behind the development of more highly mixed higher order cuprates was exactly the same as that which led to the development of the lower order analogs R_tR_rCuLi, that is, to conserve potentially valuable organolithium reagents, two equivalents of which normally go to form $R_2Cu(CN)Li_2$ whereas only one is transferred to the educt. Extensive trials have led to several observations of general applicability. It is now appreciated that for Michael additions [57], alkyl and vinylic ligands transfer selectively over acetylenic, thienyl, dimsyl and even a simple methyl moiety [66a]. Procedures for the last three residual ligand types are illustrated below. The simplest, of course, would involve an R_rLi out of a bottle, and in this sense only R_r = Me [i.e. to form $R_t(Me)Cu(CN)Li_2$] meets this criterion. With the advent of a 'cuprate in a bottle' (2-Th)Cu(CN)Li [66b]; Aldrich, catalog No. 32 417-5, however, mixed higher order cuprates can be easily obtained simply by adding to it an R_tLi of one's choosing, thus forming R_t(2-Th)$Cu(CN)Li_2$ [66c].

Michael Addition of a Mixed Thienyl Higher Order Cyanocuprate to an Enoate [67]

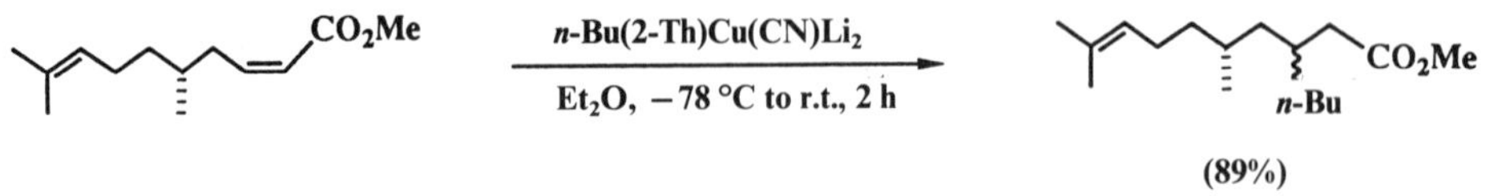

CO_2Me — n-Bu(2-Th)$Cu(CN)Li_2$ / Et_2O, −78 °C to r.t., 2 h → CO_2Me, n-Bu

(89%)

The mixed cuprate was formed using CuCN (54 mg, 0.61 mmol), 2-lithiothiophene (0.61 mmol) and *n*-BuLi (0.24 ml, 2.53 M, 0.61 mmol) in 1.4 ml of Et_2O. The unsaturated ester (121 μl, 0.55 mmol) was added to the cold (−78 °C) cuprate, the solution from which was slowly warmed to room temperature and stirred here for 2 h. Quenching with 10% aqueous ammonia–saturated aqueous NH_4Cl and extractive workup with Et_2O followed by solvent removal *in vacuo* and chromatography on SiO_2 with 5% Et_2O–pentane afforded 131 mg (89%) product: TLC, R_f (10% Et_2O–light petroleum) 0.60; IR (neat), 1745, 1165, 1020 cm^{-1}; ^1H NMR, δ 5.09 (m, 1H), 2.45–2.20 (m, 3H), 2.1–1.8 (m, 3H), 1.80–0.80 (m, 25H); mass spectrum, m/z (relative intensity, %) 268 (M^+, 2.8), 238 (1.1), 237 (5.5), 218 (4.7), 194 (35.2), 69 (100); HRMS, calculated for $C_{17}H_{32}O_2$, 268.2402; found, 268.2406.

Generation and Use of 'Me$_2$SO Cuprates,' Li$_2$[CH$_3$SOCH$_2$Cu(CN)R] [47e]

$Li_2[CH_3SOCH_2Cu(CN)\text{-}n\text{-}Bu]$; THF, −78 °C, 3 h; 0 °C, 1 h

n-Bu

(95%)

The lithio anion of DMSO was generated as a 0.2 M solution in THF by treatment of DMSO with *n*-butyllithium (1 equiv.) at 0 °C for 15 min. This was then transferred to a slurry of copper(I) cyanide (1 equiv.) in THF at −78 °C via a cannula. The mixture was warmed to 0 °C, resulting in a light green slurry which was recooled to −78 °C and *n*-butyllithium (1 equiv.) was added and allowed to warm to 0 °C to ensure cuprate formation. It was then cooled to −78 °C and a solution of 3,5,5-trimethylcyclohexen-1-one (0.45 equiv.) in THF was added with a syringe. After 3 h at −78 °C and an additional 1 h at 0 °C, the reaction was quenched with a saturated NH_4Cl solution containing 10% ammonia solution. After stirring for 15 min, it was suction filtered through Celite; the filter cake was washed with diethyl ether and the aqueous phase extracted with more diethyl ether. Analysis of the combined organic phases by VPC showed the product had formed in 95% yield; ^1H NMR ($CDCl_3$), δ 2.19–2.08 (m, 4H), 1.59 (t, 2H, J = 15 Hz), 1.49 (d, 2H, J = 14 Hz), 1.26–1.21 (m, 4H), 1.03 (s, 3H), 1.02 (s, 3H), 0.98 (s, 3H), 0.88 (t, 3H, J = 7 Hz).

Selective Ligand Transfer from a Mixed Dialkyl Higher Order Cyanocuprate to Cyclopentenone [57]

s-Bu(Me)Cu(CN)Li_2; THF, −78 °C, 30 min

(97%)

To CuCN (67.2 mg, 0.75 mmol) in 1.25 ml of cold (−78 °C) THF was added MeLi (0.47 ml, 0.75 mmol) and the mixture was warmed to 0 °C. Recooling to −78 °C and addition of *s*-BuLi (0.61 ml, 0.75 mmol) was followed by injection, with a syringe, of cyclopentenone (42 μl, 0.5 mmol) and stirring for 0.5 h. Quenching and extractive (Et_2O) work-up followed by chromatography on silica gel (40% Et_2O–pentane) afforded 67 mg (97%) of the product as a light oil: TLC, R_f [Et_2O–pentane (1 : 1)] 0.69; IR (neat), 1740 cm^{-1}; ^1H NMR, δ 0.8–1.0 (m, 6H), 1.1–2.5 (m, 10H); mass spectrum, m/z (relative intensity, %) 140.0 (5.4), 125 (20.3), 111.0 (17.9), 83.0 (47.4), 69.9 (14.9), 55.2 (100); HRMS, calculated for $C_9H_{16}O$, 140.1200; found, 140.1199.

Insofar as substitution reactions are concerned, the selectivity of transfer using R_rLi = MeLi is not satisfactory [58]. Hence it is necessary to resort to other choices, such as the thienyl-containing system $R_t(2\text{-Th})Cu(CN)Li_2$. Unhindered, monosubstituted epoxides couple very well with these mixed cuprates, even when less robust vinylic groups are undergoing transfer, as in the case of opening a chiral, non-racemic glycidol ether.

Opening of a Chiral, Non-Racemic Epoxide with a Higher Order Cyanocuprate Containing the 2-Thienyl Ligand [67]

$(2\text{-Th})Cu(CN)Li_2$; THF, 0 °C, 2.5 h; (92%)

Thiophene (88 µl, 1.1 mmol) was added to THF (1 ml) at −78 °C followed by *n*-butyllithium (0.39 ml, 1.1 mmol). The cooling bath was removed and the temperature raised to 0 °C over 5 min and stirred for an additional 30 min. The faint yellow anion was then transferred, via a cannula, into a two-necked flask containing copper (I) cyanide (89.6 mg, 1 mmol) and THF (1 ml), which was previously purged with Ar and cooled to −78 °C. Warming to 0 °C produced a light tan solution which was cooled to −78 °C and vinyllithium (0.5 ml, 1 mmol) was injected, with immediate warming to 0 °C (no visible change). It was then cooled to −78 °C and to it was added, with a cannula, a precooled solution of (2*S*)-benzyl 2-epoxypropyl ether (149 mg, 0.91 mmol) in THF (1 ml). After warming to 0 °C for 2.5 h, the reaction was quenched with 5 ml of a 90% saturated NH_4Cl–concentrated ammonia solution, extracted with diethyl ether (2 × 10 ml) and dried over sodium sulfate. Concentration followed by chromatography on silica gel (230–400 mesh) with diethyl ether–light petroleum (2:3) afforded 141 mg (92%) of a clear liquid, b.p. 90 °C (0.1 mmHg); R_f [diethyl ether–light petroleum [1:1] 0.33; $[\alpha]_D$ −2.2° ($c = 3$, $CHCl_3$); IR (neat), 3400, 3070, 3030, 1640, 1100, 740, 700 cm^{-1}; ^1H NMR ($CDCl_3$), δ 7.33 (s, 5H), 5.90–5.75 (m, 1H), 5.1 (m, 2H), 4.55 (s, 2H), 3.90 (m, 1H), 3.6–3.4 (m, 2H), 2.41 (d, 1H, $J = 3.3$ Hz), 2.26 (t, 2H, $J = 6.9$ Hz); mass spectrum, m/z (relative intensity, %) 192 (1.1, M^+), 92 (24.9), 91 (100); HRMS, calculated for $C_{12}H_{16}O_2$, 192.1150; found, 192.1161.

The diminished reactivity of higher order thienyl cuprates toward hindered (e.g. β,β-disubstituted) enones can often be overcome by simply adding $BF_3 \cdot Et_2O$ to the preformed, cold cuprate. Thus, in the case of isophorone, an otherwise sluggish coupling of a vinyl cuprate occurs at −78 °C in less than 1 h.

$BF_3 \cdot Et_2O$-Assisted 1,4-Addition of a Mixed 2-Thienyl Higher Order Cyanocuprate [$R_t(2\text{-Th})Cu(CN)Li_2$] [36]

$(2\text{-Th})Cu(CN)Li_2$; THF, Et_2O, $BF_3 \cdot Et_2O$, −78 °C, 1 h; (98%)

Thiophene (0.082 ml, 1.02 mmol) was added to THF (0.6 ml) in a 10 ml two-necked pear-shaped flask at −78 °C, followed by *n*-BuLi (0.315 ml, 1.0 mmol). Stirring was continued at the same temperature for 15 min, then at 0 °C for 30 min. The solution was transferred, with a cannula, into a slurry of CuCN (89.6 mg, 1.0 mmol) and Et_2O (0.6 ml), with a wash of 0.6 ml Et_2O. Warming to 0 °C gave a light tan solution, which was recooled to −78 °C and then vinyllithium (0.60 ml, 1.0 mmol) was introduced. After addition of neat $BF_3 \cdot Et_2O$ (0.13 ml, 1.0 mmol), isophorone (0.104 ml, 0.70 mmol) was added, neat, followed by stirring at −78 °C for 1 h and quenching. VPC analysis indicated that vinyl transfer had occurred to the extent of 98%. Chromatography on SiO_2 with Et_2O–pentane (15:85) gave pure material; TLC, R_f [pentane–Et_2O (15:85)] 0.32; IR (neat), 1710, 1635, 913 cm^{-1}; NMR ($CDCl_3$) δ 5.95–5.60, 5.10–4.85 (m, 3H), 2.58, 2.14 (2H, AB, $J = 25$ Hz), 2.15 (s, 2H), 1.67 (s, 2H), 1.14 (s, 3H), 1.07 (s, 3H), 1.00 (s, 3H); mass spectrum, m/z (relative intensity, %) 166 (M^+, 38), 151 (19), 124 (2), 123 (10), 109 (34).

4.2.4.2 Formation via Transmetallations of Organostannanes, Alanes and Zirconocenes

In a departure from prior art as a route to mixed cyanocuprates, selected higher order reagents can be formed directly by ligand-exchange processes between non-copper-containing organometallics and trivial higher order cuprates, generalized as in equation 4.6. Specifically, the chemistry applies to either vinylic or allylic stannanes, or *in situ* generated vinylic alanes and zirconocenes. Note that there is no involvement of highly basic organolithiums in this scheme.

$$RM + R'_2Cu(CN)Li_2 \longrightarrow R'M + RR'Cu(CN)Li_2 \qquad (4.6)$$

Vinylic stannanes, prepared in a variety of ways (e.g. hydrostannylation) [68], are converted to cuprates **3** on treatment with $Me_2Cu(CN)Li_2$ in THF at room temperature for *ca* 1–1.5 h (equation 4.7) [69].

$$\text{(vinyl)}SnR_3 \xrightarrow[\text{THF, 25 °C, 1 h}]{Me_2Cu(CN)Li_2} \text{(vinyl)}Cu(CN)Li_2(Me) + MeSnR_3 \qquad (4.7)$$

(3)

The trivial reagent $Me_2Cu(CN)Li_2$ can be freshly preformed (from CuCN plus 2MeLi–Et_2O in THF), or alternatively, prepared in quantity and stored in a freezer (*ca* −20 °C) for months without significant decomposition [70]. Exposure of the azeotropically dried stannane to one equivalent of $Me_2Cu(CN)Li_2$ leads to quantitative transmetallation. Once the newly formed cuprate is available, cooling to −78 °C followed by introduction of an α,β-unsaturated ketone effects the desired 1,4-addition. Some key points concerning this overall one-pot process include: (1) for BOTH vinylic and allylic stannanes (see below), the extent of transmetallation is dependent on the purity of the stannanes; (2) stannanes should *not* be stored under rubber septa, as exposure to these (rather than glass) stoppers dramatically reduces the extent of ligand exchange; (3) with vinyl stannanes, there is no obvious color change during the transmetallation; (4) the process is amenable to scale-up, having been run successfully (for the example below) on a mole scale in pilot-plant equipment [69]; (5) for substitution reactions, cuprates Me(vinylic)$Cu(CN)Li_2$ are *not* useful, since unlike their reactions with enones, they transfer the methyl ligand in preference to the vinyl group [58]. Hence the transmetallation should be conducted with Me(2-Th)$Cu(CN)Li_2$ resulting in ligand swapping to give (vinylic)(2-Th)$Cu(CN)Li_2$, which then selectively transfers the vinylic moiety [67]. Below, a procedure for the preparation of the antisecretory agent misoprostol is given, the lower side-chain being appended to the cyclopentenone using this new chemistry.

Transmetallation of a Vinylstannane with $Me_2Cu(CN)Li_2$ Followed by Michael Addition: Preparation of Misoprostol [69]

Bu_3Sn, Me, $OSiMe_3$ → 1. $Me_2Cu(CN)Li_2$, THF, r.t., 2 h; 2. $(CH_2)_6CO_2Me$, Et_3SiO; 3. $HOAc–H_2O–THF$ (3:1:1) → O, CO_2Me, Me, HO, OH (91%)

Copper cyanide (1.21 g, 13.5 mmol, flame-dried under Ar) in THF (15 ml) was treated with methyllithium (20.6 ml, 1.44 M in Et_2O, 29.7 mmol) at 0 °C. The cooling bath was removed and vinylstannane (7.65 g, 15.2 mmol) in THF (15 ml) was added. After 1.5–2 h at ambient temperature the mixture was cooled to −64 °C and the enone (3.2 g, 9.63 mmol) in THF (15 ml) was added rapidly with a cannula. The temperature rose to −35 °C. After 3 min, the mixture was quenched into a 9:1 saturated NH_4Cl–ammonia solution. Diethyl ether extraction followed by solvent removal *in vacuo* provided 11 g of residue, which was solvolyzed [acetic acid–THF–water (3:1:1), 100 ml) and chromatographed (silica gel, EtOAc–hexane eluent) to provide 3.15 g of product (8.2 mmol, 91%), which was identical in all respects with an authentic sample of misoprostol [71].

With allylic stannanes, the transmetallation is far more facile [72], not an unexpected observation judging from the order of release of ligands from tin [73]. Under standard conditions of THF, 0 °C, 30 min, higher order diallylic cuprates are formed to which can then be added various electrophiles, including vinylic triflates [74], alkyl halides [72] and epoxides (equation 4.8) [72]. 1,4-Additions under a CO atmosphere (i.e. allylic acylations) are also facile [75]. Starting with virtually colorless solutions of $Me_2Cu(CN)Li_2$ and an allylic tin, these *in situ* generated cuprates take on a bright yellow coloration. For the best results from these transmetallations, again, it is crucial that the initial stannanes be kept free from exposure to rubber septa. Also, it should be appreciated that vinylic iodides and bromides appear to be unacceptable reaction partners [74]. The former lead to (often high) percentages of the product of reduction, while the latter do not retain their stereochemical integrity (i.e. double bond geometry).

$$2\ \text{(allyl)SnBu}_3 \xrightarrow[\text{THF, 0 °C, 30 min}]{Me_2Cu(CN)Li_2} (\text{allyl})_2\text{Cu(CN)Li}_2 + 2\ MeSnBu_3 \quad (4.8)$$

Insofar as alkyl halides are concerned, since allylic cuprates are among the most reactive cuprates known, as such they can be used to displace even primary chlorides at −78 °C in minutes [72]. Thus, while bromides are also perfectly acceptable, less stable, usually light-sensitive iodides are unnecessary and are likely to afford lower yields of coupling product.

In Situ Diallylcuprate Formation from an Allylstannane: Displacement of a Primary Chloride [72]

$$\text{Cl(CH}_2)_3\text{C(OCH}_2\text{CH}_2\text{O)CH}_3 \xrightarrow[\text{THF, }-78\,^\circ\text{C, }<15\text{ min}]{(\text{CH}_2{=}\text{CHCH}_2)_2\text{Cu(CN)Li}_2} \text{CH}_2{=}\text{CH(CH}_2)_4\text{C(OCH}_2\text{CH}_2\text{O)CH}_3 \;(83\%)$$

CuCN (112 mg, 1.25 mmol) was gently flame-dried under vacuum, followed by flushing with Ar. This process was repeated twice. THF (1.5 ml) was added with a syringe and the resulting slurry was cooled to −78 °C. MeLi (1.18 M, 2 equiv, 2.11 ml, 2.5 mmol) was added dropwise, and the mixture was warmed to 0 °C and stirred until homogeneous. After recooling to −78 °C, allyltri-*n*-butylstannane (2 equiv., 0.66 g, 0.62 ml, 2.5 mmol) was added dropwise and the solution was warmed to 0 °C, stirred for 30 min (during which time it became yellow) and recooled to −78 °C. A cooled solution (0 °C) of the 5-chloropent-2-one acetal (0.8 equiv., 0.17 g, 0.15 ml, 1.0 mmol) in THF (1 ml) was transferred with a cannula to the cuprate solution. The reaction was stirred 15 min, followed by quenching with 3 ml of aqueous saturated NH_4Cl. Water (4 ml) was added and the aqueous mixture was poured into a separating funnel and extracted with diethyl ether (2 × 20 ml). The combined organic layers were washed with water (15 ml) and brine (15 ml), dried over Na_2SO_4 and the solvent was removed *in vacuo*, giving a clear liquid. Chromatography [silica gel, light petroleum–diethyl ether (13:1) as eluent] afforded 0.113 g (0.66 mmol, 83%) of a clear liquid product; TLC, R_f (10% diethyl ether in light petroleum 0.35; IR (neat), 3080, 2930, 2880, 1640, 1450, 1375, 1250, 1225, 1120, 1055, 910, 850 cm^{-1}; ^1H NMR ($CDCl_3$) δ 5.82–5.60 (m, 1H), 4.98–4.83 (m, 2H), 3.98–3.84 (m, 4H), 2.10–1.90 (br m, 2H), 1.60–1.52 (m, 2H), 1.36–1.31 (m, 4H), 1.25 (s, 3H); EI mass spectrum, *m/z* (relative intensity, %) 155 ($M^+ - CH_3$, 14), 87 (100), 55 (8), 43 (40); HRMS, calculated for $C_9H_{15}O_2$, 155.1072; found, 155.1060.

Expansion of the domain of available organometallics beyond tin which participate in ligand exchanges with higher order cyanocuprates has begun, and now includes aluminum and zirconium. Both processes known thus far begin from a terminal acetylene. With the former, an initial carboalumination gives rise to a vinylalane, which transfers its ligand from aluminum to copper on treatment with a diacetylenic higher order cyanocuprate (equation 4.9) [76]. Prior to the transmetallation step, the toluene and dichloroethane are removed *in vacuo* and exchanged for Et_2O. Only a 10–15% excess of alkyne is needed, an especially important feature when precious, enantiomerically pure materials are involved. Extension of the method to include hydroalumination/transmetallation/conjugate addition was also demonstrated. These 1,4-additions, under the conditions outlined below, are fairly rapid (*ca* 30 min), affording yields in the range 63–95%.

$$\text{R}{-}{\equiv}{-}\text{H} \xrightarrow[\text{PhMe, ClCH}_2\text{CH}_2\text{Cl}]{\text{Me}_3\text{Al, cat. Cp}_2\text{ZrCl}_2} \text{R(Me)C}{=}\text{CH}{-}\text{AlMe}_2$$

$$\xrightarrow[\text{Et}_2\text{O, THF, }-23\,^\circ\text{C}]{(\text{C}_4\text{H}_9\text{-}n{-}{\equiv})_2\text{Cu(CN)Li}_2} \text{R(Me)C}{=}\text{CH}{-}\text{Cu(CN)Li}_2(\text{C}{\equiv}\text{C}{-}\text{C}_4\text{H}_9\text{-}n) \qquad (4.9)$$

Carboalumination/Transmetallation of a 1-Alkyne to a Mixed Higher Order Cuprate Followed by 1,4-Addition [76]

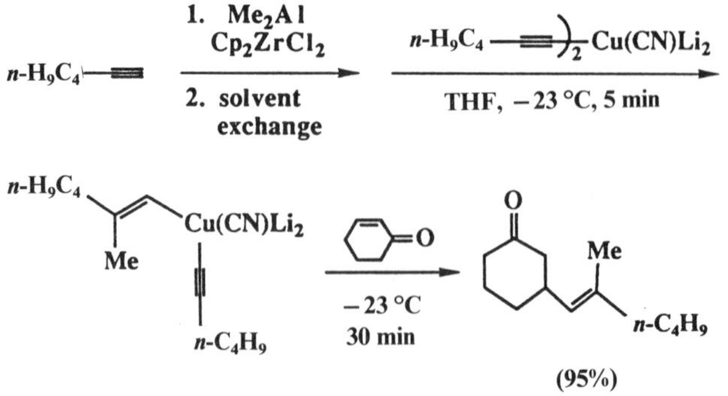

n-H_9C_4—≡ → 1. Me_2Al, Cp_2ZrCl_2; 2. solvent exchange → n-H_9C_4—≡)$_2$Cu(CN)Li_2, THF, −23 °C, 5 min → cyclohexenone, −23 °C, 30 min → (95%) (4.9)

A suspension of 53 mg (0.18 mmol) of zirconocene dichloride in 2 ml of dry 1,2-dichloroethane was treated at 0 °C with 0.78 ml (1.56 mmol) of a 2.0 M solution of trimethylaluminum in toluene, followed by addition of the solution of 46 mg (0.56 mmol) of 1-hexyne in 0.3 ml of 1,2-dichloroethane. The reaction mixture was stirred for 3 h at room temperature, the solvent was removed *in vacuo* and 3 ml of dry Et_2O were added. The ethereal solution of the vinylalane was added to the solution of 50 mg (0.56 mmol) of flame-dried CuCN in 2 ml of THF, which had previously been treated at −23 °C with 2.24 ml (1.12 mmol) of a 0.5 M solution of 1-hexynyllithium in THF–hexane (5:1). After the reaction mixture had been stirred for 5 min at −23 °C, a solution of 48 mg (0.50 mmol) of cyclohexenone in 1 ml of Et_2O was added dropwise, and stirring at −23 °C was continued for another 30 min. The mixture was quenched into a 9:1 saturated aqueous NH_4Cl–ammonia solution and extracted three times with Et_2O, and the combined organic layers were dried ($MgSO_4$), filtered through silica gel, and chromatographed [silica gel, EtOAc–hexane (1:5) as eluent] to yield 92 mg (95%) of product ketone.

Most recently, vinylzirconates have been found to serve as viable intermediates *en route* to mixed vinylic cuprates [77]. Hydrozirconation of a 1-alkyne produces the vinyl organometallic **4**, which is converted into the methylated species **5** at low temperature. Addition of MeR′Cu(CN)Li_2 presumably induces transmetallation to the desired mixed reagents **3** or **6**, which transfer the vinyl ligand in a Michael fashion to enones (Scheme

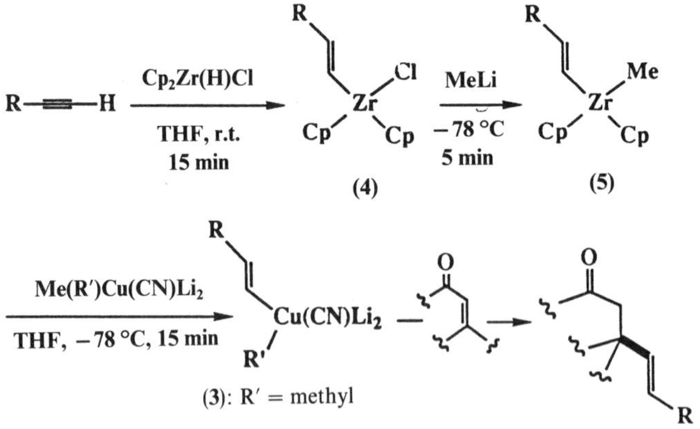

R—≡—H → $Cp_2Zr(H)Cl$, THF, r.t., 15 min → (4) → MeLi, −78 °C, 5 min → (5) → Me(R′)Cu(CN)Li_2, THF, −78 °C, 15 min → Cu(CN)Li_2 reagent

(3): R′ = methyl

(6): R′ = 2-thienyl

Scheme 4.2

4.2). The initial hydrozirconation step, normally run in aromatic hydrocarbon (e.g. benzene) as solvent [78a], is best performed here in THF, thereby speeding up the process dramatically (owing to the greater solubility of the reagent in THF) [78b]. Introduction of both the MeLi and $Me_2Cu(CN)Li_2$ to **4** should be done with precooled (to −78 °C) reagents. Their addition to yellow solutions of **4** do not lead to any noticeable differences in color during ligand exchange. The extent of transmetallation can be sensitive to the quality of the MeLi; bottles containing noticeable amounts of particulates may prove problematic. Should incomplete transmetallation be observed, use of *ca* 2 equiv. of dimethoxyethane (DME) often appears to negate the deleterious effects of dissolved extraneous lithium salts. Use of Et_2O as co-solvent, introduced along with the educt, also enhances the overall process. Final concentrations on the order of *ca* 0.1 M are recommended. The synthesis of misoprostol (in bis-silylated form) using this protocol is illustrative.

Higher Order Mixed Cyanocuprate Formation/1,4-Addition via Transmetallation of a Vinyl Zirconocene with $Me_2Cu(CN)Li_2$ [77a]

Me
OSiMe₃
1. $Cp_2Zr(H)Cl$
2. MeLi
3. $Me_2Cu(CN)Li_2$
THF, −78 °C, 15 min
4. O, $(CH_2)_6CO_2Me$, Et_3SiO
O, CO_2Me, Me, Et_3SiO, $OSiEt_3$
(92%)

A 10 ml round-bottomed flask equipped with a stir bar was charged with zirconocene chloride hydride (0.129 g, 0.50 mmol) and sealed with a septum. The flask was evacuated with a vacuum pump and purged with Ar, the process being repeated three times. THF (1.50 ml) was injected and the mixture stirred to generate a white slurry which was treated with trimethyl{[1-methyl-1-(2-propynyl)pentyl]oxy}silane (0.124 ml, 0.50 mmol). The mixture was stirred for 15 min to yield a nearly colorless solution, which was cooled to −78 °C and treated via a syringe with ethereal MeLi (0.35 ml, 0.50 mmol) to generate a bright yellow solution. Concurrently, CuCN (0.045 g, 0.50 mmol) was placed in a 5 ml round-bottomed flask equipped with a stir bar, and sealed under septum. The flask was evacuated and purged with Ar as above and diethyl ether (0.50 ml) was added via a syringe. The resulting slurry was cooled to −78 °C and treated with MeLi in diethyl ether (0.70 ml, 1.0 mmol). The mixture was warmed to yield a suspension of $Me_2Cu(CN)Li_2$, which was recooled to −78 °C and added with a cannula to the zirconium solution. The mixture was stirred for 15 min at −78 °C to yield a bright yellow solution, which was treated with a cannula with methyl 7-{5-oxo-3-[(triethylsilyl)oxy]-1-cyclopenten-1-yl}heptanoate (0.088 ml, 0.25 mmol) in diethyl ether (0.50 ml). After 10 min the mixture was quenched with 20 ml of 10% aqueous ammonia in saturated aqueous NH_4Cl. The product was extracted with 3 × 30 ml of diethyl ether and dried over Na_2SO_4. The solution was then filtered through a pad of Celite and the solvent removed *in vacuo*. The resulting residue was submitted to flash chromatography on silica gel [light petroleum–EtOAc (9:1)] to yield the protected form of misoprostol (0.132 g, 92%) as a colorless oil, which displayed spectral characteristics identical with those of authentic material [71].

Using the same vinylzirconocene intermediate **4**, two alternative procedures may be utilized which arrive at the same type of cuprate (e.g. **3**) [77a]. Thus, in the first scenario, treatment of **4** with 2 MeLi followed by (2-thienyl)Cu(CN)Li [66b] (available from Aldrich, catalog No. 32 417-5) presumably affords **6** (R = 2-Th), which then goes on to deliver the vinylic moiety in a 1,4-sense. The simplest procedure, however, is one which calls for addition of 3MeLi to **4**, to which is then added LiCl-solubilized CuCN in cold THF. Subsequent introduction of the unsaturated ketone completes this straightforward sequence.

Hydrozirconation/Transmetallation Using 2MeLi–(2-Th)Cu(CN)Li on a Vinyl Zirconocene: Michael Addition of a Mixed Cuprate to Isophorone [77a]

$Cp_2Zr(H)Cl$; THF, r.t., 15 min; Cp_2Zr, Cl; 1. 2 MeLi, −78 °C; 2. 2-ThCu(CN)Li; 3. $BF_3{\cdot}Et_2O$; 4. isophorone −78 °C, 1 h; O; (71%)

A 10 ml round-bottomed flask equipped with a stir bar was charged with zirconocene chloride hydride (0.258 g, 1.0 mmol) and sealed with a septum. The flask was evacuated with a vacuum pump and purged with Ar, the process being repeated three times. THF (3.0 ml) was injected and the mixture stirred to generate a white slurry which was treated with a syringe with phenylacetylene (0.110 ml, 1.0 mmol). The mixture was stirred for 15 min to yield a bright red solution, which was cooled to −78 °C and treated with a syringe with ethereal MeLi (1.40 ml, 2.0 mmol). Concurrently, CuCN (0.0895 g, 1.0 mmol) was placed in a 5 ml round-bottomed flask equipped with a stir bar and sealed under a septum. The flask was evacuated and purged with Ar as above and THF (1.0 ml) was added via a syringe. The resulting slurry was cooled to −78 °C and treated with a cannula with a solution of 2-thienyllithium prepared from the metallation of thiophene (0.080 ml, 1.0 mmol) with *n*-BuLi (0.43 ml, 1.0 mmol) in THF (1.50 ml) at −30 °C (25 min). The mixture was warmed to yield a suspension of (2-Th)Cu(CN)Li, which was recooled to −78 °C and added with a cannula to the vinyl zirconocene solution. The mixture was stirred for 30 min at −78 °C to yield a bright red solution, which was treated with $BF_3{\cdot}Et_2O$ (0.12 ml, 1.0 mmol) followed by the addition of isophorone (0.075 ml, 0.5 mmol). After 1 h the mixture was quenched with 10 ml of 10% aqueous ammonia in saturated aqueous NH_4Cl. The product was extracted with 3 × 50 ml of diethyl ether and dried over Na_2SO_4. The solution was then filtered through a pad of Celite and the solvent removed *in vacuo*. The resulting residue was submitted to flash chromatography on silica gel [light petroleum–ethyl acetate (9:1)], to give a 71% yield (0.086 g) of 3-(1-phenylethen-2-yl)-3,5,5-trimethylcyclohexanone as a thick yellow oil. The above procedure can alternatively be carried out with commercially available (2-Th)Cu(CN)Li (2.94 ml, 1.0 mmol), which when cooled to −78 °C can be added directly to a vinyl zirconocene solution which has been treated with 1 equiv. of MeLi (1.01 ml, 1.0 mmol) at −78 °C. TLC, R_f [light petroleum–ethyl acetate (9:1)] 0.37; IR (neat), 2960, 1955, 1885, 1812, 1705, 1600m 1450, 1280, 1230, 990, 750, 695 cm^{-1}; ^1H NMR (500 MHz, $CDCl_3$), δ 7.31–7.25 (m, 4H), 7.19–7.16 (m, 1H), 6.34 (d, 1H,J_{trans} = 16.5 Hz), 6.10 (d, 1H, J_{trans} = 16.5 Hz), 2.70 (d, 1H, J_{gem} = 13.5 Hz), 2.26 (d, 1H, J_{gem} = 13.5 Hz), 2.19 (d, 1H, J_{gem} = 13.5 Hz), 2.13 (d, 1H, J_{gem} = 13.5 Hz), 1.81 (d, 1H, J_{gem} = 14 Hz), 1.73 (d, 1H, J_{gem} = 14 Hz), 1.18 (s, 3H), 1.05 (s, 3H), 0.94 (s, 3H); EI mass spectrum, *m/z* (relative intensity, %) 242 (70), 227 (29), 185 (41), 143 (57), 129 (83), 118 (21), 115 (21), 105 (33), 91 (100), 83 (46), 77 (16), 55 (46); HRMS, calculated for $C_{17}H_{22}O$, 242.1671; found, 242.1662.

Mixed Cyanocuprate Generation from Treatment of a Vinyl Zirconocene with 3MeLi, then CuCN·2LiCl: Preparation of Protected 15-Methyl-PGE₁ Methyl Ester [77a]

1. $Cp_2Zr(H)Cl$ 2. 3MeLi 3. $CuCN{\cdot}2LiCl$; THF–Et_2O, −78 °C, 20 min (84%)

A 10 ml round-bottomed flask equipped with a stir bar was charged with zirconocene chloride hydride (0.258 g, 1.0 mmol) and sealed with a septum. The flask was evacuated with a vacuum pump and purged with Ar, the process being repeated three times. THF (0.5 ml) was injected and the mixture stirred to generate a white slurry which was treated with trimethyl {[1-methyl-1-(1-ethynyl)hexyl]oxy}silane (0.148 g, 0.70 mmol) as a solution in THF (0.5 ml). The mixture was stirred at room temperature for 30 min to yield a yellow–orange solution which was cooled to −78 °C and treated with a cannula with a diethyl ether (3.0 ml) solution of MeLi (1.91 ml, 2.10 mmol, in THF–cumene), which had been precooled to −78 °C, to generate a bright yellow solution. The mixture was stirred for 10 min. Concurrently, CuCN (0.063 g, 0.7 mmol) and LiCl (0.059 g, 1.4 mmol) were placed in a 5 ml round-bottomed flask equipped with a stir bar and sealed under a septum. The flask was evacuated and purged with Ar as above and THF (1.0 ml) was added with a syringe. The mixture was stirred for 5 min at room temperature to generate a colorless homogeneous solution, which was cooled to −78 °C and added with a cannula to the zirconocene solution. The mixture was stirred for 5 min at −78 °C to yield a bright yellow solution, which was treated with a precooled (−78 °C) solution of methyl 7-{5-oxo-3-[(triethylsilyl)oxy]-1-cyclopenten-1-yl}heptanoate (0.0883 g, 0.25 mmol) in Et_2O (0.50 ml). After 20 min the mixture was quenched with 10 ml of 10% aqueous ammonia in saturated aqueous NH_4Cl. The product was extracted with 3 × 35 ml of diethyl ether and dried over Na_2SO_4. The solution was then filtered through a pad of Celite and the solvent removed *in vacuo*. The resulting residue was submitted to flash chromatography on silica gel [light petroleum–EtOAc (9:1)] to give an 84% yield (0.120 g) of the protected form of 15-methyl-PGE₁ as a yellow oil; TLC, R_f [light petroleum–EtOAc (9:1)] 0.41; IR (neat), 2950, 1745, 1460, 1375, 1250, 1100, 1010, 840, 750 cm^{-1}; ^{1}H NMR (500 MHz, $CDCl_3$), δ 5.55 (dd, 1H, J = 15.5, 11 Hz), 5.57–5.38 (m, 1H), 4.02–3.95 (m, 1H), 2.58 (dd, 1H, J = 5.5, 18 Hz), 2.42–2.36 (m, 1H), 2.23 (t, 2H, J = 7.5 Hz), 2.14 (dd, 1H, J = 8.5, 18.5 Hz), 1.90–1.78 (m, 1H), 1.56–1.49 (m, 7H), 1.45–1.35 (m, 4H), 1.29–1.15 (m, 12H), 0.90 (t, 9H, J = 8 Hz), 0.82 (t, 3H, J = 7 Hz), 0.55 (qd, 6H, J = 1, 8 Hz), 0.06 (s, 9H); ^{13}C NMR, δ 216.03, 215.97, 174.09, 140.48, 140.39, 126.89, 126.82, 75.67, 75.59, 72.97, 72.82, 53.94, 53.82, 53.56, 53.47, 51.33, 47.67, 47.62, 43.96, 43.85, 33.97, 32.27, 32.21, 29.44, 28.90, 28.11, 28.02, 27.60, 27.52, 26.56, 26.48, 24.85, 23.81, 23.70, 22.63, 14.03, 6.73, 4.73, 2.56, 2.50; CI mass spectrum, m/z (relative intensity, %) 553 (11), 539 (10), 509 (14), 497 (100), 421 (24), 365 (100), 347 (35), 315 (43), 311 (100), 293 (44), 115 (26), 103 (78), 75 (75), 73 (76); HRMS, calculated for $C_{31}H_{60}O_5Si_2$ (M^+ + 1), 568.4100; found, 568.4041.

4.2.5 ORGANOCOPPER (RCu) AND ORGANOCOPPER·LEWIS ACID REAGENTS (RCu·BF_3)

The preparation of organocopper complexes, generally expressed as RCu, with or without an additive, traditionally follows along the lines outlined earlier for cuprate formation but with one important distinction: the stoichiometry of RLi to CuX (X = I,

Br) is 1:1, rather than 2:1 (equation 4.10).

$$\underset{(X = Br, I)}{RLi + CuX} \longrightarrow RCu + LiX \equiv R(X)CuLi \qquad (4.10)$$

Although Lewis basic additives, such as sulfides (e.g. Me_2S, see below) and phosphines (e.g. n-Bu_3P) are typically present, presumably to stabilize the 'RCu' formed (as $RCu \leftarrow L$) [2c,d], there are a number of situations where solely the species derived from metathesis are useful in their own right. When using 'RCu,' it is important to appreciate that isolated, dry reagent may be explosive, especially where R = alkyl. Hence 'RCu' is best prepared and utilized *in situ* [79].

Asymmetric Conjugate Addition of MeCu to a Chiral, Non-Racemic Vinyl Sulfoximine [80]

MeCu·LiI; Et_2O, −25 °C, 1 h; 0 °C, 1 h (75%)

To a stirred suspension of copper(I) iodide (486 mg, 2.56 mmol) in diethyl ether (12.8 ml) at −25 °C was added methyllithium (2.56 mmol). After 30 min, (*SR*, 1*S*, 2*R*)-*N*-(1-methoxyl-phenyl-2-propyl)-*S*-(1-hexenyl)-*S*-phenylsulfoximine (190 mg, 0.511 mmol) in diethyl ether (2 ml) was added, and the mixture was stirred at −25 °C for 1 h. It was then allowed to warm to 0 °C over a period of 1 h and, after an additional 1 h at 0 °C, the reaction was quenched with aqueous NH_4Cl (20 ml). The layers were separated and the ether layer was dried and concentrated *in vacuo*. Analysis of the crude reaction mixture by HPLC indicated two compounds in a ratio of 96.5:3.5. Purification of the crude material by preparative TLC [ethyl acetate–hexane (2:3)] gave the product as a colorless oil: ^1H NMR, δ 7.70–7.01 (m, 10H), 3.90 (d, 1H, J = 7.6 Hz), 3.32–2.89 (m, 2H), 3.20 (s, 3H), 2.73 (dd, 1H, J = 7.6, 14.2 Hz), 2.02 (m, 1H), 1.54–1.0 (m, 6H), 1.32 (d, 3H, J = 5.9 Hz), 0.86 (t, 3H), 0.77 (d, 3H, J = 6.8 Hz); ^{13}C NMR, δ 141.22, 138.86, 131.95, 129.15, 128.76, 128.13, 127.62, 127.08, 89.24, 63.45, 57.11, 56.13, 36.35, 28.47, 28.28, 22.51, 21.83, 19.89, 13.95; CI(methane) mass spectrum, m/z (relative intensity, %) 388 (18, $M^+ + 1$), 356 (22), 266 (79), 125 (100).

More commonly, as alluded to above, an equimolar amount of a trialkylphosphine is used in conjunction with CuI, which together afford a THF-soluble Cu(I) salt. Once the RLi (1 equiv.) has been added, the '$RCu \cdot PR'_3$' is ready for use, although the price for this additive must be considered in terms of its pyrophoric nature, effect on work-up and chromatographic separation.

Michael Addition of a Phosphine-Stabilized Organocopper Complex ($RCu \cdot PR'_3$) [81]

n-$H_{11}C_5$CH(OTHP)CH=CHCu·P(n-Bu)$_3$; Et_2O, THF, −78 °C, 1 h (84%)

Copper (I) iodide (300 mg, 1.57 mmol) was placed in a 180 ml ampoule equipped with a rubber septum. After the atmosphere had been replaced by Ar, dry THF (20 ml) followed by dry, distilled tri-*n*-butylphosphine (1.02 ml, 4.10 mmol) were added at room temperature. The suspension was stirred until a clear solution resulted. In a 30 ml test-tube equipped with a rubber septum were placed (*E*)-1-iodo-3-tetrahydropyranyloxyoctene (528 mg, 1.56 mmol) and dry diethyl ether (6 ml). After cooling to −95 °C, *tert*-butyllithium (1.68 ml, 3.12 mmol) in pentane was added to this solution, with stirring, over 1 min. The mixture was stirred at −78 °C for 2 h. The resulting white suspension was added at −78 °C, with stirring, to the above-prepared ethereal solution of the copper(I) iodide-phosphine complex through a stainless-steel cannula under a slight Ar pressure. After the mixture had been stirred at −78 °C for 10 min, to this solution was then added slowly, along the cooled wall of the reaction vessel, a solution of cyclopentenone (103 mg, 1.25 mmol) in cold (−78 °C) THF (10 ml) through a stainless-steel cannula under a slight Ar pressure over 50 min. The mixture was stirred at −78 °C for 1 h. A saturated aqueous solution of NH_4Cl (15 ml) was added at −78 °C and the mixture shaken vigorously. The organic layer was separated and the aqueous layer extracted with diethyl ether (30 ml). The combined extracts were dried over $MgSO_4$, evaporated, and chromatographed on triethylamine-treated silica gel (30 g) using hexane–ethyl acetate–triethylamine (2000 : 100 : 1) as eluent to give the adduct (310 mg, 84%, mixture of diastereomers) as a colorless oil; IR (neat) 1741 (C=O) cm^{-1}; ^1H NMR (CCl_4), δ 0.90 (t, 3H, $J = 6.5$ Hz), 1.1–3.1 (m, 21H), 3.3–3.7 (m, 1H), 3.7–4.2 (m, 2H), 4.65 (br s, 1H), 5.2–5.8 (m, 2H); HRMS, calculated for $C_{13}H_{19}O_3$ ($M^+ - C_5H_{11}$), 223.1346; found, 223.1340.

Perhaps the most reactive form of $RCu \cdot PR'_3$ is arrived at via lithium naphthalenide reduction of $CuI \cdot PR'_3$, which generates highly active $Cu \cdot PR'_3$ [82a]. Although both Ph_3P and n-Bu_3P have been employed, the trialkylphosphine complexes ultimately give reagents of higher reactivity. Preformed $Cu \cdot PR'_3$, or this species generated *in situ*, give similar results *en route* to $RCu \cdot PR'_3$. Typical procedures are given below, in these cases for the coupling with an acid chloride to generate a ketone, and a 1,4-addition to an enone. It is especially noteworthy that highly functionalized organocopper reagents can be made in this fashion, wherein carboalkoxy, cyano, halo and even epoxy groups can be tolerated. More recently, lithium naphthalenide reduction of (2-Th)Cu(CN)Li at −78 °C has led to a process which does not involve phosphines [82b]. These reactions should be carried out under Ar, since lithium naphthalenide gradually reacts with N_2.

Acylation of an Organocopper Species Prepared via Lithium Naphthalenide Reduction of CuI·PPh_3; Ketone Formation [82a]

Br–$CH_2CH_2CH_2$–CO_2Et → (1. Li(0)–naphthalene, CuI·PPh_3; 2. PhCOCl, THF, −35 °C, 1.5 h, r.t., 30 min) → PhCO–$CH_2CH_2CH_2$–CO_2Et (81%)

Lithium (70.8 mg, 10.2 mmol) and naphthalene (1.588 g, 12.39 mmol) in freshly distilled THF (10 ml) were stirred under Ar until the Li was consumed (*ca* 2 h). CuI (1.751 g, 9.194 mmol) and PPh_3 (2.919 g, 11.13 mmol) in THF (15 ml) were stirred for 30 min, giving a thick white slurry which was transferred via a cannula to the dark green solution of lithium naphthalenide at 0 °C. (Later experiments showed that slightly better results were obtained if the lithium naphthalide solution was added to the CuI·PPh_3 mixture.) The resultant reddish black solution of active copper was stirred for 20 min at 0 °C. Ethyl 4-bromobutyrate (0.3663 g, 1.888 mmol) and the GC internal standard *n*-decane (0.1566 g, 1.101 mmol) were added neat with a syringe to the active copper solution at −35 °C. The solution was stirred for 10–15 min at −35 °C, followed by addition of benzoyl chloride (0.7120 g, 5.065 mmol), neat, to the organocopper solution at −35 °C. The reaction was stirred for 90 min at −35 °C, followed by warming to room temperature for 30 min. (GC analylsis showed the reaction to be essentially complete after stirring at −35 °C.) The reaction was then worked up by pouring into saturated aqueous NH_4Cl, extracting with Et_2O and drying over anhydrous sodium sulfate. (For compounds not sensitive to base, the ether layer was also washed with 5% aqueous NaOH solution.) Silica gel chromatography (hexane, followed by mixtures of hexane and ethyl acetate) and further purification by preparative TLC (2 mm layer) provided 4-carboethoxy-1-phenyl-1-butanone in 81% isolated yield (93% GC yield after quantitation using the isolated product for the preparation of GC standards); IR (neat), 3060, 2960, 1735, 1690, 1600, 1580, 1450, 1375, 1240, 1205 cm^{-1}; ^1H NMR (200 MHz, $CDCl_3$), δ 7.90–8.04 (m, 2H), 7.36–7.64 (m, 3H), 4.14 (q, 2H, $J = 7.2$ Hz), 3.05 (t, 2H, $J = 7.2$ Hz), 2.43 (t, 2H, $J = 7.2$ Hz), 2.07 (tt, 2H, $J = 7.2$, 7.2 Hz), 1.25 (t, 3H, $J = 7.2$ Hz); ^{13}C NMR ($CDCl_3$), δ 199.2, 173.0, 136.9, 132.9, 128.5, 127.9, 60.2, 37.4, 33.4, 19.4, 14.1; EI mass spectrum, *m/z* (relative intensity, %) 220 (M^+, 1.0), 175 (8.5), 147 (10.8), 133 (2.7), 120 (9.1), 105 (100.0), 77 (47.3), 55 (7.1); HRMS, calculated for $C_{13}H_{16}O_3$, 220.1100; found, 220.1092.

1,4-Addition of an Organocopper Complex Prepared by Lithium Naphthalenide Reduction of CuI·PBu$_3$ [82a]

n-$C_8H_{17}Br$ → (1. Li(0)–naphthalene, CuI·PBu_3; 2. cyclohexenone, THF, −78 °C to r.t., 4 h) → 3-(C_8H_{17}-n)cyclohexanone (93%)

Lithium (71.2 g, 10.3 mmol) and naphthalene (1.592 g, 12.42 mmol) in freshly distilled THF (10 ml) were stirred under Ar until the Li was consumed (*ca* 2 h). A solution of CuI·PBu_3 (3.666 g, 9.333 mmol) and PBu_3 (2.89 g, 14.3 mmol) in THF (5 ml) was added with a cannula to the dark green lithium naphthalenide solution at 0 °C and the resultant reddish black active copper solution was stirred for 20 min. 1-Bromooctane (0.9032 g, 4.677 mmol) and the GC internal standard *n*-decane (0.1725 g, 1.212 mmol) in THF (5 ml) were added rapidly with a cannula to the active copper solution at −78 °C. Organocopper formation was typically complete after 20 min at −78 °C. 2-Cyclohexen-1-one (0.1875 g, 1.950 mmol) in THF (10 ml) was added slowly dropwise over 20 min to the organocopper species at −78 °C. The reaction was allowed to react at −78, −50 and −30 °C and room temperature for 1 h each. The reaction was then worked up by pouring into saturated aqueous NH_4Cl, extracting with Et_2O and drying over anhydrous sodium sulfate. Silica gel chromatography (hexane followed by mixtures of hexane and ethyl acetate) and further purification by preparative TLC provided 3-*n*-octylcyclohexanone (93% GC yield after quantitation using the isolated product for the preparation of GC standards); IR (neat), 2930, 2860, 1715, 1465, 1425, 1225 cm^{-1}; ^1H NMR (200 MHz, $CDCl_3$), δ 1.50–2.50 (m, 9H), 1.15–1.45 (very broad app. s, 14H), 0.88 (app. t, 3H, $J = 6.5$ Hz); ^{13}C NMR ($CDCl_3$), δ 211.8, 48.2, 41.5, 39.1, 36.6, 31.9, 31.4, 29.7, 29.5, 29.2, 26.6, 25.3, 22.6, 14.0; mass spectrum, *m/z* (relative intensity, %) 210 (M^+, 1.4), 167 (2.4), 97 (100.0), 83 (3.1), 69 (9.0), 55 (11.5); HRMS, calculated for $C_{14}H_{26}O$, 210.1984; found, 210.1992.

Although ethereal solvents are standard fare for reactions of 'RCu,' another solvent of increasing popularity is dimethyl sulfide (Me_2S). Historically, sulfides have served mainly as additives, especially valued for their ability to solubilize CuI and CuBr, thereby obviating the usual call for warming slurries of 2RLi + CuX to effect dissolution [83]. In some situations, greater percentages of Me_2S were involved, but always no more than as co-solvent [9]. When used to the exclusion of, e.g., Et_2O and THF, it appears that not only are these reagents more thermally stable, they are also more reactive toward enones, epoxides, acid halides, etc., than in traditional ethereal media [84]. This may be due to changes in solubility properties, or possibly differences in reagent constitution. From the experimental point of view, its low boiling point (38 °C) simplifies work-up.

Conjugate Addition of 'PhCu·LiI' in Me_2S [84]

$CuI{\cdot}SMe_2$ — 1. PhLi, Et_2O/hexane, −50 °C, 30 min; 2. cyclohexenone (=O), Me_2S, −75 to 0 °C, 2.5 h →

(*ca* 90%)

A 500 ml recovery flask was charged with 20.00 g of CuI (105.0 mmol, Alfa 'ultrapure'), which were dissolved in 40 ml of deoxygenated (Ar sparge) Me_2S (Aldrich, gold label) at 25 °C under N_2. On cooling the solution to −50 °C, a white solid precipitated; therefore, an additional 120 ml of Me_2S were added to the cold suspension in order to redissolve the CuI. A 55.5 ml volume of 1.86 M PhLi (103 mmol, 0.17 M residual base, Aldrich) solution (diethyl ether–cyclohexane) was added with a syringe over *ca* 1 h. The dark greenish yellow solution was stirred at −50 °C for 30 min. It was then cooled to −75 °C (*ca* 15 min) and 9.63 g of 2-cyclohexenone (100.2 mmol, Aldrich, freshly opened bottle) dissolved in 15 ml of Me_2S were added to the rapidly stirred solution over *ca* 3 min with a cannula from a 50 ml strawberry-shaped flask cooled in a dry-ice–2-propanol bath. After 2 h at −75 °C and 30 min at 0 °C, the reaction was complete. Work-up consisted of the addition of 100 ml of 3 M aqueous NH_4Cl, separation of phases and extraction of the organic phase with 4 × 100 ml of 3 M aqueous NH_4Cl. The combined aqueous phases were back-extracted with 100 ml of diethyl ether. The combined organic layers were dried over anhydrous Na_2SO_4 and the solvent was removed by rotary evaporation *in vacuo*. (A dry-ice trap was inserted between the rotary evaporator and the aspirator to which it was connected.) The residue was treated with 100 ml of hexane and filtered; the filter cake was washed with a total of 75 ml of fresh hexane. The hexane was removed *in vacuo* and the residue, which still contained some solid, was dissolved in 100 ml of diethyl ether, which was extracted with 2 × 100 ml of 0.5 M sodium thiosulfate. The combined thiosulfate layers were back-extracted with 100 ml of diethyl ether and the combined ether layers were dried over anhydrous Na_2SO_4. Rotary evaporation left 16.8 g of crude 3-phenylcyclohexanone (94% pure by GLC). Flash chromatography on a 60 mm × 30 mm i.d. column of basic alumina (50 g, Woelm activity I) eluted with hexane afforded 13.7 g of product in the first four 50 ml fractions. Further elution with 200 ml of diethyl ether yielded 2.3 g of product. The purity was not improved by this chromatography; therefore, 15.9 g of the chromatographed material were distilled at 0.01 mmHg. Three fractions were collected: 0.8 g (65–88 °C, 49% pure by GLC), 3.0 g (88–92 °C, 91% pure) and 8.3 g (92–94 °C, 99% pure). The main impurity was biphenyl from the commercial PhLi solution. (Little biphenyl was observed in the small-scale reactions, which employed solid PhLi, free of biphenyl.) ^1H NMR ($CDCl_3$), δ 1.82 (2H), 2.10 (2H), 2.40 (2H), 2.55 (2H), 3.01 (1H), 7.20 (3H), 7.30 (2H); ^{13}C NMR ($CDCl_3$), δ 25.5, 32.8, 41.2, 44.7, 48.9, 126.6 (2C), 126.7, 128.7 (2C), 144.3 211.0; mass spectrum (70 eV), m/z (relative intensity, %) 27 (16), 28 (7), 29 (6), 39 (38), 40 (6), 41 (22), 42 (55), 50 (14), 51 (35), 52 (11), 53 (6), 55 (10), 63 (17), 65 (19), 70 (12), 74 (6), 75 (6), 76 (9), 77 (50), 78 (57), 79 (7), 82 (5), 83 (15), 89 (9), 91 (50), 92 (7), 102 (12), 103 (52), 104 (97), 105 (20), 115 (39), 116 (10), 117 (100), 118 (33), 128 (7), 129 (7), 131 (78), 132 (11), 145 (5), 146 (6), 174 (87), 175 (11).

Switching the exposure of organocopper complexes from Lewis bases to Lewis acids, most notably $BF_3 \cdot Et_2O$, brings about (not surprisingly) some major changes not only in the chemistry observed but in the composition of the reagents themselves. Introduction of $\geqslant 1$ equiv. of $BF_3 \cdot Et_2O$ to the products of metathesis between RLi and CuI (cf. equation 4.10) leads to a reagent mixture highly prone toward both displacements of allylic leaving groups [85], as well as Michael additions to enones, enoates and unsaturated acids [86, 87]. The standard protocol involves little more than cooling the RCu·LiI formed initially and introducing the Lewis acid followed by the substrate. With the former class of substrates, complete allylic rearrangement is the norm.

Allylic Alkylation of '$RCu \cdot BF_3$' [85]

Cl — n-BuCu·BF_3, THF, −78 °C to r.t. → Bu-n

(94%)

In a 200 ml flask, equipped with a magnetic stirrer and maintained under N_2, were placed 1.9 g (10 mmol) of CuI and 20 ml of dry THF. n-Butyllithium in hexane (1.3 M, 10 mmol) was added at −30 °C, and the resulting mixture was stirred at this temperature for 5 min. The mixture was then cooled to −70 °C, and $BF_3 \cdot OEt_2$ (47%, 1.3 ml, 10 mmol) was added. After the mixture had been stirred for a few minutes, cinnamyl chloride (1.53 g, 10 mmol) was added and the mixture was allowed to warm slowly to room temperature with stirring. The product was filtered through a column of alumina using light petroleum. The olefin thus obtained in essentially pure form was distilled under reduced pressure: 1.64 g, 94%, b.p. 65–66 °C (5 mmHg).

Conjugate Addition of '$RCu \cdot BF_3$' to an α,β-Unsaturated Acid [86, 87]

COOH — n-BuCu·BF_3, Et_2O, −70 °C to r.t. → COOH, Bu-n

(81%)

In a 200 ml flask, equipped with a magnetic stirrer and maintained under N_2, were placed 60 ml of dry diethyl ether and 5.7 g (30 mmol) of purified CuI. n-Butyllithium in hexane (1.3 M, 30 mmol) was slowly added at −30 to −40 °C, and the resulting dark brown suspension was stirred for 5 min. The mixture was then cooled to −70 °C, and $BF_3 \cdot OEt_2$ (47%, 3.9 ml, 30 mmol) was slowly added. The colour changed from dark brown to black and BuCu·BF_3 seemed to be present as a precipitate. After the mixture had been stirred for a few minutes, diethyl ether solution of crotonic acid (0.86 g, 10 mmol) was added at −70 °C. The color immediately changed to deep black. The mixture was allowed to warm slowly to room temperature with stirring. Addition of water, separation and distillation yielded the desired carboxylic acid: 11.16 g, 81%, b.p. 75–76 °C (1 mmHg).

Although the RLi + CuLi + BF_3 combination was originally described as $RBF_3^-Cu^+$ and now commonly as '$RCu \cdot BF_3$,' more recent evidence has shown that the LiI present (see equation 4.10) plays a critical role [88]. In fact, these reagents clearly involve iodocuprates, perhaps of the general form RCu(I)Li or $R(I_2)Cu_2Li$, which arise as a result of the metathesis and/or by the action of BF_3 on the initially formed R(I)CuLi dimer (equation 4.11). Irrespective of these subtle changes, the reagent system '$RCu \cdot BF_3$' provides, in some instances, the only alternative for successful 1,4-additions (e.g. with unsaturated acids; see above).

$$2RCu \cdot LiI \equiv [R(I)CuLi]_2 \underset{\text{THF, } -80\,^{\circ}\text{C}}{\overset{2BF_3}{\rightleftharpoons}} RLi \cdot BF_3 + 2CuI \qquad (4.11)$$

As an alternative to $BF_3 \cdot Et_2O$ as co-reagent with 'RCu' in Michael additions, Me_3SiCl has also been found to function admirably in this regard. Usually, to drive a reaction to completion forming the product silyl enol ether, other additives such as HMPA [41] and/or basic amines (e.g. Et_3N) [40d] are deemed necessary. One recent report has appeared on '$RCu \cdot Me_3SiCl$' as a successful combination for effecting 1,4-delivery of allylic ligands to α,β-unsaturated ketones [89]. The allylic copper species derive from lithiation of an allylic stannane, followed by treatment with precooled (to −78 °C) LiX-solubilized CuI in THF, to which is then added Me_3SiCl (equation 4.12). Allylic systems examined include allyl, methallyl, crotyl and prenyl. Crotylcopper reacts virtually exclusively at its α-site, as is true for prenylcopper. However, mixes of *E*- and *Z*-isomers are to be expected from the former ($E:Z \approx 3:1$). This method gives good to excellent yields of conjugate adducts, but only moderate results with highly hindered enones.

SnBu 3 → 1. MeLi, THF, −78 °C; 2. CuI·LiX, THF −78 °C; 3. Me_3SiCl, −78 °C → Cu(X)Li•TMSCl (4.12)

1,4-Addition of Allylcopper·Me_3SiCl to an α,β-Unsaturated Ketone [89]

$SnBu_3$ → 1. MeLi, THF, −78 °C; 2. CuI·LiCl THF, −78 °C → Cu(I)Li → 1. TMSCl; 2. (4-isopropyl-2-cyclohexenone); 3. n-Bu_4NF → (87%)

CuI (0.400 g, 2.10 mmol) and dry LiCl (0.89 g, 2.10 mmol) were placed in a 10 ml round-bottomed flask equipped with a stir bar and sealed with a septum. The flask was evacuated and purged with Ar; the process was repeated three times. THF (1.5 ml) was injected, and the mixture was stirred for 5 min to yield a yellow, homogeneous solution which was then cooled to −78 °C. Concurrently, a solution of allyllithium (2.0 mmol) was prepared from allyltri-*n*-butylstannane (0.62 ml, 2.0 mmol) and MeLi (1.25 ml, 2.00 mmol) in THF (1.0 ml) at −78 °C (15 min). This solution was then transferred via a dry ice-cooled cannula to the CuI·LiCl solution (−78 °C) to yield a tan solution. TMSCl (0.17 ml, 2.1 mmol) was added followed immediately by the neat addition of 4-isopropyl-2-cyclohexenone (0.11 ml, 0.75 mmol). The reaction was allowed to proceed for 30 min before being quenched with 5 ml of saturated aqueous NH_4Cl solution. Extraction with 4 × 20 ml of diethyl ether was followed by combining the organic layers and drying over Na_2SO_4.

The solvent was then removed *in vacuo* and the resulting oil was treated with THF (5 ml) and n-Bu_4NF (2.0 ml, 20 mmol) for 15 min. The solvent was again removed *in vacuo*, and the residue was subjected to flash chromatography [light petroleum–EtOAc (9:1)] to yield 0.118 g (87%) of 3-(1-propen-3-yl)cyclohexanone as a colorless oil; TLC, R_f [light petroleum–EtOAc (9:1] 0.28; IR (neat) 3080, 2960, 1710, 1640, 1450, 990, 910 cm^{-1}; ^1H NMR (500 MHz, $CDCl_3$), δ 5.74–5.65 (m, 1H), 5.05–5.00 (m, 2H), 2.38–2.34 (m, 2H), 2.26–1.80 (m, 4H), 1.66–1.25 (m, 4H), 0.97 (d, 3H), 0.80 (d, 3H); EI mass spectrum, m/z (relative intensity, %) 180 (M^+, 6), 139 (12), 111 (13), 97 (32), 95 (20, 83 (73), 69 (62), 55 (100), 43 (51); HRMS, calculated for $C_{12}H_{20}O$ (M^+), 180.1513; found, 180.1512.

4.2.6 REACTIONS OF SILYL AND STANNYL CUPRATES ($R_3MCu{\cdot}L_n$, M = Si, Sn)

4.2.6.1 Si–Cu Reagents

Organosilicon intermediates occupy a fairly prominent position in synthetic organic chemistry, as they are extremely versatile for constructing both carbon–carbon and carbon–heteroatom bonds [90]. Aside from the simplest of these materials, which may be purchased from commercial sources, organosilanes such as those bearing vinylic and allylic appendages must be prepared prior to use. One versatile approach utilizes organocuprate chemistry, based on CuCN [91]. Starting with $PhMe_2SiCl$, the red lithiosilane ($PhMe_2SiLi$) is formed using Li(0) (lithium shot) [92]. This organometallic is stable for a few days at 0 °C, and requires titration (usually 0.9–1.2 M) to insure accuracy in the subsequent cuprate-forming step.

Preparation of $Me_2PhSiLi$ [92]

$$Me_2PhSiCl \xrightarrow[\text{THF, 0 °C, 18 h}]{\text{Li(0)}} Me_2PhSiLi$$

Chlorodimethyl(phenyl)silane (5 ml, Aldrich or laboratory-made, in both cases contaminated with about 10% of the bromide, coming from its preparation using phenylmagnesium bromide and dichlorodimethylsilane) was vigorously stirred with lithium shot (1 g) in dry THF (30 ml) under N_2 or Ar at 0 °C for 18 h to give a deep red solution of the silyl lithium reagent. In general, the red color was persistent only after 0.5–1 h. The molarity of the solution was measured by adding a 0.8 ml aliquot to water (10 ml) and titrating the resulting mixture against HCl (0.1 M) using phenolphthalein as indicator. The solution was generally found to be 0.9–1.2 M.

Alternatively, $(PhMe_2Si)_2$ can be cleaved with Na-containing lithium wire to arrive at halide-free $PhMe_2SiLi$, although the presence of LiX salts has no effect on the outcome of the cuprate reactions. Once obtained, addition to CuCN (0.5 equiv.) affords reddish solutions of the higher order disilylcuprate (equation 4.13). These reagents

$$PhMe_2SiCl \xrightarrow[\substack{\text{ThF} \\ \text{0 °C, 20 min}}]{\text{Li(0)}} PhMe_2SiLi \xrightarrow[\text{THF 0 °C, 20 min}]{\text{0.5 CuCN}} (PhMe_2Si)_2Cu(CN)Li_2 \quad (4.13)$$

smoothly add Michael-wise to α,β-unsaturated ketones [91b], readily displace allylic acetates (the regio- and stereochemistry of which is substrate and conditions dependent) [91c], and effect silylcupration of 1-alkynes to afford (E)-vinylsilanes.

Silylcupration of a 1-Alkyne Using $(PhMe_2Si)_2Cu(CN)Li_2$ [91a]

$$CH_3(CH_2)_{10}C{\equiv}CH \xrightarrow[\text{THF, Et}_2\text{O, }-78\ ^\circ\text{C, 2 h; then 0 }^\circ\text{C, 30 min}]{(Me_2PhSi)_2Cu(CN)Li_2} CH_3(CH_2)_{10}CH{=}CHSiMe_2Ph \quad (95\%)$$

The silyllithium reagent (see above) (73.3 ml of a 1.14 M solution in THF, 83.6 mmol) was added by syringe to a slurry of copper(I) cyanide (3.74 g, 41.9 mmol) in THF (50 ml) at 0 °C over 2 min. The mixture was stirred at 0 °C for 20 min and cooled to −78 °C. The color of the solution changed little, and vigorous stirring was essential for all the cyanide to react. Tridecyne (5.7 g, 31.7 mmol) was added in diethyl ether (15 ml) and the mixture stirred for 2 h, allowed to warm to 0 °C and stirred for a further 30 min, by which time the solution had turned black. The reaction was quenched with aqueous NH_4Cl solution at 0 °C, stirred for 10 min at this temperature, brought to room temperature and filtered over Celite. The layers were separated and the aqueous layer extracted with diethyl ether. The organic layers were dried (Na_2SO_4) and evaporated *in vacuo*. Chromatography (SiO_2, hexane) gave the vinylsilane (10.5 g, 95%) as an oil: TLC, R_f (hexane) 0.51; IR (film), 1610 (C=C), 1245 (SiMe), and 1110 (SiPh); ^1H NMR ($CDCl_3$), δ 7.54–7.32 (m, 5H, Ph), 6.12 (dt, 1H, J = 18.5 and 6.2 Hz, SiCH=C*H*), 5.74 (dt, 1H, J = 18.5 and 1.4 Hz, SiC*H*=CH), 2.13 (m, 2H, CH=CHCH_2), 1.40–1.20 (m, 18H, CH_2), 0.88 (t, 3H, J = 6.6 Hz, Me), 0.31 (s, 6H, $SiMe_2$); mass spectrum, m/z (relative intensity, %) 316 (M^+, 6), 301 (40), 162 (30), 161 (30), 135 (50), 121 (100); HRMS, calculated for $C_{21}H_{36}Si$, 316.2586; found, 316.2572.

Silyl ligands containing other substitution patterns, e.g. *tert*-butyldimethyl- and thexyldimethylsilyl-, can be delivered in a manner analogous to the chemistry of the $PhMe_2Si$ group. The preparation of mixed cuprates t-$BuMe_2Si(R)Cu(CN)Li_2$ and (thexyl)$Me_2Si(R)Cu(CN)Li_2$, R = Me or n-Bu, however, relies on the transmetallation of the corresponding silyltrimethylstannanes, $R'Me_2SiSnMe_3$ (R′ = thexyl or *tert*-butyl) [93]. These water-white, stable liquids are easily prepared from the corresponding silyl chlorides and Me_3SnLi. By simply mixing $R'Me_2SiSnMe_3$ with, e.g., $Bu_2Cu(CN)Li_2$ at room temperature, the desired higher order cuprate is formed directly without prior lithiation of the silane (equation 4.14). Once formed, the higher order silyl cuprates can be expected to react readily with the usual sorts of coupling agents.

$$R'Me_2SiSnMe_3 + Me_2Cu(CN)Li_2 \xrightarrow[\text{r.t., 1 h}]{\text{THF}} R'Me_2Si(Me)Cu(CN)Li_2 + Me_4Sn \qquad (4.14)$$

Preparation of Thexyldimethylsilyltrimethylstannane [94]

$$Me_3SnH \xrightarrow[\text{THF}]{\text{LDA}} Me_3SnLi \xrightarrow[0\ ^\circ\text{C, 3 h}]{\text{thexyl}SiMe_2Cl} \text{thexyl}SiMe_2{-}SnMe_3 \quad (>60\%)$$

Caution: This reaction should be performed in an efficient hood owing to the extreme toxicity and volatility of Me_3SnH. To 40 ml of THF was added diisopropylamine (3 ml, 29.7 mmol) with a syringe and the solution was cooled to −78 °C, after which was added n-BuLi in hexane (12 ml, 26.4 mmol). Warming this solution by removal of the ice-bath for 10 min and then recooling to −78 °C was followed by addition of trimethyltin hydride (7.8 ml, 25.9 mmol) [95]. The reaction mixture was warmed to 0 °C for 1 h. At this temperature thexyldimethylsilyl chloride (7.4 ml, 37.6 mmol) was added and this solution was then stirred at 0 °C for 3 h. After this time, the heterogeneous mixture was filtered using a Schlenk funnel and distilled (72 °C, 0.2 mmHg) yielding a clear, water-white liquid (>60%); TLC, R_f (hexane) 0.9; IR (neat), 1460, 1260, 780 cm^{-1}; ^1H NMR (500 MHz, $CDCl_3$), δ 1.62–1.60 (m, 1H), 0.88–0.87 (m, 12H), 0.20 (s, 6H, $^3J_{Sn} = 16$ Hz), 0.07 (s, 9H, $^2J_{Sn} = 23$ Hz); EI mass spectrum, m/z (relative intensity, %) 293 (M^+ – Me, 10), 119 (20), 73 (100), 69 (85); HREIMS, calculated for $C_{11}H_{28}Si^{120}Sn$ – Me, 293.0747; found, 293.0770.

Ring Opening of Isoprene Oxide with (Thexyl)$Me_2Si(Bu)Cu(CN)Li_2$ [93, 94]

1. $n\text{-}Bu_2Cu(CN)Li_2$, THF, 22 °C, 1.5 h
2. (isoprene oxide), –78 °C, 30 min

(74%)
(E:Z 9:1)

To a two-necked flask equipped with a two-way valve connected to a source of Ar and a vacuum pump was added CuCN (45.6 mg, 0.51 mmol). It was gently flame-dried under vacuum followed by Ar addition; this evacuation and Ar re-entry process was repeated three times. THF (1 ml) was then cooled to –78 °C and thexyldimethylsilyltrimethylstannane (0.18 ml, 0.73 mmol) was introduced neat with a syringe. The reaction was warmed to 22 °C for 1.5 h to complete the transmetallation, following which the yellow homogeneous cuprate was recooled to –78 °C and epoxide (0.04 ml, 0.404 mmol) was added also neat with a syringe. The solution was then stirred for an additional 30 min, after which it was quenched at –78 °C via addition of 10% aqueous troduced neat via a syringe. The reaction was warmed to 22 °C for 1.5 h to complete the transmetallation, following which the yellow homogeneous cuprate was recooled to –78 °C and epoxide (0.04 ml, 0.404 mmol) was added also neat via a syringe. The solution was then stirred for an additional 30 min, after which it was quenched at –78 °C via addition of 10% aqueous ammonia–saturated aqueous NH_4Cl. Work-up involved warming the reaction mixture to 22 °C, addition of Et_2O, extraction and drying of the organic layer (brine–Na_2SO_4), followed by rotary evaporation *in vacuo* and chromatography [silica gel, Et_3N–Et_2O–hexane (2:40:58)] to yield the (E) and (Z)-vinylsilanes as a clear oil (68 mg, 74%) in a 9:1 ratio (GC): TLC, R_f [Et_2O–hexane (1:1)] 0.42; IR (neat), 3325, 1465 cm^{-1}. E-Isomer: ^1H NMR (500 MHz, $CDCl_3$), δ 5.46–5.43 (t, 1H, $J = 9$ Hz), 3.97 (s, 2H), 1.62 (s, 3H), 1.61 (s, 1H), 1.49–1.47 (d, 2H, $J = 9$ Hz), 1.27 (bs, 1H), 0.86–0.81 (m, 12H), 0.05 (s, 6H); EI mass spectrum, m/z (relative intensity, %) 143 (M^+ – C_6H_{13}, <3), 125 (10), 84 (20), 75 (100), 70 (90), 59 (20). Z-Isomer: ^1H NMR, δ 5.33–5.29 (t, 1H, $J = 9$ Hz), 4.07 (s, 2H), 1.77–1.76 (s, 3H), 1.66 (s, 1H), 1.59–1.58 (d, 2H, $J = 6$ Hz), 1.27 (bs, 1H), 0.86–0.81 (m, 12H), 0.06 (s, 6H); EI mass spectrum, m/z (relative intensity, %) 143 (M^+ – C_6H_{13}, <2), 125 (25), 84 (10), 75 (100), 59 (25); HREIMS, calculated for $C_{13}H_{28}OSiC_6H_{13}$, 143.0892; found, 143.0915.

4.3.6.2 Sn–Cu Reagents

Recent advances in organotin chemistry have provided considerable incentive for development of new technologies which position a trialkylstannyl moiety into organic substrates [68, 96]. Several reagents exist from which to choose, depending on the particular transformation of interest. Equations 4.15–4.19 show some of these, along with the derivation of each.

$$Me_3SnSnMe_3 \xrightarrow[\substack{THF\\-20\,^{\circ}C,\\15\,min}]{MeLi} Me_3SnLi \xrightarrow[\substack{THF,\ -48\,^{\circ}C\\10\,min}]{CuBr \cdot SMe_2} \underset{\mathbf{(7)}}{Me_3SnCu \cdot SMe_2} \tag{4.15}$$

$$Me_3SnSnMe_3 \xrightarrow[\substack{THF\\0\,^{\circ}C,\\15\,min}]{MeLi} Me_3SnLi \xrightarrow[-20\,^{\circ}C,\ 15\,min]{PhSCu,\ THF} \underset{\mathbf{(8)}}{Me_3Sn(PhS)CuLi} \tag{4.16}$$

$$Me_3SnSnMe_3 \xrightarrow[\substack{2.\ 2MeLi,\ -20\,^{\circ}C\\50\,min}]{1.\ \text{thiophene},\ THF,\ -20\,^{\circ}C} \left[\text{2-thienyl-Li} + Me_3SnLi\right] \xrightarrow{CuCN} \underset{\mathbf{(9)}}{Me_3Sn(2\text{-}Th)Cu(CN)Li_2} \tag{4.17}$$

$$2Bu_3SnH + Bu_2Cu(CN)Li_2 \xrightarrow[10\,min]{THF,\ -78\,^{\circ}C} \underset{\mathbf{(10)}}{Bu_3Sn(Bu)Cu(CN)Li_2} \tag{4.18}$$

$$Me_3SiSiMe_3 \xrightarrow[2.\ Me_3SnCl]{1.\ MeLi,\ THF,\ HMPA} Me_3SiSnMe_3 \xrightarrow{Bu_2Cu(CN)Li_2} \underset{\mathbf{(11)}}{Me_3Sn(Bu)Cu(CN)Li_2} \tag{4.19}$$

Reagent **7**, prepared from hexamethylditin (see procedure below), is especially effective at stannylcupration of terminal unactivated acetylenes [97]. When used (2 equiv.) in tandem with a proton source (MeOH, 60 equiv.), good to excellent yields of the product 2-trimethylstannyl-1-alkenes are obtained, accompanied by small amounts of the isomeric (*E*)-1-trimethylstannyl-1-alkene. These by-products tend to be unstable on chromatographic separation, and are not easily isolated in small scale work.

Preparation of $Me_3SnCu \cdot SMe_2$ from $Me_3Sn\text{-}SnMe_3$ [98]

$$Me_3SnSnMe_3 \xrightarrow[\substack{2.\ CuBr \cdot SMe_2,\ THF\\-48\,^{\circ}C,\ 10\,min}]{\substack{1.\ MeLi,\ THF\\-20\,^{\circ}C,\ 15\,min}} Me_3SnCu \cdot SMe_2$$

Me_3SnLi [98] To a cold (−20 °C), stirred solution of hexamethylditin in dry THF (*ca* 10 ml per mmol of $Me_3SnSnMe_3$) was added a solution of methyllithium (1.0 equiv.) in diethyl ether. The mixture was stirred at −20 °C for 15 min to afford a pale yellow solution of trimethylstannyllithium.

$Me_3SnCu \cdot SMe_2$ [98] To a cold (−48 °C), stirred solution of trimethylstannyllithium (0.65 mmol) in 5 ml of dry THF was added, in one portion, solid copper(I) bromide–dimethyl sulfide complex (113 g, 0.65 mmol). The mixture was stirred at −48 °C for 10 min to give a dark red solution of the cuprate reagent.

Stannylcupration of a 1-Alkyne with $Me_3SnCu \cdot SMe_2$ [97]

To a cold (−78 °C), stirred solution of $Me_3SnCu \cdot SMe_2$ (0.4 mmol) in 3 ml of dry THF (Ar atmosphere) was added sequentially a dry THF solution (0.5 ml) of the 1-alkyne (39 mg, 0.2 mmol)

$Me_3SnCu \cdot SMe_2$; THF, −78 °C, 10 min; MeOH, −63 °C, 12 h → $SnMe_3$ (84%)

and dry MeOH (0.5 ml, 12 mmol). The dark red reaction mixture was stirred at −78 °C for 10 min and at −63 °C for 12 h. Saturated aqueous NH_4Cl (pH 8) (5 ml) and diethyl ether (20 ml) were added and the mixture was allowed to warm to room temperature with vigorous stirring. Stirring was continued until the aqueous phase became deep blue. The layers were separated and the aqueous phase was extracted with diethyl ether (2 × 5 ml). The combined organic extracts were washed with saturated aqueous NH_4Cl (pH 8) (2 × 5 ml) and dried ($MgSO_4$). Removal of the solvent afforded a crude oil that, on the basis of GC analysis, consisted of a 95:5 mixture of two products. Subjection of the crude oil to flash chromatography on silica gel [15 × 2 cm i.d. column, elution with diethyl ether–light petroleum (1:20)], followed by distillation (air-bath temperature 115–120 °C, 0.7 mmHg) of the oil thus obtained gave 59 mg (84%) of the vinylstannane as a colorless oil; IR (neat), 1140, 1120, 1080, 1040, 915, 770 cm^{-1}; ^1H NMR (80 MHz), δ 0.13 (s, 9H, $^2J_{Sn}$, H = 53 Hz), 1.1–1.9 (m, 14H), 2.1–2.4 (m, 2H), 3.2–4.0 (m, 4H), 4.5–4.7 (m, 1H), 5.12 (br d, 1H, $J = 3$ Hz, $^3J_{Sn,H} = 70$ Hz), 5.63 (dt, 1H, $J = 3$, 1.5 Hz, $^3J_{Sn,H} = 156$ Hz); HRMS, calculated for $C_{15}H_{29}O_2Sn$ (M^+ − Me), 361.1190; found, 361.1203.

The mixed Gilman cuprate **8**, prepared as illustrated below, undergoes 1,4-additions fairly easily, on occasion with assistance from *in situ* HOAc or MeOH. In addition to being an alternative to reagents **7**, they offer particular advantages in reactions with acetylenic esters, where careful control of conditions can lead to excellent *E* vs *Z* stereoselectivities in the resulting enoates. Thus, when conducted in THF containing MeOH at −78 °C for 3 h, the product of *E* stereochemistry results. Warming to −48 °C for 4 h followed by addition of MeOH reverses the isomeric ratio (Scheme 4.3).

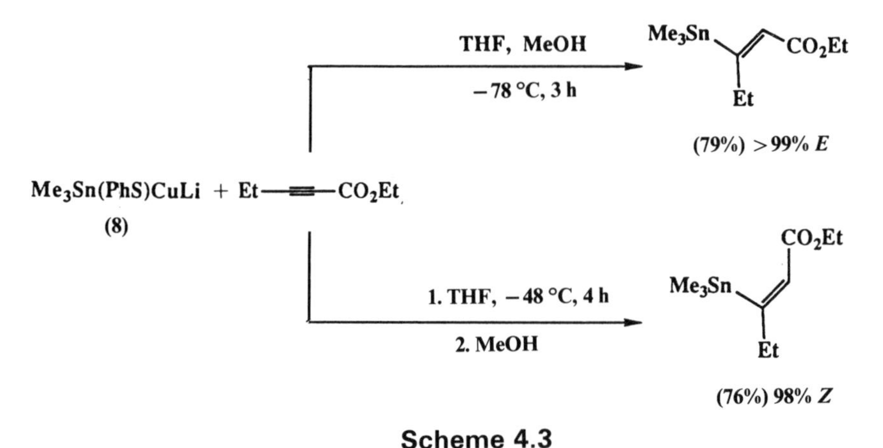

Scheme 4.3

Preparation of $Me_3Sn(SPh)CuLi$ [98]

$$Me_3SnLi \xrightarrow[\text{THF, } -20\,^\circ\text{C, 15 min}]{\text{PhSCu}} Me_3Sn(PhS)CuLi$$

To a cold (−20 °C), stirred solution of trimethylstannyllithium (0.75 mmol) in 10 ml of dry THF was added in one portion solid phenylthiocopper(I) (132 mg, 0.75 mmol). The slurry was stirred at −20 °C for 15 min to afford a dark red solution of the cuprate reagent.

Stannylcupration of an Acetylenic Ester with a Mixed Lower Order Stannylcuprate at −78 °C in the Presence of MeOH [99]

$$\mathrm{Et{-}C{\equiv}C{-}CO_2Et} \xrightarrow[\text{THF, MeOH, } -78\,^{\circ}\text{C, 3 h}]{\mathrm{Me_3Sn(PhS)CuLi}} \mathrm{Me_3Sn(Et)C{=}CH{-}CO_2Et}\ (79\%)$$

To a cold (−100 °C), stirred solution of the cuprate reagent $Me_3Sn(PhS)CuLi$ (1.0 mmol) in 10 ml of dry THF was added, dropwise, a solution of the α,β-acetylenic ester (0.5 mmol) in 0.5 ml of dry THF containing 0.85 mmol of dry MeOH. The reaction mixture was stirred at −100 °C for 15 min and at −78 °C for 3 h. MeOH (0.2 ml) and Et_2O (30 ml) were added and the mixture was allowed to warm to room temperature. The resulting slurry was filtered through a short column of silica gel (10 g, elution with 30 ml of Et_2O). The oil obtained by concentration of the combined eluate was chromatographed on silica gel (*ca* 3 g). Elution with light petroleum (*ca* 10 ml) gave $Me_3SnSnMe_3$. Further elution with Et_2O (*ca* 8 ml), followed by distillation of the material thus obtained, provided the product (79%); distillation temperature 110–125 °C (20 mmHg); IR (neat), 1715, 1598, 1175 cm^{-1}; ^1H NMR (100 MHz), δ 0.12 (s, 9H, $^2J_{Sn,H} = 54$ Hz), 0.99 (t, 3H, $J = 8$ Hz), 1.22 (t, 3H, $J = 7$ Hz), 2.85 (br q, 2H, $J = 8$ Hz), 4.11 (q, 2H, $J = 7$Hz), 5.89 (t, 1H, $J = 1.5$ Hz, $^3J_{Sn,H} = 73$ Hz); HRMS, calculated for $C_9H_{17}O_2Sn$ ($M^+ - Me$), 277.0250; found, 277.0250. Analysis: calculated for $C_{10}H_{20}O_2Sn$, C 41.28, H 6.93; found C 41.36, H 7.02%.

Stannylcupration of an Acetylenic Ester with Me₃Sn(PhS)CuLi at −48 °C [99]

$$\mathrm{Et{-}C{\equiv}C{-}CO_2Et} \xrightarrow[\text{THF, } -48\,^{\circ}\text{C, 4 h; 2. MeOH}]{\text{1. } \mathrm{Me_3Sn(PhS)CuLi}} \mathrm{Et(Me_3Sn)C{=}CH{-}CO_2Et}\ (76\%)$$

To a cold (−78 °C), stirred solution of the cuprate reagent $Me_3Sn(PhS)CuLi$ (0.39 mmol) in 5 ml of dry THF was added a solution of the α,β-acetylenic ester (0.3 mmol) in 0.5 ml of dry THF. The reaction mixture was stirred at −78 °C for 15 min and at −48 °C for 4 h. MeOH or EtOH (0.2 ml) and Et_2O (30 ml) were added and the mixture was allowed to warm to room temperature. The yellow slurry was treated with anhydrous $MgSO_4$ and was then filtered through a short column of Florisil (elution with 30 ml of Et_2O). Concentration of the combined eluate gave an oil, which was distilled directly to give the product; distillation temperature 120–125 °C (20 mmHg); IR (neat), 1701, 1601, 1195, 773 cm^{-1}; ^1H NMR (100 MHz), δ 0.12 (s, 9H, $^2J_{Sn,H} =$ 54 Hz), 0.98 (t, 3H, $J = 7.5$ Hz), 1.24 (t, 3H, $J = 7$ Hz), 2.41 (br q, 2H, $J = 7.5$ Hz), 4.15 (q, 2H, $J = 7$ (Hz), 6.34 (t, 1H, $J = 2$ Hz, $^3J_{Sn,H} = 121$ Hz). HRMS, calculated for $C_9H_{17}O_2Sn$ ($M^+ - Me$), 277.0250; found, 277.0252. Analysis: calculated for $C_{10}H_{20}O_2Sn$, C 41.28, H 6.93; found, C 41.58, H 7.10%.

Cyanocuprates **9**, **10** and **11** are of more recent vintage, and it is clear that whether a thienyl or alkyl ligand is part of their make-up, the trialkylstannyl moiety is always transferred (virtually exclusively) over the remaining ligand. All three routes (equations 4.12–4.14) involve one-pot protocols, not an insignificant feature, especially for large-scale processes. Cuprates **9** are also useful for generating vinylstannanes of general structure **12** (cf. Scheme 4.3, use of mixed cuprate **8**), and also several other product types.

(12)

Preparation of $Me_3Sn(2\text{-}Th)Cu(CN)Li_2$ [100]

$$(Me_3Sn)_2 \xrightarrow[\substack{2.\ 2\,MeLi,\ -20\,°C \\ 50\,min \\ 3.\ CuCN,\ -78\,°C \\ 10\,min \\ 4.\ -48\,°C,\ 10\,min}]{\substack{1.\ thiophene \\ THF,\ -20\,°C}} Me_3Sn(2\text{-}Th)Cu(CN)Li_2$$

To a cold (−20 °C), stirred solution of $(Me_3Sn)_2$ (164 mg, 0.5 mmol) in 10 ml of dry THF were added, successively, thiophene (42 mg, 0.5 mmol) and a solution of MeLi (1.0 mmol, low halide or LiBr complex) in Et_2O. After the pale yellow solution had been stirred at −20 °C for 50 min, it was cooled to −78 °C and CuCN (45 mg, 0.5 mmol) was added. The resulting suspension was stirred for 5 min at −78 °C and for 10 min at −48 °C to provide a bright yellow solution of the cuprate reagent. The solution was cooled to −78 °C and used immediately.

Michael Addition of the Mixed Higher Order Stannylcuprate $Me_3Sn(2\text{-}Th)Cu(CN)Li_2$ [100]

$$\xrightarrow[\text{THF, } -78 \text{ to } -20\,°C,\ 4\,h]{Me_3Sn(2\text{-}Th)Cu(CN)Li_2}$$

SnMe$_3$

(90%)

To a cold (−78 °C), stirred solution of $Me_3Sn(2\text{-}Th)Cu(CN)Li_2$ (see above) (0.5 mmol) in 10 ml of dry THF (Ar atmosphere) was added 37 mg (0.33 mmol) of the enone. After the solution had been stirrerd at −78 °C for 5 min and at −20 °C for 4 h, it was treated with saturated aqueous NH_4Cl–aqueous ammonia (pH 8) (10 ml) and Et_2O (10 ml). The vigorously stirred mixture was exposed to air and allowed to warm to room temperature. The phases were separated, and the aqueous phase was extracted with Et_2O (3 × 10 ml). The combined organic extracts were dried ($MgSO_4$) and concentrated. Flash chromatography [Et_2O–petroleum (1:4)] of the residual oil, followed by distillation (90 °C, 2.0 mmHg) of the material thus obtained gave 83 mg (90%) of the product; IR (neat), 1713, 1452, 1224, 768 cm^{-1}; ^1H NMR (300 MHz), δ 0.06, (s, 9H, $^2J_{Sn,H}$ = 50 Hz), 1.20 (s, 3H, $^3J_{Sn,H}$ = 60 Hz), 1.55–2.60 (series of m, 10H). HRMS, calculated for $C_9H_{17}OSn$ (M^+ − Me), 261.0301; found, 261.0306.

An especially simple route to Bu_3Sn-incorporated higher order cuprates involves a presumed transmetallation between Bu_3SnH and $Bu_2Cu(CN)Li_2$, which occurs on mixing of these components at −78 °C [101]. The resulting reagent behaves as a mixed cuprate (**10**) and transfers the Bu_3Sn moiety selectively to several types of educts. A 2:1 stoichiometry of tin hydride to $Bu_2Cu(CN)Li_2$ is essential, and the commercially obtained Bu_3SnH should be dry and from a relatively new bottle (or if not, then freshly distilled).

Stannylcupration of a 1-Alkyne Using $Bu_3SnCu(CN)Li_2$ Generated via Transmetallation of n-*Bu_3SnH [101]*

$$Bu_3SnH \xrightarrow[\substack{-78\,^\circ C,\ 10\ min \\ 2.\ HOCMe_2C{\equiv}CH \\ -78\,^\circ C,\ 5\ min}]{1.\ n\text{-}Bu_2Cu(CN)Li_2,\ THF} HO\text{-}CMe_2\text{-}C(=CH_2)SnBu_3 \quad (87\%)$$

CuCN (67.2 mg, 0.75 mmol) was added to a dry 10 ml round-bottomed flask equipped with a stir bar and rubber septum. The flask was evacuated with a vacuum pump and purged with Ar. This process was repeated three times. THF (2 ml) was injected and the slurry cooled to −78 °C, then *n*-BuLi (0.63 ml, 1.50 mmol) was added dropwise. The mixture was allowed to warm slightly to yield a colorless, homogeneous solution which was recooled to −78 °C where *n*-Bu_3SnH (0.40 ml, 1.5 mmol) was added via a syringe. Stirring was continued and over *ca* 10 min the solution yellowed and H_2 gas was liberated. 2-Methyl-3-butyn-2-ol (0.066 ml, 0.68 mmol) was added neat via a syringe and the reaction mixture stirred for 5 min before being quenched into a 10 ml bath of 10% aqueous ammonia–90% saturated NH_4Cl. Extraction with 3 × 20 ml of diethyl ether was followed by combining the extracts and drying over Na_2SO_4. The solvent was removed *in vacuo* and the residue chromatographed on silica gel. Elution with light petroleum–ethyl acetate (95:5) + 1% Et_3N gave 2-methyl-3(tri-*n*-butylstannyl)-3-buten-2-ol (221 mg, 87%) as a colorless oil: TLC, R_f [EtOAc–light petroleum (10:90)] 0.38; IR (neat), 3360, 2930, 1600, 1460, 1380, 990 cm^{-1}; ^1H NMR, δ 6.09–6.08 (d, 2H, J_{gem} = 2 Hz), 2.15 (s, 1H), 1.53–1.43 (m, 12H), 1.35–1.23 (m, 12H), 0.90–0.85 (t, 9H); ^{13}C NMR ($CDCl_3$), δ 155.56 (s), 122.35 (t), 30.60 (s), 29.40 (q), 29.03 (t), 27.20 (t), 13.65 (q), 9.39 (t); EI mass spectrum, *m/z* (relative intensity, %) 319 (M^+ − 57, 100), 318 (36.7), 317 (75.5), 316 (30.6), 315 (42.9), 263 (42.9), 261 (33.7), 207 (42.9), 205 (36.7), 177 (22.4), 136 (25.5), 120 (25.5), 119 (10.2), 118 (12.2), 117 (10.2), 116 (10.2), 85 (17.3), 69 (10.2), 59 (11.2); HRCIMS, calculated for $C_{17}H_{36}Sn$, (M^+ − 17), 360.1839; found, 360.1812.

Although the above transmetallation concept has been applied to the formation of the Me_3Sn analog [i.e. $Me_3Sn(Me)Cu(CN)Li_2$], another procedure which does not involve Me_3SnH [95] has been developed [102]. The *in situ* generation of $Me_3SnSiMe_3$, which uses the Me_3Si group as a bulky proton, likewise transmetallates with $R_2Cu(CN)Li_2$ to afford an Me_3Sn-containing mixed higher order cuprate (**11**) equation 4.19) [103]. Solutions of $Me_3SnSiMe_3$ in THF–HMPA are stable at room temperature for months as long as they are protected from air and moisture. All operations are conducted in a single flask, and the overall process leading to **11** appears to proceed in virtually quantitative yield.

*Stannylcupration of a 1-Alkyne Using Me_3Sn(*n*-$Bu)Cu(CN)Li_2$ Formed* in situ *from Transmetallation of $Me_3SnSiMe_3$ [102]*

$$Me_3SiSiMe_3 \xrightarrow[\substack{2.\ Me_3SnCl,\ -78\ to\ -50\,^\circ C,\ 1.5\ h \\ 3.\ n\text{-}Bu_2Cu(CN)Li_2,\ -78\ to\ -50\,^\circ C,\ 1\ h \\ 4.\ HOCH_2CH_2C{\equiv}CH,\ MeOH,\ -78\,^\circ C\ to\ r.t.}]{\substack{1.\ MeLi,\ THF,\ HMPA \\ -78\ to\ -30\,^\circ C,\ 1\ h}} HOCH_2CH_2C(=CH_2)SnMe_3 \quad (74\%)$$

Ethereal methyllithium (6.6 ml, 1.5 M) was added to a solution of hexamethyldisilane (2.0 ml, 10.0 mmol. Aldrich) in 24 ml of THF–HMPA (3:1 v.v) at −78 °C under Ar. The resulting deep red solution was stirred for 1 h while allowing it to warm to −30 °C. The reaction mixture was cooled to −78 °C, after which Me_3SnCl (1.99 g, 10.0 mmol, available from Aldrich) in 2 ml of THF was added. The reaction mixture was further stirred for 1.5 h while warming to −50 °C. In a separate vessel, $Bu_2Cu(CN)Li_2$ (10.0 mmol, prepared from 8.7 ml of 2.3 M *n*-BuLi and 0.89 g of CuCN) in 10 ml of THF was prepared at −45 °C. After stirring for 30 min, this solution was transferred with a cannula to a solution of $Me_3SnSiMe_3$ at −78 °C. The resulting lemon yellow solution was warmed to −50 °C and stirred for 1 h to ensure complete transmetallation. 3-Butyn-1-ol (0.63 g, 9.0 mmol) was then added neat with a syringe followed by 5 ml of MeOH. The reaction mixture immediately turned red. After 30 min the bath was removed and the solution warmed to room temperature. Usual work-up followed by chromatography on silica gel [hexane–ethyl acetate (8:1)] gave 1.56 g (74%0 of 4-hydroxy-2-trimethylstannyl-1-butene and 0.17 g (8%) of 4-hydroxy-1-trimethylstannyl-1-butene. GC analysis revealed a purity of >95% for both isomers. 4-Hydroxy-2-trimethylstannyl-1-butene: IR (neat), 3350 cm^{-1}; NMR (500 MHz, $CDCl_3$), δ 0.14 (s, 9H, Me_3Sn, $^2J_{Sn,H} = 54$ Hz), 1.34 (t, 1H, O*H*), 2.5 (t, 2H, $CH_2C{=}C$), 3.6 (q, 2H, OCH_2), 5.3 (d, 1H, $C{=}CH_2$, $J = 2$ Hz, $^3J_{Sn,H} = 69$ Hz), 5.7 (d, 1H, $C{=}CH_2$, $J = 2$ Hz, $^3J_{Sn,H} = 147$ Hz); EI mass spectrum, *m/z* 219 ($M^+ - 15$); HRMS, calculated for $C_6H_{13}SnO$ ($M^+ - 15$), 219.9895; found, 219.9960. 4-Hydroxy-1-trimethylstannyl-1-butene: IR (neat), 3350 cm^{-1}; NMR (500 MHz, $CDCl_3$), δ 0.10 (s, 9H, Me_3Sn, $^2J_{Sn,H} = 55$ Hz), 1.36 (t, 1H, O*H*), 2.4 (t, 2H, $CH_2C{=}C$), 3.7 (q, 2H, OCH_2), 5.9 (dt, 1H, $C{=}CH_2$, $J = 19, 6$ Hz, $^2J_{Sn,H} = 60$ Hz), 6.1 (d, 1H, $C{=}CH_2$, $J = 19$ Hz, $^3J_{Sn,H} = 32$ Hz); EI mass spectrum, *m/z* 219 ($M^+ - 15$); HRMS, calculated for $C_6H_{13}SnO$ ($M^+ - 15$), 219.9895; found, 219.9971.

4.2.7 OTHER CUPRATE AGGREGATES ($R_5Cu_3Li_2$)

The 2:1 ratio of organolithium(s) to CuX (X = halogen, CN) resulting in either R_2CuLi or $R_2Cu(CN)Li_2$ is by no means the only stoichiometry leading to discrete reagents. Higher ratios, e.g. 3RLi:CuI, give higher order cuprates R_3CuLi_2 [104, 105], while addition of 1.5 equiv. of RLi to CuI in THF leads to R_3Cu_2Li [104a,b] and 1.66 equiv. of RLi plus CuI (in Et_2O) gives $R_5Cu_3Li_2$ [104a, 106]. The higher order species Me_3CuLi_2 is in equilibrium with Me_2CuLi and free MeLi in Me_2O, which tends to limit its use [104a]. R_3Cu_2Li is relatively unreactive [104c], although in the presence of $BF_3{\cdot}Et_2O$ [34] and Me_3SiCl [107] it works well in 1,4-additions to enones. The species $R_5Cu_3Li_2$, with R = Me, is particularly effective (versus, e.g., Me_2CuLi) for conjugate methylation of α,β-unsaturated aldehydes. The quenching process is experimentally important, with best yields obtained by either rapidly adding acetic acid to the −75 °C reaction mixture, or Me_3SiCl in the presence of amines Et_3N and HMPA. A typical case involves the addition of $Me_5Cu_3Li_2$ to the β,β-disubstituted enal cyclohexyl-ideneacetaldehyde.

Conjugate Methylation of an α,β-Unsaturated Aldehyde with $Me_5Cu_3Li_2$ [106]

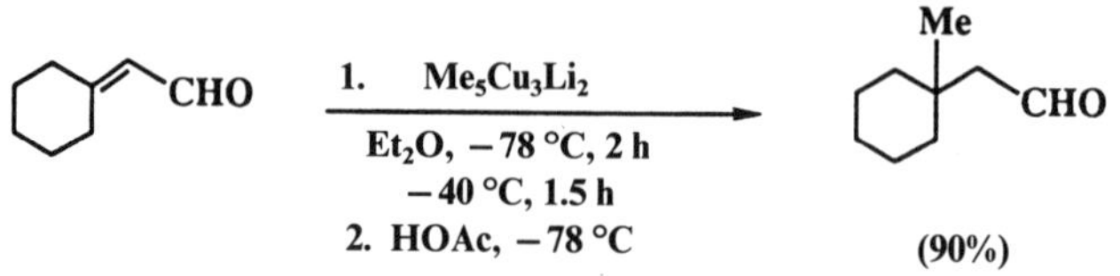

Purified copper(I) iodide (3 mmol) was placed in a dry 50 ml three-necked flask containing a magnetic stirring bar. Two necks of the flask were closed by rubber septa and the other by a vacuum take-off equipped with a stopcock. The flask was alternately evacuated and filled with N_2 (three cycles), and dry diethyl ether (10 ml) was then injected. The slurry was stirred at *ca* 0 °C (ice-bath), and commercial ethereal MeLi (*ca* 1.8 M, 5 mmol) was injected over 2–3 min. A dark yellow precipitate of methylcopper was deposited and then dissolved. Five minutes after the end of the addition the colorless (or faintly yellow) solution was cooled to −75 °C and the enal (1 mmol) in diethyl ether (1 ml plus 2 × 1 ml rinse) was added over 5 min. The reaction mixture was stirred at this temperature for 2 h. The temperature was then allowed to rise to −40 °C over 1.5 h, and the mixture was recooled to −75 °C and quenched with acetic acid. Work-up and Kugelrohr distillation (125–130 °C, 10 mmHg) gave the product (90%) of better than 97% purity (VPC, DEGS, 120 °C). The material contained 1% (VPC) of the 1,2 addition product, 1-cyclohexylidene-2-propanol; IR (film), 2715, 1710 cm^{-1}; NMR ($CDCl_3$, 200 MHz), δ 1.08 (s, 3H), 1.1–1.7 (m, 10H), 2.32 (d, 2H, $J = 3.4$ Hz), 9.83 (t, 1H, $J = 3.4$ Hz); HRMS, calculated for $C_9H_{15}O - H$, 139.1122; found, 139.1122.

4.2.8 HYDRIDO CUPRATES

The first complex metal hydride of copper, $LiCuH_2$, was reported in 1974 [108]. This species, along with several related hydrido cuprates, Li_2CuH_3, Li_3CuH_4 and Li_4CuH_5, are all prepared by $LiAlH_4$ reduction of preformed methyl cuprates resulting from various MeLi to CuI ratios (see, e.g., equation 4.2) [109]. These species have been

$$\begin{aligned} 5MeLi + CuI &\longrightarrow Li_4CuMe_5 + LiI \\ 2Li_4CuMe_5 + 5LiAlH_4 &\longrightarrow Li_4CuH_5 + 4LiAlH_2Me_2 \end{aligned} \qquad (4.20)$$

individually examined in terms of their potential to reduce haloalkanes, ketones and enones [110]. Among the many observations made was one concerning the remarkable reactivity of Li_4CuH_5 towards alkyl halides, a reagent which was shown to be more powerful than $LiAlH_4$ itself in this capacity. It is soluble in THF and is utilized at room temperature.

Preparation of Li_4CuH_5 [109]

$$\begin{aligned} 5MeLi + CuI &\longrightarrow [Li_4CuMe_5 + LiI] \\ 2Li_4CuMe_5 + 5LiAlH_4 &\longrightarrow Li_4CuH_5 + 4LiAlH_2Me_2 \end{aligned}$$

An Et_2O solution of CH_3Li (10.0 mmol) was added dropwise to a well stirred CuI (2.0 mmol) slurry in Et_2O at −78 °C. A clear solution resulted in a few minutes. To this solution was added $LiAlH_4$ (5.0 mmol) in diethyl ether and the reaction mixture was stirred at room temperature for 1 h, during which time a white crystalline solid formed. The insoluble solid was filtered, washed with diethyl ether and dried under vacuum. The product was analyzed and the X-ray powder diffraction pattern was recorded. Analysis: calculated for Li_4CuH_5, Li : Cu : H = 4.00 : 1.00 : 5.00; found, 4.10 : 1.00 : 5.09. X-ray pattern (Å): 4.05 m, 3.51 s, 2.47 m, 2.13 w, 2.02 w, 1.57 w. The white solid was stable at room temperature for over 1 week.

Another reagent for carrying out halide/sulfonate reductions [111], and also conjugate reductions of enones [112], is that derived from addition of *n*-BuLi to CuH, presumably forming LiCuH(*n*-Bu). Pure, anhydrous copper(I) hydride decomposes to hydrogen and metallic copper above −20 °C; it is indefinitely stable at −78 °C. Suspensions of copper(I) hydride in diethyl ether are relatively air insensitive; the dry solid is pyrophoric. Tri-*n*-butylphosphine and copper(I) hydride form a 1:1 complex, the high solubility of which has prevented its isolation. The procedure for pre-forming CuH from DIBAL reduction of CuBr, along with some relevant information on its properties, is given below [113]. Use of $LiAlH_4$ in place of DIBAL leads to inferior results.

Preparation of CuH [113]

$$\mathrm{CuBr} \xrightarrow[\substack{\mathrm{pyr.\,(100\,equiv.)} \\ -50\,°\mathrm{C} \\ 2.\,\mathrm{Et_2O\,(300\,equiv.)} \\ -78\,°\mathrm{C}}]{1.\,\mathrm{DIBAL}} \underset{(>90\%)}{\mathrm{CuH}}$$

Copper(I) hydride was prepared by treating 1 equiv. of copper(I) bromide dissolved in 100 equiv. of pyridine with 1.1 equiv. of diisobutylaluminum hydride (20% in heptane) at −50 °C. Vigorous mixing produced a homogeneous, dark brown solution, from which copper(I) hydride could be precipitated by dilution with *ca* 300 equiv. of diethyl ether. Centrifugation, separation of the supernatant liquid and repeated washing of the precipitate with diethyl ether, all at −78 °C, permitted isolation of copper(I) hydride as a brown solid in greater than 90% yield. The ratio of hydride to copper in this material was 0.96 ± 0.04; it contains less than 0.5% aluminum or bromine but retains *ca* 25% pyridine, based on copper.

With the CuH available, the optimum conditions for using the ate complex involve Et_2O at −40 °C for conjugate reductions, whereas halides and sulfonates usually require ambient temperature over *ca* 2 h.

Reduction of a mesylate Using Hydrido Cuprate LiCuH(n-Bu) [111]

$$\mathrm{Me(CH_2)_5\underset{|}{\overset{OSO_2Me}{C}}H(CH_2)_{10}CO_2Et} \xrightarrow[\substack{\mathrm{Et_2O,\,-40\,°C\,to\,r.t.} \\ 2\,\mathrm{h}}]{\mathrm{LiCuHBu}} \underset{(85\%)}{\mathrm{Me(CH_2)_{16}CO_2Et}}$$

CuH (6.0 mmol; see above) was prepared under Ar at −50 °C in a 50 ml round-bottomed flask, equipped with a magnetic stirrer and sealed with a rubber septum. After the CuH had been washed with four 20 ml portions of cold (−50 °C) diethyl ether, 15 ml of cold (−40 °C) diethyl ether were added with stirring and then 6.0 mmol of cold *n*-BuLi in hexane were syringed into the flask over 1 min. The resulting dark brown solution (reagent partially insoluble) was stirred for 10 min. After addition of 610 mg (1.5 mmol) of ethyl 12-mesyloxystearate in 1 ml of diethyl ether, the cooling bath was removed and stirring was continued for 2 h at room temperature. The reaction mixture was poured into aqueous saturated NH_4Cl solution, the ethereal layer was decanted and the aqueous layer was washed with 50 ml of diethyl ether. The combined ether extracts were dried, filtered and the solvents removed *in vacuo* to provide 400 mg (85%) of ethyl stearate following silicic acid chromatography.

More recently, two new methods for conjugate reduction have appeared which offer elements of convenience, efficiency, and chemoselectivity. In one, the known copper hydride hexamer $[(Ph_3P)CuH]_6$ [114], routinely used in benzene or toluene solution at room temperature, was found to be an extremely effective conjugate reductant towards α,β-unsaturated ketones and esters [115]. The reagent is a stable material, fully compatible with Me_3SiCl and insensitive to the presence of water. In fact, water is added to benzene or toluene solutions of this reagent in cases where short-lived intermediates are formed to suppress by-product formation. Isolated, unactivated olefins, however, are completely inert. The corresponding deuteride, $[(Ph_3P)CuD]_6$, is also available, with both reacting to deliver all six hydrides or deuterides per cluster to the substrate. The reagent $[(Ph_3P)CuH]_6$, a red crystalline solid, may be purchased (Aldrich, catalog No. 36 497-5) or prepared according to the procedure given below.

$$NaO\text{-}t\text{-}Bu + CuCl + PPh_3 \xrightarrow[\text{PhMe/PhH, r.t.}]{\text{1 atm } H_2} [(Ph_3P)CuH]_6 + t\text{-BuOH} + NaCl \quad (4.21)$$

Preparation of $[(Ph_3P)CuH]_6$ [116]

NaO-t-Bu Toluene (150 ml) was added with a cannula to a dry, N_2-flushed 500 ml Schlenk flask equipped with a Claisen head, a condenser topped with a gas inlet and a 50 ml pressure-equalizing addition funnel. Under positive pressure of N_2, sodium (3.52 g, 0.153 mol, cut in thin slices) was added. The toluene was heated with an oil-bath to 70–80 °C. *tert*-Butanol (45.0 ml, 35.4 g, 0.477 mol) was delivered dropwise. The vigorously stirred mixture was heated until all of the sodium had reacted (12–24 h). Under a positive N_2 flush, the condenser was replaced with a rubber septum and, for ease of subsequent cannula transfer of the resultant suspension, large clumps of the alkoxide which may be present were broken up with a glass rod.

$[(Ph_3P)CuH]_6$ Triphenylphosphine (100.3 g, 0.3825 mol) and copper(I) chloride (15.14 g, 0.1529 mol) were added to a dry, septum-capped 2 l Schlenk flask and placed under N_2. Benzene (approximately 800 ml) was added with a cannula, and the resultant suspension was stirred. The NaO-*t*-Bu–toluene suspension was transferred via a wide-bore cannula to the reaction flask, washing if necessary with additional toluene or benzene, and the yellow nearly homogeneous mixture was placed under a positive pressure (1 atm) of H_2 and stirred vigorously for 15–24 h. During this period the residual solids dissolved, the solution turned red, typically within 1 h, then dark red, and some gray or brown material precipitated. The reaction mixture was transferred under N_2 pressure through a wide-bore PTFE cannula to a large Schlenk filter containing several layers of sand and Celite. The reaction flask was rinsed with several portions of benzene, which were then passed through the filter. The very dark red solution was concentrated under vacuum to approximately one third of its volume and acetonitrile (300 ml) was layered on to the benzene, promoting crystallization of the product. The yellow–brown supernatant was removed via a cannula, and the product was washed several times with acetonitrile and dried under high vacuum to give 25.0–32.5 g (50–65%) of bright red to dark red crystals.

The degree of crystallinity and purity of the product varies somewhat with the degree of care exerted in the crystallization procedure. The major impurity present in the product was observed in the ^1H NMR spectrum as two broad resonances at $\delta \approx 7.6$ and 6.8 ppm, and has not been identified. Small amounts of this byproduct have no perceptible effect on subsequent reduction chemistry. Crystallization can also be induced by addition of hexane or pentane with no effect on product purity or yield.

Conjugate Reduction of the Wieland–Miescher Ketone with $[(Ph_3P)CuH]_6$ [115]

$[(Ph_3P)CuH]_6$ / PhH, H_2O, r.t. / 18 h

(85%)
(*cis*:*trans* 17:1)

$[(Ph_3P)CuH]_6$ (1.61 g, 0.82 mmol), weighed out under an inert atmosphere, and the keto enone (0.400 g, 2.24 mmol) were added to a 100 ml two-necked flask under positive N_2 pressure. Deoxygenated benzene (60 ml) containing 100 μl of water (deoxygenated by N_2 purge for 10 min) was added with a cannula, and the resultant red solution was stirred at room temperature until the starting material had been consumed by TLC analysis (8 h). The cloudy red–brown reaction mixture was opened to air, and stirring was continued for 1 h, during which time copper-containing decomposition products precipitated. Filtration through Celite and removal of the solvent *in vacuo* gave crude product, which was purified by flash chromatography; yield, 82% of a 17:1 mixture of *cis* and *trans* isomers, respectively.

The other new method for conjugate reduction of enones/enals entails the proposed *in situ* formation of a halo hydrido cuprate, H(X)CuLi, via transmetallation of Cl(I)CuLi (i.e., CuI·LiCl) with Bu_3SnH, according to equation 4.22 [117]. Whatever the reagent's composition, it is compatible with Me_3SiCl at low temperatures (<0 °C) and can be applied to substrates containing unprotected keto groups, esters, allylic acetates and sulfides. The procedure is best performed using excess (*ca* 2–3 equiv.) LiCl; without LiCl or with only 1 equiv. the yields generally tend to be lower.

$$\text{CuI} + \text{LiCl} \longrightarrow \text{Cl(I)CuLi} \xrightarrow[\text{THF}]{\text{Bu}_3\text{SnH}} \text{H(X)CuLi} + \text{Bu}_3\text{SnX} \quad (\text{X} = \text{Cl or I}) \tag{4.22}$$

Conjugate Reduction of an α,β-Unsaturated Aldehyde Using in situ *Generated Hydrido Cuprate H(X)CuLi [117]*

AcO … CHO → AcO … CHO

H(X)CuLi / THF, TMSCl / −60 °C to r.t. / 2 h

(quant.)

To a solution of CuI (0.1904 g, 1.00 mmol) and LiCl (0.1008 g, 2.38 mmol) in THF (4.5 ml) at −60 °C was added 8-acetoxy-2,6-dimethyl-2,6-octadienal (0.080 g, 0.391 mmol) followed by Me_3SiCl (0.27 ml, 2.09 mmol). After 10 min, Bu_3SnH (0.30 ml, 1.10 mmol) was added dropwise producing a cloudy yellow slurry. The reaction mixture was then allowed to warm to 0 °C gradually over 2 h. A concurrent darkening to a reddish brown color was observed. Quenching was carried out with 10% aqueous KF solution (3 ml), leading to an orange precipitate. The organic layer was filtered through Celite and evaporated *in vacuo*, and the residue rapidly stirred with additional amounts of 10% KF for *ca* 30 min before diluting with diethyl ether. The organic layer was then washed with saturated aqueous NaCl solution and dried (Na_2SO_4). The solvent was removed *in vacuo* and the material chromatographed on silica gel. Elution with EtOAc–

hexane (10:90) gave 82 mg (quantitative yield) of the product as a colorless oil; TLC, R_f (15% EtOAc–hexane) 0.22; IR (neat), 2960, 1740, 1460, 1380, 1240, 1030, 960 cm^{-1}; ^1H NMR ($CDCl_3$), δ 9.54 (d, 1H, $J = 1.5$ Hz), 5.27–5.24 (m, 1H), 4.51 (d, 2H, $J = 7.0$ Hz), 2.28–2.25 (m, 1H), 1.98 (d, 3H, $J = 1.0$ Hz), 1.62 (br s, 3H), 1.42–1.12 (m, 6H), 1.03 (d, 3H, $J = 6.8$ Hz); ^{13}C NMR ($CDCl_3$), δ 204.84, 170.92, 141.43, 118.57, 61.10, 45.96, 39.17, 29.78, 24.59, 20.83, 16.09, 13.16; mass spectrum, m/z (relative intensity, %), 170 (6), 153 (19), 152 (14), 135 (77), 123 (11), 109 (40), 107 (38), 97 (45), 95 (70), 94 (99), 93 (68), 83 (74), 81 (73), 79 (62), 69 (100); HRCIMS (CH_4), calculated for $C_{10}H_{18})_2$ ($M^+ - C_2H_3O$), 170.1307; found, 170.1283.

4.3 ORGANOCOPPER COMPLEXES DERIVED FROM GRIGNARD REAGENTS

Reactions of Grignard reagents under the influence of Cu(I) salts enjoy a rich history of service in organic chemistry. From the early observations of Kharasch using catalytic quantities of CuCl with RMgX [118] to the various stoichiometric recipes developed over time, few would argue today about the value of such time-honored chemistry. Undoubtedly much of the popularity stems from a combination of factors, not the least of which are the general availability of Grignard reagents and the facile nature of magnesiocopper-based addition to, e.g., α,β-unsaturated ketones and terminal acetylenes. As with lithiocopper reagents (see above), many of the same reactivity patterns are observed with Grignard-derived species, in both the catalytic and stoichiometric modes of use. In addition, most of the reaction variables and parameters to be addressed in cuprate couplings using Cu(I)–RMgX are identical with those associated with Cu(I)–RLi reagents. Thus, all reactions should be run under an inert atmosphere of N_2 or Ar, preferably the latter. Ethereal solvents are again the norm, and Grignards obtained from commercial sources should be titrated. Although early work focused on the use of CuCl as the Cu(I) salt [118], more recent developments have shifted significantly toward the usage of purified CuI, CuBr as its Me_2S complex or the Cu(II) chloride species Li_2CuCl_4 (Kochi's catalyst) [119].

4.3.1 REACTIONS USING CATALYTIC Cu(I) SALTS

Conjugate addition of a functionalized Grignard reagent, as with simpler analogs, is a popular means of setting the stage for subsequent manipulations (e.g. annulation) [2a]. When Grignards of this type (i.e. non-commercially available) are to be employed, the entire process can fortunately still be run in one pot in spite of the fact that information concerning the precise amount of Grignard present in the medium may not be known, and/or that the reagent responsible for the reaction can only be surmised. Given the truly catalytic nature of the process, it is likely that the reactive species is R_2CuMgX, formed under the conditions in a manner analogous to that of L.O. lithio cuprate generation (cf. equation 4.1). A typical procedure for preparing a Grignard reagent, for this example a protected chloroaldehyde and its eventual Michael addition to cyclohexenone, is given below. In general, with only small percentages of copper salts being necessary, work-up procedures are somewhat simplified.

Copper Bromide–Dimethyl Sulfide-Catalyzed 1,4-Addition of a Grignard Reagent to Cyclohexenone [120]

25 mol% $CuBr \cdot SMe_2$; Et_2O, −78 to 0 °C (75%)

Magnesium turnings (0.60 g, 25 mmol) were ground for a few minutes with a mortar and pestle and were immediately placed in an N_2-filled flask. A solution of 2-(3-chloropropyl)-1,3-dioxolane (1.2 ml, 8.3 mmol), 1,2-dibromoethane (0.05 ml) and THF (1.6 ml) were added at 25 °C and the mixture was stirred in a 70 °C bath, at which temperature Grignard formation began. The reaction flask was then placed in a 25 °C bath and was stirred for 30 min, diluted with additional THF (5 ml), stirred for 1.25 h and then cooled to −78 °C. A solution of copper(I) bromide–SMe_2 complex (0.41 g, 2.0 mmol) and Me_2S (4 ml) was then added dropwise and the mixture was stirred at −78 °C for 1 h. A solution of cyclohexenone (0.65 ml, 6.8 mmol) and diethyl ether (7 ml) was introduced dropwise over a 7 min period and the mixture was stirred at −78 °C for 2.5 h and then warmed in an ice–water bath. After being stirred at 0 °C for 5 min, the mixture was quenched by the addition of a saturated aqueous solution (5 ml) of NH_4Cl (adjusted to pH 8 with aqueous ammonia) and stirred at 25 °C for 1.5 h. The dark blue aqueous layer was removed and the ether layer washed with two additional 10 ml portions of water and a saturated aqueous solution (15 ml) of NaCl and dried over $MgSO_4$. Concentration by rotary evaporation *in vacuo* gave 1.28 g of the crude product, which was purified by flash chromatography [silica gel, hexanes–ethyl acetate (1 : 1)] to give 1.05 g (75%) of the product as a colorless oil. An analytical sample was obtained by bulb-to-bulb distillation [oven temperature 80 °C (0.2 mmHg)]; IR (neat), 2950, 1712 cm^{-1}; ^1H NMR ($CDCl_3$), δ 4.83 (t, 1H, $J = 4.8$ Hz), 3.90 (m, 4H), 1.15–2.55 (m, 15H). Analysis: calculated for $C_{12}H_{20}O_3$, C 67.89, H 9.50; found, C 67.76, H 9.53%.

Owing to the catalytic nature of the copper salt present, the resulting enolate must necessarily be mainly associated with a magnesiohalide counter ion (MgX^+), suggesting that *in situ* trapping with an electrophile may occur. Usually, more reactive alkylating agents work best, one example of which employs *tert*-butyl bromoacetate as the quenching agent following a copper-catalyzed 1,4 addition.

1,4-Addition of Vinylmagnesium Bromide to an Enone in the Presence of Catalytic Amounts of CuI and in situ *Trapping of the resulting Enolate [121]*

1. MgBr; CuI (3.2 mol%); THF, Me_2S
2. $BrCH_2CO_2$-*t*-Bu; HMPA, −78 to 0 °C, then r.t., 18 h

(78%)

To a slightly brownish slurry of vinylmagnesium bromide (1.5 M THF solution, 119 mmol) in Me_2S (16 ml) and THF (100 ml) was added copper(I) iodide (650 mg, 3.2 mol% vs substrate) at −78 °C and then 2-methyl-2-cyclopentenone (10.4 g, 108 mmol) in 20 ml of THF over 40 min at −50 to −60 °C to give a dark brown thick solution. After being stirred at the same temperature for 50 min, the reaction mixture was cooled to −78 °C, 47 ml (270 mmol) of HMPA were added slowly and then 44 ml (270 mmol) of *tert*-butyl bromoacetate were added. The reaction mixture was warmed very slowly to 0 °C over 6 h and stirring was continued for 18 h at room temperature, resulting in a brown solution with a white precipitate. The reaction mixture was quenched with an aqueous solution of NH_4Cl and extracted with diethyl ether three times, and the combined extracts were washed with water twice and then brine, dried over $MgSO_4$ and evaporated *in vacuo*. Fractional distillation of the residue afforded 22.2 g (114 mmol) of *tert*-butyl bromoacetate and 20.3 g (78%) of the product with 96% stereoselectivity [b.p. 80–82 °C (0.4 mmHg); IR (neat), 3060, 1730, 1639, 1151, 919 cm^{-1}; ^1H NMR (CCl_4), δ 0.77 (s, 3H), 1.40 (s, 9H), 1.6–2.5 (m, 6H), 2.7–3.2 (m, 1H), 4.9–5.2 (m, 2H), 5.5–6.1 (m, 1H)]. The stereoselectivity was calculated by the relative peak area of a singlet at δ 0.77 with that of a small singlet at δ 1.07, which was tentatively assigned to the 2-methyl protons of the *cis*-isomer.

The relative rapidity and selectivity of addition (1,4- vs 1,2-) by catalytic Cu(I)–RMgX is further manifested in the case of an α,β-unsaturated aldehyde. With substrates of this type, it is essential to use both Me_3SiCl (2 equiv.) and HMPA (2–3 equiv.), otherwise the process is slowed considerably at −70 °C, and 1,4- to 1,2-adduct ratios drop from >200:1 to 4:1. The product in all cases to be expected is the TMS enol ether, most notably of the predominantly *E* configuration.

Cu(I)-Catalyzed Conjugate Addition of RMgX to an α,β-Unsaturated Aldehyde in the Presence of Me_3SiCl_3SiCl Mixture [122]

CHO — n-$C_6H_{13}MgBr$, 3 mol% $CuBr \cdot SMe_2$ / THF, Me_3SiCl, HMPA, −78 °C → n-$H_{13}C_6$ … $OSiMe_3$

(83%)

(*E*:*Z* 94:6)

To a cooled (−78 °C) THF solution (60 ml) of *n*-hexylmagnesium bromide (prepared from 35 mmol of 1-bromohexane and 37.5 mmol of magnesium in 85–90% yield), hexamethylphosphoric triamide (10.5 ml, 60 mmol) (**Caution:** potent carcinogen) and copper(I) bromide–SMe_2 complex (257 mg, 1.25 mmol) was added dropwise a mixture of acrolein (1.67 ml, 25 mmol) and chlorotrimethylsilane (6.4 ml, 50 mmol) in 20 ml of THF over 30 min. After 3 h, triethylamine (7 ml) and hexane (100 ml) were added. The organic layer was washed with water to remove hexamethylphosphoric triamide and dried over $MgSO_4$. The product (3.86 g, 83%; 94% *E* by GLC analysis) was obtained by distillation (74 °C, 1 mmHg).

Copper-catalyzed Grignard ring openings of epoxides are commonly used mild reactions. *trans*-Diaxial opening of cyclohexene oxide, even with the usually less robust phenyl Grignard–catalytic Cu(I) mixture, takes place cleanly at −30 °C.

Opening of an Epoxide with Catalytic CuI–RMgX [123]

O + PhMgBr — 10 mol% CuI / THF, −30 °C → OH, Ph

(81%)

To 10.9 g (0.45 mol) of magnesium in 100 ml of THF was added 73.0 g (0.465 mol) of bromobenzene in 100 ml of THF over 1 h. The resulting mixture was stirred for 30 min and then 8.85 g (46.5 mmol) of copper(I) iodide was added and the mixture cooled to -30 °C. A solution of 29.45 g (0.30 mol) of cyclohexene oxide in 50 ml of THF was then added dropwise. After the addition was complete, the mixture was stirred for 3 h and then quenched by being poured into 100 ml of cold saturated aqueous NH_4Cl solution. The solution was extracted with diethyl ether and the organic layers were combined, dried and concentrated *in vacuo* to afford a liquid that was distilled at 80 °C (0.23 mmHg) to afford 43.1 g (81%) of a yellow solid which was recrystallized from pentane, m.p. 56.5–57.0 °C; IR, 3592, 3461, 2941, 2863, 1604, 1497, 1451 cm^{-1}; ^1H NMR (360 MHz), δ 7.35–7.17 (m, 5H), 3.64 (ddd, 1H, J = 5.4, 10.8, 10.8 Hz), 2.42 (ddd, 1H, J = 5.4, 10.8, 16.5 Hz), 2.11 (m, 1H), 1.84 (m, 2H), 1.76 (m, 1H), 1.62 (s, 1H), 1.53–1.25 (br m, 4H); ^{13}C NMR (90 MHz), δ 143.4 (s), 128.7 (d), 127.9 (d), 126.7 (d), 74.3 (d), 53.3 (d), 34.6 (t), 33.4 (t), 26.1 (t), 25.1 (t); mass spectrum, m/z 176 (M^+), 158, 143, 130, 117, 104, 91 (base).

Although the effect of added TMSCl is oftentimes most pronounced in conjugate addition schemes, it has a dramatic impact on additions of RMgX–catalytic Cu(I) to propargylic oxiranes [124]. As shown below, either the *syn* or *anti* isomer of the allenic products can be realized from the same educt depending on the use of this additive and the halide present in the Grignard.

Copper–Catalyzed Addition of a Grignard (RMgBr) to a Propargylic Epoxide: anti-*Allenol Formation [124]*

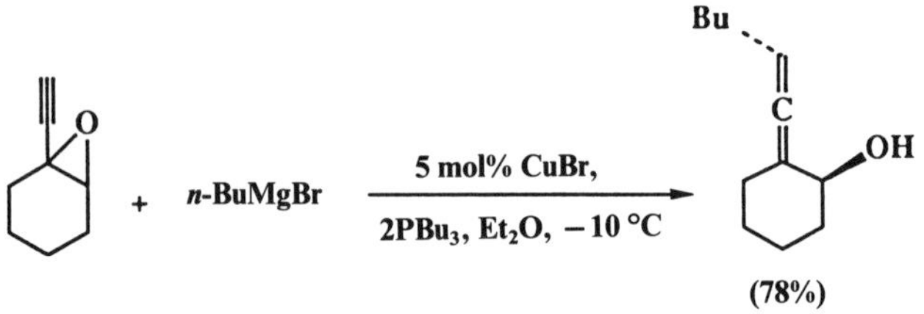

+ *n*-BuMgBr → (5 mol% CuBr, 2PBu_3, Et_2O, -10 °C) Bu, C, OH (78%)

To an ethereal solution (30 ml Et_2O) of ethynyl cyclohexene oxide (400 mg, 3.28 mmol), were successively added (1) at 0 °C a solution of CuBr·2PBu_3 (1.65 ml of a 0.1 M solution in Et_2O; 0.165 mmol; 0.05 equiv.) and (2) at -50 °C, slowly, a solution of BuMgBr (6.55 ml of a 1 M solution in Et_2O; 6.55 mmol). The stirred solution was allowed to warm to -10 °C over 30 min, then hydrolysed with 20 ml of a saturated aqueous solution of NH_4Cl admixed with 5 ml of aqueous ammonia. The aqueous phase was extracted twice with 30 ml of Et_2O, then the combined organic phases were washed with saturated aqueous NH_4Cl (20 ml), dried over $MgSO_4$ and concentrated *in vacuo*. The residue was chromatographed on silica gel [cyclohexane–EtOAc (70 : 30)]. The expected allenol (460 mg, 78% yield) was obtained as a pure diastereomer; IR, 3300, 1960 cm^{-1}; ^1H NMR, δ 5.4 (m, 1H), 4.0 (m, 1H), 2.3 (m, 3H); ^{13}C NMR, δ 197.8 (=C=), 110.4 (C=), 97.5 (—CH=), 71.4 (—CHOH), 38.6, 33.7, 32.5, 31.5, 29.5, 26.2, 24.6, 16.3.

CuBr-Catalyzed Addition of a Grignard (RMgCl) to a Propargylic Epoxide in the Presence of Me_3SiCl: syn-*Allenol Formation [124]*

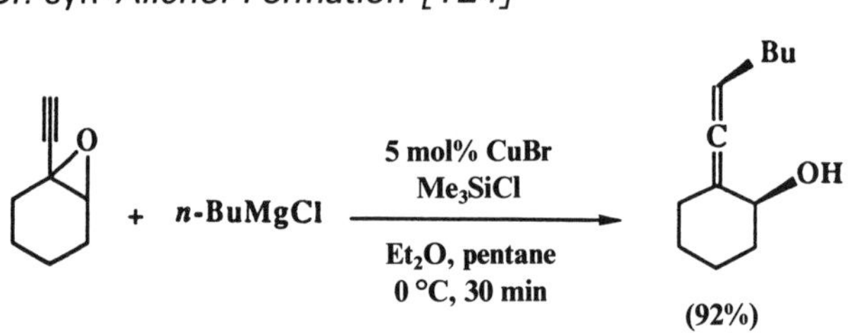

+ *n*-BuMgCl → (5 mol% CuBr, Me_3SiCl; Et_2O, pentane, 0 °C, 30 min) Bu, C, OH (92%)

To a suspension of CuBr (24 mg, 0.165 mmol) in Et_2O (20 ml) and pentane (20 ml) at 0 °C were added ethynylcyclohexene oxide (400 mg, 3.28 mmol) and then Me_3SiCl (0.4 ml, 3.28 mmol). The mixture was cooled to −50 °C and a solution of BuMgCl (6.55 ml of a 1 M solution in Et_2O; 6.55 mmol) was slowly added. The mixture was warmed to 0 °C for 30 min, then hydrolyzed and worked up as in the example above. After column chromatography, 545 mg (92% yield) of the *syn*-allenol were collected as an inseparable mixture of two diastereomers (88:12 by GC, after acetylation); IR and ^1H NMR identical with the *anti* diastereomer; ^{13}C NMR, 198 (=C=), 109.8 (C=), 97.0 (—HC=), 71.4 (—CHOH), 38.4, 33.8, 32.2, 31.5, 29.5, 25.8, 24.6, 16.3.

β-Lactones are highly susceptible to attack by RMgX modified by Cu(I), carbon–carbon bond formation occurring at the β-carbon to afford alkylated carboxylic acids [125]. The procedure is attractively straightforward and can be applied to α,β- and α,α-substituted β-propiolactones.

Catalytic CuCl–RMgX Opening of a β-Lactone: a Carboxylic Acid Synthesis [126]

β-propiolactone + *n*-BuMgBr —(2 mol% CuCl, THF, 0 °C)→ *n*-$H_{13}C_6COOH$ (90%)

n-Butylmagnesium bromide (1 M in diethyl ether, 2.4 ml, 2.4 mmol) was slowly added to a suspension of copper(I) chloride (4 mg, 0.04 mmol) in 6 ml of THF at 0 °C under Ar. β-Propiolactone (0.144 g, 2 mmol) in 2 ml of THF was next added dropwise. The mixture was stirred at 0 °C for 15 min and quenched by adding 3 M HCl. From the organic layer, heptanoic acid was extracted with 3 M NaOH solution. The alkaline solution was acidified, extracted with diethyl ether and concentrated to give pure heptanoic acid in 90% yield, b.p. 65 °C (1.0 mmHg).

Similar chemistry when performed on chiral, non-racemic (protected) α-amino β-lactones provides a quick, clean entry to α-amino acids [127]. Although excess of RMgX is present (> 5 equiv.), virtually no racemization is seen under the reaction conditions.

Ring Opening of an α-Amino-β-Propiolactone via Catalytic CuBr·SMe₂–RMgX: Synthesis of a Chiral, Non-Racemic Amino Acid [127]

Z(Bn)N-β-lactone + *i*-PrMgCl —(0.21 equiv. $CuBr\cdot SMe_2$; THF, Me_2S, −23 °C, 2 h)→ Z(Bn)N–CH(CH_2CHMe_2)–COOH (83%)

Isopropylmagnesium chloride in Et_2O (3.0 mmol, 1.0 ml) was added dropwise over 5 min to the β-lactone (180 mg, 0.578 mmol) and $CuBr\cdot SMe_2$ (25 mg, 0.122 mmol) in THF (6 ml)–SMe_2 (0.3 ml) at −23 °C. The mixture was stirred 2 h at −23 °C and quenched by addition to cold degassed 0.5 M HCl (20 ml). Extraction and washing of the ethereal phases followed by reverse phase MPLC (55% MeCN–H_2O, 3.3 ml min^{-1}) yielded 170 mg (83%) of product as an oil: $[\alpha]_D^{25}$ −44.7° (c = 2.5, $CHCl_3$); IR ($CHCl_3$), 3160 (m br), 1740 (s), 1705 (vs), 1680 (s), 1498 (m), 1468 (s), 1454 (s), 1418 (s), 1315 (s), 1240 (vs), 1208 (s), 1179 (s), 699 (vs) cm^{-1}; ^1H NMR (300 MHz, $CDCl_3$), δ 9.75 (br s, 1H, COO*H*), 7.45–7.10 (m, 10H, 2 *Ph*), 5.19 (s, 2H, PhCH_2O), 4.87–4.62 (m, 1H, C*H*), 4.60–4.30 (M, 2H, PhCH_2N), 1.90–1.20 (m, 3H, CH_2C*H*Me_2), 0.94–0.53 (m, 6H, 2 CH_3); HREIMS, calculated for $C_{21}H_{25}NO_4$, 355.1784; found, 355.1785; CIMS (NH_3), 373 ($M + NH_4$), 356 (MH^+). Analysis calculated for $C_{21}H_{25}NO_4$, C 70.97, H 7.09, N 3.94; found, C 70.68, H 7.10, N 3.87%. Optical purity analysis by GC showed no detectable *R*-isomer (i.e. ≥99.4% *ee*).

Displacement reactions of primary halides by RMgX are usually assisted by catalytic amounts of lithium tetrachlorocuprate, Li_2CuCl_4 [2d, 119]. Admixture of a Grignard reagent with this salt (1–5 mol%) leads rapidly to the Cu(I) oxidation state, which then participates in a manner similar to other Cu(I) salts, in this case effecting a substitution event. Shown below are two representative procedures, the first of which leads to the formation of a substituted butadiene [128]. Use of CuI in place of Li_2CuCl_4 affords inferior results. Other substrates tested include, e.g., 3-bromopropyl chloride, which reacts selectively at the center bearing bromine. Aryl halides also couple under roughly comparable conditions. Perhaps most noteworthy is the finding that hydroxyl, ester, ether and cyano functions present in the educt remain unaffected throughout these reactions. The second procedure, utilizing Li_2CuCl_4, demonstrates the generation of a trisubstituted aromatic derivative applicable to cannabinoid synthesis [129].

Cross-Coupling of a Vinylic Grignard with an Alkyl Halide Assisted by Catalytic Li_2CuCl_4 [128]

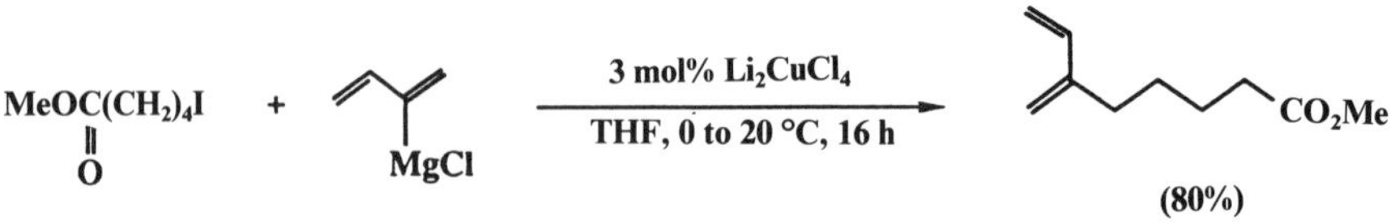

MeOC(=O)(CH$_2$)$_4$I + MgCl (vinylic Grignard) → 3 mol% Li_2CuCl_4 / THF, 0 to 20 °C, 16 h → diene–CO$_2$Me (80%)

To a mixture of halide (0.1 mol), Li_2CuCl_4 (3 mol% relative to halide) and THF (50 ml) in a 300 ml four-necked flask was added the halide (0.1 mol) in THF (100 ml) dropwise with stirring at 0 °C under an N_2 atmosphere. An exothermic reaction occurred during the addition and the color of the contents gradually changed from reddish brown to black. After the completion of the addition, stirring was continued at 20 °C for 16 h. The organic layer was separated after hydrolyzing the reaction mixture with 6 M HCl, and the aqueous layer was extracted with two portions of diethyl ether (100 ml). The combined organic extracts were washed first with 5% aqueous $NaHCO_3$ and then with water, dried (Na_2SO_4), and distilled at 71–76 °C (2 mmHg). The reaction product (80%) was identified by comparing its IR, mass and NMR spectra with the reported data. The product gave reasonable elemental analyses.

Li_2CuCl_4-Catalyzed Coupling of an Aryl Grignard with an Alkyl Halide [129]

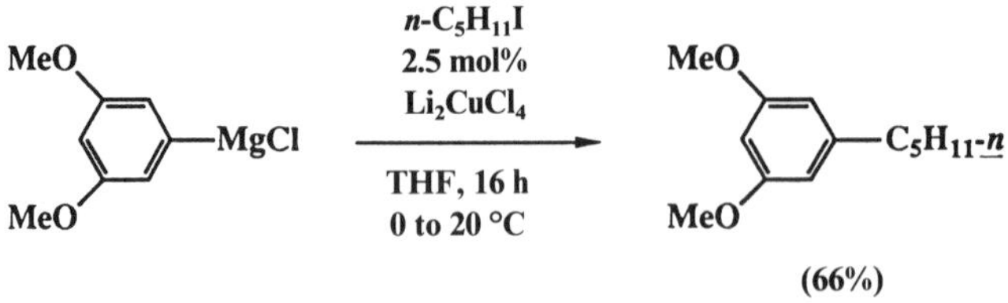

3,5-(MeO)$_2$C$_6$H$_3$–MgCl → *n*-C_5H_{11}I, 2.5 mol% Li_2CuCl_4 / THF, 16 h, 0 to 20 °C → 3,5-(MeO)$_2$C$_6$H$_3$–C_5H_{11}-*n* (66%)

Under dry N_2, 5-chloro-1,3-dimethoxybenzene (40 g, 0.23 mol), magnesium (6 g, 0.25 mol) and a small amount of 1,2-dibromoethane in THF (80 ml) were heated under reflux for 6 h. The solution was cooled in ice and a mixture of 1-iodopentane (42.6 ml, 0.325 mol) and Li_2CuCl_4 (30 ml of a

0.2 M solution in THF, 6 mmol) was added dropwise over a period of 30 min. The resulting black mixture was stirred at 0 °C for 90 min and at 20 °C for an additional 16 h. The almost solid reaction mixture was acidified with 6 M HCl (160 ml) and extracted with diethyl ether (2 × 200 ml). The organic extracts were washed with 15% aqueous ammonia (60 ml) and water (60 ml), dried with $MgSO_4$ and evaporated *in vacuo*. According to the ^1H NMR spectrum of the residual product, olivetol dimethyl ether was formed in 74% yield. Distillation afforded the pure product (31.9 g, 66%) as a colorless liquid, b.p. 152–156 °C (12 mmHg).

Displacements of primary triflates have been shown to proceed readily using catalytic CuBr–RMgBr. Even β-oxygenated educts react without complication, a rather uncommon observation, especially in cuprate chemistry. Moreover, the greater reactivity of a triflate relative to that of a tosylate is such that a double displacement can be effected in a one-pot operation given the presence of both electrophiles within the molecule. Initial substitution of the triflate requires just over 1 equiv. of the catalytic Cu(I)–RMgX system, while the follow-up coupling necessitates excess of lithiocuprate [130].

cat. CuBr
RMgX

OTf
OTs

R
R'

R'_2CuLi

Double Displacement of a Tosyl Triflate Using catalytic CuBr–RMgBr, Followed by R'_2CuLi

1. cat. CuBr, MgBr
2. $(n\text{-}C_9H_{19})_2CuLi$, THF, Et_2O, 15 °C, 3 h

OTf
OTs

C_9H_{19}-n

(58%)

3-Butenylmagnesium bromide (0.78 ml, 0.54 mmol) was added to a suspension of CuBr (14 mg, 0.1 mmol) in THF (2 ml) at 0 °C, followed by the chiral, non-racemic tosyl triflate (208 mg, 0.50 mmol) in 2.5 ml of diethyl ether, and the reaction mixture was stirred at the same temperature for 1 h. Then $(C_9H_{19})_2CuLi$, prepared from CuI (495 mg, 2.6 mmol) and $C_9H_{19}Li$ (0.82 M in diethyl ether; 6.3 ml, 5.2 mmol) in dimethyl sulfide (3 ml), was introduced at −15 °C and the reaction mixture was stirred for 3 h. The usual work-up followed by column chromatography gave 91 mg (58%) of the product as a colorless oil: TLC, R_f [hexane–diethyl ether (20:1)] 0.33; $(\alpha)_D^{22}$ + 24.6° (c = 1.78, $CHCl_3$); IR (neat), 2920, 2850, 1640, 1450, 1375, 1365, 1235, 1100, 990, 910 cm^{-1}; ^1H NMR, δ 0.88 (3H, t, J = 6.5 Hz), 1.26 (14H, s), 1.37 (6H, s), 1.4–1.7 (8H, m), 1.9–2.2 (2H, m), 3.59 (2H, m), 4.94 (1H, m), 4.99 (1H, m), 5.81 (1H, ddt, J = 17.1, 9.7, 6.5 Hz); ^{13}C NMR, δ 14.13, 22.73, 25.44, 26.17, 27.39, (× 2), 29.37, 29.59, 29.65 (× 2), 29.83, 31.97, 32.45, 33.09, 33.80, 80.84, 80.99, 107.73, 114.62, 138.41; HRMS, calculated for $C_{20}H_{38}O_2$, 310.2872; found, 310.2859.

By varying the nature of the Grignard and organolithium chosen for each step of a cuprate-mediated displacement sequence, in addition to the chirality of the bis-nucleofuge, the procedure above has been applied to syntheses of several natural products, such as (+)-*exo*-brevicomin and L-factor.

(+)-*exo*-brevicomin L-factor

Both simple and more functionalized Grignard reagents, together with a Cu(I) source, can afford products derived from couplings with allylic centers. For example, the mesitoate of a cyclohexenol, prepared specifically for stereochemical studies, clearly indicates that a CuCN-catalyzed S_N2' displacement takes place with inversion of stereochemistry [131a]. The example below not only addresses the manner in which couplings of this sort can be performed, but also highlights the use of CuCN in a catalytic role. Although this salt is not used nearly as frequently as CuI, $CuBr{\cdot}SMe_2$ or Li_2CuCl_4 for this type of cuprate reaction, it can afford completely different outcomes from the other Cu(I) salts owing to the lack of metathesis between RMgX and CuCN, as discussed previously for RLi–CuCN mixtures. Hence, with CuCN in the presence of excess RMgX, the reactive species may be either RCu(CN)MgX, or possibly the higher order magnesio cuprate $R_2Cu(CN)(MgX)_2$ [131b].

Substitution of an Allylic Carboxylate with a Grignard Reagent Catalyzed by CuCN [131a]

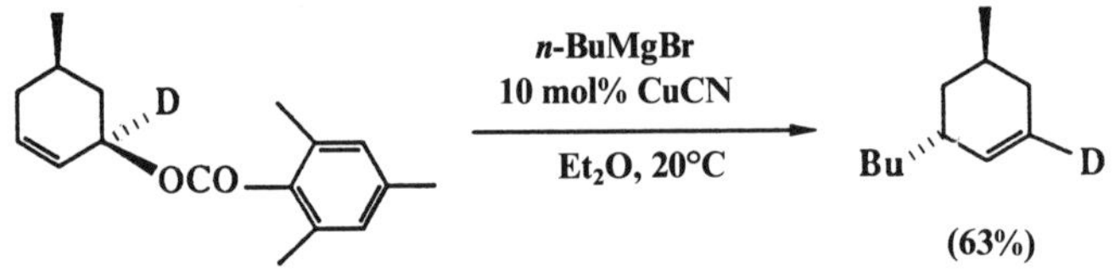

n-BuMgBr
10 mol% CuCN
Et_2O, 20°C
(63%)

A flask equipped with a magnetic stirrer and septum was charged with 54 mg (0.6 mmol) of copper(I) cyanide. After flushing with dry N_2, 2 ml of anhydrous diethyl ether were added and the suspension was chilled to −10 °C. A diethyl ether solution of *n*-butylmagnesium bromide (6 mmol, prepared from 987 mg of 1-bromobutane and 146 mg of magnesium in 8 ml of diethyl ether) was added through a cannula and, after stirring the mixture for 10 min, a solution of 778 mg (3 mmol) of α-deuterio-*cis*-5-methyl-2-cyclohexenyl mesitoate in 2 ml of diethyl ether was added. The cooling bath was removed and the mixture was stirred at room temperature for 6.5 h, after which it was quenched with 2 ml of aqueous NH_4Cl solution. The resulting mixture was filtered, the precipitate washed with diethyl ether and the ether solution was dried over $MgSO_4$. Removal of solvent by fractionation followed by column chromatography (silica gel, pentane–diethyl ether) and vacuum distillation gave 289 mg (63% yield) of a clear, mobile oil, b.p. 58–60 °C (7.4 mmHg); IR (neat), 3020, 2945, 2910, 2900, 2860, 2840, 2820, 2240, 1640, 1465, 1455, 1430, 1375, 895, 730, 710 cm^{-1}; ^1H NMR ($CDCl_3$), δ 5.63 (br s, 1H), 2.20–1.90 (br m, 2H), 1.90–1.68 (br M, 1H), 1.68–1.50 (m, 9H), 1.05–0.70 (m, 3H), 0.93 (d, 3H, $J = 7.5$ Hz); HRMS, calculated for $C_{11}H_{19}D$, 153.1622; found, 153.1628.

The Grignard of 3-chloromethylfuran, formed quantitatively and under the influence of Kochi's catalyst (Li_2CuCl_4) [119], smoothly displaces chloride ion from prenyl chloride in an S_N2 fashion almost instantaneously at 0 °C.

Coupling of an Allylic Halide with RMgX Catalyzed by Li_2CuCl_4 [132]

Cl + MgCl —(0.35 mol% Li_2CuCl_4, THF, 0 °C, 5 min)→ (85%)

To 0.104 g (4.29 mmol) of magnesium turnings covered with 3 ml of THF under Ar was added 0.5 g (4.29 mmol) of 3-chloromethylfuran in 2 ml of THF in one portion. The mixture was stirred for 30 min at room temperature, then warmed in a preheated 50 °C oil-bath for 30 min to provide a golden yellow solution. The solution was chilled in an ice–water bath and 0.448 g (4.29 mmol) of freshly distilled 1-chloro-3-methyl-2-butene in 2 ml of THF was added in one portion, followed immediately by 0.15 ml of a 0.1 M solution of Li_2CuCl_4 in THF. The resulting black suspension was stirred for 5 min at 0 °C, poured into light petroleum (50 ml), washed with 5% aqueous Na_2CO_3 solution (50 ml) and water (50 ml) and dried over Na_2SO_4. Concentration *in vacuo* provided a pale yellow liquid, which was purified by bulb-to-bulb distillation to give 0.547 g (85%) of perillene as a colorless liquid, b.p. 80 °C (20 mmHg).

Attack by a magnesiocuprate at the primary position in prenyl and related systems appears to be a general phenomenon [133]. Thus, in the case of a (*Z*)-trisubstituted allylic acetate and a Grignard reagent derived from a primary bromide, copper bromide-catalyzed coupling gave the expected product with only a small quantity of the competing S_N2' product (19 : 1). By switching to Li_2CuCl_4 as catalyst, only the product of straight substitution at carbon bearing the leaving group was afforded.

Coupling of a (Z)-Allylic Acetate with Li_2CuCl_4–RMgX [133]

AcO + BrMg ... OBn —(4 mol% Li_2CuCl_4, THF, 0 °C, 1 h)→ ... OBn (79%)

A mixture of magnesium (0.255 g, 10.5 mmol) and 1,2-dibromoethane (26 μl, 0.3 mmol) in THF (10 ml) was heated at reflux. To the activated magnesium was added a solution of 4-bromo-3-methylbutyl benzyl ether (2.57 g, 10 mmol) in THF (2 ml) at 20 °C and the mixture was refluxed for 15 min. The Grignard reagent was added dropwise at 0 °C under Ar to a mixture of the acetate (0.98 g, 5 mmol) in THF (8 ml) and a 0.1 M solution of Li_2CuCl_4 in THF (2.0 ml, 0.2 mmol). After stirring for 1 h at 0 °C, the mixture was partitioned between diethyl ether (50 ml) and saturated aqeuous NH_4Cl (50 ml). The diethyl ether layer was washed with saturated aqueous NH_4Cl (30 ml), dried with $MgSO_4$, concentrated and distilled under reduced pressure to give the product, 1.38 g (79%); b.p. 134 °C (0.2 mmHg). The product was further purified by silica gel column chromatography using hexane–diisopropyl ether (5 : 1) as eluent.

4.3.2 REACTIONS OF STOICHIOMETRIC SPECIES R_2CuMgX AND $RCu \cdot MgX_2$

Admixture of a Grignard reagent with an equivalent of a copper halide leads to a metathesis reaction affording an organocopper species in the presence of a magnesium halide salt (equation 4.23). This representation is precisely analogous to the conversion of an organolithium to $RCu \cdot LiX$, more accurately written as a halocuprate, R(X)CuLi [88]. Spectroscopic studies on the corresponding reagent $RCu \cdot MgX_2$ suggest, however, that the composition of this mixture of RMgX and CuX is not as simple as drawn (i.e. '$RCu \cdot MgX_2$') [134]. Studies of this sort are often hampered, in part, by limited reagent solubility in common ethereal solvents.

$$RMgX + CuX \xrightarrow[\text{solvent} \atop < 0\,°C]{\text{ethereal}} RCu \cdot MgX_2 \qquad (4.23)$$

When two equivalents of RMgX relative to CuX are used, the magnesium halide analog of the Gilman reagent is formed (equation 4.24). In this case, as with the 1:1 ratio, an equivalent of MgX_2 is generated relative to the lower order cuprate R_2CuMgX.

$$2RMgX + CuX \xrightarrow[< 0\,°C]{\text{THF and/or } Et_2O} R_2CuMgX + MgX_2 \qquad (4.24)$$

The chemistry of these two species, just as with the reagents themselves, is different. The lower reactivity and basicity of the neutral organocopper $RCu \cdot MgX_2$ can be used to advantage especially in carbocuprations and displacements of highly reactive electrophiles (e.g. allylic leaving groups). Magnesiocuprates, on the other hand, serve well in situations where a more robust, ate complex is needed, as in substitutions with less reactive centers and for conjugate additions.

One of the most valued reactions of $RCu \cdot MgX_2$ is their facile addition across terminal acetylenes of the elements 'R^-' and 'Cu^+,' i.e., carbocupration [135]. The regiochemistry is predictably such that the copper atom ultimately resides as the least hindered alkyne carbon, thereby forming a new organocopper$\cdot MgX_2$ complex of the vinylic type (equation 4.25). The stereochemistry is also predictable from *syn* addition. Thus, the control offered by this chemistry for stereodefined olefin preparation is particularly noteworthy, and has been utilized extensively in the area of pheromone total synthesis where even trace amounts of isomeric impurities can be detrimental to potency.

$$R{-}{\equiv}{-}H + R'Cu \cdot MgX_2 \longrightarrow (R)(R')C{=}CH(Cu \cdot MgX_2) \longrightarrow \text{etc.} \qquad (4.25)$$

Carbocupration of a 1-Alkyne Using $RCu \cdot MgX_2$ in Et_2O [136]

$$n\text{-}H_9C_4{-}{\equiv} + Me_3Si\text{CH}_2MgCl \xrightarrow[Et_2O,\ 10\,°C]{CuBr,\ LiI} (n\text{-}H_9C_4)(Me_3SiCH_2)C{=}CH_2$$

(78%)

To a suspension of copper(I) bromide (2.2 g, 15 mmol) and lithium iodide (1 M solution in diethyl ether, 20 ml, 20 mmol) in diethyl ether (50 ml) was added, at 0 °C, a solution of trimethylsilylmethylmagnesium chloride (0.9 M in diethyl ether, 17 ml, 15 mmol). The mixture first gave a yellow precipitate and then a homogeneous pale green solution, which was stirred at −5 °C for 1 h. After addition of 1-hexyne (1.0 g, 12.5 mmol), the mixture was allowed to warm to 10 °C and stirred at this temperature for 18 h (brown solution), then hydrolyzed with 100 ml of buffered ammonia solution. The mixture was filtered and decanted and then the organic layer was washed with brine (10 ml) and dried over $MgSO_4$. The solvent was evaporated under vacuum and the residue was distilled through a 10 cm Vigreux column to afford 1.8 g (78%) of pure product, b.p. 70 °C (10 mmHg); ^1H NMR (CCl_4), δ 4.8 (s, 1H), 4.6 (s, 1H), 2.0 (t, 2H), 1.8 (s, 2H), 1.4 (m, 4H), 0.98 (t, 3H), 0.05 (s, 9H). Analysis: calculated for $C_{10}H_{22}Si$, C 70.50, H 13.01; found, C 70.40, H 13.03%.

Placement of a heteroatom in the educt such that internal chelation can occur may result in a reversal of carbocupration regiochemistry. Thus, starting at the enyne stage, conjugated olefinic products of defined geometries are afforded in good yields following *syn* delivery of RCu.

Carbocupration of a Functionalized, Terminal Enyne [137]

SEt + EtCu•MgBr$_2$ $\xrightarrow[-20\,°C,\ 2\,h]{Et_2O}$ Et SEt

(82%)

To a cooled (−40 °C) suspension of purified CuBr (7.9 g, 55 mmol) or CuBr·SMe_2 complex (11.3 g, 55 mmol) in 100 ml of Et_2O was added EtMgBr (25 ml of a 2 M solution in Et_2O). After stirring for 30 min at −30 to −35 °C, a yellow–orange precipitate of ethylcopper was formed. To this suspension was added (*Z*)-1-ethylthio-1-buten-3-yne (5.6 g, 50 mmol) dissolved in 20 ml of Et_2O. The mixture was slowly warmed to −20 °C, whereupon it dissolved, and was stirred for 2 h at this temperature. The dark red solution was hydrolyzed with aqueous NH_4Cl (50 ml) admixed with 5 M HCl (30 ml), the salts were filtered off and the aqueous phase extracted once with Et_2O (50 ml). The combined organic phases were washed twice with aqueous NH_4Cl and then dried over $MgSO_4$, and the solvents were removed *in vacuo*. The residue was distilled through a 15 cm Vigreux column to afford 5.8 g (82% yield) of isomerically pure product, b.p. 87 °C (15 mmHg); IR, 3010, 1640, 1570, 970 cm^{-1}; ^1H NMR, δ 6.29 (ddt, 1H), 6.04 (t, 1H), 5.77 (d, 1H), 5.64 (dt, 1H); J_E = 14.5 Hz; J_Z = 8.5 Hz; J = 10.0 Hz.

Although a carbocupration-derived intermediate vinyl copper species (cf. equation 4.25) is readily hydrolyzed to the corresponding 1,1-disubstituted alkene, the electrophile used need not be limited to H^+. Further elaborations are indeed possible, as the following two cases demonstrate, using ethylene oxide in the former [135b] and allyl bromide in the latter [138]. Note that prior to introducing the oxirane, the vinylcopper was necessarily treated with an equivalent of pentynyllithium to form an ate species of sufficient reactivity to open even this simple coupling partner.

CuBr–RMgX-Mediated Carbocupration of an Acetylene: Trapping with an Epoxide [135b]

To a mixture of copper(I) bromide·SMe_2 (0.82 g, 4.0 mmol), diethyl ether (5 ml) and Me_2S (4 ml) at −45 °C under N_2 was added a 2.90 M solution (1.39 ml, 4.0 mmol) of methylmagnesium bromide in diethyl ether over 2 min. After 2 h, 1-octyne (0.52 ml, 3.5 mmol) was added over 1 min to the yellow–orange suspension. The mixture was stirred at −23 °C for 120 h, then the resulting dark green solution was cooled to −78 °C. A solution of 1-lithio-1-pentyne (prepared from 4.0 mmol of *n*-butyllithium and 4.0 mmol of 1-pentyne), diethyl ether (5 ml) and hexamethylphosphoric triamide (1.4 ml, 8.0 mmol) (**Caution:** potent carcinogen) was transferred to the green solution. After 1 h, ethylene oxide (0.21 ml, 4.0 mmol), which had been condensed at −45 °C, was added with a dry-ice-cooled syringe over 30 s. The resulting mixture was stirred at −78 °C for 2 h, allowed to stand at −25 °C for 24 h, quenched at 0 °C by addition of an aqueous solution (5 ml) of NH_4Cl (adjusted to pH 8 with ammonia) and then partitioned between diethyl ether and water. The crude product (90% pure by GC) was purified by column chromatography on silica gel (CH_2Cl_2) to give a colorless oil (0.44 g, 75%); IR (neat), 3300, 1669, 874 cm^{-1}; ^1H NMR, δ 5.05 (t, 1H, $J = 7$, 1 Hz), 3.55 (t, 2H, $J = 7$ Hz), 1.58 (s, 3H), 2.40–0.65 (br m, 16H); HRMS, calculated for $C_{11}H_{22}O$, 170.1667; found, 170.1691.

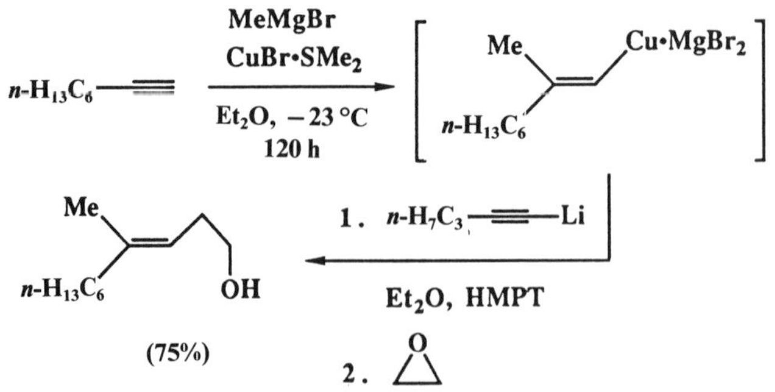

Carbocupration of Ethoxyacetylene with Br_2CuLi–RMgX Followed by Alkylation with an Allylic Halide [138]

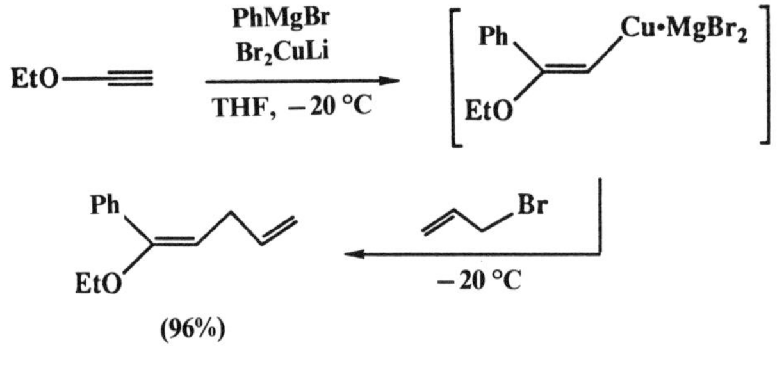

To a stirred solution of phenylcopper [prepared *in situ* by stirring phenylmagnesium bromide (0.01 mol) with 0.01 mol of the THF-soluble complex lithium dibromocuprate at −50 °C for 1 h] in THF (35 ml) was added 0.01 mol of ethoxyacetylene at −50 °C. The mixture was then stirred for 1 h at −20 °C. Subsequently, allyl bromide (0.01 mol) was added and the mixture stirred for 3 h, after which it was poured into an aqueous NH_4Cl solution (200 ml) containing NaCN (2 g) and extracted with pentane (3 × 50 ml). The combined extracts were washed with water (6 × 100 ml) to remove THF and dried over $MgSO_4$. The solvent was removed *in vacuo* and the residue purified by column chromatography, eluting with pentane, to afford the product (96%) of 95% purity by GC; IR (neat), 3080, 3060, 1645, 1600, 1495, 1238, 1128, 910, 770, 700 cm^{-1}; ^1H NMR (CCl_4), δ 7.5–7.1 (m, 5H), 5.82 (m, 1H, $J = 6.0$, 9.5, 17.5 Hz), 5.07 (br d, 1H, $J = 17.5$ Hz), 4.98 (br d, 1H, $J = 9.5$ Hz), 4.70 (t, 1H, $J = 8.0$ Hz), 3.73 (q, 2H, $J = 7.0$ Hz), 2.78 (m, 2H, $J = 6.0$, 8.0 Hz), 1.25 (t, 3H, $J = 7.0$ Hz); mass spectrum, m/z (relative intensity, %) 188 (M^+), 105 (100).

Two other valuable reactions of organocopper complexes derived from an initial carbocupration step are iodination [139] and carboxylation [140]. The vinylic iodides resulting from I_2 quenching lead to strictly defined stereochemistries (>99.9% pure), in good yields. These halides are useful precursors to the corresponding organolithiums [141] or are electrophiles for various stereospecific substitution reactions [142].

Preparation of Vinylcopper Reagents Prior to Iodination/Carboxylation [139]

$$\mathrm{RMgX} \xrightarrow[-35\,^{\circ}\mathrm{C},\ 30\ \mathrm{min}]{\mathrm{CuBr,\ Et_2O}} \xrightarrow[\mathrm{Et_2O,\ 1.5\ h;\ -35\ to\ -15\,^{\circ}C}]{\mathrm{R'{-}C{\equiv}C{-}H}} \mathrm{R'(R)C{=}CH{-}Cu{\cdot}MgX}$$

To a suspension of CuBr (50 mmol) in diethyl ether (50 ml) is added dropwise, at −35 °C, an ethereal solution of RMgBr (50 mmol). After 30 min, a yellow or brownish suspension (according to the nature of the R group) of RCu is obtained. A solution of 1-alkyne (50 mmol) in diethyl ether (30 ml) is then added dropwise and the reaction mixture is allowed to warm slowly to −15 °C. The temperature must be carefully kept for 1.5 h between −15 and −12 °C (very important!). The vinylcopper reagent is thus obtained quantitatively as a dark green solution.

To effect iodination the vinylic copper is simply treated with 1 equiv. of molecular iodine in the cold and then warmed. For large-scale (> 50 mmol) reactions, the iodine should be added portionwise. On purification of products by distillation, a small amount of copper powder added to the crude material is recommended.

Vinylic Iodides via Iodination of Vinylic Copper Reagents [139]

$$\mathrm{R(R')C{=}CH{-}Cu{\cdot}MgX} \xrightarrow[\mathrm{2.\ aq.\ NH_4Cl}]{\mathrm{1.\ I_2,\ -50\ to\ 0\,^{\circ}C,\ 1\ h}} \mathrm{R(R')C{=}CH{-}I}\quad (64\text{–}76\%)$$

Finely crushed solid iodine (50 mmol) is added at once at −50 °C to a solution of vinyl-copper reagent (50 mmol) prepared as above. The reaction mixture is then allowed to warm to 0 °C. Stirring is continued for 30 min to 1 h, until the formation of a precipitate of CuI and discoloration of the supernatant are noted. Hydrolysis is performed at −10 °C with a mixture of saturated aqueous NH_4Cl and $NaHSO_3$ (80 ml + 10 ml). Next, the precipitate is filtered off and washed with diethyl ether (2 × 50 ml). After decantation, the aqueous layer is extracted with pentane or cyclohexane. The combined organic layers are then washed with a dilute solution of $NaHSO_3$ (if necessary to eliminate free iodine) and dried over $MgSO_4$. The solvents are removed by distillation *in vacuo* and the product is isolated by distillation.

Exposure of a vinylic copper·MgX_2 species to dry CO_2 results in the nearly quantitative, stereospecific (>99.9%) conversion to α,β-unsaturated carboxylic acid salts, which ultimately give the acids on quenching with dilute HCl [140]. This particular reaction is rare in that it requires a catalytic amount of triethylphosphite [$(EtO)_3P$], but otherwise is very much akin to the follow-up procedures offered above once carbocupration has occurred.

α,β-Unsaturated Carboxylic Acids via Carboxylation of Vinylic Copper Reagents [140]

R(R')C=CH–Cu·MgX → 1. $P(OEt)_3$ (10 mol%), HMPT, −40 °C; 2. CO_2, −40 °C, 2 h then warm to r.t.; 3. 3M HCl, −30 °C → R(R')C=CH–COOH (60–96%)

To a solution of vinylcopper reagent (50 mmol) prepared as above are added, at −40 °C, HMPT (40 ml) and a catalytic amount of $P(OEt)_3$ (5 mmol). A slow stream of dried CO_2 is then bubbled into the reaction mixture for 2 h. During the carbonation, the mixture is slowly allowed to warm to room temperature. Hydrolysis is performed at −30 °C by adding 3 M HCl (80 ml). After decantation and extraction with cyclohexane (2 × 50 ml), the combined organic layers are washed with 2 M HCl (2 × 30 ml) and water (80 ml) and then dried over $MgSO_4$. The solvents are removed under vacuum and the product is isolated by distillation.

Displacements of other allylic leaving groups by organocopper·MgX_2 reagents not derived from an initial carbocupration can also be effected. The stereochemistry of the process is such that the reagent attacks the nucleofuge in an S_N2', *anti* fashion.

S_N2' Opening of a Vinylic Lactone with RCu·MgX_2 [143]

MeMgBr, CuBr·SMe_2, THF, −20 °C; Me, COOH (97%)

To a solution of copper(I) bromide·SMe_2 complex (71.0 g, 0.35 mol) in Me_2S (300 ml) and THF (700 ml) at −20 °C was added methylmagnesium bromide (125 ml, 2.85 M in THF, 0.35 mol). After stirring at −20 °C for 1 h, a solution of 2*H*-cyclopenta[*b*]furan-2-one (21.5 g, 0.18 mol) in THF (200 ml) was added dropwise via an addition funnel. The mixture was stirred at −20 °C for 5 h, poured into 1 M NaOH and stirred for 2 h. The organic layer was separated and the aqueous layer was acidified to pH ≈ 2 with 1 M HCl. After extraction with diethyl ether, the organic phase was washed with water and brine, dried over $MgSO_4$ and concentrated *in vacuo* to provide a yellow oil (23.65 g, 97%), which was characterized as the methyl ester (prepared by standard diazomethane treatment); IR ($CHCl_3$), 1730 cm^{-1}; ^1H NMR ($CDCl_3$), δ 5.65 (m, 2H), 3.65 (s, 3H), 3.14 (br m, 1H), 2.80 (br m, 1H), 2.30 (AB portion of ABX, 2H), 1.67 (m, 2H), 0.97 (d, 3H). Analysis: calculated for $C_9H_{14}O_2$, C 70.10, H, 9.15; found, C 70.01, H 9.19%.

RCu·MgX_2-based couplings with propargylic electrophiles occur readily under mild conditions to afford allenic products. For the educt below, the stereochemical outcome reflects approach of the copper species from the face of the acetylene *anti* to the departing group [144a], although the product was originally believed to be derived from cuprate *syn* delivery [144c]. This observation, therefore, is *not* an exception to the 'rule' (i.e. expected *anti* addition) [145]. Other factors, such as temperature, solvent and leaving group, can also affect the regiochemistry of these couplings [146].

Displacement of a Steroidal Propargylic Sulfinate with $RCu \cdot MgX_2$ [144a]

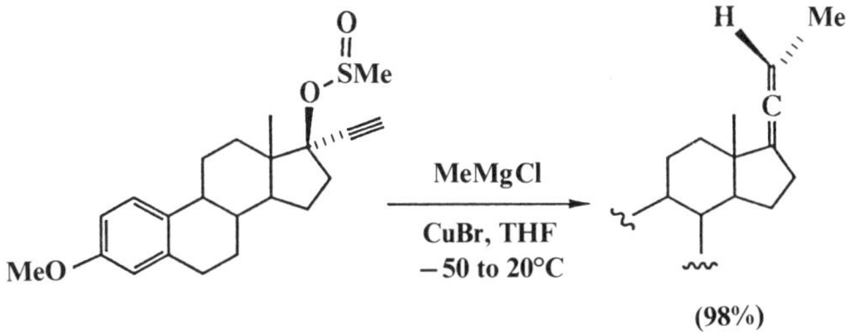

MeMgCl

CuBr, THF
−50 to 20°C

(98%)

A solution of methylmagnesuum chloride (0.03 mol) in THF (30 ml) was added cautiously to a stirred suspension of copper(I) bromide (0.03 mol) in THF (50 ml) at −50 °C and stirred at −30 °C for 30 min. 17α-Ethynyl-17β-methanesulfinyloxy-3-methoxy-1,3,5(10)-estratriene (5.58 g, 15 mmol) in THF (10 ml) was then added at −50 °C over 10 min. The temperature of the reaction mixture was raised to 20 °C within 10 min. After 45 min, it was poured into a saturated solution of NH_4Cl in water (200 ml) containing NaCN (2 g). It was then extracted with hexane (3 × 50 ml) and the combined extracts were washed with water and then dried over $MgSO_4$. Evaporation of the solvent *in vacuo* afforded the product (4.55 g, 98%), which was recrystallized from ethanol, m.p. 71.0–71.5 °C; $[\alpha]_D^{23}$ − 16.05° (CH_2Cl_2).

Magnesiocuprates, R_2CuMgX, are the reagents of choice for openings of β-propiolactones to afford carbon homologated carboxylic acids. These displacements are characterized by short reaction times, which is fortunate owing to the relative thermal instability of the cuprates. Effective stirring and cooling (internal monitoring) are essential for both reagent formation and the subsequent introduction of educt, as these are exothermic processes. A useful feature of this chemistry is the color changes associated with completion of the reaction. Thus, with R = Me, the initial slightly yellow coloration becomes a deeper yellow (also: R = *n*-Bu, gray → yellow; R = *t*-Bu, grey → greenish white; R = Ph, lemon yellow → yellow; R = vinyl, dark green → purple; R = allyl, reddish yellow → brown). Although the example below gives details for unsubstituted β-propiolactone [147], this chemistry applies to other related unsaturated substrates [148], including the six shown. In all cases, the products are those resulting from S_N2' additions.

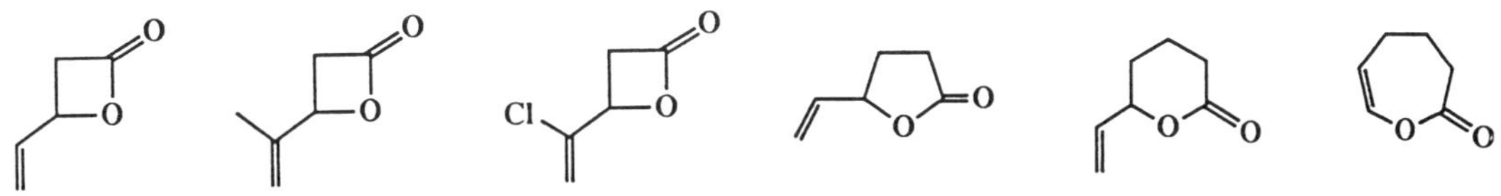

Carboxylic Acid Formation From Ring Opening of β-Propiolactone with R_2CuMgX [147]

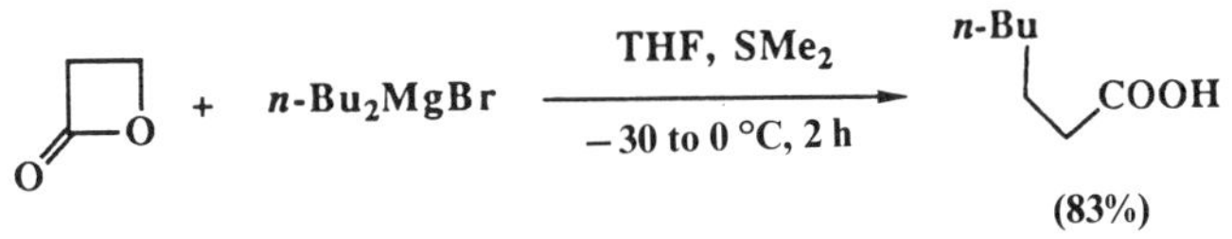

+ *n*-Bu_2MgBr —THF, SMe_2; −30 to 0 °C, 2 h→ *n*-Bu–CH_2CH_2–COOH

(83%)

A flask equipped with a magnetic stirring bar and a septum was charged with 420 mg (2.20 mmol) of CuI. After flushing with dry Ar, 10 ml of anhydrous THF and 1 ml of Me_2S were added and the solution was chilled to −30 °C. Butylmagnesium bromide (1.00 M in THF, 4.4 mmol) was slowly added to this solution and the mixture was stirred for 30 min at this temperature. Then a solution of β-propiolactone 144 mg (2.00 mmol) in 2 ml of THF was added dropwise to the flask. The mixture was stirred at −30 °C for 1 h and then allowed to warm to 0 °C for 1 h. After the reaction had been quenched by addition of 2 ml of 3 M HCl, heptanoic acid was extracted with three 5 ml portions of 3 M NaOH from the organic layer. The alkaline solution was acidified with 3 ml of 6 M HCl and then extracted with diethyl ether. The ethereal extracts were washed with brine and dried ($MgSO_4$). Concentration gave pure heptanoic acid. An analytical sample was obtained by bulb-to-bulb distillation.

Ketones are easily obtained from acid chlorides using magnesiodiorganocuprates. As with lithiocuprates, mixed magnesio reagents, generally represented as R_tR_rCuMgX, can also be expected to form and show a selectivity of transfer profile related to their lithio analogs. In the case described below, initial preparation of $(MeCu)_n$, from CuI and MeLi, can be followed by the introduction of a Grignard reagent (1 equiv.) in the usual ethereal media [149]. After appropriate manipulation of temperature to insure reagent generation (i.e. warming to obtain the MeCu into solution), the substrate is introduced. Unlike R_2CuLi or R_tR_rCuLi, Grignard-based reagents such as R(Me)CuMgX exist as suspensions at −78 °C. They do, however, qualitatively have better thermal stability.

Ketone Formation From Reaction of an Acid Halide with a Mixed Magnesiocuprate R_tR_rCuMgX [149]

Me—C₆H₄—MgBr → (PhCOCl, MeCu / THF, Et_2O, −78 to 25 °C) → 4-methylbenzophenone (79%)

In a 1 l, flame-dried, three-necked, round-bottomed flask equipped with an overhead stirrer and low-temperature thermometer, a bright yellow suspension of methylcopper was prepared by the reaction of 30 ml of a 1.73 M (51.9 mmol) diethyl ether solution of methyllithium (0.11 M in residual base) with a −78 °C suspension of 9.6 g (50.8 mmol) of CuI in 100 ml of THF. The bright yellow color characteristic of methylcopper formed when this reaction mixture was warmed to 25 °C. It was then cooled to −70 °C and 26 ml of a 1.96 M (51.0 mmol) diethyl ether solution of 4-methylphenylmagnesium bromide was added with a syringe. The resulting suspension was allowed to warm to 25 °C and, after cooling the deep purple solution to −78 °C, a solution of benzoyl chloride (13.0 ml, 112 mmol) in THF (30 ml) was added dropwise by syringe. The reaction mixture was then warmed to 25 °C and stirred for 30 min. It was quenched with 8 ml of absolute methanol and then added to 600 ml of saturated aqueous NH_4Cl solution. Stirring for 2 h dissolved the copper salts, the ethereal phase separated and the aqueous portion was washed with two 100 ml portions of diethyl ether. The combined organic fractions were washed with 100 ml of 0.1 M aqueous sodium thiosulfate, 3 × 100 ml of 1.0 M NaOH, and 200 ml saturated NaCl, then dried over patassium carbonate. The product 4-methylbenzophenone was isolated by distillation (7.8 g, 79% yield), b.p. 120–130 °C (0.6 mmHg); IR (CH_2Cl_2), 1670 cm^{-1}; ^1H NMR ($CDCl_3$), δ 7.1–7.9 (m, 9H), 2.4 (s, 3H).

Replacement of halogen by various alkyl groups at the 2-, 3- and 4-positions in the pyridine, quinoline and 1,10-phenanthroline series can be accomplished using excess of R_2CuMgX [150]. Whereas CuBr–2RMgX is recommended for methylations and ethylations, insertion of a *tert*-butyl moiety is most efficiently carried out with *t*-BuMgCl in the presence of CuCN.

Double Displacement on a Heteroaromatic System Using R_2CuMgX [150]

Cl, Cl, N, N + MeMgBr —(CuBr, THF; −78 °C, 2–3 h, then r.t., overnight)→ Me, Me, N, N

(75%)

A mixture of 8.0 g (56 mmol) of anhydrous CuBr, 250 ml of anhydrous THF and 40 ml of ethereal methylmagnesium bromide (2.9 M) was stirred under N_2 at −78 °C for 20 min. Dichlorophenanthroline (2.5 g, 7.0 mmol) was added and the reaction mixture was stirred for 2–3 h at −78 °C and then overnight at room temperature. The reaction mixture was quenched by dropwise addition of saturated aqueous ammonia and then extracted with $CHCl_3$ (3 × 75 ml). The combined extracts were stirred for 20 min with 50 ml of ethylenediamine and then 250 ml of water were added cautiously. The aqueous layer was extracted with $CHCl_3$ (3 × 75 ml) and the combined $CHCl_3$ solutions were dried ($MgSO_4$) and evaporated *in vacuo*. Column chromatography of the residue [EtOAc–CH_3OH (4:1)] gave 75% of spectroscopically pure product as a white solid, m.p. 220 °C (decomp); IR (KBr), 2940 (s), 2870 (m), 1490 (s), 1440 (m), 1387 (m) cm^{-1}; ^1H NMR (300 MHz), δ 1.95 (m, CH_2, H2, H3, H10, H11, 8H), 2.60 (s, CH_3, 6H), 2.92 (m, CH_2, H4, H9, 4H), 3.34 (m, CH_2, H1, H12, 4H), 7.90 (s, CH, 2H); ^{13}C NMR (75.5 MHz), δ 13.9, 22.8, 23.3, 27.1, 35.0, 120.9, 125.7, 129.9, 141.1, 143.5, 158.3; HRMS, calculated for $C_{22}H_{24}N_2$, 316.1956; found, 316.1930.

Michael additions based on stoichiometric magnesiocuprates are also very popular, useful processes, in particular where cyclopentenones are concerned. The initial adduct of this 1,4-event can be transformed into other derivatives, such as regiospecifically generated silyl enol ethers or α-alkylated products [151]. Both types of 1,4-additon/electrophilic trapping procedures are described below, in these cases used to provide intermediates for steroid total synthesis.

Conjugate Addition of R_2CuMgX to a Cyclopentenone with in situ *Trapping by Me_3SiCl [151]*

O, Me + MgBr —(1. CuI, THF, −40 °C; 2. Me_3SiCl, HMPA, Et_3N)→ $OSiMe_3$, Me

(89%)

To magnesium (6.07 g, 250 mmol) and one crystal of iodine in THF (100 ml) was added vinyl bromide (70.5 ml, 1 mol) in THF (60 ml) at a rate so as to maintain the reaction temperature at 45 °C. After all the magnesium had disappeared, the solution was heated at 45 °C under a stream of N_2 to remove excess vinyl bromide. The mixture was then cooled to −5 °C, CuI (25.7 g, 135 mmol) was added and the solution was stirred until it was jet black. The mixture was quickly chilled to −70 °C and 2-methylcyclopentenone (10.56 g, 110 mmol) in THF (40 ml) was added dropwise and the solution stirred at −40 °C for 45 min. After subsequent cooling to −60 °C, chlorotrimethylsilane (34 ml, 365 mmol), hexamethylphosphoric triamide (70 ml) (**Caution:** potent carcinogen) and triethylamine (50 ml) were added sequentially. The reaction mixture was allowed to warm to room temperature over a period of 2 h. Aqueous light petroleum work-up, followed by distillation, gave a colorless liquid (19.19 g, 89%); b.p. 64–66 °C (3.1 mmHg); IR (neat), 2990, 1690, 1640, 1250, 1210, 1090, 990, 840 cm^{-1}; ^1H NMR (CCl_4), δ 5.70 (overlapping (5 lines) ddd, 1H, $J = 17.5, 10, 9$ Hz), 5.00 (dd, 1H, $J = 17.5, 2.5$ Hz), 4.93 (dd, 1H, $J = 9, 2.5$ Hz), 3.00 (m, 1H), 2.5–1.4 (m, 4H), 1.47 (br s, 3H), 0.22 (s, 9H). Analysis: calculated for $C_{11}H_{20}OSi$, C 67.28, H 10.26; found, C 67.04, H 10.18%.

1,4-Addition of R_2CuMgX to a Cyclopentenone Followed by in situ *Alkylation [151]*

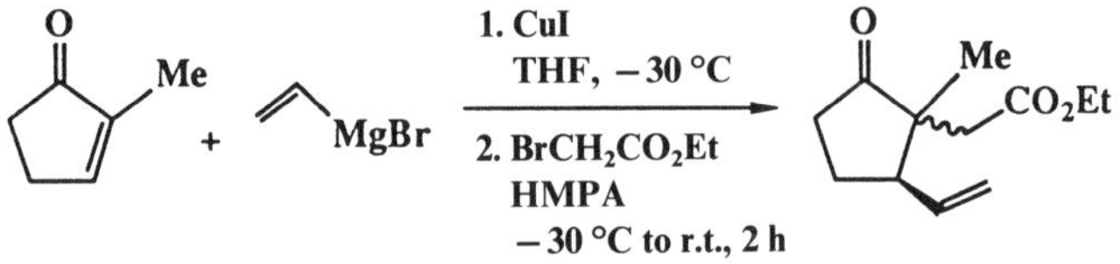

O, Me + MgBr — 1. CuI, THF, −30 °C; 2. $BrCH_2CO_2Et$, HMPA, −30 °C to r.t., 2 h → O, Me, CO_2Et

(81%)

(*cis*:*trans* 1:3)

To magnesium (2.66 g, 109 mmol) and one crystal of I_2 in THF (100 ml) was added vinyl bromide (29.5 ml, 418 mmol) in THF (60 ml) at such a rate as to maintain the reaction temperature at 45 °C. After all the magnesium had disappeared, the solution was heated at 45 °C under a stream of N_2 to remove excess vinyl bromide. The mixture was then cooled to −5 °C, copper(I) iodide (10.44 g, 54.8 mmol) was added and the solution stirred until it was jet black. The mixture was quickly chilled to −70 °C and 2-methylcyclopentenone (4.79 g, 49.9 mmol) in THF (45 ml) was added dropwise. After the addition was complete (30 min), the solution was warmed to −30 °C, stirred for 45 min and cooled to −70 °C, and HMPA (50 ml) was added, followed by ethyl bromoacetate (10 ml, 91 mmol). The solution was allowed to warm to room temperature over 90 min, then stirred for 30 min, quenched with methanol, diluted with diethyl ether, poured on to saturated NH_4Cl and stirred for another 30 min. The aqueous layer was separated and extracted twice with diethyl ether. The combined ether extracts were washed with 5% aqueous $Na_2S_2O_3$, water and brine and dried ($MgSO_4$). Evaporation of the ether left a yellow liquid which was distilled to give a colorless liquid (8.52 g, 81%), a single peak by GC and a single spot by TLC, R_f [diethyl ether–light petroleum ether (1:4)] 0.27; b.p. 85–87 °C (0.5 mmHg); IR (neat), 3100, 3000, 1740, 1645, 1470, 1210, 1040 cm^{-1}; ^1H NMR (CCl_4), δ 5.80 (m, 1H), 5.10 (m, 2H), 4.05 (q, 2H, $J = 7$ Hz), 3.23–1.53 (m, 7H), 1.23 (t, 3H, $J = 7$ Hz), 1.11 (s, 0.73H), 0.82 (s, 2.27 H); ^{13}C NMR (C_6D_6) showed two isomers, with resonances for the major isomer at δ 218.58, 170.98, 137.52, 116.63, 60.19, 49.13, 47.38, 40.07, 36.51, 24.75, 17.79, 14.05 and for the minor isomer at δ 217.97, 170.97, 137.52, 116.63, 60.19, 51.42, 49.72, 38.90, 35.58, 24.75, 22.81. 14.05; mass spectrum, m/z (relative intensity, %) 210 (M^+, 6.41), 195 (8.31), 165 (18.13), 137 (17.30), 123 (100), 95 (18.94), 81 (28.55). Analysis: calculated for $C_{12}H_{18}O_3$, C 68.54, H 8.63; found, C 68.26, H 8.44%.

With a β-halocyclopentenone, the conjugate addition of R_2CuMgX leads to reformation of the enone, a net substitution reaction. When carried out by adding the Grignard to a slurry of the chloro enone–CuI in cold THF, an excellent yield, at least with the functionalized substrate and Grignard shown below, was obtained.

Conjugate Addition–Elimination on a β-Chlorocyclopentenone Mediated by R_2CuMgX [152]

+ $BrMg(CH_2)_7O$-Si — CuI, THF, −10 °C → (95%)

The required Grignard reagent was prepared by adding a solution of 7-bromo-1-(*tert*-butyldimethylsilyloxy)heptane (6.19 g, 20 mmol) in THF (15 ml) over 1.5 h to magnesium (491 mg, 20.2 mmol) in refluxing THF (15 ml). The consumption of magnesium was complete after heating at reflux for a further 3 h. The concentration of reagent was measured by standard titration of an aliquot (1 ml) after hydrolysis. A suspension of CuI (78 mg, 0.4 mmol) in THF (2 ml) containing 3-chloro-4-(*tert*-butyldimethylsilyloxy)cyclopent-2-en-1-one (100 mg, 0.4 mmol) was stirred vigorously at −10 °C under Ar. Dropwise addition of the above Grignard reagent (0.41 M in THF, 1.85 ml, 0.76 mmol) produced a green solution which was stirred at −10 °C for 10 min. The reaction was rapidly quenched with saturated aqueous NH_4Cl solution (5 ml) and, after addition of diethyl ether (5 ml), the mixture was stirred at room temperature for 1 h before dilution with water (10 ml) and extraction with diethyl ether (5 × 10 ml). The combined extracts were washed with brine (2 × 4 ml), dried over $MgSO_4$ and evaporated. Preparative TLC [silica gel, CH_2Cl_2–methanol (50:1 v/v)] gave the product as a colorless oil (167 mg, 95%), b.p. (Kugelrohr) 135 °C (0.2 mmHg); IR, 1720 cm^{-1}; ^1H NMR, δ 0.04 (s, 6H), 0.12 (s, 3H), 0.14 (s, 3H), 0.88 (s, 9H), 0.91 (s, 9H), 1.16–1.80 (m, 10H), 2.25 (dd, 1H, J = 18.0, 3.0 Hz), 2.44 (br t, 2H, J = 8 Hz), 2.72 (dd, 1H, J = 18.0, 6.0 Hz), 3.60 (t, 2H, J = 6.0 Hz), 4.76 (dd, 1H, J = 6.0, 3.0 Hz), 5.90 (m, 1H). Analysis: calculated for $C_{24}H_{48}O_3Si_2$, C 65.40, H 11.0; found, C 65.05, H 10.8%.

4.4 MISCELLANEOUS REAGENTS

Although most organocopper complexes in use today are of the forms RCu, RR′CuM (M = Li, MgX) and $R_2Cu(CN)Li_2$, there are several alternative reagents which have also been developed. Changes in the gegenion(s) associated with, or the ligand(s) bound to, copper can play a pivotal role in controlling several reaction parameters such as (1) the regio- and stereochemistry of additions; (2) reagent reactivity; (3) stability of the reagent; (4) compatibility with functionality within the ligands on copper; and (5) tendency to form lower order vs higher order reagents. In general, it is now well established that replacement of lithium by another metal in a lower or higher order cuprate can be expected to decrease the reagent reactivity [3, 153]. While magnesiocuprates are therefore less robust than their lithio analogs, organozinc-based reagents are still less prone to react. Moving to non-associating counter ions, such as R_4N^+, depresses the reactivity still further (see above). These guidelines lend further support to the notion that many cuprate couplings are in fact electrophilically driven, with the hard Li^+ intimately involved in placing R_2CuLi or $R_2Cu(CN)Li_2$ at the top of the reactivity scale. Insight as to how to go about fine tuning the behavior of copper reagents by the judicious choice of, e.g., the counter ion(s), is now starting to accrue. As a result, significant advances have already been made, leading to powerful new methodologies to which the discussion and representative procedures that follow attest.

4.4.1 Zn-CONTAINING CUPRATES

Processes involving both catalytic and stoichiometric amounts of Cu(I) *en route* to zinc cuprates have appeared over the past few years. One case in point concerns the generation and use of a zinc homoenolate precursor, derived from a cyclopropanone ketal [154a]. Intermediates of this type are extremely versatile, participating in many different carbon–carbon bond-forming events. When exposed to catalytic quantities of a Cu(I) salt, they can be effectively utilized as Michael donors, and also as regioselective nucleophiles in allylations. The organozinc precursor is initially formed *in situ* using dried $ZnCl_2$ of high quality, which can be tolerated in slight excess. HMPA is also an essential ingredient in both reaction types, as is Me_3SiCl. DMPU can be substituted for the former additive in conjugate additions, although lower yields of 1,4-adducts are to be expected [154].

1,4-Addition of a Zinc Homoenolate Catalyzed by CuBr·SMe$_2$ [154a]

O + $Zn(CH_2CH_2CO_2Et)_2$ —cat. CuBr; TMSCl, HMPA, Et_2O, 0 °C, 3 h→ $OSiMe_3$ / CO_2Et (70–76%)

To a solution of $ZnCl_2$ (16.4–17.0 g), freshly fused under vacuum, in 500 ml of diethyl ether was added 1-(trimethylsiloxy)-1-ethoxycyclopropanone (41.80 g, 240 mmol) over 5 min. The cloudy mixture was stirred at room temperature for 1 h and then refluxed for 30 min. The clear, colorless solution of the zinc homoenolate and chlorotrimethylsilane (2 equiv. vs enone) was cooled with an ice-bath, and to this was added CuBr·SMe$_2$ (0.4 g, 2 mmol). Cyclohexenone (9.62 g, 100 mmol) and then HMPA (34.8 ml, 200 mmol) were added over 5 min. A slightly exothermic reaction occurred initially and the bath was removed after 20 min. After 3 h at room temperature, 40 g of silica gel and 300 ml of dry hexane were added while the mixture was stirred vigorously. The supernatant was decanted, and the residue was extracted twice with a mixture of diethyl ether and hexane. HMPA was collected as a low-boiling fraction (about 50–80 °C), and after about 1 g of forerun, the desired product (18.9–20.5 g, 70–76%) was obtained as a fraction boiling at 122–125 °C (2.40 mmHg); IR (neat), 1730 (s), 1655 (s), 1445, 1365, 1245, 1180 (vs), 840 (vs) cm^{-1}; ^1H NMR (CCl_4), δ 0.06 (s, 9H), 1.13 (t, 3H, $J = 7$ Hz), 1.3–2.2 (m), 3.96 (q, 2H, $J = 7$ Hz), 4.54 (br s, 1H); mass spectrum, m/z (relative intensity, %) 270 (M^+, 3), 225 (8), 219 (8), 182 (40), 169 (100), 75 (36), 73 (92); HRMS, calculated for $C_{14}H_{26}O_3Si$, 270.1652; found, 270.1659. This product contains *ca* 0.2% of the double-bond regioisomer.

CuBr·SMe$_2$-Catalyzed Allylation of a Zinc Homoenolate [154a]

Cl + $Zn(CH_2CH_2CO_2{}^iPr)_2$ —cat. CuBr; TMSCl, Et_2O, CH_3CONMe_2, 16 h, r.t.→ $CO_2{}^iPr$ + $CO_2{}^iPr$

(97%)
(S_N2' : S_N2 96 : 4)

To a mixture of cinnamyl chloride (144 μl, 1.0 mmol), $CuBr \cdot SMe_2$ (10 mg, 0.05 mmol) and HMPA (2–5 equiv.) or dimethylacetamide (4.5 ml) was added a 0.37 M ethereal solution of homoenolate (1.7 mmol, 4.5 ml; containing 2 equiv. of Me_3SiCl), and the solution was stirred for 16 h at room temperature. The reaction mixture was diluted with diethyl ether and washed five times with water. After washing with saturated NaCl, drying and concentration, the crude product was purified by chromatography (2% ethyl acetate in hexane) to obtain 148 mg (97%) of the allylated product as a 96:4 mixture of the S_N2' and the S_N2 isomers (GLC retention times on a 23 m OV-1 column at 194 °C: 5.4 and 8.1 min respectively). The isomers were separated by medium-pressure chromatography for analysis. Isopropyl 4-phenyl-5-hexenoate: b.p. 80 °C (0.3 mmHg); IR (neat), 2975 (s), 2925 (m), 1720 (s), 1490 (m), 1450 (m), 1370 (m), 1240 (m), 1175 (m), 750 (m), 700 (s) cm^{-1}; ^1H NMR (200 MHz, $CDCl_3$), δ 1.21 (d, 6H, $J = 6$ Hz), 1.91–2.33 (m, 4H), 3.26 (q, 1H, $J = 8$ Hz), 4.89–5.14 (m, 3H), 5.96 (ddd, 1H, $J = 17$, 10, 8 Hz), 7.13–7.37 (m, 5H). Elemental analysis corresponded to $C_{15}H_{20}O_2$.

Conversion of a dialkyl lithiocuprate to its zinc halide congener provides a reagent which is remarkably selective in its couplings with 4-alkoxyallylic chlorides [155a]. Products of S_N2' attack result, with the relationship between the two stereogenic centers predominantly, if not exclusively, of the *anti* configuration. Given the availability of chiral, non-racemic allylic and propargylic alcohols and the fact that cuprate attack is not limited by olefin disubstitution α- to the alkoxy moiety, these factors argue well for future applications of this chemistry to the construction of quaternary centers of defined absolute stereochemistry.

S_N2' anti *Allylation of R_2CuZnX [155a]*

OBn Cl + $Bu_2CuZnCl$ $\xrightarrow[-78\,°C,\ 3\,h]{THF}$ OBn Bu

(95%)

To a suspension of $CuBr \cdot SMe_2$ (102.8 mg, 0.5 mmol) in THF was added 1.57 M BuLi in hexane (0.57 ml, 1.00 mmol) at −70 °C. The resulting solution was warmed to −50 to −40 °C, stirred for 40 min, then cooled to −70 °C. A 1 M THF solution of fused $ZnCl_2$ (0.5 ml, 0.5 mmol) was added, turning the solution dark brown. After stirring for 10 min, 4-benzyloxy-5-methyl-2-hexenyl chloride (117 μl, 0.50 mmol) was added dropwise. After stirring for 15 h, TLC analysis indicated that the reaction was complete. The reaction mixture was then diluted with hexane (2 ml), washed with saturated $NaHCO_3$ (3 × 0.5 ml), then with saturated NaCl (3 × 0.5 ml) and dried over $MgSO_4$. Solvent was removed *in vacuo* and the residual oil was purified by column chromatography on silica gel (4 g, 2% ethyl acetate in hexane) to obtain 123.5 mg (95%) of the S_N2' product contaminated with 2% of the S_N2 product; capillary GLC (HR-1, 145 °C) analysis of the crude product indicated the absence of the diastereoisomer of the S_N2' product; ^1H NMR (200 MHz, $CDCl_3$), δ 0.78–1.00 (m, 9H involving d, $J = 7$ Hz at 0.90, and d, $J = 67$ Hz at 0.98), 1.15–1.40 (m, 6H), 1.86 (dqq, 1H, $J = 7$, 7, 7 Hz), 2.20–2.32 (m, 1H), 3.00 (dd, 1H, $J = 4$, 8 Hz), 4.58 (s, 2H), 4.98 (dd, 1H, $J = 3$, 8 Hz), 5.05 (dd, 1H, $J = 3$, 11 Hz), 5.78 (ddd, 1H, $J = 3$, 11, 18 Hz), 7.23–7.40 (m, 5H). Analysis: calculated for $C_{18}H_{28}O$, C 83.02, H 10.84; found, C 82.94, H 10.94%.

Related lower order zinc cuprates can be derived from CuCN and a preformed organozinc halide, rather than via transmetallations between lithiocuprates and ZnX_2 (see above). They enjoy a considerable level of functional group tolerance within these organometallic species owing to a less reactive carbon–zinc bond associated with the cuprate precursors RZnX. Hence, cuprates of the general stoichiometric formula RCu(CN)ZnX (X = halide) may contain within R functionality which includes esters, nitriles, halides, ketones, phosphonates, thioethers, sulfoxides, sulfones and amines. The initial phase of reagent preparation calls for generation of the organozinc iodide, using cut zinc foil (Alfa, 0.25 mm thick, 30 cm wide, 99.9% purity) or zinc dust (Aldrich, −325 mesh). A detailed procedure is given below which includes the subsequent formation of the cyanocuprate [156]. For benzylic systems, three slightly different protocols have been developed (procedures 1, 2 and 3 below) [157]. Once formed, where a yield of 90% of RZnX has been assumed, the reagent is inversely added to solubilized CuCN·2LiCl to form the lower order cyanocuprate (see procedure 4 below).

Formation of $RCH_2Cu(CN)ZnI$ via RCH_2ZnI, Prepared from RCH_2I + Zn [156]

$$RCH_2\text{-}I \xrightarrow[\text{2. CuCN·2LiCl, } -40 \text{ to } 0^\circ C]{\text{1. Zn, THF, TMSCl, 4 h; } BrCH_2CH_2Br, 35\text{–}45^\circ C} RCH_2Cu(CN)ZnI$$

A dry 100 ml, three-necked, round-bottomed flask was equipped with a magnetic stirring bar, a 50 ml pressure-equalizing addition funnel bearing a rubber septum, a three-way stopcock and a thermometer. The air in the flask was replaced with dry Ar and charged with 4.71 g (72 mmol) of cut zinc (*ca* 1.5 × 1.5 mm). The flask was flushed three times with Ar. 1,2-Dibromoethane (0.2 ml, 2.3 mmol) and 3 ml of THF were successively injected into the flask, which was then heated gently with a heat gun until ebullition of the solvent was observed. The zinc suspension was stirred for a few minutes and heated again. The process was repeated three times, after which 0.15 ml (1.2 mmol) of chlorotrimethylsilane was injected through the addition funnel. The cut zinc foil turned gray and was ready to use after 10 min of stirring. The reaction mixture was heated to 30 °C on an oil-bath and 60 mmol of the iodide dissolved in 30 ml of THF were added dropwise over 40 min. The reaction mixture was then stirred for 4 h at 35–45 °C to give a dark brown–yellow solution of the zinc reagent, the completion of which was checked by GLC analysis of an aliquot. A second dry 100 ml three-necked, round-bottomed flask was equipped with a magnetic stirring bar, a three-way stopcock and two glass stoppers. The flask was charged with 4.59 g (108 mmol) of lithium chloride, then heated on an oil-bath at 130 °C [oil-bath temperature under vacuum (0.1 mmHg)] for 2 h in order to dry the lithium chloride, cooled to 25 °C and flushed with Ar. The two glass stoppers were replaced by a low-temperature thermometer and a rubber septum and 4.84 g (54 mmol) of copper cyanide were added. The flask was flushed three times with Ar and 40 ml of freshly distilled THF were added, which led, after 15 min, to a clear yellow–green solution of the complex CuCN·2LiCl. This solution was cooled to *ca* −40 °C and the two 100 ml three necked flasks were connected via a PTFE cannula or a stainless-steel needle. The solution of the zinc reagent was then transferred to the THF solution of copper cyanide and lithium chloride. The resulting dark green solution was warmed to 0 °C within 5 min and was ready to use for the next step after 5 min of stirring at this temperature.

Preparation of Benzylic Zinc Halides from Zinc Foil (Procedure 1) [157]

$$R\text{-}C_6H_4CH_2X \xrightarrow[BrCH_2CH_2Br]{\text{Zn foil, THF}} R\text{-}C_6H_4CH_2ZnX$$

Procedure 1 Cut zinc foil (*ca* 5 mm^2 pieces, 2.30 g, 36 mmol) in 3 ml of dry THF was added to a dry, three-necked flask equipped with an Ar inlet, a thermometer and an addition funnel. 1,2-Dibromoethane (150 mg) was then added and the mixture was heated with a heat gun until evolution of soap-like bubbles of ethylene and darkening of the zinc surface indicated activation. The mixture was cooled to 0–5 °C (ice-bath) and a solution of the benzylic bromide (30 mmol) in 15 ml of THF was added dropwise (1 drop per 5–10 s). The reaction mixture was stirred at 5 °C until GLC analysis showed that the starting material was completely consumed (1–4 h).

For the Preparation of Benzylic Zinc Halides from Zinc Dust (Procedures 2 and 3) [157]

Procedure 2 Zinc dust (0.67 g, 10.5 mmol) in 1 ml of dry THF was added to a three-necked flask equipped with an Ar inlet, a thermometer and an addition funnel. The mixture was cooled in an ice-bath and a solution of the benzylic bromide (7 mmol) in 7 ml of dry THF was added dropwise (1 drop per 5–10 s). The reaction mixture was stirred at 5 °C until GLC analysis showed that the starting material was completely consumed (1–4 h).

Procedure 3 Zinc dust (1.34 g, 21 mmol) in 1.5 ml of dry THF was added to a three-necked flask equipped with an Ar inlet, a thermometer and an addition funnel. A solution of the benzylic chloride (7 mmol) in 5.5 ml of dry THF and 1.5 ml of DMSO was added dropwise at room temperature. The reaction mixture was stirred at room temperature until GLC analysis showed complete conversion of the starting material to the zinc organometallic (22–24 h).

Preparation of Benzylic Copper Derivatives from Benzylic Zinc Halides by Transmetallation (Procedure 4) [157]

R–C$_6$H$_4$–CH$_2$ZnX $\xrightarrow[\text{THF, } -40 \text{ to } -20\,°C]{CuCN\cdot 2LiCl}$ R–C$_6$H$_4$–CH$_2$Cu(CN)ZnX

Procedure 4 In a three-necked flask equipped with an Ar inlet, a thermometer and a rubber septum, heat-dried LiCl (1.14 g, 27 mmol) was combined with CuCN (1.21 g, 13.5 mmol) in 10.5 ml of dry THF. The mixture was cooled to −40 °C and the benzylic organozinc reagent (14.7 mmol, prepared as described above) was added to the copper solution by a cannula. The mixture was allowed to warm to −20 °C for 5 min, then cooled to −78 °C (dry-ice–acetone bath). The copper reagent was then ready to react with various organic electrophiles (see below).

Since only small percentages of Wurtz-like coupling occurs with, e.g., benzylic halide conversions to benzylic zinc halides, this type of organometallic can be prepared, stored at 25 °C for a few days (unlike the copper reagents derived therefrom, which must be maintained below −20 °C to avoid self-coupling) and used in many subsequent reactions. In addition to the example shown below for a Michael reaction [157], these cuprates are applicable to substitutions with allylic, trialkylstannyl and acid halides. 1,2-Additions to aldehydes readily occur in the presence of $BF_3 \cdot Et_2O$, and addition–eliminations to β-halo enones afford β-substituted, α,β-unsaturated ketones in high yields [157]. Formation of reagents containing secondary benzylic centers present no special problems, so long as precursor benzylic chlorides, rather than bromides, are used.

1,4-Addition of a Benzylic Lower Order Cyanocuprate to an Ynoate [157a]

1. CuCN·2LiCl, THF −40 to −20 °C, 5 min
2. ≡—CO_2Et −78 to −50 °C, 14 h, then −30 °C, 3 h

(13) (79%)

1,4-Diacetoxynaphthylzinc chloride was prepated from 1,4-diacetoxy-2-chloromethylnaphthalene (13 mmol) according to procedure 3 (above) with heating at 45 °C for 3 h. GLC analysis showed complete conversion with less than 3% Wurtz coupling. After the transmetallation to the corresponding copper reagent by the addition of CuCN·2LiCl (as indicated above), ethyl propiolate (1.02 ml, 10 mmol) was added at −78 °C. The reaction mixture was stirred at −50 °C for 14 h and then at −30 °C for 3 h. Work-up and purification of the residue by flash chromatography [hexane–ethyl acetate (3:1, then 2:1] afforded 0.51 g (14% yield) of the double-bond isomerized product and 2.55 g (79% yield) of the desired product **13**: IR ($CDCl_3$), 2984 (m), 1766 (s), 1715 (s), 1656 (w), 1603 (w), 1368 (s), 1277 (m), 1191 (s), 1158 (m), 1065 (m) cm^{-1}; ^1H NMR ($CDCl_3$, 300 MHz), δ 7.82 (d, 1H, $J = 1.9$ Hz), 7.54 (d, 1H, $J = 0.8$ Hz), 7.54 (m, 2H), 7.11 (s, 1H), 7.07 (dt, 1H, $J = 15.6, 6.7$ Hz), 5.86 (d, 1H, $J = 15.6$ Hz), 4.17 (q, 2H, $J = 7.1$ Hz), 3.55 (d, 2H, $J = 6.7$ Hz), 2.46 (s, 3H), 2.45 (s, 3H), 1.26 (t, 3H, $J = 7.1$ Hz); ^{13}C NMR ($CDCl_3$, 75.5 MHz), δ 168.7, 168.5, 165.4, 144.5, 141.9, 127.1, 126.9, 126.4, 126.3, 126.0, 123.0, 121.3, 60.0, 32.8, 20.5, 20.1, 13.9; EI mass spectrum, m/z (relative intensity, %) 43 (84), 77 (7), 105 (14), 115 (14), 141 (9), 152 (5), 174 (32), 191 (10), 197 (35), 226 (100), 272 (47), 314 (21), 356 (4); HRMS, calculated for $C_{20}H_{20}O_6$, 356.1260; found, 356.1254.

Unactivated primary iodides are also easily converted into the corresponding functionalized zinc iodides. These can be used to form ketones via treatment with an acid chloride [158] or α,β-unsaturated ketones in the presence of Me_3SiCl [159] and also to synthesize acetylenes by way of couplings with 1-bromoalkynes [160]. In addition, they add smoothly to nitroolefins in high yields [161].

Coupling of a Functionalized Zinc Cuprate with a Bromoacetylene [160]

NC⁀⁀Cu(CN)ZnI + Br—≡—(cyclohexenyl) —THF, −70 to −65 °C→ (79%)

The cuprate can be prepared using the general procedure above from cut zinc foil, or alternatively using zinc dust as follows. A 50 ml three-necked flask equipped with a thermometer, a 25 ml pressure-equalizing dropping funnel sealed with a septum, a vacuum/argon inlet and a magnetic stirring bar was charged with zinc dust (Aldrich, −325 mesh; 1.2 g, 18 mmol) and evacuated by means of a pump and refilled with Ar three times. 1,2-Dibromoethane (150 mg, 0.8 mmol) in 2 ml of THF was added and the suspension was heated gently to ebullition with a heat gun. The zinc suspension was stirred for *ca* 1 min and heated again. This process was repeated three times and chlorotrimethylsilane (0.1 ml, 0.8 mmol) was then added. After a few minutes, 4-iodobutyronitrile (1.66 g, 8.5 mmol) in 3.5 ml of dry THF was slowly added at 35–40 °C. After 3 h of stirring at 40 °C, GLC analysis of a hydrolyzed aliquot indicated completion of the reaction. The yield was estimated to be *ca* 85%.

The magnetic agitation was stopped and the remaining powder was allowed to settle. During this time a second 50 ml three-necked flask equipped with a magnetic stirring bar was charged with lithium chloride (*ca* 0.9 g, 21 mmol), connected to a vacuum pump and heated on an oil-bath at 130 °C for 1 h in order to dry the lithium chloride. The flask was cooled to 25 °C, flushed with Ar and equipped with a low-temperature thermometer and a rubber septum. Copper cyanide (0.9 g, 10 mmol) was added, followed by 10 ml of THF. A yellow–green solution formed after a few minutes of stirring. The reaction mixture was cooled to *ca* −40 °C and the above-prepared THF solution of 3-cyanopropylzinc iodide was added with a syringe. The reaction mixture was allowed to warm to 0 °C and was ready for use in the next step.

The solution of $NC(CH_2)_3Cu(CN)ZnI$ was cooled to −78 °C and 1-(2-bromoethynyl)cyclohexene (0.915 g, 5 mmol) was added. The reaction mixture was stirred at −70 to −65 °C for 20 h and then warmed slowly to −10 °C. No starting material was left, as indicated by GLC analysis. The reaction mixture was poured into an Erlenmeyer flask containing diethyl ether (200 ml) and a saturated aqueous solution NH_4Cl (50 ml). The insoluble copper salts were removed by vacuum filtration and the two layers were separated. The organic layer was washed with a saturated aqueous solution of NH_4Cl (2 × 50 ml) and the aqueous layer was extracted with diethyl ether (2 × 50 ml). The combined organic layers were then washed with a saturated aqueous solution of NaCl and dried over $MgSO_4$. After filtration, the solvent was removed and the resulting crude oil was purified by flash chromatography [hexane–EtOAc (6 : 1)], affording 670 mg of analytically pure 6-(1-cyclohexenyl)-5-hexyne-1-nitrile and a small second fraction of impure product (30 mg, 48% purity) in an overall yield of 79%; IR (neat), 2935 (s), 2860 (m), 2839 (s), 2248 (m), 1673 (w), 1449 (m), 1435 (m), 1359 (m), 1349 (m), 919 (m), 843 (m) cm^{-1}; ^1H NMR ($CDCl_3$, 300 MHz), δ 5.92 (br s, 1H), 2.38 (m, 4H), 1.95 (m, 4H), 1.72 (m, 2H), 1.53 (m, 4H); ^{13}C NMR ($CDCl_3$, 75.5 MHz), δ 134.6, 120.9, 119.0, 84.8, 84.6, 29.8, 26.1, 24.8, 22.8, 21.7, 18.8, 16.2; CI mass spectrum (70 eV), *m/z* (relative intensity, %) 174 (MH^+, 5), 172 (100), 158 (18), 144 (38), 130 (26), 117 (33), 105 (42), 91 (83), 79 (40), 65 (35), 51 (30); HRMS, calculated for $C_{12}H_{16}N$ (MH^+), 174.1283; found, 174.1270.

Acylation of a Functionalized Zinc Cuprate [158]

Cu(CN)ZnI —PhCOCl, THF, 0 °C, 10 h→ (80%)

To a suspension of zinc dust (activated with 1,2-dibromoethane and chlorotrimethylsilane as described above; 1.63 g, 25 mmol) in 1.5 ml of THF was slowly added 4-oxocyclohexyl 4-iodobutanoate (3.10 g, 10 mmol) in 4 ml of THF at 30 °C. After 4 h at this temperature, GLC analysis of a hydrolyzed aliquot indicated a conversion of 87%. The zinc reagent was cooled to 25 °C and added with a syringe to a solution of copper cyanide (0.67 g, 7.5 mmol) and predried lithium chloride (0.67 g, 15 mmol) in 8 ml of THF at −30 °C. The resulting milky solution was warmed to 0 °C for 5 min and cooled to −20 °C. Benzoyl chloride (0.914 g, 6.5 mmol) was added and the reaction mixture was stirred at −3 °C for 10 h. The reaction mixture was poured into a mixture of diethyl ether (200 ml) and saturated aqueous NH_4Cl solution (50 ml). After filtration and work-up as described above, the resulting crude oil was purified by flash chromatography (hexane–EtOAc), affording white crystals (m.p. 58–59 °C) of pure 4-oxocyclohexyl 5-oxo-5-phenylpentanoate (1.51 g, 80% yield); IR ($CDCl_3$), 2959 (m), 1722 (s), 1694 (s), 1595 (w), 1448 (m) cm^{-1}; ^1H NMR ($CDCl_3$, 360 MHz), δ 7.91 (d, 2H, $J = 8.1$ Hz), 7.50 (d, 1H, $J = 6.7$ Hz), 7.41 (dd, 2H, $J = 8.1$, 6.7 Hz), 2.44 (m, 4H), 2.32 (m, 2H), 2.01 (m, 6H); ^{13}C NMR ($CDCl_3$, 90.5 MHz), δ 209, 199, 172, 137, 133, 128, 127, 68, 37, 33, 30, 19; CI mass spectrum (70 eV), *m/z* (relative intensity, %) 288 (M^+, 1), 193 (15), 175 (21), 147 (13), 120 (9), 105 (100), 96 (15), 77 (27), 55 (11); HRMS, calculated for $C_{17}H_{20}O_4$ (MH^+), 289.1440; found, 289.1446.

Conjugate Addition of a Functionalized Zinc Cuprate to Cyclohexenone [159]

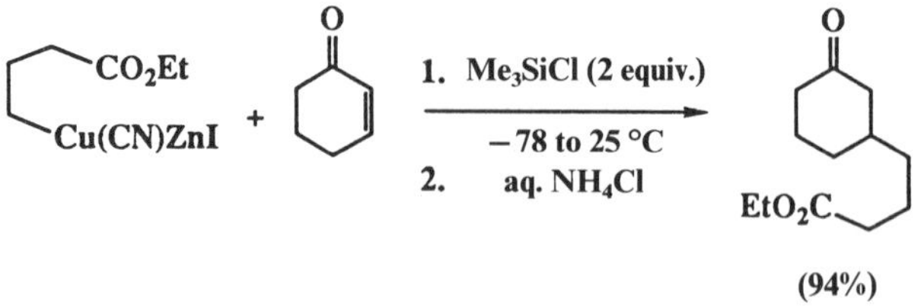

To a suspension of zinc dust (activated with 1,2-dibromoethane and chlorotrimethylsilane as described above; *ca* 3 g, 45 mmol) in 2 ml of THF was slowly added, at 45 °C, a solution of ethyl 4-iodobutyrate (4.90 g, 20 mmol) in 8 ml of THF. After 4 h at this temperature, GLC analysis of a hydrolyzed aliquot indicated over 95% conversion. The zinc reagent was cooled to 25 °C and added via a syringe to a solution of copper cyanide (1.44 g, 16 mmol) and predried lithium chloride (1.36 g, 32 mmol) in 16 ml of dry THF cooled to −30 °C. The resulting grey–greenish solutiion was warmed to 0 °C for a few minutes and cooled to −78 °C. Chlorotrimethylsilane (4.1 ml, 32 mmol) was added, followed by a solution of cyclohexenone (1.39 g, 14.5 mmol) in 5 ml of dry diethyl ether. The reaction mixture was stirred for 3 h at −78 °C, then allowed to warm to 25 °C overnight. GLC analysis of a hydrolyzed reaction aliquot showed no cyclohexenone left. The reaction mixture was poured into an Erlenmeyer flask containing diethyl ether (50 ml) and saturated aqueous NH_4Cl solution (50 ml). After 5 min of stirring, the reaction mixture was filtered over Celite and worked up as described above. Flash chromatography of the residual oil [hexane–EtOAc (85:15)] afforded pure ethyl 4-(3-oxocyclohexyl) butyrate (2.90 g, 94% yield) [162]; IR (neat), 2945 (s), 2868 (s), 1722 (s), 2433 (s), 1363 (s), 1275 (s) cm^{-1}; ^1H NMR ($CDCl_3$, 360 MHz), δ 4.2–4.0 (m, 2H), 2.5–2.2 (m, 4H), 2.1–1.8 (m, 3H), 1.8–1.6 (m, 4H), 1.5–1.1 (m, 7H); ^{13}C NMR ($CDCl_3$, 75.5 MHz), δ 211.2, 173.0, 59.9, 47.7, 41.2, 38.5, 35.7, 34.0, 30.9, 24.9, 21.8, 14.0.

Michael Addition of a Functionalized Zinc Cuprate to a Nitroolefin [161]

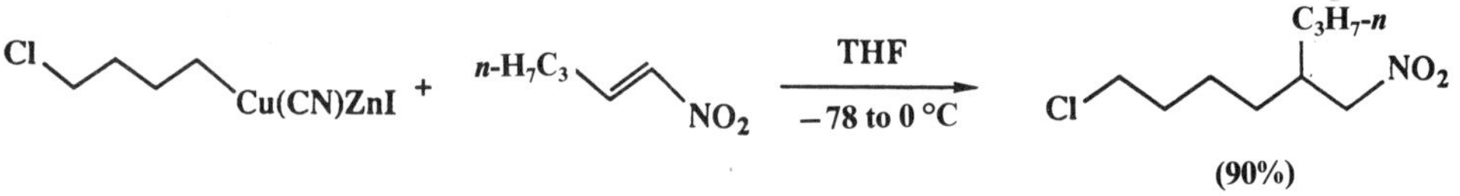

To a suspension of zinc dust (activated with 1,2-dibromoethane and chlorotrimethylsilane as described above; *ca* 2 g, 30 mmol) in 2 ml of THF was slowly added, at 40 °C, a solution of 4-chloro-1-iodobutane (2.75 g, 12.5 mmol) in 5 ml of THF. After 3 h at 40 °C, GLC analysis of a hydrolyzed aliquot indicated over 95% conversion. The zinc reagent was cooled to 25 °C and added with a syringe to a solution of copper cyanide (0.9 g, 10 mmol) and predried lithium chloride (0.85 g, 20 mmol) in 10 ml of dry THF cooled to −30 °C. The resulting greenish solution was warmed to 0 °C for a few minutes and then cooled to −78 °C. 1-Nitro-1-pentene (0.86 g, 7.5 mmol) was added dropwise at −78 °C. The cooling bath was removed and the reaction mixture was allowed to warm slowly to 0 °C. After 4 h at this temperature, no nitropentene was left as indicated by GLC analysis. The reaction mixture was cooled to −78 °C, quenched with acetic acid (2 ml in 5 ml of THF) and warmed to 0 °C. The reaction mixture was poured into an Erlenmeyer flask containing diethyl ether (200 ml) and saturated aqueous NH_4Cl solution (50 ml). The reaction mixture was filtered over Celite and worked up as described above. Flash chromatographic purification of the residual oil [hexane–diethyl ether (20:1)] afforded pure 1-chloro-5-(nitromethyl)octane (1.40 g, 90% yield); IR (neat), 2935 (s), 2872 (m), 1557 (s), 1546 (s), 1462 (m), 1446 (m), 1434 (m), 1383 (m), 1311 (w), 1210 (w), 733 (m), 650 (m) cm^{-1}; ^1H NMR ($CDCl_3$, 300 MHz), δ 4.28 (dd, 2H, $J = 7$, 2.2 Hz), 3.5 (t, 2H, $J = 6.5$ Hz), 2.18 (m, 1H), 1.73 (quint, 2H, $J = 6.5$ Hz), 1.49–1.25 (m, 8H), 0.88 (t, 3H, $J = 6$ Hz); ^{13}C NMR ($CDCl_3$, 90.5 MHz), δ 79.28,

44.58, 36.97, 33.18, 32.29, 30.36, 23.29, 19.24, 13.91; CI mass spectrum (CH_4), m/z (relative intensity, %) 208 (MH^+, 22), 177 (6), 161 (100), 125 (25), 119 (25), 105 (21), 97 (5), 89 (5), 83 (28); HRMS, calculated for $C_9H_{18}NO_2ClH$ (MH^+), 208.1104; found, 208.1118.

4.4.2 *[RCu(CN)$_2$Li[N(C$_4$H$_9$)$_4$]*

As higher order cyanocuprates are formally dianionic Cu(I) reagents, there must be either two monovalent metal ions associated with the complex, as in $R_2Cu(CN)Li_2$, or two dissimilar monovalent cations, generalized as $R_2Cu(CN)MM'$. Since the ligands on copper also need not be identical, the stoichiometry generalizes still further to $RR'Cu(CN)MM'$. Moreover, the nature of the gegenions can be chosen to reflect a cationic species which has no obligation to be of the alkali metal series. Hence, cuprates composed of less commonly employed components have started to appear which, not surprisingly, have modified properties. One case where multiple variations have been found advantageous for selected synthetic transformations concerns the higher order dicyano species derived from CuCN, n-Bu_4NCN and an organolithium (equation 4.26). The precursor, $Cu(CN)_2NBu_4$, a colorless solid, is prepared in MeOH at ambient temperature under an inert atmosphere.

$$\mathrm{CuCN} + n\text{-}\mathrm{Bu_4NCN} \xrightarrow[\text{r.t.}]{\text{MeOH}} \mathrm{Cu(CN)_2NBu_4} \xrightarrow[-40\text{ to }-25\,^\circ\mathrm{C},\ 2\,\mathrm{h}]{\text{RLi, THF}} \mathrm{RCu(CN)_2LiNBu_4} \quad (4.26)$$

Preparation of n-*Bu$_4$NCu(CN)$_2$ [163]*

To a slurry of 1.79 g (20 mmol) of copper(I) cyanide in 100 ml of anhydrous methanol at 25 °C under N_2 was added a solution of 5.36 g (20 mmol) of tetrabutylammonium cyanide in 50 ml of methanol. After stirring for 10 min at 25 °C, the resulting homogeneous solution was rotary evaporated to dryness and the resulting paste was dried by azeotropic evaporation with toluene (3 × 70 ml) at ambient temperature to furnish a white crystalline solid. The complex was dried under high vacuum at 25 °C for 24 h.

Treatment of a slurry of $Cu(CN)_2NBu_4$ in THF with 1 equiv. of an RLi (R = alkyl, phenyl, vinyl, alkenyl) leads to 'nearly homogeneous' solutions of $RCu(CN)_2LiNBu_4$, which participate in the selective delivery of the transferable ligand (R) to enones, and also in displacements of primary halides [163]. Owing to both the second nitrile group (presumably on copper) and replacement of a lithium with tetrabutylammonium as one of the two counter ions, their reactivity is diminished considerably. This is manifested in both reaction types, where conjugate additions to even unhindered α,β-unsaturated ketones are performed at temperatures above −50 °C over 1 h, and primary iodides take 4 h at −25 °C. By contrast, the corresponding cuprates, $R_2Cu(CN)Li_2$, would react at −78 °C with either functional group within minutes. Lower reactivity, however, is associated with added selectivity, and indeed the incentive for development of these higher order dicyanocuprates arose from dissatisfaction with yields obtained using other copper reagents for the substitution reaction detailed below. A second procedure highlighting the use of these cuprates as Michael donors is also given.

Displacement of a Primary Halide Using $RCu(CN)_2LiNBu_4$ [163]

1. n-BuLi, Et_2O, −40 °C, 30 min
2. n-$Bu_4NCu(CN)_2$, −25 °C, 2 h
3. I–(iodoorthoester), −20 °C, 4 h

(69%)

To a solution of 229 mg (1.0 mmol) of the stannane in 2 ml of anhydrous diethyl ether at −78 °C was added 0.404 ml of 2.50 M (1.01 mmol) *n*-BuLi. The yellow solution was stirred for 1 h at −78 °C, then for 30 min at −40 °C. The vinyllithium solution was added to a suspension of 354 mg (0.99 mmol) of tetrabutylammonium dicyanocopper(I) in 8 ml of anhydrous THF. The orange suspension was stirred for 15 min at −78 °C, then warmed to −25 °C and stirred for an additional 2 h. To the above cuprate mixture at −25 °C was added a solution of 328 mg (1.10 mmol) of the iodoorthoester in 5 ml of anhydrous THF. The mixture was stirred for 4 h at −25 °C, then quenched by the addition of 2 ml of 2 M aqueous ammonium chloride solution (pH 9.0). The crude mixture was extracted with 3 × 50 ml of pentane–diethyl ether (1 : 3), washed with brine and dried over anhydrous potassium carbonate. Evaporation of solvent *in vacuo* followed by chromatography on triethylamine-deactivated silica gel in diethyl ether–hexane (1 : 5) gave 363 mg (69%) of the product; TLC, R_f [diethyl ether–hexane (1:5)] 0.62; IR (neat), 3090 cm^{-1}; ^1H NMR ($CDCl_3$), δ 0.795 (s, 3H), 0.91 (t, 3H), 2.06 (dt, J = 7, 7 Hz), 2.75 (dd, 2H, J = 7, 7 Hz, 3.89 (s, 6H), 5.41 (m, 2H), 5.85 (d, J = 12 Hz), 6.48 (dt, J = 8, 12 Hz); mass spectrum, m/z (relative intensity, %) 527 (M^+, 2).

Typical 1,4-Addition of $R_tCu(CN)_2LiNBu_4$ to an Enone [163]

n-$BuCu(CN)_2LiNBu_4$
THF, −40 °C, 10 min

Bu-*n*

(97%)

To a suspension of 179 mg (0.5 mmol) of n-$Bu_4NCu(CN)_2$ in 4 ml of anhydrous THF at −78 °C under N_2 was added 0.202 ml of 2.5 M (0.505 mmol) *n*-BuLi. After stirring for 1 h at −40 °C, 53 mg (0.55 mmol) of 2-cyclohexenone (neat) were added and the reaction mixture was stirred for an additional 10 min at −40 °C. The reaction was quenched by the addition of 2 ml of saturated aqueous NH_4Cl and the crude product was extracted with diethyl ether. The organic phase was washed with brine and dried over anhydrous $MgSO_4$. Evaporation of solvent followed by chromatography on silica gel in diethyl ether–light petroleum afforded 74.5 mg (97%) of the product; IR (neat), 1708 cm^{-1}; ^1H NMR ($CDCl_3$), δ 0.87 (t, 3H), 2.0–2.4 (m, 4H).

4.4.3 $R_2Cu(SCN)Li_2$

Although in the term 'organocuprate' has come to imply a copper reagent arising from CuCN or a copper(I) halide salt, there is actually another precursor that leads to a unique, albeit seldomly used, species. Dating back to Gilman's original disclosure on the admixture of 2RLi with a Cu(I) salt [5], both CuI and copper(I) thiocyanate (CuSCN) were believed to arrive at the same lower order species R_2CuLi. It is now appreciated that for CuSCN, unlike copper(I) halides, the 2RLi add to form a higher order cuprate $R_2Cu(SCN)Li_2$ (**14**) [6], i.e. the thiocyano analog of $R_2Cu(CN)Li_2$ (equation 4.27). They have been reported, however, to afford results in certain contexts which are different from those obtained with other cuprates. Thus, for the displacement reaction of a configurationally defined vinylogous thioester (below), cuprates **14** were a welcomed alternative to R_2CuLi insofar as the stereoselectivities of these couplings (i.e. $E:Z$ ratios) are concerned [164]. Copper(I) thiocyanate, the precursor to **14**, is a commercially available, inexpensive, air- and light-stable, off-white solid. Given these attractive features, it is surprising that it has not been utilized to a greater extent.

$$2RLi + CuSCN \xrightarrow[\text{solvent}]{\text{ethereal}} \underset{\mathbf{(14)}}{R_2Cu(SCN)Li_2} \qquad (4.27)$$

Displacement on a Vinylogous Thioester Using $R_2Cu(SCN)Li_2$ [164a]

SMe; s-$Bu_2Cu(SCN)Li_2$; Et_2O, −78 to −58 °C, 1.5 h; s-Bu

(86%)
($E:Z$ 11:89)

To a mixture of CuSCN (0.3287 g, 2.7 mmol) in diethyl ether (20 ml) cooled to approx. −15 °C was added dropwise 4.1 ml of *sec*-butyllithium (1.33 M, 5.45 mmol). After 30 min, the light tan solution was chilled to −78 °C and the educt (0.2108 g, 1.35 mmol) was added with a syringe. After 1.5 h, the temperature had risen to −58 °C and TLC indicated that the reaction was complete. The reaction was quenched with saturated aqueous NH_4Cl and standard work-up gave 0.2357 g of a yellow oil. Purification by TLC using a preparative plate (1000 μm SiO_2, 1% v/v ethyl acetate–light petroleum) afforded the pure Z-isomer (0.1492 g, 66%), R_f 0.28, and a minor fraction [which was a mixture of the E and Z-isomers (0.0421 g), for an overall yield of 86%]. VPC analysis of the crude mixture indicated an E/Z ratio of 11:89. E-Isomer: IR, 2960 (s), 2860 (s), 1670 (s), 1620 (s), 1460 (s), 1380 (2), 1120 (s), 1060 (s) cm^{-1}; ^1H NMR, δ 0.83 (t, 3H, $J = 7$ Hz), 1.05 (d, 3H, $J = 7$ Hz), 1.09 (d, 6H, $J = 7$ Hz), 1.37 (quint, 2H, $J = 7$ Hz), 1.89–2.28 (m, 1H), 2.05 (d, 3H, $J = 1$ Hz), 2.61 (hept, 1H, $J = 7$ Hz), 6.09 (s, 1H); ^{13}C NMR, δ 205.0, 162.9, 121.5, 46.0, 41.5, 27.5, 18.8, 18.3 (2C), 155.9, 14.9. Z-Isomer: IR, 2960 (s), 2870 (m), 1687 (ss), 1610 (s), 1440 (m), 1380 (m), 1350 (m), 1210 (m), 1170 (s), 950 (m) cm^{-1}; ^1H NMR, δ 0.69–1.14 (m, 6H), 1.08 (d, 6H, $J = 7$ Hz), 1.20–1.54 (m, 2H), 1.77 (d, 3H, $J = 1$ Hz), 2.50 (hept, 1H, $J = 7$ Hz), 3.73 (sep, 1H, $J = 7$ Hz), 6.10 (s, 1H); ^{13}C NMR, δ 203.6, 161.6, 123.2, 40.9, 35.4, 27.2, 18.4 (2C), 17.9, 17.7, 11.3.

4.4.4 $R_2Cu(CN)LiMgX$

Another variation among higher order cyanocuprates involves the mixing of gegenions to reflect, e.g., the combination of an organolithium and Grignard reagent, together with CuCN, to form presumably $R_2Cu(CN)LiMgX$ (equation 4.28) [165]. The driving

$$R_rLi + R_tMgX + CuCN \xrightarrow[\text{solvent}]{\text{ethereal}} \underset{\mathbf{(15)}}{R_tR_rCu(CN)LiMgX} \qquad (4.28)$$

force behind this variation lies in the availability and ease of preparation of most Grignard reagents. When an equivalent is added to preformed (2-thienyl)Cu(CN)Li, either freshly made up or purchased in a bottle, a reagent results which is capable of effecting substitution and 1,4-additions. The solubility at low temperatures, however, of lithiomagnesiohalide cuprates is not nearly as high as that of dilithio reagents, with those derived from RMgCl (vs RMgBr) seemingly of greater tendency to dissolve in THF and Et_2O at −78 °C. The actual outcome of the reaction, however, appears not to be dependent on the halide present in the Grignard precursor (and hence in **15**). While the reactivity of cuprates **15** is lowered owing to the switch from $2Li^+$ to Li^+MgX^+, even congested enones are acceptable educts. On occasion, however, $BF_3 \cdot Et_2O$ (1 equiv.) must be present in solution (which together should not be warmed above *ca* −50 °C). With unhindered systems, the Lewis acid is not needed. THF is the best medium with respect to solubility, although the reagents often appear as greyish slurries. In Et_2O as the major or sole solvent, a far less soluble, brown, sticky material is commonly noted at the bottom of the flask.

Reaction of a Monosubstituted Epoxide with R(2-Th)Cu(CN)LiMgX [165]

n-Bu(2-Th)Cu(CN)LiMgCl; THF, −78 to 0 °C, 4 h → OH, Bu-*s* (81%)

2-Thienyllithium (0.6 mmol) was prepared from thiophene (50 μl, 0.62 mmol) and *n*-BuLi (0.25 ml, 2.40 M) in THF at −30 °C (0.4 ml, 30 min) and then added to a precooled (−78 °C) slurry of CuCN (55 mg, 0.6 mmol) in THF (1.5 ml) followed by *n*-BuMgCl (0.24 ml, 0.6 mmol). On warming to room temperature, the mixture dissolved to a green–amber solution with traces of a fluffy white precipitate. Recooled to −78 °C, the mixture thickened to a tan slurry and to it was added the neat epoxide (56 μl, 0.5 mmol). The reaction was warmed to 0 °C and stirring continued for 4 h followed by quenching and workup in the usual way. Chromatography on SiO_2 with 20% Et_2O–light petroleum afforded 63.5 mg (81%) of a clear oil; TLC, $R_f = 0.25$; identical spectroscopically with an authentic sample.

Conjugate Addition of R(2-Th)Cu(CN)LiMgX to an α,β-Unsaturated Ketone [165]

+ Cu(CN)LiMgBr; S; THF, −78 °C, 2.25 h → (85%)

CuCN (102 mg, 1.14 mmol) was placed in an oven-dried two-necked round-bottomed flask equipped with a magnetic stir bar. The salt was gently flame-dried (30 s) under vacuum and then purged with Ar. Dry THF (1.0 ml) was added and the stirred slurry was cooled to −78 °C. 2-Thienyllithium was prepared in a second two-necked round-bottomed flask from thiophene

(91 μl, 1.14 mmol) in dry THF (1.0 ml) at −30 °C, to which was added *n*-BuLi (0.47 ml, 2.44 M in hexane, 1.14 mmol) and the clear colorless solution was stirred at 0 °C for 30 min. The preformed 2-thienyllithium was added to the CuCN slurry at −78 °C and warmed to 0 °C over 30 min until the tan-brown slurry became clear. The mixture was cooled to −78 °C and the Grignard reagent-derived 2-(2-bromoethyl)-1,3-dioxane (80 μl, 1.42 M in THF, 1.14 mmol) was added dropwise and the resulting mixture was warmed to 0 °C for 2 min and cooled to −78 °C. Cyclohexenone (freshly distilled, 100 μl, 1.03 mmol) was added and the reaction was stirred at −78 °C for 2.25 h and quenched by the addition of 5 ml of a 90% aqueous NH_4Cl (saturated)-10% ammonia (concentrated) solution. After stirring at room temperature for 30 min, the solution was worked up in the usual way. Column chromatography on SiO_2 with Et_2O–light petroleum (1:1) afforded 186 mg (85%) of product as a clear liquid; IR (neat), 2960, 2940, 1712, 1408, 1380, 1145, 1008, 942, 895 cm^{-1}; ^1H NMR, δ 4.51 (t, 1H, $J = 5.1$ Hz), 4.09 (g, 2H, $J = 5.1$ Hz), 3.75 (d of t, 1H, $J = 2.4$ Hz and $J = 12.3$ Hz), 2.09 (m, 2H), 2.0–0.8 (m, 13H); mass spectrum, *m/z* (relative intensity, %) 213 ($M^+ + 1$, 18), 165 (18.4), 157 (10.5), 138 (14.7), 121 (13.2), 119 (55.6), 110 (16.5), 87 (59.2), 59 (18.3); HRMS, calculated for $C_{12}H_{20}O_3 + (M + 1)$, 213.1491; found, 213.1477.

4.4.5 RMnCl–CATALYTIC CuCl

By conversion of an RLi or RMgX into the corresponding organomanganese(II) reagents RMnCl, 1,4-additions can be effected in the presence of 1–3 mol% CuCl (equation 4.29). In some cases, yields from reactions using this combination are better than those from either the corresponding copper-catalyzed Grignard or cuprate Michael-type couplings. Notably, hindered enones (e.g. pulegone, 88%; isophorone, 95%, mesityl oxide, 94%; addition of an *n*-Bu group) are not problematic [166] and the reactions are conducted at a very convenient 0 °C over a 2–4 h period.

$$\begin{matrix} \text{RMgX} \\ \text{or} \\ \text{RLi} \end{matrix} \xrightarrow{MnCl_2} \text{RMnCl} \xrightarrow{\text{enone}} \text{1,4 adduct} \qquad (4.29)$$

Owing to the hygroscopic nature of $MnCl_2$, it must be handled quickly. It should be dried *in vacuo* before use for 2 h at 180 °C under vacuum (0.01 mmHg). A number of suppliers offer anhydrous $MnCl_2$ (purity ⩾99%) fairly inexpensively. Material supplied by Chemetall as low nickel grade flakes works especially well for this chemistry.

Preformation of RMnCl can be accomplished using either an organolithium or Grignard reagent. They must be prepared and utilized under an inert atmosphere (N_2 or Ar).

Preparation of RMnCl from RLi and $MnCl_2$ [167]

$$\underset{(\text{in } Et_2O)}{\text{RLi}} + \underset{(\text{in THF})}{MnCl_2} \xrightarrow[\text{2. 30 min to 3 h, r.t.}]{\text{1. mix at } -35 \text{ to } 0\,^\circ\text{C}} \underset{(\text{brownish solution or suspension})}{\text{RMnCl} + \text{LiCl}}$$

A solution of 50 mmol of an organolithium compound in diethyl ether or a hydrocarbon was added, with stirring, to 52 mmol of anhydrous manganese chloride in 80 ml of THF under N_2, between −35 and 0 °C. After 30 min of stirring at room temperature, the organomanganese chloride reagent was obtained quantitatively as a brownish solution (formation of an 'ate complex' with the lithium salts present in the reaction mixture).

Importantly, when using reactive organolithiums (e.g. *n*-BuLi), the reaction mixture should be kept closer to −35 °C, whereas less basic RLi (e.g. MeLi, ArLi or vinyllithiums) can be used at 0 °C. Although all RMnCl compounds have stability at ambient temperatures, those derived from *sec*-alkyl-M or *tert*-alkyl-M (M = Li or MgX) may show decomposition at room temperature should $MnCl_2$ of poor quality be used. Starting with RMgX, RMnCl is formed as follows.

Formation of RMnCl from RMgX plus $MnCl_2$ [167]

Organomanganese chloride reagents were prepared by adding, between −10 and 0 °C with stirring, a solution of 50 mmol of an organomagnesium halide in diethyl ether or, better, THF (RMgX; X = Cl, Br, I) to a suspension of 52 mmol of anhydrous manganese chloride in 80 ml of THF. The reaction was quantitative after stirring for 30 min to 3 h at room temperature. A brownish solution or suspension was thus obtained.

Once the organomanganese derivative has been prepared, finely crushed CuCl (or $CuCl_2$) is introduced prior to the substrate. In addition to efficient Michael-type processes [166], the catalytic Cu(I)–RMnCl system is highly effective for acylation reactions using acid chlorides [168]. Although it is often possible to acylate RMnCl in good yields without resorting to copper catalysis, the presence of these salts clearly accelerates the process. Moreover, with *sec*- or *tert*-alkylmanganese chlorides and MeMnCl, the yields are usually dramatically improved under the influence of Cu(I).

1,4-Addition of RMnCl Catalyzed by CuCl (or $CuCl_2$): General Procedure [166]

cat. CuCl–RMnCl

THF, 0 °C

(86–98%)

To 52 mmol of organomanganese reagent (RMnCl) in 80–100 ml of THF (prepared as above) was added, at 0 °C, 1–3% of pulverized copper chloride (CuCl or $CuCl_2$). After 3 min, 50 mmol of enone dissolved in 40 ml of anhydrous THF were added dropwise over 10 min. The reaction mixture was then allowed to warm to room temperature and, after 1.5–4 h, depending on the nature of the enone [e.g. 1.5 h for cyclohexenone and 4 h for Bu(Pr)C=CHCOBu], hydolyzed with 60 ml of a 1 M HCl. The aqueous layer was decanted and extracted with diethyl ether or cyclohexane (3 × 100 ml) and the combined organic layers were washed with dilute aqueous ammonia–NH_4Cl solution to eliminate copper salts, and then with 50 ml of aqueous Na_2CO_3 solution. After drying over $MgSO_4$, the solvents were removed *in vacuo* and the ketone was isolated by distillation in 86–98% yield.

Copper-Catalyzed Acylation of Organomanganese Reagents with Carboxylic Acid Chlorides: General Procedure [168]

$$R'C(O)Cl \xrightarrow[\text{THF, } -10\,^{\circ}\text{C to r.t., } <30\text{ min}]{\text{cat. CuCl–RMnCl}} R'C(O)R \quad (69\text{–}95\%)$$

To 52 mmol of organomanganese reagent in 80–100 ml of THF (see above for preparation) was added, at -10 °C, 1–3% of pulverized copper chloride (CuCl or $CuCl_2$). After 3 min, 50 mmol of carboxylic acid chloride dissolved in 40 ml of anhydrous THF were added to the pot dropwise over 10 min. On occasion, the yield can be improved by adding the carboxylic acid chloride at -30 °C instead of at -10 °C. The reaction mixture was then allowed to warm to room temperature and, after 20–30 min (no more!), hydrolyzed with 60 ml of 1 M HCl. The aqueous layer was decanted and extracted with diethyl ether or cyclohexane (3 × 100 ml) and the combined organic layers were washed with dilute aqueous ammonia–NH_4Cl solution to eliminate copper salts, and then with 50 ml of aqueous Na_2CO_3 solution. After drying over $MgSO_4$, the solvents were removed *in vacuo* and the ketone was isolated by distillation in a yield usually of 69–95%.

4.5 CHOOSING AN ORGANOCOPPER REAGENT

With *ca* 100 'representative' procedures presented here for forming carbon–carbon, carbon–hydrogen, carbon–tin and carbon–silicon bonds via a copper intermediate, where does one start the process of deciding upon a specific reagent for a particular transformation? Since most of the variations among the reagent pool from which to choose are for purposes of construction of C—C bonds, attention is directed toward this major subgroup rather than those composed of heteroatom or hydrido ligands.

The two main types of couplings that rely heavily on organocopper reagents are conjugate addition and substitution reactions. It is also true that carbocupration is an extremely valuable regio- and stereoselective process, but the number of different copper species which effect this reaction is relatively limited. Examples provided here include the use of lower order lithio- (R_2CuLi) and magnesiocuprates (R_2CuMgX) and Grignard-based organocopper species, $RCu{\cdot}MgX_2$, which together account for essentially all of this chemistry. Relatively few cases involving, e.g., mixed Gilman reagents have been used for carbocupration purposes, and additives such as Lewis acids do not figure in these additions to acetylenes. Further, higher order cyanocuprates bearing all-carbon ligands are excluded from contention in that they are too basic to be utilized in this context; they simply induce an acid–base reaction by abstraction of a terminal acetylene's proton (unless utilized in a intramolecular fashion).

In anticipating conditions for a particular coupling, the question of competing cuprate decomposition may arise. It was pointed out early on [2d] that, in particular, ligands containing β-hydrogens appear to be especially good candidates for *syn*-elimination of the elements of CuH [169]. A more recent study comparing a homocuprate (Bu_2CuLi) with several mixed reagents (RR'CuLi) at -50, 0 and 25 °C suggests that Gilman reagents are actually far more thermally stable that previously realized [55].

4.5.1 CONSIDERATIONS FOR CONJUGATE ADDITION REACTIONS

Focusing, then, first on the need to carry out a 1,4-addition, there are several initial, mainly visual (i.e. non-experimental) observations that can quickly narrow the field of logical choices.

1. Aside from the presence of an α,β-unsaturated carbonyl unit, what other (electrophilic or Lewis basic) functionality is located in the molecule? This question arises because, although catalytic copper-driven Grignard additions preferentially deliver in a 1,4-manner, the excess of RMgX in the pot may not be tolerated elsewhere in the molecule. On the other hand, the cornerstone of stoichiometric cuprate use is their reluctance to add, e.g., in a 1,2-fashion to isolated ketones, especially in the presence of functionalities more compatible with soft copper complexes. Further insight along these lines can be gleaned from an extensive survey on the uses of organocopper reagents in natural products-related total syntheses [2a]. What seems to come from this inspection, perhaps not unexpectedly, is that far more researchers opt for stoichiometric lithio- or magnesiocuprates that for the catalytic Cu(I)–RMgX protocol. The majority of uses of the catalytic system are with simpler substrates, where there is little room for competing modes of addition. Thus, the build-up of intermediates early in a synthetic scheme might be one of the goals. Finer control within more advanced intermediates tends to be left to the domain of stoichiometric reagents, mostly lithium based. Such a selection provides still further opportunities for taking advantage of the beneficial effects of additives such as $BF_3 \cdot Et_2O$ and Me_3SiCl (see below, point 3 on p. 373).
2. How costly is the ligand to be introduced into the substrate? When a carbon fragment of interest (R_t) is easily obtained from, say, a commercial supplier of organolithium (R_tLi) or Grignard reagents (R_tMgX), or is one step removed from readily available materials {e.g. R_tMgBr, from R_tBr, EtLi from EtBr, or (*E*)- and (*Z*)-propenyllithium from the respective halides [170]}, then it is wise to prepare the corresponding homocuprates, R_2CuLi, R_2CuMgX or $R_2Cu(CN)Li_2$. Half of the R_tM (M = Li, MgX) will be lost on work-up but the benefits associated with greater reactivity (vs that of a mixed reagent R_tR_rCuM) and the elimination of any selectivity of transfer issue far outweigh the cost of one equivalent of R_tM. Should, however, the ligand R_tM require a number of steps for preparation or involve expensive reagents (e.g. a costly protecting group) such that sacrificing one full equivalent from a cuprate coupling is less desirable or even intolerable, then it is advantageous to select a mixed cuprate (R_tR_rCuM). Since, by definition, these reagents are less active than homocuprates [171], and as the nature of the genenion M also impacts on reactivity, a lithiocuprate (rather than a Grignard-derived reagent) usually offers the best chance of success. As for the choice of R_rLi, there are many from which to choose (including, as examples, lithiated *tert*-butylacetylene, 1-pentyne, thiophene, *tert*-butoxide, thiophenoxide, trimethylsilylacetylene and 3-methoxy-3-methyl-1-butyne). At the outset, however, there is no reason to select one which is either costly or requires special handling, preparation or manipulation. Based on these stringent requirements,

thiophene and thiophenol should be seriously considered, as both easily metallate with *n*-BuLi in THF and can then be utilized directly for cuprate generation. Lithiated 3-methoxy-3-methyl-1-butyne is also an excellent choice for R_rLi, its precursor 2-methyl-3-butyn-2-ol being very inexpensive and *O*-methylated in high yield. The resulting 3-methoxy-3-methyl-1-butyne is a water-white, room temperature stable liquid which in lithiated form is extremely soluble in THF. Thus, if sulfur compounds are (for whatever reason) to be avoided, this acetylene is a good alternative as long as THF, at least in part, is in the medium.

3. Will additives such as $BF_3 \cdot Et_2O$ or Me_3SiCl help? Perhaps the best approach to answering this question is to assess the environment surrounding the chromophore. That is, although these additives do indeed have the potential to turn a completely unsuccessful 1,4-addition reaction into a remarkably efficient and rapid process, in most cases they may not be necessary and should only be considered after the results of an initial small-scale reaction are known. Their impact is most noticeable in sterically congested situations, or where the substrate is normally not a satisfactory Michael acceptor toward cuprates owing to its highly cathodic (i.e. $\leqslant -2.4$ V in DMF) reduction potential (e.g. as with unsaturated amides) [172]. For these situations, and also for even more general uses, it pays to consider '$RCu \cdot BF_3$,' as the mixture of CuI plus R_tLi (1 equiv.) followed by introduction of $BF_3 \cdot Et_2O$ in THF has been shown to undergo 1,4-additions to, e.g., α,β-unsaturated carboxylic acids. There are also small but possibly significant prices to be paid when employing these additives: with $BF_3 \cdot Et_2O$, Lewis acid-sensitive functionalities (e.g. ketals) may prohibit its use; with Me_3SiCl, the initial product is the silyl enol ether, which may require a separate 'hydrolysis' step (i.e. treatment with mild acid or fluoride ion) to arrive at the carbonyl stage. Thus, the best approach is the simplest approach: once a reagent has been decided upon, try the coupling without recourse to any other materials in the pot (including HMPA, phosphines, phosphites, $BF_3 \cdot Et_2O$ or Me_3SiX). Only when it is clear that special circumstances exist (such as with highly hindered substrates, thermally unstable cuprates, problems due to limited solubility with the substrate and/or cuprate, etc.) should one resort to these modifications.

4.5.2 CONSIDERATIONS FOR SUBSTITUTION REACTIONS

Displacements by organocopper reagents, as with conjugate additions, cover a wide range of available reagents. However, coming up with generalities for selecting one can be far more challenging in that there is a greater diversity of electrophiles associated with this aspect of organocopper chemistry. In other words, while 'Michael additions' define (in most cases) the substrate as containing an α,β-unsaturated carbonyl unit, suitable partners for substitution reactions include (a) primary, secondary, and vinylic halides and numerous sulfonate derivatives (e.g. mesylate, tosylate, triflate), (b) epoxides of the (most often) mono- and disubstituted type, (c) allylic substrates such as epoxides and acetates, (d) propargylic leaving groups and (e) miscellaneous participants, including lactones, sulfates, heteroaromatics, heteroatoms and other organometallic species, to name only a few [2a].

Of greatest interest, based solely on relative numbers, are reactions of primary leaving groups and epoxides. These two classes need be distinguished with respect to reagent choice since additives, which do not play a role in displacements of, e.g., halides, can significantly alter the chemistry of epoxides via, e.g., Lewis acid coordination. Thus, whereas the former are essentially 'push' reactions (i.e. nuclophilically driven), displacements involving oxiranes (with Li^+, MgX^+ or either in the presence of BF_3) involve a 'push–pull' type process.

For carrying out a displacement on a primary halide in THF or sulfonate in Et_2O, it is important to match the reactivity of the leaving group with the reactivity of the reagent. For simplicity, the following scale can be used as a guide, where each halide correlates with a particular lithiocuprate. Thus, if a primary iodide is chosen, a Gilman cuprate should lead to an excellent yield of the desired product at low temperatures (usually *ca* -78 °C). If, however, an iodide is of limited stability, is difficult to prepare or is light sensitive, then a bromide would suffice so long as either (a) the lower order cuprate is used at higher temperatures for longer times or (b) a more reactive higher order cyanocuprate is employed, which should do the same coupling at temperatures $\leqslant -50$ °C. At the very bottom of the reactivity scale lie the primary chlorides, which are generally unacceptable educts towards lower order cuprates and slow-reacting with most higher order cyanocuprates. By switching to allylic cyanocuprates, however, they become susceptible and can be displaced in minutes at -78 °C.

Reactivity Patterns for Cuprate Substitution Reactions at sp^3 Centers

reactivity (cuprate) ↑

(allyl)$_2$Cu(CN)Li_2 ←match→ Cl~~~

$R_2Cu(CN)Li_2$ ←match→ Br~~~

R_2CuLi ←match→ I~~~

↓ reactivity (substrate)

Although stoichiometric lithiocuprate-based substitutions are textbook reactions, strong competition comes in the form of copper-catalyzed Grignard processes. Many of the same arguments in favor of this system apply as were offered for conjugate additions (see above). Moreover, perhaps the best source of catalytic Cu(I), rarely used for 1,4-additions, is Kochi's catalyst, Li_2CuCl_4. Hence, if the starting material is not especially complex and the Grignard reagent is readily available, the catalytic Li_2CuCl_4–RMgX combination is difficult to surpass in terms of simplicity and yields can be as good as (or better than) those realized from the corresponding lithium cuprate.

Ring openings of epoxides can also be accomplished in good yields via the catalytic Cu(I)–RMgX or lithiocuprate route and, to a lesser extent, using R_2CuMgX. In general, since Et_2O is the solvent of choice over THF [in which the genenion(s) present is (are) of greater Lewis acidity], lithiocuprates will have better solubility characteristics. With monosubstituted oxiranes, assuming no likely interference from other functionalities within the substrate, essentially any of the above reagents is a valid first choice. Distinctions will merely lie in the temperatures of the couplings, which are tied not only to reagent reactivity but also reagent solubility (i.e. lithio- > magnesiocuprates in both cases). It is also well worth noting, however, that potential Schlenk equilibria associated with Grignard-based reagents may lead to halohydrin by-products (cf. Scheme 4.4). Disubstituted epoxides (1,1- or 1,2-) are more likely to be opened efficiently by lithiocuprates. When conservation of ligands (R_t) is called for, and hence reagent reactivity is decreased, the use of $\geqslant 1$ equiv. of $BF_3 \cdot Et_2O$ may prove highly beneficial. For greatest reagent reactivity without recourse to additives, higher order cyanocuprates, $R_2Cu(CN)Li_2$, in Et_2O appear to offer the best opportunities, and on occasion can be successful with trisubstituted epoxides. For activated, allylic cases, the reagent of choice is clearly the lower order cyanocuprate RCu(CN)Li.

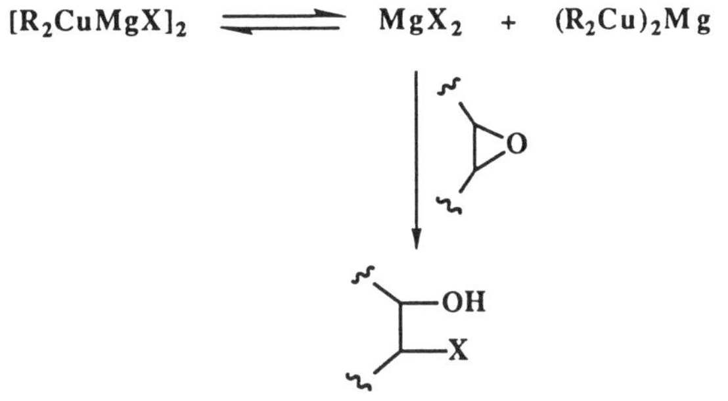

$$[R_2CuMgX]_2 \rightleftharpoons MgX_2 + (R_2Cu)_2Mg$$

Scheme 4.4

4.6 CONCLUDING REMARKS

The material compiled for this chapter not only covers an array of standard reactions in organocopper chemistry, but also traces the evolutionary developments of this extensive area of research. From the very first observations by Gilman in 1952 [5] to the most recent methodological advances for organocopper reagent generation, manipulation and use, a broad spectrum of literature procedures is contained herein. Aside from the convenience of having these collected together, the many titbits of information provided should assist researchers in quickly gaining a feel for this science. Between the experimental details and supporting comments for most of the typical reactions highlighted, it should be possible for a chemist at virtually any stage of his or her career to use organocopper reagents routinely, for not to make use of such powerful weapons would undoubtedly limit one's options in planning synthetic strategies.

4.7 ACKNOWLEDGEMENTS

It is a pleasure to acknowledge financial support for this project provided by the Lithium Division of FMC Corporation and Syntex Research (Palo Alto), and the efforts of Ms Patricia K. Ure, in typing the manuscript. Our group is indebted to Mr Otto Loeffler (Argus Chemical) for supplying the Me_2SnCl_2 used in our work on preparations of Me_3Sn-containing reagents, and to Mr Lee Kelly and Mr Jeff Sullivan (Boulder Scientific) for generously providing samples of Cp_2ZrCl_2 and $Cp_2Zr(H)Cl$ essential for the zirconium-based transmetallations discussed herein. Many of the insights offered in this chapter were provided by the groups having done the work cited. Their timely feedback is most appreciated.

4.8 REFERENCES AND NOTES

1. J. P. Collman, L. S. Hegedus, J. R. Norton, R. G. Finke, *Principles and Applications of Organotransition Metal Chemistry*, p. 682, University Science Books, Mill Valley, CA, 1987.
2. (a) B. H. Lipshutz, S. Sengupta, *Org. React.* **41** (1992), 135; (b) G. H. Posner, *An Introduction to Synthesis Using Organocopper Reagents*, Wiley, New York, 1980; (c) G. H. Posner, *Org. React.* **19** (1972) 1; (d) G. H. Posner, *Org. React.* **22** (1975) 253; (e) B. H. Lipshutz, *Synlett* (1990) 119; (f) B. H. Lipshutz, *Synthesis* (1987) 325; (g) B. H. Lipshutz, R. S. Wilhelm, J. A. Kozlowski, *Tetrahedron* **40** (1984) 5005; (h) R. J. K. Taylor, *Synthesis* (1985) 364; (i) E. Erdik, *Tedrahedron* **40** (1984) 641; (j) J. R. Normant, A. Alexakis, *Synthesis* (1981) 841; (k) T. Kauffmann. *Angew. Chem.* **86** (1974) 321.
3. C. Ouannes, G. Dressaire, Y. Langlois, *Tetrahedron Lett.* (1977) 815; G. Hallnemo, C. Ullenius, *Tetrahedron Lett.* **27** (1986) 395.
4. For a review on 1,2-additions of cuprates, see B. H. Lipshutz, in *Comprehensive Organic Synthesis*, ed. B. M. Trost, Pergamon Press, Oxford, 1991.
5. H. Gilman, R. G. Jones, L. A. Woods, *J. Org. Chem.* **17** (1952) 1630.
6. B. H. Lipshutz, J. A. Kozlowski, R. S. Wilhelm, *J. Org. Chem.* **48** (1983) 546.
7. P. G. M. Wuts, *Synth. Commun.* **11** (1981) 139; R. N. Keller, H. D. Wycoff, *Inorg. Synth.* **2** (1946) 1; A. B. Theis, C. A. Townsend, *Synth Commun.* **11** (1981) 157.
8. Based on a modification of that found in G. B. Kauffman, L. A. Tetev, *Inorg. Synth.* **7** (1963) 9; G. Linstrumelle, J. K. Krieger, G. M. Whitesides, *Org. Synth.* **55** (1976) 103.
9. H. O. House, C.-Y. Chu, J. M. Wilkins, *J. Org. Chem.* **40** (1975) 1460.
10. B. H. Lipshutz, S. Whitney, J. A. Kozlowski, C. M. Breneman, *Tetrahedrom Lett.* **27** (1986) 4273.
11. S. H. Bertz, C. P. Gibson, G. Dabbagh, *Tetrahedron Lett.* **28** (1987) 4251.
12. J. Suffert, *J. Org. Chem.* **54** (1989) 509, and references cited therein.
13. S. C. Watson, J. F. Eastham, *J. Organomet. Chem.* **9** (1967) 165.
14. H. O. House, T. V. Lee, *J. Org. Chem.* **43** (1978) 4369.
15. H. O. House, W. F. Fischer, *J. Org. Chem.* **33** (1968) 949.

16. G. M. Whitesides, W. F. Fischer, J. San Filippo, R. W. Bashe, H. O. House, *J. Am. Chem. Soc.* **91** (1969) 4871; H. Westmijze, H. Kleijn, P. Vermeer, *Tetrahedron Lett.* (1977) 2023.
17. R. H. Wollenberg, K. F. Albizati, R. Peries, *J. Am. Chem. Soc.* **99** (1977) 7365.
18. E. J. Corey, D. Enders, *Chem. Ber.* **111** (1978) 1362.
19. C. R. Johnson, G. A. Dutra, *J. Am. Chem. Soc.* **95** (1973) 7777.
20. R. K. Hill, T. F. Bradberg, *Experientia* **38** (1982) 70.
21. W. E. Willy, D. R. McKean, B. A. Garcia, *Bull. Chem. Soc. Jpn.* **49** (1976) 1989.
22. H. Gilman, F. Schulze, *J. Am. Chem. Soc.* **47** (1925) 2002.
23. C. R. Johnson, R. W. Herr, D. M. Wieland, *J. Org. Chem.* **38** (1973) 4263.
24. H. M. Sirat, E. J. Thomas, J. D. Wallis, *J. Chem. Soc. Perkin Trans 1* (1982) 2885.
25. R. G. Nelb, J. K. Stille, *J. Am. Chem. Soc.* **98** (1976) 2834.
26. J. E. McMurry, W. J. Scott, *Tetrahedron Lett.* **21** (1980) 4313.
27. J. F. Normant, *Pure Appl. Chem.* **50** (1978) 709.
28. G. Casy, M. Furber, S. Lane, R. J. K. Taylor, S. C. Burford, *Philos. Trans. R. Soc. London, Ser. A* **326** (1988) 565.
29. M. Furber, R. J. K. Taylor, S. C. Burford, *J. Chem. Soc., Perkin Trans 1* (1986) 1809.
30. C. E. Bishop, G. W. Morrow, *J. Org. Chem.* **48** (1983) 657; P. E. Sonnet, R. R. Heath, *J. Chem. Ecol.* **6** (1980) 221.
31. E. Nakamura, M. Isaka, S. Matsuzawa, *J. Am. Chem. Soc.* **110** (1988) 1297.
32. (a) M. Isaka, E. Nakamura, *J. Am. Chem. Soc.* **112** (1990) **7428**; (b) E. Nakamura, unpublished data; cf. ref. 31.
33. K. B. Baucom, G. B. Butler, *J. Org. Chem.* **37** (1972) 1730; R. Breslow, J. Pecorara, T. Sugimoto, *Org. Synth., Coll. Vol* **6** (1988) 361; D. L. Boger, C. E. Brotherton, G. I. George, *Org. Synth.* **65** (1987) 32.
34. B. H. Lipshutz, E. L. Ellsworth, T. J. Siahaan, *J. Am. Chem. Soc.* **111** (1989) 1351.
35. Y. Yamamoto, *Angew. Chem., Int. Ed. Engl.* **25** (1986) 947.
36. B. H. Lipshutz, D. A. Parker, J. A. Kozlowski, S. L. Nguyen, *Tetrahedron Lett.* **25** (1984) 5959.
37. A. B. Smith, P. J. Jerris, *J. Org. Chem.* **47** (1982) 1845.
38. M. J. Eis, B. Ganem, *Tetrahedron Lett.* **26** (1985) 1153.
39. A. Ghribi, A. Alexakis, J. F. Normant, *Tetrahedron Lett.* **25** (1984) 3083.
40. (a) E. J. Corey, N. W. Boaz, *Tetrahedron Lett.* **26** (1985) 6019; (b) see also A. Alexakis, R. Sedrani, P. Mangeney, *Tetrahedron Lett.* **31** (1990) 345; (c) A. Alexakis, J. Berlan, Y. Besace, *Tetrahedron Lett.* **27** (1986) 1047; (d) C. R. Johnson, T. J. Marren, *Tetrahedron Lett.* **28** (1987) 27.
41. Y. Horiguchi, M. Komatsu, I. Kuwajima, *Tetrahedron Lett.* **30** (1989) 7087; E. Nakamura, S. Matsuzawa, Y. Horiguchi, I. Kuwajima, *Tetrahedron Lett.* **27** (1986) 4025, 4029.
42. M. Bergdahl, E.-L. Lindstedt, M. Nilsson, T. Olsson, *Tetrahedron* **44** (1988) 2055; **45** (1989) 535.
43. S. Matsuzawa, Y. Horiguchi, E. Nakamura, I. Kuwajima, *Tetrahedron* **45** (1989) 349.
44. M. Gardette, N. Jabri, A. Alexakis, J. F. Normant, *Tetrahedron* **40** (1984) 2741. See also, N. Jabri, A. Alexakis, J. F. Normant, *Bull. Soc. Chim, Fr. II* (1983) 332.
45. G. Hallnemo, C. Ullenius, *Tetrahedron* **39** (1983) 1621.
46. C. Ullenius, unpublished data; cf. ref. 45.
47. (a) D. B. Ledlie, G. Miller, *J. Org. Chem.* **44** (1979) 1006 [3-(dimethylamino)-1-propyne]; (b) T. Tsuda, T. Yazawa, K. Watanabe, T. Fujii, T. Saegusa, *J. Org. Chem.* **46** (1981) 192

(mesitylcopper); (c) W. H. Mandeville, G. M. Whitesides, *J. Org. Chem.* **39** (1974) 400 (alkyl, aryl groups); (d) C. R. Johnson, D. S. Dhanoa, *J. Chem. Soc., Chem. Commun.* (1982) 358 (sulfonyl anion); (e) C. R. Johnson, D. S. Dhanoa *J. Org. Chem.* **52** (1987) 1885 (DMSO anion).
48. E. J. Corey, D. J. Beames, *J. Am. Chem. Soc.* **94** (1972) 7210.
49. H. O. House, M. J. Umen, *J. Org. Chem.* **38** (1973) 3893.
50. E. J. Corey, D. M. Floyd, B. H. Lipshutz, *J. Org. Chem.* **43** (1978) 3418.
51. (a) G. H. Posner, C. E. Whitten, J. J. Sterling, *J. Am. Chem. Soc.* **95** (1973) 7788; (b) G. H. Posner, D. J. Brunelle, L. Sinoway, *Synthesis* (1974) 662; (c) G. H. Posner, C. E. Whitten, *Org. Synth.* **55** (1976) 122.
52. (a) For more information about the commercially available methyl thienyl cuprate, see FMC Lithium Division Brochure, *Organometallics in Organic Synthesis*, 1990, p. 33; (b) E.-L. Lindstedt, M. Nilsson, T. Olsson, *J. Organomet, Chem.* **334** (1987) 255; E.-L. Lindstedt, M. Nilsson, *Acta Chem. Scand., Ser. B* **40** (1986) 466.
53. S. H. Bertz, G. Dabbagh, *J. Org. Chem.* **49** (1984) 1119; see also, S. H. Bertz, G. Dabbagh, G. M. Villacorta, *J. Am. Chem. Soc.* **104** (1982) 5824.
54. (a) J. P. Marino, J. C. Jaen, *J. Am. Chem. Soc.* **104** (1982) 3165; (b) R. D. Acker, *Tetrahedron Lett.* (1977) 3407; (1978), 2399; (d) L. Hamon, J. Levisalles, *Tetrahedron* **45** (1989) 489.
55. S. H. Bertz, G. Dabbagh, *J. Chem. Soc., Chem. Commun.* (1982) 1030.
56. B. H. Lipshutz, R. S. Willhelm, D. M. Floyd, *J. Am. Chem. Soc.*, **103** (1981) 7672.
57. B. H. Lipshutz, R. S. Wilhelm, J. A. Kozlowski, *J. Org. Chem.*, **49** (1984) 3938.
58. B. H. Lipshutz, R. S. Wilhelm, J. A. Kozlowski, *J. Org. Chem.* **49** (1984) 3928.
59. B. H. Lipshutz, D. Parker, J. A. Kozlowski, R. D. Miller, *J. Org. Chem.* **48** (1983) 3334.
60. B. H. Lipshutz, E. L. Ellsworth, T. J. Siahann, *J. Am. Chem. Soc.* **110** (1988) 4834.
61. J. A. Kozlowski, *PhD Thesis* University of California, Santa Barbara, 1985; cf. ref. 36.
62. M. J. Eis, J. E. Wrobel, B. Ganem, *J. Am. Chem. Soc.* **106** (1984) 3693.
63. A. Alexakis, D. Jachiet, J. F. Normant, *Tetrahedron* **42** (1986) 5607.
64. B. H. Lipshutz, E. L. Ellsworth, T. J. Siahaan, A. Shirazi, *Tetrahedron Lett.* **29** (1988) 6677.
65. R. J. Linderman, A. Godfrey, K. Horne, *Tetrahedron* **45** (1989) 495; see also R. J. Linderman, J. R. McKenzie, *J. Organomet. Chem.* **361** (1989) 31.
66. (a) For the selective transfer of a vinylic ligand over a methyl group in a Gilman cuprate, see G. H. Posner, J. J. Sterling, C. E. Whitten, C. M. Lentz, D. J. Brunelle, *J. Am. Chem. Soc.* **97** (1975) 107; see also F. Leyendecker, J. Drouin, J. J. Debesse, J. M. Conia, *Tetrahedron Lett.* (1977) 1591; (b) B. H. Lipshutz, M. Koerner, D. A. Parker, *Tetrahedron Lett.* **28** (1987) 945; (c) In using thienyl ligand-based cuprates, one common, UV-active by-product to be expected is that resulting from the oxidative coupling of this heteroaromatic group (i.e. 2,2′-dithiophene).
67. B. H. Lipshutz, J. A. Kozlowski, D. A. Parker, K. E. McCarthy, *J. Organomet. Chem.* **285** (1985) 437.
68. M. Pereyre, J.-P. Quintard, A. Rahm, in *Tin in Organic Synthesis*, Butterworths, Guildford, 1987.
69. J. R. Behling, K. A. Babiak, J. S. Ng, A. L. Campbell, R. Moretti, M. Koerner, B. H. Lipshutz, *J. Am. Chem. Soc.* **110** (1988) 2641.
70. B. H. Lipshutz, M. Koerner, unpublished data.
71. P. W. Collins, *J. Med. Chem.* **29** (1986) 437, and references cited therein.

72. B. H. Lipshutz, R. Crow, S. H. Dimock, E. L. Ellsworth, R. A. J. Smith, J. R. Behling, *J. Am. Chem. Soc.* **112** (1990) 4063.
73. D. Milstein, J. K. Stille, *J. Org. Chem.* **44** (1979) 1613; J. W. Labadie, D. Tueting, J. K. Stille, *J. Org. Chem.* **48** (1983) 4634.
74. B. H. Lipshutz, T. R. Elworthy, *J. Org. Chem.* **55** (1990) 1695.
75. B. H. Lipshutz, T. R. Elworthy, *Tetrahedron Lett.* **31** (1990) 447.
76. R. E. Ireland, P. Wipf, *J. Org. Chem.* **55** (1990) 1425.
77. (a) B. H. Lipshutz, E. L. Ellsworth, *J. Am. Chem. Soc.* **112** (1990) 7440; (b) K. A. Babiak, J. R. Behling, J. H. Dygos, K. T. McLaughlin, J. S. Ng, V. J. Kalish, S. W. Kramer, R. L. Shore, *J. Am. Chem. Soc.* **112** (1990) 7441.
78. (a) J. Schwartz, J. A. Labinger, *Angew. Chem., Int. Ed. Engl.* **15** (1976) 333; see also E. Negishi, T. Takahashi, *Synthesis* (1988) 1; *Aldrichim. Acta.* **18**, (1985) 31; E. Negishi, *Pure Appl. Chem.* **53** (1981) 2333; M. Yoshifuji, M. Loots, J. Schwartz, *Tetrahedron Lett.* (1977) 1303; (b) all hydrozirconations were performed using $Cp_2Zr(H)Cl$ obtained from Boulder Scientific, Mead, Colorado, USA.
79. G. van Koten, University of Utrecht, personal communication.
80. S. G. Pyne, *J. Org. Chem.* **57** (1986) 81.
81. M. Suzuki, T. Suzuki, T. Kawagishi, Y. Morita, R. Noyori, *Isr. J. Chem.* **24** (1984) 118.
82. (a) R. D. Rieke, R. M. Wehmeyer, T.-C. Wu, G. W. Ebert, *Tetrahedron* **45** (1989) 443; (b) R. D. Rieke, T.-C. Wu, D. E. Stinn, R. M. Wehmeyer, *Synth. Commun.* **19** (1989) 1833.
83. For representative early uses, see E. J. Corey, R. L. Carney, *J. Am. Chem. Soc.* **93** (1971) 7318; R. D. Clark, C. H. Heathcock, *Tetrahedron Lett.* (1974) 1713; G. L. Van Mourik, H. J. J. Pabon, *Tetrahedron Lett.* (1978) 2705; J.-E. Mansson, *Acta Chem. Scand., Ser. B* **32** (1978) 543.
84. S. H. Bertz, G. Dabbagh, *Tetrahedron* **45** (1989) 425.
85. K. Maruyama, Y. Yamamoto, *J. Am. Chem. Soc.* **99** (1977) 8068.
86. Y. Yamamoto, K. Maruyama, *J. Am. Chem. Soc.* **100** (1978) 3241.
87. Y. Yamamoto, S. Yamamoto, H. Yatagai, Y. Ishihara, K. Maruyama, *J. Org. Chem.* **47** (1982) 119.
88. B. H. Lipshutz, E. L. Ellsworth, S. H. Dimock, *J. Am. Chem. Soc.* **112** (1990) 5869.
89. B. H. Lipshutz, E. L. Ellsworth, S. H. Dimock, R. A. J. Smith, *J. Am. Chem. Soc.* **112** (1990) 4404.
90. E. Colvin, *Silicon in Organic Synthesis*, Butterworths, Guildford 1981; W. P. Weber, *Silicon Reagents for Organic Synthesis*, Springer, Berlin, 1983.
91. (a) I. Fleming, T. W. Newton, F. Roessler, *J. Chem. Soc., Perkin Trans. 1* (1981) 2527; (b) D. J. Ager, I. Fleming, S. K. Patel, *J. Chem. Soc., Perkin Trans. 1* (1981) 2520; (c) I. Fleming, A. P. Thomas, *J. Chem. Soc. Chem. Commun.* (1985) 411; (d) I. Fleming, M. Rowley, P. Cuadrado, A. M. Gonzalez-Nogal, F. J. Pulido, *Tetrahedron* **45** (1989) 413.
92. I. Fleming, *et al.*; c.f. M. V. George, D. J. Peterson, H. Gilman, *J. Am. Chem. Soc.* **82** (1960) 403.
93. B. H. Lipshutz, D. C. Reuter, E. L. Ellsworth, *J. Org. Chem.* **54** (1989) 4975.
94. D. C. Reuter, *PhD Thesis* University of California, Santa Barbara, 1990.
95. B. H. Lipshutz, D. C. Reuter, *Tetrahedron Lett.* **30** (1989) 4617.
96. Y. Yamamoto (ed), *Organotin Compounds in Organic Synthesis*, Tetrahedron Symposia-in-Print No. 36, 1989.

97. E. Piers, J. M. Chong, *Can. J. Chem.* **66** (1988) 1425.
98. E. Piers, H. E. Morton, J. M. Chong, *Can. J. Chem.* **65** (1987) 78.
99. E. Piers, J. M. Chong, H. E. Morton, *Tetrahedron* **45** (1989) 363; the corresponding tributyltin analogues of both reagents **7** and **8** [i.e. n-$Bu_3SnCu{\cdot}SMe_2$ and (n-Bu_3Sn)(PhS)-CuLi] have also been prepared and utilized in a similar capacity; cf. E. Piers, J. M. Chong, K. Gustafson, R. J. Andersen, *Can. J. Chem.* **62** (1984) 1.
100. E. Piers, R. D. Tillyer, *J. Org. Chem.* **53** (1988) 5366.
101. B. H. Lipshutz, E. L. Ellsworth, S. H. Dimock, D. C. Reuter, *Tetrahedron Lett.* **30** (1989) 2065; see also S. R. Gilbertson, C. A. Challener, M. E. Bos, W. D. Wulff, *Tetrahedron Lett.* **29** (1988) 4795.
102. B. H. Lipshutz, S. Sharma, D. C. Reuter, *Tetrahedron Lett.* **31** (1990), 7253.
103. See also A. C. Oehlschlager, M. W. Hutzinger, R. Aksela, S. Sharma, S. M. Singh, *Tetrahedron Lett.* **31** (1990) 165.
104. (a) E. C. Ashby, J. J. Watkins, *J. Am. Chem. Soc.* **99** (1977) 5312; (b) *J. Chem. Soc., Chem. Commun.* (1976) 784; (c) E. C. Ashby, J. J. Lin, *J. Org. Chem.* **42** (1977) 2805; (d) E. C. Ashby, J. J. Lin, J. J. Watkins, *J. Org. Chem.* **42** (1977) 1009.
105. S. H. Bertz, G. Dabbagh, *J. Am. Chem. Soc.* **110** (1988) 3668.
106. D. L. J. Clive, V. Farina, P. L. Beaulieu, *J. Org. Chem.* **47** (1982) 2572.
107. B. H. Lipshutz, E. L. Ellsworth, unpublished observations.
108. E. C. Ashby, T. F. Korenowski, R. D. Schwartz, *J. Chem. Soc., Chem. Commun.* (1974) 157.
109. E. C. Ashby, A. B. Goel, *Inorg. Chem.* **16** (1977) 3043.
110. E. C. Ashby, J.-J. Lin, A. B. Goel, *J. Org. Chem.* **43** (1978) 183.
111. S. Masamune, G. S. Bates, P. E. Georghiou, *J. Am. Chem. Soc.* **96** (1974) 3686.
112. R. K. Boeckman, R. Michalak, *J. Am. Chem. Soc.* **96** (1974) 1623.
113. G. M. Whitesides, J. San Filippo, E. R. Stredronsky, C. P. Casey, *J. Am. Chem. Soc.* **91** (1969) 6542.
114. M. R. Churchill, S. A. Bezman, J. A. Osborn, J. Wormald, *Inorg. Chem.* **11** (1972) 1818; S. A. Bezman, M. R. Churchill, J. A. Osborn, J. Wormald, *J. Am. Chem. Soc.* **93** (1971) 2063; G. V. Goeden, K. G. Caulton, *J. Am. Chem. Soc.* **103** (1981) 7354; T. H. Lemmen, K. Folting, J. C. Huffman, K. G. Caulton, *J. Am. Chem. Soc.* **107** (1985) 7774.
115. W. S. Mahoney, D. M. Brestensky, J. M. Stryker, *J. Am. Chem. Soc.* **110** (1988) 291.
116. D. M. Brestensky, D. E. Huseland, C. McGettigan, J. M. Stryker, *Tetrahedron Lett.* **29** (1988) 3749; see also J. F. Daeuble, C. McGettigan, J. M. Stryker, *Tetrahedron Lett.* **31** (1990) 2397; T. M. Koenig, J. F. Daeuble, D. M. Brestensky, J. M. Stryker, *Tetrahedron Lett.* **31** (1990) 3237.
117. B. H. Lipshutz, C. S. Ung, S. Sengupta, *Synlett* (1989) 64.
118. M. S. Kharasch, P. O. Tawney, *J. Am. Chem. Soc.* **63** (1941) 2308.
119. M. Tamura, J. K. Kochi, *J. Organomet. Chem.* **42** (1972) 205; *Synthesis* (1971) 303.
120. S. A. Bal, A. Marfat, P. Helquist, *J. Org. Chem.* **47** (1982) 5045.
121. Y. Ito, M. Nakatsuka, T. Saegusa, *J. Am. Chem. Soc.* **104** (1982) 7609.
122. Y. Horiguchi, S. Matsuzawa, E. Nakamura, I. Kuwajima, *Tetrahedron Lett.* **27** (1986) 4025.
123. J. K. Whitesell, R. M. Lawrence, H. H. Chen, *J. Org. Chem.* **51** (1986) 4779.
124. A. Alexakis, I. Marek, P. Mangeney, J. F. Normant, *Tetrahedron Lett.* **30** (1989) 2387.
125. T. Fujisawa, T. Sato, T. Kawara, K. Naruse, *Chem. Lett.* (1980) 1123; T. Fujisawa, T. Sato, T. Kawara, A. Noda, T. Obinata, *Tetrahedron Lett.* **21** (1980) 2553.
126. T. Sato, T. Kawara, M. Kawashima, T. Fujisawa, *Chem. Lett.* (1980) 571.

127. L. D. Arnold, J. C. G. Drover, J. C. Vederas, *J. Am. Chem. Soc.* **109** (1987) 4649.
128. S. Nunomoto, Y. Kawakami, Y. Yamashita, *J. Org. Chem.* **48** (1983) 1912.
129. J. Novak, C. A. Salemink, *Synthesis* (1981) 597.
130. (a) H. Kotsuki, I. Kadota, M. Ochi, *J. Org. Chem.* **55** (1990) 4417; (b) *Tetrahedron Lett.* **30** (1989) 1281, 3999.
131. (a) C. C. Tseng, S. D. Paisley, H. L. Goering, *J. Org. Chem.* **51** (1986) 2884; (b) C. C. Tseng, S.-J. Yen, H. L. Goering, *J. Org. Chem.* **51** (1986) 2892; T. L. Underiner, S. D. Paisley, J. Schmitter, L. Lesheski, H. L. Goering, *J. Org. Chem.* **54** (1989) 2369.
132. S. P. Tanis, *Tetrahedron Lett.* **23** (1982) 3115.
133. S. Suzuki, M. Shiono, Y. Fujita, *Synthesis* (1983) 804.
134. P. Four, Ph LeTri, H. Riviere, *J. Organomet. Chem.* **133** (1977) 385; E. C. Ashby, A. B. Goel, R. S. Smith, *J. Organomet. Chem.* **212** (1981) C47; E. C. Ashby, A. B. Goel, *J. Org. Chem.* **48** (1983) 2125; E. C. Ashby, R. S. Smith, A. B. Goel, *J. Org. Chem.* **46** (1981) 5133; A. B. Goel, E. C. Ashby, *Inorg. Chim. Acta* **54** (1981) L199; H. Westmijze, A. V. E. George, P. Vermeer, *Recl. Trav. Chim. Pays-Bas* **102** (1983) 322.
135. (a) M. Gardette, A. Alexakis, J. F. Normant, *Tetrahedron* **41** (1985) 5887; (b) A. Marfat, P. R. McGuirk, P. Helquist, *J. Org. Chem.* **44** (1979) 3888; (c) R. S. Iyer, P. Helquist, *Org. Synth.* **64** (1985) 1.
136. J. P. Foulon, M. B. Commercon, J. F. Normant, *Tetrahedron* **42** (1986) 1389.
137. A. Alexakis, J. F. Normant, J. Villieras, *J. Organomet. Chem.* **96** (1975) 471; see also, J. F. Normant, A. Alexakis, in *Modern Synthetic Methods*, ed. R. Scheffold, Vol. 3, p. 139, Wiley, New York, 1983.
138. P. Wijkens, P. Vermeer, *J. Organomet. Chem.* **301** (1986) 247.
139. J. F. Normant, G. Cahiez, C. Chuit, J. Villieras, *J. Organomet. Chem.* **77** (1974) 269.
140. J. F. Normant, G. Cahiez, C. Chuit, J. Villieras, *J. Organomet. Chem.* **54** (1973) C53; **77** (1974) 281.
141. G. Cahiez, D. Bernard, J. F. Normant, *Synthesis* (1976) 245.
142. J. F. Normant, A. Commercon, G. Cahiez, J. Villieras, *C. R. Acad. Sci. Ser. C* **278** (1974) 967.
143. D. P. Curran, M.-H. Chen, D. Leszczweski, R. L. Elliot, D. M. Rakiewicz, *J. Org. Chem.* **51** (1986) 1612.
144. (a) P. Vermeer, H. Westmijze, H. Kleijn, L. A. van Dyck, *Recl. Trav. Chim. Pays-Bas* **97** (1978) 56; (b) H. Westmijze, P. Vermeer, *Tetrahedron Lett.* (1979) 4101; (c) C. J. Elsevier, J. Meijer, H. Westmijze, P. Vermeer, L. A. van Dijck, *J. Chem. Soc., Chem. Commun.* (1982), 84.
145. G. Stork, A. Kreft, *J. Am. Chem. Soc.* **99** (1977) 3850; R. M. Magid, O. S. Fruchey, *J. Am. Chem. Soc.* **99** (1977) 8368; L.-I. Olsson, A. Claesson, *Acta Chem. Scand., Ser. B* **33** (1979) 679; see also I. Marek, P. Mangeney, A. Alexakis, J. F. Normant, *Tetrahedron Lett.* **27** (1985) 5499; G. Tadema, R. H. Everhardus, H. Westmijze, P. Vermeer, *Tetrahedron Lett.* (1978) 3935.
146. T. L. Macdonald, D. R. Reagan, R. S. Brinkmeyer, *J. Org. Chem.* **45** (1980) 4740.
147. T. Fujisawa, T. Mori, T. Kawara, T. Sato, *Chem. Lett.* (1982) 569.
148. T. Fujisawa, T. Sato, T. Kawara, M. Kawashima, H. Shimizu, Y. Ito, *Tetrahedron Lett.* **21** (1980) 2181; T. Fujisawa, T. Sato, T. Kawara, H. Tago, *Bull. Chem. Soc. Jpn.* **56** (1983) 345; M. Kawashima, T. Sato, T. Fujisawa, *Tetrahedron* **45** (1989) 403.
149. D. E. Bergbrieter, J. M. Killough, *J. Org. Chem.* **41** (1976) 2750.
150. T. W. Bell, L.-Y. Hu, S. V. Patel, *J. Org. Chem.* **52** (1987) 3847.

151. R. L. Funk, K. P. C. Vollhardt, *J. Am. Chem. Soc.* **102** (1980) 5253.
152. M. Gill, R. W. Rickards, *J. Chem. Soc. Perkin Trans. 1* (1981) 599.
153. B. H. Lipshutz, E. L. Ellsworth, J. R. Behling, A. L. Campbell, *Tetrahedron Lett.* **29** (1988) 893; S. H. Bertz, G. Dabbagh, *Organometallics* **7** (1988) 227.
154. (a) E. Nakamura, S. Aoki, K. Sekiya, H. Oshino, I. Kuwajima, *J. Am. Chem. Soc.* **109** (1987) 8056; (b) E. Nakamura, I. Kuwajima, *Org. Synth.* **66** (1988) 43.
155. (a) E. Nakamura, K. Sekiya, M. Arai, S. Aoki, *J. Am. Chem. Soc.* **111** (1989) 3091; (b) see also K. Sekiya, E. Nakamura, *Tetrahedron Lett.* **29** (1988) 5155.
156. M. C. P. Yeh, H. G. Chen, P. Knochel, *Org. Synth.* **70** (1991), 195.
157. (a) S. C. Berk, M. C. P. Yeh, N. Jeong, P. Knochel, *Organometallics* **9** (1990), 3053; (b) S. C. Berk, P. Knochel, M. C. P. Yeh, *J. Org. Chem.*, **53** (1988) 5789.
158. P. Knochel, M. C. P. Yeh, S. C. Berk, J. Talbert, *J. Org. Chem.* **53** (1988) 2390.
159. M. C. P. Yeh, P. Knochel, W. M. Butler, S. C. Berk, *Tetrahedron Lett.* **29** (1988) 6693.
160. M. C. P. Yeh, P. Knochel, *Tetrahedron Lett.* **30** (1989) 4799.
161. C. Retherford, M. C. P. Yeh, I. Schipor, H. G. Chen, P. Knochel, *J. Org. Chem.* **54** (1989) 5200.
162. For an alternative route to this product using organocopper chemistry, see R. M. Wehmeyer, R. D. Rieke, *J. Org. Chem.* **52** (1987) 5056.
163. E. J. Corey, K. Kyler, N. Raju, *Tetrahedron Lett.* **25** (1984) 5115.
164. (a) R. K. Dieter, L. A. Silks, *J. Org. Chem.* **51** (1986) 4687; (b) R. K. Dieter, L. A. Silks, J. R. Fishpaugh, M. E. Kastner, *J. Am. Chem. Soc.* **107** (1985) 4679; R. K. Dieter, *Tetrahedron* **42** (1986) 3029.
165. B. H. Lipshutz, D. A. Parker, S. L. Nguyen, K. E. McCarthy, J. C. Barton, S. E. Whitney, H. Kotsuki, *Tetrahedron* **42** (1986) 2873.
166. G. Cahiez, M. Alami, *Tetrahedron Lett.* **30** (1989) 3541.
167. G. Friour, G. Cahiez, J. F. Normant, *Synthesis* (1984) 37; see also G. Cahiez, *Actual. Chim.* **9** (1984) 24; G. Cahiez, M. Alami, *Tetrahedron Lett.* **27** (1986) 569.
168. G. Cahiez, B. Laboue, *Tetrahedron Lett.* **30** (1989) 7369.
169. G. M. Whitesides, E. R. Stedronsky, C. P. Casey, J. San Filippo, *J. Am. Chem. Soc.* **92** (1970) 1426.
170. G. M. Whitesides, C. P. Casey, J. K. Krieger, *J. Am. Chem. Soc.* **93** (1971) 1379.
171. W. H. Mandeville, G. M. Whitesides, *J. Org. Chem.* **39** (1974) 400.
172. H. O. House, *Acc. Chem. Res.* **9** (1976) 59.

5

Palladium in Organic Synthesis

LOUIS S. HEGEDUS
Colorado State University, Fort Collins, CO, USA

Organometallics in Synthesis—A Manual. Edited by M. Schlosser

5.1. INTRODUCTION

Although palladium complexes have been used to catalyze organic reactions for several decades, only recently has their use in the synthesis of highly functionalized, complex organic molecules been embraced by synthetic organic chemists [1]. The reasons for this tardy acceptance are both practical and psychological; practical because only recently have organometallic chemists responsible for developing the methodology demonstrated that the chemistry is indeed compatible with highly functionalized substrates, and psychological because only recently have the nether reaches of the Periodic Table, wherein resides palladium, become familiar territory to synthetic organic chemists.

Palladium complexes have a very rich organic chemistry and are among the most readily available, easily prepared and easily handled of transition metal complexes. Their real synthetic utility lies in the very wide range of organic transformations promoted by palladium catalysts, and in the specificity and functional group tolerance of most of these processes. They permit unconventional transformations and give the synthetic chemist wide latitude in his or her choice of starting materials. When utilized with skill and imagination, exceptionally efficient total syntheses can be achieved. These points are the recurring themes of this chapter.

5.2 GENERAL FEATURES

Palladium enjoys two stable oxidation states, the +2 state and the zerovalent state, and it is the facile redox interchange between these two oxidation states which is responsible for the rich reaction chemistry that palladium complexes display. Each oxidation state has its own unique chemistry.

Palladium(II) complexes are electrophilic, and tend to react with electron-rich organic compounds, particularly olefins and arenes. The most common starting material for most palladium complexes is palladium(II) chloride, $[PdCl_2]_n$, a commercially available rust red–brown chloro-bridged oligomer, insoluble in most organic solvents (Figure 5.1). The oligomeric structure is easily broken by donor ligands, resulting in monomeric $PdCl_2L_2$ complexes stable to air and soluble in most common organic solvents. Among the most useful are the nitrile complexes, $PdCl_2(RCN)_2$, prepared by stirring a suspension of $[PdCl_2]_n$ in the nitrile as solvent. These tan–golden solids are fairly soluble in organic solvents and stable to storage for years. The benzonitrile complex is most commonly used, since it is very soluble but, when used in quantity, elimination of the

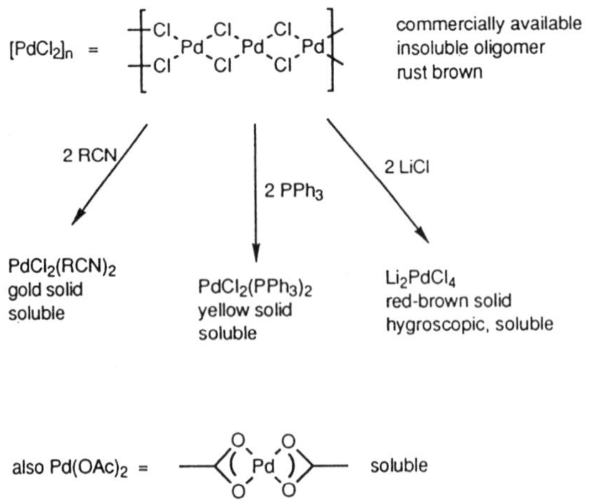

$[PdCl_2]_n$ = [Cl-bridged Pd oligomer] commercially available insoluble oligomer rust brown

2 RCN → $PdCl_2(RCN)_2$ gold solid soluble

2 PPh_3 → $PdCl_2(PPh_3)_2$ yellow solid soluble

2 LiCl → Li_2PdCl_4 red-brown solid hygroscopic, soluble

also $Pd(OAc)_2$ = soluble

Figure 5.1

odorous, water-insoluble, relatively non-volatile benzonitrile is problematic. Although slightly less soluble, the acetonitrile complex is more convenient to work with since acetonitrile is odorless, water soluble and volatile. Both nitriles are sufficiently labile to vacate coordination sites easily during reaction, making them excellent choices for catalysis. Treatment of $[PdCl_2]_n$ with triphenylphosphine produces the yellow crystalline $PdCl_2(PPh_3)_2$ complex, which again is stable and easily stored and handled. In contrast to the nitrile ligand, the phosphines are much less labile and, as a consequence, $PdCl_2(PPh_3)_2$ is infrequently used in systems requiring palladium(II) catalysis, although it is frequently the catalyst precursor of choice for palladium(0)-catalyzed processes. Even chloride ion is able to break the $[PdCl_2]_n$ oligomer, treatment of which with two equivalents of LiCl in methanol produces Li_2PdCl_4, a red–brown hygroscopic solid that is relatively soluble in organic solvents. The final commonly used palladium(II) complex is palladium(II) acetate, $Pd(OAc)_2$, again commercially available and soluble in common organic solvents. It, too, is most commonly used as a catalyst precursor for Pd(0)-catalyzed processes (see below).

Palladium(0) complexes are strong nucleophiles and strong bases, and most commonly are used to catalyze reactions involving organic halides, acetates and triflates (see below). By far the most commonly used palladium(0) complex is $Pd(PPh_3)_4$, 'tetrakis' (triphenylphosphine)palladium(0). This yellow, slightly air-sensitive (as a solid), crystalline solid is commercially available, but is expensive, and with four triphenylphosphines contributing 1048 to its molecular weight, 1 g of $Pd(PPh_3)_4$ contains very little expensive palladium and much of the inexpensive phosphine. This, combined with the tendency of this complex abruptly to lose its catalytic activity on standing, but with no accompanying change in appearance, make it advisable to prepare one's own material frequently and in small batches. Happily, it is easy to prepare, by reducing almost any palladium(II) complex in the presence of excess phosphine (Figure 5.2). In fact, in many instances palladium(0)–phosphine complexes are generated by reduction *in situ* and used

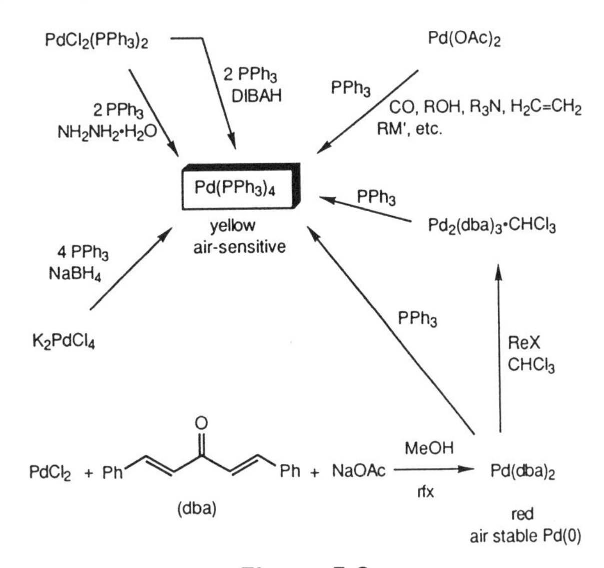

Figure 5.2

without isolation, both saving time and permitting the use of phosphine ligands other than triphenylphosphine. Another exceptionally useful palladium(0) complex is $Pd(dba)_2$, the bisdibenzylidine acetone complex. This is prepared by simply boiling palladium(II) chloride and dba together in methanol, which acts as the reducing agent (becoming oxidized to formaldehyde). A cherry red precipitate, which is $Pd(dba)_2$, forms and is easily separated by filtration; it is air stable, notwithstanding the fact that it is a palladium(0) complex. Purification (rarely necessary) by recrystallization from chloroform produces crystals of $Pd_2(dba)_3 \cdot CHCl_3$, which has the same activity as $Pd(dba)_2$. These are very useful Pd(0) catalyst precursors because they are very easily handled, can be stored without precaution for years, yet when dissolved and treated with a wide variety of phosphines, produce yellow solutions of the catalytically active PdL_n species *in situ*. Perhaps the most widely used catalyst precursor for Pd(0)-catalyzed processes is $Pd(OAc)_2$, a palladium(II) complex which, however, is very easily reduced *in situ* to palladium(0) by almost anything, including carbon monoxide, alcohols, tertiary amines, olefins, Main Group organometallics and even phosphines, all common ingredients in Pd(0)-catalyzed reactions. This has caused serious confusion amongst the uninitiated since most papers describing the use of $Pd(OAc)_2$ as a catalyst precursor carry on at great length about Pd(0) catalysis, yet every equation has $Pd(OAc)_2$ over the arrow with no comment. Rest assured, in all Pd(0)-catalyzed processes using $Pd(OAc)_2$ as the catalyst, there lurks any number of reducing agents perfectly capable of generating the requisite Pd(0) catalyst.

Hence the synthetic chemist has a large number of palladium catalysts to choose from, and the best choice is not always obvious. As is the case with traditional organic synthetic methodology, the starting point is to find as close an analogy to the desired transformations as possible, and to start with that catalyst and those conditions.

However, as it is with traditional organic chemistry, the lack of a close analogy is common, but not catastrophic; although an impediment to progress, it is also an opportunity to discover new chemistry.

5.3 PALLADIUM(II) COMPLEXES IN ORGANIC SYNTHESIS

5.3.1 NUCLEOPHILIC ATTACK ON PALLADIUM(II) OLEFIN COMPLEXES

5.3.1.1 General Features

Olefins rapidly and reversibly complex to soluble palladium(II) complexes, particularly $PdCl_2(RCN)_2$ and Li_2PdCl_4. For simple olefins, the order of complexation is $CH_2{=}CH_2$ > $RCH{=}CH_2$ > *cis*-$RCH{=}CHR$ > *trans*-$RCH{=}CHR$ $\gg$ $R_2C{=}CH_2$, $R_2C{=}CHR$; $R_2C{=}CR_2$ does not complex at all. Since palladium(II) is electrophilic, electron-rich olefins such as enol ethers and enamides complex strongly whereas electron-deficient olefins complex poorly or not at all. Although the degree of complexation is related to the degree of activation of the olefin towards further reaction, stable olefin complexes are not only unnecessary, but actually a hindrance to catalysis, and olefins which complex only very weakly can often be brought into reaction, particularly in intramolecular cases.

Once complexed, however tenuously, to palladium(II), the olefin becomes generally reactive towards nucleophilic attack, this reversal of normal reactivity (electrophilic attack) being a consequence of coordination. Under appropriate conditions, nucleophiles ranging from chloride through phenyllithium, a range of *ca* 10^{36} in basicity, can attack the metal-bound olefin. Attack usually occurs at the more substituted position of the olefin (that which would best stabilize positive charge) and from the face opposite the metal. Nucleophiles which are poor ligands for palladium(II), such as halide, carboxylates, alcohols and water, generally react without complication. Amines are substantially better ligands for palladium, and competitive attack at the metal with displacement of the olefin is sometimes a problem which can, however, often be managed. For this same reason, phosphines are rarely if ever used as spectator ligands in Pd(II)-catalyzed reactions of olefins. Low-valent sulfur species such as thiols and thioethers are catalyst poisons and cannot be used as nucleophiles. Carbanions, particularly if non-stabilized, are strong reducing agents, and reduction of Pd(II) to Pd(0) with concomitant oxidative coupling of the carbanion must be suppressed by the addition of appropriate ligands when carbanions are used as nucleophiles. Finally, remote functional groups in the substrate which strongly complex palladium(II) (e.g. phosphines, aliphatic amines and thiols) may irreversibly bind Pd(II), preventing the desired complexation and activation of the olefinic portion of the molecule.

Nucleophilic attack on the olefin produces a new carbon–nucleophile bond and a new palladium–carbon σ-bond. This (usually) unstable σ-alkylpalladium(II) intermediate, regardless of how it is produced, has a very rich reaction chemistry in its own right. Exposure of this class of complexes to subatmospheric pressures of carbon monoxide results in insertion of CO into the metal–carbon σ-bond to give a σ-acyl complex. This insertion proceeds readily at temperatures above *ca* −20 °C, and effectively competes with β-elimination (see below). The resulting σ-acyl complexes are also unstable, and it is customary to generate them in the presence of a trap such as methanol, to result in cleavage to the free organic ester, and palladium(0). Exposure of solutions of the σ-alkylpalladium(II) complex to hydrogen, again at temperatures above −20 °C, results in clean hydrogenolysis of the metal–carbon σ-bond, resulting in overall nucleophilic *addition* to the olefin. Again, Pd(0) is produced. Main Group organometallics including Sn, Zr, Zn, Cu, B, Al, MgX and Si, but not Li, readily transfer their R group to σ-alkylpalladium(II) complexes, at temperatures below *ca* −20 °C, producing dialkylpalladium(II) species which undergo reductive elimination on warming, coupling the two R groups and again producing Pd(0). If the initial σ-alkylpalladium(II) complex is warmed above −20 °C, β-hydrogen elimination rapidly ensues, regenerating the olefin (e.g. overall nucleophilic substitution) and again giving Pd(0). The stereochemistry of β-elimination requires that the metal and the β-hydrogen be *syn* coplanar, and β-elimination can often be suppressed by preventing this. However, under normal circumstances, β-elimination is very facile and is the primary competing side-reaction in processes attempting to utilize σ-alkylpalladium(II) species which have β-hydrogens. Indeed, olefin insertion, a very important process for σ-alkylpalladium(II) complexes lacking β-hydrogens (e.g. aryl or vinyl complexes; see below) can rarely be achieved with systems having β-hydrogens, since β-elimination is so facile. When β-hydrogens are lacking, simple olefins, electron-rich olefins such as enol ethers and enamides and electron-deficient olefins all readily insert into palladium(II)–carbon σ-bonds, resulting in alkylation of the olefin. The regiochemistry of the insertion is dominated by steric and not electronic effects, and alkylation occurs predominately at the less substituted olefin carbon.

Note that all of these reactions of σ-alkylpalladium(II) complexes produce Pd(0) in the final step, while Pd(II) is required to activate the olefin for the first step (nucleophilic attack). Thus, for catalysis, Pd(0) must be reoxidized to Pd(II) in the presence of substrate, nucleophile and product. A wide array of oxidants, including O_2–$CuCl_2$, $K_2S_2O_8$ and benzoquinone, are available, and many successful catalytic systems have been developed (see below). However, carrying the Pd(0) to Pd(II) redox chemistry necessary for catalysis is often the most difficult aspect of developing new Pd(II)-based systems. All of the points discussed above are summarized in Figure 5.3.

5.3.1.2 Oxygen Nucleophiles

One of the earliest palladium(II)-catalyzed reactions of olefins studied was the Wacker process, the formal oxidation of ethylene to acetaldehyde which involves, as the key step, palladium(II)-catalyzed nucleophilic addition of water to ethylene. While this particular process is of little general use in organic synthesis, the closely related palladium(II)-catalyzed oxidation of terminal olefins to methyl ketones is useful, primarily because this oxidation is selective for terminal olefins and tolerant of a wide

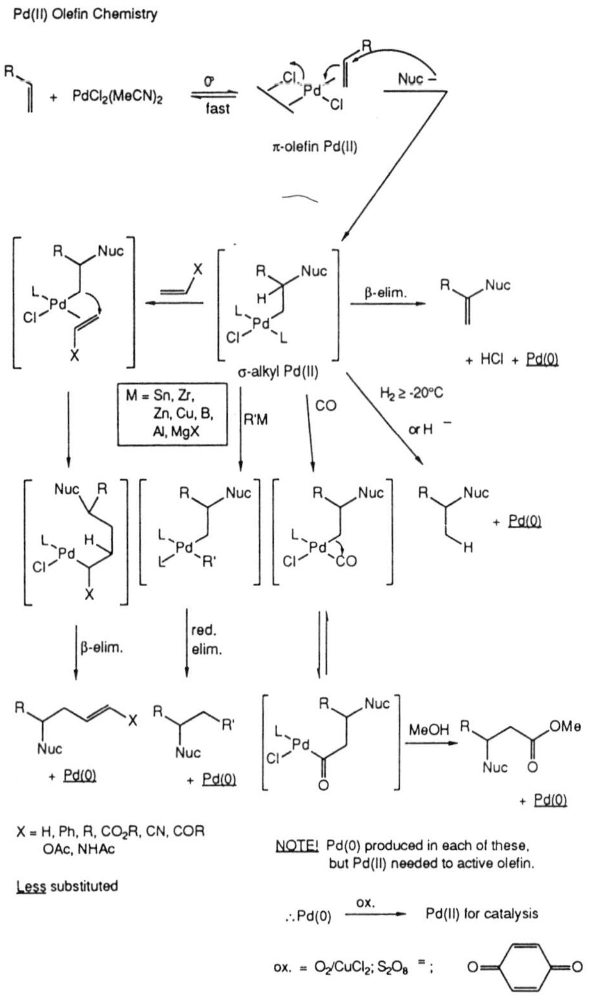

Pd(II) Olefin Chemistry

NOTE! Pd(0) produced in each of these, but Pd(II) needed to active olefin.

∴ Pd(0) $\xrightarrow{ox.}$ Pd(II) for catalysis

ox. = $O_2/CuCl_2$; $S_2O_8^{=}$;

Figure 5.3

range of functional groups [2]. The reaction probably proceeds as in Figure 5.4, and involves many of the processes discussed above.

The first step involves coordination of the olefin and, since terminal olefins complex most strongly, the reaction is selective for terminal over internal olefins. The complexed olefin then undergoes nucleophilic attack by water, regioselectively at the 2-position, ultimately leading to methyl ketones rather than aldehydes, which would result from attack at the primary carbon. β-Hydrogen elimination produces the enol, which isomerizes to the product methyl ketone, and Pd(II) + HCl. Reductive elimination of HCl gives palladium(0), which must be reoxidized to palladium(II) to re-enter the catalytic cycle.

The choice of solvent and reoxidant is critical to the success of this process. Solvent systems are particularly problematic since both the substrate olefin and water must

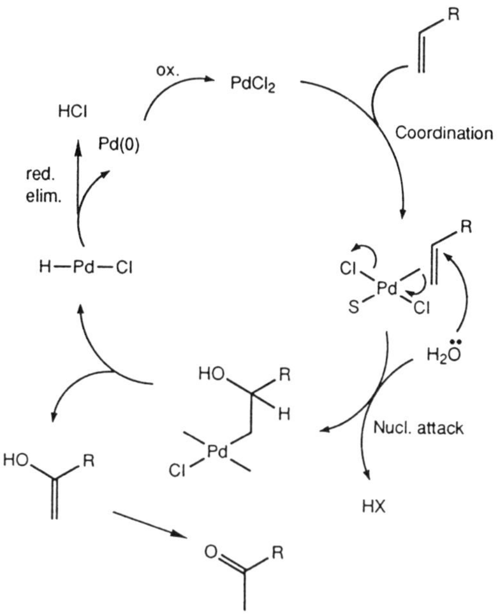

Figure 5.4

have appreciable solubility in the reaction medium. In practice, polar solvents such as dimethylformamide, *N*-methylpyrrolidinone and 3-methylsulfolane have proven most effective. Copper(I) chloride or copper(II) chloride–oxygen (1 atm) is most commonly used as the reoxidant, although a stoichiometric amount of benzoquinone provides milder conditions. Examples of the efficiency and selectivity are shown in equations 5.1–5.4.

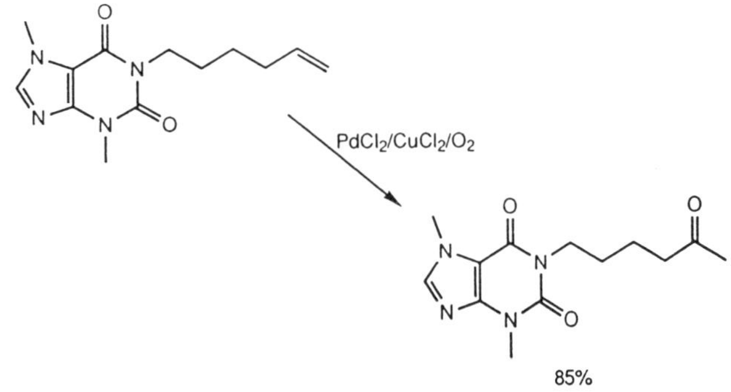

$PdCl_2/CuCl_2/O_2$ (5.1)

85%

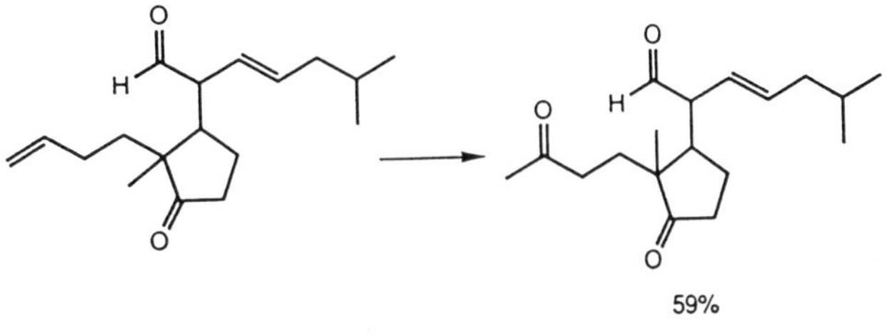

(5.2)

59%

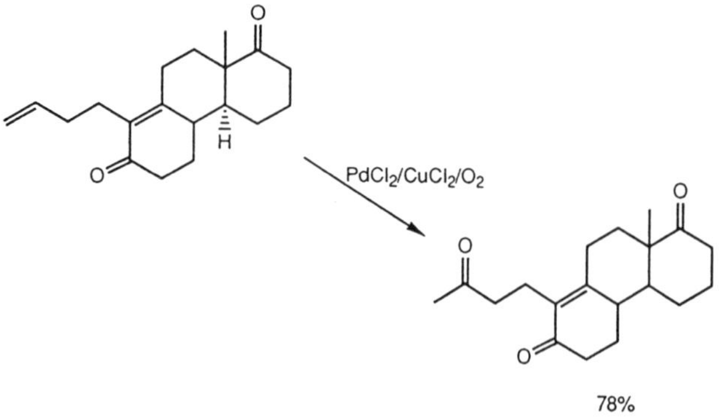

$PdCl_2/CuCl_2/O_2$

78%

(5.3)

Ph OH (OAc) → Ph OH O 87%

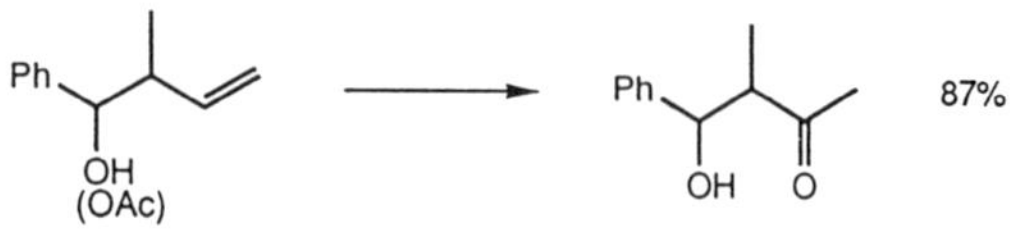

(5.4)

Internal olefins undergo reaction extremely slowly under normal reaction conditions, hence the specificity. Unfortunately, alternative conditions under which internal olefins will react have not been found. The exception to this is conjugated enones, which can be efficiently converted into β-dicarbonyl compounds under very different conditions and almost certainly by a different mechanism. However, this difference has little relevance to organic synthesis, and examples of this process, which require the use of *tert*-butyl hydroperoxide as a reoxidant, *N*-methylpyrrolidinone or 2-propanol as solvent and Na_2PdCl_4 as catalyst at 50–80 °C, are shown in equations 5.5 and 5.6.

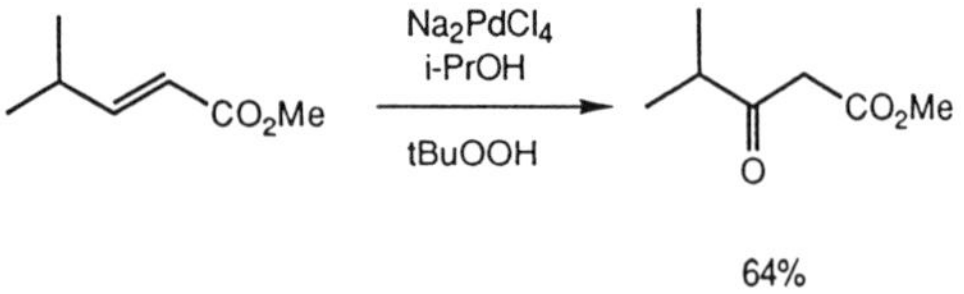

CO_2Me $\xrightarrow[\text{tBuOOH}]{Na_2PdCl_4,\ \text{i-PrOH}}$ CO_2Me

64%

(5.5)

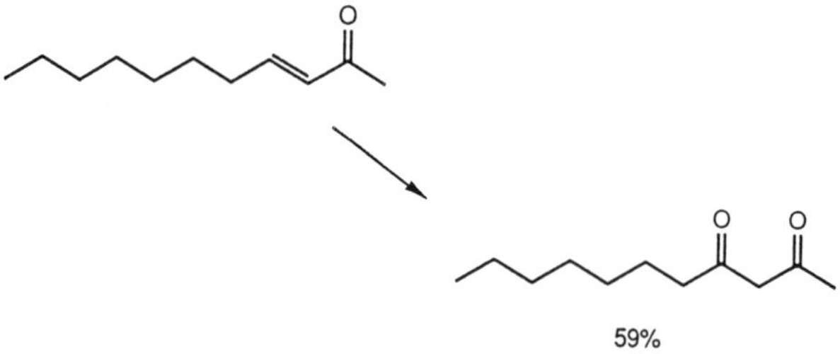

(5.6)

59%

Other oxygen nucleophiles, particularly alcohols and carboxylates, also efficiently attack olefins in the presence of palladium(II) catalysts. Intermolecular versions have been little used, but intramolecular versions, particularly with highly functionalized complex substrates, have been extensively applied. Much of this work predates the development of mild reoxidants for palladium (e.g. benzoquinone) and many literature examples were carried out using stoichiometric quantities of palladium(II), since the substrates and/or products were unstable to the relatively harsh copper–oxygen redox conditions. These cases warrant re-examination, since catalysis using milder oxidants is certainly possible. Examples of stoichiometric processes are seen in the cyclization of *o*-allylphenols to benzofurans (equation 5.7) [3] and 2-hydroxychalcones to flavones (equation 5.8) [4]. An efficient example of stoichiometric benzofuran formation is seen in the synthesis of tetracycline intermediates (equation 5.9) [5]. Even more impressive is the use of palladium(II) catalysis for the 'alkoxycarbonylation' of an olefin in the total synthesis of frenolicin (equation 5.10) [6], wherein the palladium is used twice,

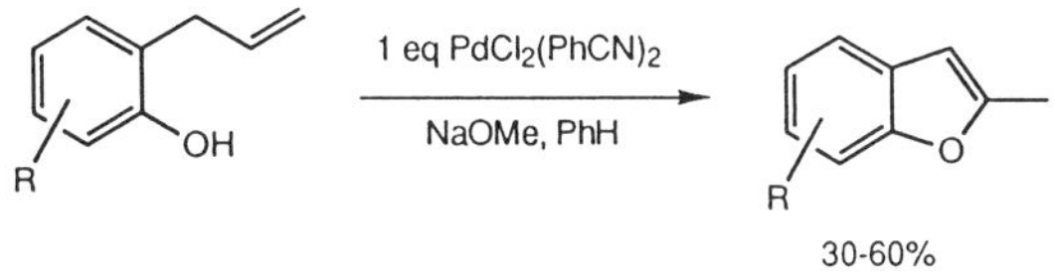

(5.7)

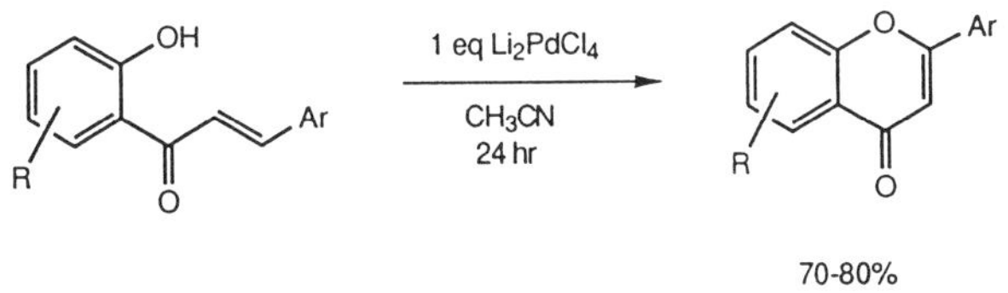

(5.8)

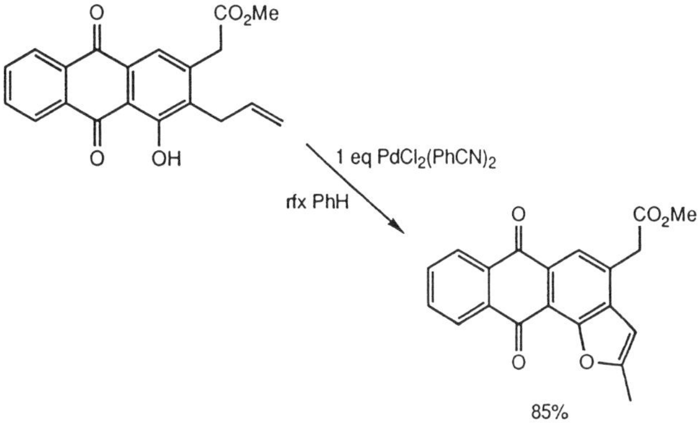

(5.9)

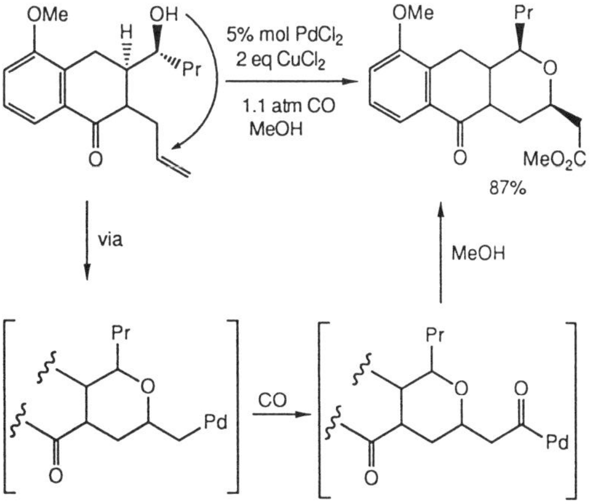

(5.10)

once to activate the olefin towards nucleophilic attack by the alcohol, and then to insert CO into the resulting σ-alkylpalladium(II) complex. This succeeded because CO insertion can compete with β-hydride elimination under appropriate conditions (see below). By using 1.5 equivalents of palladium(II) acetate and carefully controlling reaction conditions, a double insertion of two different olefins was achieved in a synthesis of a prostaglandin analog (equation 5.11) [7]. β-Elimination after the first olefin insertion was suppressed by the rigidity of the bicyclic system, which prevented the achievement of the *syn* coplanar Pd—H arrangement required for this process.

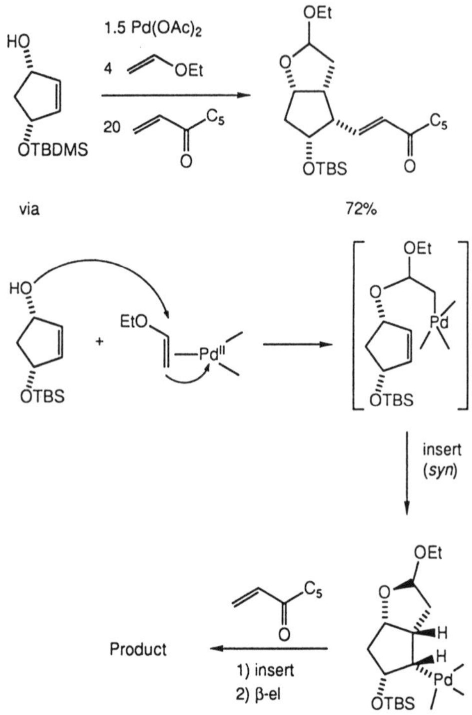

HO OTBDMS; 1.5 Pd(OAc)$_2$, 4 OEt, 20 C$_5$ O; OEt O OTBS C$_5$ O 72%

via

HO OTBS + EtO Pd^{II} → [OEt O Pd OTBS]

insert (*syn*)

OEt O H H Pd OTBS; C$_5$ O; 1) insert 2) β-el; Product

(5.11)

Carboxylic acids are also excellent nucleophiles for olefins. *o*-Allylbenzoic acids were cyclized to isocoumarins in excellent yield under both stoichiometric and catalytic conditions (equation 5.12) [8]. A useful variant of this is the palladium-catalyzed 1,4-bisacetoxylation of 1,3-dienes, which involves nucleophilic attack of acetate on the palladium complexed diene. This generates a π-allylpalladium(II) complex, also generally subject to nucleophilic attack (see Section 5.4.3). In the absence of added chloride the second acetate is delivered from the metal, resulting in overall *trans*-bisacetoxylation.

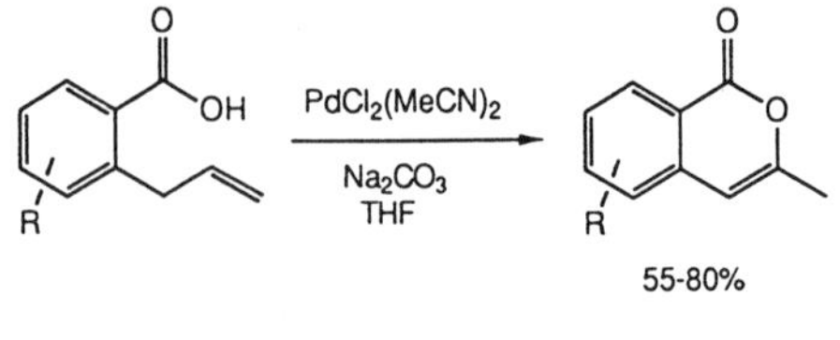

O OH R; PdCl$_2$(MeCN)$_2$, Na$_2$CO$_3$, THF; O O R 55-80%

(5.12)

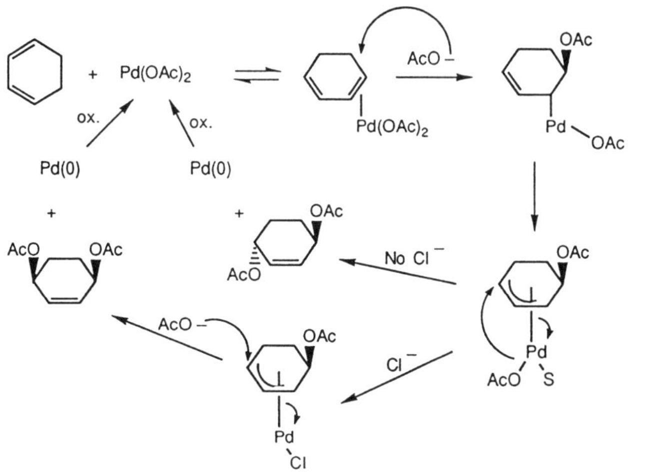
+ Pd(OAc)2
ox.
ox.
Pd(0)
Pd(0)
Pd(OAc)2
AcO
OAc
Pd
OAc
No Cl
Cl
AcO
S
Cl

Figure 5.5

Added chloride blocks this coordination site on the metal and leads to nucleophilic attack by uncomplexed acetate, from the face opposite the metal, giving the corresponding *cis*-diacetate and palladium(0) [9]. The reoxidation of Pd(0) to Pd(II) can be carried by benzoquinone–MnO_2 or more efficiently by O_2–catalytic hydroquinone–catalytic Co(salophen) [10]. Because the products of this reaction are allyl acetates, and because of the wide variety of Pd(0)-catalyzed reactions of allyl acetates (Section 5.4.3), this is a useful process. This chemistry is summarized in Figure 5.5. Note that 1,3-dienes have been catalytically dialkoxylated, using alcohols in place of acetate, under similar conditions [11]. A useful intramolecular variant of this is seen in equation 5.13 [12].

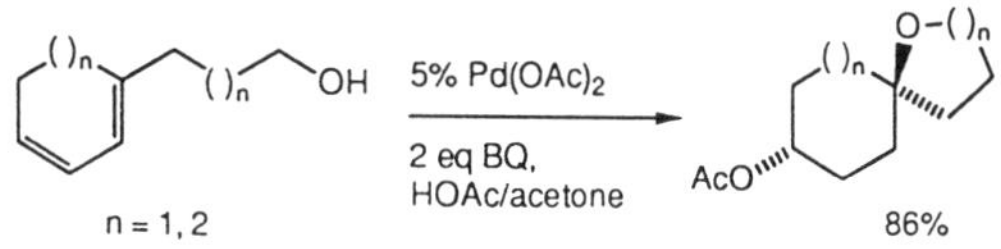
OH
5% Pd(OAc)2
2 eq BQ,
HOAc/acetone
n = 1, 2
AcO
86%

(5.13)

5.3.1.3 Nitrogen Nucleophiles

Whereas the palladium(II)-catalyzed reaction of oxygen nucleophiles with olefins is a general and efficient process, nitrogen nucleophiles pose a number of serious problems, the major one being that aliphatic amines are better ligands than olefins for palladium(II) and both the nucleophile and the product are catalyst poisons. Hence the palladium(II)-catalyzed amination of olefins by aliphatic amines has not been achieved, although a cumbersome process requiring one equivalent of Pd(II), three equivalents of a secondary amine and reductive removal of palladium has been developed [13]. The problem does indeed lie with the basicity of the amine nucleophile and, by reducing the basicity, catalysis can be achieved.

Illustrative of both the problems and successes of Pd(II)-catalyzed amination of olefins is the case of cyclization of olefinic amines (Figure 5.6). As expected, the free amines

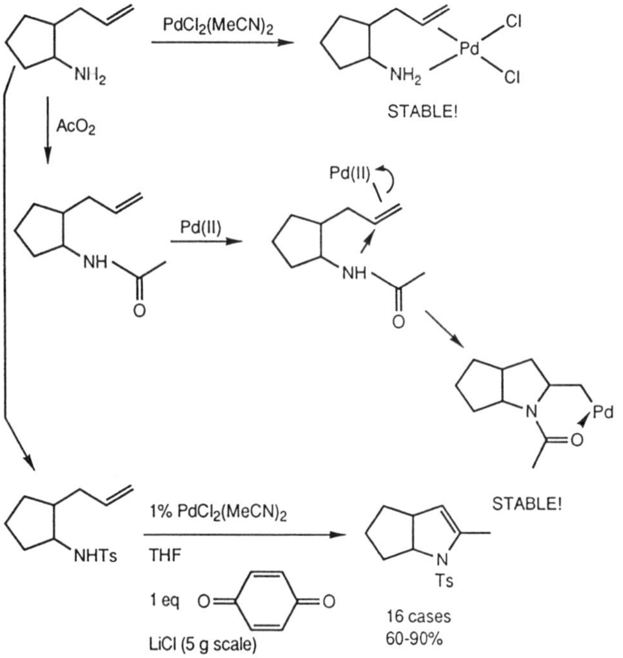

Figure 5.6

failed to cyclize when treated with either stoichiometric or catalytic amounts of palladium(II) chloride. The free amine, being a potent ligand for palladium(II), instead formed a stable complex, effectively poisoning the catalyst. *N*-Acylation of the amine sufficiently reduced the coordinating ability of the nitrogen to permit competitive olefin complexation by the Pd(II), but catalysis was again thwarted; coordination of the *N*-acyl oxygen stabilized the resulting σ-alkylpalladium(II) complex, preventing β-hydrogen elimination, and truncating the reaction. In contrast, *N*-tosylation resulted in a substrate that neither complexed the Pd(II) catalyst at nitrogen nor stabilized the intermediate σ-alkylpalladium(II) complex, and efficient catalysis was achieved [14].

In contrast to aliphatic amines, aromatic amines are *ca* 10^6 times less basic and consequently coordinate much less strongly to palladium(II) complexes. Thus *o*-allylanilines undergo catalytic cyclization to indoles without protection (deactivation) of the amine (equation 5.14) [15]. The use of the mild reoxidant benzoquinone is critical here, since copper–oxygen-based reoxidation of palladium destroys both anilines and indoles. Interestingly, both *N*-acetyl and *N*-sulfonylanilines also undergo the Pd(II)-

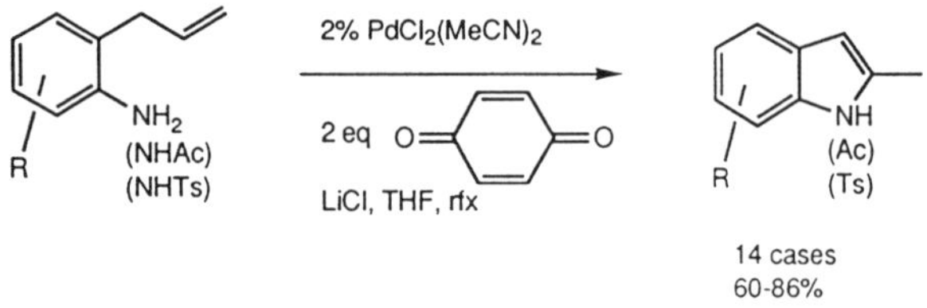

(5.14)

catalyzed cyclization. By carefully controlling the conditions to promote olefin insertion and suppress β-hydride elimination (short reaction times, polar solvents, low temperatures), the intermediate σ-alkyl palladium(II) complex could be intramolecularly trapped by an adjacent olefin, resulting in bicyclization to the pyrroloindoloquinone system (equation 5.15) [16]. Note that although this substrate has nine potential functional groups (denoted with arrows) which can coordinate to palladium(II), only coordination to one of these results in a situation where productive reaction can ensue. Provided that the other sites do not irreversibly bind the catalyst, the catalytic process proceeds smoothly.

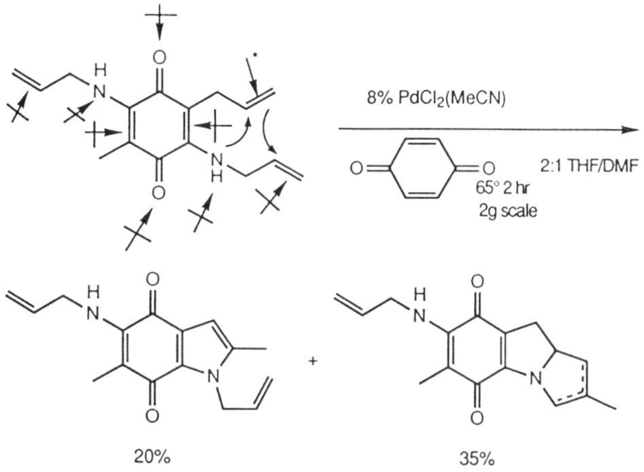

8% $PdCl_2(MeCN)$

2:1 THF/DMF

65° 2 hr

2g scale

20%

35%

(5.15)

5.3.1.4 Carbon Nucleophiles

The use of carbon nucleophiles to attack olefins coordinated to palladium(II) presents yet another set of problems. Palladium(II) salts are reasonably good oxidizing agents, and exposure of carbanions to them results in rapid reduction of Pd(II) to the metal, with concomitant oxidative coupling of the carbanion. This can be suppressed by 'protecting' the palladium(II) from direct attack by the carbanion by the addition of triethylamine to the olefin–palladium(II) complex at low temperature. Under these conditions stabilized carbanions can be directed to attack the olefin rather than the metal, and alkylation of the olefin can be achieved. Again, an unstable σ-alkylpalladium(II) complex is produced; this can β-eliminate to give the alkylated olefin, can be reduced to give the alkane or can be carbonylated to give the ester (Figure 5.7) [17]. However, this process requires a stoichiometric amount of palladium, since an oxidizing agent to reoxidize Pd(0) to which carbanions are stable has not yet been found. Hence the product must have a 'high value' before the chemistry can be utilized. Just such a case is seen in the synthesis of a relay to (+)-thienamycin (equation 5.16), wherein all of the functional groups and the absolute stereochemistry of the key stereogenic center are set in place in the initial palladium(II) assisted carboacylation step [18], with good yield and with virtually complete stereoselectivity. This reaction takes advantage of the fact that CO insertion proceeds faster than β hydrogen elimination at $-20\,^{\circ}C$,

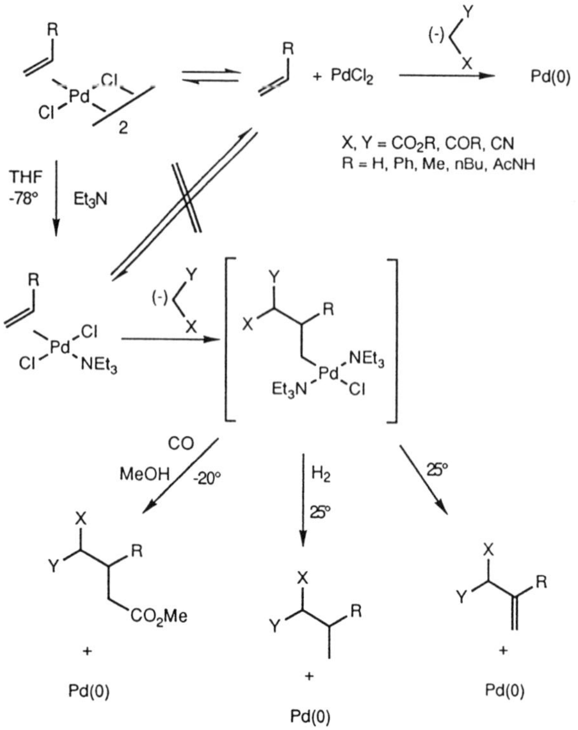

Figure 5.7

generating a σ-acyl complex which was cleaved by methanol to produce the ester. Cleavage of this same σ-acyl group with tin reagents, via transmetallation and reductive elimination (Figure 5.3), permits the introduction of a wide variety of keto groups at the former olefin terminus (equation 5.17) [19]. Another example in which the complexity of the product probably justifies the use of stoichiometric quantities of palladium is the intramolecular alkylation of olefins by TMS enol ethers (equations 5.18 and 5.19) [20]. It should be noted that palladium is not lost in these stoichiometric reactions,

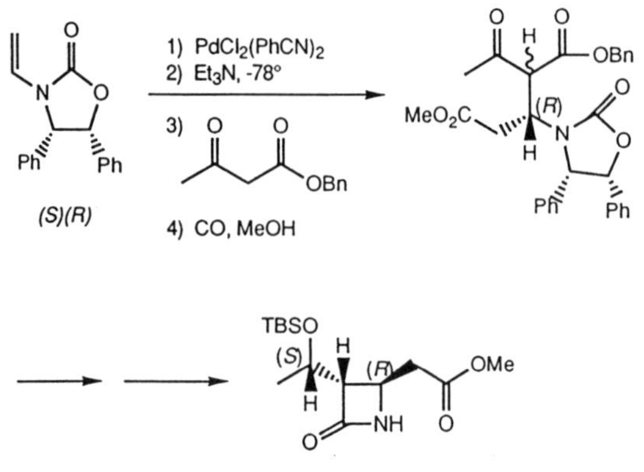

(5.16)

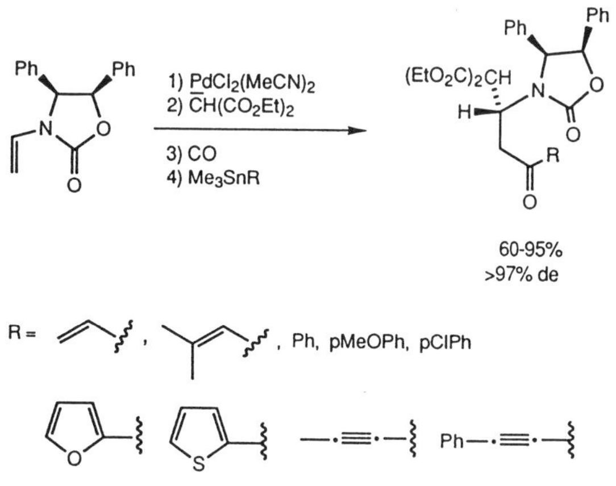

(5.17)

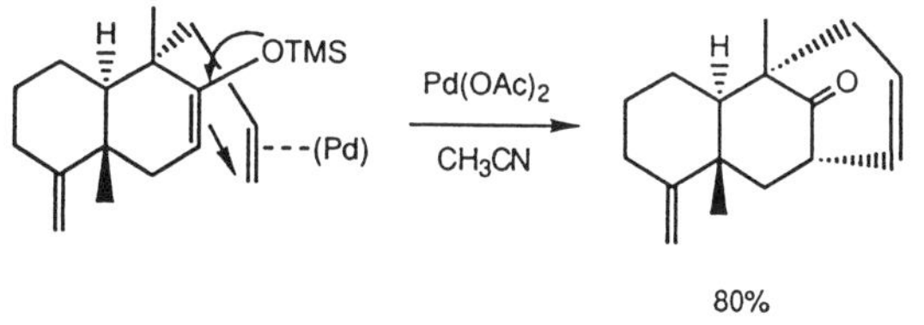

60-95%
>97% de

R = , , Ph, pMeOPh, pClPh

(5.18)

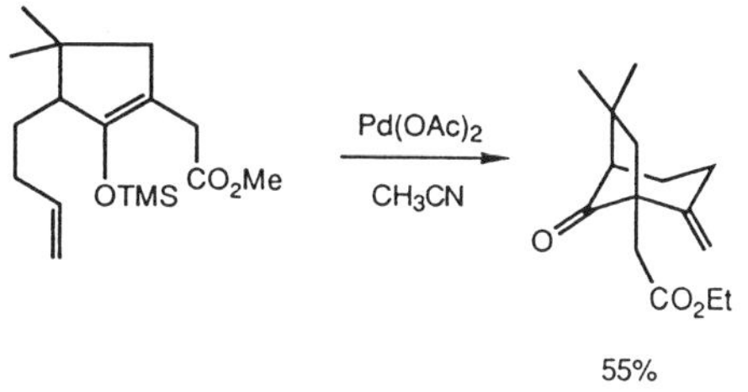

80%

(5.19)

55%

but merely reduced to palladium metal, which can be quantitatively recovered and reoxidized to palladium(II) in a separate step.

5.3.2 PALLADIUM(II)-CATALYZED REARRANGEMENTS

5.3.2.1 Rearrangements of Allylic Systems

Palladium(II) complexes are capable of catalyzing rearrangements of *all* olefinic systems to which they complex and, indeed, a competing reaction in the Pd(II)-catalyzed reactions of terminal olefins discussed above is the rearrangement of these to internal olefins, which are substantially less reactive. However, this ability to catalyze rearrangements has also been utilized in productive ways, particularly with allylic systems. The general transformation is shown in equation 5.20, along with a likely mechanism which involves coordination of the olefin to Pd(II), thereby activating it to nucleophilic attack, nucleophilic attack by Y giving a zwitterionic intermediate and collapse of the intermediate to the rearranged product with ejection of palladium(II). Note that there is no net redox here and, in a sense, Pd(II) is behaving like a very selective Lewis acid. [*Note*: palladium(0) complexes catalyze similar rearrangements but both the mechanism and the specificity are different from those for the same reactions catalyzed by palladium(II)! These will be discussed below.]

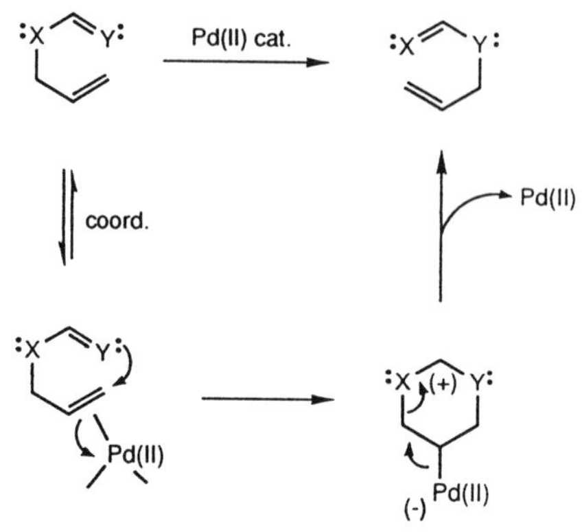

:X Y: Pd(II) cat. :X Y: coord. Pd(II) Pd(II) (+) (-) Pd(II) (5.20)

This reaction is most useful for the rearrangement of tertiary to primary allylic acetates, in that the reaction proceeds quickly at room temperature, and the selectivity and chemical yields are high (equations 5.21–5.23) [21]. When an optically active allyl acetate was rearranged, complete transfer of chirality was observed (equation 5.24) [22]. Note that the absolute stereochemistry observed at the newly formed chiral center must result from coordination of palladium to the face of the olefin opposite the acetate, and

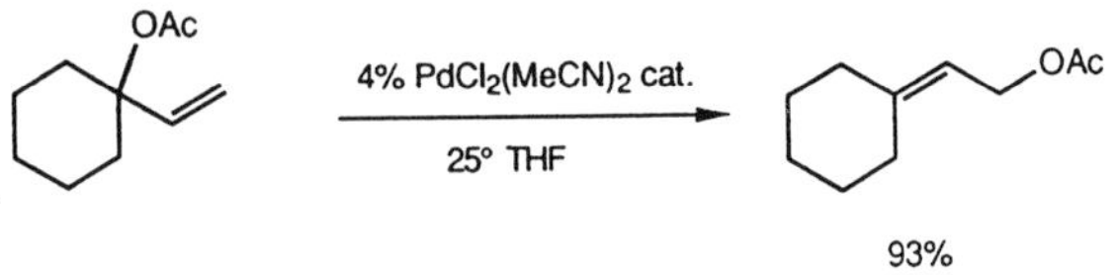

OAc 4% $PdCl_2(MeCN)_2$ cat. 25° THF OAc 93% (5.21)

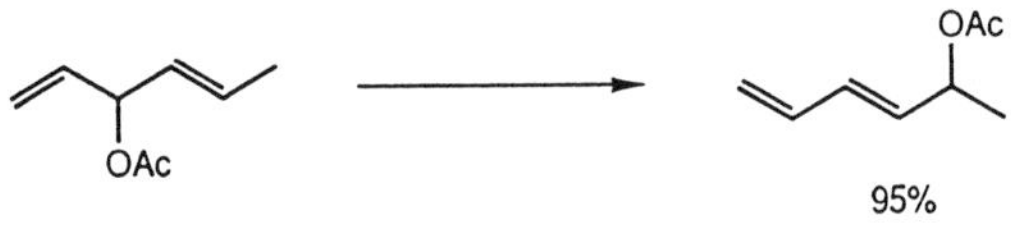

OAc OAc 95% (5.22)

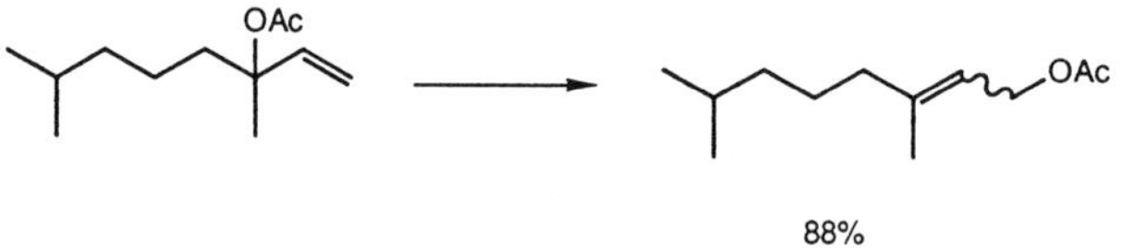

OAc OAc 88% (5.23)

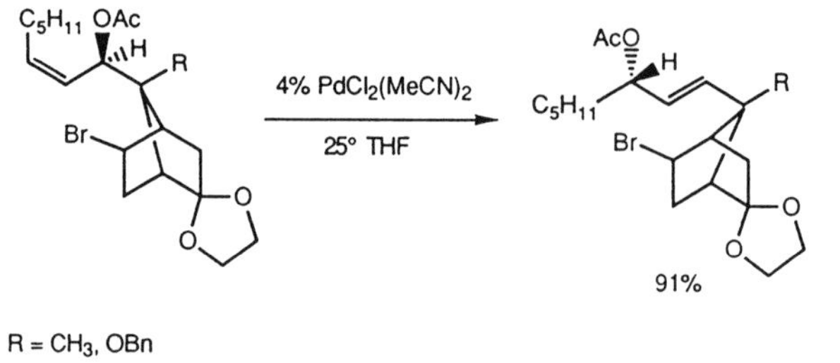

C_5H_{11} OAc H R Br O O 4% $PdCl_2(MeCN)_2$ 25° THF AcO H R C_5H_{11} Br O O 91% (5.24)

R = CH_3, OBn

attack of the olefin from the face opposite the metal, as expected for the mechanism shown in equation 5.20.

This rearrangement is not restricted to allyl acetates, but is general for allylic systems containing heteroatoms, such as allyl formimidates [23], *S*-allylthioimidates [24] and allylic thionobenzoates [25], but these have found fewer applications in synthesis. A more synthetically useful variant involves the palladium(II)-catalyzed rearrangement of oxime *o*-allyl ethers to nitrones. When carried out in the presence of a 1,3-dipolarophile, combined rearrangement and 1,3-dipolar cycloaddition can be achieved (equation 5.25) [26]. Substantially more interesting is the palladium(II)-catalyzed Cope rearrangement [21d, 27], which enjoys rate enhancements of *ca* 10^{10} over uncatalyzed processes, is more stereoselective than the thermal process and proceeds with virtually complete transfer of chirality in a chain topographic sense (equation 5.26). Although useful, this process also suffers limitations consonant with the proposed mechanism. The diene must have a substituent either at C-2 or at C-5, but not both; that is, it must have a monosubstituted olefin to complex to palladium(II), in addition to a substituent on

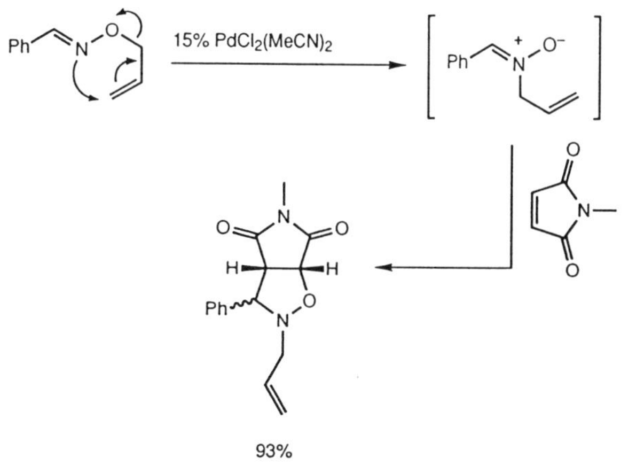

15% $PdCl_2(MeCN)_2$

93%

(5.25)

6% $PdCl_2(PhCN)_2$

25 °C, 24 hr, 87%

93 : 7

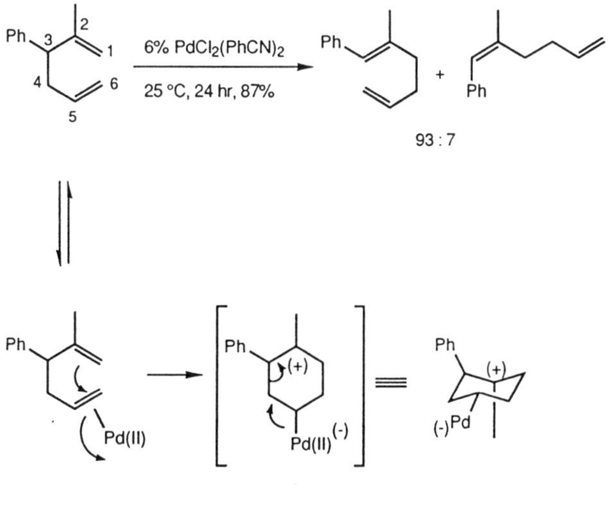

(5.26)

the other olefin to stabilize positive charge in the intermediate. Within these confines, however, the reaction is fairly useful. Oxy-Cope rearrangements have been catalyzed under similar conditions (equation 5.27) [28].

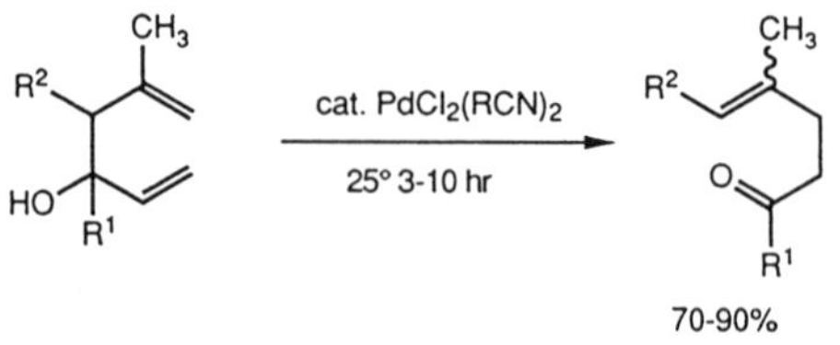

(5.27)

5.3.2.2 Palladium(II)-Catalyzed Cycloisomerization of Enynes [29]

Treatment of suitably disposed enynes with a variety of palladium(II) catalysts produces cyclic dienes in excellent yield and under mild conditions. The mechanism of this process has not yet been studied but is likely to involve hydridopalladation of the alkyne, olefin insertion and β-hydrogen elimination (equation 5.28). With palladium acetate–ligand catalysts the initial source of the requisite palladium hydride is unknown, but efficient catalysis of the same reaction by $Pd_2(dba)_3 \cdot CHCl_3$ in acetic acid, a ready source of $HPd^{II}OAc$ by oxidative addition, favors just such a mechanism. The formation of 1,3- vs 1,4-dienes is directed by the auxiliary ligands added to the catalyst, indicating that the direction of β-elimination is, as expected, ligand dependent (equation 5.29). As is typical,

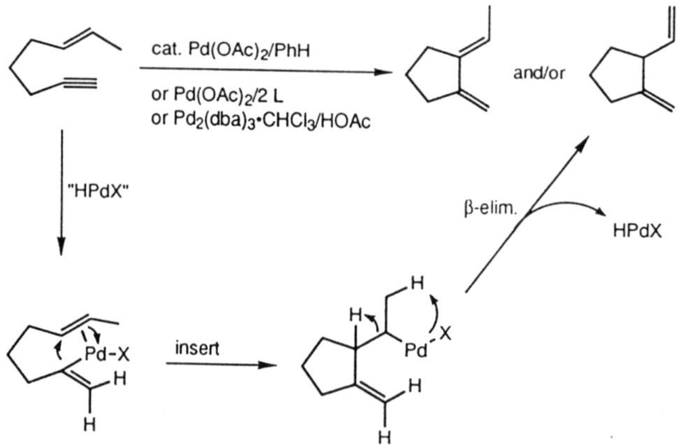

(5.28)

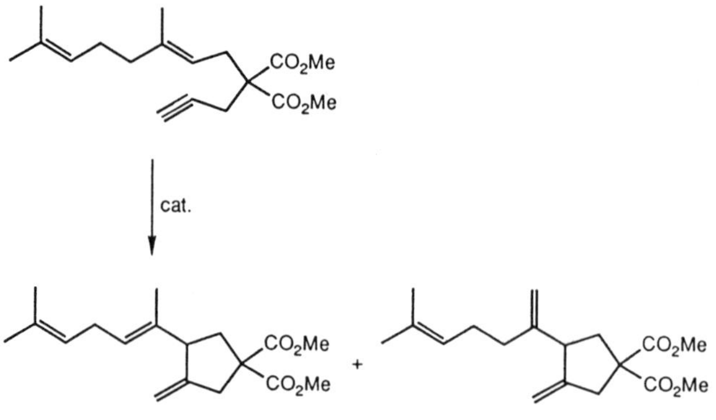

(5.29)

a wide array of functionality is tolerated, and this process has been used to synthesize a number of relatively elaborate cyclic dienes (equations 5.30–5.33).

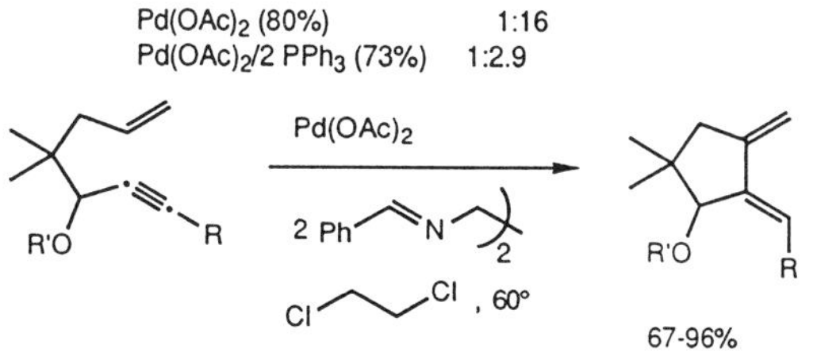

$Pd(OAc)_2$ (80%) 1:16
$Pd(OAc)_2$/2 PPh_3 (73%) 1:2.9

R'O R; $Pd(OAc)_2$, 2 Ph⁀N⁀)$_2$, Cl⁀⁀Cl, 60° → R'O R, 67-96% (5.30)

R = $COCH_3$, TMS, CH_2OCH_3, $CH_2CH_2OCH_3$, CH_3, OEt, CO_2Me

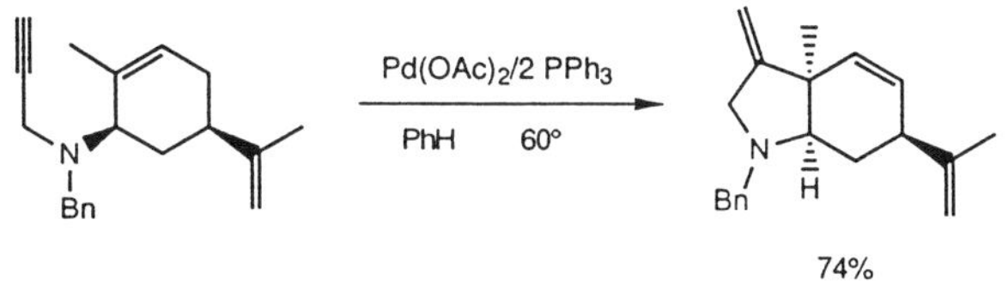

N Bn; $Pd(OAc)_2$/2 PPh_3, PhH 60° → N Bn H, 74% (5.31)

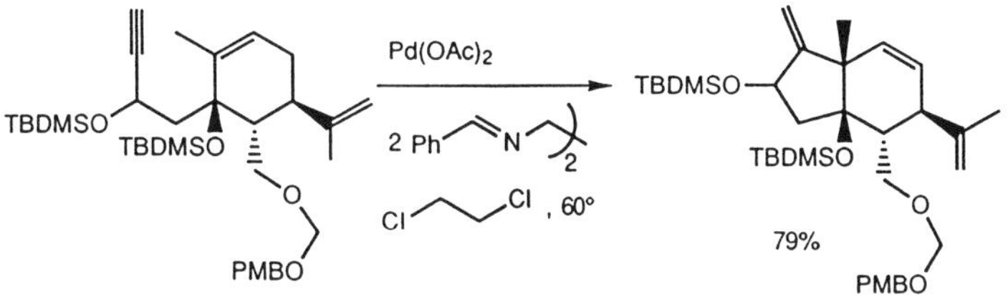

TBDMSO TBDMSO O PMBO; $Pd(OAc)_2$, 2 Ph⁀N⁀)$_2$, Cl⁀⁀Cl, 60° → TBDMSO TBDMSO O PMBO, 79% (5.32)

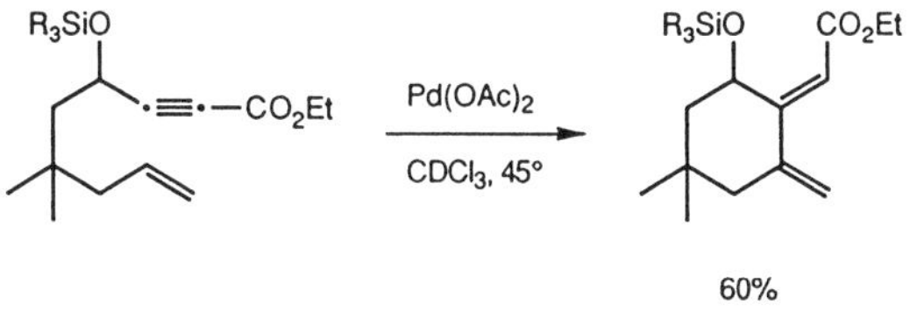

R_3SiO CO_2Et; $Pd(OAc)_2$, $CDCl_3$, 45° → R_3SiO CO_2Et, 60% (5.33)

The intermediate σ-alkylpalladium(II) species generated in the insertion step (equation 5.28) can be intercepted before β-elimination by a number of reagents. Polymethylhydrosiloxane (PMHS) reduces this complex efficiently, producing olefins rather than dienes (equation 5.34). Intramolecular trapping of this σ-alkylpalladium(II) com-

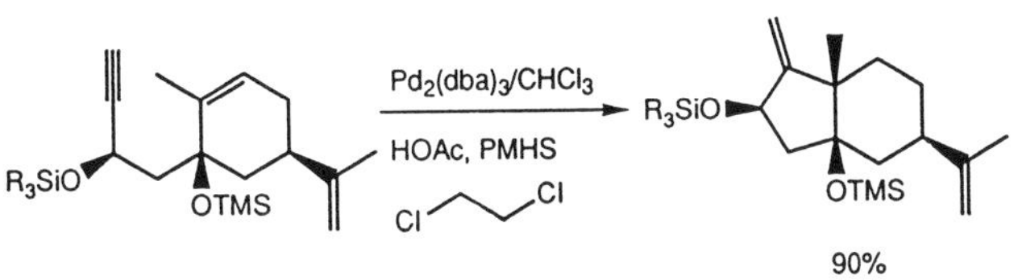

R_3SiO OTMS; $Pd_2(dba)_3/CHCl_3$, HOAc, PMHS, Cl⁀⁀Cl → R_3SiO OTMS, 90% (5.34)

plex by olefins is also efficient, leading to polycyclization (equation 5.35) [30]. This can be pushed to an extreme, with a high degree of success (equation 5.36) [30].

$Pd_2(dba)_3CHCl_3$, Ph_3P, HOAc — 57% (5.35)

$Pd_2(dba)_3 \cdot CHCl_3$, PPh_3, HOAc — 86% (5.36)

This area of chemistry is sufficiently new (1991) that it is still somewhat confused, making the best choice of catalyst and conditions difficult. However, there is already a great deal of empirical information available, and many examples. The experimental procedure is operationally simple and the transformation sufficiently useful to warrant its use in spite of this.

5.3.3 ORTHOPALLADATION

Palladium(II) salts are reasonably electrophilic and, under appropriate conditions [electron-rich arenes, $Pd(OAc)_2$, boiling acetic acid] can directly palladate arenes via an electrophilic aromatic substitution process. However, this process is neither general nor efficient, and thus has found little use in organic synthesis. Much more general is the ligand directed orthopalladation shown in equation 5.37 [31]. This process is very general, the only requirement being a lone pair of electrons in a benzylic position to precoordinate to the metal, and conditions conducive to electrophilic aromatic substitution (usually HOAc solvent). The site of palladation is always *ortho* to the directing ligand and, when the two *ortho*-positions are inequivalent, palladation results at the sterically less congested position, regardless of the electronic bias. (This is in direct

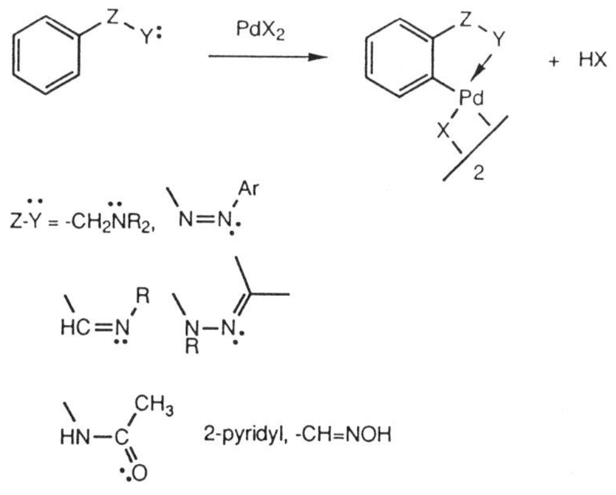

(5.37)

contrast to *ortho*-lithiation, which is electronically controlled and will usually occur at the more hindered position on rings having several electron-donating groups.) The resulting chelato-stabilized σ-arylpalladium(II) complexes are almost invariably stable, are easily isolated, purified and stored and are unreactive.

This stability is clearly due to the chelate stabilization by Y, and is the major reason why *ortho*-palladation has not yet been developed for catalytic *ortho*-functionalization of arenes. However, the reactivity intrinsic to σ-alkylpalladium(II) species (Figure 5.3) can be activated by heating the *ortho*-palladated species in the presence of excess triethylamine to disrupt the chelation by displacing Y, in the presence of an alkene. Under these conditions, facile insertion occurs (equation 5.38) [32]. Related complexes of acetanilide undergo facile carbonylation (equation 5.39) [33]. The unresolved prob-

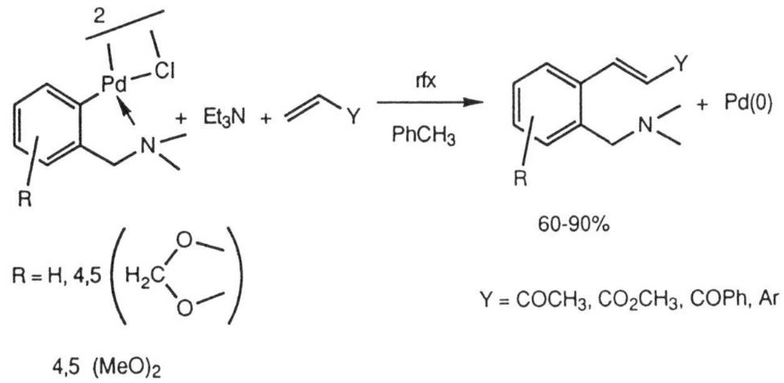

(5.38)

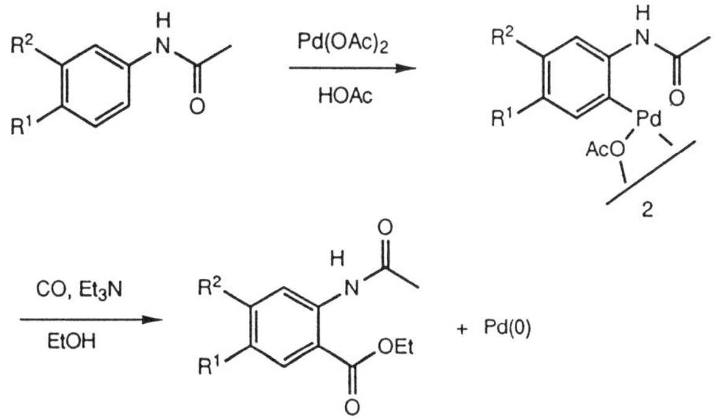

(5.39)

lems for catalysis are the incompatible conditions for the *ortho*-palladation step (HOAc) and the subsequent step (Et_3N), the stability of the *ortho*-palladated product and, as always, the reoxidation of the palladium(0) produced in the final step. Until these are solved, *ortho*-palladation will remain a stepwise stoichiometric process of synthetic interest only when its regioselectivity or specificity makes it uniquely appropriate for a desired transformation.

5.4 PALLADIUM(0) COMPLEXES IN ORGANIC SYNTHESIS

5.4.1 OXIDATIVE ADDITION–TRANSMETALLATION

5.4.1.1 General Features

The palladium(0)-catalyzed coupling of aryl and vinyl halides and triflates with Main Group organometallics via oxidative addition–transmetallation–reductive elimination sequences (Figure 5.8) has been very broadly developed and has an overwhelming amount of literature associated with it. However, relatively few fundamental principles are involved, the understanding of which will bring order out of this apparent chaos, and make the area both approachable and usable.

A very wide range of palladium catalyst precursors can be used in this system, and the choice is best made by analogy, although almost all will work reasonably well. Most often $Pd(PPh_3)_4$ or $Pd(dba)_2$ plus PPh_3 is used, although even palladium(II) catalyst precursors such as $(PPh_3)_2PdCl_2$, $Pd(OAc)_2$ or $PdCl_2(MeCN)_2$ are efficient since they are readily reduced to the catalytically active Pd(0) state by most Main Group organometallics. Recall that palladium(0) complexes are electron-rich, nucleophilic species prone to oxidation. The single most important reaction of palladium(0) complexes is their reaction with organic halides or triflates to form σ-alkylpalladium(II) complexes. This is commonly known as 'oxidative addition' because the metal is

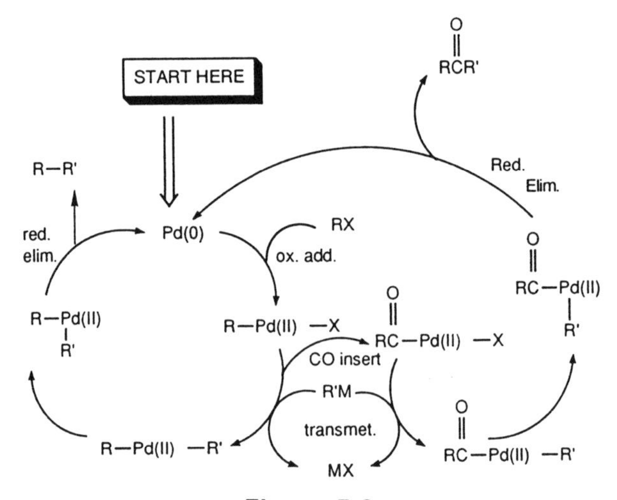

Figure 5.8

formally oxidized from Pd(0) to Pd(II) and the 'oxidizing agent,' RX, adds to the metal, hence the term oxidative addition.

The oxidative addition process has a number of general features. The order of reactivity is $I > OTf > Br \gg Cl$, such that chlorides are rarely useful in this reaction. With aryl halides, added phosphines are required for the reaction of aryl bromides, but suppress the reaction of aryl iodides, permitting easy discrimination. With most substrates, the oxidative addition step proceeds readily at room temperature. When higher temperatures are used, this is usually because some step other than the oxidative addition is sluggish at room temperature. From the standpoint of the substrate, the metal becomes oxidized, but the substrate becomes reduced. Hence electron-deficient halides are, in general, more reactive than electron-rich halides. Because β-hydrogen elimination is rapid above *ca* $-20\,°C$, the halide or triflate substrate usually cannot have β-hydrogens, restricting this process to aryl and vinyl substrates. With vinyl substrates oxidative addition occurs with retention of olefin geometry (as do all of the ensuing steps). The σ-aryl- or σ-vinylpalladium(II) complex formed by oxidative addition enjoys the rich chemistry of this class of complexes presented in Figure 5.3, including olefin and CO insertion (see Section 5.4.2), hydrogenolysis and transmetallation from Main Group organometallics.

A very wide range of Main Group organometallics 'transmetallate' to palladium(II), that is, they transfer their R group to palladium in exchange for the halide or triflate, generating a dialkylpalladium(II) complex and the Main Group metal halide or triflate. Transmetallations from Li, Mg, Zn, Zr, B, Al, Sn, Si, Ge, Hg, Tl, Cu, Ni and perhaps others have been reported, but some (e.g. Sn, B and Zn) are much more useful than others. Transmetallation is favored from more electropositive to more electronegative metals, but this is of little use in assessing reactivity since electronegativities are only crudely known and are sensitive to spectator ligands and, since, if the next step in the process is irreversible, only a small equilibrium constant for transmetallation is required for the process to proceed. Transmetallation is almost invariably the rate-limiting step, and when oxidative addition–transmetallation sequences fail, it is this step which warrants attention. [Because transmetallation is rate limiting, it is often possible to insert CO into the oxidative addition product prior to transmetallation, resulting in a carbonylative coupling (see below).] It is important that both metals involved in transmetallation benefit from the process energetically and the needs of the Main Group partner cannot be ignored. Thus, coupling of organotriflates with organotin reagents sometimes fails, since the product tin triflate is not stable, but often proceeds well with the addition of lithium chloride, permitting the production of the more stable organotin chloride rather than triflate. Similarly, organosilanes only transmetallate in the presence of added fluoride to give the very stable Si—F compounds, and organoboranes react only in the presence of added alkoxides to produce stable B—O compounds. Transmetallation occurs with retention of any stereochemistry in the organic group.

Once transmetallation has occurred, the remaining steps, rearrangement of the *trans*-dialkylpalladium(II) complex to the *cis* complex and reductive elimination, producing the coupled products and regenerating the Pd(0) catalyst, are rapid. Because of this, β-hydrogens can be present in the R group transferred from Main Group organometallic, since reductive elimination is faster than β-hydride elimination from the dialkylpalladium(II) intermediate.

5.4.1.2 Transmetallation from Li, Mg, Zn and Cu

One of the earliest examples of coupling by oxidative addition–transmetallation was the palladium-catalyzed coupling of aryl or vinyl halides with Grignard reagents [34]. This reaction is fairly general and a wide range of aromatic, heteroaromatic [35] and vinyl halides [36] couple efficiently with most Grignard reagents (even those containing β-hydrogens) under these conditions. Organolithium reagents behave in a similar manner [37], permitting the direct alkylation, vinylation and arylation of vinyl and aryl halides (equation 5.40). Because of the reactivity of Grignard and organolithium reagents, these coupling processes are limited to partners lacking much functionality.

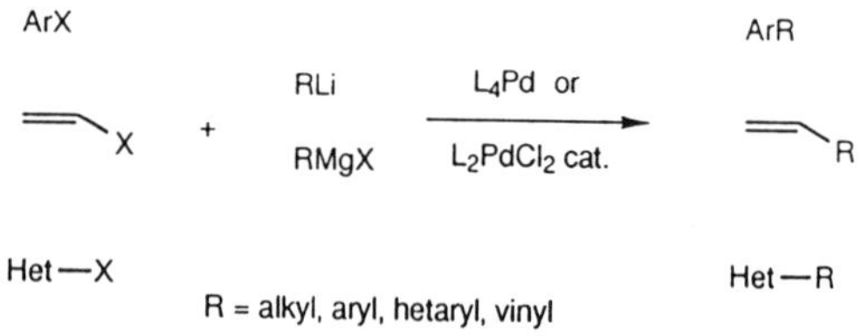

ArX

X + RLi / RMgX —(L_4Pd or L_2PdCl_2 cat.)→ R (5.40)

Het—X

ArR

Het—R

R = alkyl, aryl, hetaryl, vinyl

Much more versatile are processes involving transmetallation from zinc reagents. These can be easily prepared from organolithium (and other Main Group organometallic) species by simple addition to these of one equivalent of anhydrous zinc chloride in THF. Organozinc reagents have at least three advantages: they are among the most efficient reagents for transmetallation to palladium, they tolerate a wide range of functional groups, including carbonyls, in the substrate, and they can be prepared directly from functionalized halides (e.g. α-bromoesters), thus allowing functionality in both coupling partners (equations 5.41 [38], 5.42 [39] and 5.43 [40]).

Organocuprates transmetallate only slowly to palladium at temperatures below their decomposition points, greatly limiting their use. However, addition of one equivalent of zinc bromide promotes the coupling at *ca* $-10\,^{\circ}C$, almost certainly via the

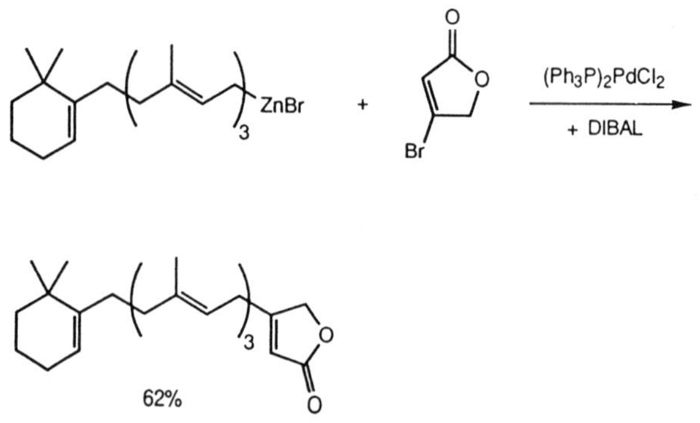

ZnBr (3) + Br, O, O —($(Ph_3P)_2PdCl_2$ + DIBAL)→ (5.41)

3, O, O

62%

OMe — 1) tBuLi, 2) $ZnCl_2$ → OMe, ZnCl — 1) ArX, R X; L_2PdCl_2/DIBAL 25°; 2) H^+ →

O, Ar or O, R

60-80%

(5.42)

OTBS, TBSO, ZnCl + OMOM, I, OMOM

L_2PdCl_2
DIBAH

OTBS, TBSO, O, OMOM, OMOM

79%

(5.43)

organozinc intermediate. With vinyl cuprates and vinyl halides (equation 5.44), retention of stereochemistry by both partners is observed, as expected [41].

($)_2$CuLi + I C_5H_{11} — 2% L_4Pd; 0°, 1eq ZnBr →

C_5H_{11}

86%
>99.5% *cis:cis*

(5.44)

5.4.1.3 Transmetallation from Zr, Al and B

Alkyl- and vinyl-zirconium, -aluminum and -boron complexes are readily available by the hydrometallation (hydrozirconation, hydroalumination, hydroboration) of alkenes and alkynes. The development of procedures that permit transmetallation from these metals to palladium has greatly expanded the range of organic starting materials which can enter into these coupling reactions.

Hydrozirconation of alkynes is tolerant of a range of functional groups, and is both regio- and stereospecific, adding *cis* to alkynes, with the Zr always occupying the least (less) substituted vinylic position [42]. These vinylzirconium species couple efficiently to aryl and vinyl halides in the presence of palladium(0) catalysts (equations 5.45 [43] and 5.46 [44]). Although internal alkynes also hydrozirconated, the resulting internal vinyl zirconium complexes transmetallated only poorly to palladium(0). However, as was the case with alkyl cuprates, addition of zinc chloride, to produce the vinyl zinc species which transmetallates more efficiently, solved the problem (equation 5.47) [45]. As is usually the case, the stereochemistry of both olefinic partners is maintained. Vinylalananes, from addition of diisobutylaluminum hydride to alkynes, undergo similar coupling reactions with similar yields and specificities [46].

THPOCH$_2$C≡CH —$Cp_2Zr(H)Cl$→ THPOCH$_2$ [Zr]

Br, CO_2Me; L_2PdCl_2/DIBAL, 4 h, 50° → THPOCH$_2$ CO_2Me 70% (5.45)

RC≡CH —$Cp_2Zr(H)Cl$→ R [Zr] —Pd(0) [from $PdCl_2(PhCN)_2$], HN, O, I, N, R', THF→ O, HN, R, O, N, R' 70-90% (5.46)

—$Cp_2Zr(H)Cl$→ [Zr]

—1) $ZnCl_2$; 2) L_4Pd, 25°; MeO_2C, Br→ CO_2Me 85% (5.47)

In a similar manner, vinyl boranes are readily available from the hydroboration of alkynes, and a vast literature with many complex examples is extant. However, transmetallation from boron to palladium was unsuccessful until the importance of generating a stable boron species in the transmetallation step was appreciated. The way to do this was to produce borates in the exchange, by using catechol borane as the hydroborating agent, and carrying out the transmetallation in the presence of alkoxide.

In this way, vinyl boranes coupled efficiently to both vinyl halides (equation 5.48) [47] and aryl halides (equation 5.49) [48], and also β-haloenones (equation 5.50) [49]. It should be noted that this process is not restricted to simple substrates, as demonstrated by the spectacular coupling shown in equation 5.51 [50].

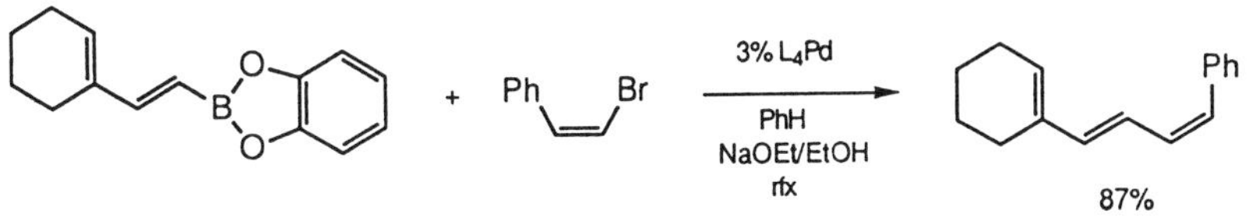

3% L_4Pd; PhH, NaOEt/EtOH, rfx; 87% (5.48)

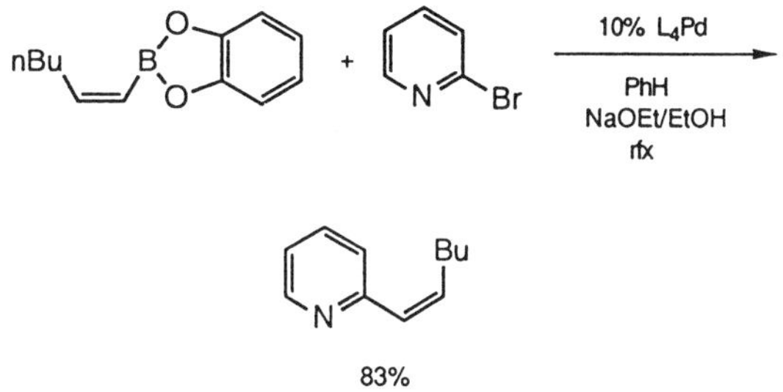

10% L_4Pd; PhH, NaOEt/EtOH, rfx; 83% (5.49)

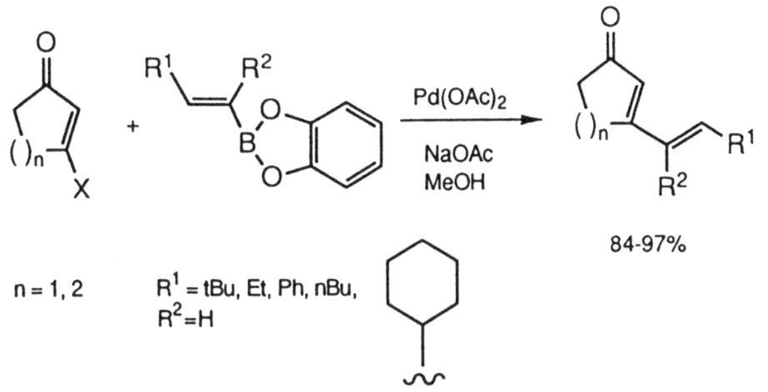

$Pd(OAc)_2$; NaOAc, MeOH; 84-97% (5.50)

n = 1, 2 R^1 = tBu, Et, Ph, nBu, R^2=H

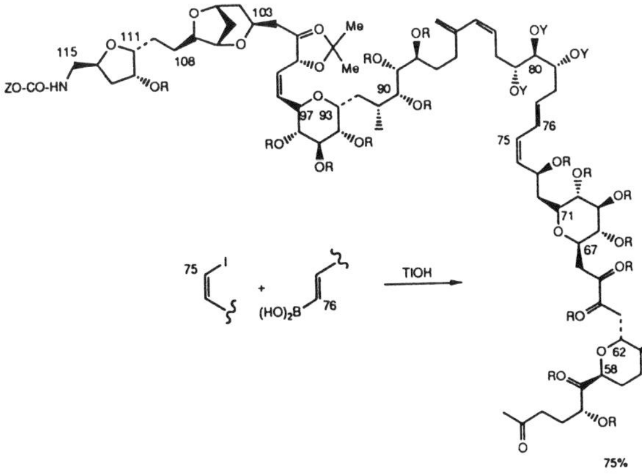

TlOH; 75% (5.51)

The utility of transmetallation from boron has been further expanded by combining it with ligand-directed *ortho*-lithiation of aromatics. In this way a variety of aryl–aryl and aryl–vinyl couplings have been achieved (equations 5.52 [51], 5.53 [52], 5.54 [53] and 5.55 [54]).

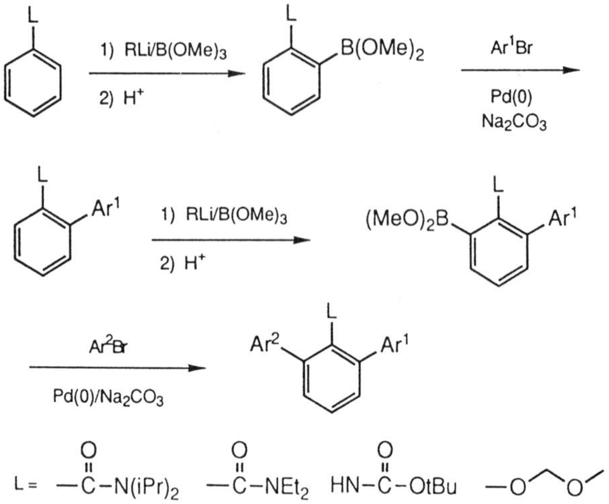

(5.52)

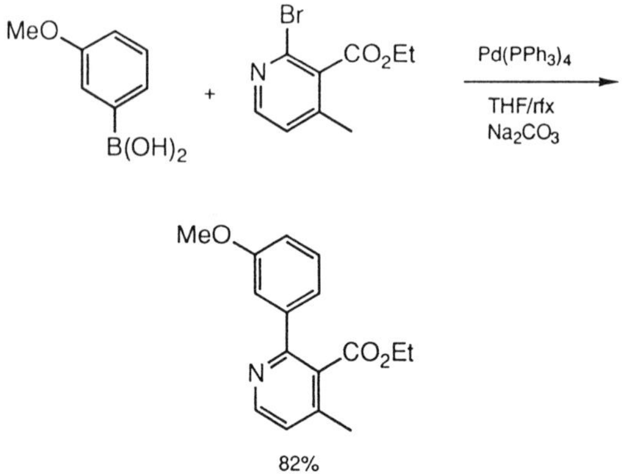

(5.53)

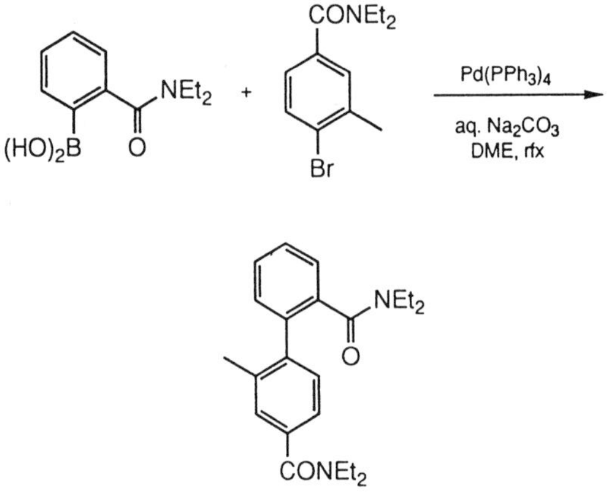

(5.54)

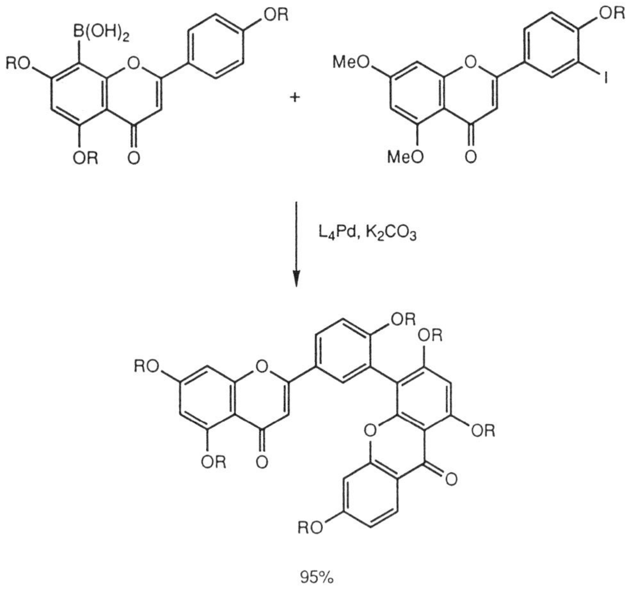

L_4Pd, K_2CO_3 (5.55)

95%

5.4.1.4 Transmetallation from Sn and Si

Transmetallation from tin to palladium is one of the most highly developed and extensively utilized processes in organopalladium chemistry [55]. The requisite organotin reagents containing a variety of useful functional groups are easily prepared, and readily stored and handled. The rate of transfer of organic groups from tin to palladium is $RC{\equiv}C > RCH{=}CH > Ar > RCH{=}CHCH_2 \approx ArCH_2 \gg C_nH_{2n+1}$, making it feasible to use trimethyl- or tributylstannanes containing the organic group to be transferred, and only transfer that single group, producing R_3SnX which is not reactive in the process. The reactions are run by adding the palladium catalyst to a solution of the substrate and the organotin reagent in a polar solvent such as DMF or THF. Coupling with vinyltin reagents occurs with retention of olefin geometry, while allyltins transfer with allylic transposition (yields and rate increases with addition of *ca* 10% $ZnCl_2$). With benzyltin reagents, inversion of the benzyl carbon is observed.

A very wide range of organic electrophiles are alkylated via oxidative addition–transmetallation from tin. The catalyst precursor of choice with acid chlorides as substrates is $(PhCH_2)Pd(PPh_3)_2Cl$ [which is reduced to the catalytically active Pd(0) species by transmetallation–reductive elimination] in HMPA or chloroform. The reaction proceeds smoothly at 65 °C, and tolerates functionality (equations 5.56 [56] and 5.57 [57]).

Vinyl [58] and aryl [59] iodides and bromides undergo coupling with a very wide range of organotin reagents at ambient temperature in DMF solvent. The catalyst precursor of choice are 1–2 mol% $PdCl_2(MeCN)_2$ or $PdCl_2(PPh_3)_2$. Specific examples abound, and more appear daily. Representative examples are shown in equations 5.58, 5.59, 5.60 [60] and 5.61 [61].

cat. $BnPd(PPh_3)_2Cl$, HMPA, 65° — Me_4Sn — 82% (5.56)

L_4Pd, $CHCl_3$, 65° — R_3SiO, Cl, Bu_3Sn, CO_2Bn — 71% (5.57)

$PdCl_2(MeCN)_2$, DMF 25° — R, X, R', SnR''_3 — 60-80% (5.58)

Ph, I, Bu, Bu_3Sn, Me_3Sn, TMS, OH

$PdCl_2(MeCN)_2$, THF or DMF — R, I, Me_3Sn—≡—R', R' (5.59)

Bu, I, Ph, O, THPO

R' = H, TMS, nBu, TMS-OCH_2, Ph, Ph O, AcO

L_4Pd — R, Br, NHTs, SnR_3 (5.60)

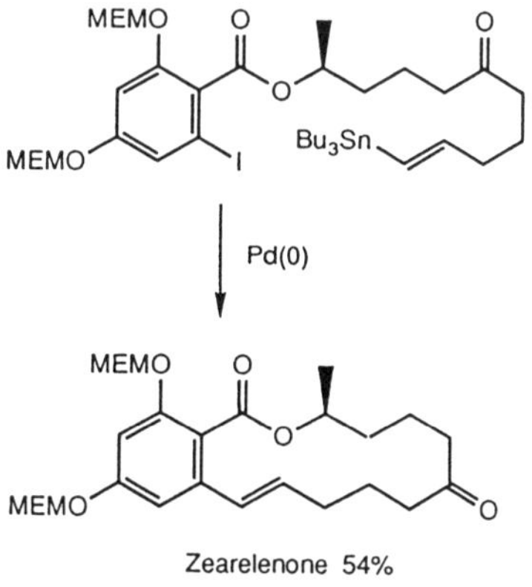

MEMO O O
MEMO I Bu$_3$Sn
Pd(0)
MEMO O
MEMO O
Zearelenone 54%

(5.61)

The utility of these couplings of vinyl and aryl halides to organotin reagents has been widened by the development of carbonylative coupling procedures (equation 5.62) wherein CO is inserted in the initial oxidative adduct prior to coupling [62] (Figure 5.8, right-hand portion). This reaction proceeds under very mild conditions (THF, 45–50 °C, 15–50 psi CO), and the catalysts of choice are $PdCl_2(PPh_3)_2$ or $(PhCH_2)Pd(Cl)(PPh_3)_2$.

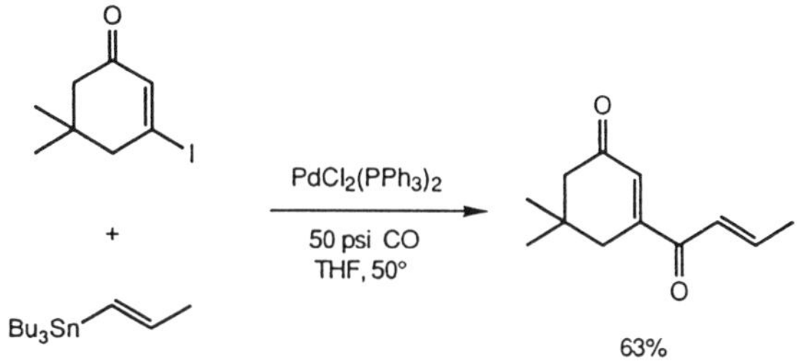

O
I
+
Bu$_3$Sn
$PdCl_2(PPh_3)_2$
50 psi CO
THF, 50°
O
O
63%

(5.62)

The discovery that vinyl triflates and aryl triflates [63] were viable coupling partners in this process if lithium chloride was added to provide a stable tin by-product (R_3SnCl vs R_3SnOTf) from the transmetallation step was a major advance. Vinyl triflates can be regioselectively generated from ketones [64], making them vinylic equivalents. Because carbonyl compounds are central to organic synthesis, highly functionalized vinyl triflates are readily available, making the triflate–tin coupling process exceptionally useful with complex substrates. The reaction proceeds under mild conditions, with $Pd(PPh_3)_4$ being the catalyst of choice, and polar solvents such as THF or DMF, which dissolve both the substrate and the required 3 equivalents of LiCl, being most efficient. Steric hindrance of the triflate has little effect, but the reaction slows if both partners are hindered. Examples of the use of this reaction in synthesis are shown in equations 5.63 [65], 5.64 [66] and 5.65 [67]. As expected, aryl [68] and vinyl triflates also participate in efficient carbonylative coupling reactions. Useful application of this methodology is seen in equations 5.66 [60] and 5.67 [70]. Vinyl [71] and aryl [72] fluorosulfonates also participate in these palladium-catalyzed coupling reactions.

F, TfO, F, O, O, OEt, N; + O, $SnBu_3$ — 2% L_2PdCl_2; 3 eq LiCl; THF, rfx → O, F, O, O, OEt, N, F; 88% (5.63)

N, O, N, O, OTf; + CO_2Et, Bu_3Sn — $Pd(PPh_3)_4$; 3 eq LiCl; THF → N, O, N, O, CO_2Et; 78% (5.64)

Ph, O, H, N, S, N, O, OTf, CO_2Bn + $R'SnBu_3$ — $Pd(dba)_2$; $ZnCl_2$; $(2\text{-furyl})_3P$; THF → (5.65)

Ph, O, H, N, S, N, O, R', CO_2Bn; 50-79%

R' = pMeOPh, H, OEt

OTf; + Me_3Sn, $SiMe_3$ — $Pd(PPh_3)_4$; CO, LiCl, THF → O, TMS; 87% (5.66)

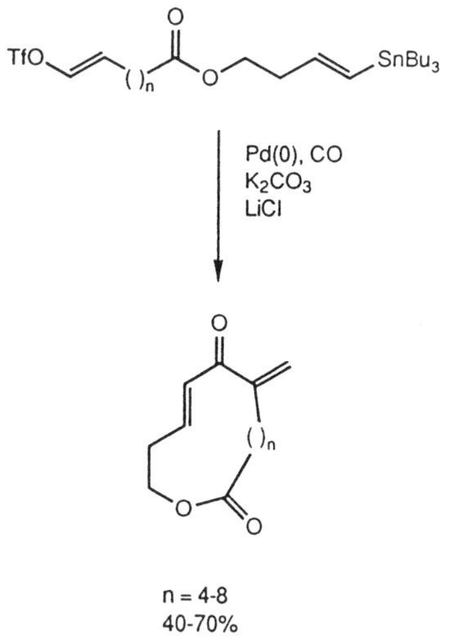

TfO ... ( )n ... O ... SnBu$_3$

Pd(0), CO
K_2CO_3
LiCl

(5.67)

n = 4-8
40-70%

In contrast to stannanes, transmetallation from silicon to palladium has only recently been developed, and to date has found little application in complex organic synthesis. Again, the problem is to provide activation of the silicon starting material and stabilization of the silicon product of the transmetallation step. This is achieved by providing a source of fluoride [73], most easily in the form of the soluble tris(diethylamino)sulfonium difluorotrimethyl silicate (TASF) [74] (equation 5.68).

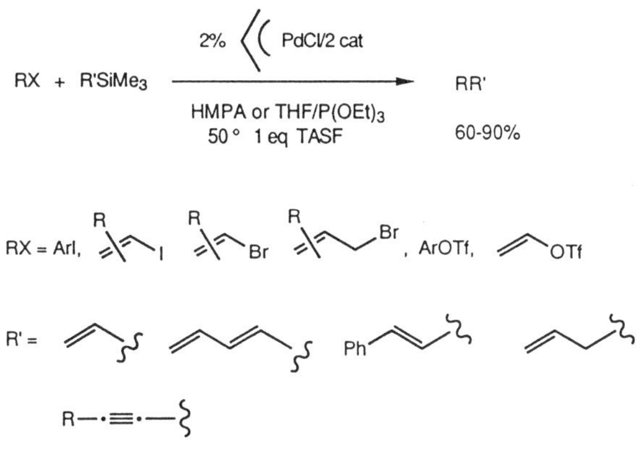

$$RX + R'SiMe_3 \xrightarrow[\text{HMPA or THF/P(OEt)}_3 \;\; 50^\circ \;\; 1\text{ eq TASF}]{2\% \text{ PdCl/2 cat}} RR' \quad (60\text{-}90\%) \tag{5.68}$$

RX = ArI, ..., ArOTf, ...OTf

R' = ..., Ph..., R—≡—

5.4.1.4 Transmetallation from Hg and Tl

Arylmercury and -thallium compounds are available by direct metallation of aromatic compounds and participate in oxidative addition–transmetallation procedures as above. The toxicity of these metals and the difficulties in removing traces of them from products have limited their use. However, they have been utilized to a greater extent in insertion chemistry, which will be discussed below.

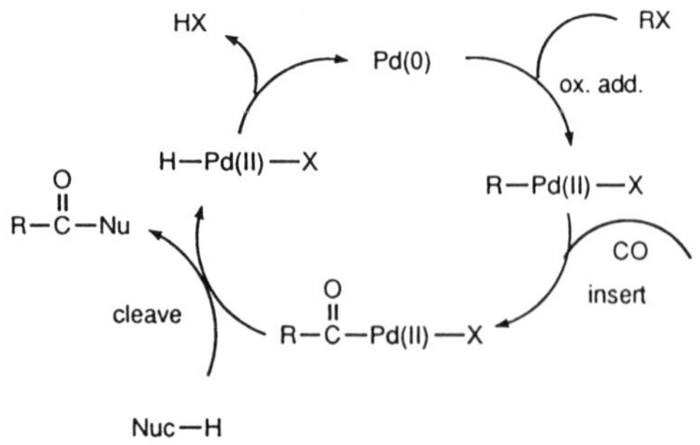

Figure 5.9

5.4.2 *PALLADIUM(0)-CATALYZED INSERTION PROCESSES*

5.4.2.1 Carbonylation

Carbon monoxide inserts very readily into palladium–carbon σ-bonds ($< -20\ ^\circ C$, < 1 atm CO), regardless of how they were formed [nucleophilic attack on olefins (Figure 5.3), *ortho*–palladation (equation 5.37), transmetallation (Figure 5.8) or oxidative addition (this section)], and the resulting σ-acyl complexes are readily cleaved by a variety of nucleophiles, making palladium-catalyzed carbonylation of organic substrates the most versatile of reactions.

By far the most commonly used palladium-catalyzed carbonylation process is that involving oxidative addition (Figure 5.9). Since oxidative addition is the most difficult step, the scope of the overall process is limited by the general scope of oxidative addition discussed above, e.g. β-hydrogens are not tolerated, aryl, vinyl, benzyl and allyl halides ($I > Br \gg Cl$) and triflates are reactive. The reaction is widely tolerant of solvents and catalysts, although some ligand to stabilize Pd(0) is usually necessary to prevent the precipitation of metallic Pd, and the nucleophile (usually methanol, to produce esters, but others such as amines also work) is normally used in excess [75]. Again, structural complexity and functionality are tolerated in the substrate, and this is probably the method of choice for the conversion of organic halides into carboxylic acid derivatives. With the development of a procedure to involve aryl [76] and vinyl triflates in palladium(0)-catalyzed oxidative additions, the scope of carbonylation was greatly broadened, since phenols, and ketones, respectively, can be carbonylated via their triflates. Because this is a recent development, only scattered applications in total synthesis have so far appeared (equations 5.69 [77] and 5.70 [78]).

Oxidative addition–carbonylation is used most frequently in an intramolecular manner, forming lactones or lactams. It works well with aryl halides (equations 5.71 [79] and 5.72 [80]) and vinyl halides (equations 5.73 [79] and 5.74 [81]). Again, this is a very general process and should be applicable to virtually any system having appropriately situated nucleophiles to trap a σ-acylpalladium complex, and compatible with the requirements of oxidative addition enunciated above. That it is usable on a

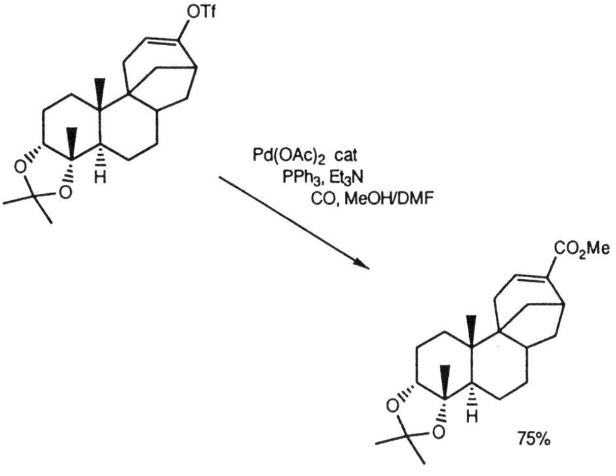

OTf
O H O
$Pd(OAc)_2$ cat
PPh_3, Et_3N
CO, MeOH/DMF
CO_2Me
O H O
75%
(5.69)

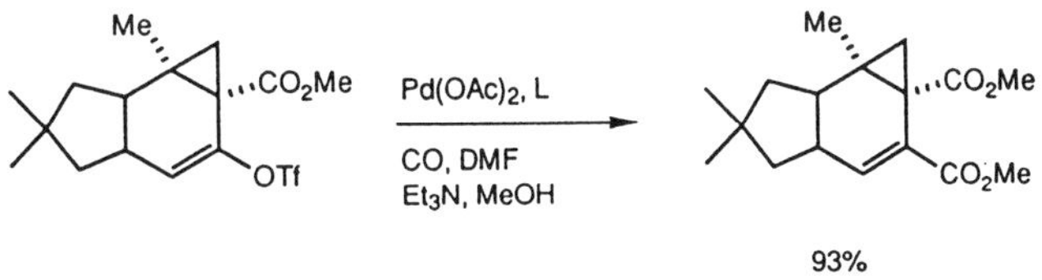

Me CO_2Me OTf
$Pd(OAc)_2$, L
CO, DMF
Et_3N, MeOH
Me CO_2Me CO_2Me
93%
(5.70)

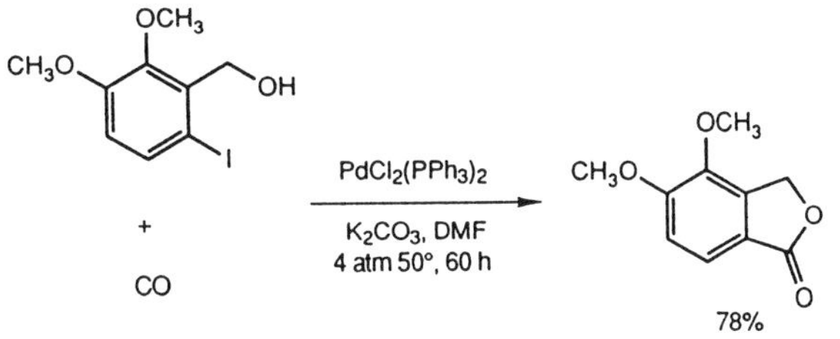

OCH_3
CH_3O OH
I
+
CO
$PdCl_2(PPh_3)_2$
K_2CO_3, DMF
4 atm 50°, 60 h
OCH_3
CH_3O O
O
78%
(5.71)

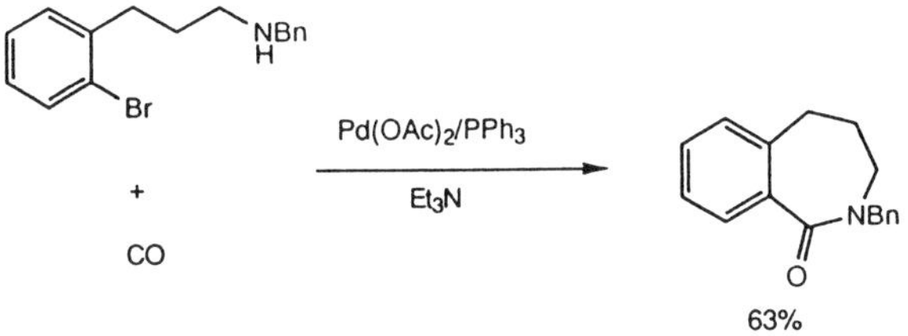

NBn
H
Br
+
CO
$Pd(OAc)_2/PPh_3$
Et_3N
NBn
O
63%
(5.72)

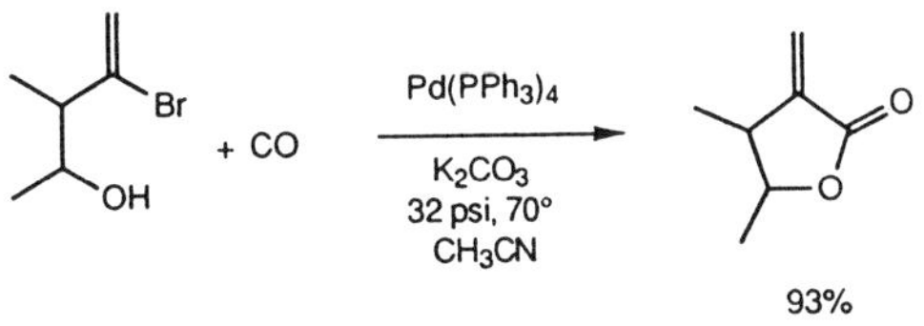

Br
OH
+ CO
$Pd(PPh_3)_4$
K_2CO_3
32 psi, 70°
CH_3CN
O
O
93%
(5.73)

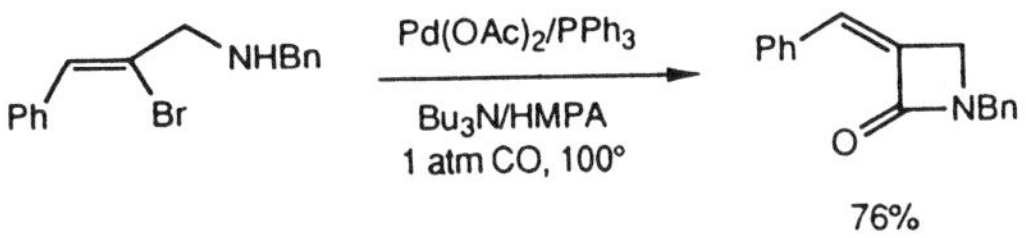

NHBn
Ph Br
$Pd(OAc)_2/PPh_3$
Bu_3N/HMPA
1 atm CO, 100°
Ph
NBn
O
76%
(5.74)

large scale is seen in equations 5.75 [82]. Under appropriate conditions, even enolate oxygens can trap the acylpalladium complex, giving isocoumarins (equation 5.76) [83].

By combining oxidative addition–CO insertion with transmetallation, the formylation of aryl and vinyl halides can be achieved (equation 5.77) [84]. In this case, as with carbonylative coupling, moderate pressures (50–90 psi) of carbon monoxide are required to ensure that CO insertion is faster than transmetallation, so that formylation, not simple reduction, occurs.

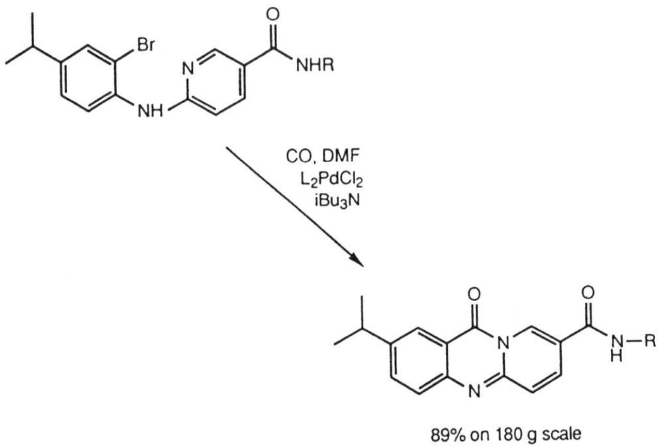

CO, DMF
L_2PdCl_2
iBu_3N

89% on 180 g scale

(5.75)

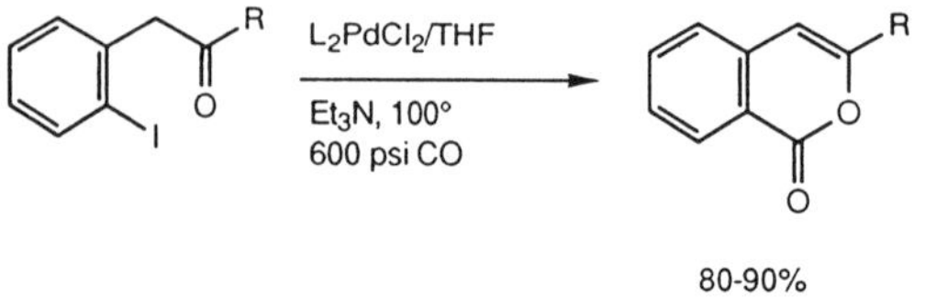

L_2PdCl_2/THF
Et_3N, 100°
600 psi CO

80-90%

(5.76)

ArI → ArCHO

Pd(0)/CO/Bu_3SnH

X = I, Br, OTf

high yields
(28 cases)

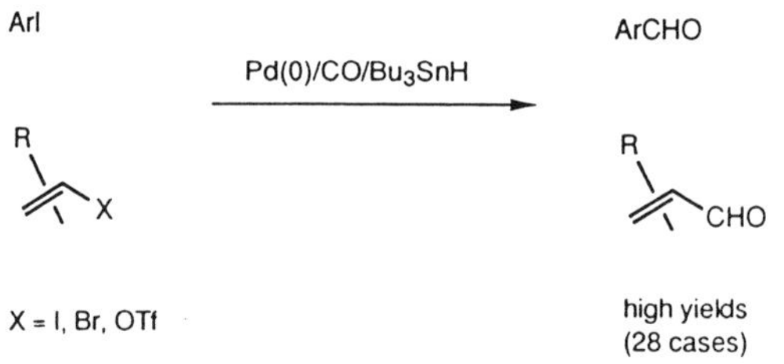

(5.77)

5.4.2.2 Palladium(0)-Catalyzed Olefin Insertion Processes

One of the most useful of Pd(0)-catalyzed reactions is the oxidative additon–olefin insertion reaction (the Heck reaction) [85] (Figure 5.10). In this case, it is the insertion step which is the most difficult, and most limitations are imposed by structural features of the olefin. The limitations on the oxidative addition step are the same, and all of the substrates discussed above are viable candidates for olefin insertion chemistry. Again, β-hydrogens are not tolerated in the substrate. A wide range of olefins undergo this

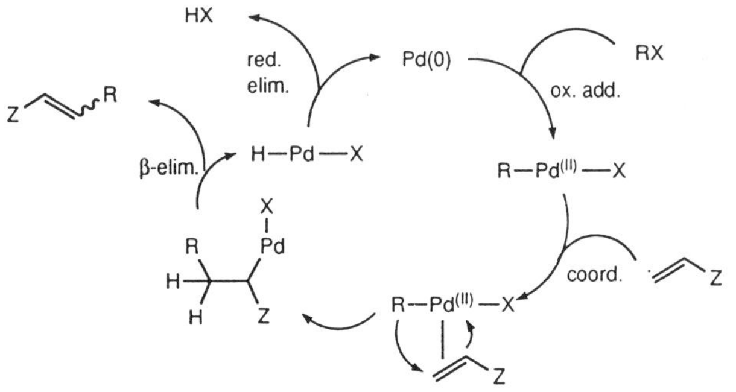

Figure 5.10

reaction, including unfunctionalized alkenes, electron-deficient alkenes such as acrylates and conjugated enones and electron-rich olefins such as enol ethers and enamides. With unfunctionalized alkenes and electron-poor alkenes, insertion occurs to place the R group at the less-substituted position of the olefin, indicating that the regioselectivity of insertion is controlled by steric, not electronic, effects with these substrates. It is a *cis* insertion, with the alkyl group being delivered to the less substituted carbon from the face of the olefin occupied by the metal. As might be expected, substitution on the olefin, particularly in the β-position, inhibits the reaction and, for intermolecular processes, β-disubstituted olefins are unreactive.

With electron-rich olefins, the regioselectivity is more complex [86]. With cyclic enol ethers and enamides, alkylation always occurs α- to the heteroatom. However, with acyclic systems, mixtures of α- and β-alkylation products are observed, the ratio of which is subject to electronic control. Electron-rich aryl halides lead to α-arylation, whereas electron-poor aryl halides lead to β-arylation of the olefin.

The classical conditions for the Heck reaction involve heating a mixture of the olefin, the halide, a catalytic amount of $Pd(OAc)_2$ and several equivalents of triethylamine in acetonitrile until the reaction is complete. {Triethylamine rapidly reduces $Pd(OAc)_2$ to the catalytically active Pd(0) state [87].} When aryl iodides are used, addition of phosphine is not required and, in fact, inhibits the reaction. (Palladium on carbon can be used to catalyze reactions of aryl iodides [88].) With bromides, added phosphine is required for the oxidative addition step. Hindered phosphines such as $(o\text{-tolyl})_3P$ are used, since triphenylphosphine reacts with the halide substrate to form quaternary phosphonium salts. However, many of these reactions can occur under much milder conditions, proceeding at room temperature when DMF is used as solvent and potassium carbonate–tetrabutylammonium chloride is added [89]. Initially, the quaternary ammonium salt was thought to promote these reactions by some unspecified solid–liquid phase-transfer process. More recent studies [90] have indicated that halide ions stabilize palladium(0) complexes during these catalytic reactions, and effect the activity of the catalyst. The quaternary ammonium halide salts provide high concentrations of soluble halides for these reactions.

The Heck reaction is one of the most widely used palladium-catalyzed processes, and a myriad of applications have been reported. Functional group compatibility is excellent. In addition to all the standard monofunctional halide–olefin couplings which abound in the literature, many complex systems have been studied. A few representative examples as shown in equations 5.78 [91], 5.79 [92] and 5.80 [93].

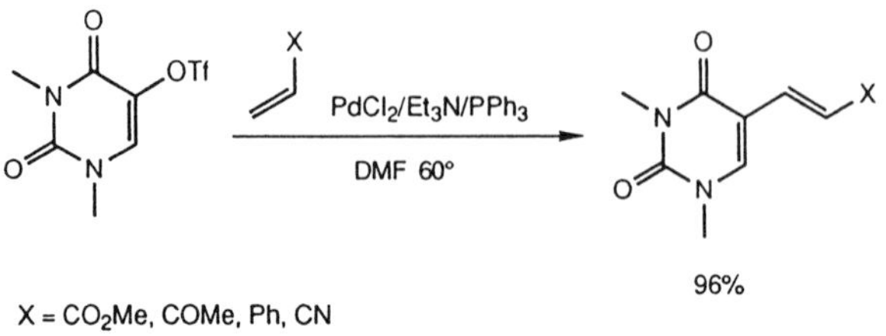

$PdCl_2/Et_3N/PPh_3$, DMF 60° — 96%

$X = CO_2Me$, COMe, Ph, CN

(5.78)

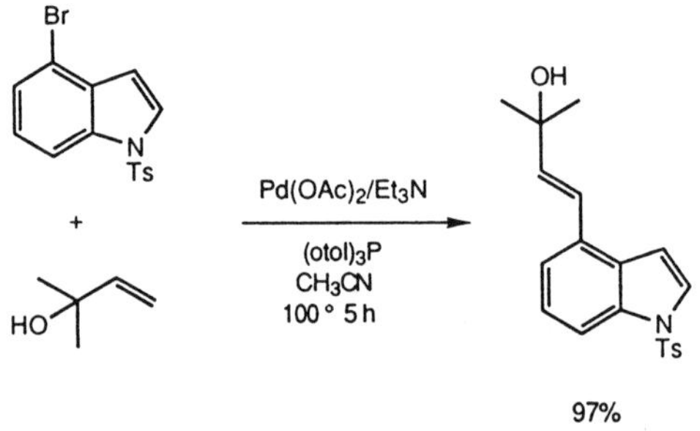

$Pd(OAc)_2/Et_3N$, $(otol)_3P$, CH_3CN, 100° 5 h — 97%

(5.79)

0.1 eq $Pd(OAc)_2$, Bu_3N, DMF, NaOAc — 85%

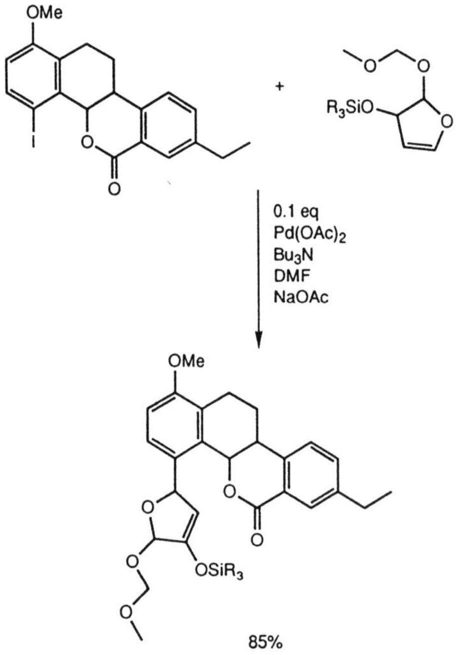

(5.80)

Intramolecular versions also abound, and this is an extremely efficient way to make both carbocyclic and heterocyclic rings. The intramolecular version is often more efficient than the intermolecular version, but regioselectivity, although usually high, is less predictable and, depending on ring size, insertion (alkylation) at the more substituted position is often observed (equations 5.81 [94], 5.82 [94] and 5.83 [95]).

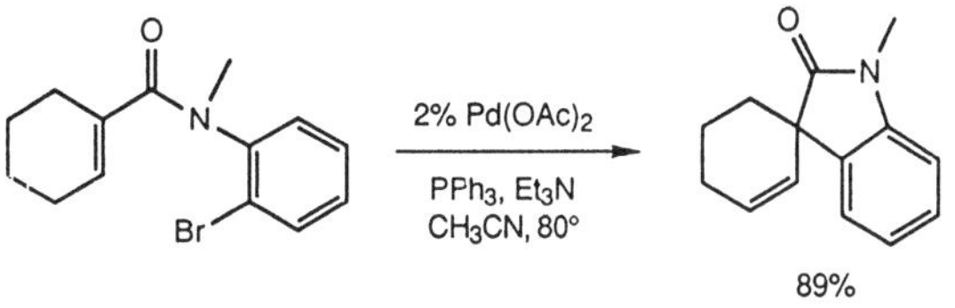
O
N
Br
2% Pd(OAc)2
PPh3, Et3N
CH3CN, 80°
O
N
89%

(5.81)

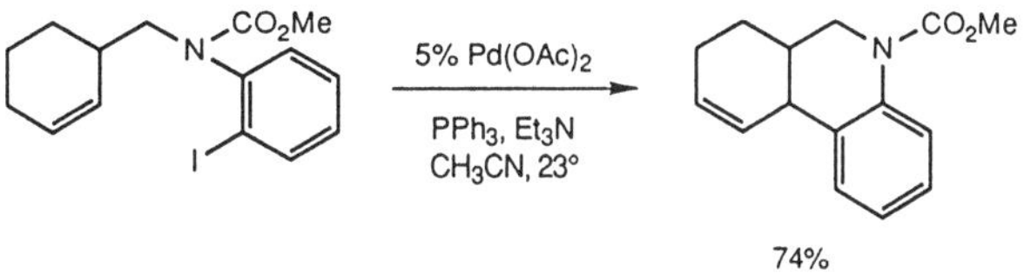
N
CO2Me
I
5% Pd(OAc)2
PPh3, Et3N
CH3CN, 23°
N
CO2Me
74%

(5.82)

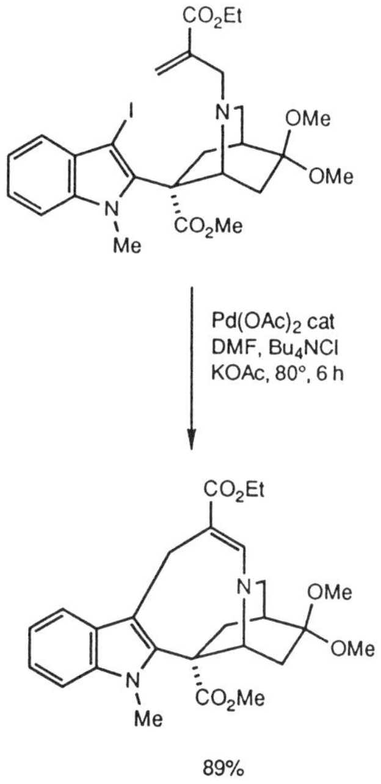
CO2Et
I
N
OMe
OMe
N
Me
CO2Me
Pd(OAc)2 cat
DMF, Bu4NCl
KOAc, 80°, 6 h
CO2Et
N
OMe
OMe
N
Me
CO2Me
89%

(5.83)

5.4.2.3 Palladium(0)-Catalyzed Multiple (Cascade) Insertion Process

As presented above, σ-aryl- and vinylpalladium(II) complexes have a very rich reaction chemistry, undergoing insertions, transmetallations, reductive eliminations and β-hydrogen eliminations. In principle, these processes can be combined in any order, and since many of them generate another σ-bonded organopalladium complex, a sequence of these processes, forming several bonds, could be envisaged. The only requirement is that the next step is more facile than reductive elimination or β-hydrogen elimination, which truncates the process. This concept is only now being developed, although forerunners can be seen in equations 5.10 and 5.15, and recent examples in equations 5.35 and 5.36. Thus, a sequence consisting of oxidative addition–CO insertion–olefin insertion–CO insertion and cleavage formed three carbon–carbon bonds per catalyst cycle (equation 5.84) [96]. This works because there are no β-hydrogens in the initial oxidative adduct, and because CO insertion is faster than olefin insertion for σ-alkylpalladium complexes, but slower for σ-acylpalladium complexes. The major trick here was to make the final CO insertion competitive with β-hydride elimination. This was achieved by using a high pressure (40 atm) of CO. This 'downhill' sequence of events is critical, as evidenced by failure when an alkyne insertion was required to precede the more facile CO insertion (equation 5.85) [97]. However, this same substrate was successful for alkyne and alkene insertion, since these processes are of comparable energy, and the requisite initial alkyne insertion is favored entropically (equation 5.86). When the olefin was also part of the same molecule, polycyclization was achieved (equations 5.87 and 5.88) [98]. The real power of this methodology can be seen by the examples in equations 5.89 [99], 5.90 [100] and 5.91 [101].

Again, any process in which σ-alkylpalladium(II) complexes participate can be used to either carry or truncate these 'cascade' processes. Thus, the reaction in equation 5.92

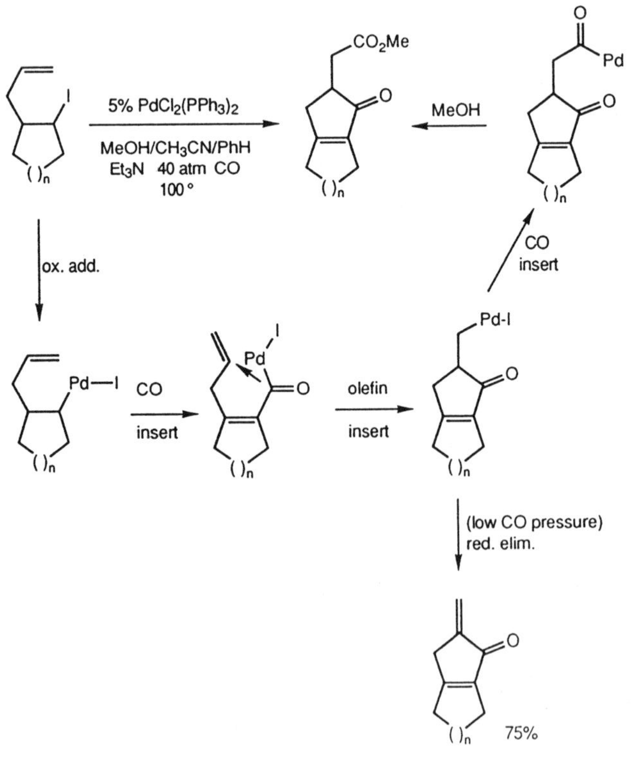

5% $PdCl_2(PPh_3)_2$; $MeOH/CH_3CN/PhH$, Et_3N, 40 atm CO, 100°; MeOH; ox. add.; CO insert; olefin insert; CO insert; (low CO pressure) red. elim.; 75%

(5.84)

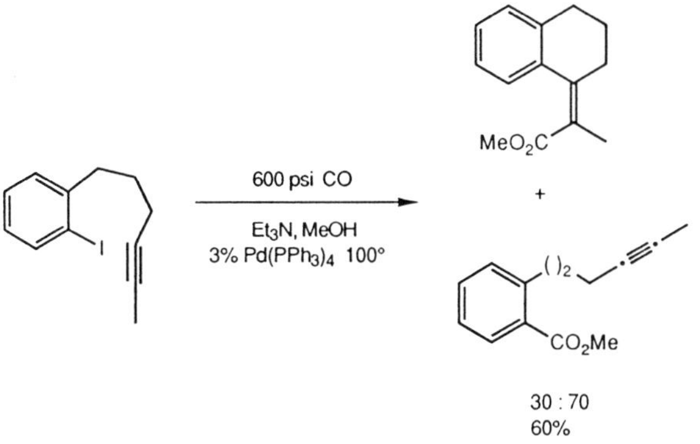

600 psi CO

Et_3N, MeOH
3% $Pd(PPh_3)_4$ 100°

MeO_2C

+

CO_2Me

30 : 70
60%

(5.85)

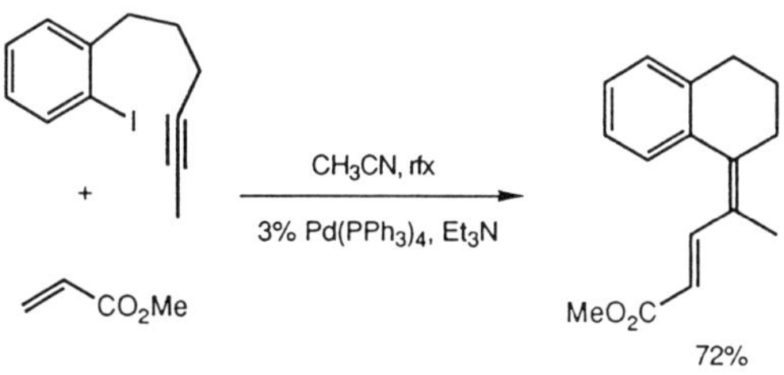

+

CO_2Me

CH_3CN, rfx

3% $Pd(PPh_3)_4$, Et_3N

MeO_2C

72%

(5.86)

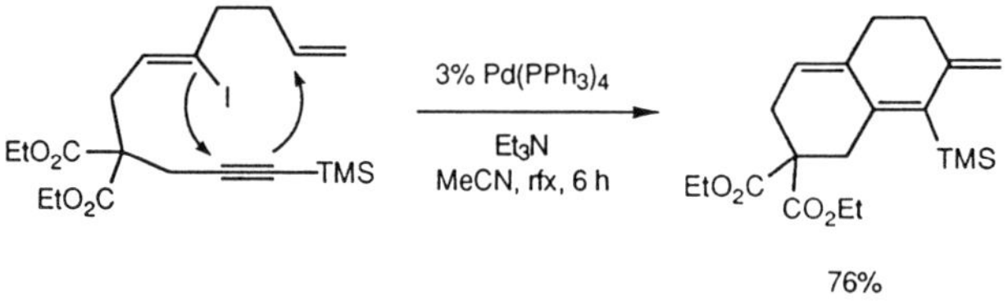

EtO_2C
EtO_2C
I
TMS

3% $Pd(PPh_3)_4$

Et_3N
MeCN, rfx, 6 h

TMS
EtO_2C
CO_2Et

76%

(5.87)

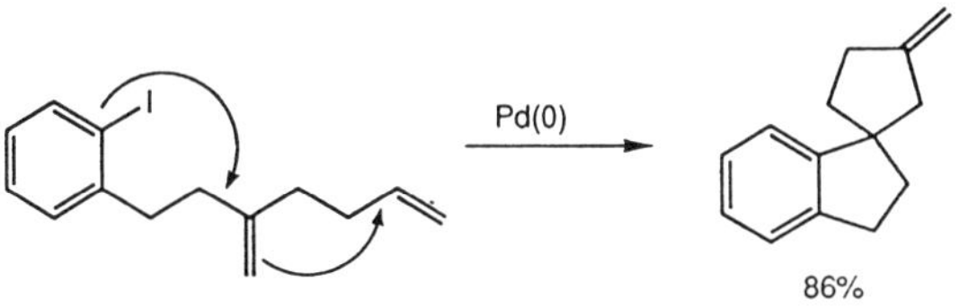

I

Pd(0)

86%

(5.88)

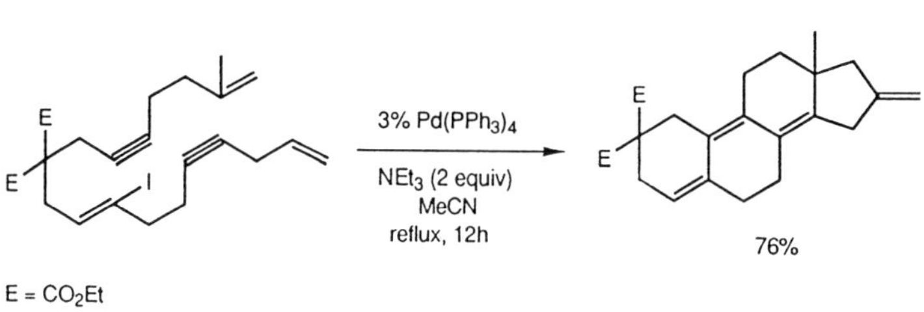

E
E
I

3% $Pd(PPh_3)_4$

NEt_3 (2 equiv)
MeCN
reflux, 12h

E
E

76%

(5.89)

E = CO_2Et

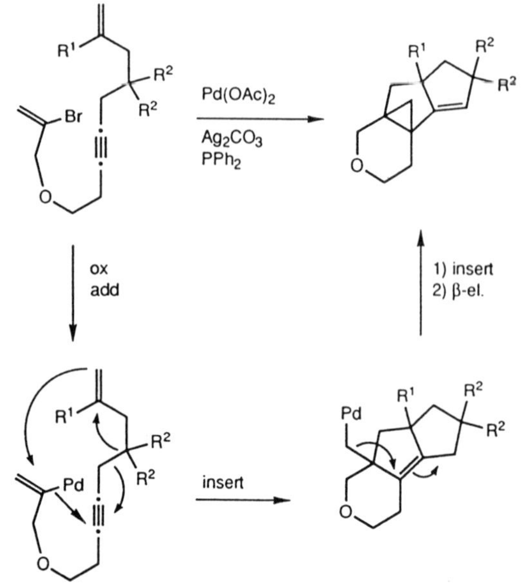

R¹ R² R² Br O Pd(OAc)$_2$ Ag$_2$CO$_3$ PPh$_2$ R¹ R² R² O

ox add 1) insert 2) β-el.

R¹ R² R² Pd O insert Pd R¹ R² R² O

(5.90)

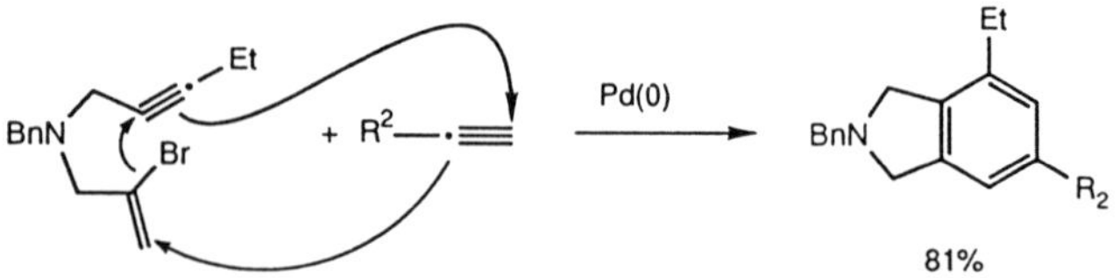

BnN Et Br + R²—≡ Pd(0) BnN Et R$_2$ 81%

(5.91)

R$_3$SiO Br OBn + R$_3$Sn—≡—OTBS L$_4$Pd R$_3$SiO OTBS OBn 86%

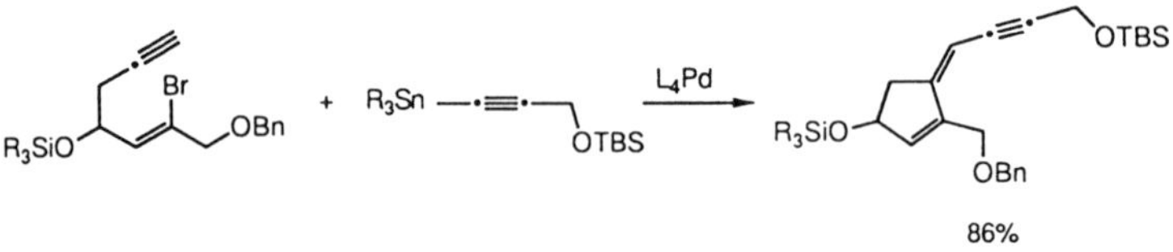

(5.92)

[102] is truncated by a transmetallation step, whereas that in equation 5.93 [103] is completed by a nucleophilic attack. There are endless variations on these themes, and it is likely that all of them will be tried.

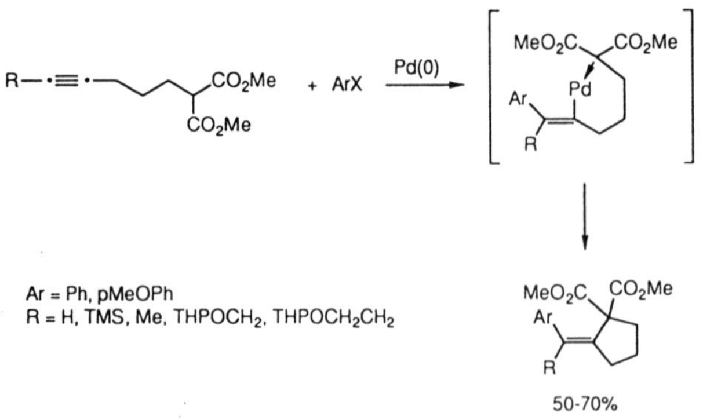

R—≡—CO$_2$Me CO$_2$Me + ArX Pd(0) [MeO$_2$C CO$_2$Me Pd Ar R]

Ar = Ph, pMeOPh
R = H, TMS, Me, THPOCH$_2$, THPOCH$_2$CH$_2$

MeO$_2$C CO$_2$Me Ar R 50-70%

(5.93)

5.4.3 PALLADIUM(0)-CATALYZED REACTION OF ALLYLIC COMPOUNDS

5.4.3.1 General Features

A very wide range of allylic substrates undergo palladium(0)-catalyzed reactions with nucleophiles, Main Group organometallics and small molecules such as olefins and carbon monoxide, resulting in a number of synthetically useful processes. Despite the disparity amongst substrates, the reactions share many common features, which are summarized in Figure 5.11.

The first step in all of these reactions is the oxidative addition of the allylic C—X system to Pd(0) (a), a step that goes with clean inversion of configuration at the allylic carbon and produces a σ-allylpalladium(II) complex. These are rarely detected (although often invoked) and rapidly collapse to the more stable π-allylpalladium(II) complex (b). These can be isolated, and are usually yellow, crystalline, air-stable solids. However, for catalysis, isolation is undesired and reactions are usually carried out under conditions which favor consumption of the π-allyl complex at a rate at least commensurate with its formation.

In the absence of added ligands, π-allylpalladium(II) complexes are relatively inert to nucleophilic attack. However, in the presence of phosphines, normally already there from the Pd(0) starting complex [most commonly $Pd(PPh_3)_4$], they undergo reaction with a wide variety of nucleophiles, most commonly amines or stabilized carbanions.

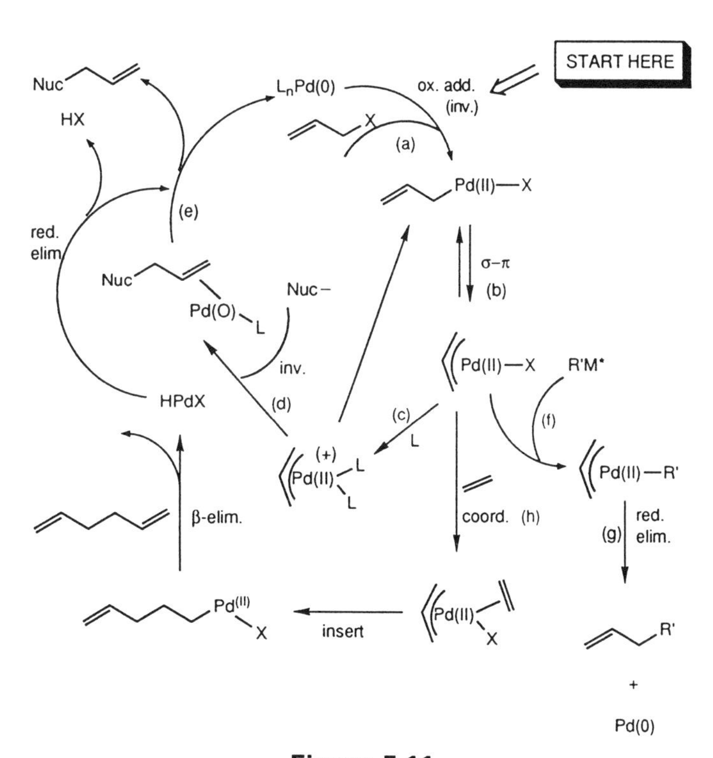

Figure 5.11

The nucleophile attacks from the face opposite the metal, with clean inversion, resulting in net retention (two inversions) for the overall process (d). With unsymmetrical allyl complexes attack usually occurs at the less substituted position, but the regioselectivity is strongly dependent on the structural features of the substrate and the conditions of the reaction. Nucleophilic attack results in a formal two-electron reduction of the metal, giving a Pd(0)–olefin complex (d) which exchanges the product olefin for ligand, regenerating the Pd(0) catalyst (e).

If, instead of a nucleophile, the process is carried out in the presence of a Main Group organometallic compound (most commonly a tin reagent), transmetallation occurs (f). This process is suppressed by added donor ligands and $PdCl_2(CH_3CN)_2$ is the most commonly used catalyst precursor. Reductive elimination of the π-allyl-σ-alkyl-palladium(II) complex results in coupling (g). Since the alkyl group is delivered from the same face of the metal, retention of configuration is observed in this process, leading to overall inversion for the total process.

Finally, π-allylpalladium(II) complexes (perhaps via their σ-form) undergo both CO and olefin insertion (h), making the carbonylation and olefin chemistry discussed above also accessible to allylic substrates.

π-Allylpalladium complexes are also produced by the addition of nucleophiles to Pd(II)–diene complexes (Figure 5.5). Once formed, these also undergo all of the reactions in Figure 5.11.

5.4.3.2 Palladium(0)-Catalyzed Allylic Alkylation

Although allylic acetates are the most commonly used substrate for this process, a very wide range of allylic substrates are subject to Pd(0)-catalyzed allylic alkylation by stabilized carbanions (equation 5.94). The reactions are usually carried out by mixing the substrate with 1–2% of $Pd(PPh_3)_4$ catalyst in THF containing an additional 5–10 equivalents (based on catalyst, not substrate) of PPh_3, generating the anion in THF in a separate vessel, adding it to the substrate–catalyst mixture and heating at reflux for several hours [104]. The reaction usually proceeds with excellent yields. There are an enormous number of examples of this process in the literature, with a very wide range of highly functionalized substrates and carbanions, so the reliability of this methodology is very well established. A few representative current examples are shown in equations 5.95 [105], 5.96 [106], 5.97 [107] and 5.98 [108].

(-) Y X + Z $\xrightarrow{L_4Pd}$ Y X

Z = Br, Cl, OAc, -OCOR, OP(OEt)$_2$, O-S(O)-R, OPh, OH, N^+R_3, NO_2, SO_2Ph, CN, (5.94)

X, Y = CO_2R, COR, SO_2Ph, CN, NO_2

OMe MeO MeO OMe OCO_2Me + SO_2Ph (-) CO_2Me

$\xrightarrow[\text{THF}]{L_4Pd}$ OMe MeO MeO OMe SO_2Ph CO_2Me (5.95)

TMS Ph OCO_2Me + O CO_2Me (-) $\xrightarrow[\text{PBu}_3\text{, THF, 25 °}]{5\%\ Pd_2(dba)_3}$ TMS Ph CO_2Me O 83% (5.96)

O O Me H H H $\xrightarrow[\text{(-)CH(CO}_2\text{Me)}_2]{Pd(PPh_3)_4\ cat.}$ HO_2C Me H MeO_2C CO_2Me H Me 90% (5.97)

AcO N N NH_2 N N + (-) PhO_2S NO_2

(5.98)

$\xrightarrow{Pd_2(dba)_3}$ NO_2 PhO_2S N N NH_2 N N

Intramolecular versions of this process are particularly useful for forming carbocyclic rings [109], and have been used to make three-, four-, five-, six-, seven-, eight-, nine-, ten- and eleven-membered and macrocyclic rings successfully. Examples are shown in equations 5.99 [110], 5.100 [111], 5.101 [112] and 5.102 [113].

Pd(PPh3)4; 2 h, 65°; NaH, THF — 70% (5.99)

NaH, Pd(dba)2; PPh3, 65°, THF — 68% (5.100)

10% Pd(PPh3)4; 10% dppp; THF, rfx — 49% (5.101)

Pd(OAc)2; (iPrO)3P — 92% (5.102)

5.4.3.3 Palladium(0)-Catalyzed Allylic Amination

A variety of nitrogen nucleophiles attack π-allylpalladium complexes and function efficiently in the palladium(0)-catalyzed amination of all of the substrate classes discussed above. These nitrogen nucleophiles include primary and secondary amines (but *not* ammonia), amides, sulfonamides, azides and some nitrogen heterocycles. The nitrogen almost invariably ends up at the less hindered terminus of the allyl system, although that might not be the initial site of attack, since Pd(0) complexes catalyze rapid allylic rearrangement of allyl amines. As with alkylation, this is experimentally a straightforward and reliable reaction which proceeds under similar conditions. Examples are shown in equations 5.103 [114], 5.104 [115] and 5.105 [116]. Intramolecular versions of aminations to form heterocycles are common and, again, the range of amenable substrates is wide (equations 5.106 [117], 5.107 [118] and 5.108 [119]).

5.4.3.4 Other Nucleophiles and Reduction/Elimination/Deprotection

Although nucleophiles other than stabilized carbanions and amines should be able to participate in Pd(0)-catalyzed reactions of allylic compounds, in fact few others have been studied. Alcohols have been used intramolecularly (equation 5.109) [120] and

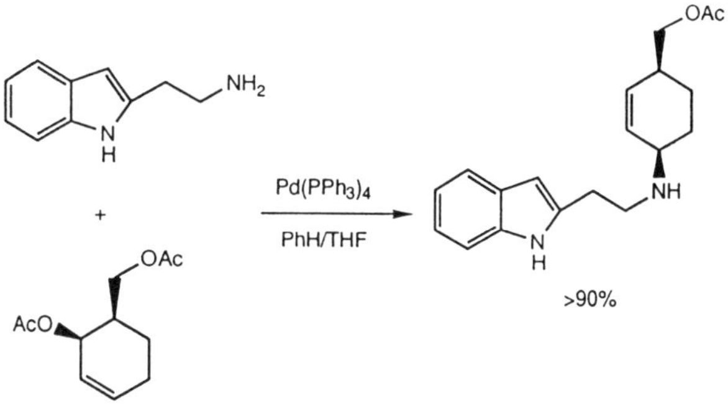

NH_2 + AcO, OAc $\xrightarrow[\text{PhH/THF}]{Pd(PPh_3)_4}$ OAc, NH, N H >90% (5.103)

+ NaN_3 $\xrightarrow[\text{Pd(dba)}_2\text{/2 PPh}_3\ 25°\ 10\ \text{min}]{\text{aq THF}}$ N_3, OH 75-85%

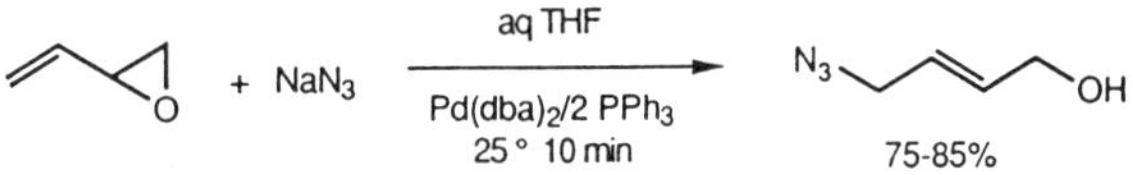

works with: (5.104)

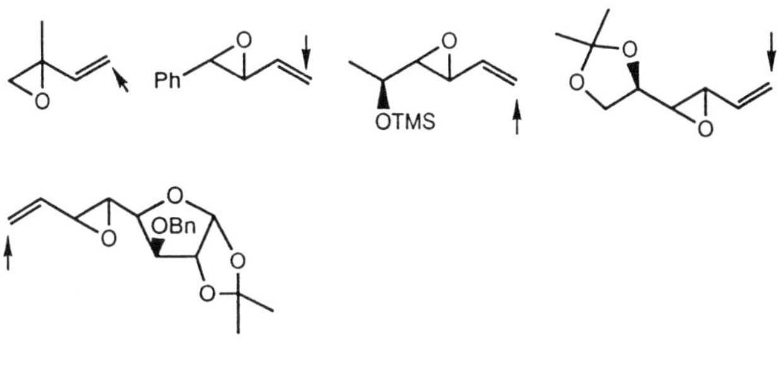

O Ph O OTMS O O O OBn O O O

OAc

MeO$_2$C CO$_2$Me

+

$(TsNH)^-$ Na^+

THF/DMSO

Pd(PPh$_3$)$_4$, 5 h
20-50 °

NHTs

MeO$_2$C CO$_2$Me

87%
(10 cases)

(5.105)

NH O

1.5% Pd$_2$(dba)$_3$•CHCl$_3$

12% P(OR)$_3$
H$_2$O/THF
65°

N OH

73%

(5.106)

O O HN AcO

5% Pd(PPh$_3$)$_4$

Et$_3$N/CH$_3$CN
70 ° 2 h

O O N

>90%

(5.107)

AcO O
H$_2$N N O
AcNH

10% Pd(PPh$_3$)$_4$
dppb, THF
70 ° high
dilution

O
HN N O
AcNH

89%

(5.108)

HO ( )$_n$ OH OAc

4% Pd$_2$(dba)$_3$•CHCl$_3$

PPh$_3$, THF, 25°, 3-6h

HO ( )$_n$ O

70-90%
6 cases

(5.109)

intermolecularly (equation 5.110) [121], but this area is either little studied or unsuccessfully studied and unreported. However, palladium(0) complexes catalyze several other useful reactions of allylic substrates.

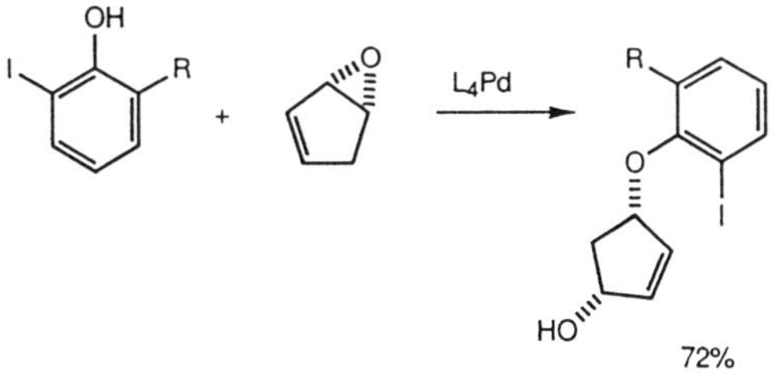

(5.110)

Treatment of allylic acetates or sulfones with Pd(0) catalysts in the presence of a hydride source (polymethylhydrosiloxane [122] is among the most efficient, although cyanoborohydride [123] and triethylborohydride [124] also work) leads to reduction to the alkene, to give a mixture of olefin isomers (equation 5.111). In the presence of a base, clean elimination to the diene is efficiently achieved (equation 5.112) [125]. Allyl groups have also been used as protecting groups for alcohols, carboxylic acids, and amines, because they are easily removed by Pd(0) catalysts via π-allyl chemistry (equations 5.113 [126], 5.114 [127] and 5.115 [128]). In these cases the allyl group is discarded with the regeneration of the catalyst.

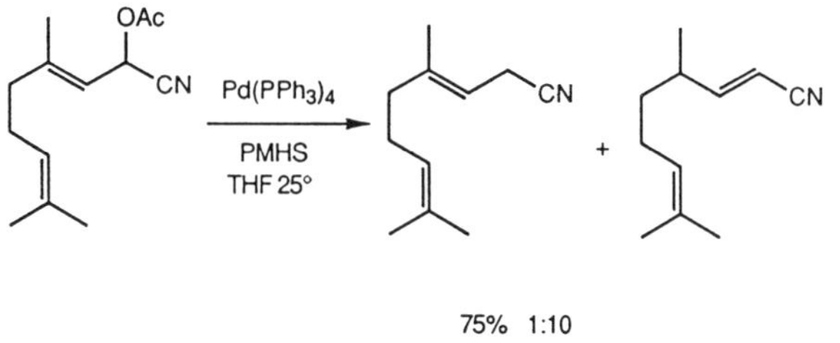

(5.111)

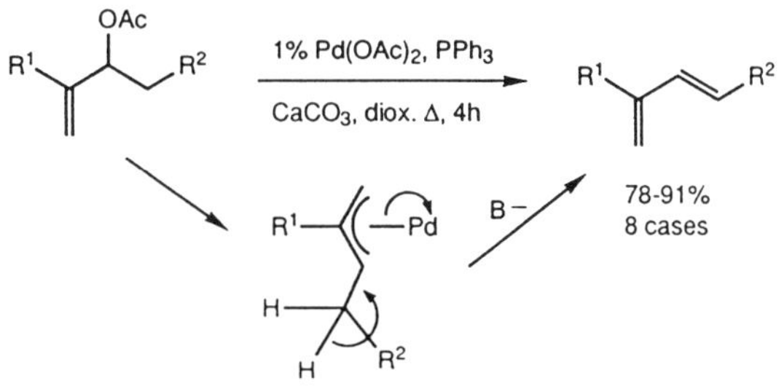

(5.112)

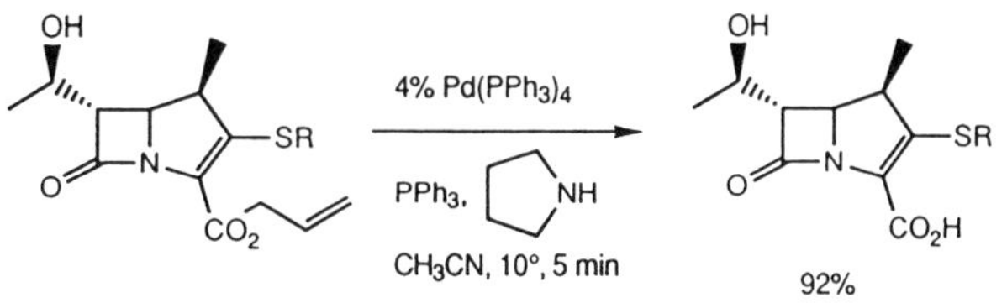

(5.113)

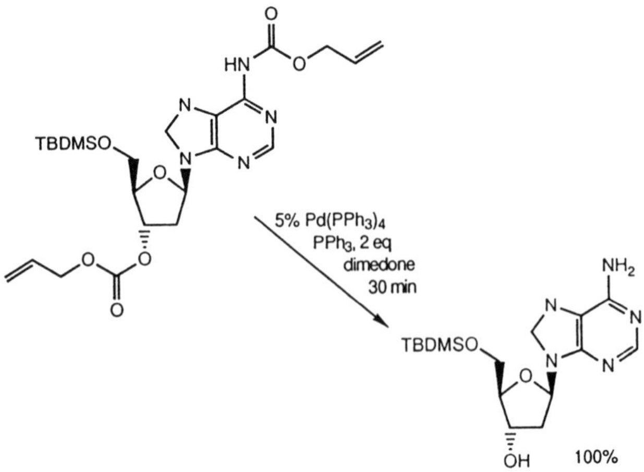

5% $Pd(PPh_3)_4$
PPh_3, 2 eq
dimedone
30 min

100%

(5.114)

$Pd(PPh_3)_4$, PPh_3
RCOOH, CH_2Cl_2, 25°
12 hr

63%

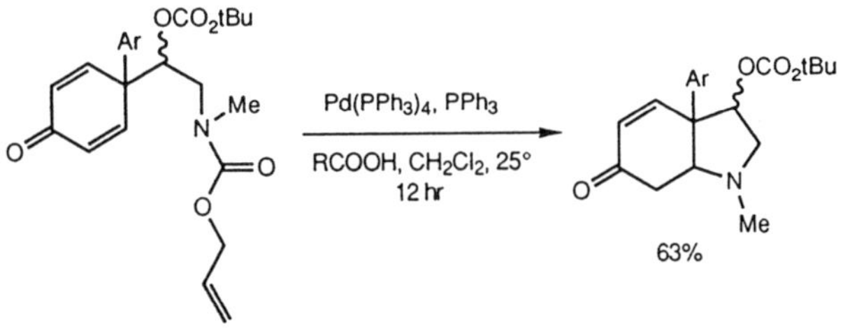

(5.115)

5.4.3.5 Palladium(0)-Catalyzed Allylic Alkylation via Transmetallation

Transmetallation has been used in allylic systems much less frequently than with aryl or vinyl systems, for several practical reasons. Allyl acetates are the most attractive class of allylic substrates, because of their ready availability from the corresponding alcohols, but they are substantially less reactive towards oxidative addition to Pd(0) than are allylic halides, and phosphine ligands are required to promote this process. Acetate coordinates fairly strongly to palladium and this, along with the presence of excess phosphine, slows down the crucial transmetallation step, already the rate-limiting step. Finally, in contrast to dialkylpalladium(II) complexes, σ-alkyl-π-allylpalladium (II) complexes under reductive elimination only slowly, compromising this final step in the catalytic cycle. It is therefore not surprising that alkylation of allylic acetates via transmetallation has been slow to develop.

However, some progress has been made. A variety of allyl acetates were alkylated by aryl- and vinyltin reagents under the special conditions of a polar (DMF) solvent, added lithium chloride (3 equivalents) to facilitate transmetallation and 3% $Pd(dba)_2$ catalyst with no added phosphine (in fact, added phosphine inhibited the reaction) (equation 5.116) [129]. Coupling occurred exclusively at the primary position of the allyl group, with clean inversion of stereochemistry. The geometry of the olefin in both the allyl substrate and vinyltin reagent was maintained, and modest functionality was tolerated

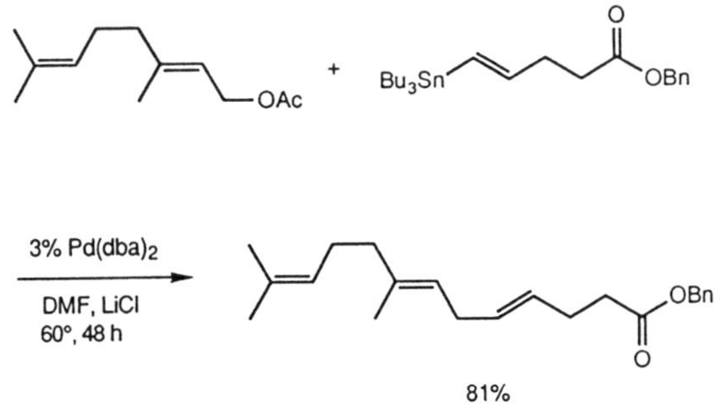

(5.116)

3% Pd(dba)2
DMF, LiCl
60°, 48 h

81%

in the tin reagent (CO_2R, OH, $OSiR_3$, OMe). 2-Ethoxy-2-propenyl diethyl phosphate also coupled to tin reagents under Pd(0) catalysis (equation 5.117) [130].

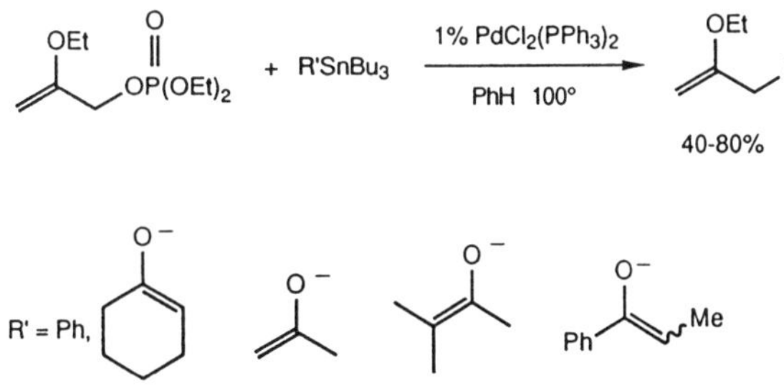

1% PdCl2(PPh3)2
PhH 100°

40-80%

(5.117)

R' = Ph,

Vinyl epoxides are more tractable substrates and undergo clean allylic alkylation by a variety of aryl- and vinylstannanes with good yields [131]. Predominantly 1,4-addition is observed, and olefin geometry is maintained in the vinyltin component but lost in the vinyl epoxide. Acetylenic, benzylic and allylic tin regents failed to transfer at all. Again, added ligands inhibited the reaction, and the best catalyst system was 3% $PdCl_2(MeCN)_2$ in DMF solvent (equations 5.118 and 5.119).

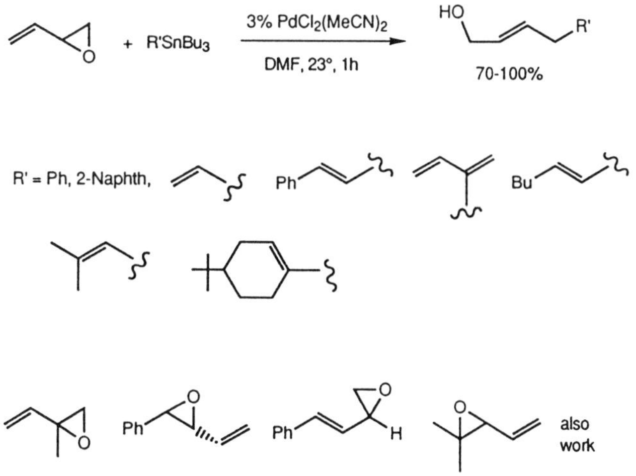

3% PdCl2(MeCN)2
DMF, 23°, 1h

70-100%

(5.118)

R' = Ph, 2-Naphth,

also work

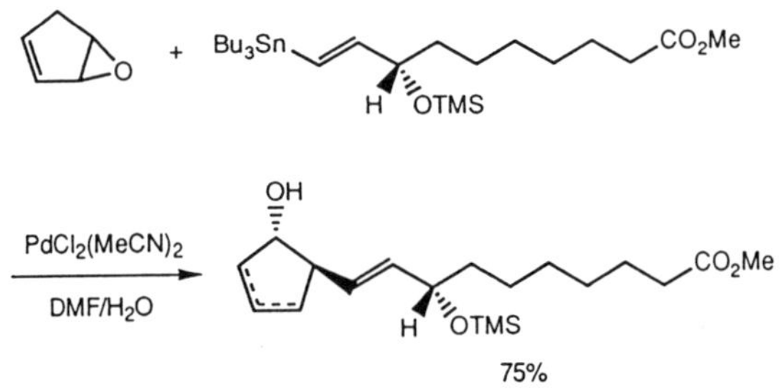

+ Bu_3Sn ... CO_2Me

H OTMS

$PdCl_2(MeCN)_2$

DMF/H_2O

OH

CO_2Me

H OTMS

75%

(5.119)

Allylic halides are much more amenable to transmetallation processes, because they both oxidatively add and transmetallate more readily than the corresponding acetates [132]. This reaction proceeds under 'normal' conditions (presence of phosphine ligands) and has been utilized in relatively complex systems (equations 5.120 [133] and 5.121)

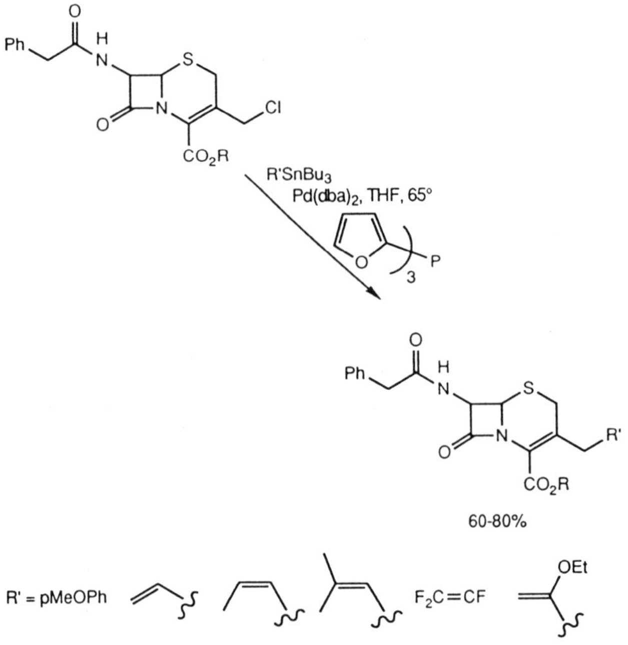

Ph O H N S N O Cl

CO_2R

R'$SnBu_3$

$Pd(dba)_2$, THF, 65°

O P 3

(5.120)

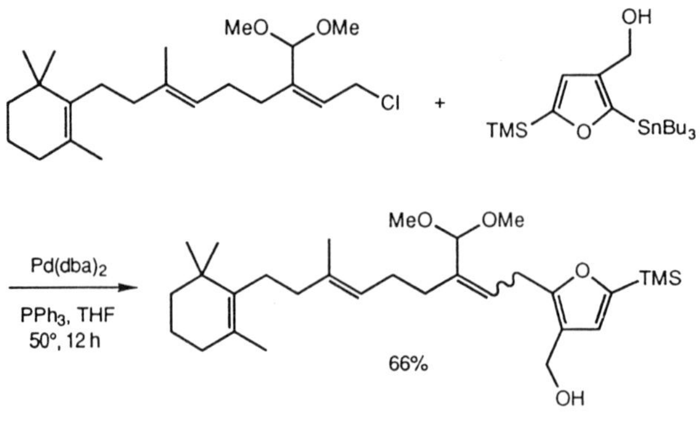

Ph O H N S N O R'

CO_2R

60-80%

R' = pMeOPh

$F_2C{=}CF$

OEt

MeO OMe

Cl

+

OH

TMS O $SnBu_3$

(5.121)

$Pd(dba)_2$

PPh_3, THF

50°, 12 h

MeO OMe

O TMS

OH

66%

[134]. Carbonylative coupling between allylic halides and aryl, vinyl [135] and allyl [136] tin reagents to give ketones has also been achieved.

5.4.3.6 Palladium(0)-Catalyzed Insertion ('Metallo-Ene') Reactions

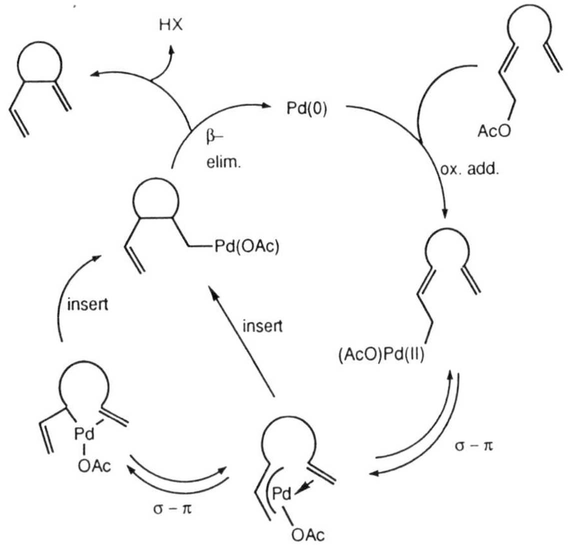

Figure 5.12

Palladium(0) complexes catalyze a number of very useful cyclization reactions of allyl acetates having remote unsaturation, via oxidative addition–insertion processes (Figure 5.12). These have been termed 'metallo-ene' reactions, although they are probably not mechanistically related to the classical organic ene reaction. The reaction proceeds under very mild conditions and is highly stereoselective. Remarkably, acetic acid promotes the reaction in an unspecified manner. The mechanism is not known, and that shown in Figure 5.12 is speculative and based on the known behavior of simpler systems. The requisite precursors for these cyclizations are almost invariably made directly using the palladium-catalyzed allylic alkylation reactions discussed in Section 5.4.3.2. A few of the already many useful examples are shown in equations 5.122–5.126 [137], along with

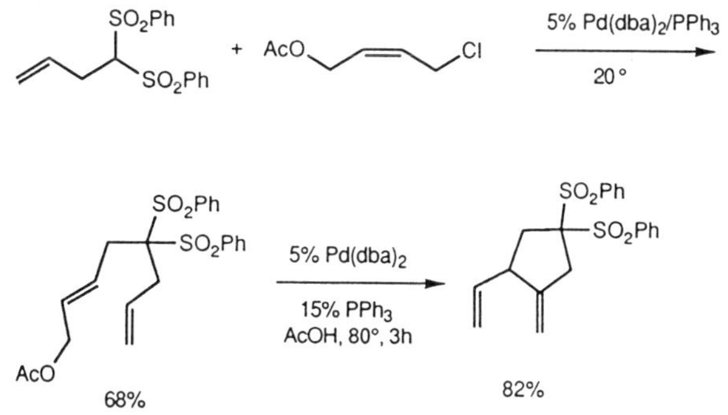

(5.122)

5% $Pd(PPh_3)_4$, AcOH, 80° 1 h — 79% (5.123)

5% $Pd(PPh_3)_4$, AcOH, 80°, 1h — 85% (5.124)

(see Figure 5.5) $Pd(PPh_3)_4$, THF (5.125)

7% $Pd(PPh_3)$, AcOH, 70°, 1h — 84%

1% $Pd(dba)_2$, 30% PPh_3, AcOH, 100°, 30 min — 77% (5.126)

the Pd(0)-catalyzed preparation of the substrates in some cases. Multiple olefin insertions can occur if the substrate is appropriately constituted (equations 5.127 [138] and 5.128 [139]). With a pendent diene rather than an olefin, insertion generates another π-allylpalladium complex, which undergoes nucleophilic attack by acetate to regenerate an allylic acetate (equations 5.129 and 5.130) [140].

L_4Pd — 70% (5.127)

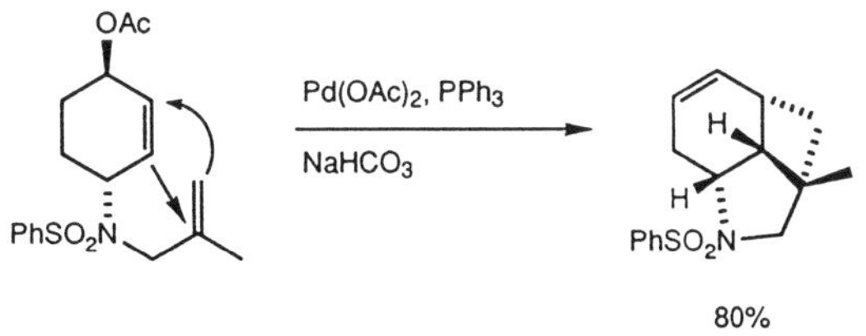

(5.128)

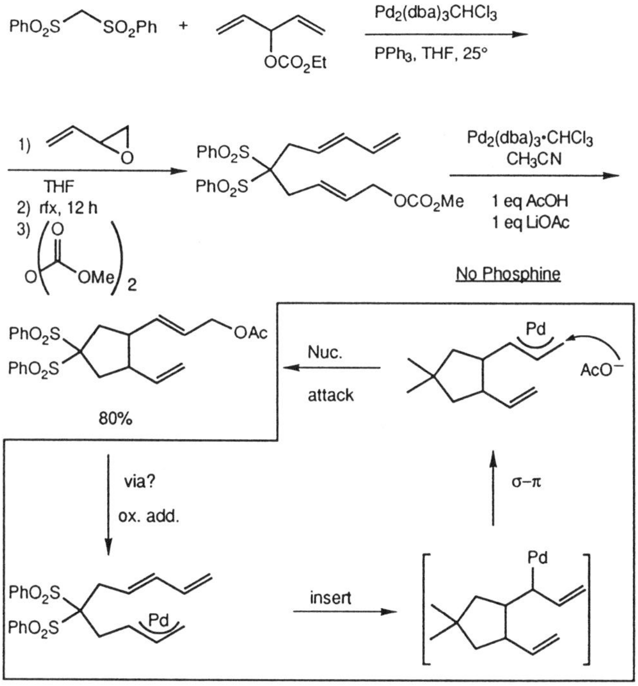

(5.129)

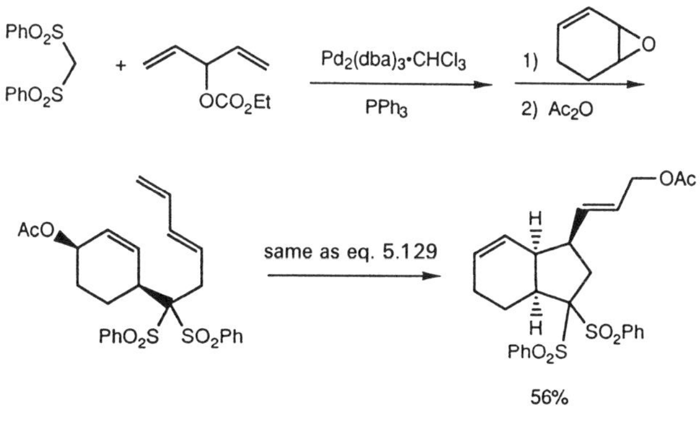

(5.130)

The initial insertion of an olefin into the π- (or σ)-allylpalladium complex produces a σ-alkylpalladium(II) complex which can insert CO competitively with β-elimination, resulting in overall cyclization–carbonylation (equation 5.131) [141]. If the olefin group of the original allyl system is sterically accessible after the cyclization–CO insertion steps, it can become involved (equation 5.132) [142]. When β-elimination is prevented by using an alkyne rather than an alkene, CO insertion followed by olefin insertion followed by another CO insertion ensues, producing many bonds in a single pot (equation 5.133) [141, 143].

Pd(dba)2, PPh3, 1 atm CO, AcOH 45° ; CO ; MeOH ; 77% (5.131)

Pd(dba)2, Bu3P, CO, AcOH, 80° ; 53% (5.132)

10% Pd(dba)2, PPh3, HOAc, 1 atm CO, MeOH, 45° ; 74% ; CO ; CO, MeOH (5.133)

5.4.3.7 Palladium(0)-Catalyzed Cycloaddition Reactions via Trimethylene Methane Intermediates

Under appropriate conditions, π-allylpalladium complexes are electrophilic at the terminal positions, and undergo attack by a wide range of nucleophiles. By devising allyl substrates having a substituted methylene group that can generate, or at least stabilize, negative charge, a 1,3-dipolar system capable of a wide range of cycloaddition reactions is generated. The most extensively studied and exploited of these is that derived from the palladium(0)-catalyzed reaction of the bifunctional silyl acetate shown in equation 5.134 [144].

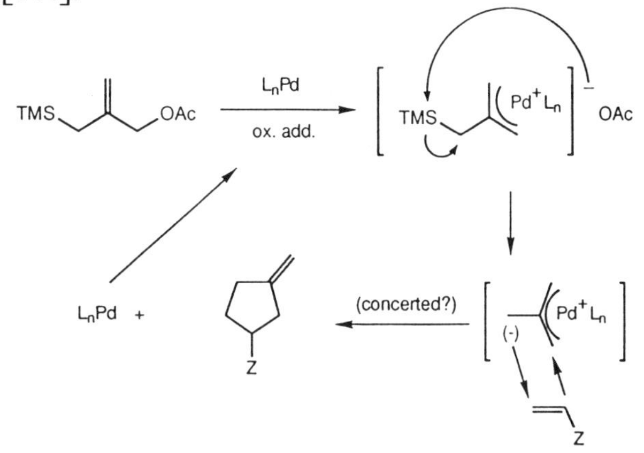

TMS OAc — L_nPd, ox. add. → [TMS Pd^+L_n] $^-$OAc → [(-) Pd^+L_n] + Z-alkene → (concerted?) → L_nPd + methylenecyclopentane-Z

(5.134)

The precise mechanism of this reaction is not known, nor is the exact nature of the intermediate complex. However, it is very useful in synthesis, proceeding with high yields under mild conditions and with excellent stereoselectivity. Thus, not only is the stereochemistry of the double bond of the acceptor maintained [there is *some* (20%) loss in certain cases], with optically active acceptors excellent diastereoselectivity is observed (equations 5.135 and 5.136) [145].

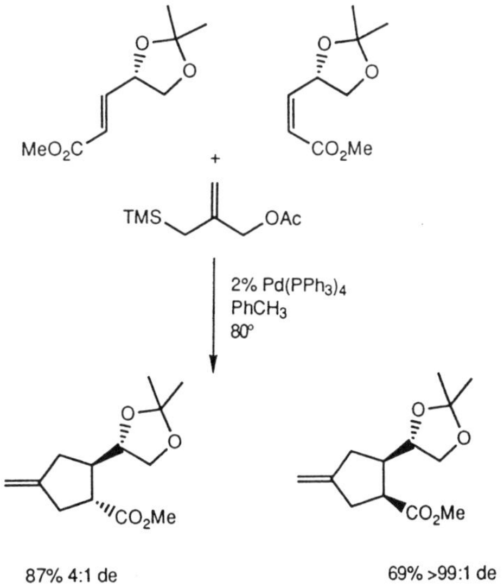

MeO_2C ... + ... CO_2Me

TMS OAc

2% $Pd(PPh_3)_4$
$PhCH_3$
80°

CO_2Me CO_2Me

87% 4:1 de   69% >99:1 de

(5.135)

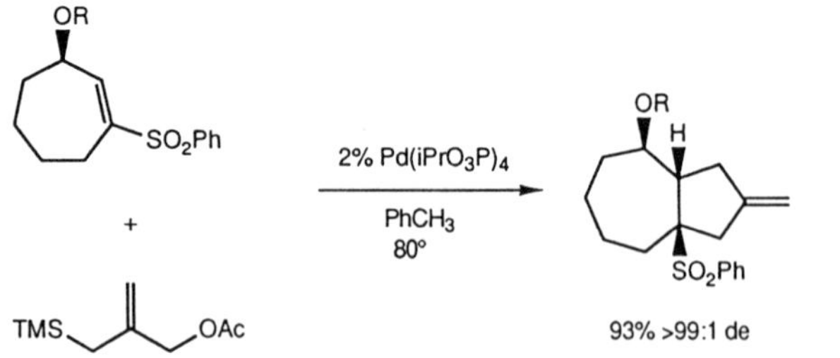

(5.136)

With unsymmetrical allylic acetate precursors, regioselective coupling occurs with both electron-withdrawing (equation 5.137) and electron-donating (equation 5.138) [146] substituents, the site of coupling always being the substituted terminus, regardless of the initial position of the substituents. This indicates that equilibration of the intermediate complex occurs at least competitively with cycloaddition. With pyrones both 3 + 2 and 3 + 4 cycloaddition occur, depending primarily on the structure of the acceptor (equations 5.139 and 5.140) [144, 147]. Tropone undergoes a 6 + 3 cycloaddition with excellent yields (equation 5.141) [147].

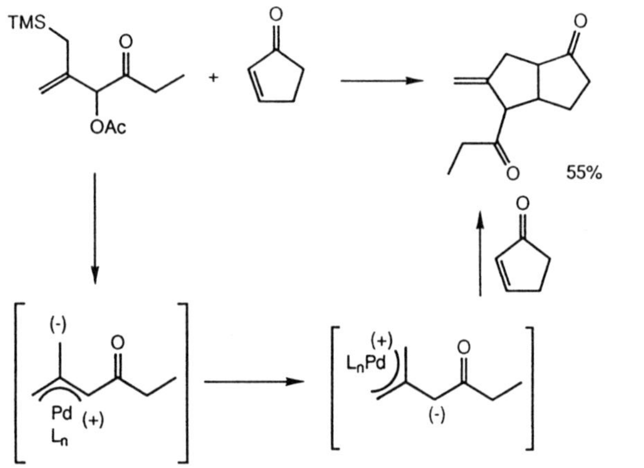

(5.137)

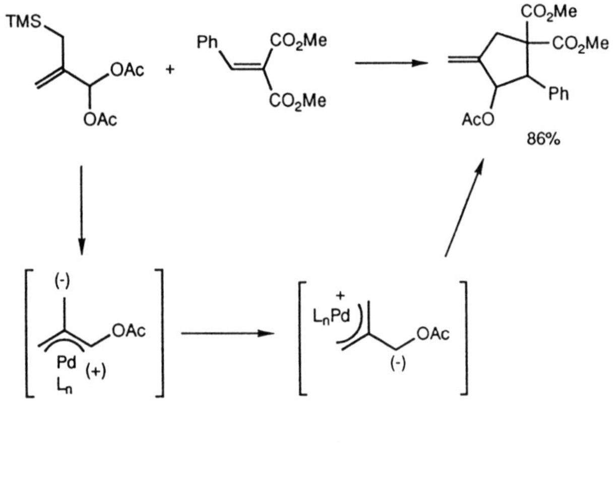

(5.138)

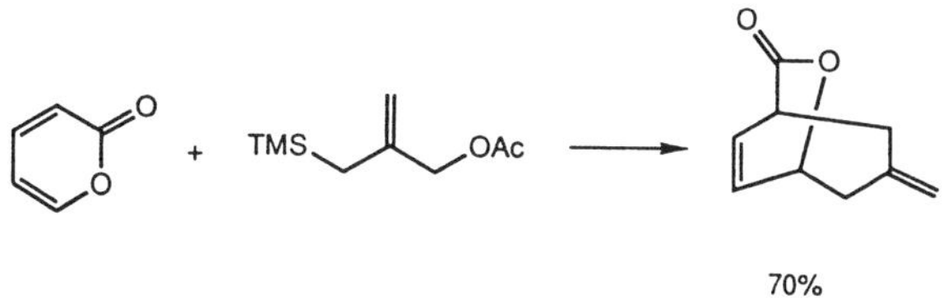

O, O + TMS, OAc → O, O (5.139)

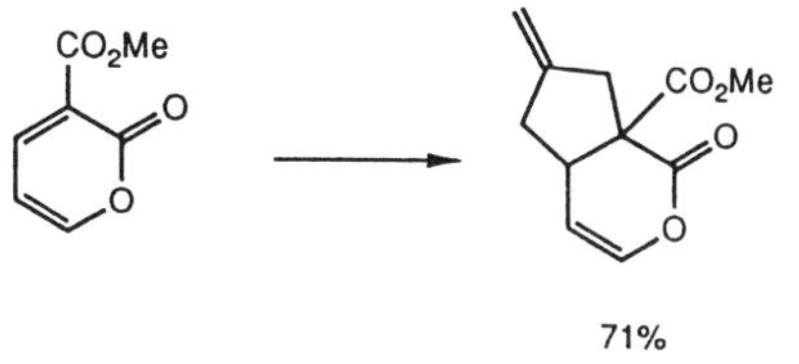

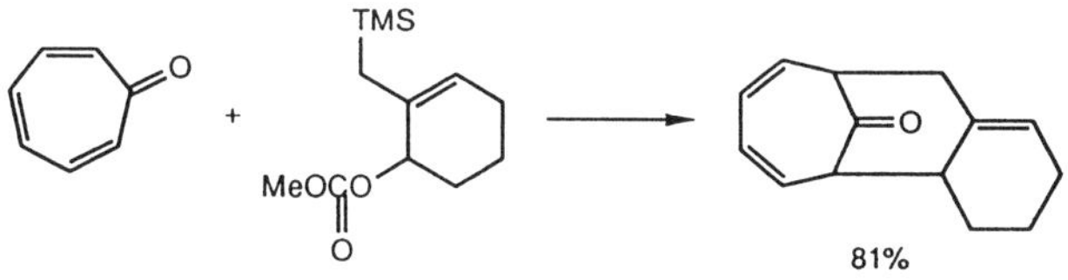

70%

CO_2Me, O, O → CO_2Me, O, O (5.140)

71%

O + TMS, MeOCO, O → O (5.141)

81%

Intramolecular versions of this 'trimethylenemethane' cycloaddition are fairly efficient for the production of polycyclic compounds (equations 5.142 [148] and 5.143 [149]).

In the presence of tin co-catalysts, particularly R_3SnOAc or R_3SnCl, cycloadditions to aldehydes occur (equation 5.144) [150]. The reaction is complex and mixtures of isomers are obtained. However, this is the subject of a full paper, and details on many specific systems are available, although generalities are difficult to enunciate.

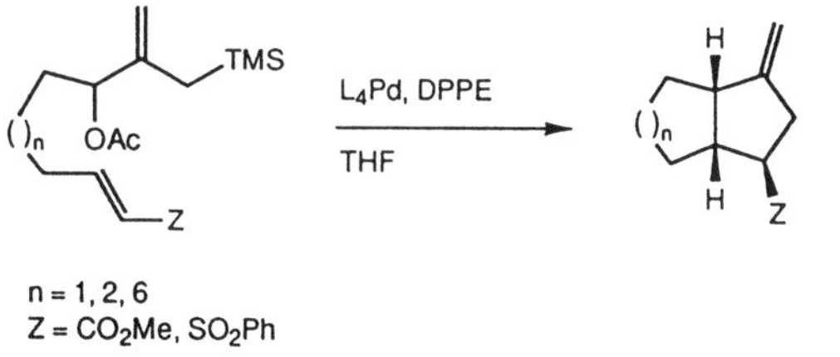

TMS, ( )n, OAc, Z —L_4Pd, DPPE / THF→ H, ( )n, H, Z (5.142)

n = 1, 2, 6
Z = CO_2Me, SO_2Ph

O, O, TMS, OAc, ( )n, SO_2Ph → O, O, ( )n, SO_2Ph (5.143)

n = 2, 3, 4, 8

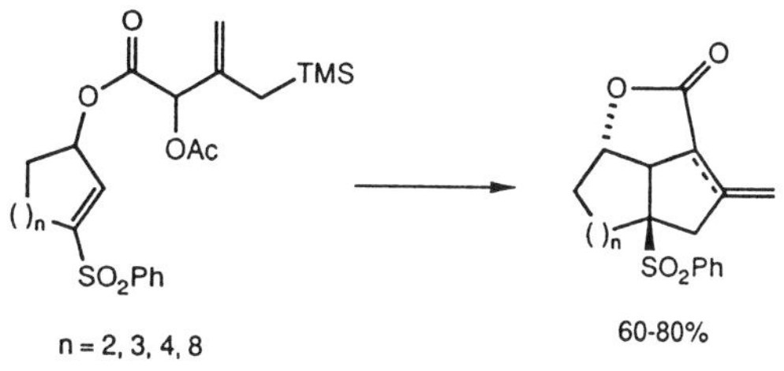

60-80%

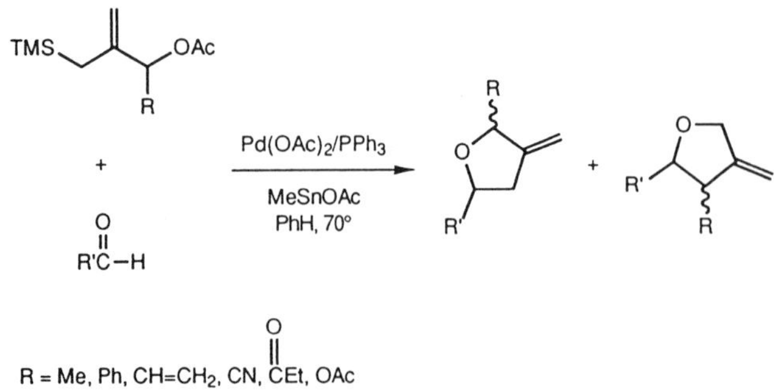

TMS OAc R + R'C(=O)H — $Pd(OAc)_2/PPh_3$, MeSnOAc, PhH, 70° → (R, R' methylenetetrahydrofurans) (5.144)

R = Me, Ph, $CH=CH_2$, CN, C(=O)Et, OAc

Methylenecyclopropanes also undergo palladium(0)-catalyzed 3 + 2 cycloadditions to electron-deficient olefins, and also a number of other cycloaddition processes. The fundamental process with relatively simple substrates has been studied in great detail [151]. It is not restricted to simple systems, however, as evidenced by equations 5.145 [152] and 5.146 [153]. Trimethylenemethane complexes from Pd(0)-catalyzed ring opening of the strained methylenecyclopropane are thought to be intermediate in these reactions.

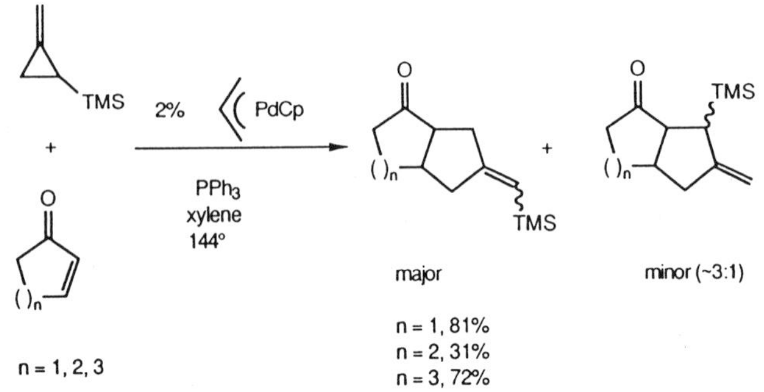

2% (allyl)PdCp, PPh_3, xylene, 144° (5.145)

n = 1, 2, 3; major; minor (~3:1)

n = 1, 81%
n = 2, 31%
n = 3, 72%

11% $Pd(dba)_2$, $(iPrO)_3P$, 110°, 48h; 79% (5.146)

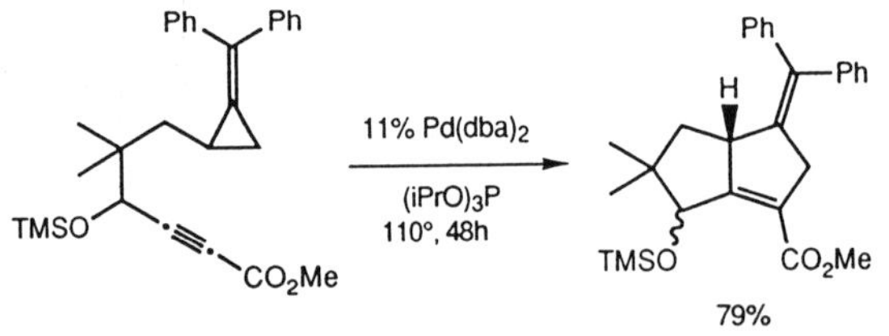

5.4.4 PALLADIUM(0)-CATALYZED TELOMERIZATION OF DIENES

Conjugated dienes combine with nucleophiles in the presence of palladium acetate–triphenylphosphine catalysts [Pd(0) generated *in situ*] to produce dimers with incorporation of one equivalent of nucleophile, in a process termed telomerization [154]. Although the mechanism of this process has not been studied in detail, it is thought to involve reductive dimerization of two diene units, followed by addition of the nucleophile to the bis(π-allyl)palladium complex thus formed (Figure 5.13). Although this

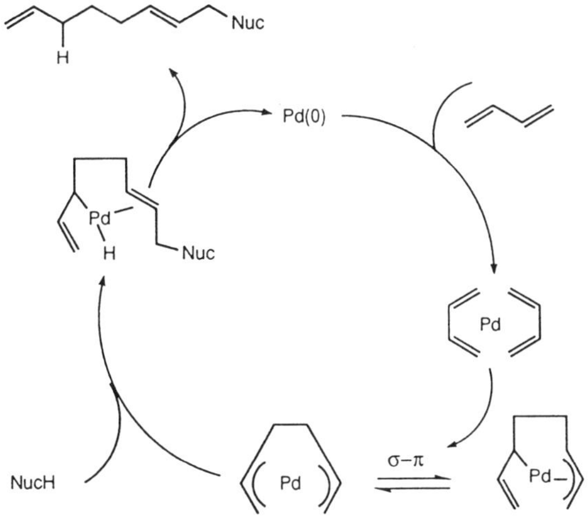

Figure 5.13

process has long been known and extensively studied, it has found little application in complex organic synthesis, except to provide very early starting materials. However, the recent development of an intramolecular version of this process promises to be of broader use (equation 5.147).

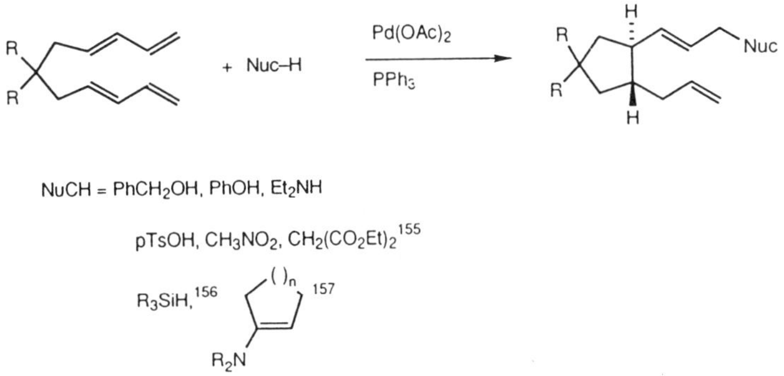

$$\text{diene} + \text{Nuc–H} \xrightarrow[\text{PPh}_3]{\text{Pd(OAc)}_2} \text{product} \qquad (5.147)$$

NuCH = $PhCH_2OH$, PhOH, Et_2NH

pTsOH, CH_3NO_2, $CH_2(CO_2Et)_2$[155]

R_3SiH,[156] enamine (R_2N)[157]

5.4.5 PALLADIUM(0)–COPPER(I)-CATALYZED COUPLING OF ARYL AND VINYL HALIDES WITH TERMINAL ALKYNES

Terminal alkynes couple with aryl and vinyl halides in the presence of palladium catalysts, copper(II) iodide and a secondary or tertiary amine [158]. The reaction conditions are very mild and consist simply of stirring the alkyne and halide for a few hours at room temperature in the amine as solvent. The most common catalysts used are $Pd(PPh_3)_4$ or $PdCl_2(PPh_3)_2$, which is rapidly reduced to Pd(0) by the amine solvent. The mechanism is not known, nor is the role of the copper iodide understood. It seems likely that copper acetylides are formed and transmetallate to the R—PdX formed by oxidative addition. However, there is no proof for this suggestion. Whatever the mechanism, the process is general and tolerant of a fair degree of functionality. The stereochemistry of the vinyl halide is maintained, as expected, making the stereospecific synthesis of enynes facile. Typical examples are seen in equations 5.148 [159], 5.149 [160], 5.150 [161], 5.151 [162], 5.152 [163] and 5.153 [164].

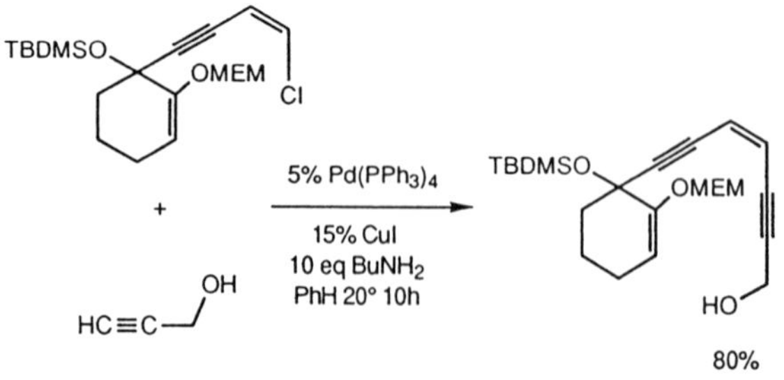

TBDMSO, OMEM, Cl + HC≡C–CH$_2$OH — 5% Pd(PPh$_3$)$_4$; 15% CuI, 10 eq BuNH$_2$, PhH 20° 10h → TBDMSO, OMEM, HO; 80% (5.148)

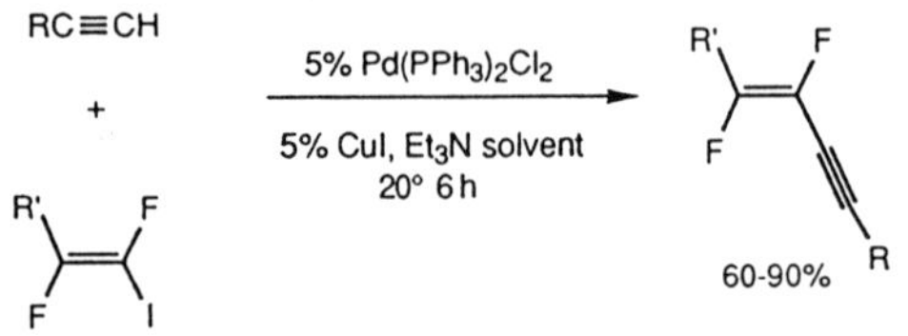

RC≡CH + R'(F)C=C(F)I — 5% Pd(PPh$_3$)$_2$Cl$_2$; 5% CuI, Et$_3$N solvent, 20° 6 h → R'(F)C=C(F)–C≡C–R; 60-90% (5.149)

R = nPr, nBu, Ph
R' = F, Ph, CF$_3$, (iPrO)$_2$P(O)

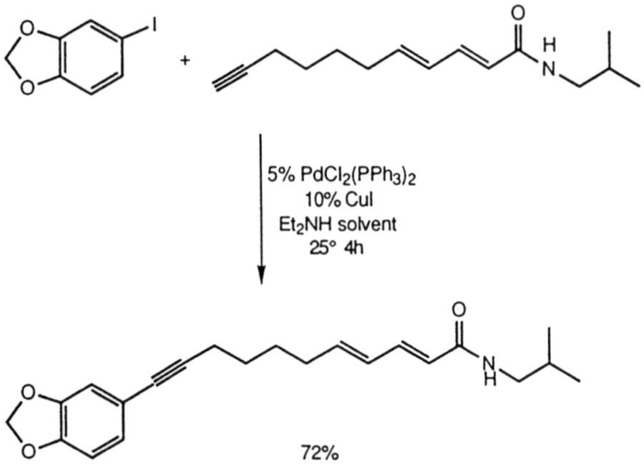

(5.150)

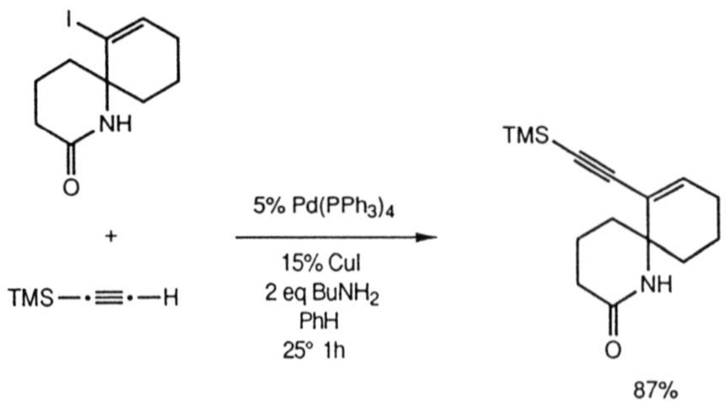

5% PdCl$_2$(PPh$_3$)$_2$
10% CuI
Et$_2$NH solvent
25° 4h

72%

NH, O, I + TMS—≡—H — 5% Pd(PPh$_3$)$_4$; 15% CuI, 2 eq BuNH$_2$, PhH, 25° 1h → TMS, NH, O; 87% (5.151)

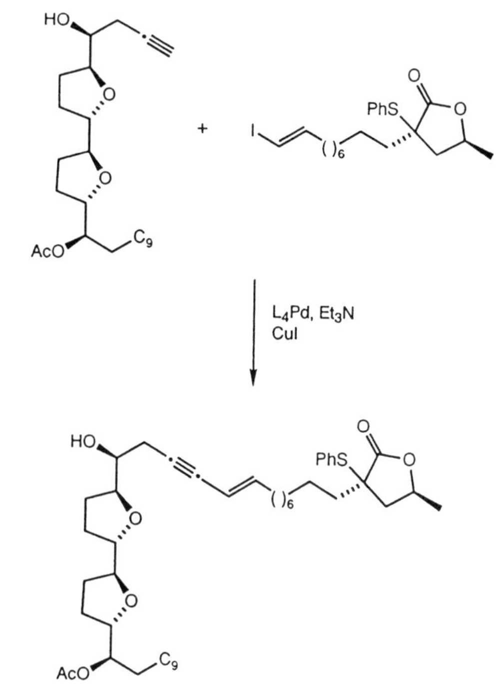

HO
O
O
AcO
C_9
+
I
$()_6$
PhS
O
O

L_4Pd, Et_3N
CuI

(5.152)

HO
O
O
AcO
C_9
PhS
O
O
$()_6$

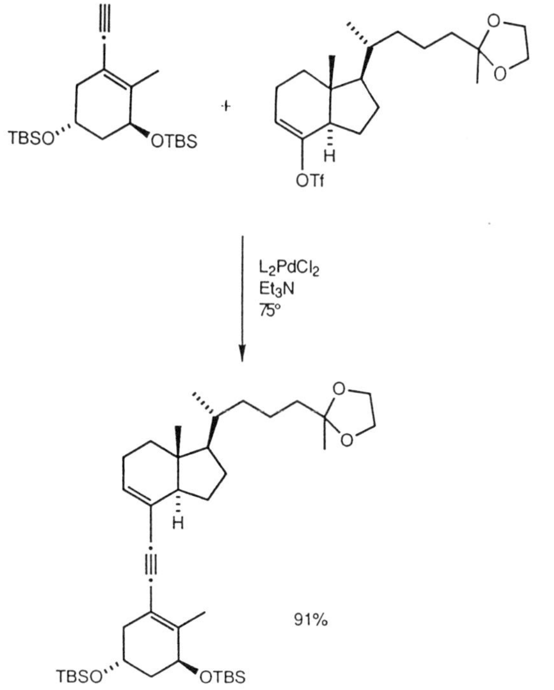

TBSO
OTBS
+
OTf
H
O
O

L_2PdCl_2
Et_3N
75°

(5.153)

O
O
H
TBSO
OTBS
91%

5.5 WORKING PROCEDURES

5.5.1 PREPARATION OF DICHLOROBIS(ACETONITRILE)PALLADIUM(II)

Anhydrous palladium(II) chloride was suspended in dry acetonitrile and the resulting dark brown slurry was stirred for 12 h at 25 °C, producing an orange–gold slurry. Filtration followed by drying on the filter gave virtually a quantitative yield of the desired compound.

5.5.2 PREPARATION OF TETRAKIS(TRIPHENYLPHOSPHINE)PALLADIUM(0) [165]

A mixture of palladium dichloride (17.72 g, 0.10 mol), triphenylphosphine (131 g, 0.50 mol) and 1200 ml of dimethyl sulfoxide was placed in a single-necked, 2 l, round-bottomed flask equipped with a magnetic stirring bar and a dual-outlet adapter. A rubber septum and a vacuum–nitrogen system were connected to the outlets. The system was then placed under nitrogen with provision made for pressure relief through a mercury bubbler. The yellow mixture was heated by means of an oil-bath with stirring until complete solution occurred (*ca* 140 °C). The bath was then removed, and the solution rapidly stirred for *ca* 15 min. Hydrazine hydrate (20 g, 0.40 mol) was rapidly added over *ca* 1 min from a hypodermic syringe. A vigorous reaction took place with evolution of nitrogen. The dark solution was then immediately cooled with a water-bath; crystallization began to occur at *ca* 125 °C. At this point the mixture was allowed to cool without external cooling. After the mixture had reached room temperature, it was filtered under nitrogen on a coarse, sintered-glass funnel. The solid was washed successively with two 50 ml portions of ethanol and two 50 ml portions of diethyl ether. The product was dried by passing a slow stream of nitrogen through the funnel overnight. The resulting slightly air-sensitive, yellow crystalline product weighed 103.5–108.5 g (90–94% yield). Although it can be handled in air for short periods of time, it must be stored under an inert atmosphere. A change in color from yellow to green indicates oxidative decomposition. This procedure can be carried out on a much smaller scale with no diminution of yield.

5.5.3 PREPARATION OF BIS(DIBENZYLIDENEACETONE)PALLADIUM(0) AND TRIS(DIBENZYLIDENEACETONE)DIPALLADIUM(0) CHLOROFORM) [166]

Palladium chloride (1.05 g, 5.92 mmol) was added to hot (*ca* 50 °C) methanol (150 ml) containing dibenzylidene acetone (4.60 g, 19.6 mmol) and sodium acetate (3.90 g, 47.5 mmol). The mixture was stirred for 4 h at 40 °C to give a reddish purple precipitate and allowed to cool to complete the precipitation. The precipitate was removed by filtration, washed successively with water and acetone and dried *in vacuo*. This product, bis(dibenzylideneacetone)palladium(0) (3.39 g), was dissolved in hot chloroform (120 ml) and filtered to give a deep violet solution, then diethyl ether (170 ml) was added slowly. Deep purple needles precipitated. These were removed by filtration, washed with diethyl ether and dried *in vacuo*. The complex, m.p. 122–124 °C, $Pd_2(dba)_3(CHCl_3)$, was obtained in 80% yield.

5.5.4 OXIDATION OF 1-(5-HEXENYL)-3,7-DIMETHYLXANTHINE WITH PALLADIUM(II) CHLORIDE AND COPPER(II) CHLORIDE (EQUATION 5.1) [2]

A solution of palladium(II) chloride (34 mg) and copper(II) chloride (330 mg) in dimethylformamide (100 ml) and water (100 ml) was prepared. To this solution, the xanthine (5 g), dissolved in dimethylformamide (50 ml) and water (50 ml), was added slowly over 4 h at 60–80 °C with passage of oxygen. After the addition, the mixture was stirred for 5 h at the same temperature. The solvent was evaporated under vacuum and the residue extracted with chloroform. After the usual work-up, the crude material was recrystallized from benzene–hexane to give pentoxifylline as crystals; yield, 4.5 g (85%); m.p. 103 °C.

5.5.5 TRANS-1,4-DIACETOXY-2-CYCLOHEXENE (FIGURE 5.5) [9]

To a stirred solution of $Pd(OAc)_2$ (700 mg, 3.1 mmol), $LiOAc \cdot 2H_2O$ (6.8 g, 66.6 mmol) and *p*-benzoquinone (1.91 g, 17.6 mmol) in acetic acid (100 ml) was added MnO_2 (6.52 g, 74.9 mmol) followed by 1,3-cyclohexadiene (5.0 g, 62.5 mmol) dissolved in pentane (200 ml). The reaction mixture, which separated into a pentane phase and an acetic acid phase, was moderately stirred at room temperature for 8 h. The pentane phase was separated and collected and the remaining acetic acid phase was diluted with saturated NaCl (100 ml) and extracted with pentane (2 × 100 ml) and pentane–diethyl ether (1 : 1) (3 × 100 ml). The combined extracts were washed with saturated NaCl (3 × 40 ml), water (3 × 10 ml) and finally 2 M NaOH (3 × 30 ml). The organic phase was dried ($MgSO_4$) and evaporated to yield 11.51 g (93%) of crystalline product (>91% *trans*). Recrystallization from hexane gave 9.87 g (80%) of isomerically pure material (>99% *trans*); m.p. 49–50 °C. IR (KBr), 1730, 1375, 1238, 1030, 1010, 920 cm^{-1}. ^1H NMR ($CDCl_3$), δ 5.89 (brs, 2H, CH=CH), 5.32 (m, 2H, CHO), 2.1 (m, 2H, CH—CH), 2.06 (s, 6H, OAc), 1.7 (m, 2H, CH—CH). Analysis: calculated for $C_{10}H_{14}O_4$, C 60.59, H 7.12, O 32.29; found, C 60.44, H 7.01; O 32.19%.

5.5.6 CIS-1,4-DIACETOXY-2-CYCLOHEXENE (FIGURE 5.5) [9]

Essentially the same procedure as for the preparation of the *trans* compound was used, but catalytic amounts of LiCl were added to a stirred solution of $Pd(OAc)_2$ (700 mg, 3.0 mmol), LiCl (521 mg, 12.3 mmol), $LiOAc \cdot 2H_2O$ (21.5 g, 211 mmol) and *p*-benzoquinone (1.6 g, 14.8 mmol) in acetic acid (100 ml), followed by MnO_2 (6.8 g, 78 mmol) and then 1,3-cyclohexadiene (5 g, 62.5 mmol) in pentane (200 ml). After 22 h, the same work-up procedure as above gave 10.6 g (86%) of essentially pure *cis*-1,4-diacetoxy-2-cyclohexene (>96% *cis*). Distillation afforded 9.8 g (79%). IR (neat) 2940, 1730, 1670, 1435, 1365, 1230, 1020 cm^{-1}. ^1H NMR ($CDCl_3$), δ 5.91 (brs, 2H, CH=CH), 5.23 (m, 2H, CHO), 2.07 (s, 6H, OAc), 1.9 (m, 4H, CH_2—CH_2).

5.5.7 4-BROMO-1-TOSYLINDOLE (EQUATION 5.14) [92]

Method A. A 200 ml, one-necked recovery flask equipped with a reflux condenser and vacuum adapter with an argon balloon was charged with 3-bromo-2-ethylaniline *p*-toluenesulfonamide (6.00 g, 17.03 mmol), *p*-benzoquinone (1.84 g, 17.03 mmol), LiCl (7.22 g, 0.17 mmol), $PdCl_2(MeCN)_2$ (0.44 g, 1.70 mmol, 10 mol%) and THF (85 ml). The orange suspension was refluxed 75 °C bath) for 18 h. The solvent was removed *in vacuo*. The residual brown, slightly gummy solid was transferred in to a Soxhlet thimble and extracted with 250 ml of hexane for 4 h. The resulting pot solution was treated with 0.5 g of charcoal and hot-gravity filtered. After cooling to room temperature, the hexane was removed *in vacuo*. The residual beige, slightly gummy solid was transferred to the top of a silica gel (10 g) column and dissolved in and eluted with hexane. Concentration of the eluent *in vacuo* afforded 4.70 g (78.8%) of the product as a colorless solid, m.p. 119–121 °C. This material may be recrystallized from hexane to afford 4.58 g (76.7%) of shiny colorless crystals, m.p. 119–121 °C.

Method B. A 12 oz pressure bottle was charged with 3-bromo-2-ethenylaniline *p*-toluenesulfonamide (6.00 g, 17.03 mmol), *p*-benzoquinone (1.84 g, 17.03 mmol), LiCl (7.13 g, 0.170 mmol), $PdCl_2(MeCN)_2$ (0.22 g, 0.85 mmol, 5 mol%) and THF (50 ml). The orange suspension was flushed with argon and then heated at 125 °C for 75 min. An identical work-up to the above afforded 4.63 g (77.7%) of the product as a colorless oil, m.p. 118–121 °C. This material may be recrystallized from hexane to afford 4.55 g (76.3%) of shiny colorless crystals, m.p. 120–122 °C. Additional recrystallizations from hexanes afforded the analytical sample as shiny colorless crystals, m.p. 117–122 °C. ^1H NMR (360 MHz, $CDCl_3$), δ 2.35 (s, 3H, CH_3), 6.73 (d, $J = 4$ Hz, 1H, indole 3H), 7.17 (t, $J = 8$ Hz, 1H, indole 6H), 7.24 (d, $J = 8$ Hz, 2H, tosyl H adjacent to methyl), 7.39 (d, $J = 8$ Hz, 1H, indole 5H), 7.62 (d, $J = 4$ Hz, 1H, indole 2H), 7.76 (d, $J = 8$ Hz, 2H, tosyl H adjacent to sulfonyl), 7.95 (d, $J = 8$ Hz, 1H, indole 7H. IR (KBr), 1598, 1568, 1472, 1374, 1358, 1170, 1132, 751, 673 cm^{-1}. Analysis: calculated for $C_{15}H_{12}BrNO_2S$ C 51.44, H 3.45, N 4.00; found, C 51.41, H 3.50, N 3.94%.

5.5.8 PALLADIUM–CATALYZED ENYNE CYCLIZATION TO A 1,3-DIENE (EQUATION 5.30, R = CH_2OTBS, R′ = PMB)

A mixture of the enyne (1.42 g, 3.53 mmol), *N*,*N*-bisbenzylideneethylenediamine (83.4 mg, 0.353 mmol) and palladium(II) acetate (39.6 mg, 0.176 mmol) in 14 ml of dry benzene in a 50 ml round-bottomed flask with a nitrogen purge was heated in an oil-bath at 80 °C for 1 h under an atmosphere of nitrogen. The brown solution was concentrated *in vacuo* and then purified via flash chromatography (150 × 25 mm i.d. silica gel column, 4% ethyl acetate in hexane as eluent) to give 1.15 g (81%) of the desired 1,3-diene product as a colorless oil. IR(neat), 2850, 1655, 1615, 1585, 1515, 830 cm^{-1}. ^{1}H NMR (400 MHz, $CDCl_3$), δ 7.25 (d, J = 9.6 Hz, 2H), 6.86 (d, J = 8.7 Hz, 2H), 6.13 (br t, J = 7.1 Hz, 1H), 5.35 (br s, 1H), 4.85 (br s, 1H), 4.46 ($d_{A,B}$, J = 11.5 Hz, 1H), 4.36 ($d_{A,B}$, J = 11.5 Hz, 1H), 4.30 (m, 2H), 3.87 (s, 1H), 3.80 (s, 3H), 2.50 ($d_{A,B}$, J = 15.6 Hz, 1H), 2.07 ($d_{A,B}$, J = 15.6 Hz, 1H), 1.15 (s, 3H), 0.90 (s, 9H), 0.87 (s, 3H), 0.06 (s, 3H), 0.05 (s, 3H). ^{13}C NMR (100 MHz, $CDCl_3$), δ 158.8, 146.7, 140.12, 131.0, 128.6, 152.2, 113.6, 104.3, 85.2, 70.1, 61.4, 55.2, 44.9, 40.6, 26.8, 26.1, 25.7, 22.3, 18.4, −5.5. Mass spectrum, *m*/*z* 402 (M^+, 1%), 345 (M^+ − *t*-Bu, 2%), 281 (M^+ − $MeC_6H_4CH_2O$, 13%), 209 (27%), 121 (100%). TLc, R_f = 0.44 (10% EtOAc in hexane; stained with anisaldehyde).

5.5.9 PALLADIUM-CATALYZED ENYNE CYCLIZATION TO A 1,4-DIENE (EQUATION 5.32)

1,2-Dichloroethane (20 ml) was added to palladium(II) acetate (180 mg, 0.80 mmol) and *N*,*N*′-bis(phenylmethylene)-1,2-ethylenediamine (208 mg, 0.88 mmol) and the mixture was stirred under nitrogen for 10 min at room temperature. The resulting solution was added to a stirred solution of enynes (9.40 g, 16.00 mmol) in 1,2-dichloroethane (15 ml). The mixture was heated at 55 °C for 24 h, then cooled to room temperature. The solvent was evaporated *in vacuo* and the residue filtered through silica gel, eluting with hexane–ethyl acetate (5 : 1). The solvent was evaporated *in vacuo*, the residue was dissolved in acetonitrile (30 ml), tetrabutylammonium chloride (0.90 g, 3.24 mmol) and potassium fluoride dihydrate (6.0 g, 63.80 mmol) were added and the mixture was heated at reflux under nitrogen for 30 h. Water (20 ml) and diethyl ether (30 ml) were added, the organic phase was separated and the aqueous layer was extracted with diethyl ether (4 × 30 ml). The combined extracts were dried ($MgSO_4$) and evaporated *in vacuo*. The residue was purified by chromatography on silica gel, eluting with hexane–ethyl acetate (3 : 2), to give the diols (5.07 g, 79%) as a 1 : 1 mixture of C-8 epimers. For characterization the 8*R*- and 8*S*-epimers could be separated by flash chromatography. The 8*R*-epimer was obtained as needles, m.p. 94–95 °C (hexane). TLc, R_f (hexane–EtOAc, 1 : 1) = 0.41; $[\alpha]_D$ − 143.1 ° (*c* 2.405, $CHCl_3$). IR ($CDCl_3$), 3588 (br), 3480 (br), 3077, 2961, 2933, 2882, 2840, 1615, 1514, 1452, 1399, 1377, 1301, 1249, 1173, 1105, 1034 cm^{-1}. ^{1}H NMR (200 MHz, $CDCl_3$), δ 7.27 (d, J = 8.6 Hz, 2H), 6.89 (d, J = 8.6 Hz, 2H), 5.42 (dd, J = 9.8, 2.4 Hz, 1H), 5.33 (d, J = 2.0 Hz, 1H), 5.2–5.1 (m, 2H), 4.9–4.7 (m, 1H), 4.86 (m, 1H), 4.81 (m, 1H), 4.7–4.75 (AB m, 2H), 4.55 (AB m, 2H), 3.92 (s, 1H), 3.81 (s, 3H), 3.9–3.6 (m, 2H), 2.50 (m, 1H), 2.22 (dd, J = 12.6, 7.7 Hz, 1H), 2.25–2.0 (m, 1H), 1.67 (s, 3H), 1.7–1.5 (m, 3H), 1.20 (s, 3H). ^{13}C NMR (50 MHz, $CDCl_3$), δ 161.8, 159.4, 145.3, 137.4, 129.5, 125.1, 113.9, 108.9, 94.6, 81.5, 72.8, 69.7, 69.3, 55.2, 50.5, 47.0, 41.9, 38.5, 21.5, 18.5. Mass spectrum (found, M − H_2O, 382.213; calculated for $C_{24}H_{30}O_4$, 382.2144), *m*/*z* 382 (0.1%), 246 (3), 216 (2), 201 (3), 173 (3), 150 (13), 137 (15), 121 (100), 91 (7), 77 (6). The 8*S*-epimer was obtained as an oil which solidified to waxy needles, m.p. 41–43 °C, TLc, R_f (hexane–EtOAc, 1 : 1) = 0.48; $[\alpha]_D$ − 161.6° (*c* 5.875, $CHCl_3$). IR ($CDCl_3$), 3453 (br), 3062, 2950, 2924, 1611, 1510, 1373, 1299, 1247, 1171, 1119, 1100, 1081, 1061, 1030 cm^{-1}. ^{1}H NMR (200 MHz, $CDCl_3$), δ 7.28 (d, J = 8.6 Hz, 2H), 6.89 (d, J = 8.6 Hz, 2H), 5.45 (s, 1H), 5.31 (dd, J = 9.8, 2.4 Hz, 1H), 5.23 (s, 1H), 5.11 (dd, J = 9.8, 2.0 Hz, 1H), 4.86 (s, 1H), 4.80 (s, 1H), 4.74 (s, 2H), 4.56 (s, 2H), 4.46 (d, J = 5.5 Hz, 1H), 3.87 (dd, J = 10.0, 5.6 Hz, 1H), 3.81 (s, 3H), 3.8–3.6 (m, 1H), 2.52 (m, 1H), 2.2–1.8 (m, 3H), 1.67 (s, 3H), 1.27 (s, 3H). ^{13}C NMR (50 MHz, $CDCl_3$), δ 162.0, 159.3, 145.2, 137.0, 129.4, 125.4, 113.8, 111.1, 94.5, 84.1, 73.2, 69.4, 68.7, 55.1, 51.4, 46.6, 40.7, 38.4, 21.6, 18.4. Mass spectrum (found, M − CH_2O, 370.2126; calculated for $C_{23}H_{30}O_4$, 370.2144), *m*/*z* 370 (0.4%), 264 (5), 246 (5), 228 (4), 216 (8), 185 (8), 150 (22), 137 (13), 121 (100).

5.5.10 (11aS)-ETHYL-3-(5,10,11,11a-TETRAHYDRO-10-METHYL-5,11-DIOXO-1H-PYRROLO[2,1-c[1,4]BENZODIAZEPIN-2-YL)PROPENOATE (EQUATION 5.64) [66]

A reaction mixture of 0.823 g (2.18 mmol) of the vinyl triflate, 0.885 g (2.27 mmol, 1.04 equiv.) of (*E*)-tributylstannyl)ethylpropenoate, 0.927 g (21.8 mmol, 10.0 equiv.) of lithium chloride, 75.8 mg (3 mol%) of tetrakis(triphenylphosphine)palladium(0) and 25 ml of THF was heated at reflux under argon overnight. The reaction mixture was worked up by extraction with chloroform and washing with water. The organic layers were combined, washed with water and brine and dried over $MgSO_4$. Filtration and evaporation of the solvent *in vacuo* left a yellow viscous oil. The residue was dissolved in acetonitrile and washed with several portions of hexane (to remove tributyltin chloride). The acetonitrile solvent was removed *in vacuo* leaving a viscous oil. The residue was further purified by column chromatography (EtOAc–hexane, 1:1) to give 0.595 g (78%) of pale yellow solid, m.p. 158–160 °C. TLc, R_f (EtOAc–hexane, 1:1) = 0.42; $[\alpha]_D^{22}$ + 620° (*c* 0.0074, $CHCl_3$). IR($CHCl_3$), 1700, 1680, 1650, 1610, 1450, 1405, 1165, 1140, 900 cm^{-1}. ^1H NMR ($CDCl_3$), δ 1.3 (t, 3H, $J = 7.1$ Hz), 2.9 (br dd, 1H), 3.45 (s, 3H), 3.8 (brd, 1H), 4.2 (q, 2H, $J = 7.1$ Hz), 4.6 (dd, 1H, $J = 3.47$, 10.8 Hz), 5.8 (d, 1H, $J = 15.57$ Hz), 7.3 (m, 3H), 7.5 (d, 1H, $J = 15.59$ Hz), 7.6 (t, 1H), 7.9 (d, 1H). ^{13}C NMR ($CDCl_3$), δ 167.38, 166.60, 162.12, 140.30, 137.03, 132.69, 130.94, 130.60, 128.27, 125.94, 123.47, 122.25, 118.66, 60.25, 57.41, 36.53, 29.57, 14.17. Analysis: calculated for $C_{18}H_{18}N_2O_4$, C 66.24, H 5.56, N 8.58; found, C 66.15, H 5.61, N 8.50%.

5.5.11 PROCEDURE FOR THE COUPLING OF CEPHAM TRIFLATE WITH (Z)-1-PROPENYLTRIBUTYLTIN (EQUATION 5.65) [67]

Triflate A (5.860 g, 0.010 mol) was dissolved in dry *N*-methylpyrrolidone (NMP) (20 ml), the solution was degassed with argon and zinc chloride (2.720 g, 0.020 mol) was added, followed by tri(2-furyl)phosphine (92 mg, 0.397 mmol) and Pd_2 $(dba)_3$ (90.8 mg, 0.198 mmol Pd). The solution was stirred for 10 min, then (*Z*)-1-propenyltributyltin (3.640 g, 0.011 mol) was added neat by syringe, rinsing with dry NMP (2 ml). The reaction mixture was stirred at room temperature for 20 h, diluted with ethyl acetate (100 ml), washed three times with water and once with brine and dried over sodium sulfate. Filtration and concentration gave a crude product that was dissolved in acetonitrile (10 ml) and washed three times with pentane (100 ml each), in order to remove the tin-containing co-products. Evaporation gave an oil that was recrystallized from warm methanol. Yield: 3.910 g (82%) of tan crystals of the product, m.p. 133–134 °C. ^1H NMR ($CDCl_3$), δ 7.4–7.2 (m, 7H), 6.84 (m, 2h), 6.1–6.0 (2 overlapping brd, $J = 12$ Hz, $J' = 9$ Hz, 2H), 5.77 (dd, $J = 9$ Hz, $J' = 4.9$ Hz, 1H), 5.62 (m, 1H), 5.12 (s, 2H), 4.95(d, $J = 4.9$ Hz, 1H), 3.78 (s, 3H), 3.60 (m, 2H), 342 (d, $J = 18$ Hz, 1H), 3.22 (d, $J = 18$ Hz, 1H), 1.51 (dd, $J = 7.1$ Hz, $J' = 1.8$ Hz, 3H). The amount of *E*-isomer present was estimated to be 2% by NMR integration, by comparison with the NMR spectrum of an authentic sample prepared according to the literature. Analysis ($C_{26}H_{26}N_2O_5S$): C, H, N, S.

5.5.12 N-[4-(1H-IMIDAZOL-1-YL)BUTYL]2-(1-METHYLETHYL)-11-OXO-11H-PYRIDO[2,1-bQUINAZOLINE-8-CARBOXAMIDE (EQUATION 75) [82]

A 1 lL glass autoclave liner was charged with 136.0 g (0.30 mol) of the starting material, 2.5 g (3.56 mmol) of bis(triphenylphosphine)palladium dichloride, 2.5 g (9.6 mmol) of triphenylphosphine, 77 ml of tributylamine and 390 ml of 1:10 aqueous DMF. The mixture was degassed with nitrogen, placed in the autoclave, swept three times with carbon monoxide and then heated for 12 h at 100 °C under 200 psi of carbon monoxide. The cooled mixture was filtered and 200 ml of glacial acetic acid was used to rinse the liner and filter pad. The filtrate was concentrated using a 70 °C water-bath. The residue was taken up in 500 ml of water, washed with 3 × 300 ml of ethyl acetate and then made alkaline by adding saturated aqueous sodium carbonate solution. The resulting precipitate was filtered, thoroughly washed with water and dried under vacuum to give 119.8 g (99.8%) of crude product as a yellow powder. The purity of material thus obtained, typically $>90\%$, was raised to $>99\%$ by serial recrystallization from acetonitrile, DMF–aqueous ammonia–water and 2-butanone with a 70–75% recovery. The product thus prepared was shown by spectral and TLC comparisons to be identical with material obtained by the route described in the literature [167].

5.5.13 4-(3-HYDROXY-3-METHYL-1-BUTEN-1-YL)-1-TOSYLINDOLE (EQUATION 5.79) [92]

A mixture of 4-bromo-1-tosylindole (0.350 g, 1.00 mmol), 2-methyl-3-buten-2-ol (0.108 g, 1.25 mmol), Et_3N (0.127 g, 1.25 mmol), $Pd(OAc)_2$ (11 mg, 0.050 mmol) and tri-*o*-tolylphosphine (61 mg, 0.20 mmol) was flushed with argon and then heated in a sealed tube at 100 °C for 5 h. After cooling to room temperature, the residue was taken up in 100 ml of CH_2Cl_2, washed with water three times, dried (Na_2SO_4), filtered and concentrated *in vacuo*.

The residue was chromatographed on silica gel (10 g) by using hexane–benzene (1 : 1) to elute impurities and benzene to elute the product. The benzene eluate was concentrated *in vacuo* to afford 0.345 g (97.3%) of the product as a colorless foam. ^1H NMR (360 MHz, $CDCl_3$), δ 1.44 (s, 6H, geminal CH_3), 2.33 (s, 3H, CH_3), 6.43 (d, $J = 16$ Hz, 1H, olefin H adjacent to aliphatic), 6.85, 6.87 (pair of d, $J = 3$, 16 Hz, 2H, olefin H and indole 3H), 7.22 (d, $J = 8$ Hz, 2H, tosyl H adjacent to methyl), 7.26–7.34 (m, 2H, indole 6 and 5H), 7.59 (d, $J = 3$ Hz, 1H, indole 2H), 7.72 (d, $J = 8$ Hz, 2H, tosyl H adjacent to sulfonyl), 7.89 (d, $J = 8$ Hz, 1H, indole 7H). IR(CCl_4), 3640–3160, 1420, 1377, 1362, 1188, 1180, 1165, 1136, 967 cm^{-1}. Analysis: calculated for $C_{20}H_{21}NO_3S$, C 67.58, H 5.96, N 3.94; found, C 67.64, H 6.09, N 3.73%.

5.5.14 PALLADIUM-CATALYZED ALKYLATION OF VINYL EPOXIDES TO FORM MACROCYCLES (EQUATION 5.102)

A solution of the palladium catalyst was prepared as follows. $Pd(OAc)_2$ (10.2 mg, 0.046 mmol) was dissolved in 10 ml of THF under nitrogen and triisopropyl phosphite (62 mg, 0.30 mmol) was added in one portion. After stirring for 15 min at room temperature, THF (15 ml) was added followed by THF (15 ml) at 0.5, 0.75, 1.0, 1.25 and 1.5 h. The catalyst must not be diluted too quickly; Pd black may form unless care is taken to dilute the Pd(0) catalyst slowly. The solution was then brought to reflux and the acyclic ester (728 mg, 0.91 mmol) in 25 ml of THF was added over a 2 h period via a syringe pump. When the addition was complete, the solvent was removed under reduced pressure and the residue was flash chromatographed (33% EtOAc in hexane) affording 667 mg (92%) of the macrocycle as a pale yellow oil. IR ($CDCl_3$, 3560, 3067, 3058, 2927, 2864, 1733, 1452, 1433, 1330, 1314, 1150, 1112, 1080, 688 cm^{-1}. ^1H NMR (200 MHz, $CDCl_3$), δ 8.00–7.82 (m, 4H), 7.72–7.30 (m, 16H), 5.95–5.41 (m, 4H), 4.04 (t, $J = 5.9$ Hz, 2H), 3.79–3.55 (m, 3H), 3.20–2.90 (m, 3H), 2.27 (t, $J = 7.3$ Hz, 2H), 1.68–1.17 (m, 21H), 1.08 (s, 9H). Analysis: calculated for $C_{53}H_{70}O_8SiS_2$, C 68.65, H 7.61; found, C 68.48, H 7.80%.

5.5.15 PREPARATION OF 1,4-DIENE DISULFONE IN EQUATION 5.122

NaH (60% in mineral oil, 0.92 g, 22.9 mmol) was added in portions to a stirred solution of allyl disulfone (7.0 g, 20.8 mmol) in dry THF (90 ml) at 0 °C under argon. After stirring the mixture for 60 min at room temperature, $Pd(dba)_2$ (0.60 g, 5 mol%) and triphenylphosphine (1.09 g, 20 mol%) were added and a solution of (*Z*)-1-acetoxy-4-chloro-2-butene (3.10 g, 20.9 mmol) in THF (20 ml) was rapidly dropped on to the reaction mixture. After stirring overnight (14 h) under argon, the reaction was quenched with water (200 ml) and the aqueous layer extracted with diethyl ether (4×100 ml). The combined organic phases were washed with water and brine (50 ml each) and dried ($MgSO_4$) and the solvent was evaporated *in vacuo*. The residue was purified by flash chromatography (silica gel) using hexane–ethyl acetate (3 : 1) as eluent. The product was obtained as yellow oil that slowly afforded white crystals on standing at room temperature (m.p. 67–68 °C) (6.35 g, 68%).

To a solution of the substrate (500 mg, 1.12 mmol) in 5 ml of HOAc (Fluka, puriss. p.a.) was added $Pd(dba)_2$ (32 mg, 5 mol%) and triphenylphosphine (43.9 mg, 15 mol%) and the mixture was stirred at 80 °C under argon for 2 h. For work-up the reaction mixture was diluted with diethyl ether (100 ml) and extracted with water (40 ml), the organic layer was separated, washed with $NaHCO_3$ (10% in water, 30 ml), water and brine (30 ml each) and dried ($MgSO_4$) and the solvent was evaporated *in vacuo*. The residue was subjected to column chromatography (silica gel) using hexane–ethyl acetate (5:1) as eluent. The product (351 mg, 81%) was obtained as a white crystalline solid (m.p. 113–114 °C); the m.p. did not change on recrystallation of the product from ethanol.

5.5.16 CIS-4-(1,1-BISCARBOMETHOXY-4-BUTENYL)-1-CYCLOHEPT-2-ENYL ACETATE

Sodium hydride (53.7 mg, 1.23 mmol) was added to a solution of allyl dimethyl malonate (215 mg, 1.25 mmol) in THF (1.5 ml) under argon and the resulting solution was stirred for 30 min. $Pd(PPh_3)_4$ (42 mg, 3 mol%) and a solution of *cis*-1-chloro-4-acetoxycyclohept-2-ene (210 mg, 1.11 mmol) in dry THF (1.5 ml) were added and the yellow suspension was stirred for 21 h. Aqueous NH_4Cl solution was added and the mixture was extracted with diethyl ether (3 × 10 ml). The organic layers were washed with brine, combined, dried ($MgSO_4$) and evaporated to give a yellow oil. Purification by flash chromatography (30 mm; pentane–diethylether, 3:1) gave a colorless oil 288 mg (80%). ^1H NMR, 1.14 (m, 1H). 1.52 (m, 1H); 1.80 (m, 3H), 2.0 (m, 1H), 2.02 (s, 3H), 2.65 (d, $J = 7.5$ Hz, 2H), 2.90 (m, 1H), 3.72 (s, 6H), 5.08 (m, 2H), 5.52 (m, 1H), 5.72 (m, 3H). ^{13}C NMR, 170.82 (s), 170.20 (s), 133.97 (d), 132.68 (d), 130.85 (d), 118.86 (t), 73.88 (d), 61.43 (s), 52.11 (q), 43.37 (d), 38.36 (t), 32.09 (t), 29.06 (t), 27.76 (t), 21.22 (q). IR, 2950, 2940, 2860, 1730, 1440, 1375, 1240, 1140, 1020 cm^{-1}. Mass spectrum, no M^+; m/z 282 (<1%), 265 (2), 251 (1), 204 (23), 172 (28), 145 (35), 110 (100).

5.5.17 (3aα,8aβ)-DIMETHYL-3-METHYLENE-1,2,3,3a,6,7,8,8a-OCTAHYDROAZULENE-1,1-DICARBOXYLATE

A solution of the *cis*-allylic acetate from the preceding procedure (140 mg, 0.43 mmol) and $Pd(PPh_3)_4$ (22.0 mg, 5 mol%) in acetic acid (1 ml) was heated at 70 °C for 1.75 h. The solution was poured into aqueous $NaHCO_3$ and extracted with diethyl ether (3 × 10 ml). The organic layers were washed with brine, combined, dried ($MgSO_4$) and evaporated to give a yellow oil. Purifaction by flash chromatography (15 mm; pentane–diethylether, 8:1) gave a colorless oil (105.6 mg, 92.5%) which slowly crystallized on standing. Crystallization from aqueous ethanol gave the title compound as white crystals (89.1 mg, 78%), m.p. 53–55 °C. IR, 3040, 3030, 2960, 2950, 2840, 1730, 1440, 1335, 1260 cm^{-1}. ^1H NMR, 1.34 (m, 2H, spin saturation at δ 2.25, NOE 12%), 1.78 (m, 1H), 2.06 (m, 1H), 2.25 (m, 2H), 2.40 (m, 1H), 2.77 (dq, $J = 17$, 3 Hz, 1H, spin saturation at δ 3.16, NOE 32%), 3.16 (dt, $J = 17$, 1 Hz, 1H, spin saturation at δ 2.77, NOE 28%, spin saturation at δ 3.30, NOE 12%); 3.30 (broad d, $J = 12$ Hz, 1H, irradiation at 2.25 broad, spin saturation at δ 3.16, NOE 12%), 3.77 (s, 6H), 5.00 (m, 2H), 6.88 (m, 2H, spin saturation at δ 2.25, NOE 9%). ^{13}C NMR, 171.90 (s), 171.69 (s), 151.44 (s), 134.71 (d), 132.48 (d), 105.83 (t), 61.52 (s), 52.33 (q), 52.02 (q), 50.30 (d), 47.41 (d), 41.01 (t), 33.13 (t), 28.55 (t), 25.91 (t). Mass spectrum, m/z 264 (M^+, 3%), 204 (22), 189 (7), 172 (9), 145 (100). High-resolution mass spectrum: found, 264.1362; $C_{15}H_{20}O_4$ requires 264.1362.

5.5.18 4(S), 1′(R),2″(R)-2,2-DIMETHYL-4-(2-CARBOMETHOXY-4-METHYLENECYCLOPENTYL)-1,3-DIOXOLANE (EQUATION 5.135) [145b]

To 160 mg (0.71 mmol) of palladium acetate in 11 ml of dry toluene under nitrogen were added 726 mg (3.48 mmol) of triisopropyl phosphite, 2.00 g (10.7 mmol) of 2-[(trimethylsilyl)methyl]allyl acetate and 2.00 g (10.7 mmol) of the unsaturated ester. The resulting solution was heated at

100 °C under nitrogen for 16 h. After cooling, the solution was concentrated. Flash chromatography (hexane–diethyl ether, 4:1) gave 2.260 g (87%) of 4(*S*),1′(*R*),2″(*R*)-2,2-dimethyl-4-(2-carbomethoxy-4-methylenecyclopentyl)-1,3-dioxolane and its 1′(*S*),2′(*S*)-diastereomer as a colorless liquid at room temperature which crystallized at −25 °C. IR($CDCl_3$), 1730, 1650, 1430, 1375, 1365 cm^{-1}. ^1H NMR (270 MHz, $CDCl_3$), δ 4.84 (br s, 2H), 4.16–3.90 (m, 2H), 3.66 (s), 3.65 (s, 3H, overall), 3.53 (m, 1H), 2.76–2.21 (m, 6H), 1.36 (s), 1.34 (s, 3H, overall), 1.30 (s), 1.28 (s, 3H, overall). ^{13}C NMR ($CDCl_3$), δ 175.5, 175.1, 148.3, 108.9, 108.6, 106.5, 106.3, 78.3, 76.7, 67.6, 67.4, 51.5, 46.9, 46.7, 45.8, 45.4, 36.9, 34.7, 33.9, 26.4, 26.1, 25.3, 25.1. Analysis: calculated for $C_{13}H_{20}O_4$, 240.1361; found, 240.1361.

5.5.19 3-[(TRIMETHYLSILYL)ETHYNYL]BENZALDEHYDE (SECTION 5.4.5) [168]

A turbid solution of 107 g (0.578 mol) of 3-bromobenzaldehyde, 92.0 g (0.939 mol) of ethynyltrimethylsilane, 1.5 g of palladium(II) acetate and 3.0 g of triphenylphosphine in 500 ml of deaerated, anhydrous triethylamine was rapidly heated to gentle reflux under argon. At *ca* 100 °C, a clear yellow solution resulted, and a white precipitate began to form after 15 min at reflux. After 4 h, the mixture was cooled and the white crystalline solid of triethylamine hydrobromide was isolated by filtration; 105 g (0.577 mol, 99.8%). The orange–brown filtrate was concentrated, mixed with 500 ml of aqueous sodium hydrogencarbonate and extracted with dichloromethane (3 × 300 ml). The organic fractions were combined, dried over magnesium sulfate and concentrated to yield an oil which was purified by distillation to yield analytically pure 3-[(trimethylsilyl)ethynyl]benzaldehyde; 93.5 g (0.463 mol, 80.2%); b.p. 120–122 °C (0.15 Torr). IR (film), 2958 (m, sharp, SiC—H), 2825 (m, sharp, H—CO), 2146 (m, sharp, C≡C), 1692 (vs. Br, C=O), 1244 (m, sharp, Si—C), 843 cm^{-1} (s, Br, Si—C bending). ^1H NMR ($CDCl_3$), δ 0.22 (s, 9H, $SiCH_3$), 7.15–9.93 (m, 4H, aromatic), 9.85 (s, 1H, CHO). Mass spectrum (70 eV), m/z (relative intensity, %) 202 (16.4 M^+), 187 (100, $M^+ - CH_3$). Analysis: calculated for $C_{12}H_{14}OSi$, C 71.24, H 6.97, Si 13.88; found, C 71.10, H 7.07, Si 14.04%.

5.6 ACKNOWLEDGEMENTS

The author thanks John Masters for collecting and organizing the material dealing with transmetallation from tin and Professors Trost and Oppolzer for providing unpublished experimental procedures.

5.7 REFERENCES

1. In 1991, over 250 papers dealing with palladium in organic synthesis appeared. For a detailed but out-of-date treatment of this topic, see R. F. Heck, *Palladium Reagents in Organic Synthesis*, Academic Press, London, 1985.
2. J. Tsuji, *Synthesis*, 369 (1984).
3. S.-I. Murahashi, T. Hosokawa, *Acc. Chem. Res.* **23** (1990) 49.
4. A. Kasahara, T. Izumi, M. Ooshima, *Bull. Chem. Soc. Jpn.* **47** (1974) 2526.
5. B. A. Pearlman, J. M. McNamara, I. Hasan, S. Hatakeyama, H. Sekiyaki, Y. Kishi, *J. Am. Chem. Soc.* **103** (1981) 4248.
6. M. F. Semmelhack, A. Zask, *J. Am. Chem. Soc.* **105** (1983) 2034.
7. R. C. Larock, N. H. Lee, *J. Am. Chem. Soc.* **113** (1991) 7815.
8. D. E. Korte, L. S. Hegedus, R. K. Wirth, *J. Org. Chem.* **42** (1977) 1329.

9. J.-E. Bäckvall, S. E. Byström, R. E. Nordberg, *J. Org. Chem.* **49** (1984) 4619.
10. (a). J.-E. Bäckvall, R. B. Hopkins, H. Grennberg, M. M. Madder, A. K. Awastä, *J. Am. Chem. Soc.* **112** (1990) 5160; (b) H. Grennberg, A. Gogoll, J.-E. Bäckvall, *J. Org. Chem.* **56** (1991) 5808.
11. J.-E. Bäckvall, J. O. Vågberg, *J. Org. Chem.* **53** (1988) 5695.
12. J.-E. Bäckvall, P. G. Andersson, *J. Org. Chem.* **56** (1991) 2274.
13. B. Åkermark, J. E. Bäckvall, L. S. Hegedus, K. Zetterberg, K. Siirala-Hansen, K. Sjöberg, *J. Organomet. Chem.* **72** (1974) 127.
14. L. S. Hegedus, J. M. McKearin, *J. Am. Chem. Soc.* **104** (1982) 2444.
15. L. S. Hegedus, G. F. Allen, J. J. Bozell, E. L. Waterman, *J. Am. Chem. Soc.* **100** (1978) 5800; for a review on the use of palladium in the synthesis and functionalization of indoles, see L. S. Hegedus, *Angew. Chem., Int. Ed. Engl.* **27** (1988) 1113.
16. P. R. Weider, L. S. Hegedus, H. Asada, S. V. D. Andrea, *J. Org. Chem.* **50** (1985) 4276.
17. (a) L. S. Hegedus, R. E. Williams, T. Hayashi, *J. Am. Chem. Soc.* **102** (1980) 4973; (b) L. S. Hegedus, W. E. Darlington, *J. Am. Chem. Soc.* **102** (1980) 4980.
18. (a) L. S. Hegedus, G. Wieber, E. Michalson, B. Åkermark, *J. Org. Chem.* **54** (1989) 4649; (b) J. Montgomery, G. Wieber, L. S. Hegedus, *J. Am. Chem. Soc.* **112** (1990) 6255.
19. J. J. Masters, L. S. Hegedus, J. Tamariz, *J. Org. Chem.* **56** (1991) 5666.
20. (a) A. S. Kende, B. Roth, P. J. Sanfilippo, T. J. Blacklock, *J. Am. Chem. Soc.* **104** (1982) 5808; (b) A. S. Kende, B. Roth, P. J. Sanfilippo, *J. Am. Chem. Soc.* **104** (1982) 1784.
21. (a) L. E. Overman, F. M. Knoll, *Tetrahedron Lett.* (1979) 321; (b) B. T. Golding, C. Pierpont, R. Aneja, *J. Chem. Soc., Chem. Commun.* (1981) 1030; (c) M. M. L. Crilley, B. T. Golding, C. Pierpont, *J. Chem. Soc., Perkin Trans. 1* (1988) 2061; (d) for a review, see L. E. Overman, *Angew. Chem., Int. Ed. Engl.* **23** (1984) 579.
22. (a) P. A. Grieco, T. Takegawa, S. L. Bongers, H. Tanaka, *J. Am. Chem. Soc.* **102** (1980) 7587; (b) P. Martes, P. Perfetti, J.-P. Zahra, B. Waegell, *Tetrahedron Lett.* **32** (1991) 765.
23. T. Ikariya, Y. Ishikawa, K. Hirai, S. Yoshikawa, *Chem. Lett.* (1982) 1815.
24. (a) Y. Tamru, M. Kagotani, Z. Yoshida, *J. Org. Chem.* **45** (1980) 5221; *Tetrahedron Lett.* (1981) 4245; (b) J. Garin, E. Melendez, F. L. Merchan, T. Tejero, S. Uriel, J. Ayestaran, *Synthesis* (1991) 147.
25. P. R. Auburn, J. Whelan, B. Bosnich, *Organometallics* **5** (1986) 1533.
26. R. Grigg, J. Markandu, *Tetrahedron Lett.* **32** (1991) 279.
27. L. E. Overman, A. F. Renaldo, *J. Am. Chem. Soc.* **112** (1990) 3945.
28. N. Bluthe, M. Malacria, J. Gore, *Tetrahedron Lett.*. **24** (1983) 1157.
29. (a) B. M. Trost, *Acc. Chem. Res.* **23** (1990) 34, and references cited therein; (b) B. M. Trost, P. A. Hipskind, J. Y. L. Chung, C. Chan, *Angew. Chem., Int. Ed. Engl.* **28** (1989) 1502.
30. B. M. Trost, Y. Shi, *J. Am. Chem. Soc.* **113** (1991) 701.
31. For reviews on orthopalladation, see M. I. Bruce, *Angew. Chem., Int. Ed. Engl.* **16** (1977) 73; J. Rehand, M. Pfeffer, *Coord. Chem. Rev.* **18** (1976) 327; I. Omae, *Chem. Rev.* **79** (1979) 287; *Coord. Chem. Rev.* **32** (1980) 235; A. D. Ryabov, *Synthesis* (1985) 233.
32. B. J. Brisdon, P. Nair, S. F. Dyke, *Tetrahedron* **37** (1981) 173.
33. H. Horino, N. Inoue, *J. Org. Chem.* **46** (1981) 4416.
34. K. Tamao, K. Sumitani, Y. Kiso, M. Zembayashi, A. Fujioka, S.-I. Kodama, I. Nakajima, A. Minato, M. Kumada, *Bull. Chem. Soc. Jpn.* **49** (1976) 1958.
35. K. Tamao, S. Kodama, I. Nakayima, M. Kumada, A. Minato, K. Suzuki, *Tetrahedron* **38** (1982) 3347.

36. H. P. Dang, G. Linstrumelle, *Tetrahedron Lett.* (1978) 191.
37. S.-I. Murahashi, M. Yamamura, K. Yanagesawa, M. Mita, K. Kondo, *J. Org. Chem.* **44** (1979) 2408.
38. (a) M. Kobayashi, E.-I. Negishi, *J. Org. Chem.* **45** (1980) 5223; (b) E.-I. Negishi, Z. Owezarczyk, *Tetrahedron Lett.* **32** (1991) 6683.
39. C. E. Russell, L. S. Hegedus, *J. Am. Chem. Soc.* **105** (1983) 943.
40. M. A. Tius, J. Gomez-Galeno, X.-Q. Gu, J. H. Zaidi, *J. Am. Chem. Soc.* **113** (1991) 5775.
41. N. Jabri, A. Alexakis, J. F. Normant, *Tetrahedron Lett.* **22** (1981) 959.
42. J. Schwartz, J. Labinger, *Angew. Chem., Int. Ed. Engl.* **15** (1976) 333.
43. N. Okukado, D. E. Van Horn, W. Dlima, E. Negishi, *Tetrahedron Lett.* (1978) 1027.
44. P. Vincent, J. P. Beaucourt, L. Pichat, *Tetrahedron Lett.* **23** (1982) 63.
45. E. Negishi, N. Okukada, A. O. King, D. E. Van Horn, B. I. Spiegel, *J. Am. Chem. Soc.* **100** (1978) 2254; E. Negishi, *Acc. Chem. Res.* **15** (1982) 340.
46. S. Baba, E. Negishi, *J. Am. Chem. Soc.* **98** (1976) 6729.
47. N. Miyaura, K. Yamada, A. Suzuki, *Tetrahedron Lett.* (1979) 3437; N. Miyaura, H. Suginome, *Tetrahedron Lett.* **22** (1981) 127.
48. N. Miyaura, A. Suzuki, *J. Chem. Soc., Chem. Commun.* (1979) 866; N. Miyaura, K. Maeda, H. Suginome, A. Suzuki, *J. Org. Chem.* **47** (1982) 2117.
49. N. Satoh, I. Ishiyama, N. Miyaura, A. Suzuki, *Bull. Chem. Soc. Jpn.* **60** (1987) 3471.
50. J.-I. Uenishi, J. M. Bean, R. W. Armstrong, Y. Kishi, *J. Am. Chem. Soc.* **109** (1987) 4756.
51. M. J. Sharp, W. Cheng, V. Snieckus, *Tetrahedron Lett.* **28** (1987) 5093; W. Cheng, V. Snieckus, *Tetrahedron Lett.* **28** (1987) 5097.
52. T. Alves, A. B. de Oliveira, V. Snieckus, *Tetrahedron Lett.* **29** (1988) 2135.
53. J.-M. Fu, M. J. Sharp, V. Snieckus, *Tetrahedron Lett.* **29** (1988) 5459; M. A. Siddiqui, V. Snieckus, *Tetrahedron Lett.* **29** (1988) 5463; M. A. Siddiqui, V. Snieckus, *Tetrahedron Lett.* **31** (1990) 1593; for a review, see V. Snieckus, *Chem. Rev.* **90** (1990) 879.
54. D. Muller, J. P. Fleury, *Tetrahedron Lett.* **32** (1991) 2229.
55. J. K. Stille, *Angew. Chem., Int. Ed. Engl.* **25** (1986) 508.
56. A. S. Kende, B. Roth, P. J. Sanfilippo, T. J. Blacklock, *J. Am. Chem. Soc.* **104** (1982) 5808.
57. J. W. Labadie, D. Tueting, J. K. Stille, *J. Org. Chem.* **48** (1983) 4634; J. W. Labadie, J. K. Stille, *J. Am. Chem. Soc.* **105** (1983) 669.
58. J. K. Stille, B. L. Groh, *J. Am. Chem. Soc.* **109** (1987) 813; J. K. Stille, J. H. Simpson, *J. Am. Chem. Soc.* **109** (1987) 2138.
59. M. E. Krolski, A. F. Renaldo, D. E. Rudisill, J. K. Stille, *J. Org. Chem.* **53** (1988) 1170; S. E. Tunney, J. K. Stille, *J. Org. Chem.* **52** (1987) 748.
60. D. Rudisill, L. A. Castonguay, J. K. Stille, *Tetrahedron Lett.* **29** (1988) 1509.
61. A. Kalivretenos, J. K. Stille, L. S. Hegedus, *J. Org. Chem.* **56** (1991) 2883.
62. W. F. Gore, M. E. Wright, P. D. Davis, S. S. Labadie, J. K. Stille, *J. Am. Chem. Soc.* **106** (1984) 6417.
63. A. M. Echavarren, J. K. Stille, *J. Am. Chem. Soc.* **109** (1987) 5478; for a review on the coupling reactions of enol triflates, see W. J. Scott, J. E. McMurray, *Acc. Chem. Res.* **21** (1988) 47.
64. J. E. McMurray, W. J. Scott, *Tetrahedron Lett.* **24** (1983) 979; G. T. Crisp, W. J. Scott, *Synthesis* (1985) 334.
65. E. Laborde, L. E. Lesheski, J. S. Kiely, *Tetrahedron Lett.* **31** (1990) 1837.
66. M. Peña, J. K. Stille, *Tetrahedron Lett.* **28** (1987) 6573; *J. Am. Chem. Soc.* **111** (1989) 5417.

67. V. Farina, S. R. Baker, C. Sapino, Jr, *Tetrahedron Lett.* **29** (1988) 6043; V. Farina, S. R. Baker, D. A. Benigni, S. I. Hauck, C. Sapino, Jr, *J. Org. Chem.* **55** (1990) 5833.
68. A. M. Echavarren, J. K. Stille, *J. Am. Chem. Soc.* **110** (1988) 1557.
69. G. T. Crisp, W. J. Scott, J. K. Stille, *J. Am. Chem. Soc.* **106** (1984) 7500.
70. J. K. Stille, D. H. Hill, P. Schneider, M. Tanaka, D. L. Morrison, L. S. Hegedus, *Organometallics* **10** (1991) 1993.
71. G. Roth, C. Sapino, *Tetrahedron Lett.* **32** (1991) 4073.
72. G. Roth, C. E. Fuller, *J. Org. Chem.* **56** (1991) 3493.
73. For an early, limited example, see J. Yoshida, K. Tamao, M. Takahashi, M. Kumada, *Tetrahedron Lett.* (1978) 2161.
74. (a) Y. Hatanaka, T. Hiyama, *J. Org. Chem.* **53** (1988) 970; (b) Y. Hatanaka, Y. Ebina, T. Hiyama, *J. Am. Chem. Soc.* **113** (1991) 7075.
75. For a review, see R. F. Heck, *Pure Appl. Chem.* **50** (1978) 691; for carbonylation of simple halides, see J. K. Stille, P. Kwan Wong, *J. Org. Chem.* **40** (1979) 532; R. F. Heck, A. Schoenberg, *J. Org. Chem.* **39** (1974) 3327.
76. (a) S. Cacchi, P. G. Ciattini, E. Moreta, G. Ortar, *Tetrahedron Lett.* **27** (1986) 3931; (b) R. E. Dolle, S. J. Schmidt, L. I. Kruse, *J. Chem. Soc., Chem. Commun.* (1987) 904.
77. C. J. Rizzo, A. B. Smith, *Tetrahedron Lett.* **29** (1988) 2793.
78. S. K. Thompson, C. H. Heathcock, *J. Org. Chem.* **55** (1990) 3004.
79. A. Cowell, J. K. Stille, *J. Am. Chem. Soc.* **102** (1980) 4193; L. D. Martin, J. K. Stille, *J. Org. Chem.* **47** (1982) 3630.
80. M. Mori, K. Chiba, Y. Ban, *J. Org. Chem.* **43** (1978) 1684.
81. M. Mori, K. Chiba, M. Okita, Y. Ban, *J. Chem. Soc., Chem. Commun.* (1979) 698; *Tetrahedron* **41** (1985) 387.
82. J. W. Tilley, D. L. Coffen, B. H. Shaer, J. Lind, *J. Org. Chem.* **52** (1987) 2469.
83. I. Shimoyama, Y. Zang, G. Wu, E.-I. Negishi, *Tetrahedron Lett.* **31** (1990) 2841.
84. V. P. Baillargeon, J. K. Stille, *J. Am. Chem. Soc.* **108** (1986) 452.
85. R. F. Heck, *Org. React.* **27** (1982) 345.
86. G. D. Daves, Jr., A. Hallberg, *Chem. Rev.* **89** (1989) 1433.
87. R. McCrindle, G. Ferguson, G. J. Arsenault, A. J. McAlees, D. K. Stephanson, *J. Chem. Res. (S)* (1984) 360.
88. C. M. Andersson, K. Karabelas, A. Hallberg, C. Andersson, *J. Org. Chem.* **50** (1985) 3891.
89. (a) T. Jeffry, *Tetrahedron Lett.* **26** (1985) 2667; *J. Chem. Soc., Chem. Commun.* (1984) 1287; *Synthesis* (1987) 70; (b) R. C. Larock, B. E. Baker, *Tetrahedron Lett.* **29** (1988) 905.
90. C. Amatore, M. Azzabi, A. Jutland, *J. Am. Chem. Soc.* **113** (1991) 8375.
91. K. Hirota, Y. Kitade, Y. Isobe, Y. Maki, *Heterocycles* **26** (1987) 355.
92. P. J. Harrington, L. S. Hegedus, *J. Org. Chem.* **49** (1984) 2657.
93. R. N. Farr, R. A. Outten, J. C.-Y. Chen, G. D. Daves, Jr, *Organometallics* **9** (1990) 3151.
94. M. M. Abelman, T. Oh, L. E. Overman, *J. Org. Chem.* **52** (1987) 4133.
95. R. J. Sundberg, R. J. Cherney, *J. Org. Chem.* **55** (1990) 6028.
96. E.-I. Negishi, H. Sawada, J. M. Tour, Y. Wei, *J. Org. Chem.* **53** (1988) 915.
97. Y. Zhang, E.-I. Negishi, *J. Am. Chem. Soc.* **111** (1989) 3454.
98. M. A. Abelman, L. E. Overman, *J. Am. Chem. Soc.* **110** (1988) 2328.
99. Y. Zhang, G.-Z. Wu, G. Agnel, E.-I. Negishi, *J. Am. Chem. Soc.* **112** (1990) 8590.
100. F. E. Meyer, P. J. Parsons, A. de Meijere, *J. Org. Chem.* **56** (1991) 6487.
101. S. Torii, H. Okumoto, A. Nishimura, *Tetrahedron Lett.* **32** (1991) 4167.

102. J. M. Nuss, B. H. Levine, R. A. Rennels, M. M. Heravi, *Tetrahedron Lett.* **32** (1991) 5243.
103. G. Fournet, G. Balme, J. Gore, *Tetrahedron* **47** (1991) 6293.
104. B. M. Trost, T. R. Verhoeven, *J. Am. Chem. Soc.* **102** (1980) 4730.
105. D. Eren, E. Keinan, *J. Am. Chem. Soc.* **110** (1988) 4356.
106. J. Tsuji, *Pure Appl. Chem.* **61** (1989) 1673.
107. B. M. Trost, T. P. Klun, *J. Am. Chem. Soc.* **103** (1981) 1864.
108. B. M. Trost, G.-H. Kuo, T. Bennecki, *J. Am. Chem. Soc.* **110** (1988) 621.
109. B. M. Trost, *Angew. Chem., Int. Ed. Engl.* **28** (1989) 1173.
110. J. P. Genet, F. Piau, *J. Org. Chem.* **46** (1981) 7414.
111. J.-E. Bäckvall, J.-O. Vågberg, K. L. Granberg, *Tetrahedron Lett.* **30** (1989) 617.
112. B. M. Trost, M. Ohmori, S. A. Boyd, H. Okawara, S. J. Brickner, *J. Am. Chem. Soc.* **111** (1989) 8281.
113. B. M. Trost, J. T. Hane, P. Metz, *Tetrahedron Lett.* **27** (1986) 5695.
114. B. M. Trost, E. Keinan, *J. Am. Chem. Soc.* **100** (1978) 7779; *J. Org. Chem.* **44** (1979) 3451.
115. A. Tenaglia, B. Waegell, *Tetrahedron Lett.* **29** (1988) 4851.
116. S. E. Bystrom, R. Arslanian, J.-E. Bäckvall, *Tetrahedron Lett.* **26** (1985) 1745.
117. B. M. Trost, T. S. Scanlan, *J. Am. Chem. Soc.* **111** (1989) 4988.
118. S. A. Godleski, J. D. Meinhardt, D. J. Miller, S. Van Wallendael, *Tetrahedron Lett.* **22** (1981) 2247.
119. B. M. Trost, J. Casey, *J. Am. Chem. Soc.* **104** (1982) 6881.
120. B. M. Trost, A. Tenaglia, *Tetrahedron Lett.* **29** (1988) 2927.
121. R. C. Larock, N. H. Lee, *J. Org. Chem.* **56** (1991) 6253.
122. E. Keinan, N. Greenspoon, *J. Org. Chem.* **48** (1983) 3545.
123. J. K. Sutherland, G. B. Tometzki, *Tetrahedron Lett.* **25** (1984) 881.
124. D. Eren, E. Keinan, *J. Am. Chem. Soc.* **110** (1988) 4356.
125. F. M. Hauser, R. Tommasi, P. Hewawasam, Y. S. Rho, *J. Org. Chem.* **53** (1988) 4886.
126. R. Deziel, *Tetrahedron Lett.* **28** (1987) 4371.
127. Y. Hayakawa, H. Kato, M. Uchiyama, H. Kajimo, R. Noyori, *J. Org. Chem.* **51** (1986) 2400.
128. S. F. Martin, S. K. Davidsen, *J. Am. Chem. Soc.* **106** (1984) 6431.
129. L. Del Valle, J. K. Stille, L. S. Hegedus, *J. Org. Chem.* **55** (1990) 3019.
130. M. Kasugi, K. Ohashi, K. Akuzawa, T. Kawazoe, H. Sano, T. Migita, *Chem. Lett.* (1987) 1237.
131. D. R. Tueting, A. M. Echavarren, J. K. Stille, *Tetrahedron* **45** (1989) 979; A. M. Echavarren, D. R. Tueting, J. K. Stille, *J. Am. Chem. Soc.* **110** (1988) 4039.
132. J. Godschalx, J. K. Stille, *Tetrahedron Lett.* **21** (1980) 2599.
133. V. Farina, S. R. Baker, D. A. Benigni, C. Sapino, Jr, *Tetrahedron Lett.* **29** (1988) 5739; see also ref. 67.
134. S. Katsumura, S. Fujiwara, S. Isoe, *Tetrahedron Lett.* **28** (1987) 1191.
135. F. K. Sheffy, J. P. Godschalx, J. K. Stille, *J. Am. Chem. Soc.* **106** (1984) 4833.
136. J. H. Merrifield, J. P. Godschalx, J. K. Stille, *Organometallics* **3** (1984) 1108.
137. For reviews, see W. Oppolzer, *Angew. Chem., Int. Ed. Engl.* **28** (1989) 38; *Pure Appl. Chem.* **62** (1990) 1941.
138. W. Oppolzer, R. J. DeVita, *J. Org. Chem.* **56** (1991) 6256.
139. R. Grigg, V. Sridharan, S. Sukirthalingam, *Tetrahedron Lett.* **32** (1991) 3855.
140. B. M. Trost, J. I. Luengo, *J. Am. Chem. Soc.* **110** (1988) 8239.
141. W. Oppolzer, J. M. Gaudin, T. N. Birkinshaw, *Tetrahedron Lett.* **29** (1988) 4705; W. Oppolzer, J. M. Gaudin, *Helv. Chim. Acta* **70** (1987) 1478.

142. W. Oppolzer, H. Bienayame, A. Genevas-Borella, *J. Am. Chem. Soc.* **113** (1991) 9660.
143. W. Oppolzer, J.-Z. Xu, C. Stone, *Helv. Chim. Acta* **74** (1991) 465.
144. For reviews, see B. M. Trost, *Angew. Chem., Int. Ed. Engl.* **25** (1986) 1; *Pure Appl. Chem.* **60** (1988) 1615.
145. (a) B. M. Trost, S. M. Mignani, *Tetrahedron Lett.* **27** (1986) 4137; (b) B. M. Trost, J. Lynch, P. Renant, D. H. Steinman, *J. Am. Chem. Soc.* **108** (1986) 284.
146. B. M. Trost, T. N. Nanninga, T. Satoh, *J. Am. Chem. Soc.* **107** (1985) 721.
147. B. M. Trost, P. R. Seoane, *J. Am. Chem. Soc.* **109** (1987) 615.
148. B. M. Trost, T. A. Grese, D. M. T. Chan, *J. Am. Chem. Soc.* **113** (1991) 7350.
149. B. M. Trost, T. A. Grese, *J. Am. Chem. Soc.* **113** (1991) 7350.
150. B. M. Trost, S. M. King, *J. Am. Chem. Soc.* **112** (1990) 408.
151. P. Binger, H. M. Büch, *Top. Curr. Chem.* **135** (1987) 77.
152. P. Binger, E. Sternberg, U. Wittig, *Chem. Ber.* **120** (1987) 1933.
153. R. T. Lewis, W. B. Motherwell, M. Shipman, *J. Chem. Soc., Chem. Commun.* (1988) 949.
154. For reviews, see J. Tsuji, *Organic Synthesis with Palladium Complexes*, Springer, Berlin, 1980; *Top. Curr. Chem.* **91** (1980) 30; *Pure Appl. Chem.* **53** (1981) 2371; **54** (1982) 197; *Acc. Chem. Res.* **6** (1973) 8; *Adv. Organomet. Chem.* **17** (1979) 141; *Ann. N.Y. Acad. Sci.* **333** (1980) 250.
155. J. M. Takacs, J. Zu, *J. Org. Chem.* **54** (1989) 5193.
156. J. M. Takacs, S. Chandramouli, *Organometallics* **9** (1990) 2877.
157. J. M. Takacs, J. Zu, *Tetrahedron Lett.* **31** (1990) 1117.
158. H. A. Dieck, R. F. Heck, *J. Organomet. Chem.* **93** (1975) 259; S. Takahashi, Y. Kuvoyama, K. Sonogashura, N. Hagihara, *Synthesis* (1980) 627; and many others.
159. P. Magnus, H. Annoura, T. Harling, *J. Org. Chem.* **55** (1990) 1709.
160. Z.-y. Yang, D. J. Burton, *Tetrahedron Lett.* **31** (1990) 1369.
161. L. Crombre, M. A. Horsham, R. J. Blade, *Tetrahedron Lett.* **28** (1987) 4879.
162. R. J. Butlin, A. B. Holmes, E. MacDonald, *Tetrahedron Lett.* **29** (1988) 2989.
163. T. R. Hoye, P. R. Hanson, A. C. Kovelesky, T. D. Ocain, Z. Zhuang, *J. Am. Chem. Soc.* **113** (1991) 9369.
164. J. Mascareñas, L. A. Savandeses, L. Castedo, A. Mouriño, *Tetrahedron* **47** (1991) 3485.
165. *Inorg. Synth.* **13** (1972) 121.
166. T. Ukai, H. Kawazawa, Y. Ishii, J. J. Bonnett, J. A. Ibers, *J. Organomet. Chem.* **65** (1974) 253.
167. *J. Med. Chem.* **30** (1987) 185.
168. W. B. Austin, N. Bilow, W. J. Kelleghan, K. S. Y. Lau, *J. Org. Chem.* **46** (1981) 2282.

6

Organoboron Chemistry

KEITH SMITH
University College of Swansea, UK

Organometallics in Synthesis—A Manual. Edited by M. Schlosser

6.1 INTRODUCTION

The discovery of the hydroboration reaction [1] in the late 1950s signalled a new era in synthetic organic chemistry by providing convenient access for the first time to organoboranes, a class of reagents which was later to prove to be of unrivalled versatility. During the 1960s, much effort, almost entirely on the part of the group of H. C. Brown, was put into the exploration of the scope of the hydroboration reaction for the synthesis of organoboranes, despite the fact that such compounds had shown few signs of being synthetically useful. This latter situation changed dramatically in the 1970s as a number of specialist organoborane research groups joined Brown's group in exploring the reactions of organoboranes. They discovered that organoboranes, despite low reactivity in traditional reactions of organometallic reagents such as Grignard reagents, were possessed of properties which gave rise to an enormous diversity and range of other synthetically useful reactions. Against this background, many non-specialist synthetic groups began in the 1980s to utilize and indeed to extend the useful reactions of organoboranes, and an article predicted that the 1990s would begin to see the use of such reactions commercially [2]. There is already evidence that this prediction will be vindicated.

Despite the evident utility of organoboranes as reagents and the inexorable trend in their utilization, however, many practising organic chemists remain reluctant to take advantage of their potential. It is in an attempt to rectify this situation that this contribution is made.

The reasons for reluctance to embrace the new opportunities may be several:

(i) fear of possible dangers associated with unfamiliar materials which might be toxic, pyrophoric or in other ways hazardous;
(ii) uncertainty about how to handle unfamiliar chemicals;
(iii) anecdotal accounts of difficulties experienced by others in inducing reactions to work as reported;
(iv) lack of knowledge of reactions which would be useful to the synthetic challenges being undertaken;
(v) insufficient confidence in ability to choose the best reagent for the job.

Safety. The sorts of boron compounds of interest for organic synthesis carry no extreme toxic hazards. Naturally, like any reactive materials, they should be treated with care and ingestion or contact should be avoided, but most of the compounds are readily decomposed into boric acid, which is a relatively minor toxic hazard. The toxicity of any more stable compounds which might be produced would most likely be unknown, but normal safety precautions associated with handling organic chemicals should be sufficient in most cases.

Handling. Most boron–carbon bonds are stable to water, although some unsaturated boron compounds, especially allylboranes, provide exceptions. However, boron–hydrogen, boron–oxygen, boron-nitrogen and boron–halogen bonds are generally fairly easily hydrolysed; since almost all precursors of organoboranes involve one or other of these types of bonds, it is usual to carry out reactions in dry solvents.

Very few organoboranes are pyrophoric. Those which are have significant volatility, e.g. trimethylborane, triethylborane and triallylborane, and even they can be handled safely enough under an inert atmosphere. Less volatile compounds, such as higher trialkylboranes, oxidize readily in air but without ignition. Increasing hindrance around the boron atom or attachment of electron-releasing substituents such as hydroxy groups to the boron atom decreases the sensitivity to oxygen, and some compounds can even be handled in air without detriment. Nevertheless, it is usual for reactions to be carried out routinely under an atmosphere of nitrogen or exceptionally argon. Mixtures are usually opened to air only after an oxidizing mixture has been admitted, in order to avoid potential side-reactions resulting from oxygen-induced radical chain reactions.

The apparatus used can be extremely simple; a dry round-bottomed flask equipped with a magnetic follower and a septum is adequate for many purposes. The system is flushed with nitrogen via syringe needles, safety vented via a paraffin oil bubbler or by use of a rubber balloon connected via a needle, charged or sampled by means of syringes and stirred magnetically. Although more sophisticated apparatus is required for reflux, filtration or other types of manipulation, there is generally no requirement for apparatus which is more complex than that used for handling other organometallic reagents.

Reproducibility. Some early reports of hydroboration reactions emphasized the *in situ* generation of borane from sodium tetrahydroborate (borohydride) and, for example, boron trifluoride etherate. Such a method is adequate for simple hydroboration–oxidation of an unsubstituted alkene, but is not recommended for anything more complex because of possible side-reactions involving the tetrahydroborate, the Lewis acid or any other component of the mixture. Unjustified use of the *in situ* procedure may have led to some disappointments, particularly in the early days. Similarly, when understanding was less than it is today there may have been attempts to carry out reactions which had little real chance of success. Such cases can lead to anecdotal accounts of difficulties which carry weight beyond their significance. However, the numerous publications by non-specialists which demonstrate successful application of organoborane reactions testify to the fact that reported reactions are reproducible. Some attention to detail, such as use of dry solvents and control of stoichiometry, may be necessary, but given such attention the reactions are just as reliable as any other organic reactions.

Selection of reaction and reagent. Whereas the range of reactions of most classes of organometallic reagents is very narrow and easily assimilated with the help of a simple mechanistic rationale, the situation for organoboranes is much more complex. First, there are many mechanistically diverse types of reaction, involving simple organic group transfers, intramolecular rearrangements, pericyclic processes, transition metal-catalysed cross-couplings, radical reactions, boron-stabilized carbanions, etc. Second, there can be up to four different organic groups attached to boron, some of which may be involved in the reaction whereas others are merely throw-away blocking groups. Alternatively, some of the organic groups may be replaced by halogen, alkoxy or other functionalities. The non-specialist is confronted with a bewildering array of possibilities.

For this reason, the following sections are highly selective and free of mechanistic discussion. Only well tried and tested reactions are included and the emphasis is on providing a simple procedure which can be immediately used by a non-specialist. The literature contains more extensive works which include further details, including mechanistic discussion [3, 4].

Commercial availability. Many simple boron compounds are available from most chemical suppliers. Aldrich Chemical has a particularly wide selection of boron compounds as a result of its own subsidiary, Aldrich-Boranes. In particular, the company supplies borane complexed with a variety of Lewis bases, a number of trialkylboranes, a number of alternative hydroborating agents such as dibromoborane–methyl sulfide complex, catecholborane and 9-BBN, a range of trialkoxyboranes and various other boron compounds. The company is able to supply bulk quantities of a number of these reagents and provides special packaging methods for the air-sensitive reagents.

6.2 SIMPLE HYDROBORATION–OXIDATION

Simple hydroboration-oxidation (e.g., equation. 1) can be carried out with any hydroborating agent [3, 4]. The most convenient and cost-effective is borane-dimethyl sulfide, which is commercially available as a neat liquid or as a solution in dichloromethane. It offers possibilities for use of various solvents and for large scale work.

6.2.1 PREPARATION OF (−)-CIS-MYRTANOL BY HYDROBORATION–OXIDATION OF (−)-β-PINENE [4]

$$3\,(\beta\text{-pinene}) + BH_3.SMe_2 \xrightarrow[0\text{–}25\,°C]{\text{hexane}} (\text{myrtanyl})_3B \xrightarrow[EtOH,\ 40°C]{NaOH, H_2O_2} 3\,(\text{cis-myrtanol}) \quad (6.1)$$

A dry, 2 l, three-necked flask fitted with a mechanical stirrer, a septum-capped pressure-equalizing dropping funnel and a reflux condenser vented via a bubbler is flushed with nitrogen and then charged with (−)-β-pinene (238 ml, 1.5 mol) and hexane (500 ml). It is cooled in an ice-bath (to dissipate the heat generated during reaction) and borane dimethyl sulfide (52.5 ml, neat liquid, 0.55 mol) is added from the dropping funnel, with stirring, over 30 min. The cooling bath is removed and the mixture is stirred for 3 h at 25 °C to ensure complete reaction, by which time the flask contains a solution of tri-*cis*-myrtanylborane.

Ethanol (500 ml) is added cautiously (on account of initial evolution of hydrogen arising from excess borane) followed by aqueous sodium hydroxide (185 ml of 3 mol l^{-1} solution). The flask is again immersed in an ice–water-bath and hydrogen peroxide (185 ml of 30% aqueous solution) is added at such a rate that the temperature does not exceed 40 °C. The cooling bath is removed and the reaction completed by heating at 50 °C for 1 h. The mixture is poured into ice–water (5 l), mixed with diethyl ether (2 l) and then separated. The organic layer is washed with water (2 × 1 l) and saturated sodium chloride (1 l), dried over potassium carbonate, filtered and evaporated to give a light yellow oil (230 g). Short-path distillation gives (−)-*cis*-myrtanol (196 g, 85%).

The above procedure is satisfactory for many alkenes. However, more hindered alkenes may require longer reaction times or more forcing conditions at either stage of the reaction. If necessary, more concentrated peroxide (50% or even 60% solution) may be used in the oxidation stage. Some alkenes react only as far as the dialkylborane (R_2BH) or monoalkylborane (RBH_2) stage, and in such cases the ratio of borane to alkene is raised from 1:3 to 1:2 or 1:1, respectively, or an excess of borane is used. (*Caution*—a much larger quantity of hydrogen will be evolved in such cases.)

If regioselectivity of hydroboration should be a problem (hydroboration of 1-hexene gives 94% attachment of B to C-1; styrene gives 80% attachment to B to C-1; internal alkenes give poor regioselectivity), the most generally useful reagent for giving high regioselectivity is the commercially available dialkylborane, 9-borabicyclo[3.3.1]nonane (9-BBN-H) [5]. This must be used at 1:1 stoichiometry with the alkene. In the case of hydroboration of alkynes, the most regioselective reagent is dimesitylborane [6], which is also commercially available. Oxidation of a vinylborane, such as is produced in the reaction of an alkyne, requires a buffered peroxide solution in order to minimize hydrolysis of the organoborane intermediate. The product is an aldehyde or ketone.

Finally, if enantioselective hydroboration is required, the reagent of choice is diisopinocampheylborane, dilongifolylborane or monoisopinocampheylborane, depending on the nature of the alkene [7]. These organoboranes are prepared freshly before use (see Section 6.6).

6.3 PREPARATION OF SIMPLE TRIALKYLBORANES AND OTHER TRIORGANYLBORANES

The most convenient and most versatile method for the synthesis of symmetrical trialkylboranes involves the hydroboration of alkenes with borane. The experimental procedure is as described for the preparation of tri-*cis*-myrtanylborane in Section 6.2. It is important that the stoichiometry be close to the theoretical 3:1 (alkene to borane) in order to avoid contamination by dialkylborane species. Alternatively, an excess of alkene may be used provided that the excess is easily removed following reaction. For most purposes, the solution of trialkylborane obtained can be used directly for further reactions, but if necessary the solvent can be removed under reduced pressure and the product can be distilled under nitrogen or under reduced pressure.

Many triorganylboranes cannot be prepared via hydroboration, including aryl-, allyl- and most tertiary-alkylboranes [8]. In such cases the most useful general method for synthesis of the organoborane involves the reaction of a reactive organometallic reagent with trifluoroborane etherate or some similar reagent (e.g. equation 6.2) [9].

6.3.1 PREPARATION OF TRIPHENYLBORANE FROM PHENYLMAGNESIUM BROMIDE AND TRIFLUOROBORANE–DIETHYL ETHERATE [9]

$$3PhMgBr + BF_3.OEt_2 \longrightarrow Ph_3B + 3MgBrF + Et_2O \qquad (6.2)$$

A dry, 4 l, three-necked flask equipped with a mechanical stirrer, a 1 l pressure-equalizing dropping funnel capped with a septum and a still-head assembled for distillation with a condenser and 2 l receiver is flushed with nitrogen, vented through a bubbler connected to the receiver adaptor. The flask is charged (double-ended needle) with trifluoroborane diethyl etherate (142 g, 1 mol) and xylene (1 l). A freshly prepared and estimated ethereal solution of phenylmagnesium bromide (3 mol in *ca* 1 l of diethyl ether) is added dropwise, via the dropping funnel, with stirring, at 25–35 °C over a period of 3 h (slow addition so as to minimize the production of tetraphenylborate). A small amount of ether distils during the addition and as soon as the addition is complete the temperature is raised to allow the remainder of the ether to distil out (stopped when the xylene begins to distil at 138 °C). The distillation set-up is replaced by a bent sinter tube attached to a 3 l, two-necked receiver vented through a bubbler and the still hot solution is forced through the sinter under a slight nitrogen pressure. The residual salts are extracted with hot (120–130 °C) xylene (2 × 500 ml) and the combined xylene extracts are distilled, without a fractionating column, under reduced pressure. After removal of the solvent and a small forerun boiling below 155 °C at 0.1 mmHg, crude triphenylborane distils at 155–166 °C (0.1 mmHg). A single recrystallization from heptane under nitrogen gives pure triphenylborane (217 g, 90%), m.p. 148 °C, which is essentially pure.

6.4 PREPARATION OF MONOORGANYLBORANES AND DIORGANYLBORANES

Hydroboration of some alkenes with borane proceeds rapidly only as far as the monoalkylborane or dialkylborane stage and more slowly thereafter [3]. Hence, by careful control of the stoichiometry and reaction temperature, it is possible, in such cases, to produce the appropriate mono- or dialklylborane cleanly. In many other cases this direct production of mono- or dialkylboranes via hydroboration is not possible.

Perhaps the most important dialkylborane of all is 9-borabicyclo[3.3.1]nonane (9-BBN-H; **1**) [10], which is commercially available as a crystalline solid. It is sufficiently stable in air to be transferred quickly without special precautions, although it should always be stored under an inert atmosphere. The preparation of this dialkylborane is slightly more complicated than that of other simple dialkylboranes and non-specialists are therefore advised to use the commercial material.

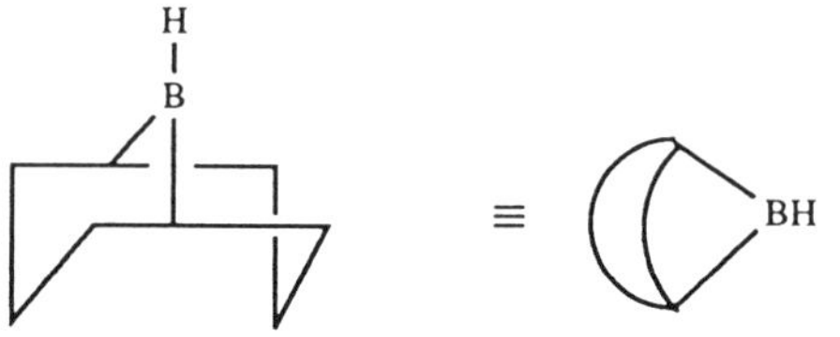

H
B
≡ BH

(1)

The types of alkenes which readily give dialkylboranes are trialkylethenes such as 2-methyl-2-butene and relatively hindered 1,2-dialkylethenes such as cyclohexene. 2-Methyl-2-butene gives the so-called disiamylborane, while cyclohexene gives dicyclohexylborane (equation 6.3). It is usual to produce them *in situ* immediately prior to their application in further reactions.

6.4.1 PREPARATION OF DICYCLOHEXYLBORANE [11]

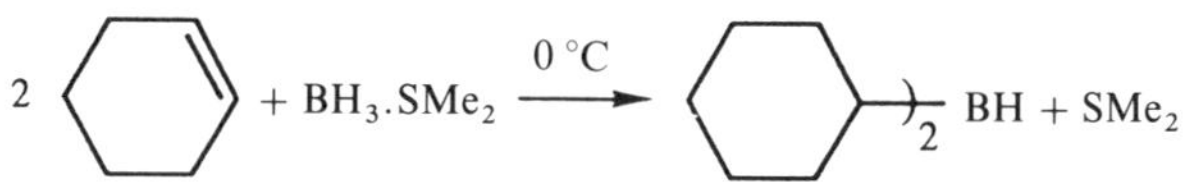

$$2\,C_6H_{10} + BH_3.SMe_2 \xrightarrow{0\,^\circ C} (C_6H_{11})_2BH + SMe_2 \qquad (6.3)$$

A 200 ml, round-bottomed, two-necked flask equipped with a pressure-equalizing dropping funnel fitted with a septum, a magnetic follower and a reflux condenser vented through a mercury bubbler is flushed with nitrogen. Cyclohexene (16.4 g, 0.2 mol) and dry diethyl ether (75 ml) are added and the mixture is cooled to 0 °C. Borane dimethyl sulfide (7.7 g, 0.1 mol) is introduced in to the dropping funnel and added dropwise to the stirred solution over 30 min. The dropping funnel is washed through with diethyl ether (25 ml) and the mixture is stirred for 3 h at 0 °C. Dicyclohexylborane (as its dimer) precipitates as white crystals and can be used directly as a suspension for most purposes. If isolation is required, the ether and dimethyl sulfide can be removed by distillation in a slow stream of nitrogen. The product can also be sublimed in vacuum. Its melting point is 103–105 °C.

The most commonly utilized monoalkylborane is thexylborane, which is not very stable over prolonged periods and therefore has to be carefully and freshly prepared (equation 6.4).

6.4.2 PREPARATION OF THEXYLBORANE [4]

$$Me_2C{=}CMe_2 + BH_3.THF \longrightarrow Me_2CHCMe_2BH_2 \qquad (6.4)$$

A dry, two-necked 200 ml flask fitted with a septum inlet, a magnetic follower and an outlet leading to a mercury bubbler is flushed with nitrogen. Borane tetrahydrofuran (100 ml of 1 mol l^{-1} solution) is added via a double-ended needle and the flask is cooled in an ice–salt bath at −10 to −15 °C. A 2 mol l^{-1} solution of 2,3-dimethyl-2-butene (50 ml) is added dropwise, with stirring, over 30 min by syringe, whilst the temperature is maintained at or below 0 °C. The mixture is then stirred for 2 h at 0 °C to complete the reaction. The clear solution so produced is used directly for further reactions.

The preparations of chiral mono- and dialkylboranes, also available by hydroboration, are described in Section 6.6. Some diorganylboranes cannot be prepared by hydroboration and in such cases organometallic reagents are generally used. An example is the preparation of dimesitylborane (equation 6.5), although this compound is also available commercially.

6.4.3 *PREPARATION OF DIMESITYLBORANE [3, 6]*

$$2\text{MesBr} \xrightarrow{2\text{Mg}} 2\text{MesMgBr} \xrightarrow{BF_3.OEt_2} \text{Mes}_2\text{BF} \xrightarrow{1/4\text{LiAlH}_4} \text{Mes}_2\text{BH} \quad (6.5)$$

(Mes = mesityl = 2,4,6-trimethylphenyl)

A dry, 250 ml, two-necked flask equipped with a septum-capped pressure-equalizing dropping funnel and a reflux condenser leading to a mercury bubbler is charged with magnesium turnings (2.73 g, 114 mmol) and then flushed with nitrogen. A solution of mesityl bromide (22.3 g, 112 mmol) in tetrahydrofuran (THF, 56 ml) is transferred into the dropping funnel by a double-ended needle and then the magnesium turnings are heated gently with an air gun. The mesityl bromide solution is added dropwise at such a rate as to give a constant reflux. After completion of the addition the reaction mixture is heated at 80–90 °C for 3 h, cooled to ambient temperature, diluted with THF (30 ml) and transferred by a double-ended needle into a 100 ml volumetric flask. The solution is made up to 100 ml with further THF and an aliquot is standardized by titration against 0.2 M HCl. The yield is *ca* 98%.

Another 250 ml, two-necked flask equipped with a septum-capped pressure-equalizing dropping funnel and a mechanical stirrer is flushed with nitrogen. The Grignard reagent prepared as above (94 ml of 1.1 mol l^{-1} solution, 104 mmol) is transferred into the funnel and the flask is charged with trifluoroborane etherate (7.44 g, 52 mmol) which has been distilled from calcium hydride. The flask is immersed in an ice–water bath and the Grignard reagent is added, at such a rate as to maintain the temperature of the reaction mixture below about 30 °C, with rapid stirring. After completion of the addition the mixture is stirred for a further 1 h at ambient temperature, the flask is disconnected and the mixture is left overnight in a deep-freeze, stoppered with a septum and under nitrogen. The supernatant liquid is transferred by a double-ended needle into another flask and the solvent is removed under reduced pressure (with protection from moisture by a drying tube). Light petroleum (b.p. 30–40 °C, 30 ml) is added to precipitate magnesium fluoride and after the mixture has settled the supernatant liquid is transferred into another nitrogen-flushed flask. The residue is washed with further light petroleum (3 × 30 ml) and the combined supernatant liquids are concentrated to *ca* 25 ml, then set aside in a deep-freeze for 18 h. The product crystallizes out and the mother liquor is removed and further concentrated to produce a second crop. Solvent is removed from the combined crystals under reduced pressure (oil pump) to give dimesitylfluoroborane (9.08 g, 70%), m.p. 70–72 °C.

A dry 500 ml flask equipped with a magnetic follower and a septum is connected via a needle to a mercury bubbler and flushed with nitrogen. Dimesitylfluoroborane (8.4 g, 50 mmol), prepared as described above, in 1,2-dimethoxyethane (50 ml) is charged via a syringe or double-ended needle and the solution is stirred whilst a solution of lithium aluminium hydride in 1,2-dimethoxyethane (63 ml of 0.2 mol l^{-1} solution, 12.6 mmol) is added dropwise through a double-ended needle. A white precipitate forms and stirring is maintained for 1 h. Dry benzene (100 ml) is added and the mixture is stirred for 30 min, then allowed to settle. The supernatant liquid is removed via a double-ended needle and the residue is washed with benzene (3 × 100 ml) in a similar manner. The combined supernatant liquids are allowed to settle overnight at ambient temperature, then filtered rapidly through a sintered-glass funnel containing a 5 cm thick layer of Celite under a stream of nitrogen. The clear filtrate is concentrated under reduced pressure (protected by a drying tube) and the crude product (9 g) is crystallized from 1,2-dimethoxyethane (*ca* 60 ml) to give colourless crystals of dimesitylborane (5.2 g, 69%), m.p. 164–166 °C.

The recent report of the preparation of 2,4,6-triisopropylphenylborane ($TripBH_2$) [12] suggests that this bulky monoarylborane may resemble the bulky monoalkylborane, thexylborane, in some of its properties.

6.5 PREPARATION OF 'MIXED' TRIORGANYLBORANES

Early attempts to synthesize 'mixed' trialkylboranes of the type $R^1R^2R^3B$ were dogged by failure, with the result that there was a view that such compounds were not stable and that they were prone to redistribution to give mixtures of the symmetrical trialkylboranes and other 'mixed' trialkylboranes (e.g. $R^1_2R^2B$). However, it is now well established that this is not the case [3, 8]. When 'redistribution' products are obtained it is the result of failure to produce the intermediate mono- and dialkylboron compounds cleanly. The key to success is therefore to ensure that the intermediate compounds are produced cleanly, and then to control the conditions carefully during their subsequent conversion into 'mixed' trialkylboranes. It is often not possible to produce a particular mono- or dialkylborane directly via hydroboration, in which case it must be made in other ways (see preparation of dimesitylborane in Section 6.4 and of other products in Section 6.8). Section 6.4 gives procedures for the clean preparation of examples of mono- and dialkylboranes which are amenable to the direct approach.

If the subsequent conversion into a 'mixed' trialkylborane involves hydroboration of a single alkene, as in the preparation of dicyclohexyl-1-octylborane by reaction of dicyclohexylborane with 1-octene (equation 6.6) or of thexyldicyclopentylborane by reaction of thexylborane with cyclopentene, the procedure is very simple. It consists simply of adding the alkene, in stoichiometric amount or slight excess, to the solution or suspension of the mono- or dialkylborane under nitrogen and allowing the mixture to stir at -10 to $+20$ °C for a period of time. Once reaction is complete, modest temperatures cease to be any problem.

6.5.1 PREPARATION OF DICYCLOHEXYL-1-OCTYLBORANE

$$(C_6H_{11})_2BH + CH_2{=}CH(CH_2)_5Me \longrightarrow (C_6H_{11})_2B-CH_2CH_2(CH_2)_5Me \qquad (6.6)$$

Dicyclohexylborane (0.1 mol) is prepared as a suspension in diethyl ether as described in Section 6.4.1. The reaction flask is cooled in ice–water to dissipate heat of reaction and the mixture is stirred whilst 1-octene (11.2 g, 100 mmol) is added dropwise. After completion of the addition the reaction mixture is stirred for 3 h at ambient temperature. (With more hindered alkenes it may be necessary to heat the mixture to 50 °C at this stage.) The suspended dicyclohexylborane dissolves during the reaction to produce a clear solution of dicyclohexyl-1-octylborane, which can be used directly in subsequent reactions.

For similar reactions using thexylborane it is necessary to keep the reaction mixture at *ca* -10 °C during addition of the alkene [13]. Even under such conditions, however, it is not possible to control the stoichiometry to produce a thexylmonoalkylborane cleanly from the least hindered alkenes such as 1-octene. Fortunately, it is possible to achieve this in the case of most internal alkenes and a further hydroboration reaction can then be used to give a fully 'mixed' trialkylborane (e.g. equation 6.7).

6.5.2 PREPARATION OF THEXYLCYCLOPENTYL(6-ACETOXYHEXYL)BORANE [13]

$$\text{Thx}{-}BH_2 \xrightarrow{\text{cyclopentene}} \text{Thx}{-}B(C_5H_9)H \xrightarrow{H_2C{=}CH(CH_2)_4OAc} \text{Thx}{-}B(C_5H_9)(CH_2)_6OAc \qquad (6.7)$$

A 200 ml flask equipped with a magnetic follower and a septum and vented by a needle leading to a paraffin oil bubbler is flushed with nitrogen. Thexylborane (100 mmol) is prepared as described in Section 6.4.2 and the reaction flask is then immersed in a cooling bath set at -10 °C. The mixture is stirred at -10 °C during the dropwise addition, by syringe, of cyclopentene (6.8 g, 100 mmol), for a further 1 h thereafter and then during the subsequent dropwise addition of 6-acetoxy-1-hexene (14.2 g, 100 mmol). The mixture is stirred for a further 1 h at -10 °C and then allowed to warm to room temperature to give a solution containing the desired organoborane ready for direct utilization in subsequent reactions.

The preceding procedure is applicable only to cases in which the first alkene is a 1,2-disubstituted ethene or a trisubstituted ethene and the second alkene is a 1-alkene, a 1,1-disubstituted ethene or a 1,2-disubstituted ethene. If other fully 'mixed' trialklyl-boranes are required a more complicated procedure is needed (e.g. equation 6.8).

6.5.3 PREPARATION OF THEXYLDECYLOCTYLBORANE [14, 15]

$$\text{Thx}{-}B(H)Cl \xrightarrow{C_6H_{13}CH{=}CH_2} \text{Thx}{-}B((CH_2)_7Me)Cl \xrightarrow[C_8H_{17}CH{=}CH_2]{K(^iPrO)_3BH} \text{Thx}{-}B((CH_2)_7Me)(CH_2)_9Me \qquad (6.8)$$

A dry, 100 ml flask equipped with a magnetic follower and a septum and vented via a needle leading to a paraffin oil bubbler is flushed with nitrogen. Thexylchloroborane dimethyl sulfide (20 mmol) in dichloromethane (20 ml) is prepared *in situ* (see Section 6.8), then cooled to 0 °C and stirred during addition of 1-octene (2.24 g, 20 mmol). The mixture is warmed to 25 °C and stirred for 2 h, then cooled to -10 °C (ice–salt bath). A solution of 1-decene (2.80 g, 20 mmol) in THF (20 ml) is added, followed by the dropwise addition of potassium triisopropoxyhydroborate (21 ml of a commercial 1 mol l^{-1} solution in THF, 21 mmol), with vigorous stirring. The mixture is stirred for 2 h at 0 °C to produce a solution of the desired product, which can be used directly.

In some cases it is either necessary or more convenient to introduce the final organic group(s) via an organometallic reagent rather than via a hydroboration reaction. The organometallic reagents generally chosen are either Grignard reagents or organolithium reagents, although others are also possible [3, 8]. The choice of the leaving group from boron can be critical. In the case of fairly hindered compounds it is reasonable to use halogenoboron compounds (e.g. equation 6.9) [3, 16]. In the case of relatively unhindered compounds, however, the triorganylborane formed may react rapidly with further organometallic reagent, resulting in the formation of some tetraorganylborate salt which can be difficult to remove and causing a lowering of the yield. In such cases it is advantageous to use a poorer leaving group, such as methoxide. The immediate product is then a methoxyborate which has to be broken down, for example by use of boron trifluoride. An example of this approach is given in Section 6.7, with the preparation of a dialkylalkynylborane.

$$\mathrm{Mes_2BF} \xrightarrow{\mathrm{EtMgBr}} \mathrm{Mes_2BEt + FMgBr} \tag{6.9}$$

6.6 PREPARATION OF CHIRAL ORGANOBORANES

The simplest way of preparing useful chiral organoboranes is via hydroboration of a chiral alkene with borane dimethyl sulfide. If the alkene is sufficiently unreactive that it can be cleanly converted into a dialkylborane (see Section 6.4), this latter species can act as a chiral hydroborating agent for the asymmetric hydroboration of prochiral alkenes. The most widely used example of such a dialkylborane is diisopinocampheylborane (Ipc_2BH) [17, 18], prepared from optically active α-pinene, which is readily available. High enantioselectivity is achieved in its reactions with (*Z*)-1,2-disubstituted ethenes such as *cis*-2-butene (equation 6.10) [3, 4]. This is therefore a very useful method for the preparation of chiral organoboranes containing the corresponding asymmetric units.

6.6.1 PREPARATION OF (R)-2-BUTYLDIISOPINOCAMPHEYLBORANE [3, 4]

$BH_3.SMe_2$ (6.10)

A 250 ml, two-necked flask equipped with a septum inlet, a magnetic follower and a distillation head leading to a condenser with a cooled (−78 °C), bubbler-vented receiver is flushed with nitrogen. The flask is charged with THF (15 ml) and neat borane dimethyl sulfide (5.05 ml, 50 mmol), cooled to 0 °C and stirred during the dropwise addition of (−)-α-pinene [15.9 ml, 100 mmol, α_D^{23} − 48.7, corresponding to 95% enantiomeric excess (ee)]. After a further 3 h at 0 °C, a mixture of the solvent and dimethyl sulfide (total 13 ml) is removed under reduced pressure (*ca* 30 mmHg; protection from moisture by a drying tube). [If α-pinene of 99% ee is used, the Ipc_2BH produced at this point is ready for direct further reaction with (*Z*)-2-butene, but with the lower purity material described here an equilibration step is now required.] The distillation set-up is rapidly replaced with a stopper and further (−)-α-pinene (2.4 ml, 15 mmol) and THF (18 ml) are added to the reaction flask. The mixture is left to equilibrate for 3 days at 0 °C to give a white suspension of diisopinocampheylborane of *ca* 99% ee. This suspension can be used directly or the excess of α-pinene can be removed (with THF) by syringe and fresh THF added.

The flask is cooled to *ca* −10 °C by immersion in an ice–salt bath and *cis*-2-butene (3.1 g, 55 mmol) is added, in solution if necessary. After stirring for 4 h at 0 °C the temperature is allowed to come to ambient and the solution contains the desired asymmetric organoborane, diisopinocampheyl-(*R*)-2-butylborane (50 mmol), available for further reaction. Oxidation (see Section 6.2) gives (*R*)-2-butanol in very high enantiomeric purity.

Diisopinocampheylborane also readily hydroborates 1,1-disubstituted ethenes, but in such cases the chiral induction is low [19]. It does not successfully hydroborate more hindered alkenes such as *trans*-1,2-disubstituted ethenes or trisubstituted ethenes because displacement of α-pinene competes with hydroboration. As a result, several different hydroborating agents are present in the mixture, each one exhibiting different stereoselectivity, which can lead overall to very low asymmetric induction. Dilongifolylborane is less prone to alkene displacement and less hindered than diisopinocampheylborane and consequently gives better asymmetric induction than the latter in reactions with the more hindered alkenes. Its preparation from optically active longifolene is straightforward, like that of dicyclohexylborane (Section 6.4).

An alternative reagent for hydroboration of relatively hindered alkenes is monoisopinocampheylborane ($IpcBH_2$, equation 6.11) [20]. This reagent presents an additional advantage because it can be reacted with an equimolar amount of such a hindered alkene to give the corresponding alkylisopinocampheylborane, which, like other dialkylboranes, tends to crystallize from the solution as its hydrogen-bridged dimer. By recrystallization, these compounds can be obtained with close to 100% ee and in substantial yield. Further, α-pinene can then be displaced by reaction with acetaldehyde to give a chiral alkyldiethoxyborane (e.g. equation 6.12). These compounds are useful for conversion into many other compounds containing one chiral organic group attached to boron [21].

6.6.2 PREPARATION OF OPTICALLY PURE MONOISOPINOCAMPHEYLBORANE

$$2Ipc_2BH \xrightarrow{\text{TMEDA}} (IpcBH_2)_2.TMEDA \xrightarrow{2BF_3.OEt_2} IpcBH_2 \qquad (6.11)$$

(−)-Diisopinocampheylborane (100 mmol) in diethyl ether (65 ml) is prepared as described above in a two-necked flask fitted with a magnetic follower, a nitrogen inlet and a reflux condenser leading to a mercury bubbler. The mixture is brought to reflux and tetramethylethylenediamine (TMEDA, 7.54 ml, 50 mmol) is added dropwise. The mixture is held at reflux for 30 min and then a small aliquot is withdrawn into a syringe and pushed back into the mixture (this process helps to induce crystallization). The mixture is allowed to cool to room temperature and then kept at 0 °C overnight, after which time the supernatant liquid is removed by means of a double-ended needle. The crystalline TMEDA complex of monoisopinocampheylborane is washed with pentane (3 × 25 ml) and the solid is dried for 1 h at 15 mmHg (drying tube needed to protect from moisture) and 2 h at 1 mmHg to give optically pure $(IpcBH_2)_2$.TMEDA (16.4 g, 79%), m.p. 140.5–141.5 °C, $[\alpha]_D^{23}$ + 69.03 °C (THF). This solid can be stored for prolonged periods and free monoisopinocampheylborane can be liberated when required, as described below.

The complex (14.6 g, 35 mmol) is charged to a 250 ml flask equipped with a magnetic follower and a septum and the flask is flushed with nitrogen. THF (50 ml) is added and the mixture is stirred until the solid dissolves. Trifluoroborane etherate (8.6 ml, 70 mmol) is added with constant stirring and the mixture is stirred for a further 1.25 h at 25 °C and then filtered under nitrogen through a sinter tube (transfer by a double-ended needle). The solid residue is washed with ice-cold THF (3 × 9 ml) and the washings are added to the original supernatant liquid. This combined solution contains monoisopinocampheylborane (typically 80–84% yield) ready for further reaction. It is advisable to check the quantity by gas titration [4].

6.6.3 PREPARATION OF OPTICALLY PURE DIETHOXY(2-PHENYLCYCLOPENTYL)BORANE [22]

1. $IpcBH_2$
2. crystallize
3. 2MeCHO

(6.12)

A 25 ml, two-necked flask equipped with a magnetic follower, a septum inlet and a tube leading to a mercury bubbler is flushed with nitrogen and then charged with a solution of mono-isopinocampheylborane [from (+)-α-pinene] in diethyl ether (52.6 ml of 0.95 mol l^{-1} solution, 50 mmol). The flask is cooled to −35 °C, 1-phenylcyclopentene (7.2 g, 50 mmol) in diethyl ether (10 ml), precooled to −35 °C, is added dropwise and with stirring and the mixture is then kept at −35 °C for 36 h without stirring. A white solid separates and the supernatant liquid is removed by a double-ended needle. The crystals are washed with cold diethyl ether (3 × 20 ml) at −35 °C. Diethyl ether (50 ml) is added to the solid and the mixture is warmed to 20 °C for 5–10 min, whereupon a homogeneous solution is obtained. It is then cooled to 0 °C for 15 h and white crystalline needles separate. The supernatant solution is removed by a double-ended needle and the crystalline solid is washed with ice-cold diethyl ether (3 × 10 ml). The solid is then dried at 15 mmHg for 1 h to yield isopinocampheyl-(1*S*,2*S*)-*trans*-2-phenylcyclopentyl)borane (10.3 g, 70%) of very high optical purity (>99% ee).

The solid dialkylborane (10.3 g, 35 mmol) is suspended in diethyl ether (30 ml), acetaldehyde (7.84 ml, 140 mmol) is added, and the mixture is stirred at room temperature for 6 h. Excess of aldehyde and diethyl ether are pumped off and the residue is distilled under reduced pressure to give diethoxy-(1*S*,2*R*)-*trans*-2-phenylcyclopentyl)borane (8.0 g, 32.5 mmol, 65% based on original monoisopinocampheylborane and alkene), b.p. 80–82 °C/0.01 mmHg. Oxidation of an aliquot of this compound in the standard way (see Section 6.2) gives (+)-*trans*-2-phenylcyclopentanol of >99% ee.

The optically active alkyldiethoxyboranes prepared as described above can be converted into other compounds by any of the other organoborane reactions described herein, ranging from reduction to the monoalkylborane and further hydroboration to direct cleavage via oxidation. They are therefore extremely important intermediates [21].

An alternative way of obtaining chiral organoboranes involves reaction of an alkyldialkoxyborane derived from the chiral diol, pinanediol, with dichloromethyllithium (e.g. equation 6.13) [23]. The α-chloroalkylboron compounds thus produced are frequently of very high enantiomeric purity and the chloride can be easily displaced in a stereodefined way to give more complex optically active species (equation 6.13).

6.6.4 PREPARATION OF (1*R*)-1-PHENYLPENTYL-[(+)-PINANEDIYLDIOXY]BORANE VIA HOMOLOGATION OF THE CORRESPONDING BUTYL DERIVATIVE [23]

1. $LiCHCl_2$, −100 °C
2. $ZnCl_2$, 20 °C
3. PhMgBr, −78 → 25 °C

(6.13)

Butyl[(+)-pinanediyldioxy]borane. (+)-Pinanediol (341 g, 2 mol) in diethyl ether (500 ml) and light petroleum (b.p. 30–40 °C, 500 ml) is stirred with boric acid (62.5 g, 1 mol), and a solution of KOH (65 g of 85%, 1 mol) is added in portions, resulting in an exothermic reaction and formation of a voluminous white precipitate, which is collected, washed with diethyl ether, dried and recrystallized twice from acetone–water (90:10). This gives potassium bis(pinanediol)borate (*ca* 45% yield) of *ca* 100% ee even from pinanediol of 92% ee. To a solution of this solid (23 g) in ice-cold water (75 ml) is added a mixture of diethyl ether and light petroleum (1:1, 150 ml), followed by ice-cold hydrochloric acid (65 ml of 2 mol l^{-1}) in small portions, with stirring. After the cloudiness in the aqueous phase has disappeared, the layers are separated and the aqueous phase is saturated with NaCl and extracted with diethyl ether (50 ml). The combined organic phases are treated with butyldihydroxyborane (butylboronic acid) (11.5 g, 113 mmol) and kept at 25 °C for 2 h, then dried over magnesium sulfate, evaporated and distilled under reduced pressure to give butyl[(+)-pinanediyldioxy]borane (*ca* 20 g, 85 mmol, 76%), b.p. 68–70 °C/0.1 mmHg.

(1R)-1-phenylpentyl[(+)-pinanediyldioxy]borane. A 250 ml flask equipped with a magnetic follower and a septum is flushed with argon, charged with dry THF (50 ml) and pure dichloromethane (3.25 ml, 60 mmol), cooled to −100 °C in a 95% ethanol–liquid nitrogen slush bath and stirred as *n*-butyllithium in hexane (25.5 ml of 2.0 mol l^{-1} solution, 51 mmol) is added dropwise by syringe down the side of the flask (so that it is already cold on reaching the reaction solution) over a period of 15 min. [The solution should remain colourless or pale yellow although a white precipitate forms; darkening is a sign of overheating and decomposition.] The solution is stirred at −100 °C for a further 15 min. The butyl(pinanediyldioxy)borane (9.98 g, 42 mmol), prepared as above, in anhydrous diethyl ether (25 ml) is injected and, under a rapid stream of argon, the septum is briefly removed so that rigorously anhydrous powdered zinc chloride (3.8 g, 28 mmol) can be admitted. The septum is replaced and the flask is flushed thoroughly with argon. The mixture and cooling bath are allowed to warm slowly to 25 °C and stirring is maintained overnight. [It is sometimes possible to use the crude solution directly at this stage, but isolation of the intermediate product is described here.] The mixture is then concentrated on a rotary evaporator (bath temperature below 30 °C) and the thick residue is stirred with light petroleum (b.p. 30–40 °C, 100 ml) and then saturated aqueous ammonium chloride (25 ml). The phases are separated, the aqueous phase is washed with light petroleum (2 × 50 ml) and the combined organic extracts are filtered through a bed of anhydrous magnesium sulfate. Concentration yields a residue which is chromatographed on silica, eluted with 4% ethyl acetate in hexane. The product at this stage is usually sufficiently pure for further reaction, but it can be distilled if required to give pure (1*S*)-1-chloropentyl[(+)-pinanediyldioxy]borane (10.3 g, 86%), b.p. 113–115 °C/0.2 mmHg.

A 100 ml flask equipped with a magnetic follower and a septum is flushed with argon and charged with dry THF (35 ml) and the chloropentyl(pinanediyldioxy)borane (2.85 g, 10 mmol), prepared as described above. The solution is cooled to −78 °C and stirred as a solution of phenylmagnesium bromide in THF or diethyl ether (10 ml of 1 mol l^{-1} solution, 10 mmol) is added by syringe. The mixture is allowed to warm to 25 °C and stirred for a further 20 h. Hydrochloric acid (10 ml of 1.2 mol l^{-1}) is added and the product is extracted into diethyl ether (3 × 25 ml). The combined ether extracts are dried over magnesium sulfate, concentrated and then distilled under reduced pressure to give (1*R*)-1-phenylpentyl)[(+)-pinanediyldioxy]borane (2.62 g, 80%), b.p. 125–128 °C/0.1 mmHg.

6.7 PREPARATION OF UNSATURATED ORGANOBORANES

Vinylboranes can in some cases be synthesized by direct hydroboration of an alkyne. This is only possible if the desired product has the right geometry (i.e. if boron and hydrogen are *cis* to each other) and if the nature of the other groups is such that the regioselectivity is appropriate (i.e. terminal alkenyl groups can be obtained from

1-alkynes, whereas symmetrical internal alkynes offer no regioselectivity problems). Even then, attempts to produce alkenylboranes from unhindered hydroborating agents such as borane dimethyl sulfide or even 9-BBN-H may be unsuccessful because of dihydroboration or polymerization. Therefore, in reality, the method is fairly restricted in scope [3, 8]. Nevertheless, for those cases where it is appropriate it is usually the simplest method, and is illustrated by the preparation of dicyclohexyl-1-octenylborane (equation 6.14).

6.7.1 PREPARATION OF DICYCLOHEXYL-(E)-1-OCTENYLBORANE

$$(C_6H_{11})_2BH + HC{\equiv}C(CH_2)_5Me \longrightarrow (C_6H_{11})_2B(H)C{=}C(H)(CH_2)_5Me \qquad (6.14)$$

Dicyclohexylborane (200 mmol) in THF (100 ml) is prepared in a 500 ml flask as described in Section 6.4 (except for solvent). The flask is immersed in an ice–salt bath until the temperature is reduced to *ca* −10 °C and a solution of 1-octyne (22.0 g, 200 mmol) in THF (20 ml) is added as rapidly as possible (to minimize any dihydroboration) whilst maintaining the temperature below 10 °C. The reaction is then allowed to warm to room temperature and stirred for 3 h to complete the hydroboration (the solid dicyclohexylborane dissolves during this process). The solution thus obtained contains the dicyclohexyloctenylborane, which may be used directly for further transformations.

For aryl-, alkynyl-, most allyl- and many alkenylboron compounds hydroboration cannot be used to put in the unsaturated group. In these cases the most versatile method is via the corresponding organolithium or organomagnesium reagents [3, 8]. The procedure is similar whichever class of unsaturated organoborane is required and is illustrated by the preparation of dicyclohexyl-1-octynylborane (equation 6.15).

6.7.2 PREPARATION OF DICYCLOHEXYL-1-OCTYNYLBORANE

$$(C_6H_{11})_2BH \xrightarrow{MeOH} (C_6H_{11})_2BOMe \xrightarrow[\text{ii, } BF_3.OEt_2]{\text{i, } LiC{\equiv}C(CH_2)_5Me} (C_6H_{11})_2B{-}C{\equiv}C(CH_2)_5Me \qquad (6.15)$$

Dicyclohexylborane (100 mmol) in THF (50 ml) is generated *in situ* (see Section 6.4) under nitrogen, in a 500 ml flask equipped with a gas inlet with stopcock, a septum inlet and a magnetic follower. Methanol (3.2 g, 100 mmol) is added, dropwise and with stirring, and the hydrogen evolved is allowed to escape through a bubbler connected via a needle through the septum. The solution is stirred for a further 1–2 h at 25 °C, during which time hydrogen evolution ceases and the dicyclohexylborane solid dissolves to form dicyclohexylmethoxyborane.

Meanwhile, a separate 250 ml flask equipped with a magnetic follower and a septum is flushed with nitrogen and then charged with 1-octyne (11 g, 100 mmol) and THF (75 ml). The mixture is cooled to $-78\ °C$ and *n*-butyllithium (40 ml of a 2.5 mol l^{-1} solution in hexane, 100 mmol) is added dropwise to give 1-lithiooctyne.

The flask containing the dicyclohexylmethoxyborane is cooled in a $-78\ °C$ bath and the 1-lithiooctyne solution is added to it by means of a double-ended needle. The 250 ml flask is rinsed with additional THF (5 ml) to complete the transfer of 1-lithiooctyne and the bulk solution is stirred for 30 min at $-78\ °C$. Trifluoroborane etherate (18.8 g, 133 mmol) is added by syringe and the mixture is stirred at $-78\ °C$ for a further 15 min and then warmed to ambient temperature. The reaction mixture can be used directly for many purposes, but if isolation is required the mixture is first concentrated under reduced pressure and then mixed with pentane. The precipitated material is removed under nitrogen by filtration through a glass sinter and the product can then be obtained from the filtrate by removal of the solvent and distillation under reduced pressure.

6.8 PREPARATION OF OTHER ORGANOBORON COMPOUNDS

For some purposes it is necessary to have a boron compound of the type R_2BX or RBX_2, where X is a methoxy or halo group. Such compounds can be readily obtained by addition of dry HX to the corresponding RBH_2 or R_2BH (see procedure for dicyclohexyl-1-octynylborane in Section 6.7, for example) or, in the case of alkoxy compounds, from the corresponding halo compounds by reaction with dry alcohol. Compounds of the type $RB(OEt)_2$ can be obtained from IpcBRH by reaction with acetaldehyde (Section 6.6) and such compounds can be chain-extended by reaction with the anion of dichloromethane (see Section 6.6) [3]. However, it is also possible to put in the organyl group directly via hydroboration using HBX_2 or H_2BX derivatives. Of these, the most conveniently available are catecholborane (equation 6.16), dibromoborane dimethyl sulfide (equation 6.17) and monochloroborane dimethyl sulfide (equation 6.18). The first two are available commercially, but are easily prepared from catechol and borane [3, 4, 24] or tribromoborane dimethyl sulfide and borane dimethyl sulfide [3] if desired.

6.8.1 PREPARATION OF B-[(E)-2-CYCLOHEXYLETHENYL]CATECHOLBORANE [24]

$$C_6H_4O_2BH + HC{\equiv}C{-}C_6H_{11} \longrightarrow C_6H_4O_2B{-}CH{=}CH{-}C_6H_{11} \qquad (6.16)$$

A 100 ml, two-necked flask equipped with a reflux condenser vented to a mercury bubbler, a PTFE stopcock capped by a silicone-rubber septum and a magnetic follower is flushed with nitrogen. Cyclohexylethyne (10.8 g, 100 mmol) and catecholborane (12.1 g, 100 mmol) are added by means of syringes and the mixture is stirred and heated at 70 °C for 1 h. After cooling, the product is ready for further reaction, but if required it can be distilled at 114 °C/2 mmHg to provide pure *B*-cyclohexylethenylcatecholborane (18.7 g, 82%).

Reactions of catecholborane, such as that described above, are intrinsically slow. Indeed, internal alkynes require longer reaction times (*ca* 4 h at 70 °C) and those of alkenes require even higher temperatures (*ca* 100 °C) [24, 25]. Fortunately, the reactions can be performed at moderate temperatures if catalysed by rhodium complexes [26] and this even allows the possibility of asymmetric hydroboration by use of homochiral catalysts [27].

6.8.2 PREPARATION OF HEXYLDIBROMOBORANE [28–30]

$$Br_3B.SMe_2 \xrightarrow{BH_3.SMe_2} Br_2BH.SMe_2 \xrightarrow{\text{1-hexene}} Br_2B(CH_2)_5CH_3.SMe_2 \xrightarrow{Br_3B} Br_2B(CH_2)_5CH_3 \quad (6.17)$$

A 100 ml flask equipped with a septum and a magnetic follower is flushed with nitrogen and then charged with dimethyl sulfide (14.9 ml, 12.4 g, 200 mmol) and pentane (50 ml). The flask is cooled in an ice-bath and fitted with a bubbler connected via a needle and the contents are stirred vigorously whilst tribromoborane (9.5 ml, 25.3 g, 100 mmol) is added dropwise by syringe (exothermic reaction). The mixture is brought to room temperature and the volatile materials are removed at the pump (protection from moisture) to leave tribromoborane dimethyl sulfide (31.5 g, 99%), m.p. 106–107 °C, as a white powder.

Dimethyl sulfide (10 ml) and borane dimethyl sulfide (4.75 ml, 47.5 mmol) are added, with stirring, and the mixture is held at 40 °C for 12 h. Excess of dimethyl sulfide is removed at the pump (protection from moisture) to leave dibromoborane dimethyl sulfide (35.1 g, *ca* 100%) as a clear, viscous liquid.

A two-necked flask equipped with a reflux condenser vented through a mercury bubbler, a PTFE stopcock capped by a silicone-rubber septum and a magnetic follower is flushed with nitrogen and charged with dichloromethane (75 ml) and 1-hexene (12.5 ml, 100 mmol) by syringe. The mixture is stirred at 25 °C during the dropwise addition of dibromoborane dimethyl sulfide (12.8 ml, 100 mmol), prepared as above, and then heated under reflux for 3 h. On cooling to 25 °C the product, hexyldibromoborane dimethyl sulfide, is ready for many further manipulations.

If required, the product can be distilled (b.p. 97–100 °C/1 mmHg) to give the pure complex (29 g, 91%). If the product free of dimethyl sulfide is needed, the mixture after the hydroboration step is cooled to 0 °C and stirred during addition of triibromoborane (10.0 ml, 105 mmol). The mixture is stirred at 25 °C for 1 h and the solvent is removed to leave a mixture of liquid hexyldibromoborane and solid tribromoborane dimethyl sulfide. This mixture is distilled directly under reduced pressure at a bath temperature less than 100 °C ($Br_3B.SMe_2$ melts at 108 °C) to give pure hexyldibromoborane (18.0 g, 71%), b.p. 56–58 °C/0.9 mmHg.

6.8.3 PREPARATION OF CHLORODICYCLOPENTYLBORANE [3, 31, 32]

$$BH_3.SMe_2 \xrightarrow{CCl_4} BH_2Cl.SMe_2 \xrightarrow{\text{cyclopentene}} (\text{cyclopentyl})_2BCl \quad (6.18)$$

A two-necked flask equipped with a septum, a reflux condenser leading to a mercury bubbler and a magnetic follower is flushed with nitrogen and charged with borane dimethyl sulfide (10.1 ml, 100 mmol). Tetrachloromethane (9.7 ml, 15.4 g, 100 mmol) is added, with stirring, and the mixture is held under reflux for 20 h, by which time the product is predominantly $BH_2Cl.SMe_2$. Estimation of an aliquot by gas titration shows the presence of 2 mole equivalents of hydrogen per mole of boron and hydrolysis and titration of the liberated HCl show 1 mole equivalent of labile chlorine per mole of boron. However, ^{11}B NMR shows the presence of equilibrium quantities (*ca* 8% each) of $Cl_2BH.SMe_2$ (δ −2.0 ppm) and $BH_3.SMe_2$ (δ −19.8 ppm) in addition to $Cl_2BH.SMe_2$ (δ −6.7 ppm).

A 200 ml flask equipped with a septum and a magnetic follower is flushed with nitrogen and charged with cyclopentene (14.7 g, 216 mmol) in pentane or diethyl ether (90 ml). The mixture is stirred at 0 °C whilst the monochloroborane dimethyl sulfide, prepared as described above, is added slowly by syringe. The mixture is then allowed to warm to 25 °C and stirred for 2 h at this temperature. The solution thus obtained can be used directly for further transformations or the solvent may be removed under reduced pressure to leave chlorodicyclopentylborane which is *ca* 93% pure (containing some $RBCl_2$ and R_3B). Distillation under reduced pressure gives the pure product (*ca* 80% yield), b.p. 69–70 °C/1.2 mmHg. Unhindered dialkylchloroboranes may retain the dimethyl sulfide during solvent removal but it is lost during distillation.

Thexylchloroborane dimethyl sulfide is prepared in a manner similar to that used for dicyclopentylchloroborane, but using monochloroborane dimethyl sulphide (100 mmol), dimethyl sulfide (1.5 ml, 20 mmol), dichloromethane (24 ml) and 2,3-dimethyl-2-butene (14.8 ml, 110 mmol). The alkene is added dropwise to the stirred mixture of the other components at 0 °C, over a period of 1 h, and the resultant clear solution is stirred for 30 min at 0 °C and 3 h at 25 °C to complete the preparation. The solution obtained can be used directly or the solvent and excess of dimethyl sulfide can be removed under reduced pressure (protection from moisture).

6.9 REPLACEMENT OF BORON BY A FUNCTIONAL GROUP

The simplest applications of organoboron compounds involve replacement of boron by a functional group such as OH, halogen or NH_2 (equation 6.19) [3, 4]. The most common application is oxidation with alkaline hydrogen peroxide, which converts alkylboron compounds into the corresponding alcohols (see Section 6.2). All stereochemical features of the organoborane are retained in the alcohol and all three groups attached to boron can be utilized.

$$\rangle B{-}R \longrightarrow R{-}X \qquad (X = OH, Cl, Br, I, NH_2) \tag{6.19}$$

In some cases the sensitivity of other functionalities present in the organic product may dictate that oxidants other than alkaline hydrogen peroxide must be used to convert an organoborane into the corresponding alcohol. In such cases, buffered hydrogen peroxide, *m*-chloroperbenzoic acid or trimethylamine *N*-oxide may be used with advantage [3]. Oxidation of vinylboranes gives aldehydes or ketones.

Replacement of boron by bromine or iodine is best carried out by use of sodium methoxide in the presence of the free halogen. In the case of iodinolysis only two of the three alkyl groups on a trialkylborane are readily converted into iodoalkane [33], whereas in the case of brominolysis all three groups can be utilized [34]. In both cases the process occurs with complete inversion of stereochemistry at the displaced carbon atom (e.g. equation 6.20) [35].

$$(\text{norbornyl})_3B + 2I_2 + 3NaOMe \longrightarrow 2\ \text{norbornyl-I} + 2NaI + Na^+\ \text{norbornyl-}\bar{B}(OMe)_3 \tag{6.20}$$

Aryldihydroxyboranes can be converted into aryl bromides [36] and vinyldihydroxyboranes can be converted into vinyl bromides, again with inversion (equation 6.21) [37].

$$\underset{R}{\overset{H}{\diagdown}}C{=}C\underset{H}{\overset{B(OH)_2}{\diagup}} \xrightarrow{Br_2,NaOH} \underset{R}{\overset{H}{\diagdown}}C{=}C\underset{Br}{\overset{H}{\diagup}} \quad (6.21)$$

The procedure for preparation of methyl 11-bromoundecanoate (equation 6.22) is representative.

6.9.1 PREPARATION OF METHYL 11-BROMOUNDECANOATE [34]

$$3CH_2{=}CH(CH_2)_8CO_2Me \xrightarrow[0\,^\circ C]{BH_3.THF} B[(CH_2)_{10}CO_2Me]_3 \xrightarrow[0\,^\circ C]{Br_2.NaOMe} 3Br(CH_2)_{10}CO_2Me \quad (6.22)$$

A two-necked flask equipped with a magnetic follower, a stopcock-protected septum and a pressure-equalizing dropping funnel connected to a bubbler is flushed with nitrogen. THF (75 ml) and methyl 10-undecenoate (29.7 g, 33.5 ml, 150 mmol) are added and the mixture is cooled to 0 °C. Borane THF (50.0 ml of a 1.0 mol l^{-1} solution in THF, 50 mmol) is added dropwise with stirring and the mixture is then stirred for 30 min at 0 °C and 30 min at 25 °C to complete the hydroboration. Methanol (1 ml) is added and the temperature is again lowered to 0 °C. Bromine (10.0 ml, 200 mmol) is added dropwise at such a rate as to maintain a temperature of 0 °C and then a freshly prepared solution of sodium methoxide in methanol (60 ml of a 4.16 mol l^{-1} solution, 250 mmol) is added dropwise and with stirring over 45 min in order that the temperature remains below 5 °C. The reaction mixture is allowed to warm to 20 °C and treated with pentane (50 ml), water (20 ml) and saturated aqueous potassium carbonate (20 ml). The pentane layer is separated and the aqueous layer is extracted with further pentane (3 × 50 ml). The combined pentane extracts are washed with water (2 × 50 ml) and then brine (50 ml), dried over K_2CO_3, filtered and evaporated under reduced pressure to give a colourless oil (41.2 g, 98%). Reduced pressure distillation at 126–128 °C/0.65 mmHg gives methyl 11-bromoundecanoate (35.4 g, 85%) contaminated by a small amount of methyl 10-bromoundecanoate.

There are several ways of converting trialkylboranes into primary amines, of which the use of hydroxylamine-*O*-sulfonic acid [38] and use of *in situ*-generated chloramine (equation 6.23) [39] are worthy of particular note, although a maximum of only two of the alkyl groups can be utilized. There are also possibilities for converting organoboranes into *N*-alkylsulfonamides [40] and other nitrogen derivatives [3].

6.9.2 PREPARATION OF METHYL 11-AMINOUNDECANOATE [39]

$$B[(CH_2)_{10}CO_2Me]_3 \xrightarrow{aq.\,NH_3,\,NaOCl} H_2N(CH_2)_{10}CO_2Me \quad (6.23)$$

Methyl 10-undecenoate (30 mmol) is hydroborated in THF solution as described in the preceding procedure. The solution is cooled to 0 °C and aqueous ammonia (4.9 ml of a 2.05 mol l^{-1} solution, 10 mmol) is added, followed by commercial bleach (15.4 ml of a 0.78 mol l^{-1} solution, 12 mmol), dropwise. A precipitate forms and the suspension is stirred for 5 min at 0 °C and then allowed to warm to room temperature. The mixture is made acidic by addition of 10% hydrochloric acid and then extracted with diethyl ether (2 × 50 ml). The aqueous layer is made alkaline with aqueous NaOH (3 mol l^{-1}) and the product is extracted into diethyl ether (2 × 75 ml). The combined ether layers are washed with brine and then dried over KOH. Removal of the solvent gives methyl 11-aminoundecanoate (1.64 g, 76% based on transformation of just one alkyl group). The method can be applied for incorporation of ^{15}N and it was also stated that the use of two mole equivalents of aqueous ammonia permits the utilization of two of the three alkyl groups [39].

Another important simple cleavage reaction is protonolysis, which is included here although there is no replacement by a functional group. With a few exceptions (such as allylboranes), boron–carbon bonds are not readily cleaved by water. Rather, carboxylic acids are more commonly used to cleave C—B bonds protonolytically, and even then it is necessary to raise the temperature to *ca* 160 °C for effective removal of all three groups (e.g. equation 6.24). The reaction is capable of maintaining all the stereochemical features of the organoborane.

6.9.3 PREPARATION OF CIS-PINANE BY PROTONOLYSIS OF TRIMYRTANYLBORANE [3, 4, 41]

B, 3; $EtCO_2H$, 160 °C, 2 h; 3 (6.24)

A two-necked flask equipped with a magnetic follower, a condenser leading to a mercury bubbler and a stopcock-controlled septum is flushed with nitrogen. Trimyrtanylborane (33 mmol) is prepared as described in Section 6.2 and then the solvent is removed under reduced pressure and replaced by diglyme (33 ml). Degassed propanoic acid (11 ml, *ca* 50% excess) is added and the mixture is stirred and heated under reflux (*ca* 160 °C) for 2 h, then cooled. Excess of aqueous NaOH (3 mol l^{-1}) is added and the diglyme phase is diluted with pentane (50 ml). The organic phase is separated, washed with ice–water (5 × 50 ml) to remove diglyme, dried over $MgSO_4$ and evaporated to give *cis*-pinane (12.4 g, 90%).

Vinylboranes are generally protonolysed more readily than alkylboranes and still yield products with complete retention of configuration [3, 4].

6.10 α-ALKYLATION OF CARBONYL COMPOUNDS VIA ORGANOBORANES

Anions derived by deprotonation of α-halocarbonyl compounds or α-halonitriles, and the chemically related α-diazocarbonyl compounds or α-diazonitriles, react with organoboranes with transfer of an organyl group from boron to the α-carbon atom. Hydrolysis produces the corresponding α-alkyl- or α-aryl-carbonyl compounds or -nitriles (equation 6.25) [3].

$$\gt B{-}R \xrightarrow[2.\ H_2O]{1.\ \bar{C}HCOY\ (X)} RCH_2COY \qquad (6.25)$$

$X = Cl, N_2^+$; $Y = OEt, CH_3$, *etc.*

Only one of the three alkyl groups of a trialkylborane is transferred and for optimum utilization of organic residues it is preferable to use organyldichloroboranes in the case of diazocarbonyl compounds (e.g. equation 6.26) [42] or *B*-alkyl-9-BBN derivatives in the case of α-halocarbonyl compounds (e.g. equation 6.27) [43, 44]. Presumably, organyldibromoboranes will behave in a similar way to organyldichloroboranes.

6.10.1 PREPARATION OF ETHYL p-CHLOROPHENYLACETATE FROM ETHYL DIAZOACETATE AND p-CHLOROPHENYLDICHLOROBORANE [42]

$$Cl{-}C_6H_4{-}BCl_2 \xrightarrow[2.\ H_2O]{1.\ N_2CHCO_2Et} Cl{-}C_6H_4{-}CH_2CO_2Et \qquad (6.26)$$

A 50 ml, two-necked flask equipped with a magnetic follower, a septum-capped stopcock and a line leading to a nitrogen supply and a vacuum pump is evacuated and filled with nitrogen (three repetitions) and then cooled to −25 °C by immersion in an acetone–tetrachloromethane–dry-ice bath. *p*-Chlorophenyldichloroborane [45] (1.94 g, 10 mmol) in THF (10 ml) is added by syringe and then stirred while ethyl diazoacetate (1.25 g, 11 mmol) in THF (10 ml) is added at a rate which allows smooth liberation of nitrogen (about 1 ml every 4 min). With the mixture still stirring at −25 °C, water (5 ml) and methanol (5 ml) are added and the cooling bath is then removed. The mixture is poured into saturated aqueous ammonium carbonate (75 ml) and extracted with diethyl ether (3 × 50 ml). The combined ether extracts are dried over magnesium sulfate and the solvent is then removed under reduced pressure. Distillation under reduced pressure gives ethyl *p*-chlorophenylacetate (1.80 g, 91%), b.p. 106–107 °C/3.5 mmHg. Alkyldichloroboranes, as opposed to aryldichloroboranes, are reacted at lower temperatures, *ca* −62 °C, and generally give lower yields.

6.10.2 PREPARATION OF A STEROID NITRILE BY α-ALKYLATION OF α-CHLOROACETONITRILE [43, 44]

(2) AcO– steroid $\xrightarrow[2.\ ClCH_2CN,\ \text{2,6-di-}t\text{-butyl-4-methylphenoxide } O^-K^+]{1.\ 9\text{-BBN-H}}$ (3) AcO– steroid–CH(CH$_3$)CH$_2$CN (6.27)

Solid 9-BBN-H (0.13 g, 1.05 mmol) is placed in a 50 ml flask equipped with a magnetic follower and a stopcock fitted with a septum and connected to a nitrogen supply and a vacuum pump.

The system is repeatedly evacuated and refilled with nitrogen and then the steroid **2** (0.35 g, 1.03 mmol) dissolved in THF (2 ml) is added by syringe, with stirring. The hydroboration step is allowed to proceed for 15 h at 20 °C and the mixture is then cooled to 0 °C. A slurry of freshly prepared potassium 2,6-di-*tert*-butyl-4-methylphenoxide in THF (2.19 ml of 0.47 mol l^{-1} slurry, 1.03 mmol) is added via a syringe fitted with a wide-bore needle, followed by chloroacetonitrile (65 μl, 1.03 mmol). The mixture is stirred for 1 h at 0 °C, then ethanol (0.4 ml) is added and the whole is stirred for 15 min at room temperature. Hexane (5 ml) is added and the solution is extracted with aqueous sodium hydroxide (3 × 25 ml of a 1 mol l^{-1} solution) and water (2 × 20 ml). The organic phase is dried over magnesium sulfate and concentrated under reduced pressure. Chromatography on silica gel gives pure **3** (0.26 g, *ca* 65% yield), m.p. 179–181 °C.

6.11 KETONES AND TERTIARY ALCOHOLS VIA REACTIONS OF ORGANOBORANES WITH ACYL CARBANION EQUIVALENTS

Acyl carbanion equivalents, or anions with two α-leaving groups, can lead to the transfer of two organyl groups from boron to carbon. Subsequent oxidation leads to the appropriate boron-free product, typically a secondary or tertiary alcohol. Of the various types of acyl carbanion equivalents available, anions of 1,1-bis(phenylthio)alkanes [46] or 2-alkyl-1,3-benzodithioles [47] are the most useful of the readily prepared materials for this application. The former anions are more hindered than the latter and are chosen for reactions with relatively unhindered organoboranes, whereas the latter are used for more hindered organoboranes (e.g. equation 6.28) [47].

6.11.1 PREPARATION OF CYCLOHEXYLDICYCLOPENTYLMETHANOL FROM THEXYLDICYCLOPENTYLBORANE AND 2-CYCLOHEXYL-1,3-BENZODITHIOLE [47]

S, S, H

1. *n*-BuLi
2. $ThxB(c\text{-}C_5H_9)_2$
3. $HgCl_2$
4. H_2O_2/OH^-

OH, C

(6.28)

A 100 ml, three-necked flask equipped with a magnetic follower, a septum-capped stopcock, an angled, rotatable side-arm and a line to a nitrogen supply and vacuum pump is charged with 2-cyclohexyl-1,3-benzodithiole (0.472 g, 2 mmol) in the flask and mercury(II) chloride (1.63 g, 6 mmol) in the side-arm and then repeatedly evacuated and refilled with nitrogen. THF (5 ml) is added to dissolve the dithiole and the solution is cooled to −30 °C. *n*-Butyllithium (1.37 ml of a 1.6 mol l^{-1} solution in hexane, 2.2 mmol) is added by syringe and the mixture is stirred for 75 min at −30 °C.

Meanwhile, in a separate flask, thexyldicyclopentylborane (2 mmol) in THF (5 ml) is prepared as described for other ‘mixed’ organoboranes in Section 6.5. This solution is transferred into the flask containing the organolithium solution by means of a syringe and further THF (2 ml) is used to transfer the last traces. The mixture is allowed to warm to room temperature over 1 h and then stirred for 3 h. [*Note*: for ketone synthesis (see below) the mixture is directly oxidized at this point.]

The mixture is cooled to −78 °C and stirred vigorously as the side-arm is rotated to release the mercury(II) chloride into the flask. The mixture is allowed to warm to room temperature and stirred overnight. [*Note*: for ketone synthesis (see below) the following oxidation procedure is carried out directly on the mixture without introduction of mercury(II) chloride.]

The mixture is cooled to 0 °C and then aqueous sodium hydroxide (10 ml of a 5 mol l^{-1} solution) and 50% aqueous hydrogen peroxide (7 ml) are successively added, the latter dropwise and with care. The mixture is stirred at room temperature for 5 h and the product is extracted into pentane (2 × 100 ml), washed with water (2 × 100 ml), dried over sodium sulfate, filtered and evaporated under reduced pressure. The syrupy residue is chromatographed on alumina (activity III), eluting successively with pentane (100 ml), dichloromethane–pentane (1 : 1, 150 ml) and dichloromethane (250 ml). The dichloromethane fraction is evaporated under reduced pressure and then pumped overnight to remove residual 2,3-dimethyl-2-butanol, leaving cyclohexyldicyclopentylmethanol (0.40 g, 80%).

If the addition of mercury(II) chloride is omitted from the above procedure only a single migration takes place. Oxidation then gives an aldehyde or ketone, depending on the 2-substituent of the benzodithiole moiety. The optimum utilization of boron-bound alkyl groups is achieved by use of *B*-alkyl-9-BBN derivatives, but to avoid unnecessary duplication of procedures, the example below uses a thexyldialkylborane (equation 6.29) [48].

6.11.2 PREPARATION OF CYCLOHEXYL CYCLOPENTYL KETONE VIA REACTION OF THEXYLDICYCLOPENTYLBORANE WITH 2-CYCLOHEXYL-1,3-BENZODITHIOLE

1. BuLi
2. $ThxB(c\text{-}C_5H_9)_2$
3. H_2O_2/OH^-

(6.29)

The procedure is the same as the preceding procedure, except that there is no need for the angled side-arm charged with mercury(II) chloride, until the point when the mercury chloride would have been added. At that point oxidation as described later in the same procedure is carried out. Isolation of the product is also very similar, but silica rather than alumina is used in the chromatography. The yield of the ketone is *ca* 76%.

The anion of dichloromethane has been used in a way similar to that described in this section for benzodithiole anions but on alkyldialkoxyboranes rather than trialkylboranes. Further, if the dialkoxyborane part is generated from an optically active diol the rearrangement step can take place in a highly stereoselective manner to give chiral products with high enantiomeric purities (see Section 6.6.4).

6.12 KETONES, TERTIARY ALCOHOLS AND CARBOXYLIC ACIDS VIA REACTIONS OF ORGANOBORANES WITH HALOFORM ANIONS

Anions derived from dichloromethyl methyl ether (DCME), chloroform or related haloforms react with trialkylboranes with spontaneous migration of all three alkyl groups from boron to carbon. Oxidation yields a tertiary alcohol in which all of the structural features of the trialkylborane have been 'riveted' into the corresponding hydroxycarbon compound (equation 6.30) [49].

$$R_3B \xrightarrow{^{-}CCl_2OMe} (Cl)(MeO)B{-}CR_3 \xrightarrow{[O]} R_3COH \qquad (6.30)$$

With small modifications the procedure can be used to convert even very hindered organoboranes into the corresponding tertiary alcohols (e.g. equation 6.31) [50].

6.12.1 PREPARATION OF CYCLOHEXYLCYCLOPENTYLTHEXYLMETHANOL

$$\text{ThexB(cyclohexyl)(cyclopentyl)} \xrightarrow[\text{3. } HOCH_2CH_2OH \;\; \text{4. } H_2O_2/OH^-]{\text{1. } CHCl_2OMe \;\; \text{2. } Et_3COLi} \text{Thex(cyclohexyl)(cyclopentyl)C{-}OH} \qquad (6.31)$$

A 500 ml, three-necked flask equipped with a magnetic follower, septum-capped stopcock, reflux condenser and mercury bubbler is flushed with nitrogen. Thexylcyclohexylcyclopentylborane (100 mmol) in THF (100 ml) is prepared in the flask by a procedure analogous to that described for other thexyldialkylboranes in Section 6.5. The solution is cooled to 0 °C and stirred whilst purified DCME (25.3 g, 0.22 mol, excess) is added, followed by a solution of lithium triethylmethoxide (111 ml of a 1.8 mol l^{-1} solution in hexane, 0.20 mol, excess) dropwise over a period of 20–30 min. The mixture is brought to room temperature and stirred for a further 30 min, during which a heavy precipitate of lithium chloride forms. Ethylene glycol (12.4 g, 0.20 mol, excess) is added and the volatile materials are removed at the pump. Ethanol (95%, 50 ml) and THF (25 ml) are added and then solid NaOH (24 g, 0.6 mol). When most of the solid has dissolved, hydrogen peroxide (30%, 50 ml) is added, cautiously and with stirring, over 2 h, whilst the temperature of the mixture is kept below 50 °C. To complete the oxidation the mixture is held at 55–60 °C for 2 h, then cooled and saturated with NaCl. The organic layer is separated and the aqueous layer is extracted with diethyl ether (3 × 50 ml). The combined organic extracts are dried over magnesium sulfate and evaporated under reduced pressure. Rapid Kugelrohr distillation (in two portions) at 0.5 mmHg gives cyclohexylcyclopentyltexylmethanol (17.4 g, 66%), b.p. 115–120 °C/0.5 mmHg.

If the same procedure is applied to dialkylmethoxyboranes the two alkyl groups are transferred from boron to carbon and oxidation gives the corresponding ketone (e.g. equation 6.32) [3, 51].

$$(C_6H_{11})_2BOMe \xrightarrow[\text{2. } Et_3COLi \;\; \text{3. } NaOH/H_2O_2]{\text{1. } CHCl_2OMe} C_6H_{11}{-}C(=O){-}C_6H_{11} \qquad (6.32)$$

The use of trichloromethyllithium as the carbanion source and 2-alkyl-1,3,2-dithiaborolanes as substrates allows the synthesis of homologated carboxylic acids (e.g. equation 6.33), but the temperature has to be kept at −100 °C during utilization of the carbanion [3, 52].

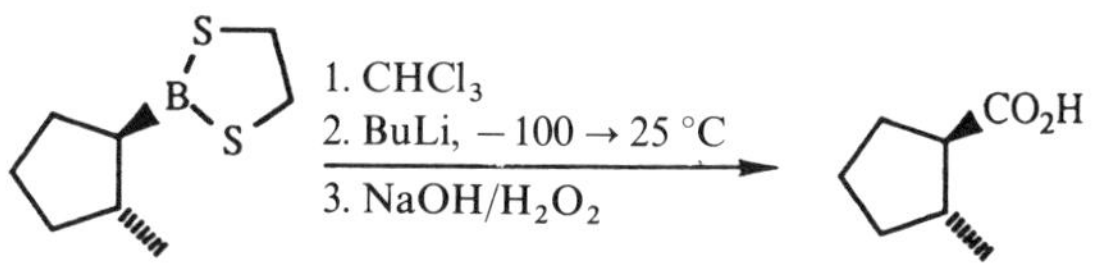

1. $CHCl_3$
2. BuLi, −100 → 25 °C
3. $NaOH/H_2O_2$

(6.33)

6.13 ALDEHYDES, KETONES AND ALCOHOLS VIA CARBONYLATION OF ORGANOBORANES

The reactions of organoboranes with carbon monoxide (carbonylation reactions) are extremely versatile [3, 53]. The reaction is slow and requires heating to *ca* 100 °C and/or the use of elevated pressures in order to achieve a reasonable rate. Oxidation of the mixture after such a process gives rise to ketones (equation 6.34) [3, 4, 54]. In practice, the use of the cyanoborate reaction (Section 6.14) achieves the same overall result under more convenient conditions, but the carbonylation reaction preceded the cyanoborate reaction and has been applied to a wider range of organoborane substrates [3, 53]. The use of thexyldialkylboranes conserves valuable alkyl groups (e.g. equation 6.34) and has the overall effect of replacement of a thexylboron unit by a carbonyl group.

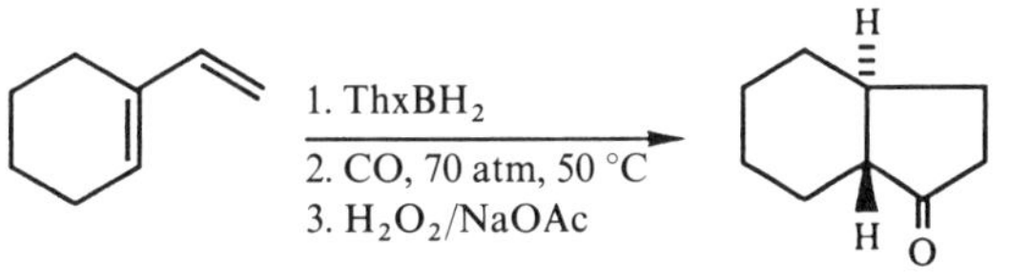

1. $ThxBH_2$
2. CO, 70 atm, 50 °C
3. $H_2O_2/NaOAc$

(6.34)

If the temperature is raised to *ca* 150 °C, especially if ethylene glycol is added, a third rearrangement may occur, giving rise to the 'riveted' product after oxidation [53]. In practice, the DCME reaction (Section 6.12) is probably the first-choice method for such a transformation. However, it is conceivable that the product from the DCME and carbonylation reactions may have different stereochemistry at the newly generated hydroxycarbon centre, so that the two methods may not always be interchangeable. A third possible method involves the cyanoborate reaction (Section 6.14) and in this case a difference has been observed with one substrate [55]. Therefore, it is of interest to have a procedure, as illustrated by the preparation of a perhydrophenalenol (equation 6.35) [56].

6.13.1 PREPARATION OF CIS,CIS,TRANS-PERHYDRO-9b-PHENALENOL FROM TRANS,TRANS,TRANS-1,5,9-CYCLODODECATRIENE [3, 4, 56]

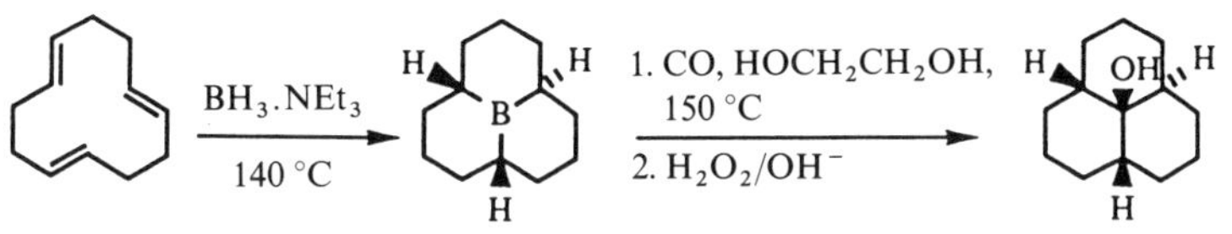

$BH_3.NEt_3$, 140 °C → (4) → 1. CO, $HOCH_2CH_2OH$, 150 °C; 2. H_2O_2/OH^-

(6.35)

A three-necked, 1 l flask equipped with a magnetic follower, a septum inlet, a mercury bubbler and a short Vigreux column connected to a distillation apparatus is flushed with nitrogen and charged with borane triethylamine (57.6 g, 0.5 mol) and diglyme (300 ml). The temperature is raised to 140 °C and a solution of *trans,trans,trans*-1,5,9-cyclododecatriene (81 g, 0.5 mol) in diglyme (100 ml) is added over 2 h by means of a syringe pump. Most of the diglyme is removed at atmospheric pressure and the residue is then heated at 200 °C (internal temperature) for 6 h (this isomerizes other organoboranes). Spinning band distillation under reduced pressure gives the perhydroboraphenalene **4** in 98% purity (b.p. 115–117 °C/10 mmHg).

Compound **4** (17.6 g, 0.1 mol), THF (50 ml) and ethylene glycol (16.8 ml, 18.6 g, 0.3 mol) are charged to an autoclave under a flow of nitrogen. The autoclave is filled with carbon monoxide at 1000 psi pressure and the temperature is raised to 150 °C for 2 h. The autoclave is cooled and opened and pentane (100 ml) is added to help remove the mixture. The solution is washed with water, dried over magnesium sulfate and evaporated.

THF (100 ml) and ethanol (95%, 100 ml) are added and the mixture is stirred during addition of aqueous NaOH (37 ml of a 6 mol l^{-1} solution, 0.22 mol, excess) and then 30% aqueous hydrogen peroxide (37 ml, excess), the latter added dropwise while the solution is maintained at a temperature below 40 °C. Once the vigorous reaction is over, the temperature is raised to 50 °C for 3 h. The mixture is then cooled to 25 °C, an equal volume of pentane is added and the organic phase is separated, washed with water (3 × 50 ml), dried over magnesium sulfate and evaporated to yield *cis,cis,trans*-perhydro-9*b*-phenalenol. Recrystallization from pentane gives the pure product (13.6 g, 70%), m.p. 78–78.5 °C.

The carbonylation reaction is at its most useful for the synthesis of aldehydes via a single migration. In order to achieve this, a hydride reducing agent is needed for the process and this also has the benefit of enhancing the rate of carbon monoxide uptake. As a result, the reaction can be carried out at 0 °C and at atmospheric pressure. The most useful hydrides are lithium trimethoxyaluminium hydride (which results in a gelatinous precipitate during work-up and can present separation difficulties) and potassium triisopropoxyborohydride (which can lead to polymerization if precautions are not taken to stop the stirring during addition of the hydride). The procedure given below is appropriate for the latter hydride. *B*-Alkyl-9-BBN derivatives allow maximum utilization of alkyl residues (e.g. equation 6.36) [3, 57].

6.13.2 PREPARATION OF CYCLOPENTANECARBOXALDEHYDE VIA HYDRIDE-INDUCED CARBONYLATION OF B-CYCLOPENTYL-9-BBN [3, 57]

$$\text{9-BBN-H} \xrightarrow{\text{cyclopentene}} \text{B-cyclopentyl-9-BBN} \xrightarrow[\text{2. } H_2O_2/OH^-]{\text{1. CO, K}(^iPrO)_3BH} \text{CHO-cyclopentane} \quad (6.36)$$

The apparatus used in this procedure is a Brown Automatic Gasimeter, which allows monitoring of the uptake of carbon monoxide. However, if this information is not specifically required the carbon monoxide can be introduced by direct feed from a cylinder or more conveniently by means of a rubber balloon connected to a syringe needle.

The Brown Automatic Gasimeter is set up for carbonylation as described elsewhere [4] and fitted with a 500 ml reaction flask equipped with a magnetic follower and a septum inlet. The entire system is flushed with nitrogen and then a solution of 9-BBN-H (40 ml of a 0.5 mol l^{-1} solution in THF, 20 mmol) and cyclopentene (1.36 g, 20 mmol) are added by syringe. The mixture is stirred for 2 h at room temperature, then cooled to room temperature. The stirrer is stopped, a solution of potassium triisopropoxyborohydride (20 ml of a 1.0 mol l^{-1} solution in THF, 20 mmol) is added and the system is then flushed with carbon monoxide (by injecting *ca* 8 ml of formic acid into the generator flask containing hot sulphuric acid). Stirring is then recommenced and uptake of carbon monoxide can be monitored. After 15 min (uptake long completed) the system is flushed with nitrogen and the mixture is stirred vigorously during addition of a pH 7 buffer (40 ml) followed by 30% hydrogen peroxide (8 ml). After the initial vigorous reaction, the cooling bath is removed and stirring is maintained for 15 min. Potassium carbonate (*ca* 50 g) is added to saturate the mixture and the organic layer is removed. The aqueous layer is extracted with diethyl ether (2 × 50 ml) and the combined organic layers are dried over potassium carbonate and then carefully evaporated to remove the solvent. Distillation at 50 mmHg gives cyclopentane-carboxaldehyde (1.84 g, 94%), b.p. 80–81 °C/50 mmHg.

6.14 KETONES AND TERTIARY ALCOHOLS VIA CYANOBORATES

Addition of solid sodium or potassium cyanide to a solution of a trialkylborane in an ether solvent results in dissolution of the solid with formation of a cyanoborate salt [58]. Addition of an electrophile induces rearrangement and acylating agents lacking α-hydrogen atoms are particularly useful. Trifluoroacetic anhydride permits reaction under mild conditions and the use of 1 mole equivalent leads to two migrations, giving a ketone after oxidation [59]. Optimum utilization of alkyl groups is achieved by use of thexyldialkylboranes and this also permits the synthesis of 'mixed' or cyclic ketones from the appropriate organoboranes (e.g. equation 6.37) [59, 60].

6.14.1 PREPARATION OF 3-*o*-ACETOXYPHENYLPROPYLIC CYCLOPENTYL KETONE VIA THE CYANOBORATE REACTION [59]

BH_2 — 1. cyclopentene; 2. OAc (o-allylphenyl acetate) → B (5) OAc

1. NaCN
2. $(CF_3CO)_2O$
3. CO_3H / Cl (m-chloroperbenzoic acid)

→ O, C, OAc (6.37)

A 100 ml, three-necked flask is fitted with a magnetic follower, a rotatable angled side-tube, a septum-capped stopcock and a line leading to a nitrogen supply and vacuum pump. The side tube is charged with dry, powdered sodium cyanide (0.54 g, 11 mmol) and the apparatus is then repeatedly evacuated and refilled with nitrogen. A solution of the trialkylborane **5** (10 mmol) in THF (*ca* 22 ml) is prepared as described for a related example in Section 6.5 by successive addition of THF, borane THF, 2,3-dimethyl-2-butene, cyclopentene and 3-*o*-acetoxyphenyl-1-propene. The mixture is then stirred at room temperature and the side-arm is rotated to introduce the sodium cyanide. Stirring is maintained for 1 h, by which time most of the cyanide has dissolved. The mixture is cooled in a bath at −78 °C, trifluoroacetic anhydride (2.53 g, 12 mmol) is added dropwise with vigorous stirring and the cooling bath is then removed. The mixture is stirred for 1 h at room temperature and then cooled to 10 °C. *m*-Chloroperbenzoic acid (3.9 g, 22.5 mmol, enough to cleave two B—C bonds) in dichloromethane (15 ml) is added. The mixture is stirred for 30 min at room temperature and the products are then extracted into pentane (150 ml). The extract is successively washed with aqueous solutions of sodium carbonate (25 ml of 1 mol l^{-1} solution), sodium thiosulfate (25 ml of 1 mol l^{-1} solution) and hydrochloric acid (25 ml of 0.01 mol l^{-1}), dried over magnesium sulfate and evaporated. The crude product is transferred into a column packed with dry silica (100 g) with the aid of a small volume of pentane and the column is eluted with pentane and then dichloromethane. The dichloromethane is removed to provide essentially pure product, which can be distilled to give pure 3-*o*-acetoxyphenylpropyl cyclopentyl ketone (2.19 g, 80%), b.p. 105 °C/5 mmHg.

By use of excess of trifluoroacetic anhydride and a period of heating the reaction can be encouraged to proceed further, resulting in a third migration and the 'riveting' of the original organoborane (e.g. equation 6.38) [61] in a manner analogous to the methods using carbon monoxide or DCME. The DCME method occurs under milder conditions, but there is a possibility of different diastereoisomeric products in some cases.

6.14.2 PREPARATION OF TRICYCLOHEXYLMETHANOL FROM TRICYCLOHEXYLBORANE VIA THE CYANOBORATE REACTION [61]

$$(C_6H_{11})_3B \xrightarrow[3.\ H_2O_2/OH^-]{1.\ KCN \quad 2.\ (CF_3CO)_2O,\ \text{excess},\ 40\ ^\circ C} (C_6H_{11})_3C{-}OH \qquad (6.38)$$

The apparatus described in the preceding procedure is set up, charged with potassium cyanide (0.72 g, 11 mmol) in the side-arm and flushed with nitrogen. Tricyclohexylborane (10 mmol) is prepared as described for other trialkylboranes in Sections 6.2 and 6.5, from cyclohexene (3.04 ml, 30 mmol) and borane THF (6.7 ml of a 1.5 mol l^{-1} solution, 10 mmol), allowing 3 h at 50 °C for completion of the hydroboration step. The THF is removed under reduced pressure and diglyme (10 ml) is added. The side-arm is rotated to introduce the cyanide and stirring is maintained for 1 h, by which time most of the solid has dissolved. The solution is cooled to 0 °C and trifluoroacetic anhydride (TFAA, 12.6 g, 60 mmol) is added. The temperature is then raised to 40 °C for 6 h. After cooling, excess of TFAA is removed under reduced pressure and the flask is immersed in an ice–water bath. Aqueous NaOH (12 ml of a 3 mol l^{-1} solution) is added, followed by hydrogen peroxide (8 ml of 50% solution), slowly and with care. Once the initial vigorous reaction has subsided, the cooling bath is removed and the oxidation is completed by stirring at 25 °C for 3 h and at 50 °C for 15 min.

The mixture is extracted with pentane (150 ml) and the extract is washed with aqueous NaOH (2 × 25 ml of a 2 mol l^{-1} solution) and water (2 × 25 ml), dried over magnesium sulfate and evaporated. The crude product is transferred with a small volume of pentane into a column filled with dry, neutralized (with triethylamine) silica gel and eluted with pentane (50 ml) and then dichloromethane (300 ml). The dichloromethane fractions are evaporated to yield tricyclohexylmethanol (2.39 g, 86%), which is pure by GC. Recrystallization from pentane gives very pure product (2.21 g, 79%), m.p. 93–94 °C.

6.15 KETONES, FUNCTIONALIZED ALKENES, ALKYNES AND DIYNES VIA ALKYNYLBORATES

Addition of alkynyllithiums to trialkylboranes gives rise to lithium trialkylalkynylborates, which are susceptible to attack by electrophiles on the triple bond. Such attack leads to rearrangement involving migration of an alkyl group from boron to the adjacent carbon atom. The exact nature of the product obtained depends on the nature of the electrophile and the method of work-up. If the electrophile is a protonic acid or an alkylating agent the intermediate is a vinylborane which is generally a mixture of *E*- and *Z*-isomers. However, both isomers give rise to the same ketone on oxidation, making this a useful synthesis of ketones (e.g. equation 6.39) [3, 62].

6.15.1 PREPARATION OF 8-ALLYL-7-TETRADECANONE VIA ALLYLATION OF LITHIUM TRIHEXYLOCTYNYLBORATE [3, 62]

$$Hex_3B \xrightarrow[\substack{2.\ CH_2{=}CHCH_2Br \\ 3.\ H_2O_2/OH^-}]{1.\ LiC{\equiv}CHex} HexCOCH(CH_2CH{=}CH_2)Hex \qquad (6.39)$$

A 100 ml, three-necked flask equipped with a magnetic follower, a septum-capped stopcock, a septum-capped pressure-equalizing dropping funnel and a line leading to a nitrogen supply–vacuum system is flushed with nitrogen. Trihexylborane (5 mmol) is prepared in the dropping funnel from borane THF (3.4 ml of a 1.47 mol l^{-1} solution, 5 mmol) and 1-hexene (1.26 g, 15 mmol), with swirling to mix the reagents thoroughly, and the mixture is left for 1 h. Meanwhile, the reaction flask is immersed in an ice–water bath and charged with 1-octyne (0.55 g, 5 mmol), light petroleum (b.p. 40–60 °C, 5 ml) and butyllithium (3.2 ml of a 1.56 mol l^{-1} solution in hexane, 5 mmol), with stirring. The cooling bath is removed and the mixture is stirred for 30 min at room temperature. The cooling bath is replaced and the mixture is stirred during addition of the trihexylborane solution from the dropping funnel. The funnel is rinsed out with diglyme (5 ml) and the mixture is stirred until all of the precipitated octynyllithium dissolves (only a few minutes). The volatiles are then removed under reduced pressure, leaving a solution of lithium trihexyl-octynylborate in diglyme.

Stirring is maintained for 15 min at 25 °C and then allyl bromide (0.61 g, 5 mmol) is added by syringe. The reaction mixture is stirred at 40 °C for 2 h, cooled in ice and then oxidized by successive addition of aqueous NaOH (2 ml of a 5 mol l^{-1} solution) and 50% hydrogen peroxide (1.5 ml of 50% solution), the latter dropwise (from the dropping funnel) and with vigorous stirring and after removal of the septa from the apparatus. After the initial exothermic reaction has subsided the cooling bath is removed and the mixture is stirred for 3 h at room temperature. The product is extracted into diethyl ether (2 × 20 ml) and the extract is washed with water, dried over magnesium sulfate and evaporated to give a syrupy residue which is transferred with a small volume of pentane onto a column packed with dry silica gel (100 g). Elution is with pentane (100 ml) and then dichloromethane (2 × 150 ml) and removal of the dichloromethane provides almost pure 8-allyl-7-tetradecanone (1.11 g, 88%) , which can be further purified by distillation if required (b.p. 96–98 °C/0.8 mmHg).

When the electrophile added to the alkynylborate is an α-bromocarbonyl compound, iodoacetonitrile or even propargyl bromide, the vinylborane intermediate is formed highly stereoselectively. Thus, by hydrolysis of the intermediate it is possible to obtain

(*Z*)-alkenyl ketones, carboxylates, nitriles and alkynes (e.g. equation 6.40) [3, 63]. Of course, oxidation of the intermediate to give a ketone (see above) is also possible.

6.15.2 PREPARATION OF (Z)-4-HEXYL-4-UNDECEN-2-ONE FROM TRIHEXYLOCTYNYLBORATE AND BROMOACETONE [3, 63]

$$Li^+Hex_3\bar{B}C{\equiv}CHex \xrightarrow[(-LiBr)]{BrCH_2COCH_3} (Hex)(Hex_2B)C{=}C(CH_2COCH_3)(Hex)$$

$$\xrightarrow{^iPrCO_2H} (Hex)(H)C{=}C(CH_2COCH_3)(Hex) \qquad (6.40)$$

Lithium trihexyloctynylborate (5 mmol) is prepared in diglyme (5 ml) exactly as described in the preceding procedure. The mixture is cooled in a -78 °C bath and stirred while bromoacetone (0.75 g, 5.5 mmol) is added by syringe. The mixture is allowed to warm to room temperature and then heated at 55 °C for 6 h to produce the intermediate vinylborane, followed by cooling to 25 °C.

Degassed 2-methylpropanoic acid (1 ml) is added and the mixture is stirred for 3 h at 25 °C. The mixture is neutralized by addition of aqueous NaOH (5 mol l^{-1}) and then a further 1.5 ml of the same NaOH solution is added. Hydrogen peroxide (3 ml of 50% solution, large excess) is added cautiously and the mixture is stirred overnight (this oxidizes residual B—C bonds). The product is extracted into diethyl ether and the extract is washed with water, dried over magnesium sulfate and evaporated. The product is transferred with the aid of a small volume of pentane onto a column of dry silica gel and successively eluted with light petroleum (b.p. 40–60 °C) and then pentane–dichloromethane (1 : 1). Evaporation of the latter fractions yields (*Z*)-4-hexyl-4-undecen-2-one (0.95 g, 75%), which can be further purified by distillation if required (b.p. 100–102 °C/1.5 mmHg).

When the electrophile added to a trialkylalkynylborate is iodine, the intermediate 2-iodoalkenylborane eliminates dialkyliodoborane to produce an alkyne. This can be a useful synthesis of unsymmetrical alkynes (e.g. equation 6.41) [64].

$$Ph_3B \xrightarrow[2.\,I_2]{1.\,LiC{\equiv}CCMe_3} PhC{\equiv}CCMe_3 \qquad (6.41)$$

Use of unsymmetrical triorganylboranes in order to minimize the wastage of potentially valuable organic residues is not very successful in most cases on account of the similarity of the migratory aptitudes of different groups in this reaction. However, alkynyl groups show significantly higher relative migratory aptitudes than secondary alkyl groups and so di-*sec*-alkyldialkynylborates give rise to conjugated diynes [65]. In order to make the reaction useful for the synthesis of unsymmetrical diynes, all that is required is a means of synthesis of the appropriate dialkyldialkynylborate. This can be achieved by addition of an alkynyllithium to an isolated dialkylalkynylborane [66], prepared as described in Section 6.7, or *in situ* by use of dicyclohexyl(methylthio)borane as the starting material (e.g. equation 6.42) [67].

6.15.3 PREPARATION OF 5,7-TETRADECADIYNE FROM A DIALKYNYLBORATE [67]

$$(C_6H_{11})_2BH \xrightarrow[\substack{2.\ LiC\equiv CHex \\ 3.\ LiC\equiv CBu \\ 4.\ I_2}]{1.\ MeSH} HexC\equiv CC\equiv CBu \quad (6.42)$$

A 100 ml, two-necked flask equipped with a magnetic follower, a septum and a septum-capped pressure equalizing dropping funnel is flushed with nitrogen. Dicyclohexylborane (5 mmol) is prepared in THF (10 ml) as described in Section 6.4 and methanethiol (0.75 ml of an 8.2 mol l^{-1} solution in THF, 6 mmol) is then added, with stirring and use of a bubbler connected via a needle to vent liberated hydrogen. The mixture is stirred for 2 h to complete the formation of the methylthioborane (the dicyclohexylborane dissolves) and the flask is briefly pumped (via a needle) to remove excess of methanethiol. If necessary, some additional THF is added to bring the volume back to *ca* 10 ml.

Meanwhile, in each of two separate, nitrogen-flushed 50 ml flasks equipped with a magnetic follower and a septum, the two alkynyllithium reagents are prepared. The flask is cooled in an ice–water bath and charged with the appropriate alkyne (5 mmol), pentane (5 ml) and butyllithium solution (3.9 ml of a 1.29 mol l^{-1} solution in hexane, 5 mmol). The mixture is stirred for 30 min at 25 °C and THF (4 ml) is then added. The solution is cooled to −78 °C before use in the next step.

The main flask, containing dicyclohexyl(methylthio)borane, is cooled in a −78 °C bath and the cooled solution of 1-lithiooctyne is added by syringe, residual traces being transferred with the aid of additional THF (2 ml). The reaction mixture is allowed to warm to room temperature over 15 min and then recooled to −78 °C for addition, in the same way, of the cooled 1-lithiohexyne solution. The mixture is again warmed to room temperature and stirred for 1 h, then most of the hydrocarbon solvent is removed under reduced pressure. If necessary, a small volume of THF is added back to bring the final volume to *ca* 10 ml. The solution is cooled to −78 °C and a solution of iodine (2.54 g, 10 mmol) in THF (20 ml) is added dropwise from the dropping funnel over 15 min. The mixture is stirred at −78 °C for a further 15 min then allowed to warm to room temperature over 1 h. Saturated aqueous sodium thiosulfate solution (5 ml) and aqueous NaOH (5 ml of a 5 mol l^{-1} solution) are added and the mixture is stirred for 1 h. The products are extracted into pentane (3 × 40 ml) and the extract is washed with water (3 × 50 ml), dried over magnesium sulfate and evaporated. The product is purified on a column of silica gel by elution with pentane and then 5% dichloromethane in pentane. The fractions containing the pure product are combined and evaporated to give 5,7-tetradecadiyne (0.584 g, 61%), which can be distilled if required (b.p. 66 °C/0.02 mmHg).

6.16 STEREOSPECIFIC SYNTHESIS OF ALKENES VIA ALKENYLBORATES

Most of the reactions of alkynylborates (Section 6.15) can probably be achieved with alkenylborates, although the products will obviously be at a lower oxidation level. Some of the reactions have been demonstrated [3], but in general the reactions have been less widely studied. The one reaction which has found extensive use is the reaction with iodine, which produces alkenes in a highly stereoselective way [3].

The alkenylborate used in such procedures can be obtained by addition of an alkenyllithium to a trialkylborane [68], but this results in the wastage of two boron-bound alkyl groups. This problem can be overcome by utilization of alkyldimethoxyboranes as substrates (e.g. equation 6.43) [69]. The stereochemistry around the double bond is inverted during the process.

$$\text{(cyclopentyl-Et)B(OMe)}_2 \xrightarrow[\text{2. } I_2\text{, NaOMe}]{\text{1. } (Z)\text{-RCH=CHLi}} \text{(cyclopentyl-Et)CH=CHR} \quad (6.43)$$

Alternatively, alkylalkenylboron compounds obtained via hydroboration reactions can be converted into borate salts by addition of methoxide and then reacted with iodine. For optimum utilization of organic residues an alkyldibromoborane can be reduced to an alkylmonobromoborane which is then used to hydroborate an alkyne (equation 6.44) [70].

$$RBBr_2.SMe_2 \xrightarrow[\text{2. } R^1C{\equiv}CR^2 \text{ 3. NaOMe 4. } I_2]{\text{1. 0.25 LiAlH}_4} (R)(R^1)C{=}C(R^2)(H) \quad (6.44)$$

The simplest procedure of all involves the direct hydroboration of an alkyne with an easily formed dialkylborane, followed by reaction with a base to generate the borate and iodine to induce the reaction (e.g. equation 6.45) [71]. This reaction is used to illustrate the procedure since the other cases differ primarily only in how the organoborane is generated.

6.16.1 PREPARATION OF (Z)-1-CYCLOHEXYL-1-HEXENE [3, 71]

$$(C_6H_{11})_2BH \xrightarrow[\text{2. NaOH, } I_2]{\text{1. HC}{\equiv}\text{C(CH}_2)_3\text{Me}} (C_6H_{11})(H)C{=}C(H)((CH_2)_3Me) \quad (6.45)$$

A 100 ml, three-necked flask equipped with a magnetic follower, a thermometer, a septum and a septum-capped pressure-equalizing dropping funnel is flushed with nitrogen. Dicyclohexylborane (25 mmol) is prepared as described in Section 6.4 from cyclohexene (4.1 g, 50 mmol) in THF (20 ml) and borane THF (13.9 ml of a 1.8 mol l^{-1} solution, 25 mmol) at 0–5 °C. The alkenyldialkylborane is generated, as described in Section 6.7 for a related example, by addition of 1-hexyne (2.05 g, 25 mmol) at -10 °C and then stirring at ambient temperature until the precipitate dissolves, followed by 1 h more. The mixture is then cooled to -10 °C and aqueous NaOH (15 ml of a 6 mol l^{-1} solution) is added, followed by the dropwise addition, with stirring, of a solution of iodine (6.35 g, 25 mmol) in THF (20 ml) over a period of 15 min. The mixture is allowed to warm to ambient temperature and excess of iodine is decomposed by addition of a small amount of aqueous sodium thiosulfate solution. The product is extracted into pentane (2 × 25 ml) and the extract is washed with water (2 × 10 ml), dried over magnesium sulfate and evaporated. Fractional distillation under reduced pressure gives (*Z*)-1-cyclohexyl-1-hexene (3.11 g, 75%), b.p. 44–45 °C/1 mmHg.

Conjugated dienes can be obtained by treatment of dialkenylalkoxyboranes with methoxide or hydroxide and iodine (e.g. equation 6.46) [72], whilst conjugated enynes can be obtained from alkenylalkynylborates [73].

$$\text{t-BuO–B(CH=CHR)}_2 \xrightarrow{\text{NaOH, I}_2} \text{RCH=CH–CH=CHR} \qquad (6.46)$$

6.17 STEREOSPECIFIC SYNTHESIS OF ALKENES VIA PALLADIUM-CATALYSED CROSS-COUPLING REACTIONS

The cross-coupling between an organic halide and an organoboron compound catalysed by palladium compounds has become a widely used and versatile method for the generation of carbon–carbon bonds [74]. At its simplest, the reaction can be represented as in equation 6.47.

$$R^1-X+R^2-BY_2 \xrightarrow{\text{PdLn}} R^1-R^2 \qquad (6.47)$$

The reaction is at its most efficient when the reacting partners are an alkenyl bromide and an alkenylboron compound, such as an alkenyldihydroxyborane (alkenylboronic acid) (e.g. equation 6.48) [75]. The reaction will tolerate many functional groups and the stereochemistry is retained in both portions of the diene.

6.17.1 PREPARATION OF BOMBYKOL VIA PALLADIUM-INDUCED COUPLING OF A BROMOALKENE WITH AN ALKENYLDIHYDROXYBORANE [75]

$$\text{HO(CH}_2)_9\text{C}{\equiv}\text{CH} \xrightarrow[\text{2. H}_2\text{O}]{\text{1. 2 catecholborane (C}_6\text{H}_4\text{O}_2\text{BH)}} \text{HO(CH}_2)_9\text{–CH=CH–B(OH)}_2$$

$$\xrightarrow[\text{Pd(PPh}_3)_4]{\text{Me(CH}_2)_2\text{–CH=CH–Br}} \text{HO(CH}_2)_9\text{–CH=CH–CH=CH–(CH}_2)_2\text{Me} \qquad (6.48)$$

A 100 ml, two-necked flask equipped with a magnetic follower, a septum inlet and a reflux condenser connected to a bubbler is flushed with nitrogen and charged with 10-undecyn-1-ol (1.51 g, 9 mmol) in THF (3 ml). Catecholborane (2.0 ml, 18 mmol) is added dropwise, with stirring, and the mixture is stirred at room temperature until hydrogen evolution ceases and then brought to reflux for 5 h. The mixture is cooled and water (60 ml) is added. The mixture is stirred for 2 h and then cooled to 0 °C. The solid is collected by filtration, washed with water (3 × 20 ml) and dried to give (*E*)-11-hydroxy-1-undecenyldihydroxyborane (1.6 g, 83%), which is used without further purification in the next stage.

A 50 ml, two-necked flask equipped with a magnetic follower, septum inlet and reflux condenser leading to a bubbler is flushed with nitrogen and charged successively with a solution of tetrakis(triphenylphosphine)palladium (0.29 g, 0.25 mmol) in benzene (20 ml), (*Z*)-1-bromo-1-pentene (0.75 g, 5 mmol), 11-hydroxy-1-undecenyldihydroxyborane (1.18 g, 5.5 mmol) and a solution of sodium ethoxide in ethanol (5 ml of a 2 mol l^{-1} solution). The solution is heated at reflux for 2.5 h, with stirring, and then cooled to room temperature and oxidized by successive addition of aqueous NaOH (0.5 ml of a 3 mol l^{-1} solution) and hydrogen peroxide (0.5 ml of 30% solution), to destroy any residual organoborane. The mixture is stirred for 1 h and the products are then extracted into diethyl ether (30 ml) and the extract is washed with saturated NaCl (2 × 15 ml), dried over magnesium sulfate and evaporated to yield crude bombykol (0.96 g, 82%), which can be further purified by Kugelrohr distillation (b.p. 125 °C/0.1 mmHg).

The reaction can be applied to aryl or alkynyl derivatives and with some modifications to *B*-alkyl-9-BBN derivatives [74]. It is even possible to couple the latter compounds to iodoalkanes in the presence of potassium phosphate and the palladium catalyst, but the yields are only moderate in such cases [76].

6.18 HOMOALLYLIC ALCOHOLS VIA ALLYLBORANES

Allylic boron compounds generally react rapidly with aldehydes, or less rapidly with ketones, to give homoallylic alcohols [3]. The reaction takes place with transposition of the allylic group (equation 6.49), but many allylic organoboranes themselves undergo allylic rearrangement to interconvert the isomers so care must be taken to ensure that the desired allylic organoborane is indeed the one undergoing reaction.

$$X_2BCHR^1CH{=}CHR^2 \xrightarrow[2.\ H_2O]{1.\ R^3COR^4} R^1CH{=}CHCHR^2\overset{\displaystyle OH}{\overset{|}{C}}R^3R^4 \qquad (6.49)$$

Organoboranes possessing more than one allylic group should be avoided since the rates of reaction will differ as each one reacts in turn, which may lead to complications. The nature of the other boron-bound groups (X in equation 6.49) has a strong influence on both the rate of allylic rearrangement of the organoborane and the rate of reaction of the organoborane with carbonyl compounds. The more powerfully electron-donating groups (dialkylamino > alkoxy > alkyl) slow the reactions substantially. Hence dialkylboryl derivatives must be prepared and reacted at low temperatures if it is necessary to avoid allylic rearrangement of the allylic organoborane prior to reaction with the carbonyl compound, whereas dialkoxyboryl derivatives are relatively stable up to ambient temperature. The latter compounds are also less reactive towards carbonyl compounds, however, and rearrangement may precede reactions with ketones. Reactions with aldehydes are usually free from such complications.

There is the possibility of substantial control over the various stereochemical features in the product homoallylic alcohols, viz. over the geometry about the double bond, the *syn* or *anti* relationship between the two newly created sp^3 centres and the absolute configuration of the product, when chiral. For example, in reactions of dialkoxy (1-methylallyl)boranes with aldehydes the $Z:E$ ratio for the double bond in the product can be varied from 3:1 to 1:2 simply by changing the alkoxy groups, bulky groups favouring the *Z*-product [77].

The *syn*:*anti* selectivity is primarily determined by the geometry of the double bond in the initial allylborane (which can also be altered via reversible allylic rearrangement). (*Z*)-Allylboranes give *syn*-alcohols (e.g. equation 6.50) [78], whereas (*E*)-allylboranes give *anti*-alcohols [3].

1. RCHO
2. $N(CH_2CH_2OH)_3$

OH R (*syn*) (6.50)

Chiral induction is achieved by having optically active groups on boron, such as two isopinocampheyl groups [79] or an optically active cyclic alkylenedioxy group [80]. By using a chiral auxiliary with the appropriate configuration, it is possible to obtain the appropriate enantiomer of the product, which, coupled with the right choice of geometry of the double bond in the allylic organoborane, allows the synthesis of the desired enantiomer of either diastereoisomer (*syn* or *anti*), rendering this a very versatile method [3, 79].

In terms of the details of how the reaction should be carried out, the only differences required involve the preparation and handling of the initial organoborane. Simple dialkylallylboranes can be prepared and purified by procedures similar to that described in Section 6.7 for a dialkylalkynylborane, but if isomerization of the allylic organoborane is to be avoided it may be necessary to use the organoborane directly as generated *in situ* and to maintain a low temperature [79]. From that point on, the reaction procedure given below for a simple case (equation 6.51) [81] is satisfactory.

6.18.1 PREPARATION OF 5,5-DIMETHYL-1-HEXEN-4-OL BY REACTION OF B-ALLYL-9-BORABICYCLO[3.3.1]NONANE WITH PIVALALDEHYDE [81]

$$\text{B}-CH_2CH{=}CH_2 \xrightarrow[\text{2. } H_2O]{\text{1. } Me_3CCHO} CH_2{=}CHCH_2{-}CH(OH){-}CMe_3 \qquad (6.51)$$

A 50 ml flask equipped with a magnetic follower and a septum connected via a needle to a mercury bubbler is flushed with nitrogen via a needle and then charged with *B*-allyl-9-BBN (3.775 g, 23.3 mmol) and purified (alkene-free) pentane (25 ml). The mixture is cooled to 0 °C (lower temperatures throughout are recommended for geometrically or positionally labile organoboranes) and stirred during dropwise addition, from a syringe, of freshly distilled pivaldehyde (2.60 ml,

2.00 g, 23.3 mmol). The mixture is allowed to warm up and stirred for 1 h at room temperature, then neat ethanolamine (1.40 ml, 1.42 g, 23.3 mmol) is added in order to free the product and precipitate the 9-BBN by-product. The slurry thus obtained is stirred for 30 min and the contents of the flask are then poured into a centrifuge tube, the residue being washed through with a small volume of pentane (10 ml). The mixture is centrifuged and the clear supernatant liquid is removed with a syringe. The precipitate is washed repeatedly with pentane (3 × 15 ml), each time with thorough mixing prior to centrifugation, and the combined organic solutions are concentrated by blowing with a stream of nitrogen (reduced pressure is avoided at this stage because of the volatility of the product). The residual oil is distilled under reduced pressure to give 5,5-dimethyl-1-hexen-4-ol (2.53 g, 85%), b.p. 55.5–56 °C/19 mmHg.

6.19 ALDOL REACTIONS OF VINYLOXYBORANES

Vinyloxyboranes (boron enolates) can be formed regiospecifically by a number of routes [82], but the recent increase in importance of these species has depended on their direct generation from their parent carbonyl compounds [3]. Regioselectivity can still be achieved by variation of the reaction parameters during preparation of the enolate. Thus, the use of dibutylboryl triflate, diisopropylethylamine and a short reaction period at −78 °C allows total conversion of 2-pentanone into its kinetic boron enolate, whereas the use of 9-BBN triflate, 2,6-lutidine and a long reaction period at −78 °C gives exclusively the thermodynamic enolate (equation 6.52) [3, 83]. Subsequent aldol reactions with aldehydes take place without loss of regiochemical integrity, but reactions with ketones are slower and preservation of regiochemical integrity can then be more of a problem.

OB(9-BBN) ← 9-BBN-OTf, 2,6-lutidine, −78 °C, 16 h — 2-pentanone — Bu_2BOTf, $^iPr_2NEt$, −78 °C, 15 min → $OBBu_2$ (6.52)

An advantage of boron enolates over most other kinds of enolates is the high diastereoselectivity displayed in their reactions with aldehydes. There is a very good correlation between the geometry of the enolate and the diastereoisomer formed in its aldol reactions. (*Z*)-Enolates give almost 100% stereoselectivity for formation of *syn*-aldols, whereas (*E*)-enolates strongly favour the *anti*-aldol products, especially if the groups on boron are relatively bulky (e.g. equation 6.53) [3, 84, 85].

syn (>97%) ← ($OBBu_2$, Et enolate) — CHO — ($OB(c\text{-}C_5H_9)_2$ enolate) → *anti* (88%) (6.53)

In order to take advantage of the stereoselectivity of the reaction it is necessary to be able to generate boron enolates with the appropriate geometry. Use of a hindered dialkylboryl triflate (e.g. dicyclopentyl) and diisopropylethylamine at 0 °C yields predominantly the *E*-isomer, whereas use of less hindered reagents (e.g. dibutyl) and low temperature (−78 °C) gives almost exclusively the *Z*-isomer (e.g. equation 6.54) [84].

6.19.1 PREPARATION OF SYN-1-HYDROXY-2-METHYL-1-PHENYL-3-PENTANONE VIA A BORON ENOLATE (VINYLOXYBORANE) [83, 84]

$$Bu_3B \xrightarrow{CF_3SO_3H} \underset{(6)}{Bu_2BOTf} \xrightarrow{EtCOEt}$$

OBBu$_2$ Et (7) $\xrightarrow{PhCHO}$ O OH Et Ph (6.54)

Dibutylboryl triflate (**6**) is prepared as follows. A 100 ml flask equipped with a magnetic follower and a septum connected via a needle to a bubbler is flushed with argon. Tributylborane (15.16 g, 83.3 mmol) is charged by syringe, followed by a small amount of trifluoromethanesulfonic acid (1.0 g). The mixture is stirred and warmed at 50 °C until evolution of butane begins (there is an induction period), and is then cooled to 25 °C. The remaining trifluoromethanesulfonic acid (11.51 g, total 83.3 mmol) is added dropwise at such a rate as to maintain a temperature between 25 and 50 °C. The mixture is then stirred for a further 3 h at 25 °C. Short-path distillation under reduced pressure in an atmosphere of argon gives pure dibutylboryl triflate (**6**) (19.15 g, 84%), b.p. 60 °C/2 mmHg.

A 100 ml flask equipped as described above is flushed with argon, charged with dry diisopropylethylamine (0.85 g, 6.6 mmol), dibutylboryl triflate (**6**) (1.81 g, 6.6 mmol) and dry diethyl ether (15 ml) and then cooled to −78 °C. 3-Pentanone (0.52 g, 6.0 mmol) is added, dropwise and with stirring, and the mixture is stirred for a further 30 min at −78 °C, during which the boron enolate **7** is formed along with a white precipitate of diisopropylethylammonium triflate.

Benzaldehyde (0.64 g, 6.0 mmol) is added dropwise and the mixture is stirred for a further 30 min at −78 °C and 1 h at 0 °C. The reaction is quenched by addition to a pH 7 phosphate buffer solution (50 ml) and the product is extracted into diethyl ether (2 × 30 ml). The combined ether extracts are washed with brine (2 × 10 ml) and concentrated under reduced pressure. The oil thus obtained is dissolved in methanol (20 ml), the solution is cooled to 0 °C, and hydrogen peroxide (6.5 ml of 30% solution) is added. The mixture is stirred at room temperature for 2 h and water (50 ml) is then added. Most of the methanol is removed under reduced pressure (note that the mixture contains peroxide and should not be evaporated to dryness) and the residue is extracted with diethyl ether (2 × 20 ml). The ether extracts are combined, washed with 5% aqueous sodium hydrogen carbonate (2 × 10 ml) and brine (10 ml), dried over magnesium sulfate and concentrated to a colourless oil (1.01 g, 88%). The product is chromatographed on silica gel at medium pressure using hexane–ethyl acetate (8 : 1) to give *syn*-1-hydroxy-2-methyl-1-phenyl-3-pentanone (0.89 g, 77%).

Introduction of chirality features into the vinyloxyborane unit allows the reaction to be extended to the synthesis of non-racemic products [3, 86]. Further, by appropriate choice of a homochiral aldehyde in reaction with a chiral vinyloxyborane it is possible to make use of 'double asymmetric induction' to maximize the enantioselectivity of the reaction [3, 87].

Dialkoxyvinyloxyboranes, available via a number of procedures [88, 89], also take part in aldol reactions. However, the reactions are slower and *syn*-products are obtained predominantly irrespective of the geometry of the enolate [90].

6.20 ALCOHOLS AND ALKENES VIA BORON-STABILIZED CARBANIONS

Although a dialkylboryl group can be expected to provide about as much stabilization as a carbonyl group to a carbanion centre, the generation of such boron-stabilized anions is much more problematic [3]. Nevertheless, methods have been developed which make the anions available for synthetic reactions. Probably the most useful method involves the direct deprotonation of a hindered organoborane such as an alkyldimesitylborane with a moderately hindered base such as mesityllithium [91]. The anions thus generated can then be reacted with various electrophiles to give the corresponding products [3].

For example, dimesitylboryl-stabilized carbanions react readily with primary alkyl bromides and iodides to give the corresponding alkylated organoboranes, which can be oxidized to yield alcohols (e.g. equation 6.55) [92]. Although reactions with secondary halides are less efficient, it is possible to make more highly branched derivatives by successive deprotonations and alkylations with dimethyl sulfate [93].

6.20.1 *PREPARATION OF 2-OCTANOL FROM ETHYLDIMESITYLBORANE AND 1-IODOHEXANE [92]*

$$\mathrm{Mes_2BCH_2Me} \xrightarrow{\mathrm{MesLi}} \underset{(\mathbf{8})}{\mathrm{Mes_2B\bar{C}HMe}} \xrightarrow{\mathrm{HexI}}$$

$$\mathrm{Mes_2B\overset{\overset{\displaystyle Hex}{|}}{C}HMe} \xrightarrow{[\mathrm{O}]} \mathrm{Hex\overset{\overset{\displaystyle OH}{|}}{C}HMe} \qquad (6.55)$$

A 50 ml flask equipped with a magnetic follower is charged with dry mesityl bromide (MesBr; Mes = 2,4,6-trimethylphenyl) (1.095 g, 5.5 mmol), then fitted with a septum connected via a needle to a bubbler and flushed with nitrogen via a second needle. Dry THF (10 ml) is added and the mixture is cooled to −78 °C for addition of *tert*-butyllithium (7.33 ml of 1.5 mol l^{-1} solution in hexane, 11 mmol), dropwise and with stirring. The mixture, which develops a cloudy yellow colour, is stirred for 15 min at −78 °C and then for 15 min at 25 °C. The bubbler is removed so that the solution of mesityllithium so obtained can be withdrawn into a wide-needle syringe when required.

Meanwhile, a 100 ml flask equipped as above is charged with ethyldimesitylborane (see Sections 6.5 and 6.7) (1.39 g, 5 mmol) and flushed with nitrogen. Dry THF (5 ml) is added and the mixture is stirred as the solution of mesityllithium is added by syringe. The mixture is stirred for 1 h at 25 °C (longer for more hindered cases) to generate the anion **8**. The contents of the flask are cooled to 0 °C for addition, by syringe, of 1-iodohexane (1.17 g, 5.5 mmol). The cooling bath is then removed and the pink mixture is stirred for 30 min at room temperature. The septum is removed and the organoborane is oxidized by successive addition of methanol (5 ml), aqueous NaOH (2.5 ml of 5 mol l^{-1} solution) and hydrogen peroxide (5 ml of 50% solution), followed by stirring overnight at 25 °C and then heating under reflux for 3 h (longer for more hindered cases).

The mixture is cooled and saturated with potassium carbonate. The organic layer is separated and the aqueous layer is extracted with diethyl ether (2 × 30 ml). The combined organic extracts are washed with 10% aqueous citric acid (30 ml) and water (20 ml), dried over magnesium sulfate and evaporated below 40 °C. The crude product is chromatographed on silica gel by gradient elution with dichloromethane through chloroform and diethyl ether. 2-Octanol (0.442 g, 68%) is obtained in the fractions containing chloroform or chloroform with a small amount of diethyl ether following removal of the solvent under reduced pressure at less than 40 °C.

A particularly useful type of reaction of anions such as **8** is that with carbonyl compounds, known as the boron–Wittig reaction [94]. In the reaction with aromatic aldehydes it is possible to vary the work-up procedure to obtain either (*E*)- or (*Z*)-alkenes (e.g. equation 6.56) [3, 95].

6.20.2 PREPARATION OF (E)- OR (Z)-1-PHENYL-1-NONENE BY THE BORON–WITTIG REACTION [3, 95]

$Mes_2B\bar{C}HHept$ (**9**)

↓ PhCHO

$Mes_2B\text{-}CHHept$ / $^{-}OCHPh$ (**10**)

(**10**) $\xrightarrow{(CF_3CO)_2O}$ (Z)-PhCH=CHHept (**12**)

(**10**) $\xrightarrow{1.\ Me_3SiCl;\ 2.\ HF}$ (E)-PhCH=CHHept (**11**) (6.56)

The anion **9** is prepared from dimesityloctylborane (3 mmol) in a manner analogous to that described for **8** in the preceding procedure. The solution is cooled to −78 °C. Meanwhile, a 10 ml Wheaton bottle equipped with a septum is flushed with argon, charged with freshly distilled benzaldehyde (0.223 g, 2.1 mmol) in THF (3 ml) and cooled to −78 °C. The benzaldehyde solution is added to the stirred solution of anion **9** via a double-ended needle and the mixture is stirred for 2 h at −78 °C to give intermediate **10**. This solution is used for preparation of either alkene, as detailed below.

(E)-1-Phenyl-1-nonene. A solution of chlorotrimethylsilane (0.337 g, 3.1 mmol) in THF (3 ml) in an argon-flushed Wheaton bottle is cooled to −78 °C and transferred via a double-ended needle in to the stirred solution of intermediate **10**. The mixture is stirred at −78 °C for 1 h and then gradually allowed to warm to 20 °C and stirred for a further 16 h. The volatile components are removed under reduced pressure and dry light petroleum (b.p. 30–40 °C, 30 ml) is added. The mixture is stirred and then allowed to settle and the petroleum solution is decanted from the precipitated solid, concentrated and chromatographed on alumina, eluting with pentane. An oil consisting of mesitylene and the trimethylsilyl ether of **10** is obtained. This oil is cooled to −78 °C and a solution of aqueous HF (1 ml of 40% solution) in HPLC-grade acetonitrile (20 ml) is added. The cooling bath is removed and the solid reaction mixture is allowed to warm to 20 °C and then stirred for a further 30 min. The mixture is poured into pentane (30 ml) and the pentane phase is separated. The aqueous phase is extracted a second time with pentane (30 ml) and the combined extracts are washed with water (3 × 20 ml), dried over magnesium sulfate and evaporated to give a crude product (1.68 g). Chromatography on silica, eluting with pentane, gives a fraction containing the product and mesitylene, which on pumping overnight to remove mesitylene leaves pure (*E*)-1-phenyl-1-nonene (**11**) (0.36 g, 84%). The product can be distilled under reduced pressure if required (b.p. 80–82 °C/0.1 mmHg).

(Z)-1-Phenyl-1-nonene. The solution of intermediate **10** is cooled to −110 °C and a precooled (−78 °C) solution of trifluoroacetic anhydride (0.55 g, 2.6 mmol) in THF (3 ml) is added via a cooled double-ended needle. The mixture is stirred for 1 h at −110 °C and 4 h at −78 °C then left to warm to room temperature overnight. Volatile materials are removed under reduced pressure and then light petroleum (b.p. 30–40 °C, 30 ml) is added. The mixture is stirred and then allowed to settle and the petroleum layer is decanted from the precipitated solid. The extract is evaporated and the residue is chromatographed on silica, eluting with pentane. The fractions containing mesitylene and the product are pumped overnight to remove mesitylene and leave pure (*Z*)-1-phenyl-1-nonene (**12**) (0.33 g, 77%). The product can be distilled if required (b.p. 79–83 °C, 0.1 mmHg).

The reaction with aliphatic aldehydes is less reliable using the procedures described above. Instead, alkenes are more reliably obtained by using an acid in admixture with the aldehyde during addition to the anion **9** [96]. The stereochemistry can be controlled by the choice of acid, strong acids such as trifluoromethanesulfonic acid or HCl giving predominantly the (*E*)-alkene and weak acids such as acetic acid generally giving predominantly the (*Z*)-alkene.

6.21 REDUCTIONS WITH TRIALKYLHYDROBORATES

There are two features of trialkylhydroborates (trialkylborohydrides) which justify their use instead of simple reagents such as sodium borohydride in certain circumstances: they are much more reactive and can therefore accomplish reactions which are very slow or low yielding with the simple reagents, and they are much more hindered and may therefore give rise to more selective reactions, particularly more stereoselective reactions, when steric factors are important [3].

Probably the most useful reaction which takes account of the high reactivity of trialkylhydroborates is the conversion of alkyl halides and tosylates into the corresponding alkanes (e.g. equation 6.57) [97]. Several trialkylhydroborates are commercially available and can be used directly. Otherwise, unhindered trialkylhydroborates are relatively easily synthesized by stirring THF solutions of the corresponding trialkylboranes with solid LiH, NaH or KH, whilst hindered trialkylboranes are easily converted into their lithium hydroborates by reaction with *tert*-butyllithium [3].

6.21.1 PREPARATION OF CYCLOOCTANE BY REDUCTION OF CYCLOOCTYL TOSYLATE WITH LITHIUM TRIETHYLHYDROBORATE [3, 97]

$$\text{cyclooctyl-OTs} \xrightarrow{LiEt_3BH} \text{cyclooctane} \quad (6.57)$$

A 300 ml, two-necked flask equipped with a magnetic follower, a reflux condenser connected to a bubbler and a septum-capped side arm is flushed with nitrogen, charged with dry THF (20 ml) and cyclooctyl toluenesulfonate (7.05 g, 25 mmol) by syringe and cooled to 0 °C. Lithium triethylhydroborate (33.3 ml of a 1.5 mol l^{-1} solution in THF, 50 mmol) (note that since commercial solutions are less concentrated, longer reaction times may be necessary with these unless the concentrations are adjusted) is added to the stirred solution, the ice-bath is removed, the mixture is stirred for 2 h at 25 °C and excess of hydride is then destroyed by cautious addition of water (hydrogen is evolved). Oxidation of the triethylborane by-product is achieved by successive addition of aqueous NaOH (20 ml of a 3 mol l^{-1} solution) and hydrogen peroxide (20 ml of a 30% solution, added dropwise) (see Section 6.2), followed by stirring at 25 °C for 1 h. The mixture is allowed to separate, the aqueous layer is extracted with pentane (2 × 20 ml) and the combined organic extracts are washed with water (4 × 15 ml), dried over magnesium sulfate and concentrated by distillation at atmospheric pressure (because of the volatility of the product). The residue is transferred in to a small distillation apparatus and distilled at atmospheric pressure to give cyclooctane (2.27 g, 81%) as a colourless oil, b.p. 142–146 °C, contaminated by about 3% of cyclooctene.

The ability of trialkylhydroborates to effect stereoselective reductions is well illustrated by the reduction of substituted cyclohexanones [98]. Use of lithium tri-*sec*-butylhydroborate (L-Selectride) at −78 °C allows the production of 90% of the less stable *cis*-isomer on reduction of 4-methylcyclohexanone and with lithium trisiamylhydroborate the proportion is 99%. With 4-*tert*-butylcyclohexanone the proportion is even higher (equation 6.58) [98]. Recently, results comparable to those achieved using lithium trisiamylhydroborate at −78 °C have been achieved at 0 °C with the even more hindered reagent lithium ethylbis(2,4,6-triisopropylphenyl)hydroborate [99]. In this case there is the additional advantage that the triorganylborane by-product is air stable and can easily be recovered and reused.

6.21.2 PREPARATION OF CIS-4-TERT-BUTYLCYCLOHEXANOL BY REDUCTION OF 4-TERT-BUTYLCYCLOHEXANONE WITH LITHIUM TRISIAMYLHYDROBORATE [3, 98]

$$\text{4-}t\text{-Bu-cyclohexanone}=O \xrightarrow[-78\ ^\circ C]{LiSia_3BH} \text{cis-4-}t\text{-Bu-cyclohexanol (OH)} \qquad (6.58)$$

A 250 ml, two-necked flask equipped with a magnetic follower, a reflux condenser connected to a bubbler and a septum-capped stopcock is flushed with nitrogen, charged with lithium trisiamylhydroborate solution (70 ml of a 0.4 mol l^{-1} solution in THF, 28 mmol) and immersed in a −78 °C cooling bath. A solution of 4-*tert*-butylcyclohexanone (3.7 g, 24 mmol) in THF (25 ml) is cooled to 0 °C and then added by syringe to the trisiamylhydroborate solution as rapidly as is consistent with keeping the reaction solution cold (for maximum selectivity). The mixture is stirred vigorously for 2 h at −78 °C and then allowed to warm to ambient temperature over 1 h. Water (4 ml) and ethanol (10 ml) are added and the trisiamylborane by-product is oxidized by successive addition of aqueous NaOH (10 ml of a 6 mol l^{-1} solution) and hydrogen peroxide (15 ml of a 30% solution), followed by warming at 40 °C for 30 min once the initial vigorous reaction has subsided (see Section 6.2). The mixture is cooled and the aqueous phase is saturated with potassium carbonate. The organic phase is separated, the aqueous phase is further extracted with diethyl ether–tetrahydrofuran (1 : 1, 2 × 20 ml) and the combined extracts are dried over magnesium sulfate. The volatile solvents and 3-methyl-2-butanol are removed under reduced pressure to leave *cis*-4-*tert*-butylcyclohexanol (3.65 g, 98%) as a white solid, m.p. 80 °C, which is at least 99.5% *cis*-isomer by GC.

6.22 STEREOSELECTIVE REDUCTIONS WITH ORGANOBORANES

Trialkylboranes generally display little reactivity towards aldehydes or ketones. However, organoboranes possessing an isopinocampheyl group are more reactive, effecting reduction of some compounds with concomitant displacement of α-pinene. Further, since such organoboranes can be obtained in non-racemic form, the reagents can be used to induce asymmetry during reduction. *B*-Isopinocampheyl-9-BBN (Alpine-borane) is one such reagent. It reduces aldehydes readily at 65 °C and if its deuterio derivative is used, or if a 1-deuterioaldehyde is used, the corresponding deuteriated primary alcohol is obtained in very high optical purity (e.g. equation 6.59) [3, 100].

6.22.1 PREPARATION OF (S)-BENZYL-1-d-ALCOHOL VIA REDUCTION OF 1-DEUTERIOBENZALDEHYDE WITH ALPINE-BORANE

$$\text{PhC(=O)D} \xrightarrow[\text{(+)-}\alpha\text{-pinene, 65 °C}]{\text{Alpine-borane from}} \text{Ph–C(OH)(H)(D)} \qquad (6.59)$$

A 1 l, two-necked flask equipped with a magnetic follower, a reflux condenser connected to a bubbler and a septum-capped stopcock is flushed with nitrogen and then charged with *B*-isopinocampheyl-9-BBN (52.9 g, 205 mmol, from (+)-α-pinene of 93% ee), in THF (400 ml) via a double-ended needle. Benzaldehyde-1-*d* (19.0 ml, 185 mmol) is added and the mixture is stirred for 10 min at room temperature and then heated at reflux for 1 h. The mixture is cooled to 20 °C and acetaldehyde (5 ml) is added to destroy residual trialkylborane. THF is removed under reduced pressure and the mixture is then pumped at oil-pump vacuum using a bath temperature of 40 °C to remove α-pinene. Nitrogen is readmitted, diethyl ether (150 ml) is added, the solution is cooled to 0 °C and 2-aminoethanol (12.5 g, 205 mmol) is added in order to precipitate the 9-BBN derivative. The precipitate is filtered off and washed with diethyl ether (2 × 20 ml). The combined organic extracts are washed with water (25 ml), dried over magnesium sulfate and evaporated to give a product which is further purified by distillation under reduced pressure. (*S*)-Benzyl-1-*d* alcohol (16.5 g, 82%) of 88% ee is collected at 110 °C/30 mmHg.

Alpine-borane reacts sluggishly with simple ketones, but reductions can still be effected with good enantioselectivity if the reactions are carried out without solvent and/or under pressure [101]. Alternatively, alkynyl ketones, which are much less sterically demanding, are reduced more readily to the corresponding alkynylmethanols, and the latter are easily converted into saturated alcohols. 'NB-Enantrane', the organoborane derived by reaction of 9-BBN-H with nopol benzyl ether, reduces alkynyl ketones with even greater enantioselectivity [102].

For enantioselective reduction of simple ketones, including aromatic ketones such as acetophenone (98% ee for the corresponding alcohol), hindered aliphatic ketones such as pinacolone (95% ee) and hindered alicyclic ketones such as 2,2-dimethylcyclopentanone (98% ee), the best reagent appears to be diisopinocampheylchloroborane (commercially available, but see Section 6.8 for the preparation of similar reagents) (e.g. equation 6.60) [103].

6.22.2 PREPARATION OF (S)-1-PHENYLETHANOL BY REDUCTION OF ACETOPHENONE WITH DIISOPINOCAMPHEYLCHLOROBORANE (103)

$$\mathrm{PhC(O)Me} \xrightarrow[\text{2. } HN(CH_2CH_2OH)_2]{\text{1. } Ipc_2BCl,\ -25\ ^\circ C,\ 7\ h} \mathrm{PhC(OH)(H)Me} \qquad (6.60)$$

A 250 ml flask equipped with a magnetic follower and a septum is flushed with nitrogen and charged with a solution of diisopinocampheylchloroborane (9.0 g, 28 mmol) [from (+-α-pinene] in THF (20 ml). The solution is cooled to −25 °C and acetophenone (3.05 ml, 26 mmol) is added by syringe, whereupon the mixture turns yellow. The mixture is stirred for 7 h at −25 °C (^{11}B NMR of a methanolysed aliquot shows the reaction to be complete) and the volatiles are then removed at aspirator pressure. α-Pinene is removed under reduced pressure (0.1 mmHg, 8 h) and the residue is dissolved in diethyl ether (100 ml). Diethanolamine (6.0 g, 57 mmol) is added and after 2 h the solid is removed by filtration. The solid is washed with pentane (2 × 30 ml) and the combined organic solutions are concentrated. The residue is distilled to give (*S*)-1-phenylethanol (2.3 g, 72%), b.p. 118 °C/22 mmHg, with 97% ee.

Hindered ketones such as pinacolone react much more sluggishly than acetophenone and the reaction requires 12 days at room temperature without solvent for completion [103].

6.23 CONCLUSION

The purpose of this chapter is to provide appropriate, well tried procedures for some of the more important synthetic methods based on organoboron compounds. In particular, it is hoped that it can provide a good starting point for anyone interested in making use of the enormous synthetic potential of organoboranes, especially those experiencing reluctance on account of lack of familiarity with the handling of such reagents. Naturally, in order to try to keep the material to a reasonable quantity, the choice of reactions for inclusion has had to be highly selective. Also, procedures may need to differ, even for the same reaction type, depending on factors such as the degree of steric hindrance in the substrate. Such variations can only be alluded to in a work of this length. For additional procedures and more extensive discussions of reactions of organoboranes, specialized monographs are available [3, 4].

6.24 REFERENCES

1. H. C. Brown, *Hydroboration*, Benjamin, New York, 1962; reprinted with Nobel Lecture, Benjamin/Cummings, Reading, MA, 1980.
2. K. Smith, *Chem. Ind. (London)* (1987) 603.
3. A. Pelter, K. Smith, H. C. Brown, *Borane Reagents*, Academic Press, London, 1988.
4. H. C. Brown, G. W. Kramer, A. B. Levy, M. M. Midland, *Organic Syntheses via Boranes*, Wiley, New York, 1975.
5. H. C. Brown, R. Liotta, C. G. Scouten, *J. Am. Chem. Soc.* **98** (1976) 5297.

6. A. Pelter, S. Singaram, H. C. Brown, *Tetrahedron Lett.* **24** (1983) 1433.
7. H. C. Brown, P. K. Jadhav, A. K. Mandal, *Tetrahedron* **37** (1981) 3547.
8. K. Smith, *Chem. Soc. Rev.* **3** (1974) 443.
9. R. Köster, P. Binger, W. Fenzl, *Inorg. Synth.* **15** (1974) 134.
10. H. C. Brown, C. F. Lane, *Heterocycles* **7** (1977) 453.
11. H. C. Brown, M. C. Desai, P. K. Jadhav, *J. Org. Chem.* **47** (1982) 5065.
12. A. Pelter, K. Smith, D. Buss, Z. Jin, *Heteroatom Chem.*, **3** (1992), 275.
13. E. Negishi, H. C. Brown, *Synthesis* (1974) 77.
14. S. U. Kulkarni, H. D. Lee, H. C. Brown, *J. Org. Chem.* **45** (1980) 4542.
15. H. C. Brown, J. A. Sikorski, S. U. Kulkarni, H. D. Lee, *J. Org. Chem.* **47** (1982) 863.
16. J. W. Wilson, *J. Organomet. Chem.* **186** (1980) 297.
17. H. C. Brown, M. C. Desai, P. K. Jadhav, *J. Org. Chem.* **47** (1982) 5065.
18. H. C. Brown, P. K. Jadhav, A. K. Mandal, *Tetrahedron* **37** (1981) 3547.
19. G. Zweifel, N. R. Ayyangar, T. Munekata, H. C. Brown, *J. Am. Chem. Soc.* **86** (1964) 1076.
20. H. C. Brown, A. K. Mandal, N. M. Yoon, B. Singaram, J. R. Schwier, P. K. Jadhav, *J. Org. Chem.* **47** (1928) 5069.
21. H. C. Brown, B. Singaram, *Pure Appl. Chem.* **59** (1987) 879; *Acc. Chem. Res.* **21** (1988) 287.
22. H. C. Brown, J. V. N. Vara Prasad, A. K. Gupta, R. K. Bakshi, *J. Org. Chem.* **52** (1987) 310.
23. D. S. Matteson, R. Ray, R. R. Rocks, D. J. Tsai, *Organometallics* **2** (1983) 1536; D. S. Matteson, K. M. Sadhu, M. L. Peterson, *J. Am. Chem. Soc.* **108** (1986) 810; for a review, see D. S. Matteson, *Acc. Chem. Res.* **21** (1988) 294.
24. H. C. Brown, S. K. Gupta, *J. Am. Chem. Soc.* **94** (1972) 4370; H. C. Brown, J. Chandrasekharan, *J. Org. Chem.* **48** (1983) 5080.
25. H. C. Brown, S. K. Gupta, *J. Am. Chem. Soc.* **93** (1971) 1816; **97** (1975) 5249.
26. D. Manning, H. Nöth, *Angew. Chem., Int. Ed. Engl.* **24** (1985) 878.
27. M. Sato, N. Miyaura, A. Suzuki, *Tetrahedron Lett.* **31** (1990) 231.
28. H. C. Brown, W. Ravindran, *Inorg. Chem.* **16** (1977) 2938.
29. H. C. Brown, N. Ravindran, S. U. Kulkarni, *J. Org. Chem.* **45** (1980) 384; H. C. Brown, U. S. Racherla, *J. Org. Chem.* **51** (1986) 895.
30. H. C. Brown, J. B. Campbell, *J. Org. Chem.* **45** (1980) 389.
31. W. E. Paget, K. Smith, *J. Chem. Soc., Chem. Commun.* (1980) 1169.
32. H. C. Brown, N. Ravindran, *J. Org. Chem.* **42** (1977) 2533.
33. N. R. De Lue, H. C. Brown, *Synthesis* (1976) 114.
34. H. C. Brown, C. F. Lane, *J. Am. Chem. Soc.* **92** (1970) 6660.
35. H. C. Brown, N. R. DeLue, G. W. Kabalka, H. C. Hedgecock, *J. Am. Chem. Soc.* **98** (1976) 1290; G. W. Kabalka, E. E. Gooch, *J. Org. Chem.* **46** (1981) 2582.
36. H. G. Kuivila, A. R. Hendrickson, *J. Am. Chem. Soc.* **74** (1952) 5068.
37. H. C. Brown, T. Hamaoka, N. Ravindran, *J. Am. Chem. Soc.* **95** (1937) 6456.
38. H. C. Brown, W. R. Heydkamp, E. Breuer, W. S. Murphy, *J. Am. Chem. Soc.* **86** (1964) 3565; M. W. Rathke, N. Inoue, K. R. Varma, H. C. Brown, *J. Am. Chem. Soc.* **88** (1966) 2870.
39. G. W. Kabalka, K. A. R. Sastry, G. W. McCollum, H. Yoshioka, *J. Org. Chem.* **46** (1981) 4296; G. W. Kabalka, K. A. R. Sastry, G. W. McCollum, C. F. Lane, *J. Chem. Soc., Chem. Commun.* (1982) 62.
40. V. B. Jigajinni, A. Pelter, K. Smith, *Tetrahedron Lett.* (1978) 181.
41. H. C. Brown, K. Murray, *J. Am. Chem. Soc.* **81** (1959) 4108; G. Zweifel, H. C. Brown, *J. Am. Chem. Soc.* **86** (1964) 393.

42. J. Hooz, J. N. Bridson, J. G. Calzada, H. C. Brown, M. M. Midland, A. B. Levy, *J. Org. Chem.* **38** (1973) 2574.
43. H. C. Brown, H. Nambu, M. M. Rogić, *J. Am. Chem. Soc.* **91** (1969) 6854.
44. M. M. Midland, Y. C. Kwon, *J. Org. Chem.* **46** (1981) 229.
45. For a description of the preparation of aryldichloroboranes, see J. Hooz, J. G. Calzada, *Org. Prep. Proced. Int.* **4** (1972) 219.
46. R. J. Hughes, S. Ncube, A. Pelter, K. Smith, E. Negishi, T. Yoshida, *J. Chem. Soc., Perkin Trans. 1* (1977) 1172.
47. S. Ncube, A. Pelter, K. Smith, *Tetrahedron Lett.* (1979) 1895.
48. S. Ncube, A. Pelter, K. Smith, *Tetrahedron Lett.* (1979) 1893.
49. H. C. Brown, B. A. Carlson, *J. Org. Chem.* **38** (1973) 2422.
50. H. C. Brown, J.-J. Katz, B. A. Carlson, *J. Org. Chem.* **38** (1973) 3968.
51. B. A. Carlson, H. C. Brown, *J. Am. Chem. Soc.* **95** (1973) 6876; *Synthesis* (1973) 776; B. A. Carlson, J.-J. Katz, H. C. Brown, *J. Organomet. Chem.* **67** (1974) C39; H. C. Brown, J.-J. Katz, B. A. Carlson, *J. Org. Chem.* **40** (1975) 813.
52. H. C. Brown, T. Imai, *J. Org. Chem.* **49** (1984) 892.
53. H. C. Brown, *Acc. Chem. Res.* **2** (1969) 65.
54. H. C. Brown, E. Negishi, *J. Am. Chem. Soc.* **89** (1967) 5477; *J. Chem. Soc., Chem. Commun.* (1968) 594.
55. A. Pelter, P. J. Maddocks, K. Smith, *J. Chem. Soc., Chem. Commun.* (1978) 805.
56. H. C. Brown, E. Negishi, *J. Am. Chem. Soc.* **89** (1967) 5478; H. C. Brown, W. C. Dickason, *J. Am. Chem. Soc.* **91** (1969) 1226.
57. H. C. Brown, J. L. Hubbard, K. Smith, *Synthesis* (1979) 701.
58. A. Pelter, M. G. Hutchings, K. Smith, *J. Chem. Soc., Chem. Commun.* (1970) 1529.
59. A. Pelter, K. Smith, M. G. Hutchings, K. Rowe, *J. Chem. Soc., Perkin Trans. 1* (1975) 129.
60. A. Pelter, M. G. Hutchings, K. Smith, *J. Chem. Soc., Chem. Commun.* (1971) 1048.
61. A. Pelter, M. G. Hutchings, K. Rowe, K. Smith, *J. Chem. Soc., Perkin Trans. 1* (1975) 138.
62. A. Pelter, T. W. Bentley, C. R. Harrison, C. Subrahmanyam, R. J. Laub, *J. Chem. Soc., Perkin Trans. 1* (1976) 2419.
63. A. Pelter, K. J. Gould, C. R. Harrison, *J. Chem. Soc., Perkin Trans. 1* (1976) 2428.
64. A. Suzuki, N. Miyaura, S. Abiko, M. Itoh, H. C. Brown, J. A. Sinclair, M. M. Midland, *J. Am. Chem. Soc.* **95** (1973) 3080; *J. Org. Chem.* **51** (1986) 4507.
65. A. Pelter, K. Smith, M. Tabata, *J. Chem. Soc., Chem. Commun.* (1975) 857.
66. J. A. Sinclair, H. C. Brown, *J. Org. Chem.* **41** (1976) 1078.
67. A. Pelter, R. J. Hughes, K. Smith, M. Tabata, *Tetrahedron* (1976) 4385.
68. K. Utimoto, K. Uchida, M. Yamaya, H. Nozaki, *Tetrahedron* **33** (1977) 1945; N. Miyaura, H. Tagami, M. Itoh, A. Suzuki, *Chem. Lett.* (1974) 1411; N. J. LaLima, A. B. Levy, *J. Org. Chem.* **43** (1978) 1279.
69. D. A. Evans, R. C. Thomas, J. A. Walker, *Tetrahedron Lett.* (1976) 1427; D. A. Evans, T. C. Crawford, R. C. Thomas, J. A. Walker, *J. Org. Chem.* **41** (1976) 3947.
70. H. C. Brown, D. Basaviah, *J. Org. Chem.* **47** (1982) 1792, 3806, 5407; S. U. Kulkarni, D. Basaviah, M. Zaidlewicz, H. C. Brown, *Organometallics* **1** (1982) 212.
71. G. Zweifel, H. Arzoumanian, C. C. Whitney, *J. Am. Chem. Soc.* **89** (1967) 3652.
72. G. Zweifel, N. L. Polston, C. C. Whitney, *J. Am. Chem. Soc.* **90** (1968) 6243.
73. E. Negishi, G. Lew, T. Yoshida, *J. Chem. Soc., Chem. Commun.* (1973) 874.
74. For a paper giving most of the important background references, see N. Miyaura, T. Ishiyama, H. Sasaki, M. Ishikawa, M. Satoh, A. Suzuki, *J. Am. Chem. Soc.* **111** (1989) 314.

75. N. Miyaura, H. Suginome, A. Suzuki, *Tetrahedron* **39** (1983) 3271; see also H. Yatagai, Y. Yamamoto, K. Maruyama, A. Sonoda, S.-I. Murahashi, *J. Chem. Soc., Chem. Commun.* (1977) 852; J. Uenishi, J. M. Beau, R. W. Armstrong, Y. Kishi, *J. Am. Chem. Soc.* **109** (1987) 4756.
76. T. Ishiyama, S. Abe, N. Miyaura, A. Suzuki, in press: the authors are thanked for providing this information in advance of publication.
77. R. W. Hoffmann, U. Weidmann, *J. Organomet. Chem.* **195** (1980) 137.
78. R. W. Hoffmann, H.-J. Zeiss, *Angew. Chem., Int. Ed. Engl.* **18** (1979) 306.
79. H. C. Brown, K. S. Bhat, *J. Am. Chem. Soc.* **108** (1986) 5919.
80. D. J. S. Tsai, D. S. Matteson, *Tetrahedron Lett.* **22** (1981) 2751.
81. G. W. Kramer, H. C. Brown, *J. Org. Chem.* **42** (1977) 2292; see also *J. Organomet. Chem.* **132** (1977) 9.
82. R. Köster, W. Fenzl, *Angew. Chem., Int. Ed. Engl.* **7** (1968) 735.
83. T. Inoue, T. Mukaiyama, *Bull. Chem. Soc. Jpn.* **53** (1980) 174.
84. D. A. Evans, J. V. Nelson, E. Vogel, T. R. Taber, *J. Am. Chem. Soc.* **103** (1981) 3099.
85. D. E. Van Horn, S. Masamune, *Tetrahedron Lett.* (1979) 2229.
86. S. Masamune, W. Choy, F. A. J. Kerdesky, B. Imperiali, *J. Am. Chem. Soc.* **103** (1981) 1566.
87. S. Masamune, W. Choy, J. S. Petersen, L. R. Sita, *Angew. Chem., Int. Ed. Engl.* **24** (1985) 1.
88. R. W. Hoffmann, K. Ditrich, *Tetrahedron Lett.* **25** (1984) 1781.
89. C. Gennari, L. Colombo, G. Poli, *Tetrahedron Lett.* **25** (1984) 2279.
90. C. Gennari, S. Cardani, L. Colombo, C. Scolastico, *Tetrahedron Lett.* **25** (1984) 2283.
91. A. Pelter, B. Singaram, L. Williams, J. W. Wilson, *Tetrahedron Lett.* **24** (1983) 623.
92. A. Pelter, L. Williams, J. W. Wilson, *Tetrahedron Lett.* **24** (1983) 627.
93. J. W. Wilson, *J. Organomet. Chem.* **186** (1980) 297.
94. A. Pelter, B. Singaram, J. W. Wilson, *Tetrahedron Lett.* **24** (1983) 635.
95. A. Pelter, D. Buss, E. Colclough, *J. Chem. Soc., Chem. Commun.* (1987) 297.
96. A. Pelter, K. Smith, S. Elgendy, M. Rowlands, *Tetrahedron Lett.* **30** (1989) 5647; 5643.
97. S. Krishnamurthy, H. C. Brown, *J. Org. Chem.* **41** (1976) 3064; **48** (1983) 3085.
98. S. Krishnamurthy, H. C. Brown, *J. Am. Chem. Soc.* **98** (1976) 3383.
99. K. Smith, A. Pelter, A. Norbury, *Tetrahedron Lett.* **32** (1991) 6243.
100. M. M. Midland, S. Greer, A. Tramontano, S. A. Zderic, *J. Am. Chem. Soc.* **101** (1979) 2352.
101. H. C. Brown, G. G. Pai, *J. Org. Chem.* **50** (1985) 1384; **48** (1983) 1784.
102. M. M. Midland, A. Kazubski, *J. Org. Chem.* **47** (1982) 2814.
103. J. Chandrasekharan, P. V. Ramachandran, H. C. Brown, *J. Org. Chem.* **50** (1985) 5446; **51** (1986) 3394.

7

Organoaluminium Compounds

HISAHI YAMAMOTO
Nagoya University, Japan

Organometallics in Synthesis—A Manual. Edited by M. Schlosser

7.1 INTRODUCTION

Although their role in the olefin-oriented world of petrochemicals is well established [1], organoaluminium reagents are relative newcomers as tools in selective organic syntheses [2]. However, organoaluminium compounds are the cheapest of the active metals, and are hence gradually replacing other organometallic derivatives as more economical reducing and alkylating agents.

The characteristic properties of aluminium reagents derive mainly from the high Lewis acidity of the organoaluminium monomers, which is directly related to the tendency of the aluminium atom to complete electron octets. Bonds from aluminium to electronegative atoms such as oxygen or the halogens are extremely strong; the energy of the Al—O bond is estimated to be 138 kcal mol^{-1}. Because of this strong bond, nearly all organoaluminium compounds are particularly reactive with oxygen and often ignite spontaneously in air. These properties, commonly identified with 'oxygenophilicity,' are of great value in the design of selective synthetic reactions.

The strong Lewis acidity of organoaluminium compounds appears to account for their tendency to form 1:1 complexes, even with neutral bases such as ethers. In fact, in contrast to lithium and magnesium derivatives, diethyl ether and related solvents may retard the reactivity of organoaluminium compounds. The relatively belated appreciation of organoaluminium chemistry may be traced to the longstanding confinement to ethereal solvents. The coordinated group may be activated or deactivated depending on the type of reaction. Further, on coordination of an organic molecule an auxiliary bond can become coupled to the reagent and promote the desired reaction.

The major difference between organoaluminium compounds and more common Lewis acids such as aluminium chloride and bromide is attributable to the structural flexibility of organoaluminium reagents. Thus, the structure of an aluminium reagent is easily modified by changing one or two of its ligands.

The following questions should be posed in seeking the most effective use of organoaluminium compounds:

1. What kind of stereochemical or electrochemical reactivity difference in substrates can we expect by coordination with the aluminium reagent?
2. Among the three ligands of organoaluminium compounds, which one has the highest reactivity?
3. What are the oxygenophilicities? What kind of reaction can we design with this unusual reactivity?
4. Organoaluminium reagent is a typical Lewis acid. However, once coordinated with a substrate, the aluminium reagent behaves as a typical ate complex and one of the ligands behaves as a nucleophile. This is an acid–base complexed-type reagent. What are the characteristic features of these reagents?
5. By changing from a trivalent organoaluminium reagent to a tetravalent ate complex, the reactivity of the reagent is changed significantly. What is the chemistry of the ate complex of an aluminium reagent in the presence of donor ligands?

We discuss here the chemistry of organoaluminium reagents in an attempt to answer some of these questions.

7.2 ORGANOALUMINIUM REACTIVITY AFTER COORDINATION WITH A SUBSTRATE

Most of the trivalent organoaluminium-promoted reaction was initiated by coordination with the substrate to the aluminium atom. For example, when the cyclohexanone **1** was treated with trimethylaluminium, the carbonyl oxygen was first coordinated with an aluminium atom [3]. If the reaction was performed in diethyl ether or tetrahydrofuran, the axial alcohol **2** was produced through a four-centred transition state. A similar complex was generated in hydrocarbon solvent with a 1:1 ratio of carbonyl compound and aluminium reagent. However, if the ratio was changed to 1:2, the reaction proceeded through a six-membered transition state (below) to yield the equatorial alcohol as the major product [4, 5]. This example illustrates a typical solvent effect of aluminium reagent-promoted reaction. In ethereal solution, the oxygen atom of the solvent molecules strongly coordinates with the aluminium atom and the self-association of the aluminium reagent becomes much weaker. The resulting dimethylaluminium alkoxide forms a stable dimer and the rate of methylation decreases significantly [6].

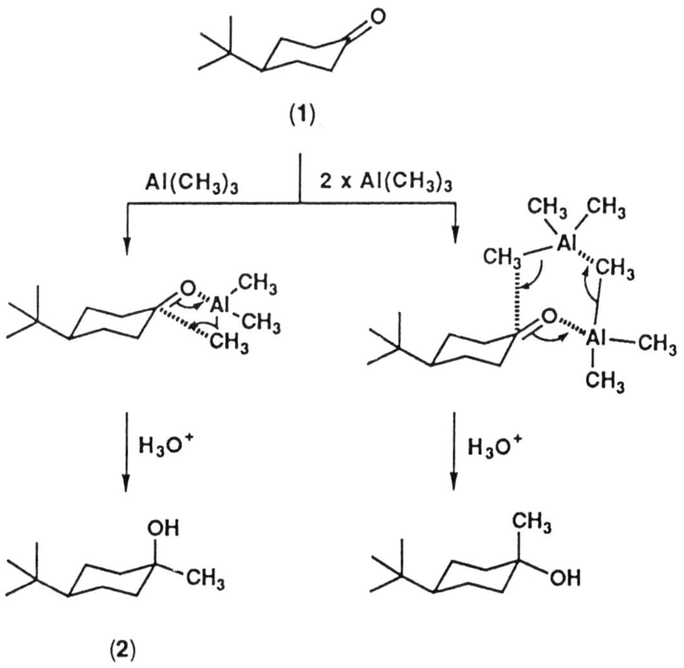

(1)

$Al(CH_3)_3$ | $2 \times Al(CH_3)_3$

H_3O^+ | H_3O^+

(2)

Pronounced solvent effects in the course of organoaluminium-induced reactions were frequently observed. Unimolecular decomposition of the 1:1 complex of citronellal–trimethylaluminium at −78 °C to room temperature yielded the acyclic compound **3** in hexane, whereas isopregol (**4**) was produced exclusively in 1,2-dichloroethane. Further, the cyclization–methylation product **5** was formed with high selectivity using excess of trimethylaluminium in dichloromethane at low temperature [7].

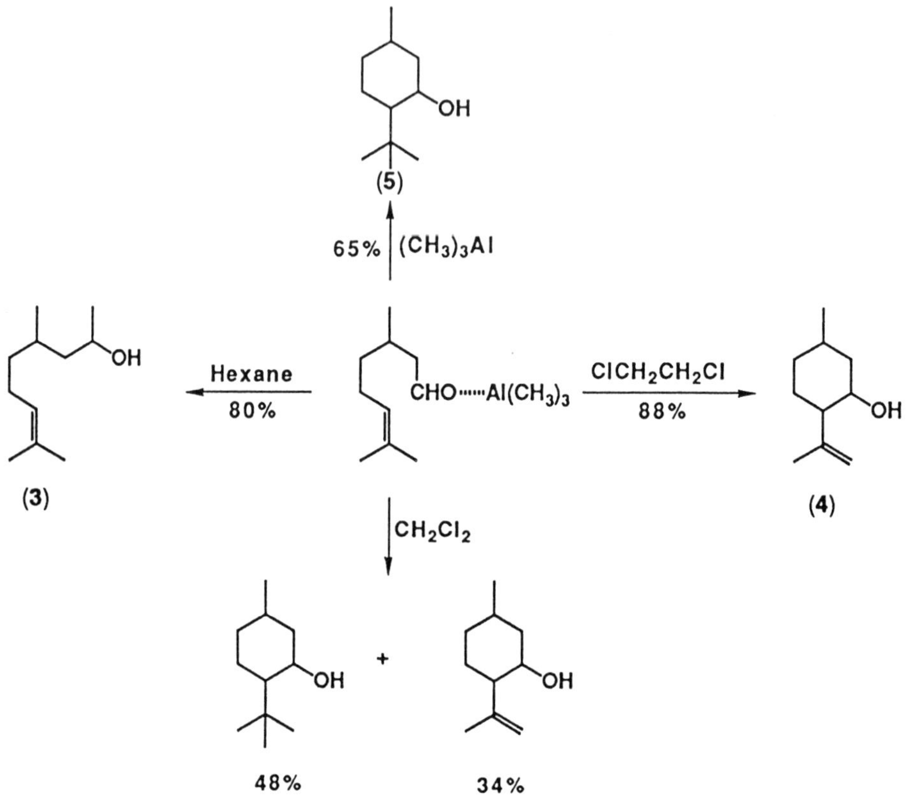

(5)

65% $(CH_3)_3Al$

Hexane 80%

$CHO\cdots Al(CH_3)_3$

$ClCH_2CH_2Cl$ 88%

(3)

(4)

CH_2Cl_2

48% 34%

Hydroalumination of acetylene using diisobutylaluminium hydride (DIBAH) is known to occur through a *cis*-addition mechanism similar to that of hydroboration [8]. If silylacetylene was used as a substrate and if the reaction was performed in hexane, the initially formed *cis*-addition product *Z*-**6** was isomerized to the *E*-isomer *E*-**6** spontaneously [9]. The olefinic π-electron may coordinate with the vacant orbital of the aluminium atom and also with the silicon atom through pπ–dπ conjugation, and these coordinations may accelerate the rotation with C—C bond [10]. On the other hand, if the reaction took place in hexane in the presence of triethylamine or in diethyl ether solvent, the vacant orbital of aluminium was occupied by these donor molecules and isomerization did not easily occur. Therefore, by choosing the solvent system, we were able to produce either a *cis*- or *trans*-olefin stereoselectively.

$RC{\equiv}CSi(CH_3)_3$ — DIBAH / ether → (*Z*-**6**) $R(H)C{=}C(Si(CH_3)_3)(Al^iBu_2)$ — 1. CH_3Li 2. R^1X → $R(H)C{=}C(Si(CH_3)_3)(R^1)$

$RC{\equiv}CSi(CH_3)_3$ — DIBAH / hexane → (*E*-**6**) $R(H)C{=}C(Al^iBu_2)(Si(CH_3)_3)$ — 1. CH_3Li 2. R^1X → $R(H)C{=}C(R^1)(Si(CH_3)_3)$

A significant solvent effect was also observed during the study of the Beckmann rearrangement-alkylation sequence [11]. For example, when the reaction of cyclopentanone oxime tosylate (**7**) with tripropylaluminium was performed in hexane at low temperature, the resulting imine was directly reduced with DIBAH and *N*-propylcyclopentylamine (**8**) was produced. Meanwhile, if the same reaction was performed in dichloromethane, the rearrangement–alkylation product, piperidine, turned out to be the major product. Further, if the reaction was conducted at 40 °C, the rearrangement–alkylation product coninine (**9**) was the sole product. It is noteworthy that no amine **8** or piperidine was produced under these latter reaction conditions [12].

(**7**) — 1. Pr_3Al, hexane, -78 °C; 2. DIBAH → (**8**)

(**7**) — 1. Pr_3Al, CH_2Cl_2, 40 °C; 2. DIBAH → (**9**)

(**7**) — 1. Pr_3Al, hexane, −78 °C; 2. DIBAH → piperidine (minor) + *N*-propylcyclopentylamine (major)

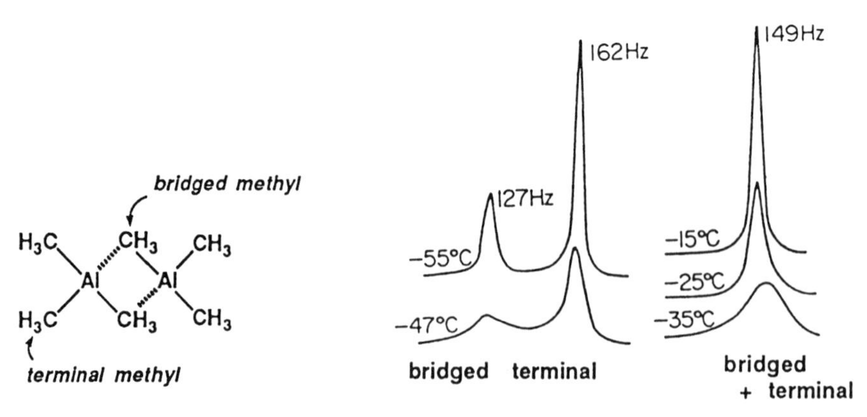

Figure 7.1. NMR spectrum of trimethylaluminium in toluene

The coordination of solvent with aluminium atom may be minimal in hydrocarbon. Thus, in these solvent systems the self-association of aluminium atoms may be more significant. For example, the association of trimethylaluminium in toluene at various temperatures is shown in Figure 7.1 [13]. At -55 °C, the ratio between bridged methyl and terminal methyl was found to be 1 : 2, so that most of the trimethylaluminium would have a dimeric structure [14]. By increasing the temperature, one of the methyl groups of trimethylaluminium was rapidly exchanged, randomizing the NMR signals of the two types of methyl groups. At low temperature, the vacant orbital of organoaluminium atom may therefore be occupied by the methyl group of other molecules and cannot be fully utilized by the reaction. At higher temperature, however, the existence of monomeric trimethylaluminium can be expected. The product distribution reflects these general features of aluminium reagent.

Bulky ligands effectively solved this association–dissociation problem. With an exceedingly bulky ligand, organoaluminium reagents no longer associate with each other but remain as monomers in solution. In other words, the aluminium reagent no longer shares the alkyl or halogen group with the adjacent molecule because of the steric bulkiness of the ligand.

2,6-Di-*tert*-butyl-4-methylphenol is a typical bulky ligand for aluminium reagents. This readily available phenol reacts with trimethylaluminium to generate methylaluminium bis (2,6-di-*tert*-butyl-4-methylphenoxide) (MAD; for the preparation procedure, see Section 7.7.2). MAD was found to be monomeric even in hydrocarbon solvents [15].

H_3C–(Ar)–OH $\xrightarrow{(CH_3)_3Al}$ H_3C–(Ar)–O–Al(CH_3)–O–(Ar)–CH_3 + $2CH_4$

MAD

Highly Lewis acidic aluminium reagents react with carbonyl compounds to form a 1:1 complex, in a kind of neutralization process. The lone pair on the carbonyl group behaves as a Lewis base during the reaction. For example, benzophenone reacts with an equivalent amount of trimethylaluminium at low temperature to generate an orange 1:1 complex (**10**) almost instantaneously [3–5]. Although similar complexes can be generated between other organometallics and carbonyl compounds, they are usually unstable and are further transformed into alcohols by successive hydride reduction or alkylation processes. With a highly Lewis acidic aluminium reagent, the complex formed is sufficiently stable and has a long enough lifetime for a variety of reactions.

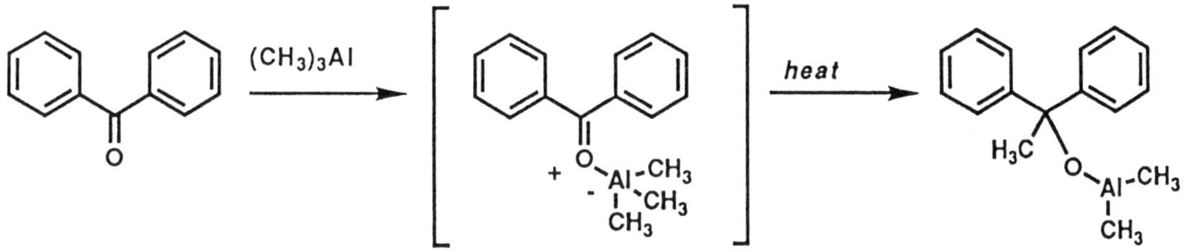

$(CH_3)_3Al$ heat

(**10**)

Bulky organoaluminium reagents were found to be of great practical use when combined with carbonyl compounds, 'MAD' [methylbis(di-2,6-*tert*-butyl-4-methylphenoxy)aluminium] and 'MAT' [methylbis(tri-2,4,6-*tert*-butylphenoxy)aluminium] belong to the most versatile members of this class of compounds. When mixed with a

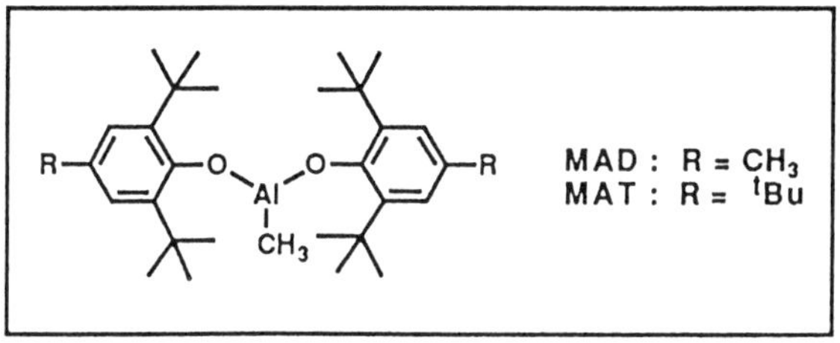

MAD : R = CH_3
MAT : R = ^tBu

carbonyl compound such as 4-*tert*-butylcyclohexanone MAD gave a stable 1:1 complex (**11**). This complex was treated with methyllithium at low temperature to yield an equatorial alcohol (**12**, overleaf), the stereochemistry of which was the opposite of that of the product from the simple reaction of cyclohexanone and methyllithium. The equatorial selectivity achieved with MAD was found to be nearly perfect, and that with MAT even superior [15].

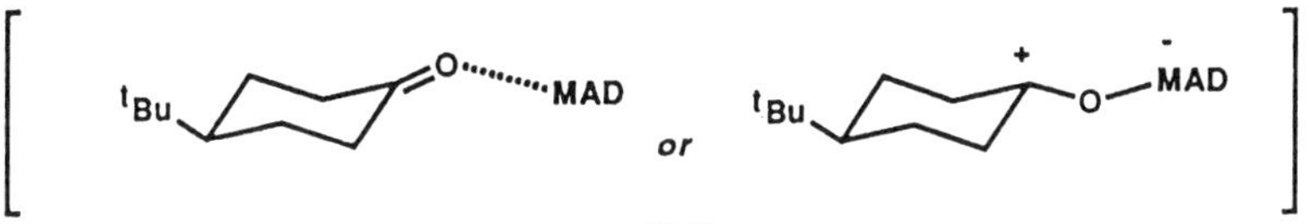

^tBu O······MAD *or* ^tBu O—MAD

(**11**)

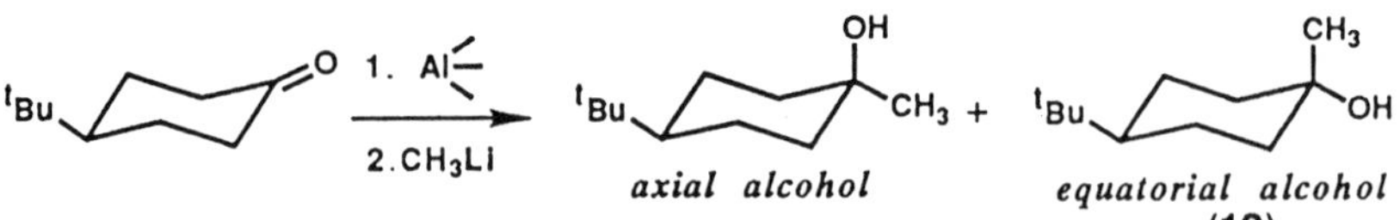

1. Al; 2. CH_3Li → axial alcohol + equatorial alcohol

(12)

CH_3Li : 85% (79 : 21)
MAD/CH_3Li : 84% (1 : 99)
MAT/CH_3Li : 92% (0.5 : 99.5)

Such complexation also allows inversion of the stereoselectivity of nucleophilic addition to chiral aldehydes. While methylmagnesium iodide, when reacting along with 2-phenylpropanal, obeys Cram's rule, the opposite mode is largely favoured in the presence of MAT.

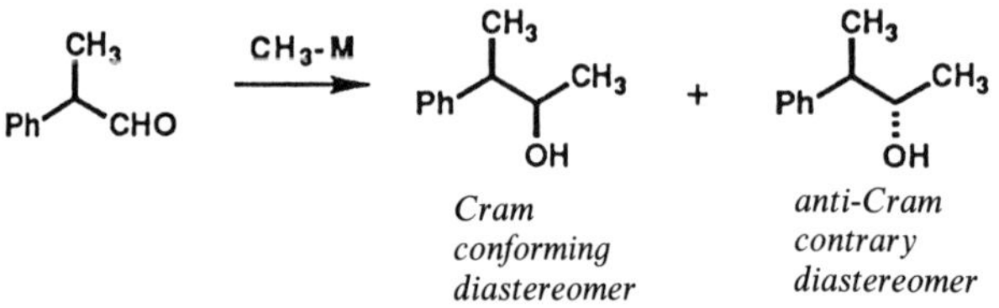

CH_3-M

Cram conforming diastereomer

anti-Cram contrary diastereomer

CH_3MgI : 64% (72 : 28)
MAT/CH_3MgI : 96% (7 : 93)

7.3 RELATIVE MOBILITY OF ORGANOALUMINIUM LIGANDS

Organoaluminium reagents are sometimes the reagent of choice for the introduction of an alkynyl group into a substrate. The alkynylaluminium reagent can be readily prepared from the corresponding lithium or magnesium compound. They smoothly undergo *anti*-periplanar addition to oxiranes and, in the resence of a nickel catalyst, 1,4-addition to enones.

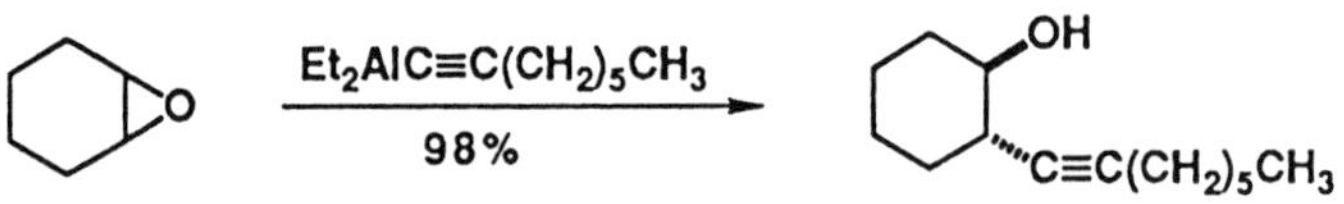

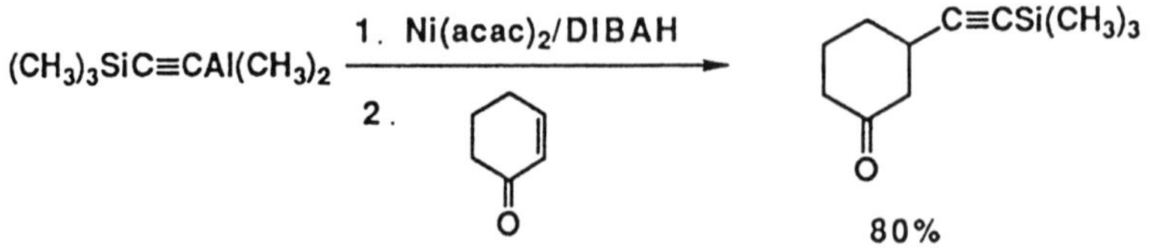

$Et_2AlC{\equiv}C(CH_2)_5CH_3$, 98% → $C{\equiv}C(CH_2)_5CH_3$

$(CH_3)_3SiC{\equiv}CAl(CH_3)_2$ 1. $Ni(acac)_2$/DIBAH 2. → $C{\equiv}CSi(CH_3)_3$ 80%

When the reactivity of these alkynylaluminium reagents was compared with that of the corresponding organocopper reagents, the reactivity trend of each ligand was found to be completely different. The acetylenic ligand attached to copper is stabilized by interaction with the d-orbital of copper and cannot be transferred easily [16]. On the other hand, such a stabilization effect with an aluminium reagent cannot be expected to occur and the less basic acetylenic ligand is transferred preferentially [17, 18].

80%

A heterosubstituted ligand of an organoaluminium compound is transferred much more easily than alkyl ligands. For example, treatment of diethylaluminium amide with methanol produced ammonia by cleaving the Al–N bond. No formation of ethane was detected in this experiment [19].

$$(CH_3CH_2)_2AlNH_2 + CH_3OH \longrightarrow (CH_3CH_2)_2AlOCH_3 + NH_3$$

Similarly, cleavage of the Al—S or Al—P bond occurred preferentially by treatment of dimethylaluminium sulphide and phosphite with methanol. In general, the R_2AlX type of reagent revealed a similar reactivity profile. This reactivity of aluminium reagents also was in sharp contrast to that of organocopper reagents, where the PhS and PhO groups were frequently utilized as non-transfer groups [16]. The reaction between diethylaluminium amide and methanol may be understood as a ligand-exchange process involving intermediates **13** to **14** which requires a much smaller activation energy than the evolution of ethane gas. This characteristic feature may be utilized broadly in organic synthesis. Table 7.1 shows several examples of reactions of AlR_2X, all of which depend on this feature of organoaluminium reagents.

$(CH_3CH_2)_2AlNH_2$ → (CH_3OH) **(13)** → **(14)** → $(CH_3CH_2)_2AlOCH_3$ + NH_3

(13) ⇢ —Al(NH_2)(CH_2CH_3)—$\overset{+}{O}$—H ↛ —Al(NH_2)—O + CH_3CH_3 ↑

Table 7.1. Organic transformations based on R_2AlX

Transformation	Reagent	Ref.
-COOR → -COX; X = NR_2, NNR^1R^2, SR, SeR	$(CH_3)_2AlX$	20, 21, 22 23
-COOR → -CN	$(CH_3)_2AlNH_2$	24
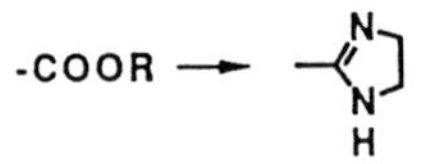 -COOR → (imidazoline, N, N, H)	$(CH_3)_2AlNHCH_2CH_2NH_2$	25
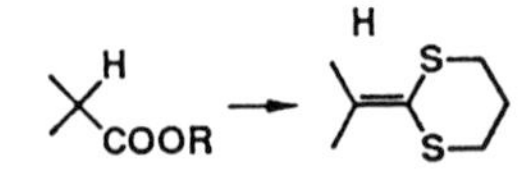 (H, COOR) → (S, S dithiane)	$((CH_3)_2AlSCH_2)_2CH_2$	26
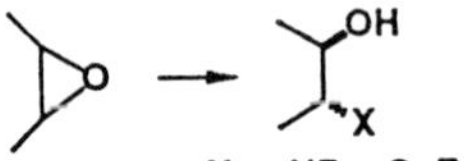 (epoxide, O) → (OH, X); X = NR_2, SeR	$(CH_3CH_2)_2AlNR_2$ $(CH_3)_2AlSeR$	27 23
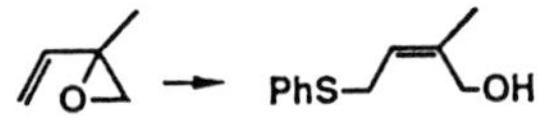 (vinyl epoxide, O) → PhS–...–OH	$(CH_3CH_2)_2AlSPh$	28
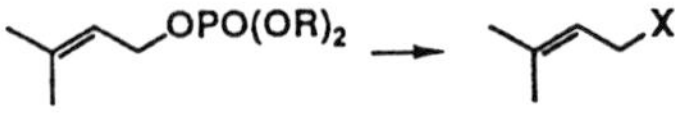 ...$OPO(OR)_2$ → ...X; X = OPh, NHPh, SR	$(CH_3)_2AlX$	29
Michael addition O=C-C=C → O=C-C-CX	$(CH_3)_2AlX$ X = SR, $SeCH_3$	23, 30
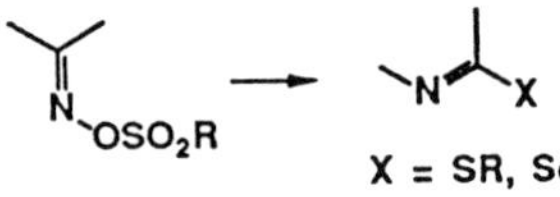 (N–OSO_2R) → (N, X); X = SR, SeR	R'_2AlX	31
RCN → R–C(=NH)–NR'R"	$CH_3Al(Cl)NR'R"$	32
 H...(epoxide, O) → ...OH	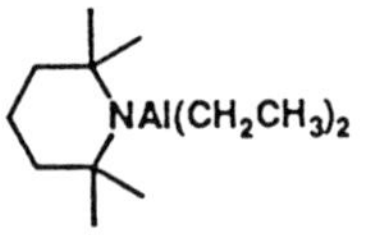 (tetramethylpiperidide) $NAl(CH_2CH_3)_2$	33
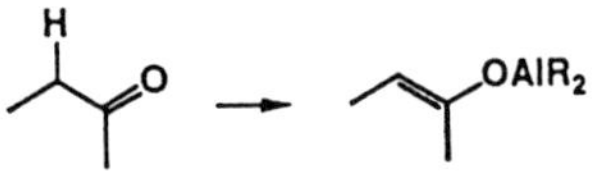 (H, ketone O) → ...$OAlR_2$	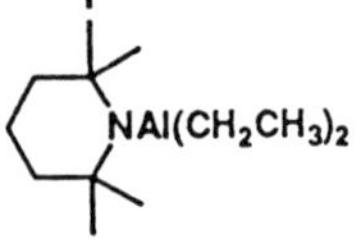 (tetramethylpiperidide) $NAl(CH_2CH_3)_2$	34

The push–pull behaviour of organic aluminium reagents can be exploited in numerous synthetically useful reactions. Table 1 summarizes representative examples.

7.4 OXYGENOPHILICITY OF ORGANOALUMINIUM REAGENTS

Organoaluminium reagents are highly reactive compounds and those with alkyl groups of four or less carbon atoms are usually pyrophoric, that is, they ignite spontaneously in air at ambient temperature. All aluminium alkyl compounds react violently with water. Increasing the molecular weight by increasing the number of carbons in the alkyl groups, or by substituting halogens for these groups, generally reduces the pyrophoric reactivity.

Whatever the detail, it seems probable by analogy with other organometallic oxygenations that the initial step in oxidation must be the formation of a peroxide (**15**), which either undergoes reaction with additional organoaluminium to form a dialkylaluminium alkoxide (**16**) or else undergoes rearrangement to methylaluminium dialkoxide [35].

$R_3Al \xrightarrow{O_2}$ (**15**) $\longrightarrow RAl(OR)_2$ (**16**) $\longrightarrow\longrightarrow Al(OR)_3$

The extraordinarily high reactivity of organoaluminium reagents must be carefully controlled if they are to be used for selective reactions; the high oxygenophilicity of the reagent should be conserved.

Combination with Cp_2TiCl_2–Me_3Al ('Tebbe's complex', **17**) resulted in an excellent reagent for the transformation of esters to vinyl ethers (Scheme 7.8). This reaction clearly utilized the oxygenophilicity of the aluminium reagent. In this particular case, the

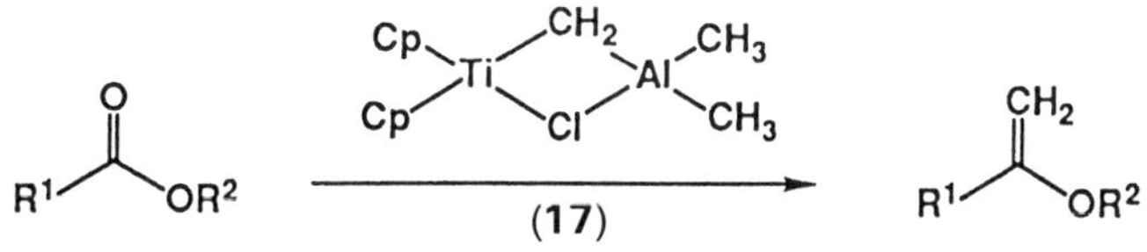

$R^1C(O)OR^2 \xrightarrow{(\mathbf{17})} R^1C(=CH_2)OR^2$

intermediate oxyanion (**18**)was eliminated before its conversion to ketone. Thus, a sufficiently oxygenophilic metal is necessary for the success of this transformation [36].

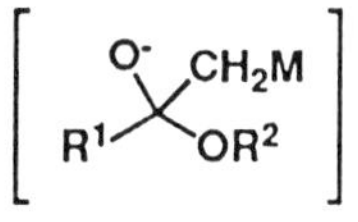

$[R^1C(O^-)(CH_2M)OR^2]$ (**18**)

Mild oxidation reactions using organoaluminium reagents [37], in particular the epoxidation of allyl alcohols and the dehydrogenation of secondary alcohols to give ketones may be carried out.

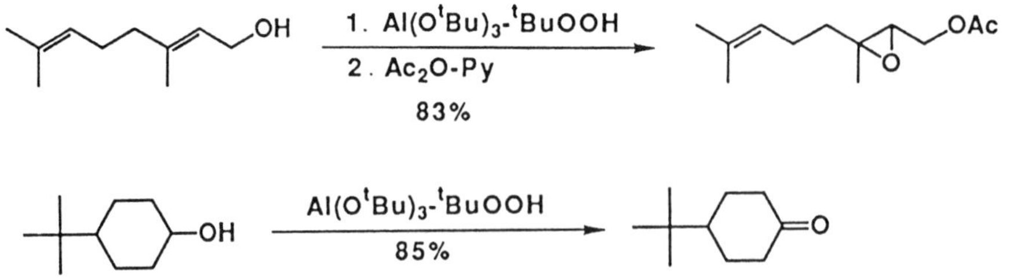

OH — 1. $Al(O^tBu)_3$-^tBuOOH; 2. Ac_2O-Py; 83% → OAc

OH — $Al(O^tBu)_3$-^tBuOOH; 85% → =O

When an aluminium reagent is treated with *tert*-butyl hydroperoxide, the intermediate **19** can be generated.

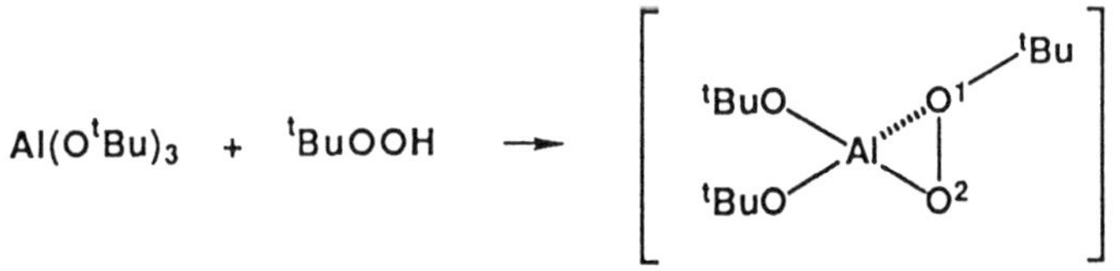

$Al(O^tBu)_3$ + ^tBuOOH →

(**19**)

Strong coordination of oxygen with an aluminium atom results in a highly polarized structure (**19**), and this oxygen turns out to be electrophilic. The first reaction above is similar to the titanium-catalysed Sharpless oxidation and the transition state should be a three-membered ring structure (**20**) [38]. If there is an appropriate olefin in the system, oxidation takes place smoothly with this oxygen. In the case of simple secondary alcohols, the oxidation produces ketone instead through the possible transition state **21**.

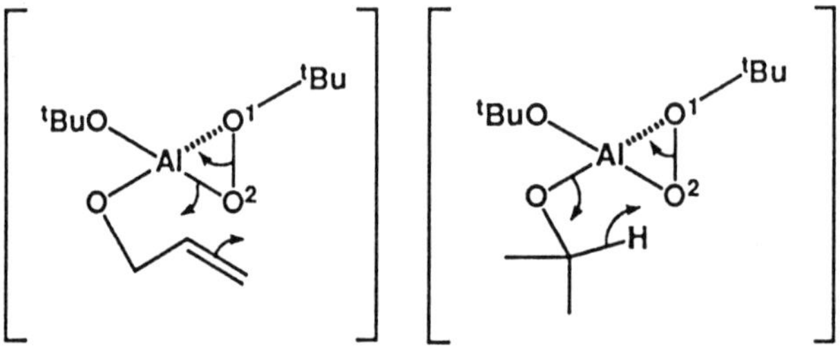

(**20**) (**21**)

Pronounced solvent and temperature effects on the course of organoaluminium-induced reactions are frequently observed. Halogenated solvents can coordinate with aluminium metal and the aluminium reagent behaves as an ate complex in these solvent systems. Such halogenophilicity of aluminium reagents can be utilized in many synthetic processes, including the following cyclopropanation reaction [39]. Thus, treatment of olefins with various organoaluminium compounds and alkylidene iodide under mild conditions produced the corresponding cyclopropanes highly efficiently. For this reaction an intermediate dialkyl(iodomethyl)aluminium species (**22**) is responsible for the cyclopropanation of olefins and it easily undergoes decomposition in the absence of olefins or with excess of trialkylaluminium. Hence the use of equimolar amounts of trialkylaluminium and methylene iodide in the presence of olefins is essential for the achievement of reproducible results.

$RCHI_2$ + R'_3Al ⟶ [I–CH(R)–AlR'_2] (**22**) —$C_{10}H_{21}$CH=CH$_2$, 84-99%→ $C_{10}H_{21}$-cyclopropane-R

R = H and CH_3
R' = CH_3, Et and i-Bu

7.5 ORGANOALUMINIUM REAGENTS AS ACID–BASE COMPLEXED SYSTEMS

Epoxides are isomerized to afford allylic alcohols under basic conditions. With a strong base, one of the protons adjacent to the epoxide is removed, resulting in a high yield of an allylic alcohol. If several protons can be removed, the product becomes a mixture.

For effective rearrangement, the reagent requires two reaction sites, a Lewis acid site to coordinate oxygen of the epoxide and a Lewis base site to remove the proton from the system. An organoaluminium amide was found to be the reagent of choice for this purpose. If aluminium metal coordinates well with oxygen (aluminium metal now acting as the centre of an ate complex!) and with a lone pair of electrons on nitrogen essential for removal of the proton, a well defined structure of the transition state **23** of this reaction could be expected. With such a rigid transition structure, a highly regioselective deprotonation occurs [33].

(**23**)

A pair of stereoisomeric epoxides was chosen for the reaction. As shown overleaf, epoxide **24** has open space on the upper side of the molecule and the epoxide **25** has open space on the lower side of the structure. With the usual reagents, no discrimination between the two structures was possible. If aluminium reagent attacks from the less hindered side of the molecule, the adjacent nitrogen would approach the proton closest to the aluminium metal. For epoxide **24**, this should be a proton from the methyl group and for epoxide **25** the methylene proton would be the one to be removed. Consequently, the two isomers would give rise to two different products from the reaction [33].

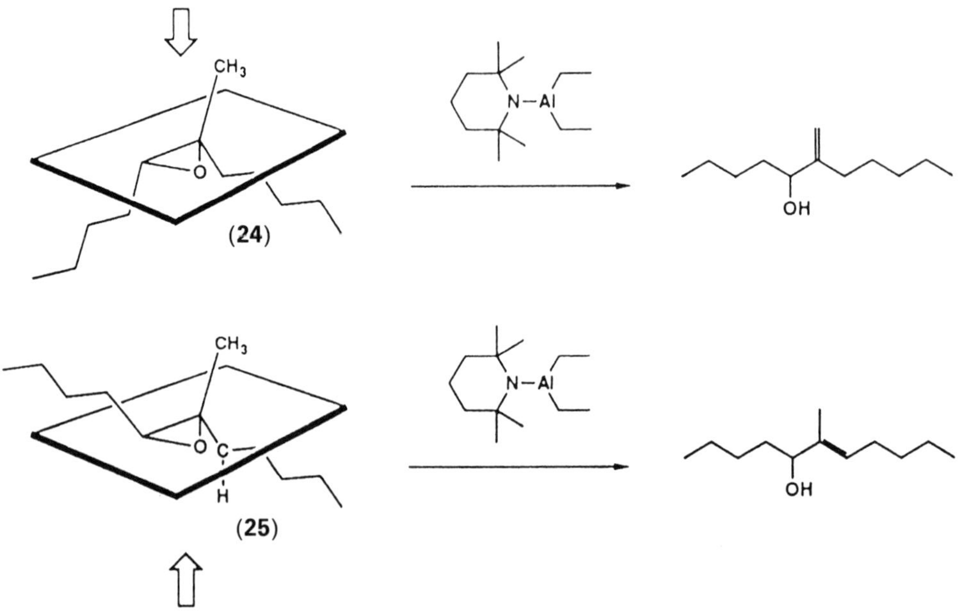

CH_3 N—Al OH (**24**)

CH_3 O C H N—Al OH (**25**)

Beckmann rearrangement involves the skeletal rearrangement of ketoximes in the presence of Brønsted or Lewis acids to give amides or lactams. The reaction has found broad application in the manufacture of synthetic polyamides. It is a preferred means of incorporating the nitrogen atom efficiently in both acyclic and alicyclic systems, thereby providing a powerful method for a variety of alkaloid syntheses. The reaction proceeds through a carbenium–nitrilium ion (**26**) after capture of the oxime oxygen by a Lewis acid. The intermediate cation adds water to generate an amide or lactam [40].

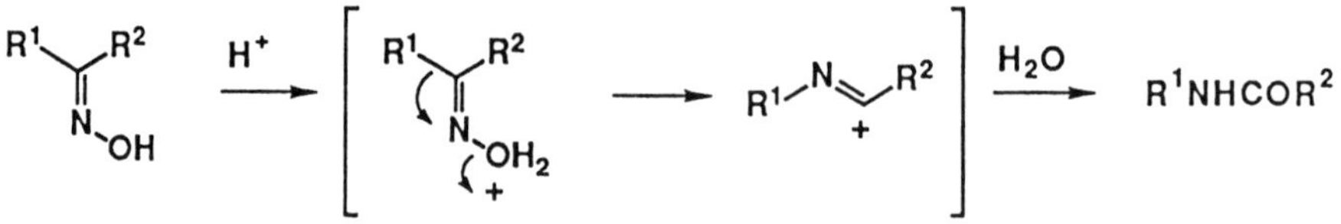

R^1 R^2 N OH $\xrightarrow{H^+}$ [R^1 R^2 N OH_2 + → R^1–N≡C–R^2 +] $\xrightarrow{H_2O}$ R^1NHCOR^2

(**26**)

If the reaction is performed in the absence of water, the resulting cation is not efficiently trapped and subsequently decomposes in various ways. With an organoaluminium reagent, a typical acid–base complex system was found to be highly useful in this case. Thus, after the rearrangement process, the resulting ate complex was able to supply the alkyl group to the cationic intermediate. The resulting imine (**24**) can be reduced to the amine with DIBAH [11, 12].

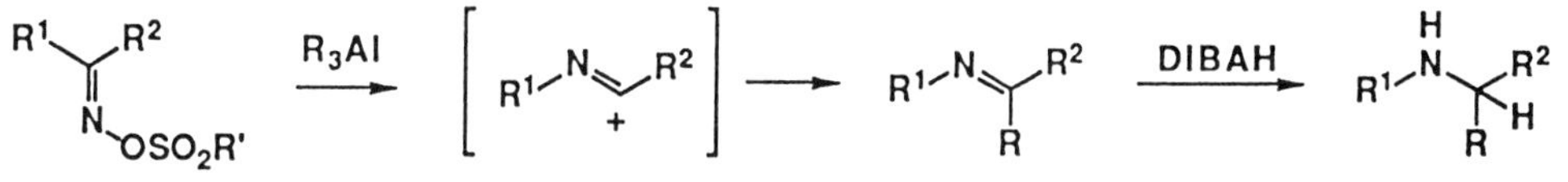

R_3Al DIBAH

(**27**)

In the reaction shown for the synthesis of pumiliotoxin C, the starting ketone was selectively prepared from the corresponding bicyclic cyclopentenone derivative **28**, and was hydrogenated in the presence of palladium black to give a *cis*-fused ring system. The saturated ketone was treated with hydroxylamine and then *p*-toluenesulphonyl chloride to give the tosylate. After recrystallization followed by a Beckmann rearrangement–alkylation process, pumiliotoxin C was obtained stereoselectively [11, 12].

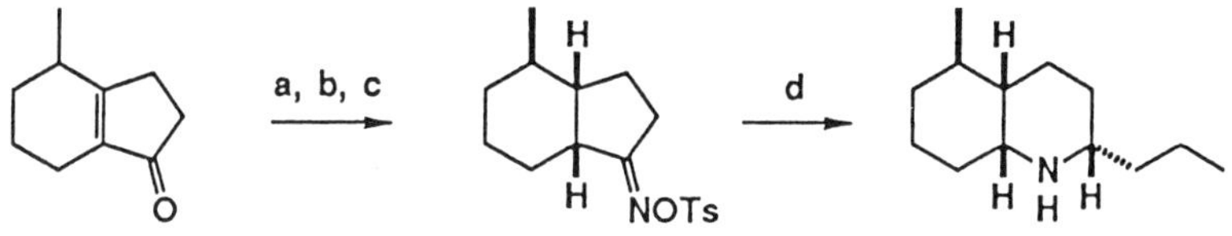

(**28**) a, b, c NOTs d Pumiliotoxin C

a: H_2/Pd - CH_3CH_2COOH
b: $H_2NOH{\cdot}HCl$ - NaOAc
c: TsCl - $N(CH_2CH_3)_3$
d: $^nPr_3Al$ - DIBAH

Claisen rearrangement of an allyl vinyl ether, such as 3-cyclohexenyl vinyl ether (**29**, overleaf), generally requires high temperatures. However, in the presence of an organoaluminium catalyst, the reaction can be effected at a much lower, ambient for example,

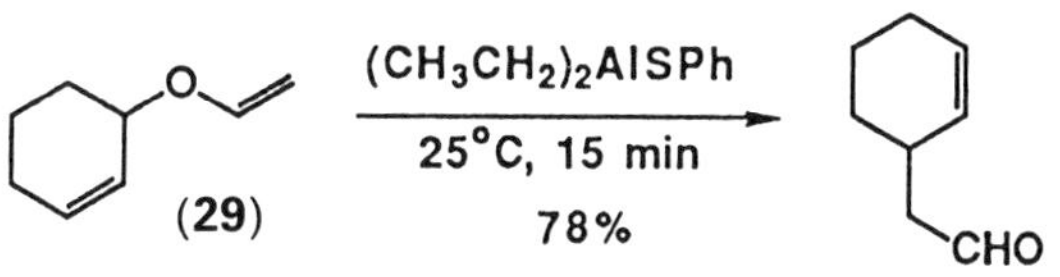

(**29**) $(CH_3CH_2)_2AlSPh$, 25°C, 15 min, 78% → CHO

temperature [41]. With trialkylaluminium reagents the resulting aldehydes may be further converted into secondary alcohols by *in situ* transfer of alkyl groups. This was observed, for example, when 1-butyl-2-propenyl vinyl ether (**30**) was treated with trimethyl- or triethylaluminium [41].

Diethylaluminium mercaptide or diethylaluminium chloride–triphenylphosphine may be used for the preparation of unsaturated aldehydes. It should be noted that other, stronger Lewis acids such as BF_3 and $TiCl_4$ could not be used for this reaction. An acid–base complex system plays an important role in this particular reaction [41].

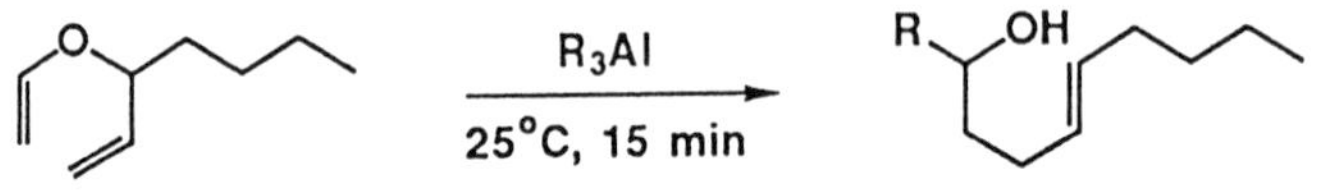

R_3Al, 25°C, 15 min

(**30**)

R = CH_3; 91% (*E/Z* = 47:53)
R = CH_2CH_3; 75% (*E/Z* = 42:58)

Bulky aluminium reagents can also be utilized for Claisen rearrangements. The bulky reagents methylaluminium bis(2,6-diphenylphenoxide) (MAPH) and methylaluminium bis(4-bromo-2,6-*tert*-butylphenoxide) (MABR) are highly effective reagents which are able to catalyse such rearrangements under very mild reaction conditions.

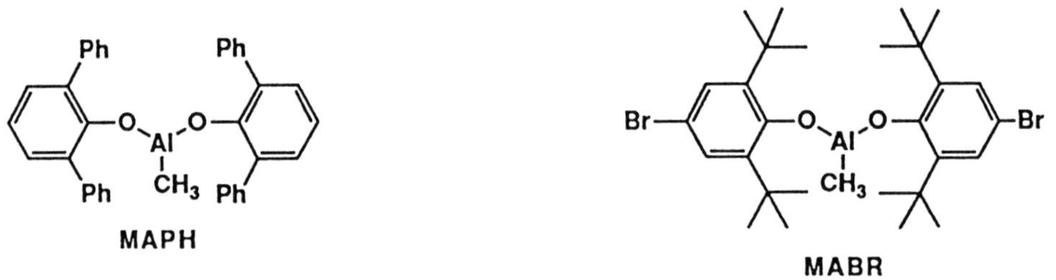

MAPH

MABR

With MABR, for example, the rearrangement takes place in a few seconds even at −78 °C. Moreover, the 4-alkenals are formed with high *cis*-selectivity [42].

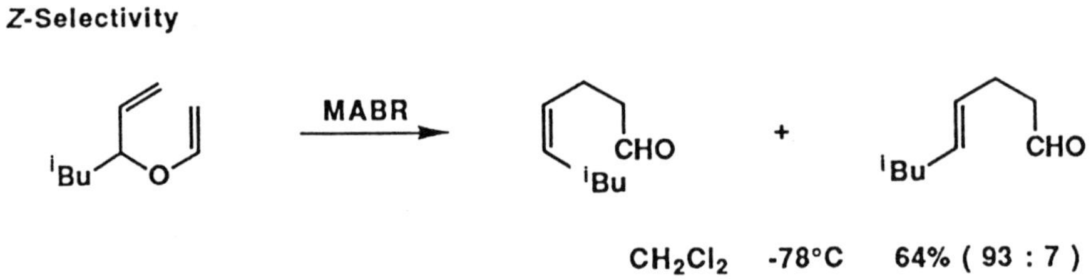

Z-Selectivity

MABR

CH_2Cl_2 -78°C 64% (93 : 7)

On the other hand, when MABR is replaced by MAPH the *trans*-isomers are obtained preferentially [42].

E-Selectivity

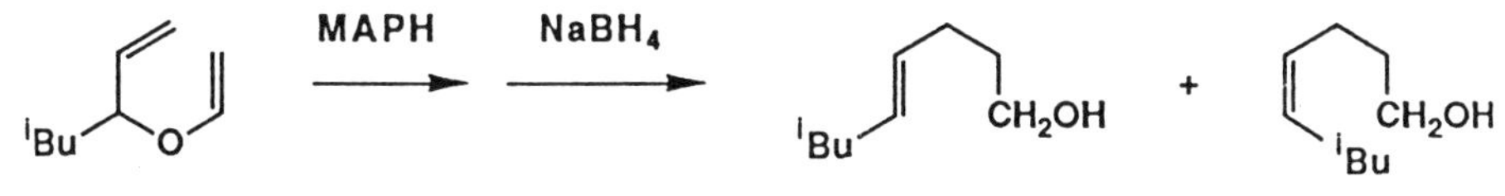

MAPH → $NaBH_4$ →

toluene	-20°C	85% (97 : 3)
thermal rearrangement		70% (92 : 8)

The observed selectivity may be due to precise steric discrimination between the substrate conformers *ax*-**31** and *eq*-**31** by the bulky aluminium reagent [42].

possible mechanism

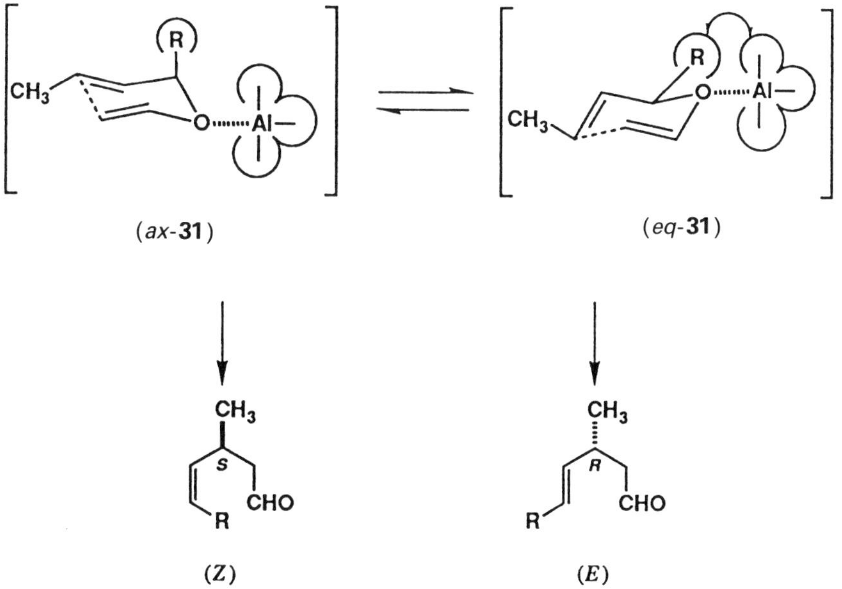

(*ax*-**31**) (*eq*-**31**)

(*Z*) (*E*)

7.6 SPECIFIC ATE COMPLEX REACTIVITY

As described previously, the initial reaction of a trivalent organoaluminium compound with organic molecules is always the coordination of the substrate with an aluminium atom. In contrast, a preformed aluminium ate complex behaves completely differently: the complex has already satisfied the octet rule and behaves as a nucleophile rather than a Lewis acid. For example, reduction of an α,β-unsaturated epoxide gave a completely different product when the aluminium reagent was changed from a neutral trivalent species to an ate complex [43]. With DIBAH, the 1,4-addition type of reduction product was found to be the major product through a possible six-membered transition state, whereas with lithium aluminium hydride, a direct reduction product was formed preferentially. This is a typical example of the difference in reactivity between trivalent aluminium reagents and ate complexes.

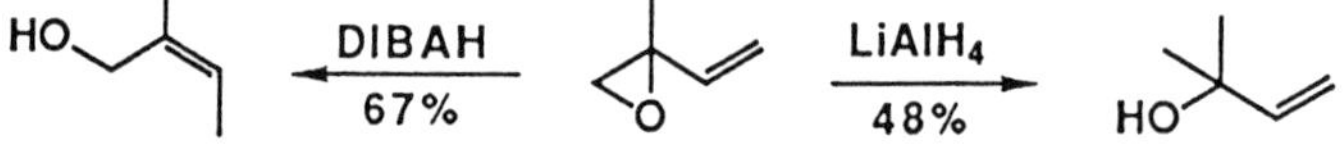

HO DIBAH 67% LiAlH$_4$ 48% HO

An alkenylaluminium reagent was used in a key synthetic process in the construction of the prostaglandin structure. The preparation of the alkenylaluminium **32** is straightforward, using simple hydroalumination of a terminal acetylene. Preparation of the ate complex **33** from this alkenylaluminium reagent is not difficult and a 1,4- rather than a 1,2-addition product can be expected from the reaction with a cyclopentenone derivative [44].

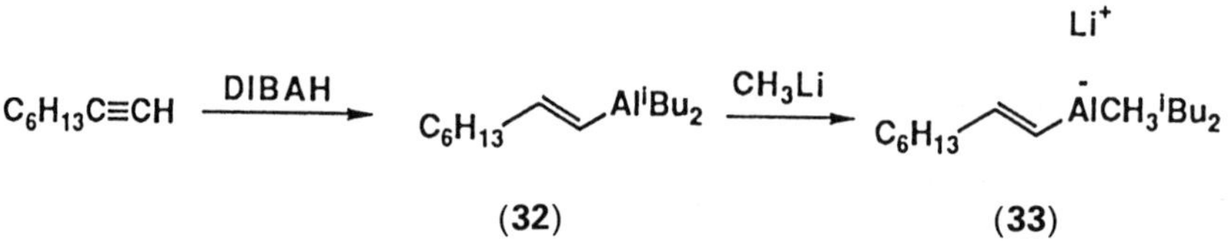

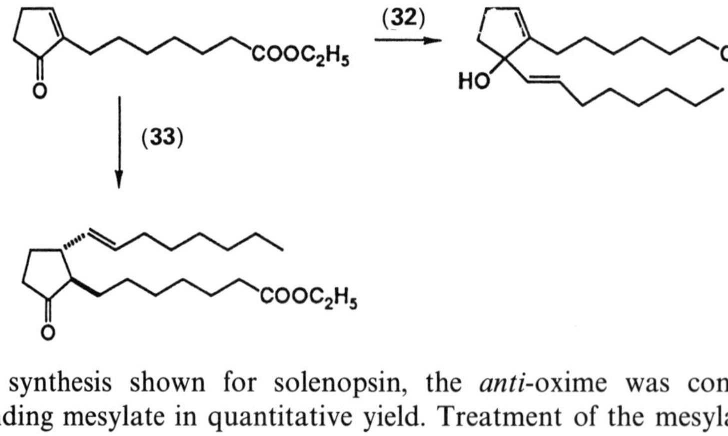

$C_6H_{13}C{\equiv}CH$ DIBAH C_6H_{13} Al^iBu_2 (32) CH_3Li C_6H_{13} Li^+ $\bar{Al}CH_3{}^iBu_2$ (33)

$COOC_2H_5$ (32) HO $COOC_2H_5$

O (33)

$COOC_2H_5$

O

In the synthesis shown for solenopsin, the *anti*-oxime was converted into the corresponding mesylate in quantitative yield. Treatment of the mesylate in methylene chloride with excess of trimethylaluminium in toluene resulted in the formation of an imine **34**. The final step of the synthesis should be simple reduction; unfortunately, however, the usual reduction with DIBAH gave the *cis*-isomer almost exclusively. Moderate selectivity for the desired *trans* isomer was observed with lithium aluminium hydride. Excellent stereoselectivity was finally attained in formation of the *trans* form lithium aluminium hydride in an ethereal solvent in the presence of an equimolar amount of trialkylaluminium at low temperature. Thus, solenopsin was obtained almost exclusively using lithium aluminium hydride–trimethylaluminium in THF at low temperature [45].

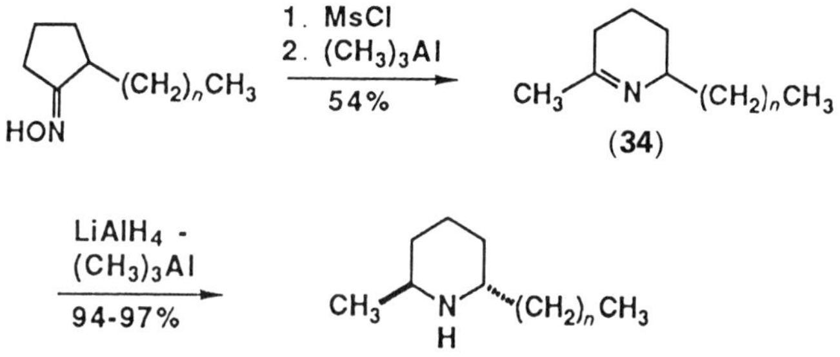

HON $(CH_2)_nCH_3$ — 1. MsCl 2. $(CH_3)_3Al$, 54% → CH_3 N $(CH_2)_nCH_3$ (**34**)

$LiAlH_4$ - $(CH_3)_3Al$, 94-97% → CH_3 N H $(CH_2)_nCH_3$

Solenopsin A (n = 10)
Solenopsin B (n = 12)

The observed selectivity can be predicted. In the hydride reduction of the imine, the underside approach of the hydride ion toward the imino π-bond is preferred by stabilization of the σ^* orbital through electron delocalization from the σ C—H bond into the σ^* orbital, producing the *cis*-isomer [46]. In the presence of trialkylaluminium

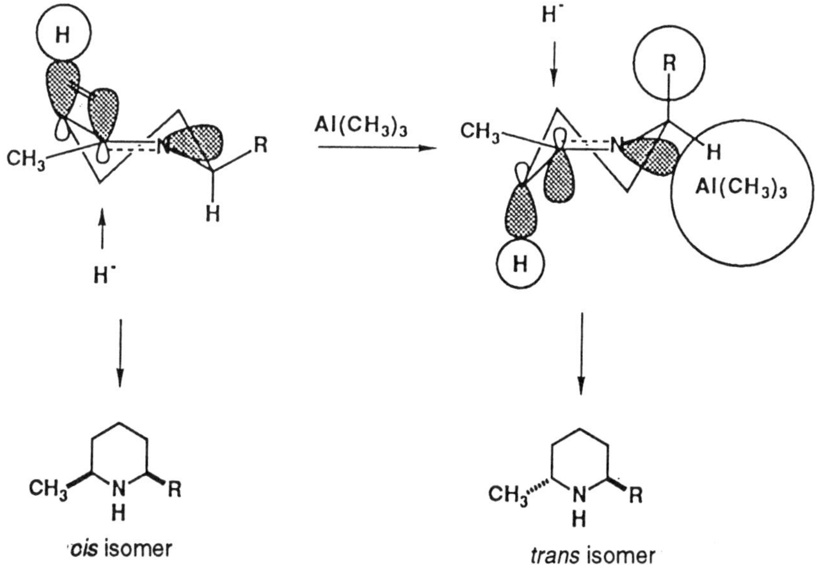

H, CH_3, R, H, H^- — $Al(CH_3)_3$ → H^-, R, CH_3, H, $Al(CH_3)_3$, H

CH_3 N H R — *cis* isomer

CH_3 N H R — *trans* isomer

as a Lewis acid, on the other hand, the alkyl group occupies the axial position because of the steric interaction between R and R_3Al coordinating the nitrogen lone pair of electrons by *A*-strain [47]. Such a conformational change facilitates the upperside approach of the hydride ion toward the imino π-bond and furnishes the desired *trans*-isomer.

Reductive cleavage of acetals with organoaluminium hydride reagents affords stereoselectively *syn* reduced products (such as **35**). The observed high diastereoselectivity was ascribed to the stereospecific coordination of the organoaluminium reagent with one of the acetal oxygens followed by hydride attack *syn* to the cleaved C—O bond [48].

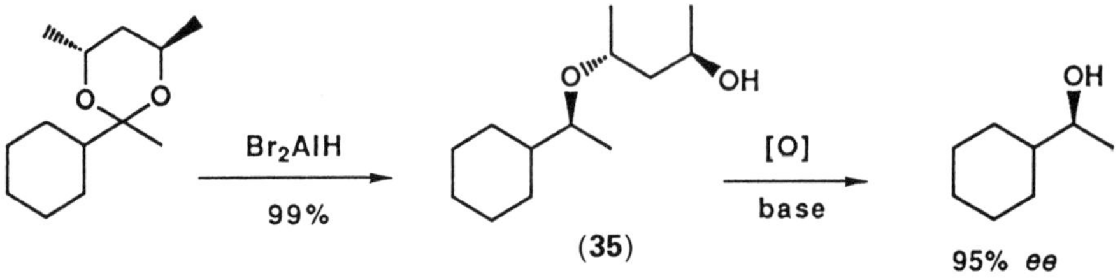

This reaction probably proceeds by a tight ion-paired S_N1-like mechanism [48].

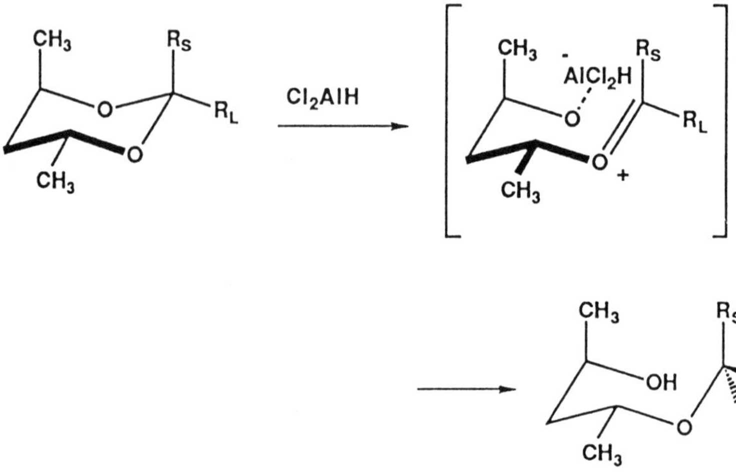

7.7 GENERAL HINTS CONCERNING THE USE OF ORGANOALUMINIUM COMPOUNDS AND WORKING PROCEDURES

Aluminium alkyls are typically clear, colourless, mobile liquids with low melting points (except trimethylaluminium, m.p. 15 °C) and low vapour pressures at ambient temperature. Aluminium alkyls are miscible in all proportions and compatible with saturated and aromatic hydrocarbons. Organoaluminium compounds are also soluble in ethers and tertiary amines, accompanied by exothermic complex formation.

Aluminium alkyls react very vigorously with air, water or other hydroxylic reagents. The violence of this reaction depends on the molecular weight of the aluminium alkyl. Dilute solutions of aluminium alkyls in hydrocarbon solvents are non-pyrophoric and could be used in place of more common organometallic reagents, such as butyllithium. Aromatic solvents such as benzene and toluene and saturated aliphatic solvents such as hexane can be used. Generally, unsaturated aliphatic and aromatic compounds should not be used. Some chlorinated or oxygenated solvents may be used, but others, such as carbon tetrachloride, may react violently with the formation of toxic gases. The non-pyrophoric concentration varies with the aluminium alkyl compound, temperature and solvent; for C_4 compounds and below concentrations up to 10–20 wt% alkyls are generally non-pyrophoric; for C_5 and higher compounds, the concentration can be increased to over 20 wt%. Hence organoaluminium reagents in hydrocarbon solvents at these concentrations can be used with equal ease as butyllithium in hexane. On the other hand, if one wishes to use the neat aluminium alkyls, extreme caution in handling is essential.

Operations with organoaluminium compounds should be carried out under an inert gas such as nitrogen or argon. This is important not only because of the pyrophoric character of the reagent, but also to eliminate losses due to hydrolysis and oxidation of the organoaluminium compounds.

The following experimental part describes typical uses of organoaluminium compounds in hydrocarbon solvents.

7.7.1 CYCLOPROPANATION USING AN ORGANOALUMINIUM REAGENT: 1-HYDROXYMETHYL-4-(1-METHYLCYCLOPROPYL)-1-CYCLOHEXENE [49]

A dry 1 l three-necked, round-bottomed flask was equipped with a gas inlet, a 50 ml pressure-equalizing dropping funnel, a rubber septum and a PTFE-coated magnetic stirring bar. The flask was flushed with argon, after which 10.7 g (0.07 mol) of (*S*)-perillyl alcohol followed by 350 ml of dichloromethane were injected through the septum into the flask. The solution was stirred and 37.3 ml (0.147 mol) of triisobutylaluminium were added from the dropping funnel over 20 min at room temperature. After the mixture had been stirred at room temperature for 20 min, 7.3 ml (0.091 mol) of diiodomethane were added dropwise with a syringe over 10 min. The mixture was stirred at room temperature for 4 h, then poured into 400 ml of ice-cold 8% aqueous sodium hydroxide. The organic layer was separated and the aqueous layer extracted twice with 100 ml portions of dichloromethane. The combined extracts were dried over anhydrous sodium sulphate and concentrated with a rotary evaporator at *ca* 20 mm. The residual oil was distilled under reduced pressure to give 11.1 g (96%) of 1-hydroxymethyl-4-(1-methylcyclopropyl)-1-cyclohexene as a colourless liquid, b.p. 132–134 °C/24 mmHg.

7.7.2 PREPARATION OF MAD [12]

To a solution of 2,6-di-*tert*-butyl-4-methylphenol (2 equiv.) in toluene was added at room temperature a 2 M hexane solution of trimethylaluminium (1 equiv.). Methane gas was immediately evolved. The resulting mixture was stirred at room temperature for 1 h and used as a solution of MAD in toluene without further purification.

7.7.3 ALKYLATION OF 3-CHOLESTANONE WITH MAD–CH_3Li SYSTEM [12]

To a solution of MAD (2 mmol) in toluene (8 ml) was added at −78 °C 3-cholestanone (258 mg, 0.67 mmol) in toluene (2 ml). After 10 min, a 1.37 M ethereal solution of methyllithium (1.46 ml, 2.0 mmol) was added at −78 °C. The mixture was stirred at −78 °C for 2 h and poured into 1 M hydrochloric acid. After extraction with diethyl ether, the combined extracts were dried, concentrated and purified by column chromatography on silica gel using diethyl ether–hexane (1:3 to 2:1) as eluent to give 3-α-methylcholestan-3β-ol (257 mg, 96%).

7.8 SPECIALIZED TOPICS

7.8.1 ZIEGLER–NATTA CATALYSIS

Although it is not the purpose of this chapter to give a detailed review of organoaluminium chemistry, it is still appropriate to mention some of the recent advances in Ziegler–Natta catalysis. As originally described [1], the Ziegler–Natta catalyst is a combination of a transition metal compound and an organometallic compound of Group I, II or III. It has been referred to as a coordination catalyst, reflecting the view that the catalysis occurs via coordination of the alkene to the active metal centre, followed by insertion into a metal–alkyl bond. A variety [50] of reagent systems have been developed and some of them are proving useful in industry. The combined use of trialkylaluminiums and cyclopentadienyltitanium catalysts is unique among them [50]. The catalyst is not particularly active, but is soluble in organic solvents and hence is popular for academic research studies. The activity of this catalyst is relatively low, but activity an order of magnitude higher is obtained when small amounts of water are added to the trialkylaluminium. Peak activities are found with between 0.2 and 0.5 mol of water per mole of aluminium alkyl [51]. Although the detailed structure of this aluminium reagent is not known, the system seems to have great potential not only for industrial use, but also for more special uses in organic synthesis. A typical combination is that of dicyclopentadienyltitanium dichloride with the methyldi-oxyaluminium species **36** or **37**.

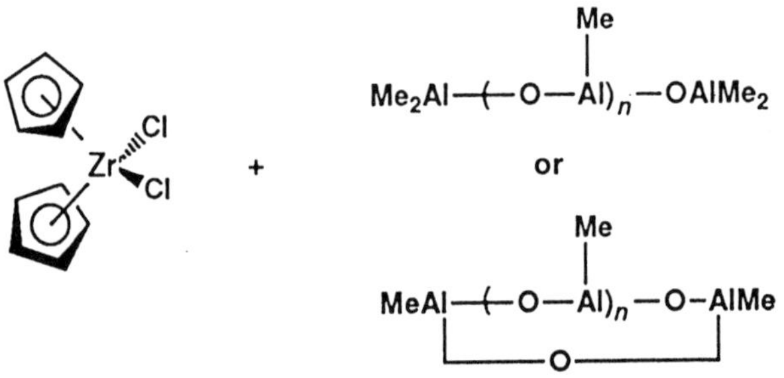

Cp_2ZrCl_2 + $Me_2Al{-}(O{-}Al(Me))_n{-}OAlMe_2$ (**36**)

or

$MeAl{-}(O{-}Al(Me))_n{-}O{-}AlMe$ (with the two terminal Al atoms bridged by O) (**37**)

7.8.2 COMPLEXATION CHROMATOGRAPHY [52]

The ready availability of different types of hindered polyphenols permits the molecular design of various polymeric organoaluminium reagents. For example, this chemistry allows the realization of complexation chromatography, i.e. the separation of heteroatom-containing solutes by complexation with stationary, insolubilized organoaluminium reagents [52]. Accordingly, treatment of the sterically hindered triphenol **38** (2 mmol) in CH_2Cl_2 with Me_3Al (3 mmol) at room temperature for 1 h gave rise to the polymeric monomethylaluminium reagent **39**. After evaporation of the solvent, the residual solid was ground to a powder and mixed with silanized silica gel (1.7 g) in an argon box. This was packed in a short-path glass column (150 mm × 10 mm i.d.) as a stationary phase and washed once with dry, degassed hexane to remove unreacted free triphenol. Then a solution of methyl 3-phenylpropyl ether and ethyl 3-phenylpropyl ether (0.5 mmol each) in degassed hexane was applied to this short-path column and eluted with hexane–diethyl ether to achieve a complete separation. This technique allows the surprisingly clean separation of structurally similar ether substrates. Ethyl 3-phenylpropyl ether and isopropyl 3-phenylpropyl ether, or the THF and THP ethers of 4-(*tert*-butyldiphenylsiloxy)-1-butanol, can be separated equally well by this short-path column chromatographic technique.

7.9 REFERENCES

1. K. Ziegler, in *Organometallic Chemistry*, ed. H. Zeiss, p. 194, Reinhold, New York, 1960; J. Boor, *Ziegler–Natta Catalysts and Polymerizations*, Academic Press, New York, 1979.
2. (a) T. Mole, E. A. Jeffery, *Organoaluminum Compounds*, Elsevier, Amsterdam, 1972; (b) G. Bruno, *The Use of Aluminum Alkyls in Organic Synthesis*, Ethyl Corporation, Baton Rouge, 1970, 1973, 1980; (c) E. Negishi, *J. Organomet. Chem. Libr.* **1** (1976) 93; (d) H. Yamamoto, H. Nozaki, *Angew. Chem., Int. Ed. Engl.* **17** (1978) 169; (e) E. Negishi, *Organometallics in Organic Synthesis*, Vol. 1, p. 286, Wiley, New York, 1980; (f) J. J. Eisch, in *Comprehensive Organometallic Chemistry*, ed. G. Wilkinson, G. G. A. Stone, E. W. Abel, Vol. 1, p. 555, Pergamon Press, Oxford, 1982; (g) J. R. Zietz, G. C. Robinson, K. L. Lindsay, in *Comprehensive Organometallic Chemistry*, ed. G. Wilkinson, G. G. A. Stone, E. W. Abel, Vol. 7, p. 365, Pergamon Press, Oxford, 1982; (h) G. Zeifel, J. A. Miller, *Org. React.* **32** (1984) 375; (i) K. Maruoka, H. Yamamoto, *Angew. Chem., Int. Ed. Engl.* **24** (1985) 668; (j) K. Maruoka, H. Yamamoto, *Tetrahedron* **44** (1988) 5001.
3. (a) E. C. Ashby, J. Laemmle, G. E. Parris, *J. Organomet. Chem.* **19** (1969) 24; (b) T. Mole, J. R. Surtees, *Aust. J. Chem.* **17** (1964) 961.
4. (a) E. C. Ashby, S. H. Yu, *J. Chem. Soc. D* (1971) 351; (b) E. C. Ashby, S. H. Yu, P. V. Roling, *J. Org. Chem.* **37** (1972) 1918; (c) J. Laemmle, E. C. Ashby, P. V. Roling, *J. Org. Chem.* **38** (1974) 2526.
5. E. C. Ashby, J. T. Laemmle, *Chem. Rev.* **75** (1975) 521.
6. T. Mole, *Aust. J. Chem.* **19** (1966) 381.
7. S. Sakane, K. Maruoka, H. Yamamoto, *J. Chem. Soc. Jpn.* (1985) 324.
8. (a) G. Wilke, H. Muller, *Chem. Ber.* **89** (1956) 444; (b) G. Wilke, H. Muller, *Liebigs Ann. Chem.* **618** (1958) 267; (c) G. Wilke, H. Muller, *Liebigs Ann. Chem.* **629** (1960) 222; (d) G. Wilke, W. Schneider, *Bull. Soc. Chim. Fr.* (1963) 1462.
9. (a) J. J. Eisch, G. A. Damasevitz, *J. Org. Chem.* **41** (1976) 2214; (b) K. Uchida, K. Utimoto, H. Nozaki, *J. Org. Chem.* **41** (1976) 2215.
10. J. J. Eisch, S. Rhee, *J. Am. Chem. Soc.* **97** (1975) 4673.
11. K. Hattori, Y. Matsumura, T. Miyazaki, K. Maruoka, H. Yamamoto, *J. Am. Chem. Soc.* **103** (1981) 7368.
12. K. Maruoka, T. Miyazaki, M. Ando, Y. Matsumura, S. Sakane, K. Hattori, H. Yamamoto, *J. Am. Chem. Soc.* **105** (1983) 2831.
13. N. S. Ham, T. Mole, *Prog. Nucl. Magn. Reson. Spectrosc.* **4** (1961) 91.
14. (a) K. C. Ramey, J. F. O. Brien, I. Hasegawa, A. E. Borchert, *J. Phys. Chem.* **69** (1965) 3418; (b) K. C. Williams, T. L. Brown, *J. Am. Chem. Soc.* **88** (1966) 5460; (c) E. A. Jeffery, T. Mole, *Aust. J. Chem.* **22** (1969) 1129.
15. K. Maruoka, T. Itoh, M. Sakurai, K. Nonoshita, H. Yamamoto, *J. Am. Chem. Soc.* **110** (1988) 3588.
16. (a) E. J. Corey, D. J. Beames, *J. Am. Chem. Soc.* **94** (1972) 7210; (b) P. W. Collins, E. Z. Dajani, M. S. Bruhn, C. H. Brown, J. R. Palmer, R. Pappo, *Tetrahedron Lett.* **13** (1975) 4217; (c) G. H. Posner, C. E. Whitten, J. Sterling, *J. Am. Chem. Soc.* **95** (1973) 7788; (d) G. H. Posner, D. J. Brunelle, L. Sinoway, *Synthesis* (1974) 662.
17. (a) J. Fried, C. Lin, S. H. Ford, *Tetrahedron Lett.* **7** (1969) 1379; (b) J. Fried, C. Lin, M. Mehra, W. Kao, P. Dalven, *Ann. N.Y. Acad. Sci.* **180** (1971) 36.
18. R. T. Hanse, D. B. Carr, J. Schwartz, *J. Am. Chem. Soc.* **100** (1978) 2244.

19. K. Gosling, J. D. Smith, D. H. W. Wharmby, *J. Chem. Soc. A* (1969) 1738.
20. (a) A. Basha, M. Lipton, S. M. Winreb, *Tetrahedron Lett.* **15** (1977) 4171; (b) E. J. Corey, D. J. Beames, *J. Am. Chem. Soc.* **95** (1973) 5829.
21. A. Benderly, S. Stavchansky, *Tetrahedron Lett.* **29** (1988) 739.
22. R. P. Hatch, S. M. Weinreb, *J. Org. Chem.* **42** (1977) 3960.
23. A. P. Kozikowski, A. Ames, *J. Org. Chem.* **43** (1978) 2735.
24. J. L. Wood, N. A. Khatri, S. M. Weinreb, *Tetrahedron Lett.* **17** (1979) 4907.
25. G. Neef, U. Eder, G. Sauer, *J. Org. Chem.* **46** (1981) 2824.
26. E. J. Corey, A. P. Kozikowski, *Tetrahedron Lett.* **13** (1975) 925.
27. L. E. Overman, L. A. Flippin, *Tetrahedron Lett.* **22** (1981) 195.
28. A. Yasuda, M. Takahashi, H. Takaya, *Tetrahedron Lett.* **22** (1981) 2413.
29. Y. Kitagawa, S. Hashimoto, S. Iemura, H. Yamamoto, H. Nozaki, *J. Am. Chem. Soc.* **98** (1976) 5030.
30. A. Itoh, S. Ozawa, K. Oshima, H. Nozaki, *Tetrahedron Lett.* **21** (1980) 361.
31. K. Maruoka, T. Miyazaki, M. Ando, Y. Matsumura, S. Sakane, K. Hattori, H. Yamamoto, *J. Am. Chem. Soc.* **105** (1983) 2831.
32. R. S. Garigipati, *Tetrahedron Lett.* **31** (1990) 1969.
33. (a) A. Yasuda, S. Tanaka, K. Oshima, H. Yamamoto, H. Nozaki, *J. Am. Chem. Soc.* **96** (1974) 6513; (b) S. Tanaka, A. Yasuda, H. Yamamoto, H. Nozaki, *J. Am. Chem. Soc.* **97** (1975) 3252.
34. H. Nozaki, K. Oshima, K. Takai, S. Ozawa, *Chem. Lett.* (1979) 379.
35. (a) A. G. Davies, C. D. Hall, *J. Chem. Soc.* (1963) 1192; (b) H. Hock, H. Kropf, F. Ernst, *Angew. Chem.* **71** (1959) 541.
36. S. H. Pine, R. Zahler, D. A. Evans, R. H. Grubbs, *J. Am. Chem. Soc.* **102** (1980) 3270.
37. K. Takai, K. Oshima, H. Nozaki, *Tetrahedron Lett.* **21** (1980) 1657.
38. (a) T. Katsuki, K. B. Sharpless, *J. Am. Chem. Soc.* **102** (1980) 5974; (b) B. E. Rossiter, T. Katsuki, K. B. Sharpless, *J. Am. Chem. Soc.* **103** (1981) 464; (c) V. S. Matin, S. S. Woodard, T. Katsuki, Y. Yamada, M. Ikeda, K. B. Sharples, *J. Am. Chem. Soc.* **103** (1981) 6237.
39. K. Maruoka, Y. Fukutani, H. Yamamoto, *J. Org. Chem.* **50** (1985) 4412.
40. C. G. McCarty, in *Chemistry of the Carbon–Nitrogen Double Bond*, ed. S. Patai, p. 408, Wiley, Chichester, 1970.
41. K. Takai, I. Mori, K. Oshima, H. Nozaki, *Tetrahedron Lett.* **22** (1981) 3985.
42. K. Nonoshita, H. Banno, K. Maruoka, H. Yamamoto, *J. Am. Chem. Soc.* **112** (1990) 316.
43. R. S. Lenox, J. A. Katzenellenbogen, *J. Am. Chem. Soc.* **95** (1973) 957.
44. K. F. Bernady, M. B. Floyd, J. F. Poletto, M. J. Weiss, *J. Org. Chem.* **44** (1979) 1438.
45. Y. Matsumura, K. Maruoka, H. Yamamoto, *Tetrahedron Lett.* **23** (1982) 1929.
46. A. S. Cieplak, *J. Am. Chem. Soc.* **103** (1981) 4540.
47. A. S. Narula, *Tetrahedron Lett.* **22** (1981) 2017.
48. K. Ishihara, A. Mori, H. Yamamoto, *Tetrahedron* **46** (1990) 4595.
49. K. Maruoka, S. Sakane, H. Yamamoto, *Org. Synth.* **67** (1989) 176.
50. J. R. Young and J. R. Stille, *J. Am. Chem. Soc.* **114** (1992) 4936, and references cited therein.
51. H. Sinn, W. Kaminsky, H.-J. Vollmer, R. Woldt, *Angew. Chem., Int. Ed. Engl.* **19** (1980) 390.
52. K. Maruoka, S. Nagahara, H. Yamamoto, *J. Am. Chem. Soc.* **112** (1990) 6115.

8

Organotin Chemistry

HITOSI NOZAKI
Okayama University of Science, Japan

Organometallics in Synthesis—A Manual. Edited by M. Schlosser

8.1 INTRODUCTION

Organotin compounds have been produced commercially as poly(vinyl chloride) stabilizers and as an active ingredients of antifouling paints, etc., since the early 1950s [1]. A recent topic is marine pollution with hexabutyldistannoxane ($R_3SnOSnR_3$, R = butyl). Synthetic applications as reagents or catalysts started slightly later than for the silicon counterparts. This is exemplified by the fact that Colvin's book [2] on organosilicon reagents was published in 1980, whereas the Pereyre's book [3] on tin compounds appeared in 1987. A Tetrahedron Symposium in this field [4] appeared in 1989. Certainly, this is one of the most rapidly expanding branches of chemistry and refs 3 and 4 are extremely useful in understanding the state of the art [5]. This chapter is intended to give a brief outline of tin reagents in selective organic synthesis for those who are not familiar with this type of organometallic compounds. Literature citations are illustrative rather than exhaustive and the selection has been made on the basis of synthetic usefulness.

8.2 TIN–HYDROGEN BOND REACTIVITIES AND RADICAL CYCLIZATIONS

The reagent used here is mostly tributylstannane, but trimethyl- and triphenylstannane are also used. The tin–hydrogen bond cleavage proceeds via radical chains. The displacement of halogen attached to carbon by hydrogen may be of little synthetic importance in general but represents the basic reactivity of the tin–hydrogen bond. The reaction proceeds as follows:

$$R_3SnH + \cdot Y \longrightarrow R_3Sn\cdot + HY$$

$$R_3Sn\cdot + XR' \longrightarrow R_3SnX + \cdot R'$$

$$R_3SnH + \cdot R' \longrightarrow R_3Sn\cdot + HR'$$

The reaction is initiated by the radical · arising from initiators such as azoisobutyronitrile or by irradiation and the important chain carrier is a trisubstituted stannyl radical. This type of radical reaction finds important applications in carbon–carbon bond formation as described below.

8.2.1 RADICAL CYANOETHYLATION AND RING CLOSURE

The cyanoethylation sequence comprises several steps:

$$R_3SnH + \cdot X \longrightarrow R_3Sn\cdot + HX;$$

$$R'Br + \cdot SnR_3 \longrightarrow R'\cdot + BrSnR_3$$

$$R'\cdot + CH_2{=}CHCN \longrightarrow R'CH_2\dot{C}HCN;$$

$$R_3SnH + R'CH_2\dot{C}HCN \longrightarrow R_3Sn\cdot + R'CH_2CH_2CN$$

where X· represents the radical from a chain initiator, R = butyl and R'· = the tetraacetylated 1α-glucopyranosyl radical. Cyanoethylation of tetra-*O*-acetyl-α-D-glucopyranosyl bromide produces specifically the cyanoethylation product **1** with retention of configuration at C-1. The experimental details are described in a recent issue of *Organic Syntheses* [6], and are outlined under Working Procedures (Section 8.9). Cyanoethylation takes place exclusively at the α-side of the sugar molecule and this selectivity is explained on the basis of the pseudo-axial attack (from below) on the skew-boat type pyranose radical **2**.

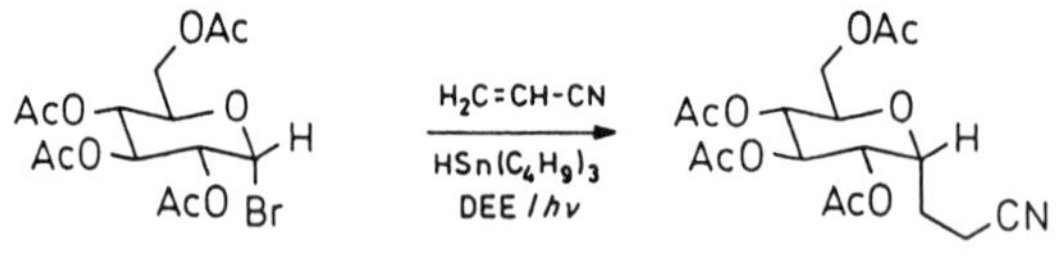

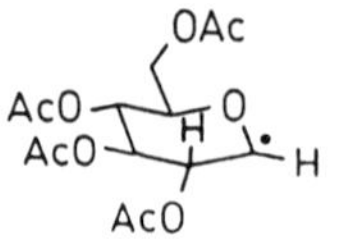

The 5-hexenyl radical cyclizes fairly easily to the cyclopentylmethyl radical, which may again be trapped with acrylonitrile as in the previous case. This type of cyclization is performed generally under high dilution (0.02 M) and in the presence of azoisobutyronitrile as a chain initiator.

Stork [7] and others [8] have utilized such cyclization in the synthesis of cyclopentane derivatives. Synthetically very useful are reactions of 2-bromo-1,6-heptadiene (**3**) and 1-hepten-6-yne (**4** and **5**) type substrates leading to methylenecyclopentanes [9–11]. The extension of this principle to oxygen-containing five-membered ring closure has been recorded [12, 13]. The cyclization of the α,α-disubstituted malonate **5** has been described in *Organic Syntheses* [14]. Details are given in Section 8.9.

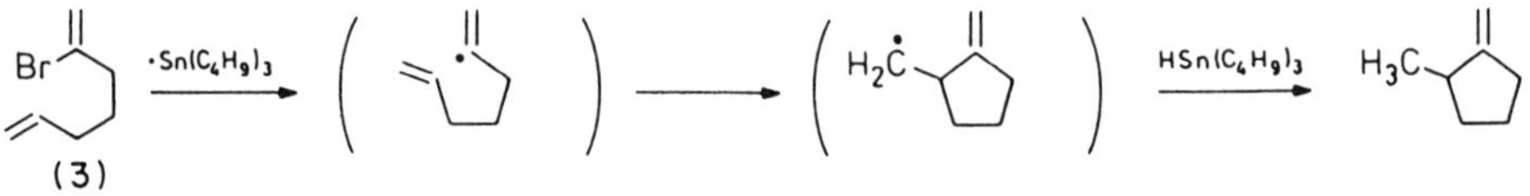

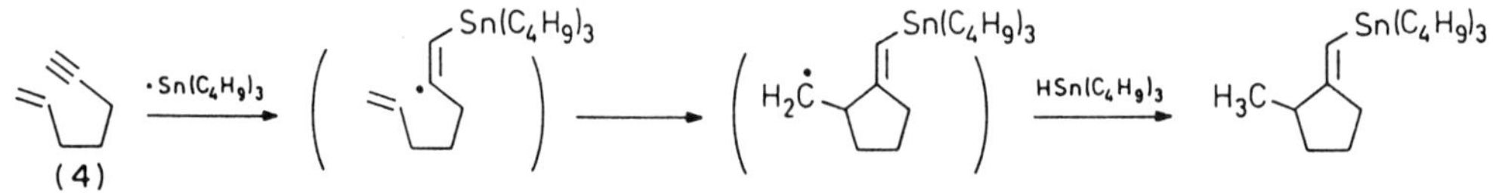

Radicals are produced smoothly from triethylborane in the presence of a trace of atmospheric oxygen. Oshima and co-workers [15] have utilized this fact and succeeded in carrying out the hydrostannylation and subsequent cyclization of the enyne **6**. The stannylmethylenecyclopentane obtained was converted into a mixture of dehydroiso-iridodiol isomers (**7**).

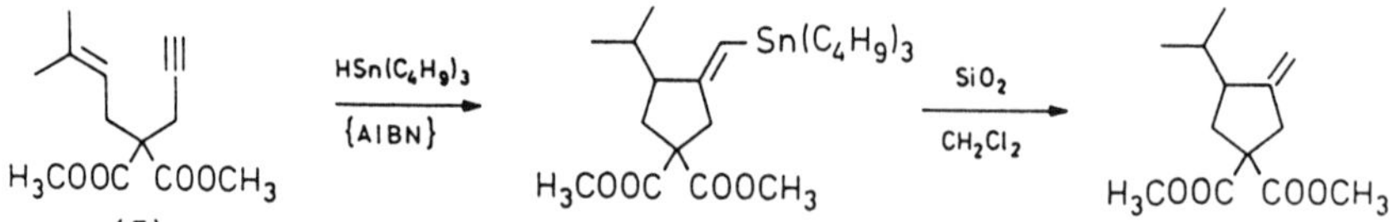

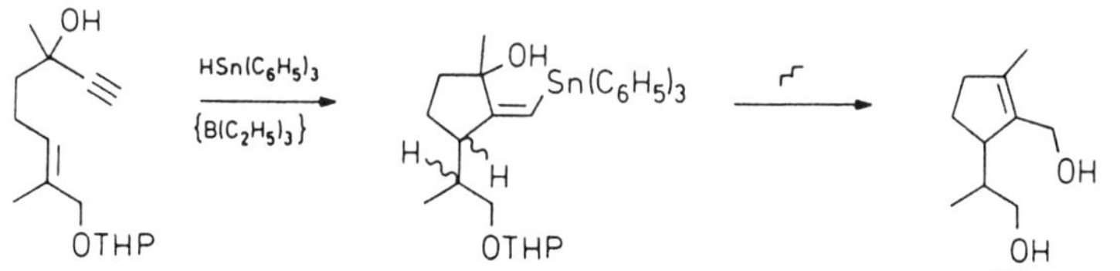

Palladium catalysis [16] and sonochemical initiation [17] of hydrostannylation have been recorded.

8.2.2 CYCLIZATION OF 2-CYCLOALKENOL-DERIVED BROMOETHANAL ACETALS

Cyclization of bromoethanal acetals (**8**) of 2-cycloalkenols provides a method of introducing a formylmethyl group at position 2 of an allyl alcohol the double bond of which is reduced [8]. This sort of cyclization proceeds equally well with five- and six-membered cyclic ring substrates. The acetal moiety serves as a detachable link.

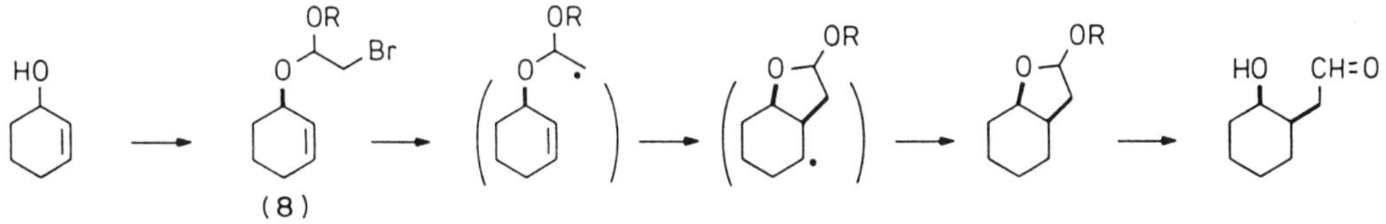

Synthetically it would, of course, be even more attractive to introduce a functional chain on C-3 rather than just a hydrogen atom. This goal has been achieved by carrying out the cyclization of the iodide **9** in the presence of *tert*-butyl isocyanide, a good radical trap. In this way, a cyano group was introduced at C-3 from the desired β-site. The liberated *tert*-butyl radical abstracts hydrogen from tributylstannane to form tributylstannyl radical, which is the chain carrier. This method introduces useful functional groups both on C-2 and C-3 of the olefinic linkage of the initial 2-cyclopentene-1,4-diol. The 3-cyano group can subsequently be transformed into a formyl and ultimately 3*S*-(*E*)-3-hydroxy-1-octenyl moiety, whereas the formylmethyl group on C-2 can be converted into a *cis*-olefin by the Wittig reaction. The prostaglandin skeleton is thus constructed stereospecifically.

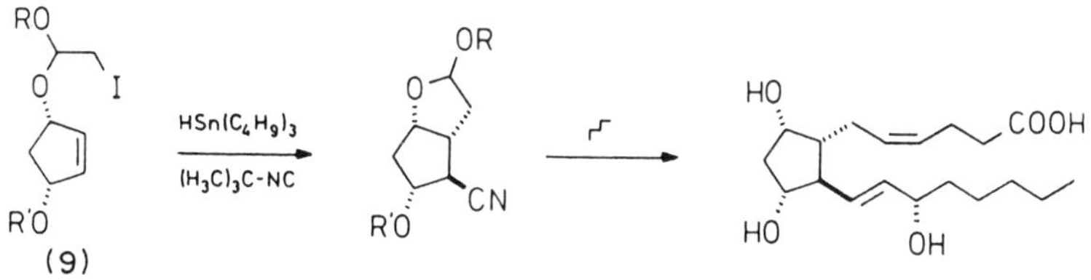

The same type of reaction enables one to prepare a calcitriol intermediate starting from 3-methyl-2-cyclohexenol. Elaboration of the cyano function into an acetylmethyl group followed by aldol-type ring closure produces the functionalized carbocyclic ring **10** having the desired stereochemistry. The introduction of the cyano group occurs as required from the α-face, since the transient radical **11** reacts only selectively in this sense.

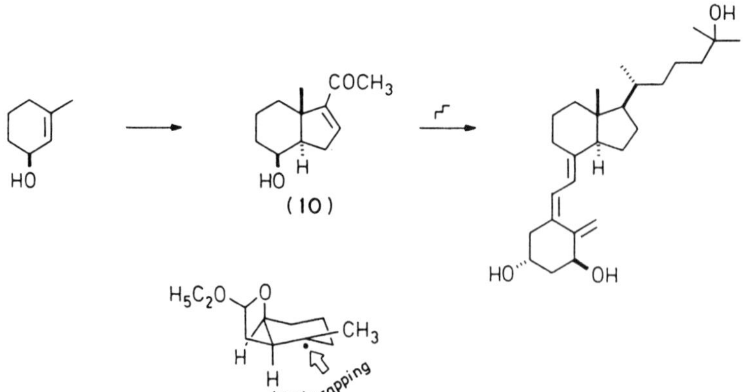

Bromomethylsilyl ether such as **12** provide a means of creating the *trans* junction of the bicyclo[4.3.0]nonane skeleton. The hydrogen atom provided by tin hydride becomes attached to the remote olefinic carbon from the opposite side of the allylic oxygen. Thus, the cyclization product **13** emerges in the *trans* configuration.

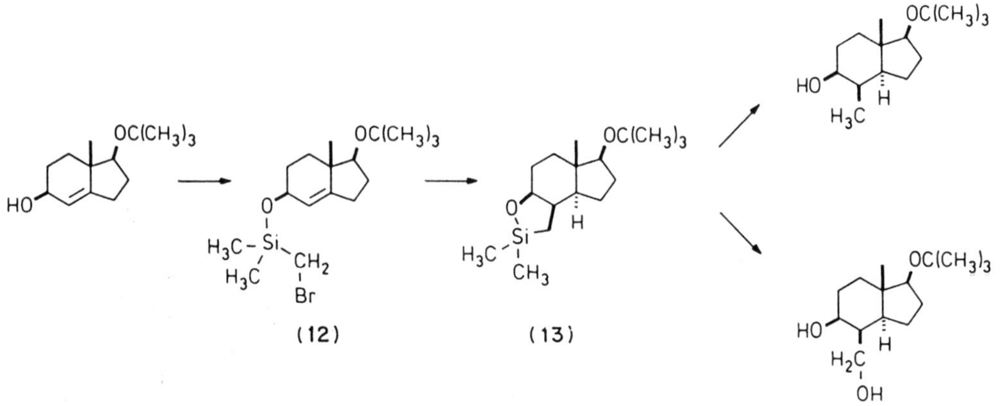

What type of ring junction would be expected if we started from the epimer having the allylic oxygen on the opposite α-site? The ring junction should naturally have a *cis* configuration. In this way we have provided a logical handle controlling the *cis*, *trans* geometry of the bicyclo[4.3.0]nonane system [11]. The cyclization method has been successfully applied to the synthesis of reserpinol and should find more synthetic applications in the future.

8.2.3 RELATED REACTIONS

The selenium–carbonyl bond in the selenyl carbonate **14** is easily cleaved by tributylstannane. The cyclization of the carbon radical thus generated affords an α-methylene-γ-lactone [18].

A pyrrolidine ring synthesis [19] is based on the stannane reduction of an allyl(2,2,2-trichloroethyl)amine such as **15**. The transient radical **16** prefers the *syn* rather than the *anti* conformation, thus minimizing the interaction between chlorine and phenyl substituents. This is at the origin of the *cis* arrangement of ethyl and phenyl groups in the cyclized product. The dichloromethylene group of the latter can be reduced to a methylene group with tin hydride [20].

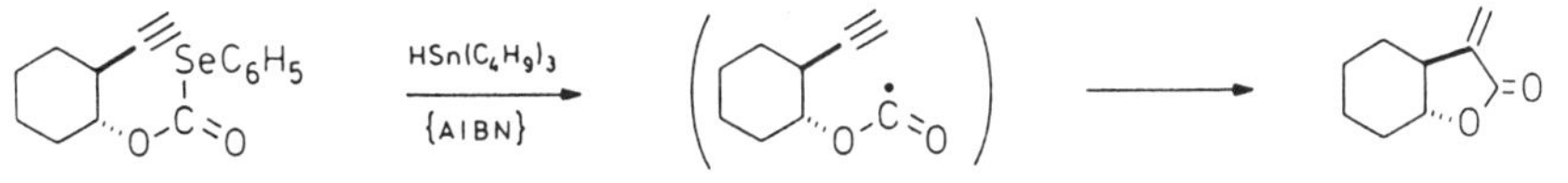

(14)

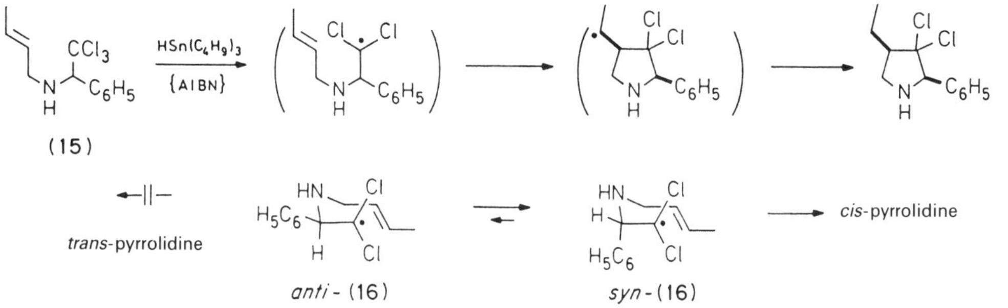

(15)

trans-pyrrolidine

cis-pyrrolidine

anti-(16) syn-(16)

Iminyl radical intermediates resulting from tributylstannane cleavage of *S*-phenylsulphenylimines readily cyclize on to suitably located double bonds intramolecularly. The intermediate carbon radical thus produced is trapped intermolecularly by a variety of electron-poor olefins. The sequence of radical reactions proceeding in one pot provides access to pyrroline derivatives [21].

The deoxygenation of an alcohol ROH into the corresponding hydrocarbon RH via the xanthate ROCSSR′ by the action of tributylstannane is preparatively useful and is now known as the Barton reaction [22]. Facile reduction of the xanthates and of halides by means of the tributylstannane–triethylborane system has been reported [23, 24]. The key intermediate is the tributylstannyl radical. The attack may give either a mixture $\cdot C(=S)OR + R'SSnBu_3$ or alternatively $R\cdot + O{=}C(SR')SSnBu_3$. This alternative has been controversively discussed [24, 25]. The deoxygenation of a glucose derivative (**17**) by means of such a reaction sequence has been described in *Organic Synthesis* [22]. The working procedure is given in Section 8.9.

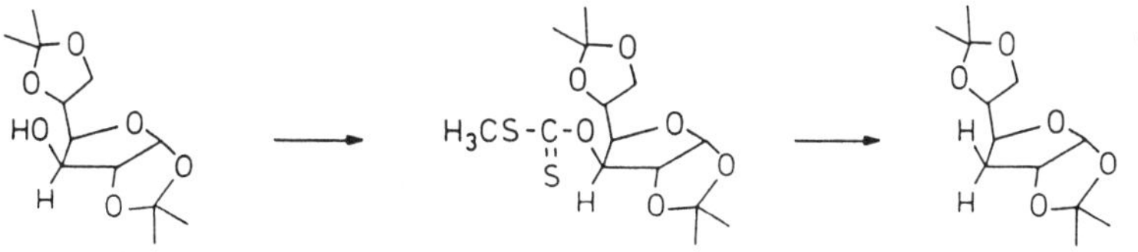

(17)

Yamaguchi *et al.* [26] have reported on the tributylstannane-promoted debromination of the *N*-aroyldihydroisoquinoline **18**, which produced a nitrogen-containing tetracycle. As evidenced by the high yield, the tributylstannyl-mediated halogen abstraction from an aromatic carbon atom proceeds equally well as from an olefinic site.

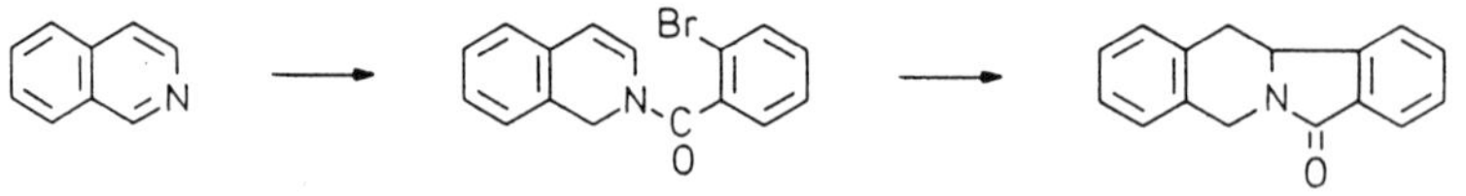

Br
N C
O
N
O
(18)

Similar radical cyclization has been utilized in the synthesis of oxindoles [27, 28]. A synthesis of the diquinane framework [29] can be analogously accomplished by the cyclization bromomethylsilyl propargyl ether **19**, followed by the intramolecular attack of the resulting radicals **20**, **21** and, after intermolecular combination with acrylonitrile, **22**, on the olefinic bonds.

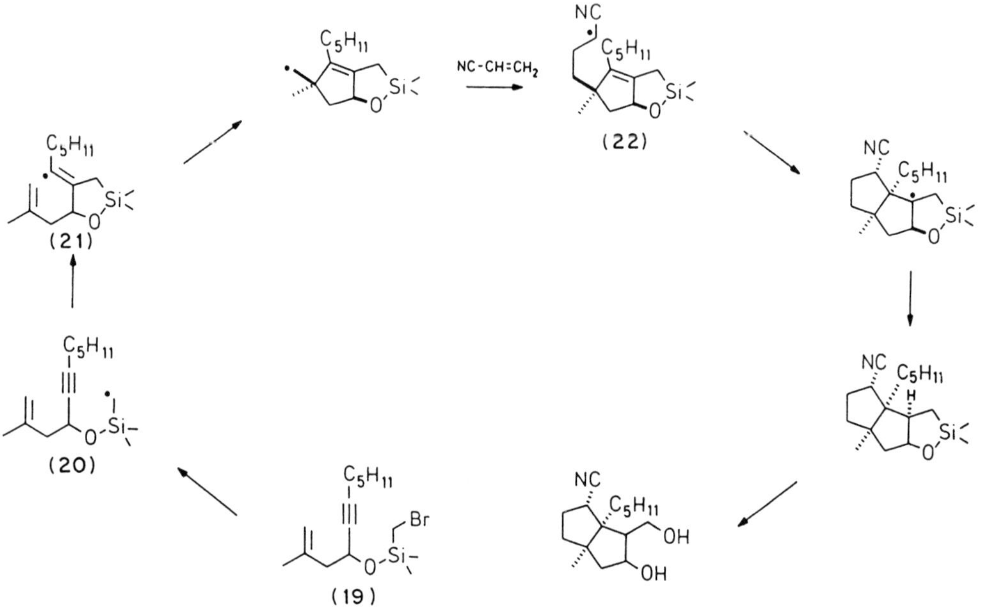

NC
C_5H_{11}
Si
O
NC-CH=CH_2
C_5H_{11}
Si
O
(22)
C_5H_{11}
Si
O
(21)
NC
C_5H_{11}
Si
O
C_5H_{11}
O-Si-
(20)
NC C_5H_{11}
H
Si
O
C_5H_{11}
Br
O-Si-
(19)
NC
C_5H_{11}
OH
OH

Stannyl-promoted reactions may also find applications in chain extension. The reductive ring expansion of the β-bromoketone **23** occurs through radical transfer to afford a γ-keto ester [30].

O
CH_2Br
$COOCH_3$
(23)
O
$\dot{C}H_2$
$COOCH_3$
O
$COOCH_3$
O
$COOCH_3$
H

Such a kind of radical transfer is obviously also involved in the coenzyme B_{12}-mediated isomerization of methylmalonyl coenzyme A (**24**) into succinyl coenzyme A (**25**). In order to elucidate further details, Wollowitz and Halpern [31] studied the reaction of tributylstannane with β-bromo esters $BrCH_2C(CH_3)XCOOR$, where X is acetyl, ethylthiocarbonyl, etc., particularly the radical isomerization $\cdot CH_2C(CH_3)XCOOR \rightarrow XCH_2\dot{C}(CH_3)COOR$ and competition with tin hydride trapping.

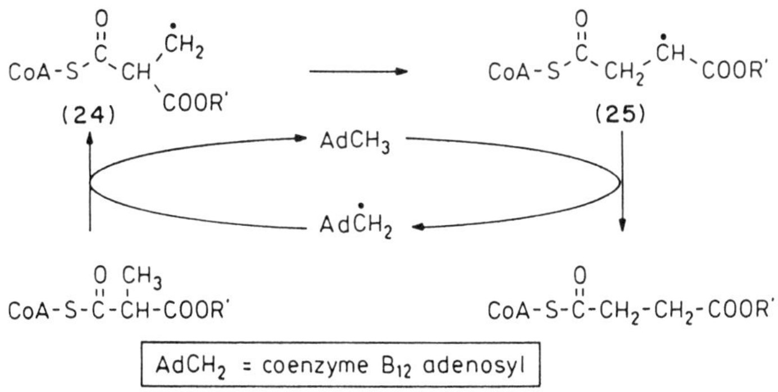

CoA-S-C(=O)-CH($\dot{C}H_2$)-COOR' (24) ⟶ CoA-S-C(=O)-CH_2-$\dot{C}H$-COOR' (25)

$AdCH_3$; $Ad\dot{C}H_2$

CoA-S-C(=O)-CH(CH_3)-COOR' ; CoA-S-C(=O)-CH_2-CH_2-COOR'

$AdCH_2$ = coenzyme B_{12} adenosyl

Finally boron enolates may form through radical intermediates [32]. When α-bromocarbonyl substrates $RCOCH_2Br$ are treated with triphenylstannane in the presence of a slight excess of triethylborane they produce boron enolates, $RC(OBEt_2)=CH_2$, one of the ethyl groups being split off as ethane. The enolates thus obtained can undergo aldol addition reactions. Further, treatment of methyl vinyl ketone with an alkyl iodide RI and tributylstannane in the presence of a small excess of triethylborane produces an enolate $RCH_2CH=C(CH_3)OBEt_2$. Quenching with methanol gives a methyl ketone $RCH_2CH_2COCH_3$, whereas combination of the enolate with an aldehyde R′CHO produces the aldol adduct $RCH_2CH[CH(OH)R']COCH_3$.

8.3 TIN–METAL BOND REAGENTS: REACTIVITIES AND SYNTHETIC APPLICATIONS

Reagents having a tin–lithium linkage were described as early as 1963 [33, 34]. Recent interest has focused on tin–silicon [35, 36] and tin–aluminium [37] bond reactivities. The Piers' tin–copper reagents are discussed in Section 8.5.1. Addition of these tin–metal (Sn–M) reagents to an acetylenic triple bond proceeds in the presence of transition metal catalysts; subsequent protonolysis leads to alkenylstannanes which are formally identical with those obtained by *cis* hydrostannylation of acetylenes. Naturally such carbon to metal and carbon to tin bonds react differently with electrophiles and thus allow simultaneous and specific activation of both acetylenic carbon atoms.

Yamamoto and co-workers [38] reported the cross-coupling of allylic bromides with a functionalized allylic sulphide (**26**) mediated by hexamethyldistannane under

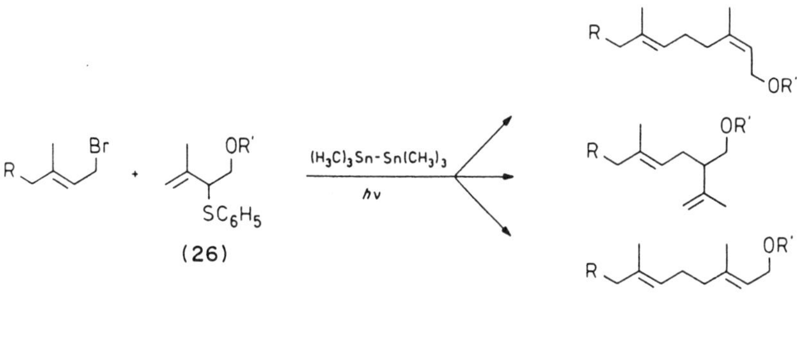

$(H_3C)_3Sn-Sn(CH_3)_3$, $h\nu$

SC_6H_5 (26) OR' R Br

irradiation, affording biallyl-type products. The reaction presumably involves a trimethylstannyl radical as a key intermediate [39].

8.3.1 ADDITION OF TIN–METAL BOND REAGENTS TO ACETYLENIC BONDS AND SYNTHETIC APPLICATIONS OF ALKENYLSTANNANES

A review on alkenylstannanes was published by Kosugi and Migita [40]. Hydrostannylation of 1-alkynes and the cyclization of the stannylalkenyl radicals often proceed with limited regio- and stereoselectivities and, therefore, have found few synthetic applications. Oshima and co-workers have accumulated knowledge on silicon–magnesium, silicon–aluminium [41] silicon–zinc [42, 43] and silicon–manganese [44] bond reactivities and extended it to the chemistry of tin–metal reagents, improving the regioselectivity of the hydrostannylation of acetylenes. As the results summarized in Table 8.1 reveal, the catalyst employed plays a major role [45, 46].

No addition takes place with internal acetylenes. The reaction proceeds in a *cis* fashion with respect to both 1- and 2-stannylated olefins in the reaction of phenylacetylene with the diethyl(tributylstannyl)aluminium reagent ($Bu_3Sn–AlEt_2$) catalyzed by copper(I) cyanide [45].

Treatment of enol triflates (**27**) or 1-iodoalkenes (**28**) with the system methyl(trimethylstannyl)magnesium–copper(I) cyanide also gives rise to 2-stannyl-1-alkenes and 1-stannylated isomers, respectively [47]. Tin–metal exchange reactions to produce more reactive species are discussed later (Section 8.5).

Species having tin–silicon bonds were studied by Chernard and Van Zyl [48]. The reaction of the respectively Sn–Si reagent with 1-hexyne in the presence of tetrakis(triphenylphosphine)palladium gives 1,2-difunctionalized 1-alkenes such as **29a** or **29b**. The addition proceeds in a *cis* fashion and affords 1-silyl-2-stannyl products only.

The intermediates thus formed can be further converted by translithiation with butyllithium and subsequent addition to carbonyl compounds or by Friedel–Crafts-type reactions with acetyl chloride and aluminium chloride.

Table 8.1. Transition metal-catalysed stannylmetallation of 1-alkynes RC≡CH[a]

R	Reagent	Catalyst	Yield (%)	Ratio[b]
$C_6H_5CH_2OCH_2CH_2$	$(C_4H_9)_3SnMgCH_3$	CuCN	88	100:2
	$(C_4H_9)_3SnAlCH_2CH_3$	CuCN	86	81:19
	$[(C_4H_9)_3Sn]_2Zn$	$Pd[P(C_6H_5)_3]_4$	81	14:86
C_6H_5	$(C_4H_9)_3SnMgCH_3$	CuCN	89	>95:5
	$[(C_4H_9)_3Sn]_2Zn$	$PdCl_2[P(C_6H_5)_3]_4$	89	>95:5
$C_{10}H_{21}$	$[(C_4H_9)_3Sn]_2Zn$	$Pd[P(C_6H_5)_3]_4$	70	<5:95

[a] The reaction of the 1-alkynes with the stannylmetal reagent was carried out in the presence of the respective catalyst and the carbon–metal bond in resulting products was preferentially hydrolised at 0 °C to give the stannylalkene isomers.
[b] Isomer ratio of (*E*)-RCH=CHSn$(C_4H_9)_3$ to RCSn$(C_4H_9)_3$=CH_2.

$R{-}C{\equiv}C{-}H \rightarrow$ [(*Z*)-28] $\rightarrow$ R–CH=CH–$Sn(CH_3)_3$ (*Z*)

$R{-}C{\equiv}C{-}H \rightarrow$ [(*E*)-28] $\rightarrow$ R–CH=CH–$Sn(CH_3)_3$ (*E*)

$R{-}C{\equiv}C{-}H \rightarrow$ R–C(OTf)=CH_2 (27) $\rightarrow$ R–C($Sn(CH_3)_3$)=CH_2

$OTf = OSO_2CF_3$

$H_9C_4{-}C{\equiv}C{-}H \rightarrow (H_3C)_3Sn(H_9C_4)C{=}C(H)Si(CH_3)_2C(CH_3)_3$ (29a) $\rightarrow$ $H_3C{-}C(=O)(H_9C_4)C{=}C(H)Si(CH_3)_2C(CH_3)_3$; $H_5C_6{-}CH(OH)(H_9C_4)C{=}C(H)Si(CH_3)_2C(CH_3)_3$

$H_5C_6{-}C{\equiv}C{-}H \rightarrow (H_9C_4)_3Sn(H_5C_6)C{=}C(H)Si(CH_3)_3$ (29b) $\rightarrow$ $(H_9C_4)_3Sn(H_5C_6)C{=}C(H)C(=O)CH_3$

It should be noted that a palladium(0) catalyst such as tetrakis(triphenylphosphin)palladium catalyses the dismutation reaction of the $(H_3C)_3Sn{-}Si(CH_3)_3$ reagent into hexamethylyldistannane and hexamethyldisilane. Reagents carrying larger alkyl groups than methyl do not undergo this reaction and, therefore, the tin–silicon reagents shown are fully utilized in forming both carbon–tin and carbon–silicon bonds on acetylenic carbons.

8.3.2 PREPARATION AND SUBSEQUENT CYCLIZATION OF 1-TRIALKYLSTANNYL-1,4-PENTADIEN-3-ONES

The Nazarov cyclization of 1,4-pentadien-3-one compounds to cyclopentenones is not a trivial process owing to the highly acidic nature of the reaction media [49]. Attempts have been made to overcome some of the problems by stabilizing the transient carbocations with β-silyl or β-stannyl substituents. Chernard *et al.* [50a] and Denmark and Jones [50b, 51] have carried out the cyclization of the *β*-silylated enones **30** by means of boron trifluoride ethereate to obtain a *cis*-4-silylated and the corresponding desilylated *cis*-hydroindenones in 37% and 30% yields, respectively. In contrast, the cyclization of the structurally closely related precursor **31** produced a *trans*-4,5-disubstituted cyclopentenone derivative in 80% yield.

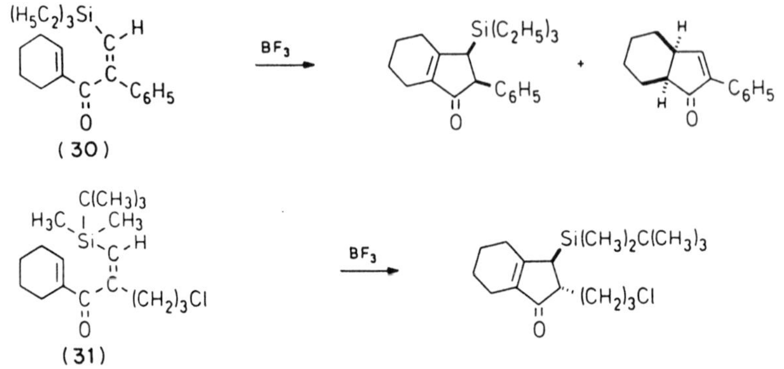

Similar reaction of stannylated dienones of type **32** [52] gives tin-free cyclopentenones. The required starting material, 1,2-bis(tributylstannyl)ethene, is prepared by hydrostannylating tributylstannylethyne with tributylstannane (see Section 8.5.2), to give 1,2-bis(tributylstannyl)ethene followed by treatment with an acyl chloride in the presence of aluminium trichloride.

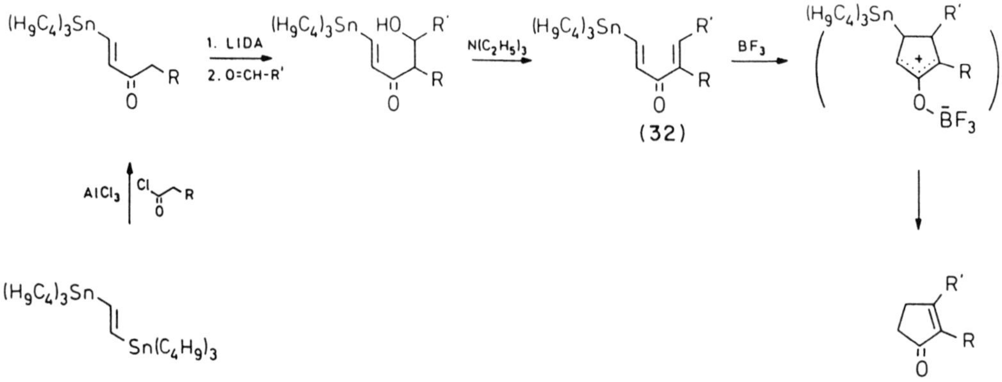

8.3.3 ALUMINIUM ENOLATE FORMATION AND CARBONYL ADDITION

Like boron enolates [32], aluminium enolates (**33**) are valuable intermediates in organic syntheses. They can be conveniently prepared by treatment of α-bromo ketones with stannylalanes [53] such as $Bu_3Sn–AlEt_2$ (which can be prepared from tributylstannyllithium and diethylaluminium chloride) or $F_2ClSn–AlEt_2$ [from tin(II) fluoride and diethylaluminium chloride). The aluminium centre having strong Lewis acid character will first coordinate with the carbonyl oxygen and thus facilitate the reduction of the

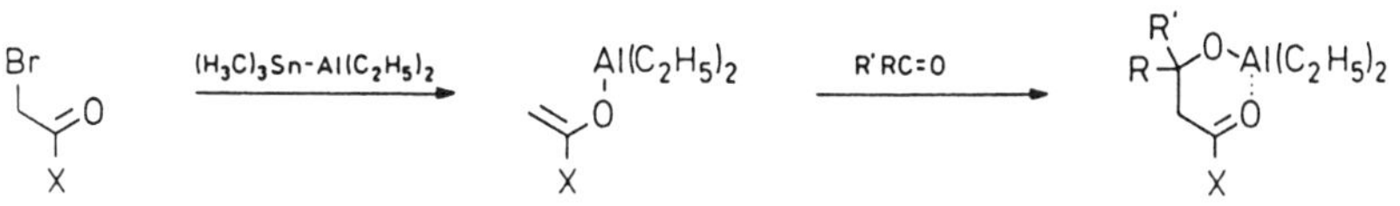

Br—C bond by the tin moiety, affording the aluminium enolate **33** and bromotributylstannane.

The subsequent reaction with a carbonyl component R′RCO produces the addition product shown.

The formation of aluminium enolates by zinc reduction in the presence of diethylaluminium chloride was previously reported by Yamamoto and co-workers [54]. Tin–aluminium reagents are soluble in reaction solvents and therefore are free from problems with the insufficiently active metal surface often encountered with metals such as zinc or magnesium.

Posner *et al.* [55] have reported a case of conjugate addition of tributylstannyllithium reagent to 2-cyclohexenone. The resulting lithium enolate is treated with a vinyl ketone, $CH_2{=}CHCOR$, and then with an aldehyde R′CHO successively to afford the four-component adduct **34**, whose oxidation gives a medium-ring lactone.

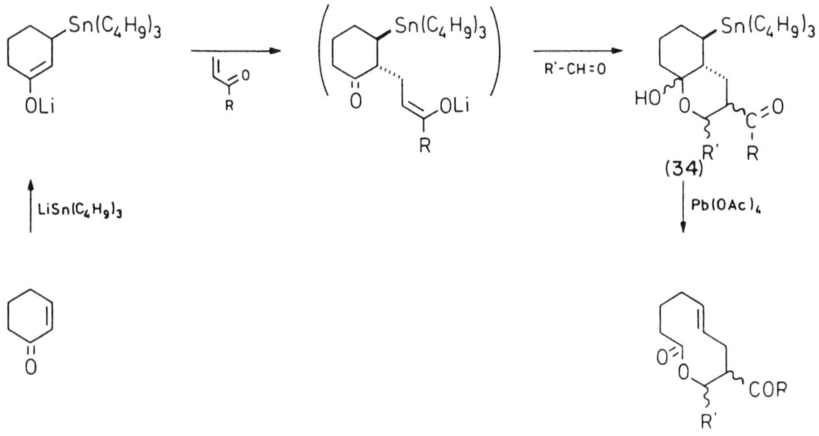

8.4 ADDITION OF ALLYLSTANNANES TO CARBONYL COMPOUNDS

The carbon bond to tin is fairly resistant to hydrolysis and is not nucleophilic enough to perform carbonyl addition, the reaction which is commonly observed with organomagnesium and organolithium reagents. Exceptionally, however, allylstannanes do react with aldehydes when heated to 100–150 °C or when treated with boron trifluoride ethereate in dichloromethane at −78 °C. The reaction involves electrophilic attack at the metal-free allyl terminus and double-bond shift and produces homoallyl alcohols.

8.4.1 METHODS FOR THE PREPARATION OF ALLYLSTANNANES

The stability of the tin–carbon bond towards aqueous media is exemplified by the reaction of benzyl chloride with metallic tin powder suspended in boiling water, which produces tribenzylchlorostannane, as reported by Sisido *et al.* [56] in 1961. A similar reaction with allyl bromide produces diallyldibromostannane [57]. Nokami and co-workers [58, 60] allowed the resulting stannane to react *in situ* with ketones (the Barbier conditions) and obtained homoallylic alcohols in the aqueous suspension. Water obviously accelerates the addition and the presence of aluminium powder also proves to be helpful [61–63]. The reaction has been applied to the synthesis of five- and six-membered carbocycles and α-methylene-γ-butyrolactones [64–66]. Similar reactions are carried out by means of chromium(II) salts instead of elemental tin in aprotic solvents [67–69].

Several new methods have been reported for the preparation of allylstannanes, all proceeding under anhydrous conditions. The tin–aluminium reagents of Matsubara *et al.* [70] (Section 8.3.3) react with allylic phosphates to afford allylstannanes. The reagents thus produced react with carbonyl components in the absence of boron trifluoride ethereate to give homoallylic alcohols. Clearly the phosphorylation of allylic alcohols is much easier than their conversion into allylic bromides. Another procedure involves the reaction of allylic alcohols with tin(II) chloride under palladium(0) catalysis. Addition of the resulting allyltin(IV) species to aldehydes is reported [71]. Other entries to allylstannanes are based on the electrochemical coupling of allyl halides with tributylchlorostannane [72] and on the thermochemical or photochemical stannylation of allyl sulphides, sulphonates and carbamates [73].

Desponds and Schlosser [74] have recently devised a new method for the preparation of pure allyl- and benzyl-type organometallics using tin compounds as key intermediates. Unsaturated hydrocarbons are first metallated with their superbase [75] and the products are treated with trimethylstannyl or tributylstannyl chloride. The resulting allylstannanes are purified and treated with trimethylsilylmethylpotassium [76]. An immediate metal–metal exchange takes place. The simultaneously formed tributylstannyl(trimethylsilyl)methane is removed by evaporation in order to avoid secondary reactions. The organopotassium species can be converted into other organometallic compounds, e.g. magnesium or lithium derivatives [77]. The manipulations are easy to perform (see Section 8.9) and satisfactory yields are obtained on electrophilic quenching even when the 'double reaction' test [78] is applied.

8.4.2 STEREOCHEMISTRY OF CARBONYL ADDITION IN THE PRESENCE OF LEWIS ACIDS

Yamamoto, one of the major contributors in this field, has published several reviews [79–81]. Allyltins, like allylchromiums [68, 69], react with aldehydes via six-membered transition states. Hence (*E*)-2-butenylmetal (*trans*-crotylmetal) species give homoallylic alcohols having the *anti* configuration (*anti*-**35**), whereas the *cis* isomers afford the corresponding *syn* isomers (*syn*-**35**). In the presence of boron trifluoride etherate, however, exclusively *syn*-alcohols are produced from both *trans*- and *cis*-crotyltin reagents. The same selectivities are observed with crotylsilane reagents also. Reactive species in this kind of carbonyl addition have been identified by means of ^{13}C NMR spectroscopy [82, 83]. This selectivity has been exploited in the transformation of 4-methoxycarbonyl-2-methylpentanal to the adduct **36**. The latter was then converted into the Prelog–Djerassi lactone [79].

syn-(35)

anti-(35)

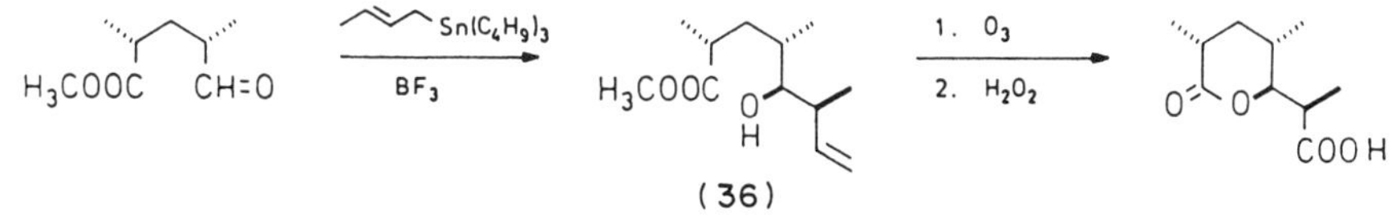

(36)

8.4.3 RELATED REACTIONS

Umani-Ronchi and co-workers [84] have succeeded in the preparation of allylstannane reagents **37** carrying chiral substituents by oxidative addition of tin(II) species to allyl bromides. Subsequent reaction with aldehydes gave adducts with 40–60% enantiomeric excess.

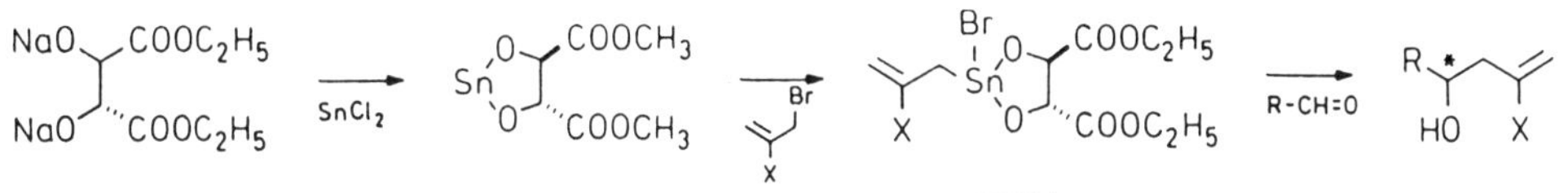

(37)

The chiral allylstannanes **38** and **39** have been prepared starting with (*R*)-glyceraldehyde acetonide and (*S*)-2-benzyloxypropanal. The procedure involved a Wittig–Horner

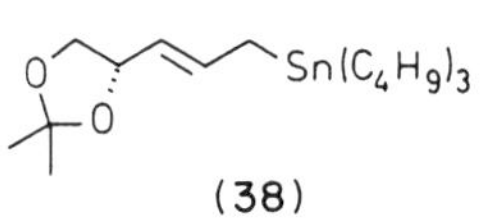

(38)

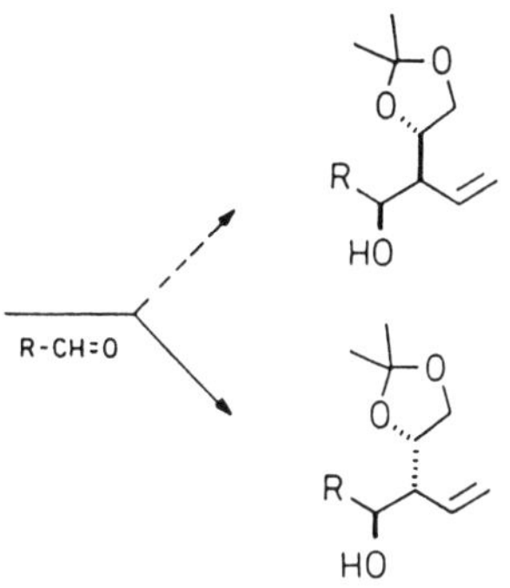

reaction, reduction of the unsaturated esters thus obtained to allylic alcohols, transformation to the xanthates, allylic rearrangement of the latter to the isomeric dithiocarbonate, and final stannylation. When the allylstannanes are heated in the presence of an aldehyde without solvent to 150 °C, an enantioselective addition reaction occurs [85].

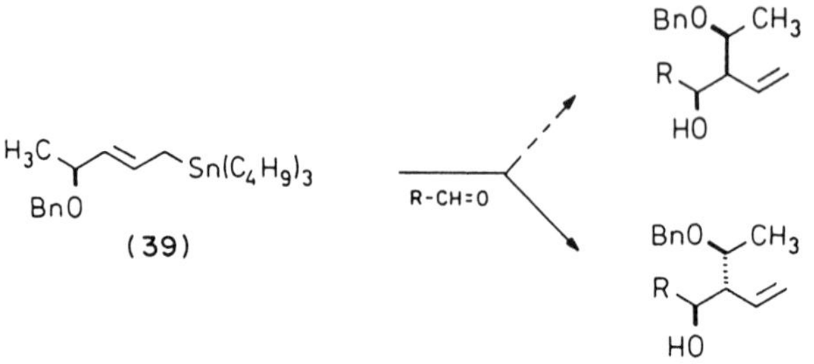

(39) —R-CH=O→

p-Bromobenzaldehyde reacts with a *cis–trans* isomeric mixture of (1-ethoxy-2-butenyl)tributylstannane in the presence of boron trifluoride ethereate in methylene chloride at −78 °C to afford the α-(hydroxylalkyl)allyl ether **40** and the γ-(hydroxy)ene ether, both components existing as *syn* and *anti* diastereoisomers [86]. The formation of the products of type **40** can be explained by assuming a rapid equilibration between α- and γ-tributyl-substituted crotyl ethyl ether.

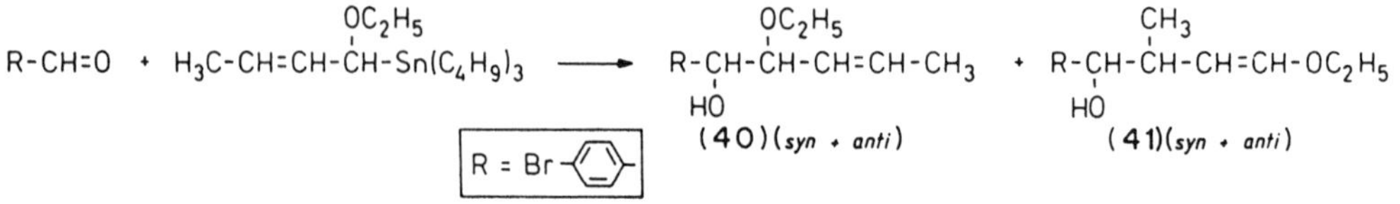

R-CH=O + H_3C-CH=CH-CH(OC_2H_5)-$Sn(C_4H_9)_3$ → R-CH(HO)-CH(OC_2H_5)-CH=CH-CH_3 (40)(*syn* + *anti*) + R-CH(HO)-CH(CH_3)-CH=CH-OC_2H_5 (41)(*syn* + *anti*)

R = Br-C_6H_4-

The intramolecular carbonyl addition of the α-alkoxyallylstannane **42** has been applied to form a twelve-membered ring [87]. This reaction product serves as an intermediate for the synthesis of cembranolides.

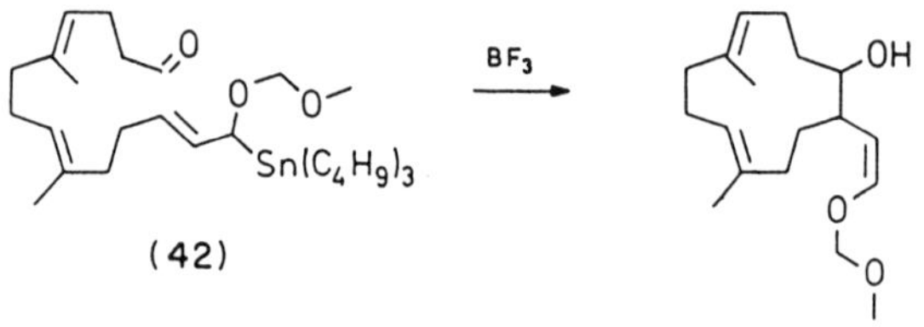

(42) —BF_3→

Carbenium–immonium ions are readily produced by acid-catalysed condensation of amines RR′NH with formaldehyde. They react with allylstannanes to afford homoallylic products **43** and **44** in virtually quantitative yield [88].

N-(2,4-Alkadienoyl)isoquinolinium salts react with allylstannanes to afford products of type **45** (see Section 8.2.3) [89]. The subsequent intramolecular Diels–Alder cycloaddition gives rise to a tetracyclic alkaloid skeleton.

C_6H_5-CH_2-NH_2 → C_6H_5-CH_2-$N(CH_2CH_2CH{=}CH_2)_2$ (43)

+ $H_2C{=}CH$-CH_2-$Sn(C_4H_9)_3$

+ $CH_2{=}O$ {F_3CCOOH}

C_6H_5-CH_2-NH-CH_3 → C_6H_5-CH_2-$N(CH_3)$-$CH_2CH_2CH{=}CH_2$ (44)

Cl⁻ {$ClSn(C_4H_9)_3$} $Sn(C_4H_9)_3$ H N O ΔT

(45)

The reaction of alkynylstannanes with 3-acyl-1-methoxycarbonylpyridinium salts affords a mixture of alkynylated products **46** and **47**. The reaction is believed to proceed through a tin-bridged carbocation, which is then attacked by the nucleophile [90].

CO-R′ R-$C{\equiv}C$-$Sn(CH_3)_3$ / Cl-$COOCH_3$ → CO-R′, $C{\equiv}C$-R, N, H, $COOCH_3$ (46) + R-$C{\equiv}C$, H, N, CO-R′, $COOCH_3$ (47)

Otsuji and co-workers [91] described a three-component coupling leading to products **48**, with *El* being a cyano or methoxy carbonyl electrophilic substituent, R alkyl or allyl, R′ hydrogen or methyl and R″ aryl. Attack of the radical R on the electron-deficient olefin initiates the chain process. When the resulting stabilized radical reacts with allylstannane, tributylstannyl radical is regenerated.

$(H_9C_4)_3Sn$ (R″) + *El*, *El* C=CH-R′ + I-R {AIBN} or *hν* → R′, R, R″ *El* *El* (48)

Intramolecular cyclization of 2-(6-tributylstannyl-4-hexenyl)oxiranes was carried out in the presence of a Lewis acid catalyst such as titanium(IV) chloride, tin(IV) chloride or trimethylsilyl triflate [92].

Finally, a novel type of [2 + 3] ring closure relying on an acyl-iron complex **49** as the key intermediate has been reported [93]. The electron-deficient, terminal olefinic carbon is equivalent to the β-olefinic carbon of an acrylate. When this enolate is attacked by the nucleophilic allylstannane, a new carbon–carbon bond is created. Ultimately ring closure to afford a cyclopentane ring derivative occurs.

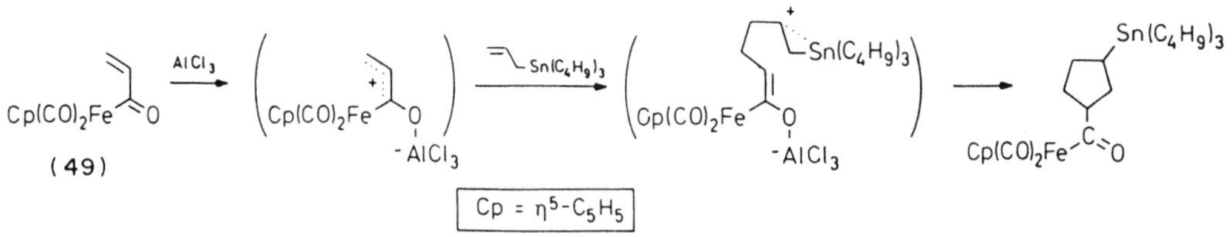

(49)

$Cp = \eta^5\text{-}C_5H_5$

The effect of the tin–carbon bond in stabilizing a positive charge on the β-carbon has been studied by Lambert and Wang [94].

8.5 SYNTHETIC REACTIONS MEDIATED BY TRANSMETALLATION OF ORGANOTINS

Useful reagents are derived from tin compounds by converting the tin–carbon into the more reactive lithium–carbon bond. Further, catalytic palladium transmetallation of the tin–carbon bond in the presence of organometallics affords organopalladium species whose reductive elimination affords carbon–carbon coupling products with regeneration of the palladium(0) catalyst. This section deals with this type of metal exchange of tin–carbon bonds and the synthetic applications. This subject has already been reviewed [95].

8.5.1 TRANSMETALLATION OF TIN–CARBON BOND INTO LITHIUM–CARBON BOND

Both 1- and 2-trialkylstannyl-1-alkenes (see Section 8.3.1) undergo a metal–metalloid exchange with butyllithium generating the corresponding lithium reagents. The tin moiety is converted into a non-reactive tetraalkylated stannane which need not be removed from the reaction mixture.

Chernard [96] reported the conjugate addition of trialkylstannyllithium reagents to 2-cyclohexenone and the subsequent silylation of the resulting enolate to afford the 1-siloxy-3-stannylcyclohexene intermediate (**50**). Tin–lithium exchange increases the nucleophilicity of the allyl moiety. Addition to an aldehyde takes place rapidly and preferentially at C-3. Protection of the hydroxyl function with methyl iodide, desilylation

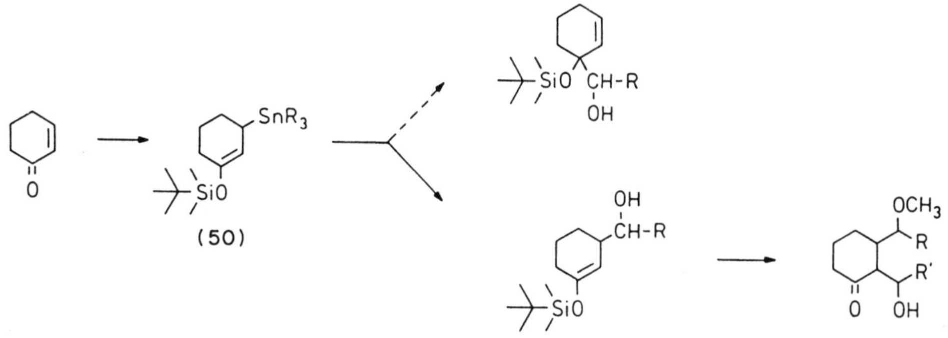

(50)

and subsequent aldol-type addition of a second aldehyde at C-2 result in the formation of a product which carries an α-hydroxyalkyl and an α-methoxy-alkyl moiety attached on both olefinic carbons of cyclohexenone.

Noyori's three-component coupling route to prostaglandins is well known [97]. Johnson and Penning [98] have modified the access to the key nucleophile, an organocopper species (**51**; M = CuCNLi). This reagent was prepared from the respective 1-alkyne derivative and tributylstannane under irradiation and the resulting 1-alkenylstannane (**51**; M = tributylstannyl) was then treated consecutively with butyllithium and copper(I) cyanide in tetrahydrofuran (−78 °C). 1,4-Addition to the cyclopentenone derivative was followed by alkylation with the *cis*-allylic iodide component in hexamethylphosphoramide at −30 °C (3 h, 53% yield) [99].

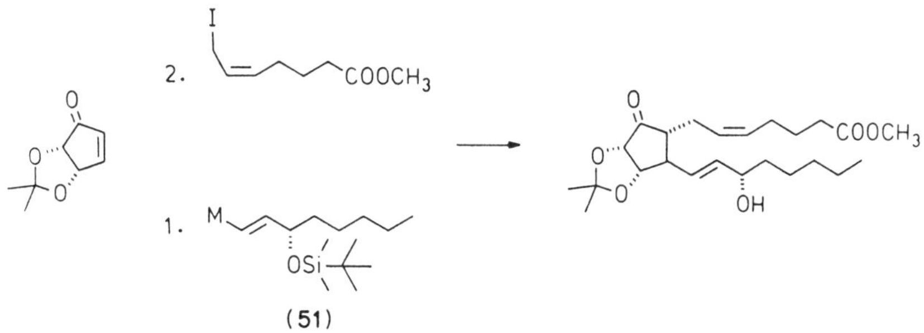

2. I–CH₂CH=CH(CH₂)₃COOCH₃

1. (51)

The synthesis of 2-trialkylstannyl-1-alkenes based on the addition of tin–copper reagents to 1-alkyne derivatives has been studied by Piers and Karunaratne [100]. The addition reaction was applied to the synthesis of such a unique reagent as 2-lithio-4-chloro-1-butene via 3-tributylstannyl-3-buten-1-ol and 4-chloro-2-tributylstannyl-1-butene.

[2,3]-Sigmatropic Wittig rearrangements in organic synthesis have been reviewed by Nakai and Mikami [101]; an allyl stannylmethyl ether rearranges into a homoallylic alcohol after treatment with butyllithium. Obviously, the tin–carbon bond is initially transformed into a lithium–carbon bond and then the rearrangement proceeds [102, 103].

The tin–lithium exchange at oxygen-bearing carbon atoms was first described by Still and has recently found a variety of synthetic applications. Hanessian *et al.* [104] were able to stannylate the C—H bond of a dihydropyrane-type enol ether by using Schlosser's super base [75] for metallation before adding tributylstannyl chloride. Subsequently, the tin–carbon–oxygen species **52** is transformed into an organolithium reagent before being methylated to afford the desired final product.

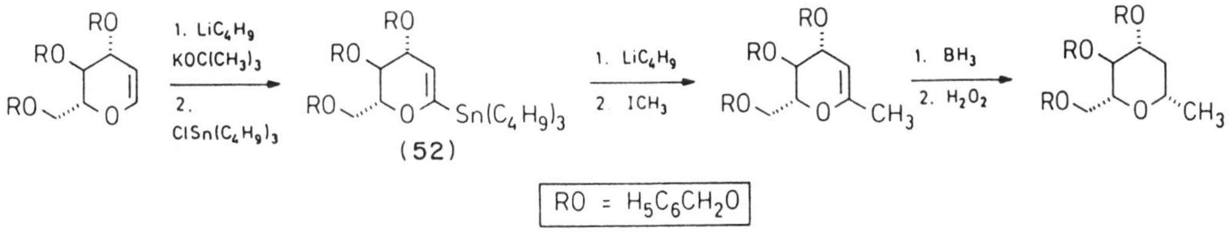

1. LiC_4H_9 / $KOC(CH_3)_3$; 2. $ClSn(C_4H_9)_3$ → $Sn(C_4H_9)_3$ (52) → 1. LiC_4H_9; 2. ICH_3 → CH_3 → 1. BH_3; 2. H_2O_2 → CH_3

RO = $H_5C_6CH_2O$

Tributyl(methoxymethoxymethyl)stannane, $(C_4H_9)_3SnCH_2OCH_2OCH_3$, has been prepared by the reaction of tributylstannyllithium with formaldehyde (paraformaldehyde), followed by etherification with chloromethyl methyl ether. This reagent can be used for attaching the masked hydroxymethyl moiety as a nucleophile to the β-carbon atom of conjugated enones. For activation, the trialkyltin group has to be consecutively replaced by lithium and copper [105, 106]. The resulting adducts (**53**) can be cyclized to give oxygen-containing five-membered ring skeletons [107].

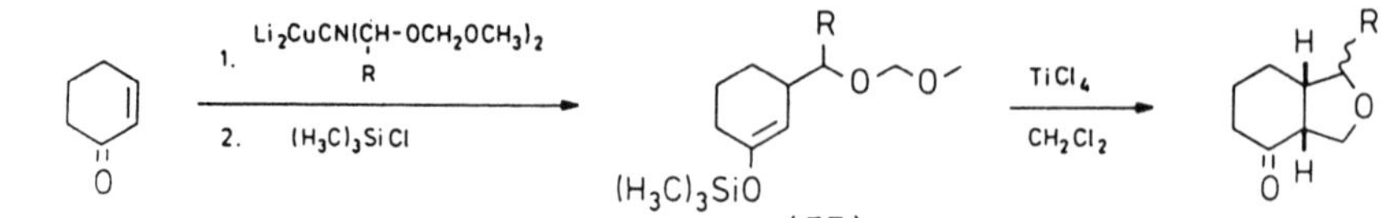

1. $Li_2CuCN(CH(R)-OCH_2OCH_3)_2$
2. $(H_3C)_3SiCl$

$(H_3C)_3SiO$ (53)

$TiCl_4$, CH_2Cl_2

Conjugated addition to α-enones of α-alkoxyalkylcopper reagents obtained from the corresponding stannanes serves to produce *C*-nucleosides stereospecifically [108]. Chong and co-workers [109] and Hutchinson and Fuchs [110] have prepared the enantiomerically enriched α-alkoxystannanes by reduction of acylstannanes with BINAL-H (2,2′-dihydroxy-1,1′-binaphthyl-modified lithium aluminium hydride) and subsequent methoxymethylation. They utilized the reactive intermediates obtained by translithiation to the synthesis of *endo*-brevicomin, γ-lactones, α-hydroxy acid derivatives.

Organolithium species can be readily converted into organotin compounds by the action of trialkylstannyl chloride. An application of such a transformation was utilized in the preparation of 2-aryloxazolines by Dondoni *et al.* [111]. The crucial organolithium intermediate **54** are first trapped with trimethylstannyl chloride before being treated with aryl halides in the presence of palladium(0) catalyst (see also Section 8.5.2).

Allylic and benzylic tin compounds prepared from organolithium precursors have been employed in terpenoid synthesis [112].

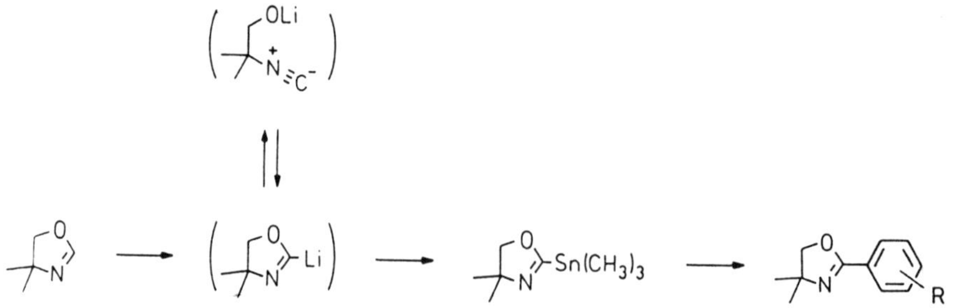

(54)

$Sn(CH_3)_3$

8.5.2 TIN–PALLADIUM EXCHANGE IN CATALYSED CROSS-COUPLING

The cross-coupling of Grignard reagents with organic halides can be mediated by nickel catalysts. The reaction was first described by Kumada and co-workers [113, 114] and Corriu and Masse [115]. The technique was subsequently extended to other systems

[116] containing palladium(0) catalysts and organotin compounds [117, 118]. Such efforts have considerably extended the scope of the coupling reaction [119–121]. When the reaction is carried out under a carbon monoxide atmosphere, two organic groups are connected with a carbonyl moiety to produce ketones, which can also be prepared by similar coupling of organotin reagents with acid chlorides [122, 123]. This reaction finds useful applications in the synthesis of divinyl ketones [124, 125] and also other classes of compounds [126, 127]. The detailed preparation of tributylstannylacetylene, (*E*)-1,2-bis(tributylstannyl)ethylene, ethyl (*E*)-3-tributylstannylpropenoate (**55**) and ethyl (*E*)-4-(4-nitrophenyl)-4-oxo-2-butenoate by this method has been described in *Organic Syntheses* [123] and the working procedure is given in Section 8.9.

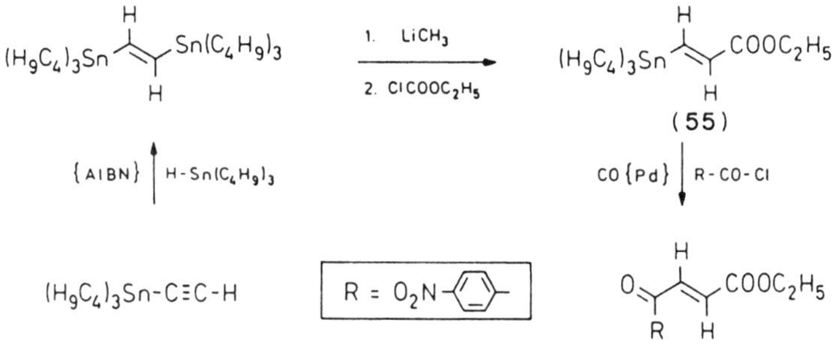

Kosugi *et al.* [128] published a new ketone synthesis, which consists in the reaction of 1-ethoxy-1-tributylstannylethene with aryl halides catalysed by tetrakis(triphenylphosphine)palladium in toluene solution. The acid hydrolysis of the reaction mixture gives the aryl methyl ketones. When the aryl halide is *p*-bromoacetophenone, *p*-diacetylbenzene is obtained in 89% yield, and analogously *p*-nitroacetophenone is prepared in 91% yield.

A ring-closure reaction of compounds of type **56** carrying both 1-alkenylstannane and enol triflate functional groups was reported by Stille and Tanaka [129]. Large-ring lactones have been obtained under high dilution conditions (0.001 M) in *ca* 60% yield irrespective of the ring size. The *E*-configuration of the olefinic linkage is retained. Developments in this field continue [130, 131].

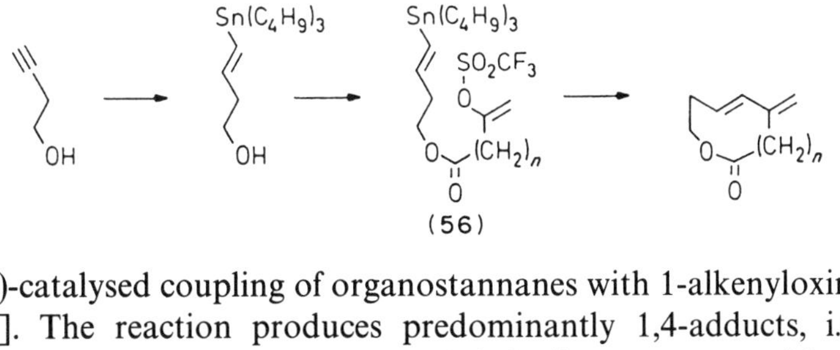

Palladium(0)-catalysed coupling of organostannanes with 1-alkenyloxiranes has been reported [132]. The reaction produces predominantly 1,4-adducts, i.e. 4-alkylated 2-buten-1-ols. Interestingly, lithium 1-alkenyltrialkylborates react with trialkylstannyl chloride to give dialkyl[1-alkyl-2-stannyl-1-(*E*)-alkenyl]boranes, in which the boron and tin atoms occupy a *cis* position [133].

Tunney and Stille [134] have further extended this reaction to the synthesis of triarylphosphines. The substrates are diphenyl(trimethylstannyl)phosphine, diphenyl (trimethylsilyl)phosphine, etc. When aryl iodides are allowed to react with them in the presence of bis(triphenylphosphine)palladium(II) dichloride catalyst in benzene, the desired aryldiphenyl phosphines are formed with good yields.

8.6 SYNTHETIC REACTIONS WITH TIN(II) SALTS

The main topics in this section are concerned with tin(II) enolates mediating aldol-type reactions and the Michael additions. Mukaiyama and co-workers have made major contributions in this field, hence particular attention is drawn to their research accounts [135–137].

8.6.1 TIN(II) ENOLATES AND ALDOL-TYPE ADDITION

Mukaiyama and co-workers have generated tin(II) enolates in the presence of the chiral auxiliary (*S*)-1-methyl-2-(piperidinomethyl)pyrrolidine. Subsequent aldol-type addition with carbonyl compounds gave the adducts **57** and **58** with high enantioselectivities (80–90% ee). Aldehydes and ketones react with tin enolates to afford *syn* diastereomers preferentially. The 1,3-thiazolidine-2-thione group of the aldol-type adducts **58** is easily convertible into the methyl or benzyl ester and also into an aldehyde group in high yields. This method can also be applied to 3-acyl-1,3-oxazolidin-2-ones [138].

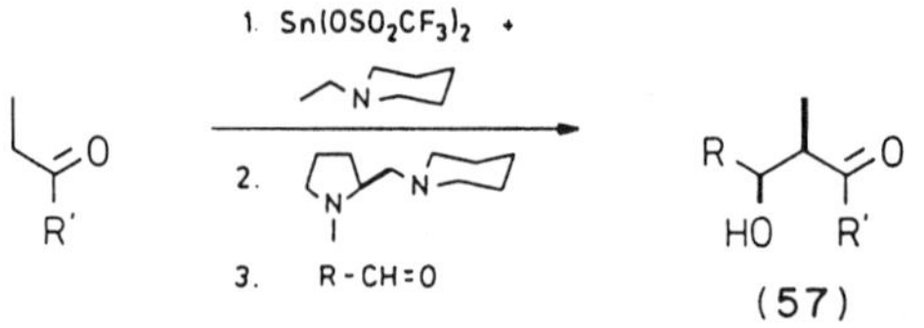

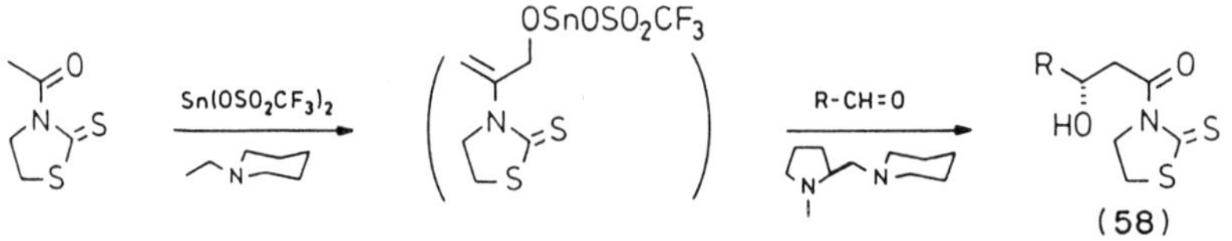

One might expect tin(IV) enolates to be formed when α-bromo ketones are allowed to react with tin(II) salts ('oxidative' addition as mentioned in Section 8.3.3). However, the products obtained after the reaction with aldehydes are in reality bromohydrins, indicating that tin(II) α-bromoenolates (**59**) had been formed by simple deprotonation of the bromo ketones. The bromohydrins are shown to have *syn* stereochemistry exclusively. They can be converted into *cis*-3-substituted oxiranylcarbaldehydes (**60**; X = H) and oxiranylcarboxylic acids (**60**; X = OH).

An efficient catalytic asymmetric aldol-type reaction has been reported [139]. A variety of aldehydes react with silyl enol ether of *S*-ethyl propanethioate in propionitrile

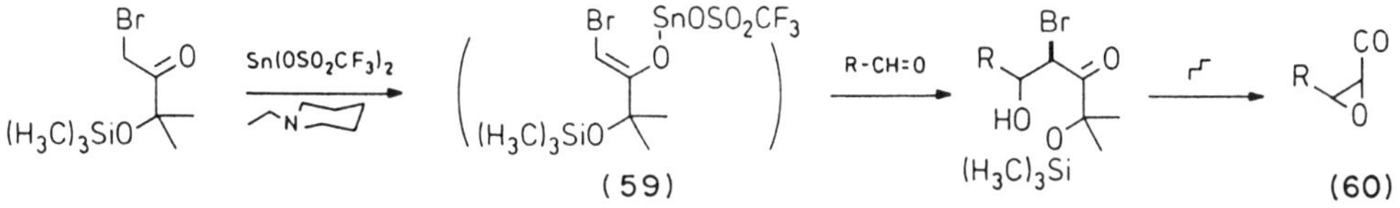

(59) (60)

to afford the aldol adducts with high stereochemical control by the use of tin(II) triflate coordinated with another proline-derived diamine as a chiral auxiliary.

8.6.2 TIN(II) SALT CATALYSIS IN 1,4-ADDITION

Mukaiyama *et al.* have investigated the 1,4-addition of ethylthiotrimethylsilane to conjugated enones and the subsequent aldol reaction with aldehydes to give the *syn*-adducts **61**, which can be elaborated to β-hydroxy-α-methylenealkanones. The first steps of the sequence appear to be catalysed by 10 mol% of tin(II) triflate ($H_5C_2SSnOTf$).

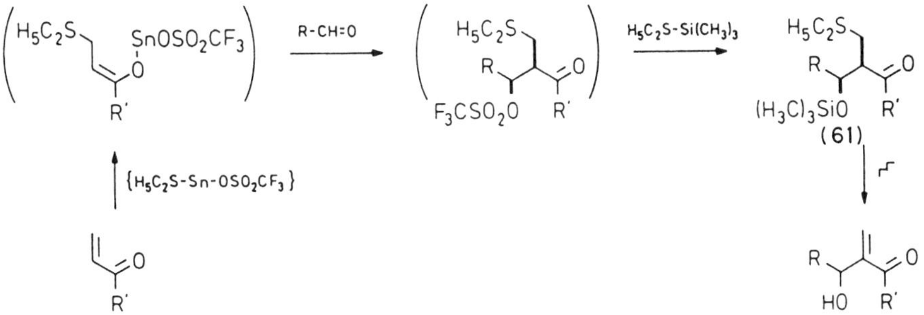

(61)

3-Acyl-1,3-oxazolidin-2-one-derived tin(II) enolates **62** react with conjugated enones only in the presence of chlorotrimethylsilane or chlorodimethylsilane. Results of asymmetric synthesis by means of the Michael addition of tin(II) enolates have been described.

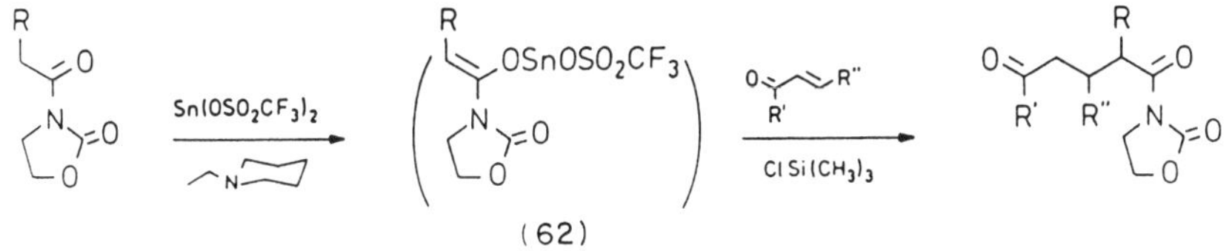

(62)

Further reports deal with enantioselective Michael reaction of tin(II) enolates [140] and enethiolates [141]. Alkylation of ketene dithioacetals with α,β-unsaturated orthoesters is catalysed by tin(II) chloride and other activators. The final products of type **63** are formed on hydrolysis [142].

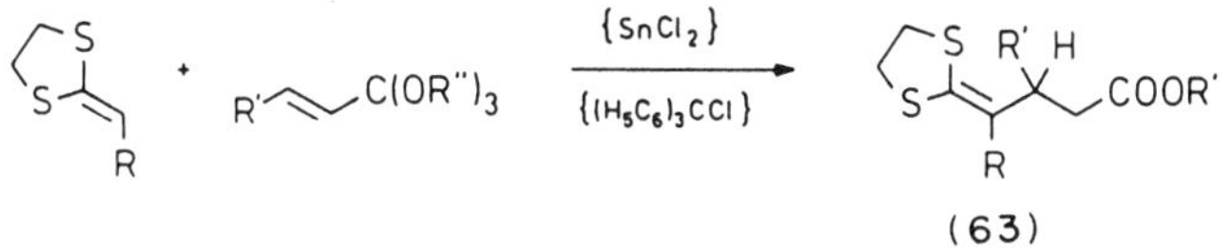

(63)

8.6.3 *OTHER REACTIONS OF TIN(II) SALTS*

The oxidative addition of tin(II) fluoride to an allylic iodide was also reported by Mukaiyama *et al.* [142]. The reaction of the bifunctional reagent **64** with 1,2-diones affords *cis*-diols stereoselectively. Thus, a new type of [3 + 2] cycloaddition leading to the methylenecyclopentane skeleton has been established [143].

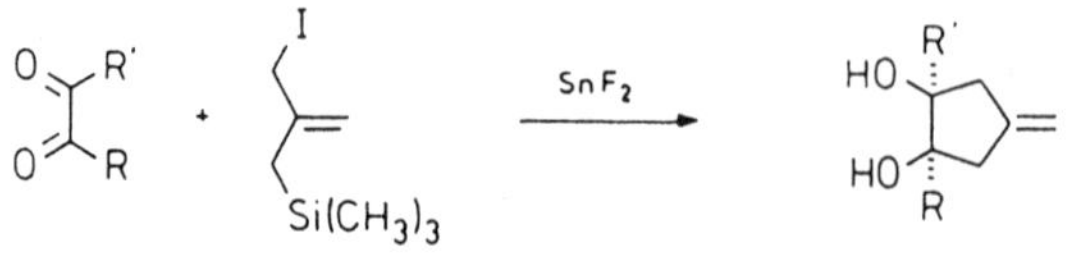

I

SnF_2

$Si(CH_3)_3$

(64)

A stereoselective glycosidation is based on the reaction of a 1α-glycosyl chloride with alcohols. This condensation is mediated by tin(II) triflate as a promoter and tetramethylurea as a base [144]. The 1,4-addition of bis(trimethylsilyl)stannylene to enones immobilized in the *s-cis* conformation gives rise to stanna-cyclic products of type **65** [145].

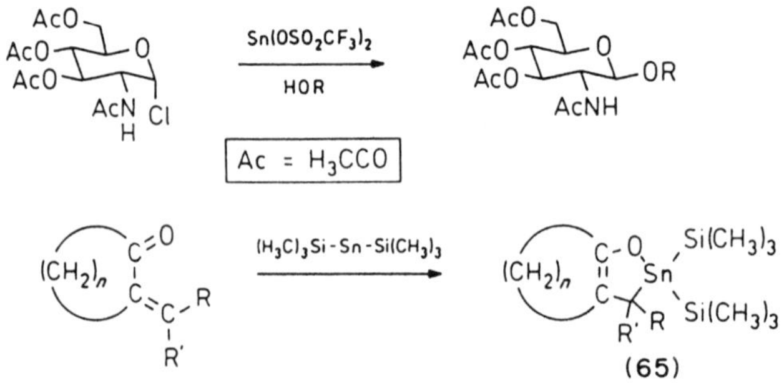

$Sn(OSO_2CF_3)_2$

HOR

Ac = H_3CCO

$(H_3C)_3Si-Sn-Si(CH_3)_3$

(65)

8.7 TIN–HETEROATOM BOND REACTIVITIES

The first part of this section deals with the selective transformation of carbonyl functional groups as effected by the catalysis of distannoxanes having the formula X—$SnBu_2OSnBu_2$—Y, where X and Y are hydroxyl, halo or isothiocyanato. One of these groups is exchanged into an alkoxyl group by the action of an alcohol as a solvent and the resulting alkoxyl group attached to pentacoordinated tin proves to be nucleophilic enough to attack the carbonyl carbon, whose electron density is decreased by coordination of the carbonyl oxygen on the tin atom of the catalyst. The reactions discussed here first are transesterifications and large-ring lactonizations [146].

The second part of this section covers the reactivities of tin–sulphur reagents in the presence of a Lewis acid as applied to the preparation of monothioacetals and vinylogues thereof and also their synthetic transformations. Further, the conversion of carbinols, obtained by the nucleophilic addition of α-(phenylthio)methoxymethyllithium to aldehydes, into 1,3- and 1,4-dicarbonyl compounds are described and some examples of the selective activation of hydroxyl and carboxyl groups by means of tin–sulfur reagents are presented. Finally, the unique properties of dibutyltin ditriflate as a Lewis acid are reported.

8.7.1 DISTANNOXANE-CATALYSED TRANSESTERIFICATION, LACTONIZATION AND RELATED REACTIONS

Distannoxanes such as compound **66** are known to have the tightly dimeric ladder structures [147, 148]. All the tin–oxygen bonds have approximately the same length and therefore similar bond orders. All tin atoms are pentacoordinated and occupy the centre of a slightly distorted trigonal bipyramid [149]. Further coordination with the oxygen atom of carbonyl groups still increases the already strong nucleophilicity of the alkoxyl group attached to the tin atoms. Finally, the dimeric tetrabutyldistannoxanes are remarkably soluble even in paraffinic solvents, despite the polarity of the tin–heteroelement bonds. This unique solubility is ascribed to the hydrophobic attraction on the eight butyl groups.

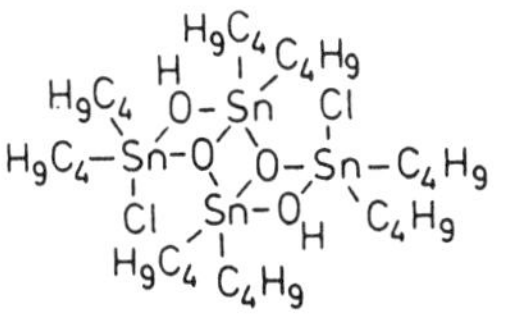

(66)

Distannoxanes are excellent catalysts for transesterification reactions [150, 151]. The presumed mechanism is illustrated by the transition states **67a** and **67b**: a dioxa-distanna core acts as a turntable that mediates the departure of the ester alkoxy group OR′, its replacement by a catalyst alkoxy group OR″ and the restitution of the latter by an entering alcohol molecule present in the solution.

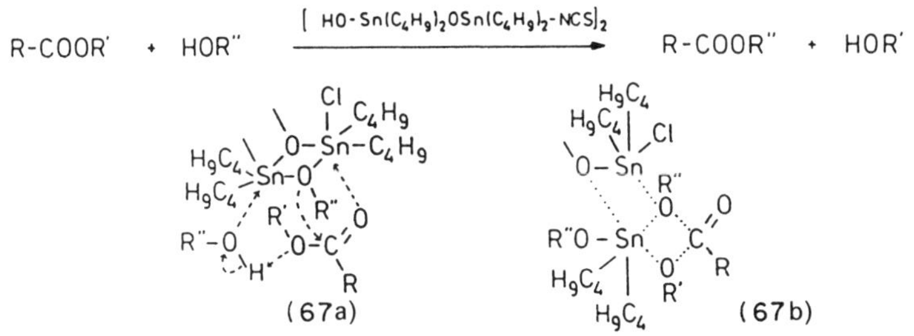

$$\text{R-COOR}' + \text{HOR}'' \xrightarrow{[\text{HO-Sn}(C_4H_9)_2\text{OSn}(C_4H_9)_2\text{-NCS}]_2} \text{R-COOR}'' + \text{HOR}'$$

(67a) (67b)

The ester exchange proceeds smoothly under neutral conditions in alcohol-containing non-polar solvents including aliphatic and aromatic hydrocarbons and carbon tetrachloride, but only sluggishly in polar solvents such as 1,4-dioxane, acetonitrile, and diethylene glycol dimethyl ether [152]. Obviously, the polar distannoxane catalyst effectively coordinates the reactants, if in a non-polar solvent, but not in a polar medium. As the data given in Table 2 suggest, the transesterification process is sensitive to steric hindrance next to the carbonyl group and not in the alkoxy moiety. Consequently, Sn-OR bond formation must occur in a fast, rather than rate-limiting step.

Table 8.2. Solvent effect and steric hindrance in transesterification of RCOOR′ with HOR″R^{3}OH in the presence of $SCNSn(C_4H_9)_2OSn(C_4H_9)_2OH$ (10 mol%)

R	R′	R″	Solvent[a]	Yield of RCOOR″ (%)
C_3H_7	CH_3	$C_6H_5CH_2$	Benzene	95
C_3H_7	CH_3	$C_6H_5CH_2$	Toluene	100
C_3H_7	CH_3	$C_6H_5CH_2$	Hexane	99
C_3H_7	CH_3	$C_6H_5CH_2$	Carbon tetrachloride	100
C_3H_7	CH_3	$C_6H_5CH_2$	Acetonitrile	7
C_3H_7	CH_3	$C_6H_5CH_2$	1,4-Dioxane	15
C_3H_7	CH_3	$C_6H_5CH_2$	Diethylene glycol dimethyl ether	2
C_3H_7	CH_3	C_4H_9	Benzene	99
C_3H_7	CH_3	$(CH_3)_3CCH_2$	Benzene	89
C_6H_5	CH_3	$C_6H_5CH_2$	Benzene[b]	92
$(CH_3)_2CH$	CH_3	$C_6H_5CH_2$	Benzene	77
$(CH_3)_3CCH_2$	CH_3	$C_6H_5CH_2$	Benzene	19
CH_3	$CH_2C(CH_3)_3$	$C_6H_5CH_2$	Benzene	38

[a] The mixture was heated at reflux for 20 h.
[b] Heating for 15 h.

Optional *monoacylation* of the β-methylglucoside derivatives **68** at position 2 or *diacylation* at positions 2 and 6 is possible by the appropriate choice of the tin reagent: dibutyltin oxide and hexabutylstannoxane, respectively. The other hydroxy groups can be left unprotected; they are not affected [153].

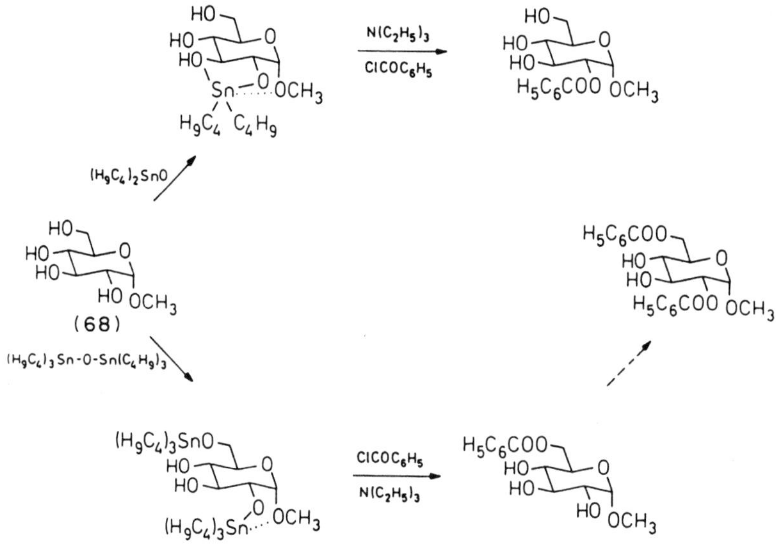

The high nucleophilicity of oxygen attached to a pentavalent tin atom has found useful applications also outside sugar chemistry [154, 155]. Schreiber and co-workers [156] applied it to the preparation of the β-iodoacrylate **69** of a multifunctionalized, complex secondary alcohol having three asymmetry carbon atoms including the one carrying the hydroxyl group. The final ring closure step to brefeldin C relies on chromium chemistry (as already discussed in Section 8.4.1) [67–69].

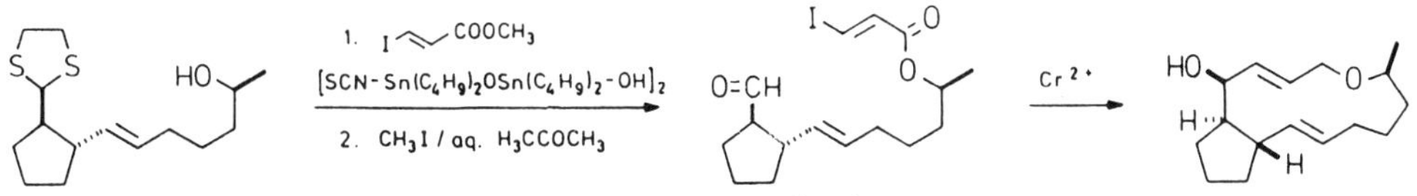

1. $I\text{-}CH{=}CH\text{-}COOCH_3$ / $[SCN\text{-}Sn(C_4H_9)_2OSn(C_4H_9)_2\text{-}OH]_2$
2. CH_3I / aq. H_3CCOCH_3

Cr^{2+}

(69)

The transesterification technique also finds application in the transformation of diaceates $CH_3COO(CH_2)_nOCOCH_3$ to the respectively monoacetates $HO(CH_2)_nOCOCH_3$. High yields are obtained under mild and essentially neutral conditions [157].

Lactonization of ω-hydroxycarboxylic acids giving large rings proceeds successfully in the presence of the distannoxanes dissolved in decane [158]. Typical results are summarized in Table 8.3. This novel method requires neither high dilution nor a Dean–Stark trap. Related lactonization techniques utilizing dibultyltin oxide give less satisfactory yields [159–161].

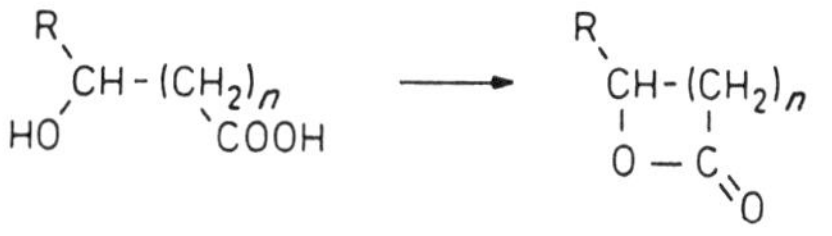

$$R\text{-}CH(OH)\text{-}(CH_2)_n\text{-}COOH \longrightarrow R\text{-}CH\text{-}(CH_2)_n \text{ (ring closed via } O\text{-}C{=}O)$$

Supposedly, the substrate is first attached to the catalyst molecule to form a tin alkoxide (**70**). The repulsion of the solvent with the polar groups favours the hairpin-like conformation of the substrate and thereby saves entropic work. Alkyl groups bigger than butyl attached to the hydroxy-carrying carbon atom apparently improve the steric conditions of cyclization.

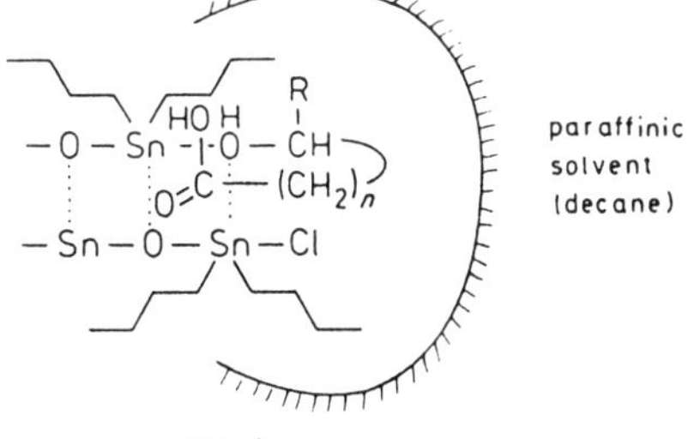

paraffinic solvent (decane)

(70)

Dibutyldichlorostannane-catalyzed esterifications of carboxylic acids have been reported [162]. The active catalyst may well be again a distannoxane. The results of a

Table 8.3. Macrolactonization of hydroxy-substituted carboxylic acids $HOCHR(CH_2)_nCOOH$ catalyzed by distannoxanes $HOSn(C_4H_9)_2OSn(C_4H_9)_2X$ yields of lactones isolated after 24 h of reflux in decane

n	R	Catalyst X	Yield (%)[a]
14	H	Cl	81 (60)
13	H	Cl	78 (43)
12	H	NCS	19
12	C_6H_{13}	NCS	48
11	H	NCS	20
11	CH_3	NCS	–[b]
11	C_3H_7	NCS	–[b]
11	C_4H_9	NCS	63 (0)
11	$(CH_3)_2CHCH_2$	NCS	56
11	C_6H_{13}	NCS	64[c]
10	H	Cl	–[b,d] (22)
10	C_6H_{13}	Cl	90

[a] In parentheses are given the yields obtained with dibutyltin oxide as the catalyst [104].
[b] No monolide obtained.
[c] Plus 19% diolide.
[d] Only diolide (82%) found.

systematic study of distannoxane-catalyzed esterifications are compiled in Table 8.4 [163].

Finally, desilylation of siloxyl bond, deacetalization and acetalization have also been observed in the presence of distannoxane catalysts [164].

Table 8.4. Distannoxane-catalyzed esterification of typical carboxylic acids RCOOH with alcohols R′OH

R	R′	Yield (%)[a]
C_3H_7	C_4H_9	97
C_3H_7	$(CH_3)_3CCH_2$[b]	95
$(CH_3)_2CH$	C_4H_9	97
Cyclohexyl	C_4H_9	88
$(CH_3)_3CCH_2$	C_4H_9	20

[a] The reaction was carried out with 1 mmol of RCOOH dissolved in 3 ml of R′OH at 80 °C for 20 h in the presence of 10 mol% of $HOSn(C_4H_9)_2OSn(C_4H_9)_2NCS$.
[b] The alcoholic component was employed in a 1.1 molar ratio.

8.7.2 TIN–SULPHUR BOND REACTIVITIES IN THE PRESENCE OF LEWIS ACIDS: THE CHEMISTRY OF MONOTHIOACETALS AND ITS VINYLOGUES

Synthetic applications of aldehyde monothioacetals have been reviewed [165]. Treatment of acetals, conveniently prepared by the method of Noyori and co-workers [166], of both aldehydes and ketones with tributyl(phenylthio)stannane or dibutylbis(phenylthio)stannane in the presence of boron trifluoride etherate in toluene below 0 °C has turned out to give good yields of the desired monothioacetals starting materials **71** [167].

$$R'RC(OR'')_2 \xrightarrow[BF_3]{(H_9C_4)_nSn(SC_6H_5)_{4-n}} R'RC(OR'')(SC_6H_5) \quad (71)$$

Alternatively, formaldehyde monothioacetal can be alkylated at the active methylene group. The resulting monothioacetals can be converted into homoallyl (or homopropargyl) ethers (**72**) by reaction with allyltributylstannane in the presence of boron trifluoride etherate. If, however, the Lewis acid is titanium(IV) chloride instead, the same organotin reagents gave homoallyl (or homopropargyl) thioethers (**73**). In this way the formaldehyde monothioacetal serves as either a methoxycarbene or phenylthiocarbene synthon connecting organic halides nucleophilically on one side and organotin reagent electrophilically on the other side [168].

$$H_2C(OCH_3)(SC_6H_5) \xrightarrow[2.\ R-X]{1.\ LiC(CH_3)_3} R-CH(OCH_3)(SC_6H_5) \xrightarrow{H_2C{=}CH-CH_2-Sn(C_4H_9)_3}$$

+ {BF_3} → R–CH(OCH_3)–CH_2–CH=CH_2 (72), with H_3CO

+ {$TiCl_4$} → R–CH(SC_6H_5)–CH_2–CH=CH_2 (73), with H_5C_6S

α-Enal-derived acetals can be readily converted into 3-phenylthio-1-alkenyl ethers (**74**) (in general mixtures of *Z*- and *E*-isomers), which are versatile intermediates [169, 170]. After deprotonation with *tert*-butyllithium they may be submitted to electrophilic substitution at the 3-position, suitable electrophiles being alkyl halides, aldehydes and oxiranes. Oxidative desulphurization and hydrolysis afford α-enals (again, in general, as a mixture of *Z*- and *E*-isomers). This type of transformation was first explored by Kondo and Tunemoto [171] with sodium phenylsulphinate as a key reagent.

$$\text{R,R'-enal} \longrightarrow \text{R,R'-enal acetal (OR'')}_2 \xrightarrow{(H_9C_4)_2Sn(SC_6H_5)_2} (74)\ [H_5C_6S,\ OR''] \xrightarrow[2.\ El-X \quad 3.\ NaIO_4]{1.\ LiC(CH_3)_3} \text{El-substituted enal (R, R')}$$

This method has been applied to the synthesis of a few natural products. The D, L-nuciferal (**75**) thus prepared was 100% pure *E*-isomer, whereas the analogously obtained citral **76** was found to be a 3:7 mixture of *Z*- and *E*-isomers.

H_5C_6S OCH$_3$ — 1. LiC(CH$_3$)$_3$; 2. (aryl iodide) → H_5C_6S OCH$_3$ — NaIO$_4$, H$_2$O → (75)

H_5C_6S OCH$_3$ — 1. LiC(CH$_3$)$_3$; 2. $(H_3C)_2C{=}CH{-}(CH_2)_3{-}CH_2I$ → OCH$_3$ SC$_6$H$_5$ — NaIO$_4$, H$_2$O → (76)

The reaction of the lithiated intermediates **74** with aldehydes leads, after oxidative elimination of the phenylthio group, to γ-hydroxy-α-enals **77**, while that with oxiranes leads to δ-hydroxy-α-enals **78** [170]. All these products have mainly, if not exclusively, the *E*-configuration.

H_5C_6S OCH$_3$ R — 1. LiC(CH$_3$)$_3$; 2. R'-CH=O → H_5C_6S OCH$_3$ R' HO R — NaIO$_4$ → R' HO R (77)

— 1. LiC(CH$_3$)$_3$; 2. R'-oxirane → HO S(C$_6$H$_5$) OCH$_3$ R' R — NaIO$_4$ → HO R' R (78)

When lithiated formaldehyde monothioacetal is combined with aldehydes and the resulting adducts are treated with methanesulphonyl chloride, highly reactive intermediates (**79**) are obtained. Aqueous hydrolysis causes the migration of the phenylthio group and affords α-phenylthio-substituted aldehydes by this very convenient modification of the de Groot–Jansen sequence. These aldehydes may be submitted to a Wittig reaction employing α-alkoxyxcarbonyl-substituted ylides and the resulting α,β-unsaturated esters may be converted into γ-oxo esters (**80**). Alternatively, the intermediate mesylates **79** may be treated with enol silyl ethers in the presence of tin(IV) tetrachloride to produce adducts, which, depending on the conditions of manipulation and hydrolysis, can be transformed into 1,4- or 1,3-dicarbonyl compounds (**81** or **82**) [173, 174].

When a tetrahydropyranyl ether of a common alcohol is treated with a tin–sulphur reagent in the presence of a Lewis acid, the alcohol moiety is converted into the tin alkoxide **83**, the phenylthio group always being picked up by the tetrahydropyranyl group to form a monothioacetal [176]. The oxygen atom of tin alkoxides is more nucleophilic than that of a free hydroxyl group, so that the tin—oxygen bond is preferentially hydrolysed under conditions in which other protecting groups of hydroxyl remain untouched. The tin alkoxides thus produced are easily alkylated [177], acylated or selectively oxidized (with pyridinium chlorochromate).

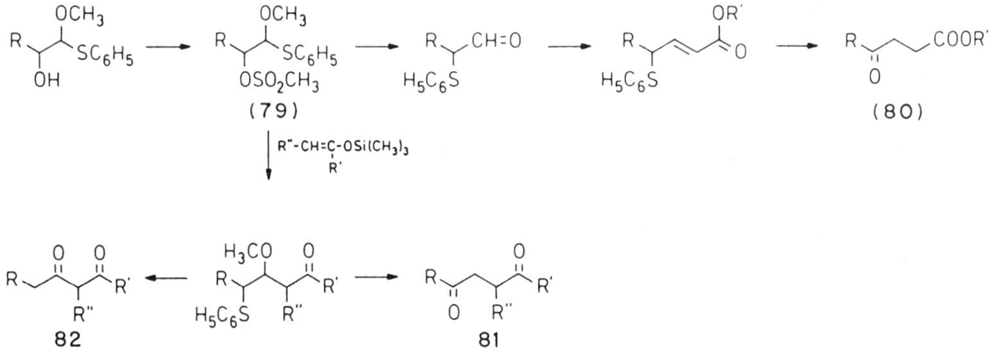

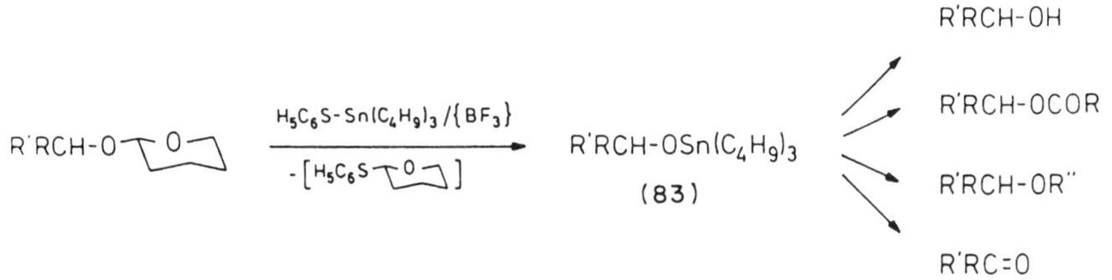

Accordingly, tetrahydropyranylation is not just a protection of a hydroxyl group, but may serve as a means of selective activation. This may provide a means for the preferential transformation of some of the hydroxyls of polyol compounds. The technique was further extended to carboxyl group manipulation [178]. Dimethylbis (methylthio)stannane is a distillable and easily purified liquid. It was found to react in the presence of boron trifluoride with α-methylcinnammates (**84**) to afford cinnamyl sulphides **85** and tin carboxylates (**86**) selectively. The compounds can undergo hydrolysis to give carboxylic acids or alkylation to give esters. Other protective groups of

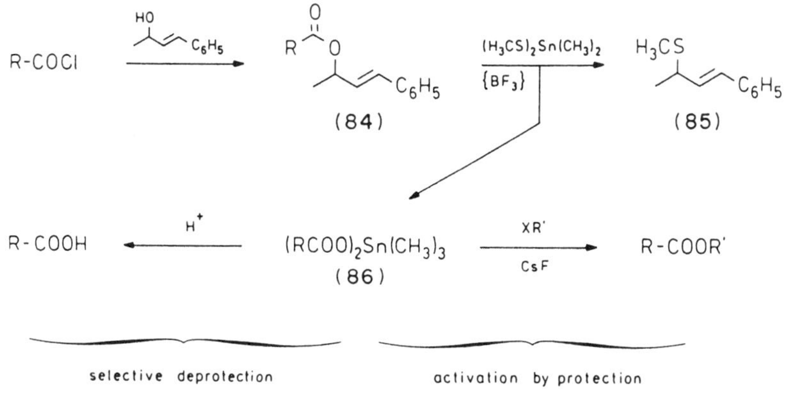

selective deprotection

activation by protection

hydroxyl or carboxyl remain intact on the treatment with these tin–sulphur reagents. This method provides a means for the selective transformation of a particular carboxyl group in preference to others.

8.7.3 ORGANOTIN TRIFLATE AS A UNIQUE LEWIS ACID

1,3-Dithianes can be prepared from aldehydes or acetals by transacetalization using a novel Lewis acid, dibutyltin ditriflate, $(H_9C_4)_2Sn(OTf)_2$. Typical examples are compounds **87–90** [179]. A number of competitive experiments have established the following order of reactivities: aromatic aldehyde acetals > aliphatic aldehydes > aromatic aldehydes > aliphatic aldehyde acetals. Hence acetalization increases the reactivities of aromatic aldehydes but deactivates those of aliphatic aldehydes. The reactivity sequence may be explained on the basis of the rate-determining formation of alkoxylated carbocations from acetals or of the formation of stannyloxy-substituted carbocation from unprotected aldehydes.

H_9C_4-CH=O + C_6H_5-CH=O; $(H_9C_4)_2Sn(S(CH_2)_3S)$ {$(H_9C_4)_2Sn(OSO_2CF_3)_2$}; fast → H_9C_4-CH(S(CH_2)_3S) (87); slow → C_6H_5-CH(S(CH_2)_3S) (88)

H_9C_4-CH$(OCH_3)_2$ + C_6H_5-CH$(OCH_3)_2$; $(H_9C_4)_2Sn(S(CH_2)_3S)$ {$(H_9C_4)_2Sn(OSO_2CF_3)_2$}; slow → H_9C_4-CH(S(CH_2)_3S) (89); fast → C_6H_5-CH(S(CH_2)_3S) (90)

The further utility of dibultyltin ditriflate in the reaction of acetals with silylated nucleophiles has been reported [180]. The new technique is outlined by the competitive experiments of acetals of ketones and acetals in the introduction of silylated nucleophiles such as hydride, cyanide, ally and enolate. As the product composition **91** vs **92** in

R_3Si-X {$(H_9C_4)_2Sn(OSO_2CF_3)_2$}

$(H_3C)(H_{13}C_6)C(OCH_3)_2$ —fast→ $(H_3C)(H_{13}C_6)C(OCH_3)(X)$ (91)

$H_{17}C_8$-CH$(OCH_3)_2$ —slow→ $H_{17}C_8$-CH$(OCH_3)(X)$ (92)

Table 8.5. Competition of ketone and aldehyde acetals toward silylated nucleophiles

Si–Nu	Combined yield (**91** + **92**) (%)	Ratio **91** : **92**
$(C_2H_5)_3Si—H$	92	95 : 5
$(CH_3)_3Si—CN$	75	95 : 5
$Cl_3Si—CH_2CH{=}CH_2$	48	98 : 2
$(CH_3)_3Si—OC(OCH_3){=}C(CH_3)_2$	67	94 : 6

competition experiments has revealed, ketone-derived acetals are far more reactive than aldehyde-derived acetals under such conditions (see Table 8.5).

A similar selectivity behaviour was observed when other silyl enolates were allowed to react under Mukaiyama conditions. The ketone-derived acetals (affording **93**) were consumed far more rapidly than their aldehyde-derived counterparts (affording **94**). The product ratios found are listed in Table 8.6.

$R'\text{-}CH{=}C(OSi(CH_3)_3)R$ + $\{(H_9C_4)_2Sn(OSO_2CF_3)_2\}$

$(H_3C)(H_9C_4)C(OCH_3)_2$ —fast→ $H_9C_4\text{-}C(CH_3)(OCH_3)\text{-}CH(R')\text{-}C(=O)R$ (**93**)

$H_{15}C_7\text{-}CH(OCH_3)_2$ - - slow - -→ $H_{15}C_7\text{-}CH(OCH_3)\text{-}CH(R')\text{-}C(=O)R$ (**94**)

These new methods will certainly find future synthetic applications in the transformation of polyfunctional carbonyl compounds. Further, application of the Michael addition of silyl enolates to methyl vinyl ketone itself has been carried out by means of this Lewis acid catalyst, dibutyltin ditriflate, with good preparative yields [181–182].

Table 8.6. Lewis acids in the Mukaiyama reaction of acetals with *O*-silylene ethers

R	R′	Lewis acid[b]	Combined yield (**93** + **94**)	Ratio **93** : **94**
H	C_6H_5	$(C_4H_9)_2Sn(OTf)_2$	79	99 : 1
CH_3	C_2H_5	$(C_4H_9)_2Sn(OTf)_2$	81	99 : 1
H	$(CH_3)_3C$	$(C_4H_9)_2Sn(OTf)_2$	80	100 : 0
H	$(CH_3)_3C$	$(CH_3)_3SiOTf$	59	85 : 15
H	$(CH_3)_3C$	$(C_6H_5)_3CClO_4$	56	89 : 11
H	$(CH_3)_3C$	$SnCl_2 + (CH_3)_3SiCl$	50	74 : 26
H	$(CH_3)_3C$	$HOSO_2CF_3$	100	81 : 19

[a] Unprotected ketones gave no aldol products, whereas aldehydes reacted normally to produce aldols.
[b] OTf = OSO_2CF_3.

8.8 CONCLUSION

The mechanistic concepts postulated in this chapter are still speculative and await physical or physicochemical verification. Nevertheless, it is impressive to find that simple tin reagents such as distannoxanes in neutral organic solvents behave almost like enzymes. The key metal atoms are effective 'binding sites' and the nucleophiles, being activated by coordination of the substrate, are 'active sites.' It should be possible in the future to design a variety of such 'combined acid–base' catalysts [146] or reagents by drawing still more inspiration from inorganic chemistry. One has to learn more about the 'higher order' structures of the reacting systems which involve various kinds of key metal atoms in order to build up the basic idea of organic synthesis in the 21st century.

8.9 WORKING PROCEDURES

8.9.1 CYANOETHYLATION FOR 2,3,4,5-TETRA-O-ACETYL-1α-BROMOGLUCOPYRANOSIDE

A mixture of 0.60 g (50 mmol) of 2,3,4,5-tetra-*O*-acetyl-1α-glucopyranosyl bromide in 100 ml of anhydrous diethyl ether is brought to reflux in a nitrogen atmosphere. To this clear solution 13.5 g (0.25 mol) of acrylonitrile and 161.5 g (55 mmol) of tributylstannane are added. The solution is irradiated with a sunlamp, which maintains a vigorous reflux for 4 h. The precipitated solids are removed by filtration, the solution is mixed with an additional 6.6 g (0.120 mol) of acrylonitrile and 5.8 g (20 mmol) of tributylstannane, and irradiation is continued for another 4 h. The cooled mixture is filtered and the filtrate A is saved. All the solids are combined, extracted with 250 ml of hot 2-propanol and the extract is filtered. The hot filtrate is concentrated to a total volume of 75 ml and allowed to cool. Filtration gives 7.8 g (40%) of the pure α-cyanoethylated product, 1-deoxy-2,3,4,6-tetra-*O*-acetyl-1-(2-cyanoethyl)-α-D-glucopyranose (**1**), m.p. 121–122 °C, and the mother liquor B. The filtrate A is concentrated, taken up in 50 ml of acetonitrile and extracted three times with 50 ml portions of pentane. The acetonitrile phase is combined with the mother liquor B and concentrated. Flash chromatography of the resulting syrup on silica gel using ethyl acetate–hexane (1 : 1) as eluent affords an addition 2.2 g of **1**, which brings the total yield to 53% [6a].

8.9.2 CYCLIZATION OF (3-METHYL-2-BUTENYL)PROPARGYLPROPANEDIOIC ACID DIMETHYL ESTER (5) [14]

A mixture of the disubstituted malonate **5** (24 g, 0.10 mol), tributylstannane (30 g, 0.10 mol) and 40 mg (0.25 mmol) of azoisobutyronitrile (AIBN) is heated in an oil-bath at 75–85 °C with stirring. After an induction period of less than 30 min, an exothermic reaction takes place to produce evolution of a small amount of gas and a rise in the oil-bath temperature. At this point the reaction is virtually completed and the product is subjected to the subsequent protodestannylation. The reaction mixture is transferred in to a 2 l Erlenmeyer flask which contains 1 l of dichloromethane, 350 g of silica and a stirring bar. After the flask has been stoppered, the mixture is stirred for 36 h, filtered and the filtrate concentrated. Short-path distillation of the residual oil gives 20 g (84% yield) of 3-methylene-4-isopropyl-1,1-cyclopentanedicarboxylic acid dimethyl ester, b.p. 80–85 °C/0.2 mmHg.

8.9.3 DEOXYGENATION OF 1,2:5,6-DI-O-ISOPROPYLIDENE-α-D-GLUCOFURANOSE (17)

A solution of 26 g (0.10 mol) of the glucofuranose diacetonide **17** and 25 mg of imidazole dissolved in 400 ml of tetrahydrofuran is treated under nitrogen with 7.2 g (0.150 mol) of a 50% sodium hydride dispersion in portions, and then with 23 g (0.30 mol) of carbon disulphide. After stirring the mixture for 30 min, 25 g (0.177 mol) of methyl iodide are added and the solution is stirred for another 15 min. The excess of sodium hydride is destroyed with 5.0 ml of glacial acetic acid. The solution is filtered and the filtrate is concentrated. The semi-solid residue is extracted with three 100 ml portions of diethyl ether and the combined ether extracts are washed with saturated sodium hydrogencarbonate solution and with water. Concentration followed by distillation (Kugelrohr) gives 32 g (92%) of the xanthate, b.p. 153–160 °C/1.0 mmHg. A mixture of 500 ml of toluene, 25 g (0.085 mol) of tributylstannane and 19.2 g (0.055 mol) of the xanthate is heated at reflux under a nitrogen atmosphere for 4–7 h until the colour changes from yellow to nearly colourless and thin-layer chromatography gives one spot. Concentration of the toluene solution gives a thick, oily residue that is partitioned between 250 ml portions of hexane and acetonitrile. The nitrile layer is separated and washed with three 100 ml portions of hexane and concentrated in a rotary evaporator. The residual oil is taken up in hexane–ethyl acetate (10:1) and filtered through a pad of silica gel. The filtrate is concentrated and distillation of the residue gives 10.0 g (75%) of 3-deoxy-1,2:5,6-di-*O*-isopropylidene-α-D-ribohexofuranose as a colourless syrup, b.p. 72–73 °C/0.2 mmHg; n_D^{25} 1.4474 [22].

8.9.4 TRIBUTYL(METHOXYMETHOXYMETHYL)STANNANE: PREPARATION AND CONSECUTIVE TREATMENT WITH BUTYLLITHIUM AND A CARBONYL COMPOUND

To a solution of diisopropylamine (9.9 g, 13.7 ml, 98 mmol) in tetrahydrofuran (125 ml) at 0 °C (ice-bath) is added dropwise butyllithium (1.56 M in hexane, 57 ml, 90 mmol). The resulting solution is stirred at 0 °C for 15 min and mixed with tributylstannane (22 g, 20 ml, 75 mmol), added dropwise. After 15 min at 0 °C solid paraformaldehyde (2.7 g, 90 mmol) is added and the ice-bath is removed. After 1.5 h at 25 °C the reaction mixture is poured into hexane (400 ml) and washed with water and brine. Drying over magnesium sulphate and concentration *in vacuo* gives crude tributylstannylethanol as a pale yellow liquid, which is subjected to etherification directly. A mixture of dimethoxymethane (114 g, 133 ml, 1.50 mol), chloroform (124 ml) and phosphorus pentoxide (25 g) is stirred for 5 min and the crude tributylstannylmethanol obtained above and phosphorus pentoxide (25 g) are added. The reaction mixture is stirred vigorously for 15 min and then poured into 300 ml of saturated aqueous sodium carbonate and the residual phosphorus pentoxide is washed with hexane (400 ml). The combined organic layers are washed with brine, dried over magnesium sulphate and concentrated *in vacuo*. Vacuum distillation gives tributyl(methoxymethoxymethyl)stannane as a clear liquid, b.p. 84–87 °C/0.05 mmHg (16.4 g, 50%).

To a solution of tributyl(methoxymethoxymethyl)stannane (1.75 g, 4.8 mmol) in 15 ml of tetrahydrofuran at −78 °C is added butyllithium (2.5 M in hexane, 1.84 ml, 4.6 mmol) over a period of 2 min while the temperature is maintained below −65 °C. Stirring is continued 5 min and the carbonyl component (4.0 mmol) is added neat. After the mixture has been stirred for 15 min at −78 °C, the reaction is quenched by the addition of 20 ml of saturated aqueous ammonium chloride. The mixture is extracted with ethyl acetate and the combined organic layers are washed consecutively with water and brine and dried over anhydrous magnesium sulphate. Concentration followed by flash chromatography over silica gel generally gives >94% yield of the carbonyl adducts, i.e. the 2-(methoxymethoxy)ethanols [106].

8.9.5 ALLYL- OR BENZYL-TYPE ORGANOTIN COMPOUNDS: PREPARATION AND ISOLATION

A precooled solution of an olefin or alkylbenzene (40 mmol) and potassium *tert*-butoxide (4.5 g, 40 mmol) in tetrahydrofuran (60 ml) is added to butyllithium (40 mmol in tetrahydrofuran (20 ml), kept in a dry-ice–methanol bath (−75 °C). After 2 h at −50 °C, the homogeneous mixture is cooled to −75 °C and trimethylstannyl chloride (8.0 g, 40 mmol) in tetrahydrofuran (20 ml) is added. The solvent is evaporated and the volatile component of the residue is distilled under reduced pressure [74].

8.9.6 CONVERSION OF AN ORGANOTIN INTO THE CORRESPONDING ORGANOPOTASSIUM COMPOUND FOLLOWED BY ELECTROPHILIC TRAPPING

A solution of trimethylsilylmethylpotassium (10 mmol) in tetrahydrofuran (40 ml), from which the amalgam has been removed by centrifugation [183], is cooled to −100 °C. A precooled (−75 °C) solution of 3-trimethylstannyl-2-phenylpropene (2.8 g, 10 mmol) is added. After being kept for 2 h at −100 °C and 1 h at −75 °C, the mixture is quenched with an appropriate electrophilic reagent [74].

8.9.7 TRIBUTYLSTANNYLETHYNE AND (E)-BIS(TRIBUTYLSTANNYL)ETHENE: PREPARATION AND APPLICATION

Under a nitrogen atmosphere a mixture of 24 g (0.26 mol) of lithium acetylide–ethylenediamine complex (Aldrich) and 800 ml of tetrahydrofuran is cooled in an ice–water bath, 71 g (0.22 mol) of tributylchlorostannane are added dropwise and the mixture is stirred for 18 h at room temperature. The excess of acetylide is destroyed with 10 ml of water under cooling with an ice-bath. The reaction mixture is concentrated under reduced pressure and the residue is washed with hexane (3 × 50 ml). The organic layers are combined and dried over anhydrous magnesium sulphate. Evaporation under reduced pressure gives a colourless oil. Distillation yields 21 g (32% yield) of tributylstannylethyne, b.p. 90–94 °C/0.5 mmHg in 31% yield. A mixture of 21 g (0.066 mol) of tributylstannylethyne, 23 g (0.079 mol) of tributylstannane and 0.25 g (0.9016 mol) of azoisobutyronitrile is heated at 90 °C with magnetic stirring for 6 h. Distillation gives (*E*)-1,2-bis(tributylstannyl)ethene, b.p. 180–218 °C/0.5 mmHg, in 88% yield. A solution of 27 g (0.044 mol) of (*E*)-1,2-bis(tributylstannyl)ethene in 100 ml of tetrahydrofuran is cooled in a dry-ice–acetone bath. By means of a double-ended needle a dropping funnel is charged with 44.8 ml of a 1.2 M-ethereal solution of methyllithium, which is added dropwise during 40 min with cooling. Stirring is continued for an additional 2 h at −78 °C. Another dry flask is charged with 5.8 g (0.53 mol) of ethyl chloroformate and 150 ml of tetrahydrofuran and the solution is cooled to −78 °C. Under a gentle nitrogen pressure, the lithiated reagent is transferred to the ethyl chloroformate solution by means of a double-ended needle at −78 °C. After an additional 30 min at −78 °C, 20 ml of methanol are added. Hexane extraction followed by silica gel column chromatography gives 10.2 g (59% yield) of ethyl (*E*)-3-tributylstannylpropenoate (**55**) as a yellow oil. Attempted purification by vacuum distillation accompanies 7–8% isomerization to ethyl (*Z*)-3-(tributylstannyl)propenoate. A bright yellow solution is made from 3.2 g (17.2 mmol) of *p*-nitrobenzoyl chloride, 0.08 g (0.10 mmol) of benzyl(chloro)(triphenylphosphine)palladium(II) (Aldrich) and 30 ml of chloroform. The bright yellow solution is evacuated and refilled with carbon monoxide (three cycles) utilizing a gas bag. After an additional 10 min at room temperature a solution of 8.0 g (21 mmol) of ethyl (*E*)-3-(tributylstannyl)propenoate in 5 ml of chloroform is added to the flask by syringe. After heating at 50 °C for 12 h under 1 atm CO pressure (gas bag) with stirring, the mixture is treated with 18 ml of a 1.2 M solution of pyridinium poly(hydrogen fluoride) and 10 ml pyridine. The reaction mixture is stirred at room temperature overnight and then transferred in to a 250 ml separating funnel containing 75 ml of water. After addition of 30 ml of chloroform, the organic layer is washed successively with 10% hydrochloric acid (3 × 20 ml), saturated sodium hydrogencarbonate solution (3 × 20 ml), water (25 ml) and brine

(25 ml). The organic layer is dried over anhydrous sodium sulphate, filtered and concentrated. The product is dissolved in chloroform and 15 g of silica gel are added to the solution. Concentration gives a brown powder of silica coated with product. This is placed at the top of a silica gel column (50 g) and elution is carried out with ethyl acetate. Repeated column chromatography gives 3.4 g (80% yield) of yellow green crystals of ethyl (*E*)-4-(4-nitrophenyl)-4-oxo-2-butenoate m.p. 69–71 °C [123].

8.9.8 1-HYDROXY-3-(ISOTHIOCYANATO)TETRABUTYLDISTANNOXANE

1,3-Dichlorotetrabutyldistannoxane is available from Aldrich; 1-hydroxy-3-isothiocyanotetrabutyldistannoxane is prepared as follows. A mixture of dibutyltin oxide (14.9 g, 60 mmol) and dibutyltin diisothiocyanate [151] (21 g, 60 mmol) in 95% ethanol (100 ml) is heated at reflux for 6 h and concentrated. The residual solid is finely powdered and allowed to stand in the atmospheric moisture, which removes the ethoxydistannoxane contamination. Recrystallization from hexane at 0 °C gives the product which decomposes in the temperature range 120–130 °C (15.1 g, 59%). Both are almost equally effective catalysts [150, 152].

8.9.9 DISTANNOXANE-CATALYSED TRANSESTERIFICATION OF (1R,2S,5R)-METHYL ACETOACETATE

A toluene (50 ml) solution of ethyl acetoacetate (0.64 ml, 5.0 mmol) (−)-methanol (3.9 g, 25 mmol) and 1-hydroxy-3-isothiocyanatotetrabutyldistannoxane (0.3 g, 0.5 mmol) is heated at reflux for 24 h. The mixture is concentrated and the residue is subjected to silica gel column chromatography with hexane–ethyl acetate (80:20) as the eluent to give the product (0.94 g, 78%) [150, 152].

8.10 REFERENCES AND NOTES

1. (a) A. G. Davies and P. J. Smith, in *Comprehensive Organometallic Chemistry*, ed. G. Wilkinson, F. G. A. Stone and E. W. Abel, Section 11 (Tin), p. 519, especially see p. 608, Pergamon Press, Oxford, 1982; (b) I. Omae, *Journal of Organometallic Chemistry Library, Vol. 21*, *Organotin Chemistry*, Elsevier, Amsterdam, 1989; (c) J. Otera, *Kagaku* **42** (1987) 675; *Chem. Abstr.* **109** (1988) 141860m; (d) P. G. Harrison (ed.), *Chemistry of Tin*, Blackie, Glasgow and London, and Chapman and Hall, New York, 1989.
2. (a) E. W. Colvin, *Silicon in Organic Synthesis*, Butterworths, Guildford, 1981; (b) E. W. Colvin (ed.), *Silicon Reagents*, Academic Press, New York, 1988.
3. W. Pereyre, J.-P. Quintard, A. Rahm, *Tin in Organic Synthesis*, Butterworths, Guildford, 1987.
4. Y. Yamamoto (ed.), Tetrahedron Symposia-in-Print No. 36, *Tetrahedron* **45** (1989) 909.
5. Review on tributylstannyllithium and tributylstannylmethyllithium: T. Sato, *Synthesis* (1990) 259.
6. (a) B. Giese, J. Dupuis, M. Nix, *Org. Synth.* **65** (1987) 236; (b) for similar reaction of tris(trimethylsilyl)silane, see B. Giese, B. Kopping, C. Chatgilialoglu, *Tetrahedron Lett.* **30** (1989) 681; (c) Approximate rate constant for addition of alkyl radicals to allylstannanes: D. P. Curran, P. A. van Elburg, B. Giese, S. Gilges, *Tetrahedron Lett.* **31** (1990) 2861.
7. G. Stork, in *Current Trends in Organic Synthesis*, ed. H. Nozaki, p. 359, Pergamon Press, Oxford, 1983.

8. (a) This type of cyclization was independently discovered: Y. Ueno, K. Chino, M. Watanabe, O. Moriya, M. Okawara, *J. Am. Chem. Soc.* **104** (1982) 5564; (b) see also B. Giese, *Radicals in Organic Synthesis: Formation of Carbon–Carbon Bonds*, Pergamon Press, Oxford, 1986; (c) D. P. Curran, *Synthesis* (1988) 417, 489.
9. G. Stork, R. Mook, *J. Am. Chem. Soc.* **109** (1987) 2829.
10. G. Stork, *Bull. Chem. Soc. Jpn.* **88** (1988) 149.
11. G. Stork, R. Mah, *Tetrahedron Lett.* **30** (1989) 3613.
12. G. Stork, R. Mook, S. A. Biller, S. D. Rychnovsky, *J. Am. Chem. Soc.* **105** (1983) 3741.
13. J. L. Belletire, N. O. Mahmoodi, *Tetrahedron Lett.* **30** (1989) 4363.
14. R. Mook, P. M. Sher, *Org. Synth.* **66** (1987) 75.
15. K. Nozaki, K. Oshima, K. Utimoto, *J. Am. Chem. Soc.* **109** (1987) 2547; *Tetrahedron* **45** (1989) 923.
16. H. Miyake, K. Yamamura, *Chem. Lett.* (1989) 981.
17. E. Nakamura, D. Machii, T. Inubushi, *J. Am. Chem. Soc.* **111** (1989) 6849.
18. (a) M. D. Bachi, E. Bosch, *Tetrahedron Lett.* **29** (1988) 2581; *J. Org. Chem.* **54** (1989) 1234; (b) cyclization radical species arising from selenoesters: D. L. Boger, R. J. Mathvink, *J. Am. Chem. Soc.* **112** (1990) 4008.
19. Y. Watanabe, Y. Ueno, C. Tanaka, M. Okawara, T. Endo, *Tetrahedron Lett.* **28** (1987) 3953.
20. Tributylstannane in dehalogenation reactions: W. P. Neumann, *Synthesis* (1987) 665; triethylborane-induced hydrodehalogenation of organic halides with tin hydride: K. Miura, Y. Ichinose, K. Nozaki, K. Fugami, K. Oshima, K. Utimoto, *Bull. Chem. Soc. Jpn.* **62** (1989) 143.
21. Pyrroline synthesis via iminyl radical cyclization: J. Boivin, E. Fouquet, S. Z. Zard, *Tetrahedron Lett.* **31** (1990) 3545.
22. S. Iacono, J. R. Rasmussen, *Org. Synth.* **64** (1985) 63.
23. (a) K. Nozaki, K. Oshima, K. Utimoto, *Tetrahedron Lett.* **29** (1988) 6125, 6127; *Bull. Chem. Soc. Jpn.* **63** (1990) 2578; (b) K. Miura, Y. Ichinose, K. Nozaki, K. Fugami, K. Oshima, K. Utimoto, *Bull. Chem. Soc. Jpn.* **62** (1989) 143; (c) D. H. R. Barton, D. O. Jang, J. C. Jaszberenyi, *Tetrahedron Lett.* **31** (1990) 4681.
24. Radical chain reaction of xanthic anhydrides mediated by organotin hydride: J. E. Forbes, S. Z. Zard, *Tetrahedron Lett.* **30** (1989) 4367.
25. (a) M. D. Bachi, E. Bosch, *Tetrahedron Lett.* **29** (1988) 2581; (b) D. Crich, *Tetrahedron Lett.* **29** (1988) 5805.
26. R. Yamaguchi, T. Hamasaki, K. Utimoto, *Chem. Lett.* (1988) 913.
27. W. R. Bowman, H. Heaney, B. M. Jordan, *Tetrahedron Lett.* **29** (1988) 6657.
28. Facile cyclization of 5-hexenyl radical is well known, but the presence of a carbonyl group in the competing position produces the cyclized oxygen radical instead of cyclopentylmethyl radical: R. Tsang, J. K. Dickson, H. Pak, R. Walton, B. Fraser-Reid, *J. Am. Chem. Soc.* **109** (1987) 3484.
29. (a) Diquinane framework: M. Journet, W. Smadja, M. Malacria, *Synlett* (1990) 320; (b) stannane-mediated alkylation and spiroannulation of 3-bromo-2-cycloalkenones with olefins: E. Lee, D.-S. Lee, *Tetrahedron Lett.* **31** (1990) 4341; (c) halogen atom transfer mediated by hexabutyldistannane under irradiation: D. L. Flynn, D. L. Zabrowski, *J. Org. Chem.* **55** (1990) 3673.
30. P. Dowd, S.-C. Choi, *J. Am. Chem. Soc.* **109** (1987) 3493.
31. S. Wollowitz, J. Halpern, *J. Am. Chem. Soc.* **110** (1988) 3112.

32. K. Nozaki, K. Oshima, K. Utimoto, *Tetrahedron Lett.* **29** (1988) 1041.
33. C. Tomborski, F. E. Ford, E. J. Soloski, *J. Org. Chem.* **28** (1963) 237.
34. W. C. Still, *J. Am. Chem. Soc.* **100** (1978) 1481.
35. H. Azizian, C. Eaborn, A. Pidcock, *J. Organomet. Chem.* **215** (1981) 49.
36. M. Kosugi, T. Ohya, T. Migita, *Bull. Chem. Soc. Jpn.* **56** (1983) 3539.
37. S. Matsubara, N. Tsuboniwa, Y. Morizawa, K. Oshima, H. Nozaki, *Bull. Chem. Soc. Jpn.* **57** (1984) 3242.
38. A. Yanagisawa, Y. Noritake, H. Yamamoto, *Chem. Lett.* (1988) 1899.
39. Generation of stannyl radicals from distannanes and benzylic stannanes: W. P. Neumann, H. Hillgärtner, K. M. Baines, R. Dicke, K. Vorspohl, U. Kobs, U. Nussbeutel, *Tetrahedron* **45** (1989) 951.
40. M. Kosugi, T. Migita, *J. Synth. Org. Chem. Jpn.* **38** (1980) 1142.
41. H. Hayami, M. Sato, S. Kanemoto, Y. Morizawa, K. Oshima, H. Nozaki, *J. Am. Chem. Soc.* **105** (1983) 4491.
42. Y. Okuda, K. Wakamatsu, W. Tückmantel, K. Oshima, H. Nozaki, *Tetrahedron Lett.* **26** (1935) 4629.
43. K. Wakamatsu, T. Nonaka, Y. Okada, W. Tückmantel, K. Oshima, H. Nozaki, *Tetrahedron* **42** (1986) 4427.
44. J. Hibino, S. Nakatsukasa, K. Fugami, S. Matsubara, K. Oshima, H. Nozaki, *J. Am. Chem. Soc.* **107** (1985) 6416.
45. J. Hibino, S. Matsubara, Y. Morizawa, K. Oshima, H. Nozaki, *Tetrahedron Lett.* **25** (1984) 2151.
46. S. Sharma, A. C. Oehlschlager, *J. Org. Chem.* **54** (1989) 5064.
47. (a) S. Matsubara, J. Hibino, Y. Morizawa, K. Oshima, H. Nozaki, *J. Organomet. Chem.* **285** (1985) 163; (b) T. Nonaka, Y. Okuda, S. Matsubara, K. Oshima, K. Utimoto, H. Nozaki, *J. Org. Chem.* **51** (1986) 4716.
48. B. L. Chernard, C. M. Van Zyl, *J. Org. Chem.* **51** (1986) 3561.
49. T. Hiyama, M. Shinoda, H. Saimoto, H. Nozaki, *Bull. Chem. Soc. Jpn.* **54** (1981) 2747.
50. (a) B. L. Chernard, C. M. Van Zyl, D. R. Sanderson, *Tetrahedron Lett.* **27** (1986) 2801; (b) S. E. Denmark, J. K. Jones, *J. Am. Chem. Soc.* **104** (1982) 2642.
51. S. E. Denmark, J. K. Jones, *J. Am. Chem. Soc.* **104** (1982) 2642.
52. M. R. Peel, C. R. Johnson, *Tetrahedron Lett.* **27** (1986) 5947.
53. N. Tsuboniwa, S. Matsubara, Y. Morizawa, K. Oshima, H. Nozaki, *Tetrahedron Lett.* **25** (1984) 2569; (b) S. Matsubara, N. Tsuboniwa, Y. Morizawa, K. Oshima, H. Nozaki, *Bull. Chem. Soc. Jpn.* **57** (1984) 3242.
54. (a) K. Maruoka, S. Hashimoto, Y. Kitagawa, H. Yamamoto, H. Nozaki, *J. Am. Chem. Soc.* **99** (1977) 7705; (b) K. Maruoka, S. Hashimoto, Y. Kitagawa, H. Yamamoto, H. Nozaki, *Bull. Chem. Soc. Jpn.* **53** (1980) 3301.
55. G. H. Posner, E. Asirvatham, K. S. Webb, S.-S. Jew, *Tetrahedron Lett.* **28** (1987) 5071.
56. K. Sisido, Y. Takeda, Z. Kinugawa, *J. Am. Chem. Soc.* **83** (1961) 538.
57. K. Sisido and Y. Takeda, *J. Org. Chem.* **26** (1961) 2301.
58. J. Nokami, J. Otera, T. Sudo, R. Okawara, *Organometallics* **2** (1983) 191; see also T. Mukaiyama, T. Harada, S. Shoda, *Chem. Lett.* (1980) 1507.
59. T. Mandai, J. Nokami, T. Yano, Y. Yoshinaga, J. Otera, *J. Org. Chem.* **49** (1984) 172.
60. Homoallyl alcohol synthesis with pentacoordinate allyl silicates in hydroxylic media: M. Kira, K. Sato, H. Sakurai, *J. Am. Chem. Soc.* **112** (1990) 257.

61. Reactions of cinnamyl chloride with aldehydes under a variety of conditions: J. M. Coxon, S. J. van Eyk, P. J. Steel, *Tetrahedron* **45** (1989) 1029.
62. Effect of water on carbonyl allylation with allylic tin reagents: D. Furlani, D. Marton, G. Tagliavini, M. Zordan, *J. Organomet. Chem.* **341** (1988) 345; see also ref. 71c.
63. J. Nokami, S. Watanabe, R. Okawara, *Chem. Lett.* (1984) 869.
64. J. Nokami, T. Tamaoka, H. Ogawa, S. Wakabayashi, *Chem. Lett* (1986) 541.
65. J. E. Baldwin, R. M. Adlington, J. B. Sweeney, *Tetrahedron Lett.* **27** (1986) 5423.
66. K. Tanaka, H. Yoda, Y. Isobe, A. Kaji, *J. Org. Chem.* **51** (1986) 1856.
67. Lactone synthesis by means of organochromium intermediates: Y. Okuda, S. Nakatsukasa, K. Oshima, H. Nozaki, *Chem. Lett.* (1985) 481.
68. Homoallyl alcohol synthesis with chromium(II): T. Hiyama, *J. Synth. Org. Chem. Jpn.* **39** (1981) 81; *Chem. Abstr.* **97** (1982) 109178t.
69. Recent advances in chromium(II)-mediated organic synthesis: (a) K. Takai, K. Utimoto, *J. Synth. Org. Chem. Jpn.* **46** (1988) 66; *Chem. Abstr.* **109** (1988) 22505q; (b) P. Cintas, *Synthesis* (1992) 248; (c) N. A. Saccomano, in *Comprehensive Organic Synthesis* (B. M. Trost, I. Fleming, eds.), vol. 1, p. 173, Pergamon, Oxford, 1991.
70. S. Matsubara, K. Wakamatsu, Y. Morizawa, N. Tsuboniwa, K. Oshima, H. Nozaki, *Bull. Chem. Soc. Jpn.* **58** (1985) 1196.
71. (a) Y. Masuyama, J. P. Takahara, Y. Kurusu, *J. Am. Chem. Soc.* **110** (1988) 4473; (b) Y. Masuyama, R. Hayashi, K. Otake, Y. Kurusu, *J. Chem. Soc., Chem. Commun.* (1988) 44; (c) Y. Masuyama, J. P. Takahara, Y. Kurusu, *Tetrahedron Lett.* **30** (1989) 3437.
72. J. Yoshida, H. Funahashi, H. Iwasaki, N. Kawabata, *Tetrahedron Lett.* **27** (1986) 4469.
73. Y. Ueno, S. Aoki, M. Okawara, *J. Am. Chem. Soc.* **101** (1979) 1979.
74. O. Desponds, M. Schlosser, *J. Organomet. Chem.* **409** (1991) 93.
75. M. Schlosser, *J. Organomet. Chem.* **8** (1967) 9; *Pure Appl. Chem.* **60** (1988) 1627.
76. J. Hartmann, M. Schlosser, *Helv. Chim. Acta* **59** (1976) 453.
77. M. Schlosser, J. Hartmann, *Angew. Chem.* **85** (1973) 544; *Angew. Chem., Int. Ed. Engl.* **12** (1973) 439.
78. M. Schlosser, H. Bossert, *Tetrahedron* **47** (1991) 6287.
79. Y. Yamamoto, *Aldrichim. Acta* **20** (1987) 45; *Chem. Abstr.* **107** (1987) 23373r.
80. Y. Yamamoto, *Acc. Chem. Res.* **20** (1987) 243.
81. Reactions of 1-methyl-2-butenylstannanes and aldehydes: C. Hull, S. V. Mortlock, E. J. Thomas, *Tetrahedron* **45** (1989) 1007.
82. (a) S. E. Denmark, E. J. Weber, *J. Am. Chem. Soc.* **106** (1984) 7970; (b) S. E. Denmark, B. R. Henke, E. Weber, *J. Am. Chem. Soc.* **109** (1987) 2512; (c) S. E. Denmark, T. Wilson, T. M. Wilson, *J. Am. Chem. Soc.* **110** (1988) 984; (d) S. E. Denmark, E. J. Weber, T. M. Wilson, *Tetrahedron* **45** (1989) 1053.
83. See also: Y. Naruta, Y. Nishigaichi, K. Maruyama, *Tetrahedron* **45** (1989) 1067.
84. G. P. Boldrini, L. Lodi, E. Tagliavini, C. Tarasco, C. Trombini, A. Umani-Ronchi, *J. Org. Chem.* **52** (1987) 5447.
85. S. V. Mortlock, E. J. Thomas, *Tetrahedron Lett.* **29** (1988) 2479.
86. (a) G. Dumartin, M. Pereyre, J.-P. Quintard, *Tetrahedron Lett.* **28** (1987) 3935; (b) J.-P. Quintard, G. Dumartin, B. Ellissondo, A. Rahm, M. Pereyre, *Tetrahedron* **45** (1989) 1017.
87. (a) J. A. Marshall, B. S. DeHoff, S. L. Crooks, *Tetrahedron Lett.* **28** (1987) 527; (b) J. A. Marshall, W. Y. Gung, *Tetrahedron Lett.* **30** (1989) 309; (c) J. A. Marshall, W. Y. Gung, *Tetrahedron* **45** (1989) 1043.
88. P. A. Grieco, A. Bahsas, *J. Org. Chem.* **52** (1987) 1378.

89. R. Yamaguchi, A. Otsuji, K. Utimoto, *J. Am. Chem. Soc.* **110** (1988) 2186.
90. (a) R. Yamaguchi, E. Hata, K. Utimoto, *Tetrahedron Lett.* **20** (1988) 1785; (b) R. Yamaguchi, T. Hamasaki, K. Utimoto, *Chem. Lett.* (1988) 913; (c) R. Yamaguchi, T. Hamasaki, K. Utimoto, S. Kozima, H. Takaya, *Chem. Lett.* (1990) 2161; (d) R. Yamaguchi, *J. Synth. Org. Chem. Jpn.* **49** (1991) 128.
91. K. Mizuno, M. Ikeda, S. Toda, Y. Otsuji, *J. Am. Chem. Soc.* **110** (1988) 1288.
92. M. Yoshitake, M. Yamamoto, S. Kohmoto, K. Yamada, *J. Chem. Soc., Perkin Trans. 1* (1990) 1226.
93. J. H. Herndon, *J. Am. Chem. Soc.* **109** (1987) 3165.
94. J. Lambert, G.-T. Wang, *Tetrahedron Lett.* **29** (1988) 1785.
95. M. Yamamoto, S. Komoto, K. Yamada, *J. Synth. Org. Chem. Jpn.* **46** (1988) 1134; *Chem. Abstr.* **111** (1989) 232879g.
96. B. L. Chernard, *Tetrahedron Lett.* **27** (1986) 2805.
97. (a) R. Noyori, M. Suzuki, *Angew. Chem.* **96** (1984) 854; *Angew. Chem., Int. Ed. Engl.* **23** (1984) 847; (b) M. Suzuki, A. Yanagisawa, R. Noyori, *J. Am. Chem. Soc.* **110** (1988) 4718.
98. C. R. Johnson, T. D. Penning, *J. Am. Chem. Soc.* **108** (1986) 5655.
99. Tetrahydrofuran synthesis by means of 4-[(*tert*-butyldiphenylsilyl)oxy]-2-(*E*)-buten-1-ol: A. G. M. Barrett, T. E. Barta, J. A. Flygare, *J. Org. Chem.* **54** (1989) 4246.
100. E. Piers, V. Karunaratne, *Tetrahedron* **45** (1989) 1089.
101. T. Nakai, K. Mikami, *Chem. Rev.* **86** (1988) 885; see also B. Kruse, R. Brückner, *Tetrahedron Lett.* **31** (1990) 4425; J. A Marshall, E. D. Robinson, *J. Org. Chem.* **55** (1990) 3450.
102. M. M. Midland, Y. C. Kwon, *Tetrahedron Lett.* **26** (1985) 5013.
103. B. Fraser-Reid, R. D. Dawe, D. B. Tulshian, *Can. J. Chem.* **57** (1979) 746.
104. S. Hanessian, M. Martin, R. C. Desai, *J. Chem. Soc., Chem. Commun.* (1986) 926.
105. J. S. Sawyer, A. Kucerovy, T. L. Macdonald, G. J. McGarvey, *J. Am. Chem. Soc.* **110** (1988) 842.
106. (a) C. R. Johnson, J. R. Medlich, *J. Org. Chem.* **53** (1988) 4131; (b) L. A. Paquette, J. Reagan, S. L. Schreiber, C. A. Teleha, *J. Am. Chem. Soc.* **111** (1989) 2331; (c) functionalized zinc and copper organometallics at the α-position to an oxygen: T.-S. Chou, P. Knochel, *J. Org. Chem.* **55** (1990) 4791.
107. R. J. Linderman, A. Godfrey, *J. Am. Chem. Soc.* **110** (1988) 6249.
108. (a) P. C. M. Chan, J. M. Chong, *J. Org. Chem.* **53** (1988) 5586; (b) J. M. Chong, E. K. Mar, *Tetrahedron* **45** (1989) 7709.
109. (a) J. M. Chong, E. K. Mar, *Tetrahedron Lett.* **31** (1990) 1981; (b) P. C. M. Chan, J. M. Chong, *Tetrahedron Lett.* **23** (1990) 1985.
110. D. K. Hutchinson, P. L. Fuchs, *J. Am. Chem. Soc.* **109** (1987) 4930.
111. A. Dondoni, M. Fogagnolo, G. Fantin, A. Medici, P. Pedrini, *Tetrahedron Lett.* **27** (1986) 5269.
112. M. Andrianome, K. Häberle, B. Delmond, *Tetrahedron* **45** (1989) 1079.
113. M. Kumada, *Pure Appl. Chem.* **52** (1980) 669.
114. Application in asymmetric synthesis: T. Hayashi, M. Konishi, Y. Kobori, M. Kumada, T. Higuchi, K. Hirotsu, *J. Am. Chem. Soc.* **106** (1984) 158.
115. R. J. Corriu, J. P. Masse, *J. Chem. Soc., Chem. Commun.* (1927) 144.
116. E. Negishi, *Acc. Chem. Res.* **15** (1982) 340.
117. M. Kosugi, Y. Shimizu, T. Migita, *Chem. Lett.* (1977) 1423: *J. Organomet. Chem.* **129** (1977) C36.

118. J. K. Stille, *Angew. Chem.* **98** (1986) 504; *Angew. Chem., Int. Ed. Engl.* **25** (1986) 508.
119. Diene synthesis related to cephalosporin chemistry: V. Farina, S. R. Baker, C. Sapino, *Tetrahedron Lett.* **29** (1988) 6043.
120. Pd(0)-catalysed coupling of aryl halides and 1-alkenylstannanes: W. Tao, S. Nesbitt, R. F. Heck, *J. Org. Chem.* **55** (1990) 63; E. Laborde, L. E. Lesheski, J. S. Kiely, *Tetrahedron Lett.* **31** (1990) 1837; α-lithiated cyclic enol ethers: P. Kocienski, S. Wadman, *J. Am. Chem. Soc.* **111** (1989) 2363; see also ref. 157.
121. Coupling of 2,5-dihydro-2-thienylstannanes with vinyl iodides: H. Takayama, T. Suzuki, *J. Chem. Soc., Chem. Commun.* (1988) 1044.
122. D. Milstein, J. K. Stille, *J. Org. Chem.* **44** (1979) 1613.
123. A. F. Renaldo, J. W. Labadie, J. K. Stille, *Org. Synth.* **67** (1988) 86.
124. Y. Ito, M. Inouye, M. Murakami, *Tetrahedron Lett.* **29** (1988) 47.
125. Enol triflates in divinyl ketone synthesis: W. J. Scott and J. E. McMurry, *Acc. Chem. Res.* **21** (1988) 47.
126. Preparation of *O*-methyl-3-acyltetronic acids via direct acylation of stannyl tetronates: S. V. Ley, D. J. Wadsworth, *Tetrahedron Lett.* **30** (1989) 1001.
127. Stereospecific coupling of nucleophilic and electrophilic chiral carbons: D. S. Matteson, P. B. Triphathy, A. Sarkar, K. M. Sadhu, *J. Am. Chem. Soc.* **111** (1989) 4399.
128. M. Kosugi, T. Sumiya, Y. Obara, M. Suzuki, H. Sano, T. Migita, *Bull. Chem. Soc. Jpn.* **60** (1987) 767.
129. J. K. Stille, M. Tanaka, *J. Am. Chem. Soc.* **109** (1987) 3785.
130. B. Burns, R. Grigg, P. Ratananukul, V. Sreidharan, P. Stevenson, S. Sukirthalingam, T. Worakun, *Tetrahedron Lett.* **29** (1988) 5565.
131. Jatrophane synthesis through Pd-catalysed carbonylative coupling: A. C. Gyorkos, J. K. Stille, paper presented at the 198th ACS Annual Meeting, Miami, FL, September 10–15, 1989, ORGN 183.
132. D. R. Tueting, A. M. Echavarren, J. K. Stille, *Tetrahedron* **45** (1989) 979.
133. K. K. Wang, K.-H. Chu, Y. Lind, J.-H. Chen, *Tetrahedron* **45** (1989) 1105.
134. S. E. Tunney, J. K. Stille, *J. Org. Chem.* **52** (1987) 748.
135. Aldol-type reactions: T. Mukaiyama, N. Iwasawa, R. W. Stevens, T. Haga, *Tetrahedron* **40** (1984) 1381; N. Iwasawa, T. Yura, T. Mukaiyama, *Tetrahedron* **45** (1989) 1197.
136. Michael reactions: T. Mukaiyama, N. Iwasawa, T. Yura, R. S. J. Clark, *Tetrahedron* **43** (1987) 5003.
137. T. Mukaiyama, *Yuuki-Gosei-Hanno* (*Organic Synthetic Reactions*, in Japanese), Tokyo Kagaku Dojin, 1987; an English translation will be published by Oxford University Press.
138. T. Yura, N. Iwasawa, T. Mukaiyama, *Chem. Lett.* (1987) 791.
139. S. Kobayashi, Y. Fujishita, T. Mukaiyama, *Chem. Lett.* (1990) 1455.
140. T. Yura, N. Iwasawa, T. Mukaiyama, *Chem. Lett.* (1987) 1021.
141. T. Yura, N. Iwasawa, K. Narasaka, T. Mukaiyama, *Chem. Lett.* (1988) 1025.
142. T. Mukaiyama, H. Sugumi, H. Uchiro, S. Kobayashi, *Chem. Lett.* (1988) 1291.
143. G. A. Molander, D. C. Shubert, *J. Am. Chem. Soc.* **108** (1986) 4683.
144. A. Libineau, J. deGallic, A. Malleron, *Tetrahedron Lett.* **28** (1987) 5041.
145. K. Hillner, W. P. Neumann, *Tetrahedron Lett.* **27** (1986) 5347.
146. Review: H. Nozaki, J. Otera, in *Organometallics in Organic Synthesis*, ed. A. de Meijere, H. tom Dieck, p. 169, Springer, Berlin, 1987.

147. R. Okawara, N. Kasai, K. Yasuda, in *Proceedings of the 2nd International Symposium on Organometallic Chemistry, Wisconsin, USA, 1965*, p. 128.
148. H. Puff, I. Bung, E. Friedrichs, A. Janzen, *J. Organomet. Chem.* **253** (1983) 23.
149. Increased electrophilicity of pentacoordinated silicates: M. Kira, M. Kobayashi, H. Sakurai, *Tetrahedron Lett.* **28** (1987) 4081; A. Hosomi, S. Kohra, Y. Tominaga, *J. Chem. Soc., Chem. Commun.* (1987) 1517; A. Hosomi, S. Kohra, Y. Tominaga, *Chem. Pharm. Bull.* **31** (1987) 2155; T. Hayashi, Y. Matsumoto, T. Kiyoi, Y. Ito, S. Kohra, Y. Tominaga, A. Hosomi, *Tetrahedron Lett.* **29** (1988) 5667.
150. J. Otera, T. Yano, A. Kawabata, H. Nozaki, *Tetrahedron Lett.* **27** (1986) 2383.
151. Distannoxane synthesis: M. Wada, M. Nishino, R. Okawara, *J. Organomet. Chem.* **3** (1965) 70.
152. J. Otera, S. Ioka, H. Nozaki, *J. Org. Chem.* **54** (1989) 4013.
153. Y. Tsuda, M. E. Haque, K. Yoshitomo, *Chem. Pharm. Bull.* **31** (1983) 1612.
154. (a) D. Wagner, J. P. H. Verheyden, J. G. Moffat, *J. Org. Chem.* **39** (1974) 24; (b) S. Danishefsky, R. Hungate, *J. Am. Chem. Soc.* **108** (1986) 2486.
155. Increased nucleophilicity of oxygen appears to be also crucial for the cycloaddition of trimethylenemethane with aldehydes: B. M. Trost, S. A. King, T. Schmidt, *J. Am. Chem. Soc.* **111** (1989) 5902.
156. (a) S. L. Schreiber, H. V. Meyers, *J. Am. Chem. Soc.* **110** (1988) 5198; (b) S. L. Schreiber, D. Desmaele, J. A. Porco, *Tetrahedron Lett.* **20** (1988) 6689.
157. J. Otera, N. Dan-oh, H. Nozaki, *J. Chem. Soc., Chem. Commun.* (1991) 1742; *Tetrahedron* **49** (1993), 3065.
158. J. Otera, T. Yano, Y. Himeno, H. Nozaki, *Tetrahedron Lett.* **27** (1986) 4501.
159. K. Steliou, M.-A. Poupart, *J. Am. Chem. Soc.* **105** (1983) 7130.
160. A. Shanzer, N. Mayer-Schochet, F. Frolow, D. Rabinovich, *J. Org. Chem.* **46** (1981) 4662.
161. Y. Tor, J. Libman, F. Frolow, H. E. Gottlieb, R. Lazar, A. Shanzer, *J. Org. Chem.* **50** (1985) 5476.
162. A. K. Kumar, T. K. Chattopadhyay, *Tetrahedron Lett.* **28** (1987) 3713.
163. J. Otera, N. Dan-oh, H. Nozaki, *J. Org. Chem.* **56** (1991) 5307.
164. (a) J. Otera, H. Nozaki, *Tetrahedron Lett.* **27** (1986) 5743; (b) J. Otera, T. Mizutani, H. Nozaki, *Organometallics* **8** (1989) 2063; (c) J. Otera, N. Dan-oh, H. Nozaki, *Tetrahedron* **48** (1992) 1449.
165. J. Otera, *Synthesis* (1988) 85.
166. T. Tsunoda, M. Suzuki, R. Noyori, *Tetrahedron Lett.* **25** (1984) 5913.
167. T. Sato, T. Kobayashi, T. Gojo, E. Yoshida, J. Otera, H. Nozaki, *Chem. Lett.* (1987) 1661.
168. T. Sato, S. Okura, J. Otera, H. Nozaki, *Tetrahedron Lett.* **28** (1987) 6299.
169. T. Sato, H. Okazaki, J. Otera, H. Nozaki, *Tetrahedron Lett.* **29** (1988) 2979.
170. T. Sato, J. Otera, H. Nozaki, to be published.
171. K. Kondo, D. Tunemoto, *Tetrahedron Lett.* (1975) 1005.
172. T. Sato, J. Otera, H. Nozaki, *J. Org. Chem.* **54** (1989) 2779.
173. T. Sato, H. Ozaki, J. Otera, H. Nozaki, *J. Am. Chem. Soc.* **110** (1988) 5209.
174. T. Sato, M. Inoue, S. Kobara, J. Otera, H. Nozaki, *Tetrahedron Lett.* **30** (1989) 91.
175. T. Sato, Y. Hiramura, J. Otera, H. Nozaki, *Tetrahedron Lett.* **30** (1989) 2821.
176. (a) 2-Tetrahydropyranyl ethers: T. Sato, T. Tada, J. Otera, H. Nozaki, *Tetrahedron Lett.* **30** (1989) 1665; *J. Org. Chem.* **55** (1990) 4770; (b) Thio- and selenoglycosides: T. Sato, Y. Fujita, J. Otera, H. Nozaki, *Tetrahedron Lett.* **33** (1992) 239.

177. N. Nagashima, M. Ohno, *Chem. Lett.* (1987) 141.
178. T. Sato, J. Otera, H. Nozaki, *Tetrahedron Lett.* **30** (1989) 2959; see also T. Sato, J. Otera, H. Nozaki, *J. Org. Chem.* **57** (1992) 57.
179. (a) 1,3-thianes: T. Sato, E. Yoshida, T. Kobayashi, J. Otera, H. Nozaki, *Tetrahedron Lett.* **29** (1988) 3977; 1,3-dithiopropenes: T. Sato, J. Otera, H. Nozaki, *Synlett* (1991) 903; *J. Org. Chem.* **58** (1993), 4971.
180. T. Sato, J. Otera, H. Nozaki, *J. Am. Chem. Soc.* **112** (1990) 901.
181. T. Sato, Y. Wakahara, J. Otera, H. Nozaki, *Tetrahedron Lett.* **31** (1990) 1581.
182. (a) T. Sato, Y. Wakahara, J. Otera, H. Nozaki, S. Fukuzumi, *J. Am. Chem. Soc.* **113** (1991) 4028; (b) T. Sato, Y. Wakahara, J. Otera, H. Nozaki, *Tetrahedron* **47** (1991) 9773; (c) J. Otera, Y. Wakahara, H. Kamei, T. Sato, H. Nozaki, S. Fukuzumi, *Tetrahedron Lett.* **32** (1992) 2405.
183. M. Stähle, M. Schlosser, *J. Organomet. Chem.* **220** (1981) 277.

Subject Index

Formula Index

The gross formulas of the organometallic reagents and intermediates listed below are ordered according to these principles:

- Only the elements of the organic moiety to be transferred are counted. For example, both methyltitanium triisopropoxide and tetramethyltitanium appear under the heading CH_3.
- The metal attached to this transferable organic group is shown in parentheses. The metal priority decreases when moving in the Periodic Table from left to right and, within a row, from top to bottom (Li > Na > K > Rb > Cs > Be > Mg > Ca > Ba > Ti > Pd > Cu > B > Al > Sn). There is, however, one inconsistency: Zn directly follows Mg.